second edition

COLLEGE ALGEBRA:
A PROBLEM-SOLVING APPROACH

Arnold L. Steffensen
L. Murphy Johnson
Northern Arizona University

ADDISON-WESLEY

An Imprint of Addison Wesley Longman, Inc

Reading, Massachusetts • Menlo Park, California • New York • Harlow, England
Don Mills, Ontario • Sydney • Mexico City • Madrid • Amsterdam

Manager of Addison Wesley Longman Custom Books: Caralee Woods
Production Administrator: Liz Faerm
Cover Design: John Callahan

COLLEGE ALGEBRA:
A Problem-Solving Approach
Second Edition

Copyright © 1997 Addison-Wesley Educational Publishers Inc.

All rights reserved. Printed in the United States of America. No part of this book may be used or reproduced in any manner whatsoever without written permission from the publisher except testing materials, which may be copied for classroom use. For information, address Addison Wesley Educational Publishers Inc., 10 East 53rd Street, New York, NY 10022.

ISBN: 0-201-30149-0

97 98 99 3 2 1

PREFACE

College Algebra: A Problem-Solving Approach, Second Edition, is designed to provide comprehensive coverage of the standard topics in algebra needed by students for later courses in mathematics, engineering, business, statistics, or the natural sciences. The emphasis on real-world applications also makes the text suitable for liberal arts courses. Students with two years of high school algebra or a background in intermediate algebra at the college level should have the necessary prerequisite skills.

The text is organized for maximum instructional flexibility. Chapter 1 provides a review of basic algebra, which some classes may cover quickly or skip altogether. Chapters 2 through 5 present the major topics in college algebra, including equations and inequalities, functions and their graphs, the theory of polynomials, and polynomial, rational, exponential, and logarithmic functions and their many applications. For added flexibility, systems of equations and inequalities and matrices and determinants have been presented in Chapters 6 and 7. A thorough treatment of conics and nonlinear systems is given in Chapter 8, and the text concludes with a variety of topics including sequences, series, mathematical induction, permutations, combinations, the binomial theorem, and probability.

FEATURES

STUDENT GUIDEPOSTS These lead students through each section, specifying important terms, definitions, and procedures. The guideposts are designed to help students locate important concepts while they begin studying each section, as well as to provide direction during their review of the material for examinations.

EXAMPLES More than 500 carefully selected examples consist of detailed step-by-step solutions that include helpful color annotations.

PRACTICE EXERCISES These parallel each example to reinforce the concepts presented. Space for working each practice exercise is provided, with answers immediately following.

CAUTIONS This feature calls the attention of students to common mistakes and special problems.

COLOR is used pedagogically to point out important steps and to emphasize methods and terminology. Key definitions, rules, and procedures are set off in colored boxes for increased emphasis. Figures and graphs use color to clarify the concepts presented, and examples present important steps and helpful side comments in color.

EXERCISES Two parallel exercise sets (A and B) and a collection of extension exercises (C) follow each section and offer a wealth of practice for students and flexibility for instructors. There are more than 4300 exercises, including about 650 applied problems, which are identified by discipline.

Exercises A This set of exercises includes space for working the problems and provides answers immediately following. Containing a blend of drill and comprehension problems, the exercises can be used as in-class practice or assigned homework.

Exercises B This set parallels set A, problem for problem, but are presented without work space or answers. These exercises can be used as homework problems, examples in class, or additional practice.

Exercises C This set is designed to give students an extra challenge. These problems, which extend the concepts of the section, either are more demanding than exercises in set A or B or require extensive use of a calculator. Answers or hints are given for selected exercises.

FOR REVIEW This heading identifies exercises that reinforce material from earlier sections of the chapter and review special topics necessary for success in the following section. These exercises are located at the end of most A and B exercise sets.

END-OF-CHAPTER MATERIAL Each chapter concludes with an **Extension Project**, **Chapter Review**, and a **Chapter Test**. The extension projects often involve use of a calculator or personal computer and provide interesting topics for further study, such as square root approximation, how to use a computer to solve a system of equations, or how to perform operations on matrices. The review contains a glossary of key terms and key concepts, along with review exercises. Part of the review exercises are referenced to sections in the chapter, and the rest are mixed, requiring that students recognize types of problems without the aid of a section reference.

FINAL REVIEW EXERCISES referenced to each chapter, with answers supplied, are located at the back of the text.

USE OF CALCULATOR It is assumed that most students have a scientific calculator. Steps for calculator solutions have been provided for such topics as powers, radicals, logarithms, and antilogarithms. However, no specific emphasis has been given to the use of a calculator in many of the exercises. In working exercises that involve dividing decimals, approximating irrational numbers, or logarithms, a calculator can be an invaluable tool, and it is important for students to learn without specific instruction when a calculator should or should not be used. Some exercises, however, do specifically refer to calculator use.

NEW IN THIS EDITION

- Explanations have been streamlined where appropriate, making greater use of figures and illustrations.
- Exercise sets have been reviewed relative to grading and balance of coverage with an increase in the number and variety of practical applications. Additional challenging exercises and a greater emphasis on the use of a calculator have been incorporated in Exercises C. Practice exercises that parallel examples also have been provided.
- Student guideposts are used to help students find their way through each section and to call attention to important topics or procedures.
- Additional "For Review" exercises have been added to review topics in preparation for the next section.
- Key Words given in the Chapter Review include brief definitions along with section references.
- Greater emphasis has been given to the use of figures and illustrations in examples and exercises, particularly applications.
- Instructions for using a calculator have been expanded, especially in the areas of exponents, radicals, and logarithms.
- Real-world applications are used in the chapter introductions to motivate the use of the skills to be discussed.
- Graphing in the coordinate plane is again presented early in the text, and a greater emphasis has been given to graphing various types of functions throughout the text.
- Much emphasis has been given to the important concept of problem solving, and a section on mathematical modeling has been added to Chapter 3.

- Complex numbers now are introduced early in Chapter 2, allowing for a more thorough discussion of quadratic equations.
- An expanded treatment of polynomials and rational functions has been added, and coverage of the solutions to polynomial (quadratic) inequalities and rational inequalities has been moved to Chapter 4 to take advantage of knowledge about graphs.
- The use of a calculator to study logarithms replaces the previous dual presentation involving both the calculator and logarithmic tables. Logarithmic tables and interpolation are still available in Appendix A.
- Chapter 6 has been expanded to include linear programming and partial fractions, while the treatment of matrices in Chapter 7 has been updated to provide more recognition of important applications.

INSTRUCTIONAL FLEXIBILITY

College Algebra: A Problem-Solving Approach, Second Edition, offers proven flexibility for a variety of teaching situations such as individualized instruction, lab instruction, lecture classes, or a combination of methods.

Material in each section of the book is presented in a well-paced, easy-to-follow sequence. Students in an individualized or lab instruction setting, aided by the student guideposts, can work through a section completely by reading the explanation, following the detailed steps in the examples, working the practice exercises, and then doing the exercises in set A.

The book also can serve as the basis for, or as a supplement to, classroom lectures. The straightforward presentation of material, numerous examples, practice exercises, and three sets of exercises offer the traditional lecture class an alternative approach within the convenient workbook format.

SUPPLEMENTS

The **INSTRUCTOR'S TESTING MANUAL** contains a placement test, student information sheet, six different but equivalent tests for each chapter (four open-response and two multiple-choice), and two final examinations, with all answers supplied in a convenient format.

The **INSTRUCTOR'S SOLUTIONS MANUAL** contains complete, worked-out solutions to every exercise in sets B and C.

The **STUDENT'S SOLUTIONS MANUAL** contains complete, worked-out solutions to each practice exercise in the text, to every exercise in set A, and to all the chapter review exercises and chapter tests.

The **HARPERCOLLINS TEST GENERATOR FOR MATHEMATICS,** available in IBM and Macintosh formats, enables instructors to select questions for any section in the text or to use a ready-made test for each chapter. Instructors may generate tests in multiple-choice or open-response formats, scramble the order of questions while printing, and produce 25 versions of each test. The system features printed graphics and accurate mathematical symbols. The program also allows instructors to choose problems randomly from a section or problem type or to choose questions manually while viewing them on the screen, with the option to regenerate variables. The editing feature allows instructors to customize the chapter data disks by adding their own problems.

The **INTERACTIVE TUTORIAL SOFTWARE,** based on a mastery learning approach, has been developed for this text. Available for Apple and IBM computers, each tutorial module covers one of the important topics in *College Algebra*. Each module presents a tutorial lesson, followed by drill and practice problems that are randomly generated. Each module also contains one or more computer "tools" that assist students with problem solving or demonstrate essential concepts in *College Algebra*.

ACKNOWLEDGMENTS

We extend our gratitude to the instructors and students who used the first edition of this book and offered suggestions for improvement. Special thanks go to the instructors at Northern Arizona University. In particular, the assistance given over the years by Joseph Mutter, James Kirk, and Michael Ratliff is most appreciated. Also, we sincerely appreciate the support and encouragement of the Northern Arizona University administration, especially President Eugene M. Hughes. It is a pleasure and privilege to serve on the faculty of a university that recognizes quality teaching as its primary role.

We also express our thanks to the following users and reviewers for their countless beneficial suggestions at various stages of the book's development:

Leonard Andrusaitis, *University of Lowell*
Patricia Bradley, *College of DuPage*
Dennis C. Ebersole, *Northampton Community College*
James Hamilton, *Vernon Regional Junior College*
Scott Higinbotham, *Middlesex Community College*
Robert Krenz, *Northwest College*
Philip R. Montgomery, *University of Kansas*
Thomas Rourke, *North Shore Community College*
Dorothy Sulock, *University of North Carolina–Asheville*
Ray T. Treadway, *Bennett College*
Jamie Whitehead, *Texarkana College.*

To everyone at HarperCollins we are greatly indebted. Special thanks go to Jack Pritchard, Anne Kelly, Kathy Richmond, Randee Wire, Sarah Joseph, and Carol Leon.

Finally, we appreciate the support of our families and in particular our wives, Barbara and Barbara, whose encouragement over the years cannot be measured.

Arnold R. Steffensen
L. Murphy Johnson

CONTENTS

To the Student   x

A Word About Calculators   xii

1 REVIEW OF FUNDAMENTAL CONCEPTS

1.1   The Real Number System   2
1.2   Integer Exponents and Scientific Notation   12
1.3   Algebraic Expressions and Polynomials   21
1.4   Factoring Polynomials   32
1.5   Rational Expressions   42
1.6   Radical Expressions   56
1.7   Rational Exponents   68
      CHAPTER 1 EXTENSION PROJECT   76
      CHAPTER 1 REVIEW   76
      CHAPTER 1 TEST   82

2 EQUATIONS, INEQUALITIES, AND PROBLEM SOLVING

2.1   Linear and Absolute Value Equations   85
2.2   Problem Solving and Applications of Linear Equations   98
2.3   Complex Numbers   113
2.4   Quadratic Equations   121
2.5   Equations that Result in Quadratic Equations   135
2.6   Problem Solving and Applications of Quadratic Equations   141
2.7   Linear and Absolute Value Inequalities   154
      CHAPTER 2 EXTENSION PROJECT   166
      CHAPTER 2 REVIEW   166
      CHAPTER 2 TEST   172

3 RELATIONS, FUNCTIONS, AND GRAPHS

3.1   The Rectangular Coordinate System   175
3.2   Linear Equations   183
3.3   Relations and Functions   196
3.4   Properties of Functions and Transformations   204
3.5   Composite and Inverse Functions   218
3.6   Quadratic Functions   228

3.7 Mathematical Modeling   238
3.8 Variation   245
   CHAPTER 3 EXTENSION PROJECT   252
   CHAPTER 3 REVIEW   253
   CHAPTER 3 TEST   261

4 POLYNOMIAL AND RATIONAL FUNCTIONS

4.1 Polynomials and Synthetic Division   264
4.2 The Remainder and Factor Theorems   272
4.3 More Theorems Involving Polynomials   278
4.4 Bounds and The Rational Root Theorem   286
4.5 Graphing Polynomial Functions   295
4.6 Rational Functions   307
4.7 Polynomial and Rational Inequalities   319
   CHAPTER 4 EXTENSION PROJECT   330
   CHAPTER 4 REVIEW   330
   CHAPTER 4 TEST   335

5 EXPONENTIAL AND LOGARITHMIC FUNCTIONS

5.1 Introduction to Logarithms   339
5.2 Exponential and Logarithmic Functions   345
5.3 Properties of Logarithms   355
5.4 Common and Natural Logarithms   364
5.5 Exponential and Logarithmic Equations   374
5.6 More Applications of Exponentials and Logarithms   382
   CHAPTER 5 EXTENSION PROJECT   392
   CHAPTER 5 REVIEW   393
   CHAPTER 5 TEST   398

6 SYSTEMS OF EQUATIONS AND INEQUALITIES

6.1 Linear Systems in Two Variables   401
6.2 Linear Systems in More than Two Variables   410
6.3 Problem Solving Using Systems of Equations   419
6.4 Linear Systems of Inequalities   429
6.5 Linear Programming   440
6.6 Partial Fractions   449
   CHAPTER 6 EXTENSION PROJECT   456
   CHAPTER 6 REVIEW   457
   CHAPTER 6 TEST   462

7 MATRICES AND DETERMINANTS

7.1 Matrices   465
7.2 Matrix Multiplication   474
7.3 Solving Systems of Equations Using Matrices   484
7.4 The Inverse of a Square Matrix   493
7.5 Determinants and Cramer's Rule   505

7.6 More on Determinants 517
   CHAPTER 7 EXTENSION PROJECT 526
   CHAPTER 7 REVIEW 528
   CHAPTER 7 TEST 533

8 TOPICS IN ANALYTIC GEOMETRY

8.1 Conic Sections and the Circle 536
8.2 The Ellipse 543
8.3 The Hyperbola 554
8.4 The Parabola 564
8.5 Nonlinear Systems 574
   CHAPTER 8 EXTENSION PROJECT 583
   CHAPTER 8 REVIEW 584
   CHAPTER 8 TEST 589

9 SEQUENCES, SERIES, AND PROBABILITY

9.1 Sequences and Series 592
9.2 Arithmetic Sequences and Series 601
9.3 Geometric Sequences and Series 608
9.4 Infinite Geometric Sequences and Series 615
9.5 Mathematical Induction 622
9.6 Permutations and Combinations 630
9.7 The Binomial Theorem 642
9.8 Probability 649
   CHAPTER 9 EXTENSION PROJECT 661
   CHAPTER 9 REVIEW 662
   CHAPTER 9 TEST 667

Final Review Exercises 669

Answers to Chapter Tests 679

Appendix   Logarithmic Tables and Interpolation 681

Table of Common Logarithms 686

Index 688

TO THE STUDENT

During the past several years we have taught college algebra to more than 1500 students having a variety of career choices. Some were taking mathematics to satisfy graduation requirements, while others were preparing for more advanced courses in mathematics, science, business, or engineering. Regardless of your educational goals, this text has been written with you, the student, in mind. The material is introduced gradually, building from basic to more advanced skills. We have tried to demonstrate the relevance and usefulness of mathematics throughout the text by including practical everyday applications. As you begin this course, keep in mind these guidelines that are both necessary and helpful.

GENERAL GUIDELINES

1. Mastering algebra requires motivation and dedication. Just as an athlete does not improve without commitment to his or her goal, an algebra student must be prepared to work hard and spend time studying.

2. Algebra is not learned simply by watching, listening, or reading; *it is learned by doing*. Use your pencil and practice. When your thoughts are organized and written in a neat and orderly way, you have taken a giant step toward success. Be complete and write out all details. The following are samples of two students' work on an applied problem. Can you tell which one was more successful in the course?

STUDENT A

Let n = number of units produced during the month
C = total cost per month
$C = 10n^2 - 100n - 2000$
$10{,}000 = 10n^2 - 100n - 2000$
$0 = 10n^2 - 100n - 12{,}000$
$0 = n^2 - 10n - 1200$
$0 = (n-40)(n+30)$
$n - 40 = 0$ or $n + 30 = 0$
$n = 40$ or $n = -30$
Thus, 40 units were produced.

STUDENT F

$10\,(\cancel{10{,}000})^2 - 100\,(10{,}000) - 2000$
$=$
$n =$ units
$n = 10n^2 - 100n - 2000$
$0 = 10n^2 - 101n - 2000$

3. A calculator is useful in any course in algebra. Become familiar with your calculator by consulting your owner's manual. Use the calculator as a time-saving device for work with decimals or complicated functions, but do not become so dependent that you use it for simple calculations that can be done mentally. Learn when to use and when not to use your calculator. See "A Word About Calculators" for more information about how calculators can be used with this text.

SPECIFIC GUIDELINES

1. As you begin each section, look through the material for a preview of what is coming.
2. Return to the beginning of the section and start reading slowly. The STUDENT GUIDEPOSTS will help you find important concepts as you progress or review.
3. Read through each EXAMPLE and make sure you understand every step. The side comments in color will help you if something is not clear.
4. After reading each example, work the parallel PRACTICE EXERCISE. This will reinforce what you have just read and start the process of practice. (*Note:* Complete solutions to the Practice Exercises are available in the *Student's Solutions Manual.*)
5. Periodically you will encounter a CAUTION or a NOTE. The Cautions warn you of common mistakes and special problems to avoid. The Notes provide pertinent information or additional explanations.
6. After you finish the material in the body of the section, you need to practice, practice, practice! Begin with the exercises in set A. Answers to all of these problems are at the end of the set for easy reference. To practice more, do the exercises in set B; to challenge yourself, try the exercises in set C. (*Note:* Solutions to all problems in set A are available in the *Student's Solutions Manual.*)
7. After you have completed all of the sections in the chapter, read the CHAPTER REVIEW that contains the definitions of Key Terms. The Review Exercises provide more practice before you take the CHAPTER TEST. Answers to the tests are at the back of the book. Solutions are provided in the *Student's Solutions Manual.*
8. To help you study for your final examination, we have concluded the book with a comprehensive set of FINAL REVIEW EXERCISES.

If you follow these steps and work closely with your instructor, you will greatly improve your chances for success in algebra.

Best of luck, and remember that you can do it!

A WORD ABOUT CALCULATORS

It is assumed that most students will have a hand calculator in this course. Although it is not absolutely essential, your work will be easier if you use a calculator. The major difference between the types of calculators is in the way they perform various operations. The type that uses Algebraic Logic (ALG) is more desirable at this level, since the order of operations is the same as in algebra. The alternative system, Reverse Polish Notation (RPN), is preferred by many mathematicians and professionals, however. Each system, with its advantages and disadvantages, will perform the calculations necessary in this course. Throughout the text we will illustrate both systems using ALG for Algebraic Logic and RPN for Reverse Polish Notation. As an example, we show the sequence of steps used in each system to compute

$$\frac{(2)(4.5) - (1.3)^2}{5\sqrt{3}}.$$

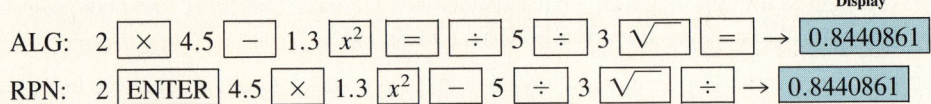

ALG: 2 × 4.5 − 1.3 x^2 = ÷ 5 ÷ 3 √ = → Display 0.8440861

RPN: 2 ENTER 4.5 × 1.3 x^2 − 5 ÷ 3 √ ÷ → 0.8440861

Notice that RPN calculators use an ENTER key instead of the = key found on ALG calculators. This is an essential difference between the two operating systems. Other variations in the types of keys are strictly notational. For example, to change the sign of a number (for entering negative numbers), some calculators have a +/− key, while others have a CHS key. Also, one calculator uses the STO key to place a number in memory, while another has an M key. We will try to point out some of the differences that arise as we consider various computations. However, since it is impossible to mention all of these differences, the best advice is to read your owner's manual.

With calculators, slight variations in accuracy due to rounding differences are bound to occur. Most of these will appear in the seventh or eighth decimal place and should not be of much concern. Throughout the text we have not rounded results until the final step, holding calculated values in memory. Even with this agreement, small variations due to individual calculator differences may arise. Don't panic if your calculator gives an answer that disagrees slightly with what we have shown.

Finally, keep in mind that a calculator is a tool for doing complicated computations; it does not think and only reacts to your input. Do not become so dependent on your calculator that you reach for it to make simple computations that can be made mentally. You must learn when a calculator should and should not be used and when your results are reasonable and appropriate. Some problems for which a calculator may be especially helpful are marked with a calculator symbol 🖩

CHAPTER 1

Review of Fundamental Concepts

AN APPLIED PROBLEM IN RECREATION

Linda Youngman hiked to the bottom of the Grand Canyon at a rate of 4 mph and returned to the top at a rate of 2 mph. To the nearest tenth, what was her average rate for the hike?

Analysis

At first glance we might assume that the average rate is 3 mph, the average of 4 and 2. However, upon further analysis, it is clear that her average rate must be closer to 2 mph than 4 mph since she hiked for a longer period of time at this slower rate. The basic distance formula, $d = rt$ (distance = rate · time), can be used after first solving for t,

$$t = \frac{d}{r}.$$

Tabulation

Let D = the distance hiked down into the canyon (also the distance hiked out of the canyon),

$\dfrac{D}{4}$ = time it took to hike down into the canyon,

$\dfrac{D}{2}$ = time it took to hike out of the canyon.

Translation

Then the average rate is: $\dfrac{\text{total distance hiked}}{\text{total time of the hike}} = \dfrac{2D}{\dfrac{D}{4} + \dfrac{D}{2}}.$

By simplifying this complex fraction, we can find the average rate.

A knowledge of algebra gives not only a foundation for the study of more advanced mathematics but also the tools for solving a variety of applied problems in business, science, and engineering. The application at the left is solved completely in Example 10 in Section 1.5. The material in this chapter should be a review of topics you have already studied in high school or in intermediate algebra. If your background in algebra is fresh, you might be able to skip over much of this material. The Chapter Test at the end of the section could be used as a pretest of your knowledge. Chapter 1 presents a review of the basic properties of the real number system, and briefly covers algebraic expressions, radicals, and exponents.

1.1 THE REAL NUMBER SYSTEM

STUDENT GUIDEPOSTS

1. Basic Sets of Numbers
2. The Set of Real Numbers
3. Order and the Axioms of Equality
4. Absolute Value
5. Operations on Real Numbers
6. Properties of Real Numbers

1 BASIC SETS OF NUMBERS

Two terms used in mathematics are *set* and *element*. A **set** is a collection of objects, and each object is called an **element** of the set. The elements that belong to a particular set are sometimes listed within braces { } with a capital letter used to identify the set. Primarily we are interested in sets of numbers. The most basic set of numbers is the set of *natural* or *counting numbers*.

> Natural (counting) numbers: $\{1, 2, 3, \ldots\}$

The three dots mean that the sequence or pattern continues on in the same manner. When zero is included with the set of natural numbers, we have the *whole numbers*.

> Whole numbers: $\{0, 1, 2, 3, \ldots\}$

If the negatives of all the natural numbers are combined with the whole numbers, the result is the set of *integers*.

> Integers: $\{\ldots, -3, -2, -1, 0, 1, 2, 3, \ldots\}$

We often refer to the set of **negative integers**, $\{\ldots, -3, -2, -1\}$ and to the set of **positive integers** $\{1, 2, 3, \ldots\}$. The set of *rational numbers* includes the set of integers together with all quotients of integers (remember that division by zero is excluded).

> Rational numbers: $\left\{\frac{a}{b} \text{ such that } a \text{ and } b \text{ are integers, } b \neq 0\right\}$

An alternative way to look at rational numbers is in decimal form. For example, the rational number 3/8 can be written as 0.375 (by dividing 3 by 8), and 3/11 as 0.2727 . . . (dividing 3 by 11). The decimal 0.375 is a **terminating decimal** because the sequence of digits comes to an end, while 0.2727 . . . is a **repeating decimal** because the block of digits 27 repeats indefinitely. Repeating decimals are often written with a bar over the block of digits that repeats. For example,

$$\frac{3}{11} = 0.\overline{27} \quad \text{and} \quad \frac{1}{3} = 0.\overline{3}.$$

Every rational number has a decimal form that either terminates or repeats. This property is sometimes used to define the set of rational numbers. Numbers that are not rational are called *irrational numbers*.

Irrational numbers: {x such that x is not rational}

Thus, we can characterize the irrational numbers as numbers whose decimal form neither terminates nor repeats. One of the best known irrational numbers is π, the ratio of the circumference of any circle to its diameter. Numbers such as $\sqrt{2}$ and $\sqrt{26}$, square roots of positive integers that are not perfect squares, are also irrational. In practical situations, irrational numbers are often approximated by rational numbers. For example, we might use 3.14 as an approximation for π, or 1.414 as an approximation for $\sqrt{2}$. Another example of an irrational number is

$$0.1010010001000010000010000001\ldots$$

Notice that although there is a pattern to the digits, there is no block of digits that will ever repeat.

EXAMPLE 1   RECOGNIZING TYPES OF NUMBERS

Consider the numbers -2, 1/4, 2.5, 0, 15, $\sqrt{5}$, $-10/7$, and 2π.

(a) Which of these numbers are natural numbers? The only natural number in the list is 15.

(b) Which of these numbers are whole numbers? The whole numbers are 0 and 15.

(c) Which of these numbers are integers? The integers are -2, 0, and 15.

(d) Which of these numbers are rational numbers? The rational numbers are -2, 1/4, 2.5, 0, 15, and $-10/7$.

(e) Which of these numbers are irrational numbers? Only $\sqrt{5}$ and 2π are irrational.

PRACTICE EXERCISE 1

Consider the numbers: 0.01, -23, $\sqrt{13}$, -1.09, $\pi/2$, and 500.

(a) Which of these are natural numbers?

(b) Which of these are whole numbers?

(c) Which of these are integers?

(d) Which of these are rational?

(e) Which of these are irrational?

Answers: (a) 500   (b) 500   (c) -23 and 500   (d) 0.01, -23, -1.09, and 500   (e) $\sqrt{13}$ and $\pi/2$

② THE SET OF REAL NUMBERS

When the rational numbers and the irrational numbers are combined, the result is the set of *real numbers*.

Real numbers: {x such that x is rational or x is irrational}

Thus, all of the numbers in Example 1 are real numbers. The relationships among the sets of numbers we have discussed is displayed in Figure 1.1.

4 CHAPTER 1 REVIEW OF FUNDAMENTAL CONCEPTS

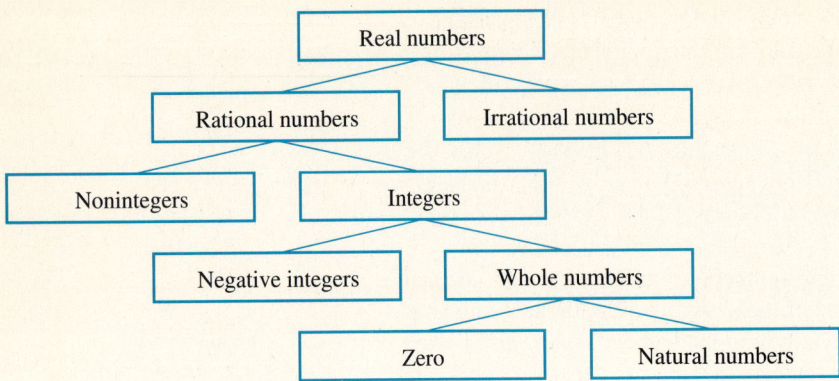

Figure 1.1 The Real Number System

An excellent means of displaying numbers and showing some of their important properties is by use of a **number line,** as shown in Figure 1.2. The **origin** is labeled zero and unit lengths in both directions are marked off. Points to the right of zero are identified with positive numbers, while points to the left of zero correspond to negative numbers.

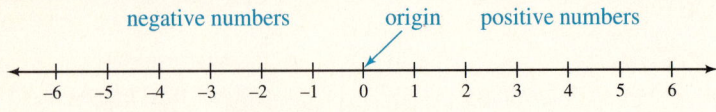

Figure 1.2 Number Line

Every real number can be identified with exactly one point on a number line, and every point on a number line corresponds to exactly one real number. Figure 1.3 shows a number line with points corresponding to several real numbers plotted on it.

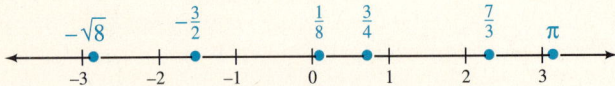

Figure 1.3 Points on a Number Line

3 ORDER AND THE AXIOMS OF EQUALITY

Numbers that are **equal** ($a = b$) correspond to the same point on the number line. If a is to the left of b, we say a is **less than** b and write $a < b$. We also say that b is **greater than** a and write $b > a$. More formally, $a < b$ or $b > a$ if $b - a$ is a positive number ($b - a > 0$).

If a and b are any real numbers the **trichotomy property** states that exactly one of the following holds.

$$a > b, \quad a = b, \quad \text{or} \quad a < b \qquad \text{Trichotomy property}$$

The **transitive property** for inequalities states that

$$\text{if } a < b \text{ and } b < c, \text{ then } a < c. \qquad \text{Transitive property}$$

Both the trichotomy and transitive properties are **axioms,** properties that are accepted without verification.

The axioms of equality that are used in solving equations and simplifying algebraic expressions are summarized next.

Axioms of Equality

Let a, b, and c be real numbers.

Reflexive property $a = a$

Symmetric property   If $a = b$, then $b = a$.

Transitive property   If $a = b$ and $b = c$, then $a = c$.

Substitution property   If $a = b$, then either may replace the other in any statement without affecting the truth of the statement.

EXAMPLE 2   USING ORDER AND EQUALITY AXIOMS

State the property illustrated.

(a) If $x = 5$, then $5 = x$.   Symmetric property

(b) $-2 < -\frac{1}{2}$ and $-\frac{1}{2} < 3$ implies $-2 < 3$.   Transitive property for $<$

(c) If $x = 7$ and $y + x = 6$, then $y + 7 = 6$.   Substitution property

(d) If $y = 6$ and $6 = z$, then $y = z$.   Transitive property for $=$

PRACTICE EXERCISE 2

Complete each statement using the property given.

(a) Reflexive property: $x + 1 =$ _____

(b) Transitive property: If $y < 5$ and $5 < w$ then _____.

(c) Substitution property: If $a = 3$ then $a - 5 =$ _____.

(d) Trichotomy property: If z is a real number, then $z < 0$, $z > 0$, or _____.

Answers: (a) $x + 1$ (b) $y < w$ (c) $3 - 5$ or -2 (d) $z = 0$

4 ABSOLUTE VALUE

The number of units that a number a is from zero on the number line is called the *absolute value* of a and is denoted $|a|$. See Figure 1.4.

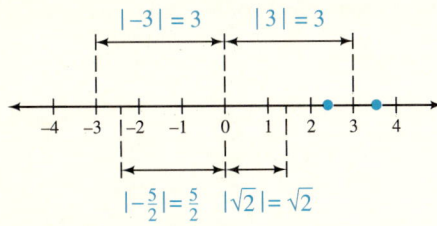

Figure 1.4 Finding Absolute Values

Later in this section we will give a more precise and useable definition of absolute value.

EXAMPLE 3   FINDING ABSOLUTE VALUES

Find the absolute value.

(a) $|-9|$ The distance -9 is from 0 on a number line is 9, so $|-9| = 9$.

(b) $|0|$ Since 0 is 0 units from 0 on a number line, $|0| = 0$.

PRACTICE EXERCISE 3

Find the absolute value.

(a) $|\sqrt{3}|$

(b) $|-1/4|$

Answers: (a) $\sqrt{3}$ (b) $1/4$

5 OPERATIONS ON REAL NUMBERS

The basic operations on the real numbers are *addition* and *multiplication*. We will assume that these operations on nonnegative numbers are well understood, and briefly review what happens when negative numbers are involved. These summaries are justified by theorems that follow later in this section.

Addition of Real Numbers

1. To add two numbers with the same signs, add their absolute values and attach the common sign.
2. To add two numbers with different signs, subtract their absolute values and attach the sign of the number with the larger absolute value. If the absolute values are the same, the sum is 0.

EXAMPLE 4 ADDITION OF REAL NUMBERS

Add.

(a) $-4 + (-7) = -(7 + 4) = -11$

(b) $4.8 + (-2.3) = 4.8 - 2.3 = 2.5$

(c) $-\dfrac{1}{2} + \dfrac{1}{4} = -\left(\dfrac{1}{2} - \dfrac{1}{4}\right) = -\dfrac{1}{4}$

PRACTICE EXERCISE 4

Add.

(a) $-1.2 + (-3.5)$

(b) $13 + (-25)$

(c) $(-2\pi) + \pi$

Answers: (a) -4.7 (b) -12 (c) $-\pi$

Multiplication of Real Numbers

1. To multiply two numbers with the same signs, multiply their absolute values. The product is positive.
2. To multiply two numbers with different signs, multiply their absolute values. The product is negative.
3. If one (or both) of the numbers is zero, the product is zero.

EXAMPLE 5 MULTIPLICATION OF REAL NUMBERS

Multiply.

(a) $(4)(-2) = -(4)(2) = -8$

(b) $\left(-\dfrac{1}{2}\right)\left(-\dfrac{1}{3}\right) = \left(\dfrac{1}{2}\right)\left(\dfrac{1}{3}\right) = \dfrac{1}{6}$

(c) $(-\pi)(0) = 0$

PRACTICE EXERCISE 5

Multiply.

(a) $(-1.3)(2.5)$

(b) $(-8)(-3)$

(c) $(0)(-\sqrt{5})$

Answers: (a) -3.25 (b) 24 (c) 0

6 PROPERTIES OF REAL NUMBERS

The set of real numbers satisfies a number of properties under the operations of addition and multiplication. These properties are reviewed below.

Axioms of the Real Numbers

Let a, b, and c be real numbers.

Closure properties	$a + b$ is a real number
	ab is a real number
Commutative properties	$a + b = b + a$
	$ab = ba$
Associative properties	$a + (b + c) = (a + b) + c$
	$a(bc) = (ab)c$
0 is the additive identity	$a + 0 = a = 0 + a$
1 is the multiplicative identity	$a \cdot 1 = a = 1 \cdot a$
Negatives (additive inverses)	$a + (-a) = 0 = (-a) + a$
Reciprocals (multiplicative inverses)	$a\left(\dfrac{1}{a}\right) = 1 = \left(\dfrac{1}{a}\right)a \quad (a \neq 0)$
Distributive property	$a(b + c) = ab + ac$

EXAMPLE 6 IDENTIFYING PROPERTIES OF REAL NUMBERS

State the property illustrated.

(a) $3(x + y) = 3x + 3y$ Distributive property

(b) $7 + (-7) = 0$ Existence of negatives

(c) $5(xy) = (5x)y$ Associative property of multiplication

PRACTICE EXERCISE 6

Complete each statement using the property given.

(a) Commutative property of multiplication:
$2(a + b) = \underline{\quad}$

(b) Additive identity:
$(2x) + 0 = \underline{\quad}$

(c) Reciprocals:
$6(\underline{\quad}) = 1$

Answers: (a) $(a + b)2$ (b) $2x$ (c) $1/6$

The operations of *subtraction* and *division* are defined in terms of addition and multiplication, respectively.

Subtraction and Division of Real Numbers

If a and b are real numbers,

$$a - b = a + (-b) \quad \text{and} \quad a \div b = a/b = a(1/b) \quad (b \neq 0).$$

Since 0 has no reciprocal, remember that division by 0 is not defined. Also, division of signed numbers has the same properties as multiplication: quotients of numbers with the same signs are positive, and quotients of numbers with different signs are negative.

EXAMPLE 7   SUBTRACTION AND DIVISION OF REAL NUMBERS

Perform the indicated operations.

(a) $(-4) - (-9) = (-4) + 9 = 9 - 4 = 5$

(b) $0 - (-2) = 0 + 2 = 2$

(c) $(-9) \div 3 = (-9)\left(\dfrac{1}{3}\right) = -(9)\left(\dfrac{1}{3}\right)$
$= -\left(\dfrac{9}{3}\right) = -3$

(d) $|(-4) \div (-2)| = \left|(-4)\left(-\dfrac{1}{2}\right)\right|$
$= \left|\dfrac{4}{2}\right| = |2| = 2$

PRACTICE EXERCISE 7

Perform the indicated operations.

(a) $6 - (-5)$

(b) $-\sqrt{2} - 0$

(c) $-\dfrac{1}{2} \div \left(-\dfrac{1}{4}\right)$

(d) $\left|6 \div \left(-\dfrac{1}{5}\right)\right|$

Answers: (a) 11   (b) $-\sqrt{2}$ (c) 2   (d) 30

Properties that can be proved using axioms and definitions are called **theorems**. We present a proof of the first of the following theorems leaving some of the remaining proofs as Exercises C.

Theorems Involving Zero

If a is any real number,

1. $a \cdot 0 = 0 \cdot a = 0$.
2. $a - 0 = a$ and $0 - a = -a$.
3. $0 \div a = \dfrac{0}{a} = 0$ $(a \neq 0)$. $\left(\dfrac{a}{0} \text{ is undefined.}\right)$

Proof of 1

$0 = a + (-a)$	Additive inverse
$= a \cdot 1 + (-a)$	Multiplicative identity
$= a \cdot (0 + 1) + (-a)$	Additive identity
$= (a \cdot 0 + a \cdot 1) + (-a)$	Distributive property
$= (a \cdot 0 + a) + (-a)$	Multiplicative identity
$= a \cdot 0 + (a + (-a))$	Associative property of addition
$= a \cdot 0 + 0$	Additive inverse
$= a \cdot 0$	Additive identity

Therefore, by the transitive property of equality, $0 = a \cdot 0$. By the symmetric property of equality, $a \cdot 0 = 0$. Since $0 \cdot a = a \cdot 0$ by the commutative property of multiplication, by the substitution property of equality, $0 \cdot a = 0$.

The following theorems, easily proved using the axioms of equality, are important for solving equations in Chapter 2.

Theorems of Equality

Let a, b, and c be real numbers.

1. If $a = b$ then $a + c = b + c$.   Addition property
2. If $a = b$ then $ac = bc$.   Multiplication property

1.1 THE REAL NUMBER SYSTEM

The proofs of several of the following properties of negatives follow from the axioms and definitions and are left as Exercises C.

Theorems Involving Negatives

Let a and b be real numbers.

1. $-(-a) = a$
2. $(-a)b = -(ab) = a(-b)$
3. $(-a)(-b) = ab$
4. $(-1)a = -a$
5. $-(a + b) = -a - b$
6. $-(a - b) = -a + b$
7. $\dfrac{-a}{b} = -\dfrac{a}{b} = \dfrac{a}{-b}$ $(b \neq 0)$
8. $\dfrac{-a}{-b} = -\dfrac{-a}{b} = -\dfrac{a}{-b} = \dfrac{a}{b}$ $(b \neq 0)$

Notice that parts of the above theorem justify the rules for operating with signed numbers. For example, part 2 shows why the product of numbers with different signs is negative, and part 3 shows why the product of two negative numbers is positive.

EXAMPLE 8 PROPERTIES OF ZERO AND NEGATIVES

Use the theorems involving zero and negatives.

(a) $3 - (x - y) = 3 - x + y$

(b) $0 - 8y = -8y$

(c) $(4x + 2y) \cdot 0 = 0$

(d) $\dfrac{x - y}{y - x} = \dfrac{x - y}{-(x - y)} = -1$

PRACTICE EXERCISE 8

Use the theorems involving zero and negatives.

(a) $1 - (-a)$

(b) $0 \div (1 + z)$

(c) $(x + y) - (-x - y)$

(d) $\dfrac{-a - (-b)}{b - a}$

Answers: (a) $1 + a$ (b) 0 (c) $2x + 2y$ (d) 1

We can now give a more formal definition of absolute value.

Let a be any real number. The **absolute value** of a is denoted $|a|$ and

$$|a| = \begin{cases} a, & \text{when } a \geq 0; \\ -a, & \text{when } a < 0. \end{cases}$$

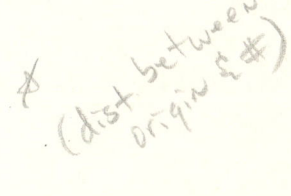
(dist between origin & #)

The symbol $\geq$ is read "greater than or equal to." Similarly we read $\leq$ as "less than or equal to." Notice that for any real number a, $|a| \geq 0$ and that $|-a| = |a|$.

1.1 EXERCISES A

Answer true *or* false *in Exercises 1–12. If the statement is false, explain why.*

1. 3 is a rational number.
2. -5 is a natural number.
3. $\sqrt{5}$ is a rational number.
4. $0.\overline{65}$ is a real number.

CHAPTER 1 REVIEW OF FUNDAMENTAL CONCEPTS

5. Every integer is a rational number.

6. Some irrational numbers are integers.

7. The set of whole numbers is closed with respect to addition.

8. The set of whole numbers is closed with respect to division.

9. For every real number a, $|a| \geq 0$.

10. For all real numbers a and b, $|a - b| = |b - a|$.

11. For all real numbers x and y, $x - y = -(y - x)$.

12. For every real number x, $1/x$ is a real number.

In Exercises 13–16 state the illustrated property.

13. $x + 6 = 6 + x$

14. If $x = 7$, then $7 = x$.

15. $(5x + y) + 0 = 5x + y$

16. 4π is a real number.

In Exercises 17–20 complete each statement using the specified law.

17. Associative property:
$x + (3 + 7) = $ _____

18. Transitive property:
If $a < b$ and $b < 9$ then _____

19. Existence of negatives:
$15 + $ _____ $= 0$

20. Additive identity:
_____ $+ 0 = \pi$

In Exercises 21–24 use the properties of negatives to simplify each expression.

21. $-(-(-3))$

22. $6 - (x - 4)$

23. $-[-(1 - x)]$

24. $-[x - (x + y)]$

Perform the indicated operations in Exercises 25–33.

25. $(-14) - (-3)$

26. $17 + (-5)$

27. $-\frac{1}{9} + \left(-\frac{1}{3}\right)$

28. $\frac{2}{3} - \left(-\frac{3}{4}\right)$

29. $\left(-\frac{2}{3}\right)\left(\frac{3}{4}\right)$

30. $-\frac{2}{3} \div \left(-\frac{3}{4}\right)$

31. $0 \div (-5\pi)$

32. $\frac{7}{8} \div 0$

33. $-2.5 - 0.3$

In Exercises 34–39 rewrite each expression without using absolute value notation.

34. $\left|-\frac{3}{4}\right|$

35. $|x - y|$ if $x < y$

36. $|-7 + 4|$

37. $|-5| - |-8|$

38. $\frac{|-12|}{|3|}$

39. $|(-3)(-2) - (-4)|$

*If a and b are real numbers, we define the **distance between a and b** to be $|a - b|$. Use this definition in Exercises 40–42 to find the distance between the given numbers.*

40. -3 and 2

41. 7 and y if $y > 7$

42. x and y if $x < 0$ and $y > 0$

43. Find real numbers a, b, and c such that $(a - b) - c \neq a - (b - c)$. This shows that subtraction is not associative.

44. Show by example that division is not commutative.

ANSWERS: 1. true 2. false 3. false 4. true 5. true 6. false 7. true 8. false 9. true 10. true 11. true 12. false 13. commutative property of addition 14. symmetric property of equality 15. additive identity 16. closure property of multiplication 17. $(x + 3) + 7$ 18. $a < 9$ 19. -15 20. π 21. -3 22. $10 - x$ 23. $1 - x$ 24. y 25. -11 26. 12 27. $-4/9$ 28. $17/12$ 29. $-1/2$ 30. $8/9$ 31. 0 32. undefined 33. -2.8 34. $3/4$ 35. $y - x$ 36. 3 37. -3 38. 4 39. 10 40. 5 41. $y - 7$ 42. $y - x$ 43. One set of numbers is $a = 1$, $b = 2$, and $c = 3$. 44. One example is $4 \div 2 = 2$, but $2 \div 4 = 1/2$.

1.1 EXERCISES B

Answer true *or* false *in Exercises 1–12. If the statement is false, explain why.*

1. $3/2$ is a rational number.
2. 0 is a natural number.
3. $0.\overline{281}$ is a rational number.
4. $\sqrt{3} - 1$ is a real number.
5. Every rational number is an integer.
6. Some rational numbers are integers.
7. The set of whole numbers is closed with respect to multiplication.
8. The set of whole numbers is closed with respect to subtraction.
9. For every real number a, $|a| > 0$.
10. For all real numbers a and b $|a| + |b| = |a + b|$.
11. For all real numbers x and y, $-x + y = -(x - y)$.
12. For every real number x, $-x$ is a negative number.

In Exercises 13–16 state the illustrated property.

13. $3(xy) = (3x)y$
14. $y < 3$ or $y = 3$ or $y > 3$
15. $(x + 1)\left(\dfrac{1}{x + 1}\right) = 1$
16. $\sqrt{7} = \sqrt{7}$

In Exercises 17–20 complete each statement using the specified law.

17. Distributive property:
$3a + 3x = $ _____
18. Transitive property:
If $x = 4$ and $4 = w$ then _____
19. Multiplicative identity:
$1 \cdot 28 = $ _____
20. Commutative property:
$3 + (x + y) = $ _____

In Exercises 21–24 use the properties of negatives to simplify each expression.

21. $-[-(-(-5))]$
22. $8 - (-a + 7)$
23. $-[-(y - 2)]$
24. $-[-z - (z - y)]$

Perform the indicated operations in Exercises 25–33.

25. $(-2) - (-8)$
26. $4 + (-13)$
27. $-\dfrac{2}{3} + \dfrac{1}{9}$
28. $\dfrac{3}{5} - \left(-\dfrac{2}{5}\right)$
29. $\dfrac{1}{5}\left(-\dfrac{5}{3}\right)$
30. $(-2.7) \div (-0.9)$
31. $0 \div \sqrt{7}$
32. $5\pi \div 0$
33. $\dfrac{4}{7} - \dfrac{5}{7}$

12 CHAPTER 1 REVIEW OF FUNDAMENTAL CONCEPTS

In Exercises 34–39 rewrite each expression without using absolute value notation.

34. $-|-2|$

35. $|x + y|$ if $x < 0$ and $y < 0$

36. $|-7| + |4|$

37. $|-5 - (-8)|$

38. $\left|\dfrac{-12}{3}\right|$

39. $|(-2)(4) - (-5)|$

*If a and b are real numbers, we define the **distance between a and b** to be $|a - b|$. Use this definition in Exercises 40–42 to find the distance between the given numbers.*

40. $-1/2$ and $-1/3$

41. -6 and x if $x < -6$

42. x and y if $x > y$

43. Find real numbers a, b, and c such that $a + (b \cdot c) \neq (a + b) \cdot (a + c)$. This shows that addition does not distribute over multiplication.

44. Show by example that subtraction is not commutative.

1.1 EXERCISES C

Use the properties of real numbers to prove the following where a, b, and c are real numbers.

1. $0 - a = -a$

2. If $a = b$ then $a + c = b + c$.

3. $-(-a) = a$

4. $0 \div a = 0, \quad a \neq 0$

5. If $a = b$ then $ac = bc$.

6. $(-1)a = -a$

7. $-(a + b) = -a - b$

8. $-(a - b) = -a + b$

Converting a rational number in fraction form to a decimal, either terminating or repeating, is a matter of dividing the numerator by the denominator. A terminating decimal can be converted to a fraction by using the definition of place value. For example, 0.5 is really $\frac{5}{10} = \frac{1}{2}$. However, converting a repeating decimal to a fraction is a bit more challenging. For example, to convert $2.353535\ldots = 2.\overline{35}$ consider the following. Let $n = 2.353535\ldots$.

$$100n = 235.353535\ldots$$
$$n = 2.353535\ldots$$
$$99n = 233$$
$$n = \frac{233}{99}$$

Follow this example and convert the repeating decimals in Exercises 9–12 to fractions.

9. $1.\overline{47}$

10. $1.\overline{9}$
[Answer: 2]

11. $23.\overline{456}$
$\left[\text{Answer: } \dfrac{7811}{333}\right]$

12. $8.1\overline{32}$

1.2 INTEGER EXPONENTS AND SCIENTIFIC NOTATION

STUDENT GUIDEPOSTS

1. Exponential Notation
2. Properties of Exponents
3. Scientific Notation
4. Significant Digits

1 EXPONENTIAL NOTATION

To multiply a number or expression by itself several times, we can use **exponential notation,** and avoid long strings of factors. For example,

$$\underbrace{3 \cdot 3 \cdot 3 \cdot 3}_{4 \text{ factors}} = 3^{\underset{\uparrow}{4}}. \leftarrow \text{exponent}$$
$$\text{base}$$

Exponential Notation

If a is any real number and n is a positive integer,

$$a^n = \underbrace{a \cdot a \cdot a \ldots a}_{n \text{ factors}},$$

where a is the **base,** n is the **exponent,** and a^n is the **exponential expression.**

CAUTION

An exponent applies only to the factor next to it unless parentheses are used. For example,

$$2y^3 = 2yyy \quad \text{and} \quad (2y)^3 = (2y)(2y)(2y) = 8y^3.$$

Also, $-x^2 \neq (-x)^2$. Thus,

$$-4^2 = -16 \quad \text{but} \quad (-4)^2 = 16.$$

2 PROPERTIES OF EXPONENTS

When we perform operations involving exponential expressions, our work can be simplified by using the basic properties of exponents. For example,

$$a^3 a^4 = \underbrace{(a \cdot a \cdot a)}_{3 \text{ factors}} \underbrace{(a \cdot a \cdot a \cdot a)}_{4 \text{ factors}} = \underbrace{a \cdot a \cdot a \cdot a \cdot a \cdot a \cdot a}_{7 \text{ factors}} = a^7.$$

Thus, in general, if m and n are positive integers,

$$a^m a^n = \underbrace{a \cdot a \ldots a}_{m \text{ factors}} \cdot \underbrace{a \cdot a \ldots a}_{n \text{ factors}} = \underbrace{a \cdot a \ldots a}_{m + n \text{ factors}} = a^{m+n}.$$

We have just proved the theorem

$$a^m a^n = a^{m+n}. \quad \text{Product rule}$$

Similarly, we can show that if m and n are positive integers and $m - n$ is positive,

$$\frac{a^m}{a^n} = a^{m-n}. \quad \text{Quotient rule}$$

When raising a power to a power, we multiply exponents. For example,

$$(a^2)^3 = \underbrace{(a^2)(a^2)(a^2)}_{3 \text{ factors}} = \underbrace{a \cdot a \cdot a \cdot a \cdot a \cdot a}_{6 \text{ factors}} = a^6.$$

Thus,

$$(a^m)^n = a^{mn}. \quad \text{Power rule}$$

EXAMPLE 1 Using the Rules of Exponents

Simplify.

(a) $2x^6 x^3 = 2x^{6+3} = 2x^9 \qquad a^m a^n = a^{m+n}$

(b) $\dfrac{b^5}{b^2} = b^{5-2} = b^3 \qquad \dfrac{a^m}{a^n} = a^{m-n}$

(c) $(3^2)^3 = 3^{2 \cdot 3} = 3^6 \qquad (a^m)^n = a^{mn}$

PRACTICE EXERCISE 1

Simplify.

(a) $3y^2 y^4 y^5$

(b) $\dfrac{2x^8 y^4}{x^3 y}$

(c) $(z^4)^5$

Answers: (a) $3y^{11}$ (b) $2x^5 y^3$ (c) z^{20}

Rules also exist for raising products or quotients to a power. For example,

$$(4x)^3 = (4x)(4x)(4x) = 4 \cdot 4 \cdot 4 \cdot x \cdot x \cdot x = 4^3 x^3$$

and

$$\left(\dfrac{4}{x}\right)^3 = \left(\dfrac{4}{x}\right)\left(\dfrac{4}{x}\right)\left(\dfrac{4}{x}\right) = \dfrac{4 \cdot 4 \cdot 4}{x \cdot x \cdot x} = \dfrac{4^3}{x^3}.$$

Thus, for n a positive integer,

$$(ab)^n = a^n b^n \quad \text{and} \quad \left(\dfrac{a}{b}\right)^n = \dfrac{a^n}{b^n} \quad (b \neq 0).$$

EXAMPLE 2 Using the Rules of Exponents

Simplify.

(a) $(3x^4 y^3)^4 = 3^4 \cdot (x^4)^4 (y^3)^4 = 3^4 x^{16} y^{12}$

(b) $\left(\dfrac{2w^2}{z^4}\right)^3 = \dfrac{(2w^2)^3}{(z^4)^3} \qquad \left(\dfrac{a}{b}\right)^n = \dfrac{a^n}{b^n}$

$\qquad = \dfrac{2^3 (w^2)^3}{(z^4)^3} \qquad (ab)^n = a^n b^n$

$\qquad = \dfrac{8w^6}{z^{12}} \qquad (a^m)^n = a^{mn}$

PRACTICE EXERCISE 2

Simplify.

(a) $(-2a^2 b^5)^3$

(b) $\left(\dfrac{-3z^2}{u^3}\right)^4$

Answers: (a) $-8a^6 b^{15}$ (b) $\dfrac{81z^8}{u^{12}}$

In order to extend the quotient rule to the case where $m = n$, consider

$$\dfrac{a^m}{a^m} = a^{m-m} = a^0.$$

Since $a^m/a^m = 1$, this suggests the following definition. If a is any real number except zero,

$$a^0 = 1. \quad \text{Zero exponent}$$

To come to a definition for negative exponents, consider extending the quotient rule to $n > m$. For example, let $m = 2$ and $n = 5$.

$$\dfrac{a^m}{a^n} = \dfrac{a^2}{a^5} = a^{2-5} = a^{-3}$$

Also $\qquad \dfrac{a^m}{a^n} = \dfrac{a^2}{a^5} = \dfrac{\cancel{a} \cdot \cancel{a}}{\cancel{a} \cdot \cancel{a} \cdot a \cdot a \cdot a} = \dfrac{1}{a \cdot a \cdot a} = \dfrac{1}{a^3}.$

This suggests that $a^{-3} = 1/a^3$. Thus, if $a \neq 0$ and n is a positive integer, then

$$a^{-n} = \frac{1}{a^n}. \quad \text{Negative exponents}$$

EXAMPLE 3 Using the Rules of Exponents

Simplify and write without using negative exponents.

(a) $8^0 = 1$
(b) $(2x^2y^2)^0 = 1 \quad (x \neq 0 \text{ and } y \neq 0)$
(c) $(-2)^{-3} = \dfrac{1}{(-2)^3} = -\dfrac{1}{8}$
(d) $\dfrac{1}{3^{-2}} = \dfrac{1}{\frac{1}{3^2}} = 3^2 = 9$

Practice Exercise 3

Simplify and write without using negative exponents.

(a) $(-5)^0$
(b) $(-3)^{-2}$
(c) $(2x)^{-3}$
(d) $\dfrac{1}{x^{-5}}$

Answers: (a) 1 (b) $\frac{1}{9}$ (c) $\frac{1}{8x^3}$ (d) x^5

CAUTION

Do not write 5^{-2} as -5^2 or $(-2)(5)$. These are all different numbers.

$$5^{-2} = \frac{1}{5^2} = \frac{1}{25}, \quad -5^2 = -25, \quad (-2)(5) = -10$$

We can extend the rules developed for positive integer exponents to negative integers. These rules are now summarized for all integer exponents.

Rules of Exponents

Let a and b be any real numbers, m and n any integers.

1. $a^m a^n = a^{m+n}$
2. $\dfrac{a^m}{a^n} = a^{m-n} \quad (a \neq 0)$
3. $(a^m)^n = a^{mn}$
4. $(ab)^n = a^n b^n$
5. $\left(\dfrac{a}{b}\right)^n = \dfrac{a^n}{b^n} \quad (b \neq 0)$
6. $a^0 = 1 \quad (a \neq 0)$
7. $a^{-n} = \dfrac{1}{a^n} \quad (a \neq 0)$
8. $\dfrac{1}{a^{-n}} = a^n \quad (a \neq 0)$

EXAMPLE 4 Using the Rules of Exponents

Simplify and write without negative exponents.

(a) $(3x)^{-1} = \dfrac{1}{(3x)^1} = \dfrac{1}{3x} \quad a^{-n} = \dfrac{1}{a^n}$

(b) $3x^{-1} = 3\left(\dfrac{1}{x}\right) = \dfrac{3}{x} \quad$ Compare with (a)

Practice Exercise 4

Simplify and write without negative exponents.

(a) $(-2w)^{-2}$
(b) $-2w^{-2}$

16 CHAPTER 1 REVIEW OF FUNDAMENTAL CONCEPTS

(c) $\left(\dfrac{x^2 y^3}{x y^4}\right)^{-2} = (x^{2-1} y^{3-4})^{-2}$ $\quad \dfrac{a^m}{a^n} = a^{m-n}$

$\qquad = (xy^{-1})^{-2}$

$\qquad = x^{-2} y^2 \quad\quad (ab)^n = a^n b^n$ and $(a^m)^n = a^{mn}$

$\qquad = \dfrac{y^2}{x^2} \quad\quad a^{-n} = \dfrac{1}{a^n}$

(d) $\dfrac{2^0}{a^{-2} + b^{-2}} = \dfrac{1}{\dfrac{1}{a^2} + \dfrac{1}{b^2}}$ $\quad a^0 = 1$ and $a^{-n} = \dfrac{1}{a^n}$

$\qquad = \dfrac{1}{\dfrac{b^2}{a^2 b^2} + \dfrac{a^2}{a^2 b^2}} \quad$ Find a common denominator

$\qquad = \dfrac{1}{\dfrac{b^2 + a^2}{a^2 b^2}} = \dfrac{a^2 b^2}{b^2 + a^2}$

(e) $\left[\left(\dfrac{3 x^6 y^{-6}}{x^{-5} y^{-3}}\right)^3\right]^{-1} = \left(\dfrac{3 x^6 y^{-6}}{x^{-5} y^{-3}}\right)^{-3}$ $\quad (a^m)^n = a^{mn}$

$\qquad = (3 x^{11} y^{-3})^{-3} \quad \dfrac{a^m}{a^n} = a^{m-n}$

$\qquad = 3^{-3} x^{-33} y^9 \quad (ab)^n = a^n b^n$

$\qquad = \dfrac{y^9}{27 x^{33}} \quad\quad a^{-n} = \dfrac{1}{a^n}$

(c) $\left(\dfrac{a^5}{3 x^{-3}}\right)^{-2}$

(d) $\dfrac{1}{a^{-1} + b^{-1}}$

(e) $\left[\left(\dfrac{5 u^{-3} v^3}{u^{-1} v^{-2}}\right)^{-2}\right]^{-1}$

Answers: (a) $\dfrac{1}{4w^2}$ (b) $-\dfrac{2}{w^y}$ (c) $\dfrac{9}{a^{10} x^6}$ (d) $\dfrac{ab}{a+b}$ (e) $\dfrac{25 v^{10}}{u^4}$

③ SCIENTIFIC NOTATION

An important application of integer exponents involves *scientific notation*. Numbers that are very large or very small can be written in a more compact way using this notation. For example, a plant cell might contain 208,000,000,000,000 molecules. This can be written as 2.08×10^{14}. Also, one estimate for the weight of an oxygen molecule is 0.000000000000000000000053, which can be written as 5.3×10^{-23}.

> **Scientific Notation**
>
> A number is written in **scientific notation** when it is expressed in the form
>
> $a \times 10^n$ where $1 \leq a < 10$ and n is an integer.

Numbers that are greater than 10 will have positive integer exponents in scientific notation, while numbers less than 1 will show a negative power of 10. The exponent can be determined by counting decimal places as shown in the following example.

EXAMPLE 5 Using Scientific Notation

Write each number in scientific notation.

(a) $93,\!000,\!000 = 9.3 \times 10^7$

    7 places

(b) $0.000000093 = 9.3 \times 10^{-8}$

    8 places

(c) $10 = 1 \times 10^1 = 1 \times 10$

    1 place

(d) $0.1 = 1 \times 10^{-1}$

    1 place

Practice Exercise 5

Write in scientific notation.

(a) 142,000,000

(b) 0.0000000142

(c) 52

(d) 0.003

Answers: (a) 1.42×10^8
(b) 1.42×10^{-8} (c) 5.2×10
(d) 3×10^{-3}

Calculations in scientific notation can be made using the rules of exponents. For example,

$$\frac{(3.2 \times 10^{-8})(8.4 \times 10^4)}{2.1 \times 10^{16}} = \frac{(3.2)(8.4)}{(2.1)} \times \frac{10^{-8} 10^4}{10^{16}}$$

$$= 12.8 \times 10^{-8+4-16}$$

$$= 12.8 \times 10^{-20}$$

$$= 1.28 \times 10^1 \times 10^{-20}$$

$$= 1.28 \times 10^{-19}$$

Very large or very small numbers must be written in scientific notation in order to be entered into a calculator. The $\boxed{\text{EE}}$, $\boxed{\text{EXP}}$, or $\boxed{\text{EEX}}$ key allows us to enter numbers in scientific notation. The calculation above would be carried out as follows:

ALG: 3.2 $\boxed{\text{EE}}$ 8 $\boxed{+/-}$ $\boxed{\times}$ 8.4 $\boxed{\text{EE}}$ 4 $\boxed{\div}$ 2.1 $\boxed{\text{EE}}$ 16 $\boxed{=}$ $\boxed{1.28 \qquad -19}$ Calculator display

RPN: 3.2 $\boxed{\text{EEX}}$ 8 $\boxed{\text{CHS}}$ $\boxed{\text{ENTER}}$ 8.4 $\boxed{\text{EEX}}$ 4 $\boxed{\times}$ 2.1 $\boxed{\text{EEX}}$ 16 $\boxed{\div}$ $\boxed{1.28 \qquad -19}$ Calculator display

The answer in scientific notation is 1.28×10^{-19}.

EXAMPLE 6 Calculations in Scientific Notation

Use scientific notation and a calculator to perform the indicated operations.

(a) $\dfrac{(0.000\ 036)(69,\!000,\!000)}{0.000\ 000\ 012} = \dfrac{(3.6 \times 10^{-5})(6.9 \times 10^7)}{1.2 \times 10^{-8}}$

ALG: 3.6 $\boxed{\text{EE}}$ 5 $\boxed{+/-}$ $\boxed{\times}$ 6.9 $\boxed{\text{EE}}$ 7 $\boxed{\div}$

1.2 $\boxed{\text{EE}}$ 8 $\boxed{+/-}$ $\boxed{=}$ $\boxed{2.07 \qquad 11}$

RPN: 3.6 $\boxed{\text{EEX}}$ 5 $\boxed{\text{CHS}}$ $\boxed{\text{ENTER}}$ 6.9 $\boxed{\text{EEX}}$ 7 $\boxed{\times}$

1.2 $\boxed{\text{EEX}}$ 8 $\boxed{\text{CHS}}$ $\boxed{\div}$ $\boxed{2.07 \qquad 11}$

Thus, the result in scientific notation is 2.07×10^{11}.

Practice Exercise 6

Use scientific notation and a calculator to perform the indicated operations.

(a) $\dfrac{(0.000\ 000\ 004\ 2)^3}{(24,\!000,\!000)(0.000\ 000\ 35)}$

(b) $\dfrac{(3.25 \times 10^9)^2}{(7.86 \times 10^{26})(5.82 \times 10^{-10})}$

ALG: 3.25 [EE] 9 [x²] [÷] 7.86 [EE] 26 [÷]
5.82 [EE] 10 [+/−] [=] | 2.309   01 | Calculator display

RPN: 3.25 [EEX] 9 [ENTER] [x²] 7.86 [EEX] 26 [÷]
5.82 [EEX] 10 [CHS] [÷] | 2.309   01 | Calculator display

Thus, rounded to two decimal places, the answer is 2.31×10^1.

(b) $\dfrac{(1.24 \times 10^{10})(3.45 \times 10^{-9})}{(7.18 \times 10^{-7})^2}$

Answers: (a) 8.82×10^{-27}
(b) 8.30×10^{13}

4 SIGNIFICANT DIGITS

Notice that we rounded the answer in Example 6(b) to the same number of decimal places as the numbers in the original problem. Rounding is necessary when approximate measurement numbers are used in a calculation, since it would be inappropriate to give an answer with a higher degree of accuracy than the values used to compute it. The idea of *significant digits* is often used to describe approximate values. Suppose that a number x has been written in scientific notation as $a \times 10^n$ where a has been rounded to k decimal places. In this case, x is said to be accurate to $k + 1$ **significant digits.** For example, given that $x = 12.3456$, we have the following:

Approximation of x	12.346	12.35	12.3	12	10
Number of significant digits	5	4	3	2	1

1.2 EXERCISES A

Write in exponential notation in Exercises 1–3.

1. $4 \cdot 4yyyyy$

2. $(7a)(7a)(7a)$

3. $(x + y)(x + y)(x + y)(x + y)$

Write without exponents in Exercises 4–6.

4. $a^2b^3c^4$

5. $x^2 + y^2$

6. $5b^5$

In Exercises 7–24, simplify and write without negative exponents.

7. $a^2a^{-3}a^4$

8. $3x^3x^{-2}$

9. $\dfrac{x^2x^{-5}}{x^3}$

10. $(b^{-3})^{-2}$

11. $\dfrac{a^3}{b^2}$

12. 5^0

13. 0^0

14. $(3y)^{-1}$

15. $3y^{-1}$

16. $\dfrac{2x^2}{x^{-2}}$

17. $\left(\dfrac{2x^2}{y^{-3}}\right)^{-2}$

18. $\left(\dfrac{2x^2}{y^{-3}}\right)^2$

19. $\dfrac{1}{x^{-1}+y^{-1}}$ 20. $\dfrac{1}{(x+y)^{-2}}$ 21. $\dfrac{3x^3y^{-5}}{12x^{-4}y^{-1}}$

22. $\left(\dfrac{6^0 x^6 y}{3x^{-2}y^{-1}}\right)^3$ 23. $\left(\dfrac{2x^{-2}y^{-3}}{x^4 y^{-1}}\right)^{-2}$ 24. $\left[\left(\dfrac{2a^{-2}b^{-3}}{a^{-4}b^2}\right)^{-2}\right]^{-1}$

In Exercises 25–28 write each number in scientific notation.

25. 456,000,000   26. 0.000 000 000 003 21   27. 0.01   28. 100

In Exercises 29–32 write each number in decimal notation.

29. 2.35×10^8 30. 8.62×10^2 31. 4.17×10^{-8} 32. 3.2×10^{-1}

Use your calculator in Exercises 33–36 to perform the indicated operations. Round to three significant digits and give answers in scientific notation.

33. $(0.000\ 273)(428,000)$ 34. $(0.000\ 000\ 018\ 2)(0.007\ 26)$

35. $\dfrac{(86,000)^2(0.000\ 000\ 392)}{(0.001\ 57)^2}$ 36. $\dfrac{(0.000\ 005\ 62)^2(369,000)^2}{183,000,000,000}$

Solve using scientific notation in Exercises 37–40.

37. **ECONOMICS** The national debt of a small country is $6,250,000,000 and the population is 2,120,000. What is the amount of debt per person?

38. **BUSINESS** A company produced 925,000 hand mixers in one year and made a profit of $8,620,000. What was the profit on each mixer?

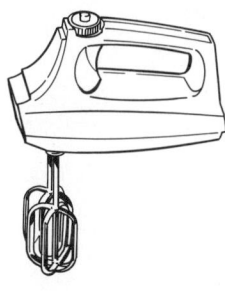

39. **ASTRONOMY** The earth is approximately 92,900,000 miles from the sun. If 1 mi = 1.61×10^3 m, what is the distance to the sun in meters?

40. **PHYSICS** If the speed of light is 3.00×10^8 m/sec, use the answer to Exercise 39 to estimate how long it takes light to get from the sun to the earth.

20   CHAPTER 1   REVIEW OF FUNDAMENTAL CONCEPTS

FOR REVIEW

Answer true *or* false *in Exercises 41–44. If the statement is false, explain why.*

41. Every natural number is a whole number.

42. The set of integers is closed with respect to division.

43. For all real numbers x, $|x| = |-x|$.

44. If $x > y$, then $|x - y| = x - y$.

ANSWERS: 1. 4^2y^5 2. $(7a)^3$ 3. $(x+y)^4$ 4. $aabbcccc$ 5. $xx + yy$ 6. $5bbbbb$ 7. a^3 8. $3x$ 9. $\frac{1}{x^6}$ 10. b^6
11. $\frac{a^3}{b^2}$ 12. 1 13. undefined 14. $\frac{1}{3y}$ 15. $\frac{3}{y}$ 16. $2x^4$ 17. $\frac{1}{4x^4y^6}$ 18. $4x^4y^6$ 19. $\frac{xy}{x+y}$ 20. $(x+y)^2$ 21. $\frac{x^7}{4y^4}$
22. $\frac{x^{24}y^6}{27}$ 23. $\frac{x^{12}y^4}{4}$ 24. $\frac{4a^4}{b^{10}}$ 25. 4.56×10^8 26. 3.21×10^{-12} 27. 1×10^{-2} 28. 1×10^2 29. 235,000,000 30. 862
31. 0.000 000 041 7 32. 0.32 33. 1.17×10^2 34. 1.32×10^{-10} 35. 1.18×10^9 36. 2.35×10^{-11}
37. $\$2.95 \times 10^3 = \$2950$ 38. $\$9.32 \times 10^0 = \$9.32$ 39. 1.50×10^{11} m 40. 5.00×10^2 sec = 8.33 min 41. true
42. false 43. true 44. true

1.2 EXERCISES B

Write in exponential notation in Exercises 1–3.

1. $3 \cdot 3 \cdot 3xxxx$

2. $(-3y)(-3y)(-3y)(-3y)$

3. $(a-b)(a-b)(a-b)$

Write without exponents in Exercises 4–6.

4. $x^3y^2z^4$

5. $a^3 - b^3$

6. $3z^4$

In Exercises 7–24 simplify and write without negative exponents.

7. x^4x^3

8. $6x^3x^{-2}$

9. $\frac{y^3y^{-7}}{y^2}$

10. $(b^2)^{-3}$

11. a^4b^3

12. $(5^0)^3$

13. $\frac{3}{2^0}$

14. $(8x)^{-1}$

15. $8x^{-1}$

16. $\frac{9y^{-3}}{3y^3}$

17. $\left(\frac{6x^2}{y^{-4}}\right)^{-2}$

18. $\left(\frac{6x^2}{y^{-4}}\right)^2$

19. $\frac{1}{x^{-2} + y^{-2}}$

20. $\frac{1}{(x+y)^{-1}}$

21. $\frac{8a^{-5}b^8}{2a^7b^{-3}}$

22. $\left(\frac{8^0a^{-5}b^8}{2a^7b^{-3}}\right)^4$

23. $\left(\frac{3x^4y^{-2}}{x^{-4}y^3}\right)^{-3}$

24. $\left[\left(\frac{x^{-3}y^4}{2x^{-5}y^{-1}}\right)^{-2}\right]^3$

In Exercises 25–28 write each number in scientific notation.

25. 29,300,000

26. 0.000 000 000 539

27. 0.0001

28. 500

In Exercises 29–32 write each number in decimal notation.

29. 3.87×10^9

30. 6.42×10^3

31. 5.91×10^{-11}

32. 8.7×10^{-2}

Use your calculator in Exercises 33–36 to perform the indicated operations. Round to three significant digits and give answers in scientific notation.

33. (0.000 000 042 1)(6,500,000)

34. (0.000 000 392)(0.000 421)

35. $\dfrac{(24,000)^2(0.000\ 065\ 4)}{(0.000\ 000\ 033\ 9)^2}$

36. $\dfrac{(0.000\ 005\ 56)^2(0.000\ 691)^2}{423,000,000}$

Solve using scientific notation in Exercises 37–40.

37. **ECONOMICS** The annual operating budget for a state is $4,350,000,000 and the population of the state is 1,850,000. What amount is budgeted per resident?

38. **BUSINESS** A manufacturing firm produced 1,320,000 items during a recent year and made a profit of $14,300,000. What was the profit on each item?

39. **ASTRONOMY** The earth is approximately 93,000,000 miles from the sun. If 1 mi = 5280 ft, what is the distance to the sun in feet?

40. **PHYSICS** If the speed of light is 9.84×10^8 ft/sec, use the answer to Exercise 39 to estimate how long it takes light to get from the sun to the earth.

FOR REVIEW

Answer true or false in Exercises 41–44. If the statement is false, explain why.

41. Every integer is a rational number.

42. The set of whole numbers is closed with respect to subtraction.

43. For all real numbers w, $|-w| \geq 0$.

44. For all real numbers x and y, $|y - x| = |x - y|$.

1.2 EXERCISES C

Find the error that was made in simplifying each expression in Exercises 1–3. What is the correct answer?

1. $a^{-2}(a^3)^{-1} = a^{-2+3-1}$
   $= a^0$
   $= 1$

2. $\dfrac{u^3 v^{-3}}{u^{-2} v^2} = u^{3-2} v^{-3+2}$
   $= u^1 v^{-1}$
   $= \dfrac{u}{v}$

3. $(2x^3 y^{-1})^3 = 2(x^3)^3 (y^{-1})^3$
   $= 2x^9 y^{-3}$
   $= \dfrac{2x^9}{y^3}$

   [Answer: The factor of 2 must also be raised to the third power. The correct answer is $8x^9/y^3$.]

1.3 ALGEBRAIC EXPRESSIONS AND POLYNOMIALS

STUDENT GUIDEPOSTS

1. Algebraic Expressions
2. Order of Operations
3. Polynomials
4. Addition and Subtraction of Polynomials
5. Multiplication of Polynomials
6. Division of Polynomials

1 ALGEBRAIC EXPRESSIONS

Before beginning our work with polynomials, we review several definitions. Remember that a **variable** is a symbol or letter used to represent an arbitrary element of a set containing more than one element. When a symbol is used to represent a specific number, it is called a **constant**. An **algebraic expression** involves sums, differences, products, and quotients of numbers and variables raised to various powers. A **term** of an algebraic expression includes only the product of a number or **numerical coefficient**, and variables raised to various powers. For example,

$$3x^2y - 2xy^2, \quad \sqrt{xy} + 2x + 5y, \quad \text{and} \quad \frac{x+y}{x-y} - \frac{2xy}{3x-y}$$

are three algebraic expressions, and $3x^2y$, $2x$, and $-\dfrac{2xy}{3x-y}$ are terms.

2 ORDER OF OPERATIONS

We are often interested in *evaluating an algebraic expression* when specific numerical values for the variable(s) are given. For example, suppose we evaluate $3 + 2x$ when $x = 5$. Substitute 5 for x.

$$3 + 2x = 3 + 2 \cdot 5 \quad \text{Replace } x \text{ with } 5$$

There can be some confusion as to the result unless we agree to a particular order for performing the operations. We could obtain 25 if we added first then multiplied, or 13 if we multiplied first then added. Of course, you probably realize that the second procedure is the correct one to use since you are familiar with the *order of operations* summarized below.

Order of Operations

1. Evaluate within grouping symbols first (parentheses or brackets), beginning with the innermost set if more than one set is used.
2. Evaluate all powers and roots.
3. Perform all multiplications and divisions in order from left to right.
4. Perform all additions and subtractions in order from left to right.

EXAMPLE 1 Evaluating Algebraic Expressions

Evaluate when $a = -2$ and $b = -1$.

(a) $2a^2 + 1 = 2(-2)^2 + 1$ Substitute -2 for a
$\qquad\quad\; = 2(4) + 1$ Square first
$\qquad\quad\; = 8 + 1$ Multiply next
$\qquad\quad\; = 9$ Add last

(b) $2 - [a - (b - a^2)]$
$\quad = 2 - [(-2) - ((-1) - (-2)^2)]$
$\quad = 2 - [(-2) - (-1 - 4)]$ Work inside parentheses first, squaring
$\quad = 2 - [(-2) - (-5)]$ Subtract inside parentheses
$\quad = 2 - [3]$ Subtract inside brackets
$\quad = -1$ Subtract last

PRACTICE EXERCISE 1

Evaluate when $x = 4$ and $y = -3$.

(a) $-2y^3 + 3x$

(b) $5[(x + y)^3 - 2]$

1.3 ALGEBRAIC EXPRESSIONS AND POLYNOMIALS

(c) $b + \sqrt{-4a - b}$
$= (-1) + \sqrt{-4(-2) - (-1)}$
$= -1 + \sqrt{8 - (-1)}$ The radical acts like a grouping symbol; multiply first
$= -1 + \sqrt{9}$ Subtract next
$= -1 + 3$ Root next
$= 2$ Add last

(c) $\dfrac{x^2 - 2y^2 + 2}{2(1 - y)}$

[*Hint:* A fraction bar acts like a grouping symbol too.]

Answers: (a) 66   (b) -5 (c) 0

CAUTION

Mistakes are often made evaluating expressions such as $-2a^2$ and $-a^2$. Remember to take powers first then multiply the result by -2 and -1, respectively. For example, when $a = -3$, $-2a^2 = -2(-3)^2 = -2(9) = -18$ *NOT* 18, and $-a^2 = -(-3)^2 = -(9) = -9$ *NOT* 9. It helps to use parentheses around the variables at the substitution step as shown in Example 1.

Many applied problems involve evaluating an algebraic expression or formula.

EXAMPLE 2   A Physics Problem

The height h (in feet) of a rocket t seconds after firing is given by the equation

$$h = -16t^2 + 180t.$$

What is the height of the rocket 4 seconds into its flight?
Substitute 4 for t in the formula to find the value of h.

$h = -16t^2 + 180t = -16(4)^2 + 180(4)$
$= -16(16) + 180(4)$ Power first
$= -256 + 720$ Multiply next
$= 464$ Add last

The height of the rocket after 4 seconds is 464 ft. This would seem to be a reasonable answer to this type of problem. Had we come up with a negative number or a much larger positive number, we might suspect that we had made an error.

Practice Exercise 2

Use the formula in Example 2 and find the height of the rocket 11.25 seconds into the flight.

Answer: 0 ft (After 11.25 sec, the rocket has returned to the ground. If you substitute a value for t greater than 11.25, what kind of value do you get for h? Does such a value make sense in this problem?)

Make sure when you solve an applied problem that your answer is reasonable. Before finding the actual value, make an estimate and try to determine what kind of answers will be appropriate.

❸ POLYNOMIALS

We now review some of the properties of some special algebraic expressions, *polynomials*.

Polynomials

A **polynomial** is an algebraic expression with terms that are products of numbers and variables raised to whole-number exponents.

Of the three algebraic expressions given at the beginning of this section, only $3x^2y - 2xy^2$ is a polynomial. In the following table we give examples of polynomials, their terms, and their coefficients.

Polynomial	Terms	Coefficients
$3x^4 - 2x^2 + x - 8$	$3x^4, -2x^2, x, -8$	$3, -2, 1, -8$
$4x^2y - 3xy^2 - xy$	$4x^2y, -3xy^2, -xy$	$4, -3, -1$
$\sqrt{2}x^4y^4 - \frac{3}{4}$	$\sqrt{2}x^4y^4, -\frac{3}{4}$	$\sqrt{2}, -\frac{3}{4}$
6	6	6

Expressions such as $\sqrt{x} + 5$ and $\frac{3}{xy} + y^2$ are not polynomials since variables are not raised to whole-number powers. Also, the expression 3^x is not a polynomial since the exponent on 3 is a variable.

A **monomial** is a polynomial with one term, while a **binomial** has two terms and a **trinomial** has three terms. A polynomial with more than three terms has no special name; it is simply a polynomial. Polynomials are further classified by degree. First we define the **degree of a term** to be the sum of the powers on the variables in the term. The **degree of the polynomial** is the same as that of the term with highest degree. A polynomial which is a constant, such as 5, has degree zero since it can be written $5x^0$. However, the zero polynomial, 0, has no degree. We illustrate the notion of degree in the following table.

Polynomial	Type	Degree
$8x^4$	Monomial	4
$3x^4 - 6x^7 + 5x$	Trinomial	7
16	Monomial	0
$3x^3y^2 - 7xy^6$	Binomial	7
$8a^2b^2c^2 - 6a^5b - 7b^4c^4$	Trinomial	8

Before we begin the study of operations on polynomials, we must understand the expression *like terms*. Two terms are **like terms** if they contain the same variables raised to the same powers. The following pairs of terms are like terms.

$$7x^3y \text{ and } -5x^3y$$
$$-6a^2b^2c \text{ and } 15a^2b^2c$$
$$5 \text{ and } -8$$
$$5a^7x^6y^3 \text{ and } 10a^7x^6y^3$$

When a polynomial contains like terms, we can collect or **combine like terms** using the distributive law. For example, consider the polynomial $2x^2 + 3 - x + x^2 - 5x$. It has two pairs of like terms: $2x^2, x^2$ and $-x, -5x$. We can simplify the polynomial by collecting these like terms.

$2x^2 + 3 - x + x^2 - 5x = 2x^2 + x^2 - x - 5x + 3$ Use commutative law to place the like terms together

$\qquad = (2 + 1)x^2 + (-1 - 5)x + 3$ Distributive law

$\qquad = 3x^2 - 6x + 3$

4 ADDITION AND SUBTRACTION OF POLYNOMIALS

To add or subtract polynomials, we simply indicate the operation and collect or combine like terms.

EXAMPLE 3 Adding Polynomials

Add.

(a) $3x^3 + 2x - 8$ and $x^3 - x^2 - 7x + 3$

$(3x^3 + 2x - 8) + (x^3 - x^2 - 7x + 3)$ Indicate addition
$= 3x^3 + 2x - 8 + x^3 - x^2 - 7x + 3$ Remove parentheses
$= 3x^3 + x^3 - x^2 + 2x - 7x - 8 + 3$ Commute
$= (3 + 1)x^3 - x^2 + (2 - 7)x + (-8 + 3)$ Distributive law
$= 4x^3 - x^2 - 5x - 5$ Like terms are collected; this is the desired sum

(b) $x^4y - 5x^3y^2 - 3xy^3 + 7y$ and $-5x^4y + 8x^2y + 4xy^3 + 3$

$(x^4y - 5x^3y^2 - 3xy^3 + 7y) + (-5x^4y + 8x^2y + 4xy^3 + 3)$ Indicate addition
$= x^4y - 5x^3y^2 - 3xy^3 + 7y - 5x^4y + 8x^2y + 4xy^3 + 3$ Remove parentheses
$= (1 - 5)x^4y - 5x^3y^2 + 8x^2y + (-3 + 4)xy^3 + 7y + 3$ Collect like terms
$= -4x^4y - 5x^3y^2 + 8x^2y + xy^3 + 7y + 3$

PRACTICE EXERCISE 3

Add.

(a) $y^4 - 5y^3 + 3y - 9$ and $7y^4 + y^3 - 4y^2 - y + 12$

(b) $a^2b^2 + 3a^3b - 2ab^3 - ab$ and $4a^2b^2 - a^3b + 4ab^3 + ab$

Answers: (a) $8y^4 - 4y^3 - 4y^2 + 2y + 3$ (b) $5a^2b^2 + 2a^3b + 2ab^3$

When subtracting polynomials, we must be careful that all signs within parentheses that follow a minus sign are changed when the parentheses are removed.

EXAMPLE 4 Subtracting Polynomials

(a) Subtract $-8y^3 + 6y^2 - 2y - 3$ from $-4y^3 - 6y^2 + 5y - 1$.

$(-4y^3 - 6y^2 + 5y - 1) - (-8y^3 + 6y^2 - 2y - 3)$ Be sure to subtract in the correct order

$= -4y^3 - 6y^2 + 5y - 1 + 8y^3 - 6y^2 + 2y + 3$ Remove parentheses; watch the signs

$= (-4 + 8)y^3 + (-6 - 6)y^2 + (5 + 2)y + (-1 + 3)$ Combine like terms
$= 4y^3 - 12y^2 + 7y + 2$

(b) Perform the indicated operations.

$(5x^3y^3 - 3x^2y) + (-7x^3y^3 + 4xy) - (9x^2y + 8xy)$
$= 5x^3y^3 - 3x^2y - 7x^3y^3 + 4xy - 9x^2y - 8xy$
$= (5 - 7)x^3y^3 + (-3 - 9)x^2y + (4 - 8)xy$
$= -2x^3y^3 - 12x^2y - 4xy$

PRACTICE EXERCISE 4

(a) Subtract $-3a^2b^3 + 9ab^2 - 5$ from $4a^2b^3 + 7ab^2 - 3b^2 + 8$.

(b) Perform the indicated operations.

$(3u^2v^2 - 2uv^2) - (uv + 2u^2v^2) - (4uv^2 - uv - 7u^2v^2)$

Answers: (a) $7a^2b^3 - 2ab^2 - 3b^2 + 13$ (b) $8u^2v^2 - 6uv^2$

5 MULTIPLICATION OF POLYNOMIALS

Multiplication of polynomials was introduced in Section 1.2 when we multiplied monomials using the product rule for exponents. To multiply a monomial by a polynomial with two or more terms, use the distributive property. To multiply

26 CHAPTER 1 REVIEW OF FUNDAMENTAL CONCEPTS

two polynomials, neither of which is a monomial, multiply each term of one by each term of the other and collect like terms. When multiplying two binomials we may use the FOIL method (F stands for First terms, O for Outside, I for Inside, and L for Last).

EXAMPLE 5 Multiplying Polynomials

Multiply.

(a) $-4a^3b^2(2a^2b^3 - 3a) = (-4a^3b^2)(2a^2b^3) - (-4a^3b^2)(3a)$ Distributive property

$= (-4)(2)(a^3a^2)(b^2b^3) - (-4)(3)(a^3a)b^2$

$= -8a^5b^5 + 12a^4b^2$

(b) $(5a + 2b)(3a - 7b) = (5a)(3a) + (5a)(-7b) + (2b)(3a) + (2b)(-7b)$

$= 15a^2 - 35ab + 6ab - 14b^2$

$= 15a^2 - 29ab - 14b^2$

Practice Exercise 5

Multiply.

(a) $-2x^2y(3xy^3 - 5xy)$

(b) $(2x - 3y)(x + 5y)$

Answers: (a) $-6x^3y^4 + 10x^3y^2$
(b) $2x^2 + 7xy - 15y^2$

When polynomials with three or more terms are multiplied, we often find it convenient to write one above the other and arrange like terms of the product in vertical columns.

EXAMPLE 6 Using the Column Method

Multiply.

$3x^2 - 5xy + 2y^2$
$7x - 8y$
$\overline{21x^3 - 35x^2y + 14xy^2}$ $7x$ multiplies each term of the top polynomial
$\quad\quad - 24x^2y + 40xy^2 - 16y^3$
$\overline{21x^3 - 59x^2y + 54xy^2 - 16y^3}$ $-8y$ multiplies each term of the top polynomial

Practice Exercise 6

Multiply.

$2a^3 - 3a^2 + a - 5$
$a - 1$

Answer: $2a^4 - 5a^3 + 4a^2 - 6a + 5$

Several special products, useful when multiplying, are important for factoring.

Product Formulas

$(a + b)(a - b) = a^2 - b^2$
$(a + b)(a + b) = (a + b)^2 = a^2 + 2ab + b^2$
$(a - b)(a - b) = (a - b)^2 = a^2 - 2ab + b^2$

EXAMPLE 7 Special Product Formulas

Multiply using the special product formulas.

(a) $(x + 5)(x - 5) = x^2 - 5^2$ Using $(a + b)(a - b) = a^2 - b^2$

$= x^2 - 25$

Practice Exercise 7

Multiply using the special product formulas.

(a) $(a - 3b)(a + 3b)$

(b) $(x + 5)(x + 5) = (x + 5)^2$
$= x^2 + 2(x)(5) + 5^2$ Using $(a + b)^2 = a^2 + 2ab + b^2$, not $a^2 + b^2$
$= x^2 + 10x + 25$

(c) $(3a^2 - 2b)(3a^2 + 2b) = (3a^2)^2 - (2b)^2$
$= 3^2(a^2)^2 - 2^2 b^2$ $(ab)^n = a^n b^n$
$= 9a^4 - 4b^2$ $(a^m)^n = a^{mn}$

(d) $(3a^2 - 2b)(3a^2 - 2b) = (3a^2 - 2b)^2$
$= (3a^2)^2 - 2(3a^2)(2b) + (2b)^2$ Using $(a - b)^2 = a^2 - 2ab + b^2$
$= 9a^4 - 12a^2 b + 4b^2$

(b) $(u + 2v)(u + 2v)$

(c) $(2x^3 + y)(2x^3 - y)$

(d) $(2x^3 - y)(2x^3 - y)$

Answers: (a) $a^2 - 9b^2$
(b) $u^2 + 4uv + 4v^2$ (c) $4x^6 - y^2$
(d) $4x^6 - 4x^3 y + y^2$

6 DIVISION OF POLYNOMIALS

To divide a polynomial by a monomial, divide each term of the polynomial by the monomial as illustrated in the following example. With practice, you will be able to skip the middle step and proceed directly to the final answer.

EXAMPLE 8   DIVIDING BY A MONOMIAL

Divide.

(a) $\dfrac{5x^3 + 10x^2 - 20x}{5x} = \dfrac{5x^3}{5x} + \dfrac{10x^2}{5x} - \dfrac{20x}{5x} = x^2 + 2x - 4$

(b) $\dfrac{6a^4 b^3 + 9a^2 b^5 - 18ab^7 + 6a^2}{6a^2 b^3} = \dfrac{6a^4 b^3}{6a^2 b^3} + \dfrac{9a^2 b^5}{6a^2 b^3} - \dfrac{18ab^7}{6a^2 b^3} + \dfrac{6a^2}{6a^2 b^3}$
$= a^2 + \dfrac{3b^2}{2} - \dfrac{3b^4}{a} + \dfrac{1}{b^3}$

PRACTICE EXERCISE 8

Divide.

(a) $\dfrac{-7y^5 - 14y^3 + 21y}{-7y^2}$

(b) $\dfrac{3a^3 b^4 + 6a^2 b^3 - 9ab^5 + 3ab}{3a^2 b^2}$

Answers: (a) $y^3 + 2y - \dfrac{3}{y}$
(b) $ab^2 + 2b - \dfrac{3b^3}{a} + \dfrac{1}{ab}$

To divide a polynomial by a binomial, we follow the procedure used in long division of numbers. The next example reviews this technique.

EXAMPLE 9   DIVIDING BY A BINOMIAL

(a) $x - 3 \overline{\smash{\big)}\, x^2 + 4x - 25}$
      quotient: $x + 7$
      $\underline{x^2 - 3x}$ $x - 3$ times x, the first term of the quotient
      $7x - 25$ Subtract and bring down -25; note that $4x - (-3x) = 7x$
      $\underline{7x - 21}$ $x - 3$ times 7, the new term of the quotient
      -4 Subtract $7x - 21$ from $7x - 25$ to obtain the remainder, -4

The answer is $x + 7$ remainder -4 or $x + 7 - \dfrac{4}{x - 3}$.

PRACTICE EXERCISE 9

Divide.

(a) $y - 3 \overline{\smash{\big)}\, y^2 + 2y - 15}$

(b)
$$y^2 + 3 \overline{)\begin{array}{l} y^2 - 4y + 3 \\ y^4 - 4y^3 + 6y^2 + 8 \\ \underline{y^4 + 3y^2} \\ -4y^3 + 3y^2 \\ \underline{-4y^3 - 12y} \\ 3y^2 + 12y + 8 \\ \underline{3y^2 + 9} \\ 12y - 1 \end{array}}$$

The answer is $y^2 - 4y + 3 + \dfrac{12y - 1}{y^2 + 3}$.

(b) Divide $a^3 + 8$ by $a + 2$.

Answers: (a) $y + 5$
(b) $a^2 - 2a + 4$ (Remainders are 0.)

CAUTION

When dividing a polynomial by a binomial, be sure to arrange the terms in descending order and leave space for any missing terms. Also, remember to subtract at each step before bringing down the next term.

1.3 EXERCISES A

In Exercises 1–12 evaluate each expression when $a = -2$, $b = 3$, and $c = -1$.

1. $2a^2$
2. $(2a)^2$
3. $-2a^2$
4. $(-2a)^2$
5. $(a - c)^{-3}$
6. $(c - a)^3$
7. $-a^2 + b$
8. $3[a - (b + c)]$
9. $\dfrac{a^2 - b^2}{c}$
10. $\sqrt{6a^2 + 1} - b$
11. $\sqrt{(a + b)^2}$
12. $b + 3[c - (4 - a)]$

In Exercises 13–15 simplify each expression and collect like terms.

13. $2y - 3 + 5y + 7$
14. $2x - [x - 2(x - 1)]$
15. $-3[4w - (1 + w)]$

In Exercises 16–18 state whether each polynomial is a monomial, binomial, or trinomial and give the degree of each.

16. $x^3 - 6x$
17. $a^3b^3 - 6a^2bc^4 + 8ab$
18. -2

Add the polynomials in Exercises 19–20.

19. $3x^2 + 2x - 5$ and $-2x^2 + 5$
20. $6a^2b^2 - 3ab + 2$ and $4a^2b^2 + 8ab - 5$

Subtract the polynomials in Exercises 21–22.

21. $-2x^2 + 3x - 4$ from $6x^2 - 4x + 1$
22. $-7a^2b^2 + 6ab - 3$ from $2a^2b^2 - 6ab - 3$

Perform the indicated operations in Exercises 23–26.

23. $(a^2 + 3a - 4) + (a^3 - a^2 - a + 4)$

24. $(y^5 + 4y^2 - 2) - (2y^2 - y^5 + 6)$

25. $(5a^4b^2 - 7a^2b^4 + 3) - (6a^4b^2 + 6a^2b^4 - 7)$

26. $(4a^2b^2 - 2ab + 3) + (2a^2b^2 + ab - 7) - (-a^2b^2 - ab - 6)$

Find each product in Exercises 27–42. Assume n is a positive integer.

27. $(-6x^2y^3)(-3x^4y)$

28. $4ab^3(-3a^2b^2 + 5ab)$

29. $(a - b)(a + 2b)$

30. $(5xy + 3)(2xy - 7)$

31. $(x - 2y)(x + 2y)$

32. $(x + 2y)^2$

33. $(x - 2y)^2$

34. $(3a^2 - b^2)(3a^2 + b^2)$

35. $(6x - 7y + 2)(2x - 3y)$

36. $(3a - 7b)(4a^2 - 2ab + 7b^2)$

37. $(x + 2y - 5)^2$

38. $[(x - 2y) + 5][(x - 2y) - 5]$

39. $x^{2n}y^n(x^n - y^n)$

40. $(x^n - y^n)(x^n + y^n)$

41. $(x^{n+1} + y^{n+1})^2$

42. $(x^2 - 3xy + 2y^2)(2x^2 + 4xy - 3y^2)$

Find each quotient in Exercises 43–48.

43. $\dfrac{5y^5 - 30y^4 + 25y^3 - 5}{5y^2}$

44. $(15x^4y^3 - 5x^3y^4 + 10x^5y^5) \div (-5x^2y^3)$

45. $a - 1 \overline{\smash{)}a^3 \quad\quad + a - 3}$

46. $2y - 1 \overline{\smash{)}8y^3 \quad\quad\quad -1}$

47. $(2x^2 + 5xy - 3y^2) \div (2x - y)$

48. $\dfrac{x^5 - x^4 - 2x^3 + 4x^2 - 15x + 5}{x^2 - 5}$

49. Find a value for m so that the polynomial
$$x^3 - x^2 + x + m$$
is divisible by $x + 1$ with remainder zero.

50. Find a value for m so that when the polynomial
$$2x^3 + x^2 - 4x + m$$
is divided by $2x + 1$, the remainder is 8.

51. METEOROLOGY The temperature measured in degrees Celsius C can be obtained from degrees Fahrenheit F by using

$$C = \frac{5}{9}(F - 32).$$

Find C when F is 5°.

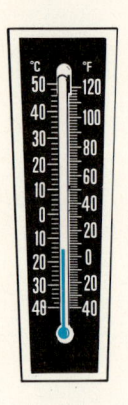

52. CONSUMER A French perfume sells for $52.50 per cubic centimeter (cc). To the nearest dollar, what will a bottle of this perfume cost if it is in the shape of a cylinder with radius 2.1 cm and height 5.6 cm? Use $V = \pi r^2 h$ and 3.14 for π.

53. ECONOMICS In a business the manufacturing cost in dollars is given by $M = 4u^2 - u + 5$, and the wholesale cost in dollars is given by $W = 2u^2 + u + 6$, where u is the number of units produced.
(a) Find the polynomial that represents the total cost C of the operation, if $C = M + W$.
(b) Find the total cost when 15 units are produced and sold.

FOR REVIEW

In Exercises 54–56, simplify and write without negative exponents.

54. $\left(\dfrac{3x^3}{y^{-2}}\right)^{-2}$

55. $\dfrac{4x^5 y^{-2}}{10x^{-2} y^{-4}}$

56. $\left(\dfrac{2^0 x^4 y^{-9}}{3(xy)^{-6}}\right)^{-2}$

ANSWERS: 1. 8 2. 16 3. -8 4. 16 5. -1 6. 1 7. -1 8. -12 9. 5 10. 2 11. 1 12. -18 13. $7y + 4$ 14. $3x - 2$ 15. $-9w + 3$ 16. binomial; 3 17. trinomial; 7 18. monomial; 0 19. $x^2 + 2x$ 20. $10a^2b^2 + 5ab - 3$ 21. $8x^2 - 7x + 5$ 22. $9a^2b^2 - 12ab$ 23. $a^3 + 2a$ 24. $2y^5 + 2y^2 - 8$ 25. $-a^4b^2 - 13a^2b^4 + 10$ 26. $7a^2b^2 + 2$ 27. $18x^6y^4$ 28. $-12a^3b^5 + 20a^2b^4$ 29. $a^2 + ab - 2b^2$ 30. $10x^2y^2 - 29xy - 21$ 31. $x^2 - 4y^2$ 32. $x^2 + 4xy + 4y^2$ 33. $x^2 - 4xy + 4y^2$ 34. $9a^4 - b^4$ 35. $12x^2 - 32xy + 21y^2 + 4x - 6y$ 36. $12a^3 - 34a^2b + 35ab^2 - 49b^3$ 37. $x^2 + 4xy + 4y^2 - 10x - 20y + 25$ 38. $x^2 - 4xy + 4y^2 - 25$ 39. $x^{3n}y^n - x^{2n}y^{2n}$ 40. $x^{2n} - y^{2n}$ 41. $x^{2n+2} + 2x^{n+1}y^{n+1} + y^{2n+2}$ 42. $2x^4 - 2x^3y - 11x^2y^2 + 17xy^3 - 6y^4$ 43. $y^3 - 6y^2 + 5y - \frac{1}{y^2}$ 44. $-3x^2 + xy - 2x^3y^2$ 45. $a^2 + a + 2 - \frac{1}{a-1}$ 46. $4y^2 + 2y + 1$ 47. $x + 3y$ 48. $x^3 - x^2 + 3x - 1$ 49. $m = 3$ 50. $m = 6$ 51. $-15°C$ 52. $4071 53. (a) $C = 6u^2 + 11$ (b) $1361 54. $\frac{1}{9x^6y^4}$ 55. $\frac{2x^7y^2}{5}$ 56. $\frac{9y^6}{x^{20}}$

1.3 EXERCISES B

In Exercises 1–12, evaluate each expression when $x = -3$, $y = 2$, and $z = -2$.

1. $3z^2$

2. $(3z)^2$

3. $-3z^2$

4. $(-3z)^2$

5. $(x - z)^{-2}$

6. $(z - x)^2$

7. $-x^2 + y$

8. $4[x - (y + z)]$

9. $\dfrac{y^2 - z^2}{x}$

10. $\sqrt{1 - 5x} - z$

11. $\sqrt{(x + z)^2}$

12. $x - 2[y - (1 - z)]$

In Exercises 13–15, simplify each expression and collect like terms.

13. $5x - 8 - 2x + 7$

14. $2a - [a - (3a - 4)]$

15. $-4[2u - (u - 3)]$

In Exercises 16–18, indicate whether each polynomial is a monomial, binomial, or trinomial and give the degree of each.

16. $a^4 - 3a^3$

17. $-3a^2b^3$

18. $x^3y^3 - 2xy^4z^2 - 3xy$

Add the polynomials in Exercises 19–20.

19. $-8x^3 + 2x^2 - 1$ and $5x^3 + x - 5$

20. $8a^3b^2 - 3ab + 9$ and $-4a^3b^2 + 4ab - 3$

Subtract the polynomials in Exercises 21–22.

21. $8x^3 + 4x - 5$ from $-9x^3 + x^2 + 9$

22. $4a^4b - 6ab^4 + 2$ from $2a^4b + ab^4 - 7$

Perform the indicated operations in Exercises 23–26.

23. $(x^3 + 4x^4 - 2) + (x^4 - x^3 - x + 1)$

24. $(3z^2 + 2z - 8) - (2z^3 - z^2 - z - 2)$

25. $(5x^2y^2z^2 - 6xyz) - (-2x^2y^2z^2 + xyz + 1)$

26. $(-4a^3b^2 - 6a^2b^3 + a^2b^2) - (-8a^2b^3 + 3a^2b^2 - 5) - (7a^3b^2 + a^2b^3 + 8)$

Find each product in Exercises 27–42. Assume that n is a positive integer.

27. $(5a^2b)(-7a^3b^3)$

28. $-6x^2y(2xy - 8x^3)$

29. $(a - b)(a - 2b)$

30. $(2ab - c)(3ab + 2c)$

31. $(7a + 2b)(7a - 2b)$

32. $(7a + 2b)^2$

33. $(7a - 2b)^2$

34. $(3a^2 - b^2)(3a^2 - b^2)$

35. $(-5x + 8y - 4)(4x + 2y)$

36. $(4a + 5b)(8a^2 + ab - 3b^2)$

37. $(x - 2y + 5)^2$

38. $[x - (2y - 5)][x + (2y - 5)]$

39. $x^{n+2}y^2(x^n + y^n)$

40. $(a^{2n} + b^{2n})(a^{2n} - b^{2n})$

41. $(x^ny^n - z^n)^2$

42. $(2a^2 - 5ab + 2b^2)(3a^2 - 6ab - b^2)$

Find each quotient in Exercises 43–48.

43. $\dfrac{-22x^4 + 11x^3 - 33x^2}{-11x}$

44. $(18a^5b^3 - 24a^2b^5 + 6ab) \div (-6a^3b^3)$

45. $a - 2 \overline{)-a^3 + a^2 \quad\quad + 1}$

46. $3a + 1 \overline{)27a^3 \quad\quad\quad +1}$

47. $(6a^3 + 5a^2b + 4ab^2 + b^3) \div (3a + b)$

48. $\dfrac{y^5 - y^4 + y^2 + 3y + 2}{y^2 + y + 1}$

49. Find a value for m so that the polynomial $y^3 + y^2 + 2y + m$ is divisible by $y - 1$ with remainder zero.

50. Find a value for m so that when the polynomial $3x^3 + x^2 + 4x + m$ is divided by $3x - 2$, the remainder is -3.

51. BANKING Use the formula $A = P(1 + r)^t$ to find the amount of money A in an account at the end of 2 years t if a principal P of $2000 is invested at 9% interest r, compounded annually.

52. GEOMETRY The perimeter of a rectangle of length l and width w is given by $P = 2l + 2w$. What is the perimeter of a rectangle of length 15 cm and width 10 cm?

53. ECONOMICS In a business the manufacturing cost in dollars is given by $C = 6u^2 + 11$ and the total revenue in dollars from the operation by $R = 7u^2 - u + 3$, where u is the number of units produced. **(a)** Find the polynomial that represents the profit P, if $P = R - C$. **(b)** Find the profit when 8 units are produced and sold.

FOR REVIEW

In Exercises 54–56, simplify and write without negative exponents.

54. $\left(\dfrac{2a^{-4}}{b^6}\right)^2$

55. $\dfrac{9a^4 b^{-6}}{18a^{-2} b^8}$

56. $\dfrac{9(a^2 b)^{-3}}{3^0 (a^{-4} b^{-2})^{-1}}$

1.3 EXERCISES C

Use a calculator to perform the operations in Exercises 1–2.

1. $(3.51a + 2.036b)(4.05a - 1.96b)$

2. $y + 2.1 \overline{)y^3 - 4.2y^2 + 15.85y + 61.068}$
[Answer: $y^2 - 6.3y + 29.08$]

Find the error that was made in the work in Exercises 3–5. What is the correct answer?

3. $(x + 2y)^2 = x^2 + 4y^2$
[Answer: The middle term has been omitted. The correct answer is $x^2 + 4xy + 4y^2$.]

4. $(a - 3b)(a - 3b) = a^2 - 9b^2$

5. $x - [2x + 3y] = x - 2x + 3y = -x + 3y$

6. $(2a + b)(a - 2b) = 2a^2 - 2b^2$

7. GEOMETRY The early Greeks gave geometric proofs of many of the formulas we use in algebra today. For example, to show that $(a + b)(a - b) = a^2 - b^2$, for the case when $a > b$, consider the figures below. Show that one expression for the unshaded area is $a^2 - b^2$ and that a second expression for the same area is $(a + b)(a - b)$.

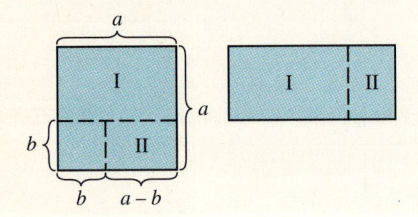

1.4 FACTORING POLYNOMIALS

STUDENT GUIDEPOSTS

1. Removing Common Factors
2. Factoring by Grouping
3. Factoring Trinomials
4. Special Factoring Formulas

① REMOVING COMMON FACTORS

Factoring a polynomial involves writing the polynomial as a product; as such, it is the reverse of multiplying. Suppose we multiply $4x^2 y(2x - 3y)$ to obtain the polynomial $8x^3y - 12x^2y^2$. If, however, we had started with the polynomial $8x^3y - 12x^2y^2$, we could use the distributive law in reverse and *factor* the polynomial obtaining $4x^2y(2x - 3y)$. This illustrates the technique of factoring by **removing the greatest common factor**. Notice that the polynomial $8x^3y - 12x^2y^2$ could be factored in several other ways such as $4(2x^3y - 3x^2y^2)$ or $2xy(4x^2 - 6xy)$. However, in these cases we have not factored out the *greatest* common factor and the factorization is not *complete*. When we consider any factoring problem, we will always **factor completely**.

EXAMPLE 1 Removing the Greatest Common Factor

Factor.

(a) $3x^2 - 9x = (3x)x - (3x)3$
$ = 3x(x - 3)$

(b) $6a^3b^2 + 2ab^2 - 10a^2b^4$
$ = (2ab^2)(3a^2) + (2ab^2)(1) + (2ab^2)(-5ab^2)$
$ = 2ab^2(3a^2 + 1 - 5ab^2)$

Practice Exercise 1

Factor.

(a) $5y^3 - 15y$

(b) $12x^2y^4 + 4x^4y^3 - 8xy^3$

Answers: (a) $5y(y^2 - 3)$
(b) $4xy^3(3xy + x^3 - 2)$

2 FACTORING BY GROUPING

We can also use the distributive property to remove common binomial factors. For example,

$3a(2a + b) - 2b(2a + b) = (3a - 2b)(2a + b)$.   *$2a + b$ is a common factor*

In fact this is the last step in a process called **factoring by grouping.** We group terms, factor each group, and then, if possible, remove any common binomial factors from the result. For example, had we been given

$$6a^2 + 3ab - 4ab - 2b^2$$

to factor, by factoring $3a$ from the first two terms and $-2b$ from the second two terms, we would have gotten the expression above, which was then factored by removing the common factor $2a + b$.

EXAMPLE 2 Factoring by Grouping

Factor.

(a) $3x^2y - 6xy^2 + 4x - 8y = 3xy(x - 2y) + 4(x - 2y)$
$ = (3xy + 4)(x - 2y)$ *$x - 2y$ is a common factor*

(b) $7ab^3 - 14a^2b^2 - b + 2a = 7ab^2(b - 2a) - (b - 2a)$ *Watch signs*
$ = (7ab^2 - 1)(b - 2a)$ *$b - 2a$ is a common factor*

Practice Exercise 2

Factor.

(a) $3x^2 + 6xy + 5x + 10y$

(b) $6ab^2 - 3ab - 14b + 7$

Answers: (a) $(3x + 5)(x + 2y)$
(b) $(3ab - 7)(2b - 1)$

3 FACTORING TRINOMIALS

We now turn our attention to factoring a trinomial of the form $ax^2 + bxy + cy^2$. First recall the patterns involved in multiplying binomials using FOIL.

34 CHAPTER 1 REVIEW OF FUNDAMENTAL CONCEPTS

$$(5x - 3y)(2x + 7y) = (5x)\cdot(2x) + (5x)\cdot(7y) + (-3y)\cdot(2x) + (-3y)\cdot(7y)$$
$$= (5\cdot 2)x^2 + (5\cdot 7)xy + (-3)(2)xy + (-3)\cdot(7)y^2$$
$$= 10x^2 + 35xy - 6xy - 21y^2$$
$$= 10x^2 + 29xy - 21y^2$$

(with circled letters F, O, I, L labeling the FOIL pattern: F = First, O = Outer, I = Inner, L = Last, and F + O + I + L)

The first two terms of the binomials multiply Ⓕ to give $10x^2$, the last two terms multiply Ⓛ to give $-21y^2$, and $29xy$ comes from the products Ⓞ and Ⓘ, $35xy - 6xy$. Thus, to factor $10x^2 + 29xy - 21y^2$ we need to factor $10x^2$, factor $-21y^2$, and tell which products forming the middle terms will add to give $29xy$. To factor $ax^2 + bxy + cy^2$ we need to fill the blanks in

Factors of c
$$ax^2 + bxy + cy^2 = (\_x + \_y)(\_x + \_y).$$
Factors of a

We must choose the factors so that the Ⓞ and Ⓘ terms add to give bxy. In the case where $a = 1$, the procedure is greatly simplified because then we are looking for the factors of c that add to give b.

EXAMPLE 3   FACTORING $x^2 + bxy + cy^2$ ($a = 1$)

Factor.

(a) $x^2 + 5xy + 6y^2 = (\_x + \_y)(\_x + \_y)$

In this case, since $a = 1$, both coefficients of x must be 1.
$$x^2 + 5xy + 6y^2 = (x + \_y)(x + \_y)$$

Also, since all the signs in the trinomial are $+$, only positive factors of 6 ($c = 6$) need to be tried. Since the factors of 6 are 1,6 and 2,3, and the pair 2,3 add to give 5 ($b = 5$), the factors are:
$$x^2 + 5xy + 6y^2 = (x + 2y)(x + 3y).$$

(b) $x^2 - 5xy + 6y^2 = (\_x + \_y)(\_x + \_y)$
$$= (x + \_y)(x + \_y)$$

Since c is positive ($c = 6$) and b is negative ($b = -5$), the factors of c must both be negative to add to the negative number, -5. The factors are:
$$x^2 - 5xy + 6y^2 = (x - 2y)(x - 3y).$$

(c) $x^2 + xy - 6y^2 = (x + \_y)(x + \_y)$

Since c is negative ($c = -6$), the factors of c must be opposite in sign. The factors are:
$$x^2 + xy - 6y^2 = (x - 2y)(x + 3y).$$

(d) $x^2 - xy - 6y^2 = (x + \_y)(x + \_y)$

This time again the factors of c must have opposite signs.
$$x^2 - xy - 6y^2 = (x + 2y)(x - 3y)$$

PRACTICE EXERCISE 3

Factor.

(a) $u^2 + 7u + 12$

(b) $u^2 - 7u + 12$

(c) $u^2 + u - 12$

(d) $u^2 - u - 12$

Answers: (a) $(u + 4)(u + 3)$
(b) $(u - 4)(u - 3)$
(c) $(u + 4)(u - 3)$
(d) $(u - 4)(u + 3)$

The four parts of Example 3 show the different possibilities for the signs of b and c when factoring $x^2 + bxy + cy^2$. Recognizing these cases can help to cut down on the number of trials used to find the correct factorization. Next we consider the situation where a is not 1.

EXAMPLE 4   FACTORING $ax^2 + bxy + cy^2$ ($a \neq 1$)

Factor.

(a) $2x^2 + 11xy + 12y^2$ ($a = 2$, $b = 11$, $c = 12$)

$$2x^2 + 11xy + 12y^2 = (\_x + \_y)(\_x + \_y)$$

with Factors of 2 on the x terms and Factors of 12 on the y terms.

Since all terms of the trinomial are positive we use only positive factors. Factors of c must be tried in both orders with the factors of a.

Factors of a	Factors of c
1, 2	1, 12 or 2, 6 or 3, 4

$2x^2 + 11xy + 12y^2$
$= (x + \_y)(2x + \_y)$ The only factors of a are 1 and 2
$\stackrel{?}{=} (x + 12y)(2x + y)$ Does not work since $xy + 24xy = 25xy$
$\stackrel{?}{=} (x + y)(2x + 12y)$ Does not work since $12xy + 2xy = 14xy$
$\stackrel{?}{=} (x + 3y)(2x + 4y)$ Does not work since $4xy + 6xy = 10xy$
$\stackrel{?}{=} (x + 4y)(2x + 3y)$ This works since $3xy + 8xy = 11xy$

Thus, $2x^2 + 11xy + 12y^2 = (x + 4y)(2x + 3y)$.

(b) $-12u^2 - 14uv + 40v^2$

First remove the common factor and make the leading coefficient $a > 0$.

$$-12u^2 - 14uv + 40v^2 = -2(6u^2 + 7uv - 20v^2)$$

Now factor $6u^2 + 7uv - 20v^2$ with $a = 6$, $b = 7$, and $c = -20$. Since $c = -20$, we need to consider both positive and negative factors of c.

$6u^2 + 7uv - 20v^2 \stackrel{?}{=} (u + 20v)(6u - v)$ Does not work since
$-uv + 120uv = 119uv$

$\stackrel{?}{=} (2u - 5v)(3u + 4v)$ Does not work since
$8uv - 15uv = -7uv$

$\stackrel{?}{=} (2u + 5v)(3u - 4v)$ This works since
$-8uv + 15uv = 7uv$

Thus, $-12u^2 - 14uv + 40v^2 = -2(2u + 5v)(3u - 4v)$.

PRACTICE EXERCISE 4

Factor.

(a) $6x^2 - 19x + 10$

(b) $-24u^2 + 6u + 9$

Answers: (a) $(3x - 2)(2x - 5)$
(b) $-3(2u + 1)(4u - 3)$

CAUTION

You should never give a wrong answer to a factoring problem since you can always check if the factors you give are correct by multiplying them to see if their product is the original polynomial. Also, always remove any common factors first, as shown in Example 4(b), and remember to give the common factor as part of your final answer.

The method we have illustrated above is somewhat of a trial-and-error method. An alternative method preferred by some students involves factoring by grouping. Consider the following.

$$2x^2 + 11xy + 12y^2 = 2x^2 + 8xy + 3xy + 12y^2 \quad \text{11xy = 8xy + 3xy}$$
$$= (2x^2 + 8xy) + (3xy + 12y^2) \quad \text{Group terms}$$
$$= 2x(x + 4y) + 3y(x + 4y) \quad \text{Factor groups}$$
$$= (2x + 3y)(x + 4y) \quad \text{Factor out the common factor } x + 4y$$

We have factored $2x^2 + 11xy + 12y^2$ by grouping the appropriate terms. But how did we decide to write $11xy = 8xy + 3xy$? Notice that 8 and 3 are factors of $ac = (2)(12) = 24$.

To Factor $ax^2 + bxy + cy^2$ by Grouping

1. Find the product ac.
2. List the factors of ac until a pair is found which add to give b.
3. Write bxy as a sum using these factors as coefficients of xy.
4. Factor the result by grouping.

EXAMPLE 5 FACTORING TRINOMIALS BY GROUPING

Factor by grouping.

(a) $3x^2 - 8xy + 5y^2$

Factors of $ac = (3)(5) = 15$	Sum of factors
15, 1	$15 + 1 = 16$
$-15, -1$	$-15 + (-1) = -16$
5, 3	$5 + 3 = 8$
$-5, -3$	$-5 + (-3) = -8$

$$3x^2 - 8xy + 5y^2 = 3x^2 - 5xy - 3xy + 5y^2$$
$$= x(3x - 5y) - y(3x - 5y) \quad \text{Factor out } x \text{ and } -y$$
$$= (x - y)(3x - 5y) \quad \text{Common factor is } 3x - 5y$$

Check this by multiplying.

(b) $7u^2 - 33uv - 10v^2$

We try those factors of $ac = 7(-10) = -70$ which seem most likely to add to $b = -33$.

Factors of $ac = -70$	Sum of factors
35, -2	$35 + (-2) = 33$
$-35, 2$	$-35 + 2 = -33$

$$7u^2 - 33uv - 10v^2 = 7u^2 - 35uv + 2uv - 10v^2$$
$$= 7u(u - 5v) + 2v(u - 5v) \quad \text{Factor groups}$$
$$= (7u + 2v)(u - 5v) \quad \text{Common factor is } u - 5v$$

Check this by multiplying.

PRACTICE EXERCISE 5

Factor by grouping.

(a) $12x^2 + 17x + 6$

(b) $-12y^2 + 22y - 6$

Answers: (a) $(3x + 2)(4x + 3)$
(b) $-2(2y - 3)(3y - 1)$

④ SPECIAL FACTORING FORMULAS

The three special product formulas presented in Section 1.3 can be used in reverse to factor certain polynomials. In addition, there are two other formulas for factoring a sum or a difference of two cubes. These forms, listed below in a summary, can be verified by multiplication, and should be memorized.

Factoring Formulas

1. $a^2 - b^2 = (a + b)(a - b)$ **Difference of squares**
2. $a^2 + 2ab + b^2 = (a + b)^2$ **Perfect square**
3. $a^2 - 2ab + b^2 = (a - b)^2$ **Perfect square**
4. $a^3 + b^3 = (a + b)(a^2 - ab + b^2)$ **Sum of cubes**
5. $a^3 - b^3 = (a - b)(a^2 + ab + b^2)$ **Difference of cubes**

Of course, a trinomial that is a perfect square can be factored using the general method for trinomials. However, finding the factors is often quicker if we recognize the special form of two terms that are perfect squares with the middle term twice the product of their square roots. The next example illustrates, in order, the formulas given above.

EXAMPLE 6   FACTORING USING SPECIAL FORMULAS

Factor.

(a) $u^2 - 25v^2$
$= u^2 - (5v)^2$
$= (u + 5v)(u - 5v)$

(b) $u^2 + 14uv + 49v^2$
$= u^2 + (2)(u)(7v) + (7v)^2$
$= (u + 7v)^2$

(c) $49x^4 - 70x^2y + 25y^2$
$= (7x^2)^2 - (2)(7x^2)(5y) + (5y)^2$
$= (7x^2 - 5y)^2$

(d) $u^3 + 125v^3$
$= u^3 + (5v)^3$
$= (u + 5v)(u^2 - 5uv + 25v^2)$

(e) $16x^3 - 54y^6$
$= 2[8x^3 - 27y^6]$
$= 2[(2x)^3 - (3y^2)^3]$
$= 2[(2x - 3y^2)(4x^2 + 6xy^2 + 9y^4)]$
$= 2(2x - 3y^2)(4x^2 + 6xy^2 + 9y^4)$

PRACTICE EXERCISE 6

Factor.

(a) $18x^6 - 8y^4$
[*Hint:* Don't forget to remove the common factor first.]

(b) $9x^2 + 30xy + 25y^2$

(c) $x^2 - 12x + 36$

(d) $2x^3 + 16y^3$

(e) $2u^3 - 250v^3$

Answers: (a) $2(3x^3 + 2y^2)(3x^3 - 2y^2)$
(b) $(3x + 5y)^2$ (c) $(x - 6)^2$
(d) $2(x + 2y)(x^2 - 2xy + 4y^2)$
(e) $2(u - 5v)(u^2 + 5uv + 25v^2)$

We conclude this section with a summary of the factoring techniques we have reviewed.

Factoring Polynomials

1. Factor out any common factors first. It is also helpful when factoring $ax^2 + bxy + cy^2$ to factor out (-1) if a is negative.
2. If the polynomial has two terms try to use

   $a^2 - b^2 = (a + b)(a - b)$ Difference of squares
   $a^3 + b^3 = (a + b)(a^2 - ab + b^2)$ Sum of cubes
   or $a^3 - b^3 = (a - b)(a^2 + ab + b^2)$.   Difference of cubes

   The sum of squares, $a^2 + b^2$, cannot be factored.
3. If the polynomial has three terms, use the trial-and-error method or the special technique by grouping. Also, be aware of perfect squares and use

   $a^2 + 2ab + b^2 = (a + b)^2$ Perfect square
   or $a^2 - 2ab + b^2 = (a - b)^2$.   Perfect square
4. If the polynomial has four terms, try to factor by grouping.

1.4 EXERCISES A

Factor each polynomial in Exercises 1–33. Check all answers by multiplying.

1. $6x^2y^2 - 3xy$
2. $8x^3y^2 + 4x^2y^3 - 12x^2y^2$
3. $5a^2b - 10ab + 3a - 6$
4. $8a^3b^3 - 2a^2b^2 - 12abc^2 + 3c^2$
5. $x^2 + 8x + 7$
6. $x^2 - 8x + 7$
7. $x^2 + 6x - 7$
8. $x^2 - 6x - 7$
9. $a^2 + 9a + 20$
10. $x^2 - 2x - 35$
11. $3x^2 - 5x - 2$
12. $a^2 - 4ab - 5b^2$
13. $16x^2 + 10xy - 21y^2$
14. $4x^2 - 9y^2$
15. $4x^2 + 12xy + 9y^2$
16. $4x^2 - 12xy + 9y^2$
17. $27u^3 - v^3$
18. $27u^3 + v^3$

19. $28x^2 - 58xy - 30y^2$

20. $6x^2 - 6y^2$

21. $25x^2 + 10xy + y^2$

22. $-4u^2 - 34u - 70$

23. $5u^2v^2 - 11uv + 2$

24. $3a^2 - 60ab + 300b^2$

25. $32x^3 + 4y^9$

26. $x^2 + 2xy + 2$

27. $-5u^2 + 70uv - 240v^2$

28. $2x^2 + 2y^2$

29. $(u - v)^2 - 16$

30. $(u - v)^2 - 4(u - v) + 4$

31. $9u^6 - 30u^3v^2 + 25v^4$

32. $u^6 + 2u^3v^3 + v^6$

33. $x^6 + 7x^3y^3 - 8y^6$

In Exercises 34–36, factor by grouping using the special formulas.

34. $4x^2 - 9y^2 - 2x - 3y$

35. $x^2 - y^2 - 4y - 4x$

36. $x^3 - y^3 - x^2y + xy^2$

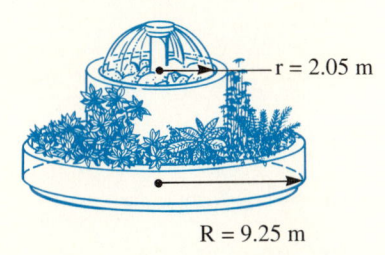

37. AGRICULTURE A circular garden of radius R has a circular fountain in the center with radius r.

$r = 2.05$ m

$R = 9.25$ m

(a) Find a polynomial which represents the area of the garden that can be planted.
(b) Factor this polynomial.
(c) Find the area, correct to two decimal places, when $R = 9.25$ m and $r = 2.05$ m. Use 3.14 for π.
(d) How many plants should be purchased if each plant requires approximately 0.75 m² to grow properly?

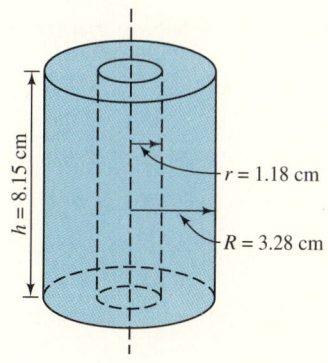

38. ENGINEERING A circular cylinder of radius R has its center drilled out to a radius of r. The cylinder is h units high.

$h = 8.15$ cm
$r = 1.18$ cm
$R = 3.28$ cm

(a) Find a polynomial that represents the volume of the remaining metal.
(b) Factor this polynomial.
(c) Find the volume, correct to two decimal places, when $R = 3.28$ cm, $r = 1.18$ cm, and $h = 8.15$ cm. Use 3.14 for π.

FOR REVIEW

39. Add the polynomials.
$2y^3 - y^2 + 3$ and $4y^3 - 6y - 3$

40. Subtract $3a^2 - a + 1$ from $6a^2 - a + 3$.

41. Find the product.
$(3u - v)^2$

42. Divide.
$$\frac{4x^3 - 2x^2 + 6x}{-2x}$$

ANSWERS: 1. $3xy(2xy - 1)$ 2. $4x^2y^2(2x + y - 3)$ 3. $(5ab + 3)(a - 2)$ 4. $(2a^2b^2 - 3c^2)(4ab - 1)$ 5. $(x + 1)(x + 7)$
6. $(x - 1)(x - 7)$ 7. $(x - 1)(x + 7)$ 8. $(x + 1)(x - 7)$ 9. $(a + 4)(a + 5)$ 10. $(x - 7)(x + 5)$ 11. $(3x + 1)(x - 2)$
12. $(a + b)(a - 5b)$ 13. $(2x + 3y)(8x - 7y)$ 14. $(2x - 3y)(2x + 3y)$ 15. $(2x + 3y)^2$ 16. $(2x - 3y)^2$
17. $(3u - v)(9u^2 + 3uv + v^2)$ 18. $(3u + v)(9u^2 - 3uv + v^2)$ 19. $2(2x - 5y)(7x + 3y)$ 20. $6(x + y)(x - y)$
21. $(5x + y)^2$ 22. $-2(2u + 7)(u + 5)$ 23. $(uv - 2)(5uv - 1)$ 24. $3(a - 10b)^2$ 25. $4(2x + y^3)(4x^2 - 2xy^3 + y^6)$
26. cannot be factored 27. $-5(u - 6v)(u - 8v)$ 28. $2(x^2 + y^2)$ 29. $(u - v + 4)(u - v - 4)$ 30. $(u - v - 2)^2$
31. $(3u^3 - 5v^2)^2$ 32. $(u + v)^2(u^2 - uv + v^2)^2$ 33. $(x - y)(x^2 + xy + y^2)(x + 2y)(x^2 - 2xy + 4y^2)$
34. $(2x + 3y)(2x - 3y - 1)$ 35. $(x + y)(x - y - 4)$ 36. $(x - y)(x^2 + y^2)$ 37. (a) $\pi R^2 - \pi r^2$ (b) $\pi(R + r)(R - r)$
(c) 255.47 m^2 (d) approximately 341 38. (a) $\pi R^2 h - \pi r^2 h$ (b) $\pi h(R + r)(R - r)$ (c) 239.69 cm^3 39. $6y^3 - y^2 - 6y$
40. $3a^2 + 2$ 41. $9u^2 - 6uv + v^2$ 42. $-2x^2 + x - 3$

1.4 EXERCISES B

Factor each polynomial in Exercises 1–33. Check all answers by multiplying.

1. $3a^2b^4 - 6ab$
2. $9a^4b^5c^2 - 18a^4b^3c + 81a^2b^2c^2$
3. $3xy^2 - 15xy - 4y + 20$
4. $10x^3y^3 + 25x^2y^2z - 12xyz - 30z^2$
5. $x^2 + x - 30$
6. $x^2 + 11x + 30$
7. $x^2 - x - 30$
8. $x^2 - 11x + 30$
9. $y^2 - 9y + 20$
10. $z^2 + z - 42$
11. $4x^2 + 4x - 15$
12. $x^2 - 4xy - 21y^2$
13. $5u^2 + 24uv - 5v^2$
14. $25u^2 - 16v^2$
15. $16x^2 + 8xy + y^2$
16. $16x^2 - 8xy + y^2$
17. $8x^3 - 27y^3$
18. $8x^3 + 27y^3$
19. $6x^2 + 21xy - 12y^2$
20. $5u^2 - 5v^2$
21. $49x^2 + 14xy + y^2$
22. $-3x^2 - 12xy + 96y^2$
23. $6x^2y^2 + 4xy - 2$
24. $5a^2 - 35ab - 150b^2$
25. $16a^3 - 2y^6$
26. $x^2 - 2xy + 2$
27. $-6x^2 + 12xy + 90y^2$
28. $7a^2 + 7b^2$
29. $(x + y)^2 - (u + v)^2$
30. $(x + 2y)^2 - (x - 2y)^2$
31. $x^4 + 2x^2y^2 - 15y^4$
32. $8u^6 - 7u^3v^3 - v^6$
33. $u^4 - 10u^2v^2 + 21v^4$

In Exercises 34–36, factor by grouping using special formulas.

34. $x^2 - 4y^2 + 6x - 12y$ **35.** $u^2 + 10uv + 25v^2 - w^2$ **36.** $x^3 + y^3 + 2x^2y + 2xy^2$

37. RECREATION   A circular exercise track for horses with radius R has a circular grassy area in the center with radius r.

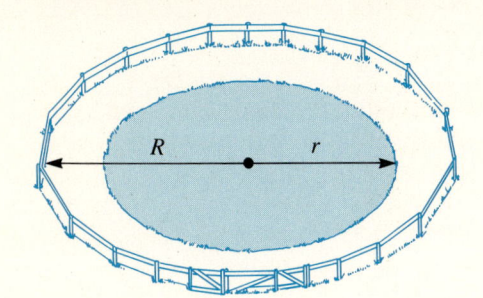

(a) Find a polynomial that represents the area of the track.
(b) Factor the polynomial.
(c) Find the area, correct to two decimal places, when $R = 50.5$ yd and $r = 48.25$ yd. Use 3.14 for π.
(d) To the nearest cubic yard, how many cubic yards of sand should be purchased to cover the track if 1 yd^3 will cover 12 yd^2 of the track?

38. ENGINEERING   A metal machine part in the shape of a cube with edge E has a rectangular solid removed from the center as shown.

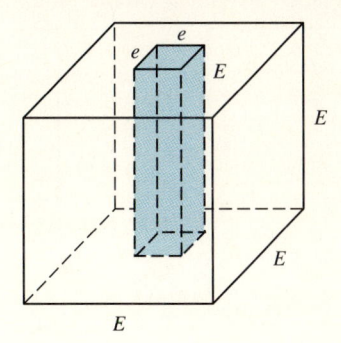

(a) Find a polynomial that represents the volume of the remaining metal.
(b) Find the factored form of this polynomial.
(c) Find the volume when $E = 12$ cm and $e = 3$ cm.

FOR REVIEW

Perform the indicated operations.

39. Add the polynomials.
$6z^3 - 3z^2 + z - 5$ and $2z^3 - 7z + 3$.

40. Subtract $x^3 - x + 3$ from $7x^3 - x + 2$.

41. Find the product.
$(3u + v)^2$

42. Divide.
$$\frac{11a^3 - 22a^2 + 55a}{-11a}$$

1.4 EXERCISES C

Factor each polynomial in Exercises 1–6. Assume n is a positive integer.

1. $4x^{2n} - y^{2n}$

2. $27x^{3n} + 8y^{3n}$
[Answer: $(3x^n + 2y^n)(9x^{2n} - 6x^ny^n + 4y^{2n})$]

3. $27x^{3n} - 8y^{3n}$

4. $x^{2n} - 4y^{6n}$
[Answer: $(x^n + 2y^{3n})(x^n - 2y^{3n})$]

5. $8x^{3n} + y^{3n}$

6. $125x^{3n} - 64y^{3n}$

42 CHAPTER 1 REVIEW OF FUNDAMENTAL CONCEPTS

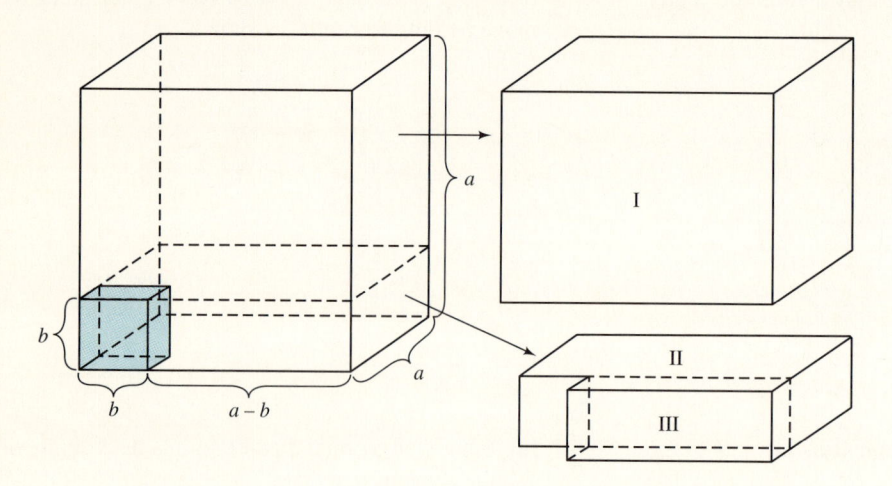

7. **GEOMETRY** The early Greeks gave geometric proofs of many of the formulas we use in algebra today. For example, to show that $a^3 - b^3 = (a - b)(a^2 + ab + b^2)$, for the case when $a > b$, consider the figures below. Show that the volume of the unshaded region in the left figure is $a^3 - b^3$. Then show that the volume of the box labeled I is $(a - b)a^2$, the volume of the box labeled II is $(a - b)ab$, and the volume of the box labeled III is $(a - b)b^2$. What happens when the volumes of I, II, and III are added together?

1.5 RATIONAL EXPRESSIONS

STUDENT GUIDEPOSTS

1. Equivalent Fractions
2. Reducing to Lowest Terms
3. Multiplication and Division of Rational Expressions
4. Least Common Denominator (LCD)
5. Addition and Subtraction of Rational Expressions
6. Complex Fractions

1 EQUIVALENT FRACTIONS

An **algebraic fraction** is a quotient of algebraic expressions. This section looks at **rational expressions** which are quotients of two polynomials. For example,

$$\frac{x^2 + 4}{x - 2}, \quad \frac{x + 3y}{2x - 5y}, \quad \frac{u^2 + 3v^2}{(x + y)^2}, \quad \frac{3x(x + 3)}{(x - 5)(x + 3)}$$

are all rational expressions. Since division by zero is undefined, when working with rational expressions we have to be careful to exclude replacements for the variable(s) that make the denominator zero.

> **To Find the Values to Exclude in a Rational Expression**
> 1. Set the denominator equal to zero.
> 2. Solve the resulting equation; any solution must be excluded from consideration.

EXAMPLE 1 FINDING VALUES TO EXCLUDE

Find the values of the variable(s) that must be excluded.

(a) $\dfrac{3x}{x-5}$

Set $x - 5$ equal to zero and solve.
$$x - 5 = 0$$
$$x = 5$$

Thus 5 must be excluded since when x is 5, the denominator is 0.

(b) $\dfrac{2a - 3b + 1}{a^2 - b^2}$

Setting $a^2 - b^2 = 0$ and factoring, we obtain $(a - b)(a + b) = 0$. Then if $a - b = 0$, that is, $a = b$, or if $a + b = 0$, that is $a = -b$, the rational expression will not be defined. So we have to exclude all values of a and b when they are either equal or negatives.

PRACTICE EXERCISE 1

Find the values of the variable(s) that must be excluded.

(a) $\dfrac{y + 3}{y + 1}$

(b) $\dfrac{x}{x^2 + 3x - 10}$

Answers: (a) $y = -1$ (b) $x = 2$ and $x = -5$

Knowing the values to exclude in a rational expression plays a major role in determining when two fractions are *equivalent*.

> **Fundamental Principle of Fractions**
>
> If the numerator and denominator of an algebraic fraction are multiplied or divided by the same nonzero expression, the resulting fraction is **equivalent** to the original. That is,
>
> $$\frac{a}{b} = \frac{ac}{bc} \quad \text{and} \quad \frac{ac}{bc} = \frac{a}{b} \quad (\text{when } b \neq 0 \text{ and } c \neq 0).$$

We can verify this theorem by multiplying the fraction $\dfrac{a}{b}$ by 1 written in the form $\dfrac{c}{c}$.

$$\frac{a}{b} = \frac{a}{b} \cdot 1 = \frac{a}{b} \cdot \frac{c}{c} = \frac{ac}{bc}$$

EXAMPLE 2 EQUIVALENT FRACTIONS

Are the fractions equivalent?

$$\frac{x}{x+1} \quad \text{and} \quad \frac{x(x-1)}{(x+1)(x-1)}$$

The fractions are equivalent (as long as $x \neq 1$) because the second has been obtained from the first by multiplying the numerator and denominator by the nonzero expression $x - 1$. Also, the first can be thought of as resulting from the second by dividing the numerator and denominator by the nonzero expression $x - 1$.

PRACTICE EXERCISE 2

Are the fractions equivalent?

$$\frac{2a}{3y} \quad \text{and} \quad \frac{2a + 5}{3y + 5}$$

Answer: no

❷ REDUCING TO LOWEST TERMS

The Fundamental Principle of Fractions is used to *reduce a rational expression to lowest terms*. A rational expression has been **reduced to lowest terms** when 1 (or -1) is the only polynomial that will divide both the numerator and the denominator.

> **To Reduce a Rational Expression to Lowest Terms**
> 1. Factor both the numerator and denominator.
> 2. Divide out all common factors.

For example, consider the rational expression

$$\frac{3x^2 + 9x}{x^2 - 2x - 15} = \frac{3x(x + 3)}{(x - 5)(x + 3)}.$$

The common factor $(x + 3)$ can be divided out provided $x \neq -3$, and we obtain the fraction $\frac{3x}{x-5}$, reduced to lowest terms. This is usually done by **canceling** the common factor(s) as shown below.

$$\frac{3x^2 + 9x}{x^2 - 2x - 15} = \frac{3x\cancel{(x+3)}}{(x - 5)\cancel{(x+3)}} = \frac{3x}{x - 5}.$$

Remember that *canceling* is simply a convenient way of *dividing out* common factors, but you should use this technique with care.

EXAMPLE 3 REDUCING TO LOWEST TERMS

Reduce the rational expression to lowest terms.

$$\begin{aligned}
\frac{2a^3 + a^2b - ab^2}{a^3 - ab^2} &= \frac{a(2a^2 + ab - b^2)}{a(a^2 - b^2)} \\
&= \frac{\cancel{a}\cancel{(a+b)}(2a - b)}{\cancel{a}\cancel{(a+b)}(a - b)} \quad \text{Cancel common factors} \\
&= \frac{2a - b}{a - b} \quad \text{This is reduced to lowest terms}
\end{aligned}$$

PRACTICE EXERCISE 3

Reduce the rational expression to lowest terms.

$$\frac{8x^4 + 2x}{2x}$$

Answer: $4x^3 + 1$ [Note that both factors in denominator divide out (cancel) leaving 1.]

CAUTION

Cancel only factors, not terms. For example,

$$\frac{2a - b}{a - b} \quad \text{is not} \quad \frac{2\cancel{a} - \cancel{b}}{\cancel{a} - \cancel{b}}.$$

③ MULTIPLICATION AND DIVISION OF RATIONAL EXPRESSIONS

If a/b and c/d are rational expressions, with $b \neq 0$ and $d \neq 0$, then

$$\frac{a}{b} \cdot \frac{c}{d} = \frac{ac}{bd}.$$ Multiplication of rational expressions

Thus, to multiply, we multiply numerators and multiply denominators. However, it is better to reduce the fractions before finding any products.

To Multiply Two or More Rational Expressions

1. Factor all numerators and denominators and place the indicated product of all numerator factors over the product of all denominator factors.
2. Divide out (cancel) all common factors.
3. Multiply the remaining numerator factors, then multiply the remaining denominator factors. The result is the product, reduced to lowest terms.

EXAMPLE 4 MULTIPLYING RATIONAL EXPRESSIONS

Multiply.

$$\frac{a^2 - 4b^2}{49a^3 - 4ab^2} \cdot \frac{7a^2 - 2ab}{a^2 + 2ab} = \frac{(a - 2b)(a + 2b)}{a(7a - 2b)(7a + 2b)} \cdot \frac{a(7a - 2b)}{a(a + 2b)}$$

$$= \frac{(a - 2b)\cancel{(a + 2b)}\cancel{(7a - 2b)}}{a\cancel{(7a - 2b)}(7a + 2b)\cancel{(a + 2b)}}$$

$$= \frac{a - 2b}{a(7a + 2b)} = \frac{a - 2b}{7a^2 + 2ab}$$

Either $\dfrac{a - 2b}{a(7a + 2b)}$ or $\dfrac{a - 2b}{7a^2 + 2ab}$ is an acceptable answer.

PRACTICE EXERCISE 4

Multiply.

$$\frac{x^2 - 5x + 6}{x^2 + x - 6} \cdot \frac{x + 3}{x - 2}$$

Answer: $\frac{x - 3}{x - 2}$

Remember that we defined division as multiplication by the multiplicative inverse or reciprocal of the divisor. Thus, if a/b and c/d are two rational expressions, for which values of the variable(s) have been excluded so that b, c, and $d \neq 0$,

$$\frac{a}{b} \div \frac{c}{d} = \frac{a}{b} \cdot \frac{d}{c} = \frac{ad}{bc} \qquad \frac{d}{c} \text{ is the reciprocal of } \frac{c}{d}$$

As was the case with multiplication, it is better to factor and reduce the fractions before multiplying in the last step.

To Divide One Rational Expression by Another

1. Replace the divisor by its reciprocal and multiply.
2. Follow the same procedure as for multiplying rational expressions.

EXAMPLE 5   Dividing Rational Expressions

Perform the indicated operations.

(a) $\dfrac{a^2 - b^2}{a + 2b} \div \dfrac{a - b}{a^2 + 3ab + 2b^2} = \dfrac{a^2 - b^2}{a + 2b} \cdot \dfrac{a^2 + 3ab + 2b^2}{a - b}$

$= \dfrac{(a - b)(a + b)}{(a + 2b)} \cdot \dfrac{(a + 2b)(a + b)}{(a - b)}$

$= \dfrac{\cancel{(a - b)}(a + b)\cancel{(a + 2b)}(a + b)}{\cancel{(a + 2b)}\cancel{(a - b)}}$

$= (a + b)(a + b) = (a + b)^2$

(b) $\dfrac{a^2 - 10ab + 25b^2}{2a^2 - 7ab - 4b^2} \cdot \dfrac{3a^2 - 17ab + 20b^2}{a^2 - 25b^2} \div \dfrac{3a^2 - 20ab + 25b^2}{2a^2 - ab - b^2}$

$= \dfrac{a^2 - 10ab + 25b^2}{2a^2 - 7ab - 4b^2} \cdot \dfrac{3a^2 - 17ab + 20b^2}{a^2 - 25b^2} \cdot \dfrac{2a^2 - ab - b^2}{3a^2 - 20ab + 25b^2}$

$= \dfrac{\cancel{(a - 5b)}(a - 5b)\cancel{(3a - 5b)}(a - 4b)(2a + b)(a - b)}{\cancel{(a - 4b)}\cancel{(2a + b)}\cancel{(a - 5b)}(a + 5b)\cancel{(3a - 5b)}\cancel{(a - 5b)}} = \dfrac{a - b}{a + 5b}$

Practice Exercise 5

Perform the indicated operations.

(a) $\dfrac{x^2}{x + 2} \div \dfrac{x}{x^2 + 4x + 4}$

(b) $\dfrac{a^2 - b^2}{a^2 + a - 2} \cdot \dfrac{a^2 + 2a}{a^2 + 2ab + b^2}$

$\div \dfrac{a^3 - b^3}{a^2 - a + ba - b}$

Answers: (a) $x(x + 2)$ (b) $\dfrac{a}{a^2 + ab + b^2}$

4   LEAST COMMON DENOMINATOR (LCD)

Adding or subtracting rational expressions with a common denominator is defined as follows where $c \neq 0$:

$\dfrac{a}{c} + \dfrac{b}{c} = \dfrac{a + b}{c}$ and $\dfrac{a}{c} - \dfrac{b}{c} = \dfrac{a - b}{c}$ Addition and subtraction

If the denominators are not the same, we need to convert the fractions into equivalent fractions that have a common denominator. With rational expressions this is best done by finding the **least common denominator (LCD).**

> **To Find the LCD of Two or More Fractions**
> 1. Factor all denominators and reduce each fraction, if possible.
> 2. Each factor of a denominator appears in the LCD as many times as it appears in the denominator where it is found the greatest number of times.

EXAMPLE 6   Finding the LCD

Find the least common denominator (LCD).

(a) $\dfrac{ab}{a^2 - 5ab + 6b^2}$ and $\dfrac{a + 2b}{a^3 - ab^2}$

$\dfrac{ab}{a^2 - 5ab + 6b^2} = \dfrac{ab}{(a - 2b)(a - 3b)}$ Denominator: one $(a - 2b)$ one $(a - 3b)$

$\dfrac{a + 2b}{a^3 - 4ab} = \dfrac{\cancel{a + 2b}}{a\cancel{(a + 2b)}(a - 2b)} = \dfrac{1}{a(a - 2b)}$ Denominator: one a one $(a - 2b)$

Thus the factors of the LCD are one $(a - 2b)$, one $(a - 3b)$, and one a.
The LCD is $a(a - 2b)(a - 3b)$.

Practice Exercise 6

Find the least common denominator (LCD).

(a) $\dfrac{a + 1}{a^3 + 2a^2 + a}$ and $\dfrac{5 - a}{a^2 - 4a - 5}$

$\left[\text{Hint: } \dfrac{5 - a}{a - 5} = -1 \right]$

(b) $\dfrac{3x+y}{x^2-6xy+9y^2}$, $\dfrac{xy}{x^2-9y^2}$, and $\dfrac{x-y}{4x^5-4x^4y+x^3y^2}$

$\dfrac{3x+y}{x^2-6xy+9y^2} = \dfrac{3x+y}{(x-3y)^2}$ Denominator: two $(x-3y)$

$\dfrac{xy}{x^2-9y^2} = \dfrac{xy}{(x+3y)(x-3y)}$ Denominator: one $(x+3y)$
   one $(x-3y)$

$\dfrac{x-y}{4x^5-4x^4y+x^3y^2} = \dfrac{x-y}{x^3(2x-y)^2}$ Denominator: three x
   two $(2x-y)$

Thus, the factors of the LCD include two $(x-3y)$ factors, one $(x+3y)$ factor, three x factors, and two $(2x-y)$ factors. The LCD is

$$x^3(x+3y)(x-3y)^2(2x-y)^2.$$

(b) $\dfrac{3x+2}{6x^4-12x^3+6x^2}$, $\dfrac{4x}{x^2-7x+6}$, and $\dfrac{5x^2+1}{8(x^5-12x^4+36x^3)}$

Answers: (a) $a(a+1)$
(b) $24x^3(x-1)^2(x-6)^2$

5 ADDITION AND SUBTRACTION OF RATIONAL EXPRESSIONS

To add or subtract rational expressions, we need to change each fraction to an equivalent fraction with the LCD as its denominator using the Fundamental Principle of Fractions.

> **To Add or Subtract Rational Expressions**
> 1. Express denominators in factored form and reduce all fractions.
> 2. Find the LCD of the fractions.
> 3. Multiply the numerator and denominator of each fraction by all factors present in the LCD but missing in the denominator of the particular fraction, so the fractions all have the same denominator, the LCD.
> 4. Indicate the sum or difference of all numerators, using parentheses (especially in subtraction), and place the result over the LCD.
> 5. Simplify and reduce the resulting fraction.

EXAMPLE 7 ADDING AND SUBTRACTING FRACTIONS

Perform the indicated operations.

(a) $\dfrac{2}{3x-21} + \dfrac{x}{49-x^2} = \dfrac{2}{3(x-7)} + \dfrac{x}{(7-x)(7+x)}$

$= \dfrac{2}{3(x-7)} + \dfrac{(-1)x}{(-1)(7-x)(7+x)}$ $x-7$ and $7-x$ are negatives

$= \dfrac{2}{3(x-7)} + \dfrac{-x}{(x-7)(7+x)}$ LCD $= 3(x-7)(x+7)$

$= \dfrac{2(x+7)}{3(x-7)(x+7)} + \dfrac{3(-x)}{3(x-7)(x+7)}$ Supply missing factors

$= \dfrac{2(x+7)-3x}{3(x-7)(x+7)}$

$= \dfrac{2x+14-3x}{3(x-7)(x+7)} = \dfrac{-x+14}{3(x-7)(x+7)}$

PRACTICE EXERCISE 7

Perform the indicated operations.

(a) $\dfrac{3x+y}{x^2+4xy-21y^2} + \dfrac{x+y}{9y^2-x^2}$

48 CHAPTER 1 REVIEW OF FUNDAMENTAL CONCEPTS

(b) $\dfrac{2xy}{x^2 - y^2} - \dfrac{y}{x + y} + 2$

$= \dfrac{2xy}{(x + y)(x - y)} - \dfrac{y}{x + y} + \dfrac{2}{1} \qquad 2 = \dfrac{2}{1}$

$= \dfrac{2xy}{(x + y)(x - y)} - \dfrac{y(x - y)}{(x + y)(x - y)} + \dfrac{2(x + y)(x - y)}{(x + y)(x - y)} \qquad \text{LCD} = (x + y)(x - y)$

$= \dfrac{2xy - y(x - y) + 2(x + y)(x - y)}{(x + y)(x - y)}$

$= \dfrac{2xy - yx + y^2 + 2(x^2 - y^2)}{(x + y)(x - y)} = \dfrac{2xy - xy + y^2 + 2x^2 - 2y^2}{(x + y)(x - y)}$

$= \dfrac{xy - y^2 + 2x^2}{(x + y)(x - y)} = \dfrac{2x^2 + xy - y^2}{(x + y)(x - y)}$

$= \dfrac{(2x - y)\cancel{(x + y)}}{\cancel{(x + y)}(x - y)} \qquad$ This can be simplified by canceling common factors in the numerator and denominator

$= \dfrac{2x - y}{x - y}$

(b) $\dfrac{y^2 - 1}{y^3 - 1} - \dfrac{y}{2y^2 + 2y + 2}$

Answers: (a) $\dfrac{2(x + 2y)(x - y)}{(x - 3y)(x + 3y)(x + 7y)}$
(b) $\dfrac{y + 2}{2(y^2 + y + 1)}$

///////////// **CAUTION** /////////////

Remember that a fraction bar acts like a grouping symbol. Thus, when subtracting rational expressions, be sure to enclose the numerator of the fraction being subtracted in parentheses to avoid a common sign error. For example,

$\dfrac{x + 1}{x - 5} - \dfrac{x - 2}{x - 5} = \dfrac{x + 1 - (x - 2)}{x - 5} = \dfrac{x + 1 - x + 2}{x - 5} = \dfrac{3}{x - 5}$

NOT $= \dfrac{x + 1 - x - 2}{x - 5} = \dfrac{-1}{x - 5}.$

/////////

---

⑥ COMPLEX FRACTIONS

An algebraic fraction which has at least one fraction in its numerator or denominator is called a **complex fraction.** The following are complex fractions:

$\dfrac{\tfrac{3}{13}}{\tfrac{7}{5}}, \quad \dfrac{\tfrac{x + y}{y}}{\tfrac{x - y}{x}}, \quad \dfrac{\tfrac{1}{a} + \tfrac{1}{b}}{a - b}, \quad \dfrac{\tfrac{1}{(x + h)^4} - \tfrac{1}{x^4}}{h}.$

Complex fractions occur frequently in mathematical models of real-world applications. For example, the focal length of a simple convex lens f is given by

$$f = \dfrac{1}{\dfrac{1}{d_1} + \dfrac{1}{d_2}}$$

where d_1 is the distance from an object to the lens and d_2 is the distance from the lens to the image of the object (see Figure 1.5).

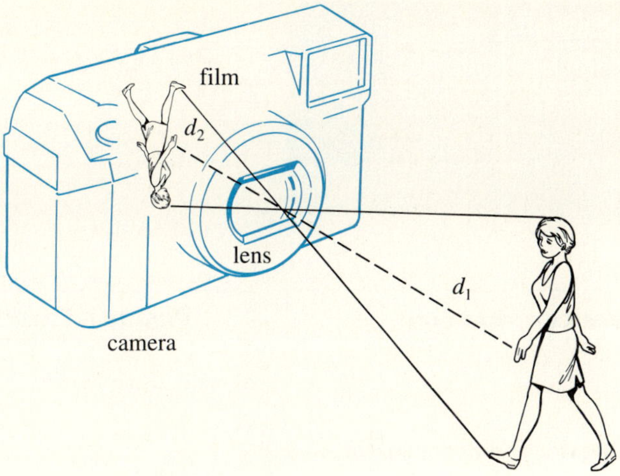

Figure 1.5

It is best to *simplify* a complex fraction before working with it in an application. A complex fraction has been simplified when all of its component fractions have been eliminated and a simple fraction obtained. There are two methods to use for simplifying a complex fraction.

To Simplify a Complex Fraction (Method 1)

1. Change the numerator and denominator to single fractions.
2. Divide the two fractions and reduce to lowest terms.

EXAMPLE 8   SIMPLIFYING A COMPLEX FRACTION

Simplify using Method 1.

$$\frac{\frac{1}{a} - \frac{1}{b}}{1 - \frac{1}{ab}} = \frac{\frac{b}{ab} - \frac{a}{ab}}{\frac{ab}{ab} - \frac{1}{ab}} \qquad \text{The common denominator for all fractions is } ab$$

$$= \frac{\frac{b-a}{ab}}{\frac{ab-1}{ab}} \qquad \text{Change numerator and denominator to single fractions}$$

$$= \frac{b-a}{ab} \cdot \frac{ab}{ab-1} \qquad \text{Divide numerator by denominator}$$

$$= \frac{b-a}{ab-1} \qquad \text{Reduce to lowest terms}$$

PRACTICE EXERCISE 8

Simplify using Method 1.

$$\frac{\frac{b}{a+b} - 1}{\frac{b}{a-b} + 1}$$

Answer: $\frac{b-a}{a+b}$

An alternative method is easier to use on some complex fractions.

> **To Simplify a Complex Fraction (Method 2)**
>
> 1. Find the LCD of all fractions within the complex fraction.
> 2. Multiply numerator and denominator of the complex fraction by the LCD to obtain an equivalent fraction. Reduce the result to lowest terms.

EXAMPLE 9 SIMPLIFYING A COMPLEX FRACTION

Simplify using Method 2.

$$\frac{y-x}{\frac{1}{y^3}-\frac{1}{x^3}} = \frac{(y-x)x^3y^3}{\left(\frac{1}{y^3}-\frac{1}{x^3}\right)x^3y^3}$$ Multiply numerator and denominator by x^3y^3

$$= \frac{(y-x)x^3y^3}{\frac{x^3y^3}{y^3}-\frac{x^3y^3}{x^3}}$$ Cancel common factors

$$= \frac{(y-x)x^3y^3}{x^3-y^3} = \frac{(y-x)x^3y^3}{(x-y)(x^2+xy+y^2)}$$

$$= \frac{-(x-y)x^3y^3}{(x-y)(x^2+xy+y^2)} = \frac{-x^3y^3}{x^2+xy+y^2}$$

PRACTICE EXERCISE 9

Simplify using Method 2.

$$\frac{a-\dfrac{4}{a}}{\dfrac{16}{a^3}-a}$$

Answer: $\dfrac{-a^2}{4+a^2}$

We can now solve the applied problem given in the chapter introduction, which illustrates the method of simplifying a complex fraction.

EXAMPLE 10 A RECREATION PROBLEM

Linda Youngman hiked to the bottom of the Grand Canyon at a rate of 4 mph and returned to the top at a rate of 2 mph. To the nearest tenth, what was her average rate for the hike?

At first glance we might assume that the average rate is 3 mph, the average of 4 and 2. However, upon further analysis, it is clear that her rate must be closer to 2 mph than 4 mph since she hiked for a longer period of time at this slower rate. The basic distance formula, $d = rt$ (distance = rate · time), can be used after first solving for t, $t = d/r$.

Let D = the distance hiked down into the canyon (also the distance hiked out of the canyon),

$\dfrac{D}{4}$ = time it took to hike down into the canyon,

$\dfrac{D}{2}$ = time it took to hike out of the canyon.

Then the average rate is:

$$\frac{\text{total distance hiked}}{\text{total time of the hike}} = \frac{2D}{\dfrac{D}{4}+\dfrac{D}{2}}$$

PRACTICE EXERCISE 10

Kathy Richmond took the same hike as Linda in Example 10. However, she hiked down the canyon at a rate of 5 mph, and she hiked out at a rate of 1.5 mph. To the nearest tenth, what was Kathy's average rate on the hike?

By simplifying this complex fraction, we can find the average rate.

$$\frac{2D}{\frac{D}{4}+\frac{D}{2}} = \frac{2D}{\frac{D}{4}+\frac{2D}{4}} = \frac{2D}{\frac{D+2D}{4}}$$

$$= \frac{2D}{\frac{3D}{4}} = \frac{2D}{1} \cdot \frac{4}{3D}$$

$$= \frac{8D}{3D} = \frac{8}{3} \approx 2.7$$

Thus, the average rate on Linda's hike was about 2.7 mph.

Answer: 2.3 mph

1.5 EXERCISES A

In Exercises 1–3 give the values of the variable that must be excluded.

1. $\dfrac{3}{x-5}$

2. $\dfrac{(y-3)(x+2)}{(y^2-5y+6)(x+7)}$

3. $\dfrac{6a+b}{a^2+ab}$

In Exercises 4–6, are the given fractions equivalent where they are defined?

4. $\dfrac{x^2}{5x}, \dfrac{x}{5}$

5. $\dfrac{2}{a^3}, \dfrac{2+b}{a^3+b}$

6. $\dfrac{a-b}{a^2-b^2}, \dfrac{1}{a+b}$

Reduce each fraction to lowest terms in Exercises 7–9.

7. $\dfrac{77x^2y^4}{33x^3y^3}$

8. $\dfrac{a+b}{a^2+b^2}$

9. $\dfrac{a^3-b^3}{a-b}$

In Exercises 10–19 perform the indicated operations and simplify.

10. $\dfrac{8x^3}{9x} \cdot \dfrac{45}{16}$

11. $\dfrac{3(x+y)}{6x^2y^2} \cdot x^3y^3$

12. $\dfrac{2a+2}{a-3} \cdot \dfrac{a^2-9}{4a+4}$

13. $\dfrac{(a-3)^2}{35(a+3)} \cdot \dfrac{5(a^2+6a+9)}{a^2-9}$

14. $\dfrac{a^2-4}{a^2-4a+4} \cdot \dfrac{a^2-9a+14}{a^3+2a^2}$

15. $\dfrac{x^2-x-12}{x^2-9} \div \dfrac{x^2-16}{5x-5}$

16. $\dfrac{a^3+64}{2a^2+18a+40} \div \dfrac{a^2-4a+16}{a^2+4a}$

17. $\dfrac{uv-uw+xv-xw}{v-w} \div \dfrac{v^2-2vw+w^2}{xv-xw}$

18. $\dfrac{y^2-1}{2y+2} \cdot \dfrac{y^2-8y+12}{y^2-4y+4} \cdot \dfrac{y+2}{y-6}$

19. $\dfrac{x^2-y^2}{x^2-xy+y^2} \cdot \dfrac{2x^2-3xy-2y^2}{x^2+2xy+y^2} \div \dfrac{2x^2-xy-y^2}{x^3+y^3}$

Find the LCD of the fractions given in Exercises 20–22.

20. $\dfrac{a}{35}$ and $\dfrac{a^2}{50}$

21. $\dfrac{a+2}{a^2 - 5a + 6}$ and $\dfrac{a}{a^2 + a - 12}$

22. $\dfrac{x+7}{x^5 - 27x^2}$ and $\dfrac{x-3}{5x^3 + 15x^2 + 45x}$

Perform the indicated operations in Exercises 23–30.

23. $\dfrac{6}{5x - 10} + \dfrac{3}{x + 2}$

24. $\dfrac{3a}{a - 4} + \dfrac{5a}{4 - a}$

25. $\dfrac{y}{y^2 - 4} - \dfrac{2}{y + 2}$

26. $\dfrac{a - 3}{a^2 - 9} + \dfrac{1}{(a + 3)^2}$

27. $\dfrac{5x}{x^3 - 16x} - \dfrac{4}{x - 4}$

28. $\dfrac{6a}{a^3 - 27} - \dfrac{4}{2(a^2 + 3a + 9)}$

29. $\dfrac{2ab}{a^2 - b^2} - \dfrac{b}{a - b} + 5$

30. $\dfrac{x^2}{x^2 - 2xy + y^2} + \dfrac{1}{x^2 - xy} - \dfrac{x}{x - y}$

Simplify the complex fractions in Exercises 31–40.

31. $\dfrac{\dfrac{1}{2y} - \dfrac{1}{3y}}{1 + \dfrac{1}{4y}}$

32. $\dfrac{\dfrac{1}{a} + \dfrac{2}{a^2}}{\dfrac{1}{4} - \dfrac{1}{a^2}}$

33. $\dfrac{x + 6 + \dfrac{8}{x}}{x + 4 + \dfrac{4}{x}}$

34. $\dfrac{1 + \dfrac{5}{a} + \dfrac{4}{a^2}}{1 + \dfrac{1}{a + 3}}$

35. $\dfrac{1}{1 + \dfrac{1}{1 + \dfrac{1}{x}}}$

36. $a - \dfrac{a}{1 - \dfrac{a}{1 - a}}$

37. $\dfrac{\dfrac{1}{x+h} - \dfrac{1}{x}}{h}$

38. $\dfrac{\dfrac{1}{xy} + \dfrac{1}{yz} + \dfrac{1}{xz}}{\dfrac{x+y+z}{xyz}}$

39. $\dfrac{x^{-1} + y^{-1}}{x^{-1}y^{-1}}$

40. $\dfrac{a^{-1} - b^{-1}}{a^{-1} + b^{-1}}$

41. Remember the rules for order of operations and simplify.

$$\dfrac{x}{x^2-y^2} - \dfrac{y}{x^2-y^2} \cdot \dfrac{x+y}{x-y} + \dfrac{y(x+y)}{x^2-3xy-4y^2} \div \dfrac{(x+y)^2}{x-4y}$$

42. **TRANSPORTATION** Finding an average rate of speed results in a complex fraction. Suppose a trucker transports a heavy load from Denver to Chicago and returns empty. With the load he travels at a rate of 50 mph and returning he travels 65 mph. What was his average rate for the trip? At first glance we might assume that the average rate is 57.5 mph, the average of 50 and 65. However, since he traveled for a longer period of time at 50 mph, the actual average must be closer to 50 mph than 65 mph. Suppose we let D represent the distance traveled in one direction. Then using the distance formula, $d = rt$, solving for t, we have $t = \dfrac{d}{r}$. Then

$$\text{time with load} = \dfrac{\text{distance with load}}{\text{rate with load}} = \dfrac{D}{50}$$

$$\text{time empty} = \dfrac{\text{distance empty}}{\text{rate empty}} = \dfrac{D}{65}$$

$$\text{average rate} = \dfrac{\text{total distance}}{\text{total time}} = \dfrac{2D}{\dfrac{D}{50} + \dfrac{D}{65}}.$$

Simplify the complex fraction and find the average rate for the trip, correct to the nearest tenth of a mph.

43. **ENGINEERING** In an electronic circuit the frequency is given by an expression of the form

$$\dfrac{\dfrac{1}{p}\left(1 + \dfrac{p}{q}\right)}{1 + (1+p)\dfrac{q}{p}}.$$

Simplify the complex fraction.

FOR REVIEW

Factor the polynomials in Exercises 44–46.

44. $10x^2 - 1000y^2$

45. $6u^2 + uv - 35v^2$

46. $27u^3 - 64v^3$

Answer true or false in Exercises 47–50. If the statement is false, explain.

47. -18 is an integer.

48. $\sqrt{8}$ is an irrational number.

49. Some integers are whole numbers.

50. The set of whole numbers is closed with respect to subtraction.

1.5 EXERCISES C

1. Simplify the complex fraction.

$$1 - \cfrac{1}{1 - \cfrac{1}{1 - \cfrac{1}{1 - \cfrac{1}{x}}}}$$

$\left[\text{Answer: } \dfrac{x-1}{x}\right]$

2. Find the error in the work below. What is the correct answer?

$$\frac{x}{x+1} - \frac{2}{x-1} = \frac{x(x-1)}{(x+1)(x-1)} - \frac{2(x+1)}{(x+1)(x-1)}$$

$$= \frac{x^2 - x - 2x + 2}{(x+1)(x-1)}$$

$$= \frac{x^2 - 3x + 2}{(x+1)(x-1)}$$

$$= \frac{(x-1)(x-2)}{(x+1)(x-1)}$$

$$= \frac{x-2}{x+1}$$

[Answer: The error was made when $2(x+1)$ was subtracted from $x(x-1)$. The correct answer is $\dfrac{x^2 - 3x - 2}{(x+1)(x-1)}$.]

1.6 RADICAL EXPRESSIONS

STUDENT GUIDEPOSTS

1 Roots and Radicals
2 Simplifying Radicals
3 Rationalizing Denominators and Numerators
4 Approximating Radicals

1 ROOTS AND RADICALS

If $b^2 = a$, then b is a **square root** of a, and if $b^3 = a$, then b is a **cube root** of a. For example, 3 is a square root of 9 since $3^2 = 9$, and -3 is a square root of 9 since $(-3)^2 = 9$. Also, 2 is a cube root of 8 since $2^3 = 8$. Notice that since $(-2)^3 = -8$, -2 is not a cube root of 8. However, -2 is a cube root of -8. We can generalize this notion to include higher order roots.

1.6 RADICAL EXPRESSIONS

kth Root of a Number

Let k be a positive integer greater than 1. If
$$b^k = a,$$
then b is a **kth root of a**.

Notice that we use the terms *square root* instead of 2nd root and *cube root* instead of 3rd root. The number of kth roots of a real number a depends on k and on a. We saw that 9 has *two* real square roots, -3 and 3, but 8 has *only one* cube root, 2. In general, the following is true:

Number of Real kth Roots of a Real Number

Let a be a real number, with k a positive integer greater than 1.

1. If k is even and $a > 0$, then a has **two** real kth roots, one positive and one negative.
2. If k is even and $a = 0$, then a has **one** real kth root, 0.
3. If k is even and $a < 0$, then a has **no** real kth roots.
4. If k is odd and a is any real number, then a has **one** real kth root with the same sign as a.

A **radical** $\sqrt{}$ is the symbol used to denote roots of a number. For example, $\sqrt{9}$ is the nonnegative square root of 9, 3. Since -3 is also a square root of 9, there could be some confusion if we were to allow $\sqrt{9}$ to represent -3 also. Thus, we agree that with square roots,

$\sqrt{a}$ **represents the nonnegative square root of a,**

called the **principal square root** of a. To represent the negative square root we use $-\sqrt{a}$. For example,
$$\sqrt{25} = 5 \quad \text{and} \quad -\sqrt{25} = -5.$$

To denote kth roots when k is greater than 2, we use an *index* on the radical. For example, the cube root of a is denoted by $\sqrt[3]{a}$, the principal fourth root of a (the nonnegative root) by $\sqrt[4]{a}$, and in general, $\sqrt[k]{a}$ denotes the kth root of a. The number under the radical is called the **radicand,** and k is called the **index** on the radical. For convenience, we will often refer to a radical expression such as $\sqrt{x}$ simply as a *radical* rather than a *radical expression*.

One way to be sure that we always use the principal root (nonnegative root) when k is even is to use the following definition, motivated by our remarks above.

Definition of $\sqrt[k]{a^k}$

1. If k is even, then $\sqrt[k]{a^k} = |a|$.
2. If k is odd, then $\sqrt[k]{a^k} = a$.

EXAMPLE 1 EVALUATING ROOTS

Evaluate.

(a) $\sqrt{(-7)^2} = |-7| = 7$ *k is even*

(b) $\sqrt[3]{(-7)^3} = -7$ *k is odd*

PRACTICE EXERCISE 1

Evaluate.

(a) $\sqrt{(-5)^2}$

(b) $\sqrt[5]{(-5)^5}$

(c) $\sqrt[4]{-16}$ is not a real number

(d) $-\sqrt[6]{64} = -\sqrt[6]{2^6} = -|2| = -2$

(e) $\sqrt{25x^2} = \sqrt{(5x)^2} = |5x| = 5|x|$

(c) $\sqrt{-25}$
(d) $-\sqrt{(6)^2}$
(e) $\sqrt[3]{27y^3}$
Answers: (a) $|-5| = 5$ (b) -5
(c) not a real number (d) -6
(e) $3y$

CAUTION

Remember that when the index k is even, $\sqrt[k]{a^k}$ is never negative, even if a itself is negative (see Example 1(a)). To represent the negative kth root when k is even we have to use $-\sqrt[k]{a^k}$.

Notice in Example 1(e) we need to leave the answer as $5|x|$ since we do not know if x is positive or negative. Thus, when a radicand involves variables in addition to specific known real numbers, there can be some confusion applying the definition of a kth root. To avoid this problem, we will agree to the following.

Basic Assumption about Radicals

For the remainder of this section, unless specifically stated otherwise, we will assume that *variables and algebraic expressions under even indexed radicals represent nonnegative real numbers*. This eliminates the need for absolute values.

For example, we will now write $\sqrt{x^2} = x$ (assuming x is nonnegative) instead of $|x|$.

❷ SIMPLIFYING RADICALS

In Example 1 we *simplified* several radicals that had radicands that were perfect kth powers. When a radicand is not a perfect kth power, often we can still simplify the radical by removing factors that are perfect powers using the following.

Product Rule for Radicals

If a and b are any real numbers or expressions, where a and b are nonnegative when k is even, then

$$\sqrt[k]{ab} = \sqrt[k]{a}\sqrt[k]{b}.$$

The kth root of a product is the product of the kth roots.

CAUTION

When using the product rule on radicals that include a variable, for example $\sqrt{x-2}$, we are assuming that $x - 2 \geq 0$, or that $x \geq 2$, for otherwise the radical is not defined. Of course, $\sqrt[3]{x-2}$ is defined for all values of x.

1.6 RADICAL EXPRESSIONS

To use the product rule for radicals, look for perfect kth powers as factors of the radicand, and *remove them* from under the radical, as in the next example.

EXAMPLE 2 Using the Product Rule

Simplify.

(a) $\sqrt{32x^5} = \sqrt{2 \cdot 2^4 \cdot x \cdot x^4}$ 2^4 and x^4 are perfect squares
$= \sqrt{2^4 x^4 \cdot 2x}$
$= \sqrt{2^4 x^4}\sqrt{2x}$
$= 2^2 x^2 \sqrt{2x} = 4x^2\sqrt{2x}$

(b) $\sqrt[5]{9a^2b^3}\sqrt[5]{81a^3b^4} = \sqrt[5]{3^6 a^5 b^7}$ $\sqrt[5]{a}\sqrt[5]{b} = \sqrt[5]{ab}$
$= \sqrt[5]{3^5 a^5 b^5 \cdot 3b^2}$ Look for perfect fifth powers
$= \sqrt[5]{3^5 a^5 b^5}\sqrt[5]{3b^2}$
$= 3ab\sqrt[5]{3b^2}$

Notice in the first step that we do not multiply the numbers. Rather, we factor them, looking for perfect fifth powers.

PRACTICE EXERCISE 2

Simplify.

(a) $\sqrt[3]{24y^5}$

(b) $\sqrt{50x^3y}\sqrt{6x^5y^3}$

Answers: (a) $2y\sqrt[3]{3y^2}$
(b) $10x^4y^2\sqrt{3}$

When a radical includes a quotient, the following simplifying rule may be applied. To use it, look for perfect powers in both the numerator and denominator.

Quotient Rule for Radicals

If a and b are any real numbers or expressions ($b \neq 0$), where a is nonnegative and b is positive when k is even, then

$$\sqrt[k]{\frac{a}{b}} = \frac{\sqrt[k]{a}}{\sqrt[k]{b}}$$

The kth root of a quotient is the quotient of the kth roots.

EXAMPLE 3 Using the Quotient Rule

Simplify.

(a) $\dfrac{\sqrt{27u^3}}{\sqrt{3u}} = \sqrt{\dfrac{27u^3}{3u}}$ $\dfrac{\sqrt{a}}{\sqrt{b}} = \sqrt{\dfrac{a}{b}}$
$= \sqrt{9u^2}$
$= 3u$ Since we assume $u > 0$, $|u| = u$

(b) $\sqrt[3]{\dfrac{7a^3b^{-1}}{56a^2b^2}} = \sqrt[3]{\dfrac{a}{8b^3}}$ Reduce fraction
$= \dfrac{\sqrt[3]{a}}{\sqrt[3]{8b^3}}$ Quotient rule
$= \dfrac{\sqrt[3]{a}}{2b}$ Product rule in the denominator

PRACTICE EXERCISE 3

Simplify.

(a) $\dfrac{\sqrt[3]{40x^4}}{\sqrt[3]{5x}}$

(b) $\sqrt[4]{\dfrac{48u^2v^5}{3u^{-2}v^{-1}}}$

Answers: (a) $2x\sqrt[3]{x}$ (b) $2uv\sqrt[4]{v}$

When a radicand includes a radical, the next simplifying rule may apply.

> ### Index Rule for Radicals
> If a is any real number or expression for which the following radicals are defined,
> $$\sqrt[m]{\sqrt[n]{a}} = \sqrt[mn]{a}.$$

EXAMPLE 4   Using the Index Rule for Radicals

Simplify.

$$\sqrt[3]{\sqrt{64x^7y^{13}}} = \sqrt[6]{64x^7y^{13}} \qquad \sqrt[3]{\sqrt{a}} = \sqrt[6]{a}$$
$$= \sqrt[6]{2^6 x^6 y^{12} \cdot xy}$$
$$= \sqrt[6]{2^6 x^6 (y^2)^6} \sqrt[6]{xy}$$
$$= 2xy^2 \sqrt[6]{xy}$$

Practice Exercise 4

Simplify.

$$\sqrt[4]{\sqrt{256a^8b^{11}}}$$

Answer: $2ab\sqrt[8]{b^3}$

The simplifying rules we have reviewed can all be proved as theorems using the properties of rational exponents presented in the next section. In fact, as we shall see, it is probably better to use rational exponents rather than the index rule.

The simplifying rules for products and quotients are sometimes used to simplify expressions that involve addition and subtraction of radicals. Remember that there are no rules for addition and subtraction of radicals similar to those for multiplication and division. For example,

$$5 = \sqrt{25} = \sqrt{9+16} \neq \sqrt{9} + \sqrt{16} = 3 + 4 = 7 \qquad 5 \neq 7$$
$$4 = \sqrt{16} = \sqrt{25-9} \neq \sqrt{25} - \sqrt{9} = 5 - 3 = 2 \qquad 4 \neq 2$$

Therefore, $\sqrt{a+b} \neq \sqrt{a} + \sqrt{b}$ and $\sqrt{a-b} \neq \sqrt{a} - \sqrt{b}$. Even without rules like these, many sums or differences of radical expressions can be simplified by combining *like radicals* using the distributive law; it is similar to combining like terms in polynomials. Two radicals are **like radicals** if they have the same radicand and the same index. For example,

$$3\sqrt{5x} + 2\sqrt{5x} - 6\sqrt{5x} = (3 + 2 - 6)\sqrt{5x} = -\sqrt{5x}.$$

Even when an expression contains radicals that are not like radicals, they can sometimes be transformed into like radicals using the product or quotient rule first.

$$\sqrt[3]{16x^4} + 3x\sqrt[3]{2x} = \sqrt[3]{8x^3 \cdot 2x} + 3x\sqrt[3]{2x}$$
$$= 2x\sqrt[3]{2x} + 3x\sqrt[3]{2x}$$
$$= (2x + 3x)\sqrt[3]{2x} = 5x\sqrt[3]{2x}$$

EXAMPLE 5   Adding and Subtracting Like Radicals

Add or subtract.

(a) $6\sqrt{125} + 3\sqrt{20} = 6\sqrt{25 \cdot 5} + 3\sqrt{4 \cdot 5}$
$$= 6\sqrt{25}\sqrt{5} + 3\sqrt{4}\sqrt{5} \qquad \sqrt{ab} = \sqrt{a}\sqrt{b}$$
$$= 6 \cdot 5\sqrt{5} + 3 \cdot 2\sqrt{5}$$
$$= 30\sqrt{5} + 6\sqrt{5} = 36\sqrt{5}$$

Practice Exercise 5

Add or subtract.

(a) $5\sqrt{147} - 9\sqrt{75}$

(b) $4\sqrt[3]{8x^4y^5} - 3\sqrt[3]{27x^4y^5} = 4\sqrt[3]{2^3x^3y^3 \cdot xy^2} - 3\sqrt[3]{3^3x^3y^3 \cdot xy^2}$
$= 4(2xy)\sqrt[3]{xy^2} - 3(3xy)\sqrt[3]{xy^2}$
$= 8xy\sqrt[3]{xy^2} - 9xy\sqrt[3]{xy^2}$
$= -xy\sqrt[3]{xy^2}$

(b) $-3\sqrt{125a^5b^2} + 5a\sqrt{80a^3b^2}$

Answers: (a) $-10\sqrt{3}$
(b) $5a^2b\sqrt{5a}$

③ RATIONALIZING DENOMINATORS AND NUMERATORS

For uniformity of answers to a problem needing radical simplification, we usually *rationalize the denominator*. For example, consider the following.

$$\frac{2a}{\sqrt{5}} = \frac{2a\sqrt{5}}{\sqrt{5}\sqrt{5}} = \frac{2a\sqrt{5}}{\sqrt{5^2}} = \frac{2a\sqrt{5}}{5}$$

and $\quad \dfrac{3}{\sqrt[3]{x}} = \dfrac{3\sqrt[3]{x^2}}{\sqrt[3]{x}\sqrt[3]{x^2}} = \dfrac{3\sqrt[3]{x^2}}{\sqrt[3]{x^3}} = \dfrac{3\sqrt[3]{x^2}}{x}$

Notice that in both cases the radical in the denominator was eliminated by multiplying both the numerator and denominator of the fraction by a radical that makes the radicand in the denominator a perfect kth power. The process of doing this is called **rationalizing the denominator.**

If a denominator has a binomial (two-term) factor of the form $c\sqrt{x} + d\sqrt{y}$ then we multiply both numerator and denominator by $c\sqrt{x} - d\sqrt{y}$ to *rationalize a binomial denominator*. Notice that this procedure uses the fact that $(a + b)(a - b) = a^2 - b^2$, so that if a or b or both are square root radicals, by squaring them the radical(s) are eliminated. The expressions $c\sqrt{x} + d\sqrt{y}$ and $c\sqrt{x} - d\sqrt{y}$ are **conjugates** of each other.

EXAMPLE 6 RATIONALIZING DENOMINATORS

Rationalize the denominator.

(a) $\dfrac{\sqrt{7x}}{\sqrt{8y}} = \dfrac{\sqrt{7x}}{\sqrt{4 \cdot 2y}}$ $\quad x \geq 0$ and $y > 0$

$= \dfrac{\sqrt{7x}}{2\sqrt{2y}} \cdot \dfrac{\sqrt{2y}}{\sqrt{2y}}$ Multiply by $\dfrac{\sqrt{2y}}{\sqrt{2y}}$

$= \dfrac{\sqrt{7x}\sqrt{2y}}{2(\sqrt{2y})^2}$

$= \dfrac{\sqrt{14xy}}{2(2y)}$ Since $y > 0$, $|2y| = 2y$

$= \dfrac{\sqrt{14xy}}{4y}$

(b) $\dfrac{\sqrt[3]{5a^4b}}{\sqrt[3]{9ab^2}} = \sqrt[3]{\dfrac{5a^4b}{9ab^2}}$

$= \sqrt[3]{\dfrac{5a^3}{9b}}$ Simplify the fraction

$= \sqrt[3]{\dfrac{5a^3}{9b} \cdot \dfrac{3b^2}{3b^2}}$ $9b \cdot 3b^2 = 27b^3$, a perfect cube

$= \dfrac{\sqrt[3]{15a^3b^2}}{\sqrt[3]{27b^3}} = \dfrac{a\sqrt[3]{15b^2}}{3b} = \dfrac{\sqrt[3]{a^3 \cdot 15b^2}}{\sqrt[3]{(3b)^3}}$

PRACTICE EXERCISE 6

Rationalize the denominator.

(a) $\sqrt{\dfrac{36x^2}{y}}$

(b) $\dfrac{\sqrt[5]{2y}}{\sqrt[5]{x^2}}$

(c) $\dfrac{\sqrt{5}}{2\sqrt{7}+\sqrt{5}} = \dfrac{\sqrt{5}(2\sqrt{7}-\sqrt{5})}{(2\sqrt{7}+\sqrt{5})(2\sqrt{7}-\sqrt{5})}$ Since the denominator is $2\sqrt{7}+\sqrt{5}$, multiply numerator and denominator by $2\sqrt{7}-\sqrt{5}$

$= \dfrac{\sqrt{5}(2\sqrt{7}-\sqrt{5})}{(2\sqrt{7})^2-(\sqrt{5})^2}$ $(a+b)(a-b) = a^2 - b^2$

$= \dfrac{2\sqrt{35}-5}{4(7)-5}$

$= \dfrac{2\sqrt{35}-5}{23}$

(c) $\dfrac{x-y}{\sqrt{x}-\sqrt{y}}$

Answers: (a) $\dfrac{6x\sqrt{y}}{y}$ (b) $\dfrac{\sqrt[5]{2x^3y}}{x}$ (c) $\sqrt{x}+\sqrt{y}$

In higher mathematics it is sometimes necessary to rationalize a numerator. The technique is the same as rationalizing a denominator. For example, to rationalize the numerator in the fraction below, multiply both the numerator and the denominator by the conjugate of $\sqrt{x}+\sqrt{y}$, which is $\sqrt{x}-\sqrt{y}$.

$\dfrac{\sqrt{x}+\sqrt{y}}{x-y} = \dfrac{(\sqrt{x}+\sqrt{y})(\sqrt{x}-\sqrt{y})}{(x-y)(\sqrt{x}-\sqrt{y})} = \dfrac{x-y}{(x-y)(\sqrt{x}-\sqrt{y})} = \dfrac{1}{\sqrt{x}-\sqrt{y}}$

⚠️ CAUTION

The method shown above for rationalizing a binomial denominator or numerator works only when the radicals are *square* roots. The technique would not apply, for example, to a denominator of the form $\sqrt[3]{x}-\sqrt[3]{y}$.

The rules reviewed in this section are all used to change a radical expression into *simplest form*.

Simplest Form of a Radical Expression

A radical expression is in **simplest form** provided:

1. The radicand contains no factor raised to a power greater than or equal to the index.
2. There are no fractions in a radicand nor radicals in a denominator.
3. The index on the radical and exponents on all factors in the radicand have no common factors greater than 1.

For example, $\sqrt{x^3}$ is not in simplest form since it violates property 1, $\sqrt{\dfrac{1}{2}}$ and $\dfrac{1}{\sqrt{2}}$ are not in simplest form since they violate property 2, and $\sqrt[4]{a^2}$ is not in simplest form since it violates property 3 because $\sqrt[4]{a^2}$ can be written as $\sqrt{\sqrt{a^2}}$ which is $\sqrt{a}$ using the index rule.

4 APPROXIMATING RADICALS

In an applied problem, an answer with a radical that represents an irrational number is often approximated by a decimal rational number. For example, if the answer to a problem is $\sqrt{71}$ feet, most of us would have a better idea of the size of this measurement if we used 8.4 feet. When an approximation for a radical is needed, the $\boxed{\sqrt{}}$ key on a calculator can be used. For example, the steps to use to approximate $\sqrt{71}$ are given below.

ALG and RPN: 71 $\boxed{\sqrt{}}$ $\boxed{8.426149773}$

EXAMPLE 7 A Physics Problem

The time in seconds T it takes for a pendulum to make one complete swing is related to the length of the pendulum in feet L by the formula

$$T = 2\pi\sqrt{\frac{L}{32}}.$$

If a pendulum is 17 feet long, find the time it takes to make one swing, to the nearest tenth of a second.

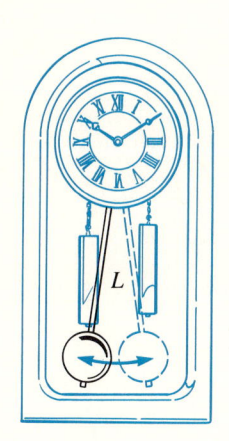

Figure 1.6

Substitute 17 for L and evaluate.

$$T = 2\pi\sqrt{\frac{17}{32}}$$

The steps to use on a calculator are:

ALG: 17 $\boxed{\div}$ 32 $\boxed{=}$ $\boxed{\sqrt{}}$ $\boxed{\times}$ 2 $\boxed{\times}$ $\boxed{\pi}$ $\boxed{=}$ $\boxed{4.579618909}$
RPN: 17 $\boxed{\text{ENTER}}$ 32 $\boxed{\div}$ $\boxed{\sqrt{}}$ 2 $\boxed{\times}$ $\boxed{\pi}$ $\boxed{\times}$ $\boxed{4.579618911}$

Thus, rounded to the nearest tenth, the time for one complete swing of the pendulum is approximately 4.6 seconds.

Practice Exercise 7

Use the formula in Example 7 to find the time for one swing of a pendulum that is 39 feet long. Give the answer correct to the nearest tenth.

Answer: 6.9 sec

Notice in Example 7 that we used the $\boxed{\pi}$ key in our calculation. If your calculator does not have this key, you can use an approximation such as 3.14 for π. Answers may vary a bit, but they should be close enough for practical purposes. You might also note that the two results shown above differ in the eighth and ninth decimal places. This too can happen when different calculators are used, as was the case here.

64 CHAPTER 1 REVIEW OF FUNDAMENTAL CONCEPTS

CAUTION

Do not get in the habit of reaching for your calculator when you see a radical problem. You should use a calculator only in an applied problem that requires an approximate answer. **Never use a calculator** when you are simplifying radical expressions similar to those we studied earlier in this section.

1.6 EXERCISES A

Assume all variables and algebraic expressions under even-indexed radicals are positive. Simplify each expression in Exercises 1–39. Rationalize all denominators.

1. $\sqrt{36}$
2. $\sqrt[3]{-27}$
3. $-\sqrt{121}$
4. $-\sqrt[5]{-32}$
5. $\sqrt{4a^8}$
6. $\sqrt[3]{8x^6}$
7. $\sqrt[5]{-32x^5y^{10}}$
8. $\sqrt[4]{81x^8y^4}$
9. $\sqrt{x^2 - 10x + 25}$
10. $\sqrt{4x^2 + 12x + 9}$
11. $2\sqrt{2}\sqrt{6}$
12. $4\sqrt{5}\sqrt{15}\sqrt{3}$
13. $4\sqrt[3]{24}$
14. $3\sqrt{72}$
15. $\sqrt[3]{3xy^2}\sqrt[3]{9x^2y^4}$
16. $\sqrt[4]{4x^3y}\sqrt[4]{4xy^5}$
17. $\dfrac{\sqrt{8x^3z}}{\sqrt{2xz^3}}$
18. $\dfrac{\sqrt[3]{16w^4z^2}}{\sqrt[3]{2wz^5}}$
19. $\dfrac{\sqrt[3]{2^{-1}x^{-3}y^{-2}}}{\sqrt[3]{16^{-1}x^{-4}y^{-8}}}$
20. $\sqrt[3]{\sqrt[4]{a^{24}b^{48}}}$
21. $\sqrt[3]{\sqrt[3]{11u^{25}v^{19}}}$
22. $3\sqrt{27} - 4\sqrt{12}$
23. $2\sqrt[3]{16} + 4\sqrt[3]{54}$
24. $5\sqrt[4]{32} - 3\sqrt[4]{2}$
25. $2\sqrt{\dfrac{25}{4}} - 5\dfrac{\sqrt{8}}{\sqrt{12}}$
26. $7\sqrt{27a^2b} - 3a\sqrt{3b}$
27. $\sqrt{5a^2 + 10a + 5} + \sqrt{20a^2 + 40a + 20}$
28. $\dfrac{2 + \sqrt{8}}{2}$
29. $\dfrac{5 - \sqrt{50}}{10}$
30. $\sqrt{\dfrac{125}{45}}$
31. $\dfrac{\sqrt[4]{5}}{\sqrt[4]{2}}$
32. $\dfrac{\sqrt{3x}}{\sqrt{6z}}$
33. $\sqrt{\dfrac{4a^3}{ab}}$
34. $\dfrac{\sqrt[4]{32a^5}}{\sqrt[4]{2b^3}}$
35. $\sqrt[3]{\dfrac{72x^2}{3xy}}$
36. $\dfrac{\sqrt{5}}{\sqrt{3} - \sqrt{5}}$
37. $\dfrac{\sqrt{7} + \sqrt{2}}{\sqrt{7} - \sqrt{2}}$
38. $\dfrac{\sqrt{x} + 1}{\sqrt{x} - 1}$
39. $2\sqrt{18x} + \dfrac{3\sqrt{x}}{\sqrt{2}}$

In Exercises 40–42 rationalize each numerator.

40. $\dfrac{1 + \sqrt{2}}{1 - \sqrt{2}}$

41. $\dfrac{\sqrt{a} + \sqrt{b}}{\sqrt{ab}}$

42. $\dfrac{\sqrt{a+b} + \sqrt{a}}{b}$

43. PHYSICS The distance in miles that can be viewed on the surface of the ocean can be approximated by $d = 1.4\sqrt{h}$ where h is the height of the viewer in feet above the surface of the water. From a helicopter at an altitude of 3000 ft, is it possible to view an island at a distance of 90 mi?

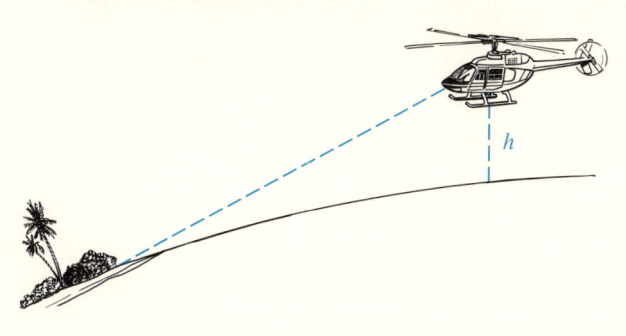

44. PHYSICS If an object is dropped from a height of h ft, the time in seconds required for the object to hit the ground is given by $t = \dfrac{\sqrt{h}}{4}$. Find the time it would take, to the nearest tenth of a second, for a rock dropped from a cliff 250 ft high to reach the ground.

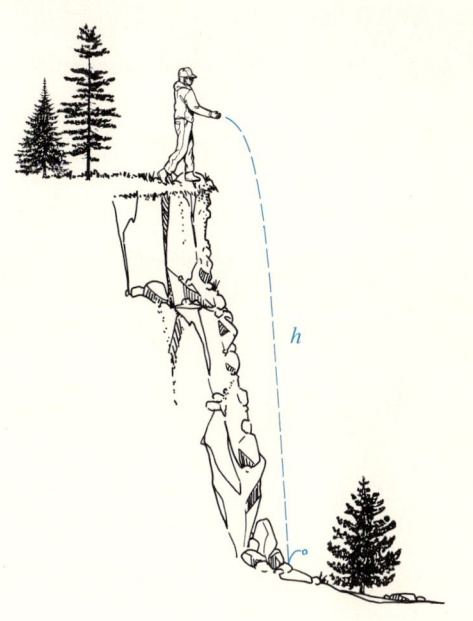

45. GEOMETRY If a and b are the lengths of the legs and c the length of the hypotenuse in a right triangle, the Pythagorean theorem states that $c^2 = a^2 + b^2$. Finding a side of a right triangle involves evaluating a radical. A baseball diamond is a square with sides 90 ft in length. To the nearest tenth of a foot, what is the distance from home plate to second base?

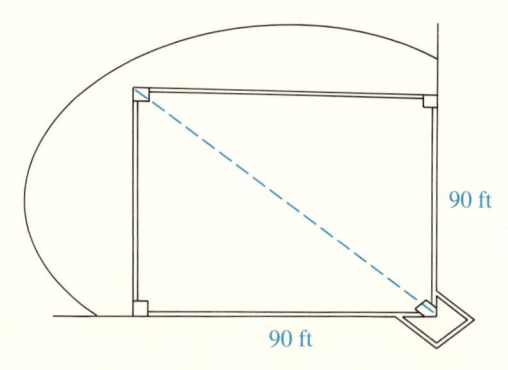

CHAPTER 1 REVIEW OF FUNDAMENTAL CONCEPTS

FOR REVIEW

Perform the indicated operations in Exercises 46–49.

46. $\dfrac{6y^2 - xy - x^2}{2y^2 - 3xy + x^2} \div \dfrac{3y^2 - 5xy - 2x^2}{x^2 - y^2}$

47. $\dfrac{3}{x+1} + \dfrac{4-2x}{x^2-1}$

48. $(2x^2y^2 - 5xy^2 + 8xy) - (3x^2y^2 - 2xy^2 + 4xy)$

49. $\dfrac{u^{-1} + v^{-1}}{1 - (uv)^{-1}}$

50. Find the error in the work shown below. What is the correct answer?

$$\dfrac{2}{x^2-1} - \dfrac{1}{x-1} = \dfrac{2}{(x-1)(x+1)} - \dfrac{1}{x-1}$$
$$= \dfrac{2}{(x-1)(x+1)} - \dfrac{x+1}{(x-1)(x+1)}$$
$$= \dfrac{2-x+1}{(x-1)(x+1)}$$
$$= \dfrac{3-x}{(x-1)(x+1)}$$

ANSWERS: 1. 6 2. −3 3. −11 4. 2 5. $2a^4$ 6. $2x^2$ 7. $-2xy^2$ 8. $3x^2y$ 9. $x-5$ 10. $2x+3$ 11. $4\sqrt{3}$ 12. 60 13. $8\sqrt[3]{3}$ 14. $18\sqrt{2}$ 15. $3xy^2$ 16. $2xy\sqrt{y}$ 17. $\frac{2x}{z}$ 18. $\frac{2w}{z}$ 19. $2y^2\sqrt[3]{x}$ 20. a^2b^4 21. $u^2v^2\sqrt[9]{11u^7v}$ 22. $\sqrt{3}$ 23. $16\sqrt[3]{2}$ 24. $7\sqrt[3]{2}$ 25. $\frac{15-5\sqrt{6}}{3}$ 26. $18a\sqrt{3b}$ 27. $3(a+1)\sqrt{5}$ 28. $1+\sqrt{2}$ 29. $\frac{1-\sqrt{2}}{2}$ 30. $\frac{5}{3}$ 31. $\frac{\sqrt[4]{40}}{2}$ 32. $\frac{\sqrt{2xz}}{2z}$ 33. $\frac{2a\sqrt{b}}{b}$ 34. $\frac{2a\sqrt[4]{ab}}{b}$ 35. $\frac{2\sqrt[3]{3xy^2}}{y}$ 36. $\frac{\sqrt{15}+5}{-2}$ 37. $\frac{9+2\sqrt{14}}{5}$ 38. $\frac{x+2\sqrt{x}+1}{x-1}$ 39. $\frac{15\sqrt{2x}}{2}$ 40. $\frac{1}{2\sqrt{2}-3}$ 41. $\frac{a-b}{a\sqrt{b}-b\sqrt{a}}$ 42. $\frac{1}{\sqrt{a+b}-\sqrt{a}}$ 43. No; the viewing distance is about 77 mi. 44. 4.0 sec 45. 127.3 ft 46. $\frac{x+y}{2x-y}$ 47. $\frac{1}{x-1}$ 48. $-x^2y^2 - 3xy^2 + 4xy$ 49. $\frac{u+v}{uv-1}$ 50. The error was made when $x+1$ was subtracted from 2 without using parentheses. The numerator should be $2 - (x+1) = 2 - x - 1$. The correct answer is $-\frac{1}{x+1}$. (Note that $1-x$ and $x-1$ are negatives.)

1.6 EXERCISES B

Assume all variables and algebraic expressions under even-indexed radicals are positive. Simplify each expression in Exercises 1–39.

1. $\sqrt{81}$
2. $\sqrt[3]{-64}$
3. $-\sqrt[4]{625}$
4. $-\sqrt[3]{-27}$
5. $\sqrt[4]{16a^{12}}$
6. $\sqrt[3]{27x^{12}}$
7. $\sqrt[5]{-243a^5b^{10}}$
8. $\sqrt[4]{625x^{16}y^8}$
9. $\sqrt{x^2+10x+25}$
10. $\sqrt{4x^2+8x+4}$
11. $4\sqrt{5}\sqrt{125}$
12. $3\sqrt{7}\sqrt{49}\sqrt{7}$
13. $7\sqrt[3]{128}$
14. $8\sqrt{108}$
15. $\sqrt[3]{2xy^2}\sqrt[3]{4x^4y^2}$
16. $\sqrt[4]{8u^3v^7}\sqrt[4]{8u^2v^8}$
17. $\dfrac{\sqrt{125x^3y}}{\sqrt{5xy^3}}$
18. $\dfrac{\sqrt[3]{81a^4b^5}}{\sqrt[3]{3ab}}$
19. $\dfrac{\sqrt[3]{25x^3y^2}}{\sqrt[4]{5^{-2}x^{-1}y^{-2}}}$
20. $\sqrt{\sqrt[3]{x^{12}y^{18}}}$
21. $\sqrt{\sqrt{16x^6y^9}}$
22. $6\sqrt{147} - 3\sqrt{75}$
23. $4\sqrt[3]{81} + 5\sqrt[3]{24}$
24. $3\sqrt[4]{48} + 5\sqrt[4]{243}$

25. $5\sqrt{\dfrac{16}{9}} - 3\dfrac{\sqrt{50}}{\sqrt{32}}$

26. $6\sqrt{8x^3y} + 2x\sqrt{8xy}$

27. $2\sqrt[3]{27x^4y^6} - 4y\sqrt[3]{125x^4y^3}$

28. $\dfrac{8 + \sqrt{48}}{4}$

29. $\dfrac{9 - \sqrt{63}}{6}$

30. $\sqrt{\dfrac{250}{27}}$

31. $\dfrac{\sqrt[3]{6}}{\sqrt[3]{7}}$

32. $\dfrac{\sqrt{8x}}{\sqrt{6y}}$

33. $\sqrt{\dfrac{9x^3}{2xy}}$

34. $\dfrac{\sqrt[3]{3xy}}{\sqrt[3]{2y^2}}$

35. $\sqrt{\dfrac{3a^3b^2}{150ab^3}}$

36. $\dfrac{2}{\sqrt{3} - \sqrt{5}}$

37. $\dfrac{3\sqrt{5} - \sqrt{3}}{\sqrt{5} + \sqrt{3}}$

38. $\dfrac{\sqrt{a} + 2\sqrt{b}}{\sqrt{a} - 2\sqrt{b}}$

39. $3\sqrt{20a} - \dfrac{2\sqrt{a}}{\sqrt{5}}$

In Exercises 40–42 rationalize each numerator.

40. $\dfrac{1 + \sqrt{3}}{1 - \sqrt{3}}$

41. $\dfrac{\sqrt{x} - \sqrt{y}}{\sqrt{xy}}$

42. $\dfrac{\sqrt{x + h} - \sqrt{x}}{h}$

43. GEOMETRY The radius of a circle inscribed in a triangle with sides a, b, and c is given by

$$r = \dfrac{\sqrt{s - a}\sqrt{s - b}\sqrt{s - c}}{\sqrt{s}}$$

where $s = \tfrac{1}{2}(a + b + c)$. Express r as a single radical and use the result to find the radius of a circle inscribed in a triangle with sides 12.0 in, 14.0 in, and 20.0 in.

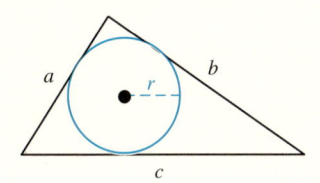

44. PHYSICS The relationship between the distance in feet d required to reach a velocity v in ft/sec when the acceleration a in ft/sec² is given can be approximated by $v = \sqrt{2ad}$. Find v to the nearest tenth if a is 88 ft/sec² and d is 60 ft.

45. A boat leaves an island and sails due east 20 miles then due south 35 miles. To the nearest tenth of a mile, how far is the boat from the island? [*Hint:* Use the Pythagorean theorem.]

FOR REVIEW

Perform the indicated operations in Exercises 46–49.

46. $\dfrac{x^2 - 3xy - 40y^2}{x^2 - 25y^2} \cdot \dfrac{x - 5y}{x^2 - 16xy + 64y^2}$

47. $\dfrac{a}{a^3 - 9a} - \dfrac{3}{a^2 - 7a + 12}$

48. $\dfrac{\dfrac{1}{x + y} - \dfrac{1}{x}}{\dfrac{1}{x + y} - \dfrac{1}{y}}$

49. $(3u - 2v + 1)^2$

50. Find the error in the work shown below. What is the correct answer?

$$(x + y) \div \left(\dfrac{1}{x} + \dfrac{1}{y}\right) = (x + y) \cdot \left(\dfrac{x}{1} + \dfrac{y}{1}\right) = (x + y)(x + y) = (x + y)^2$$

1.6 EXERCISES C

1. Rationalize the denominator.

$$\frac{1}{\sqrt[3]{a} + \sqrt[3]{b}}$$

   [*Hint:* Multiply the numerator and denominator by $\sqrt[3]{a^2} - \sqrt[3]{a}\sqrt[3]{b} + \sqrt[3]{b^2}$.]

2. Rationalize the numerator.

$$\frac{\sqrt[3]{x+h} - \sqrt[3]{x}}{h}$$

3. **GEOMETRY** Triangle *ABC*, shown here, is an *equilateral triangle* with the three equal sides of length *x*. Construct an altitude in the triangle and find a formula for the area of the triangle in terms of *x*. Suppose that *D*, *E*, and *F* are midpoints of the three sides of triangle *ABC*. Find a formula for the area of triangle *DEF* in terms of *x*.

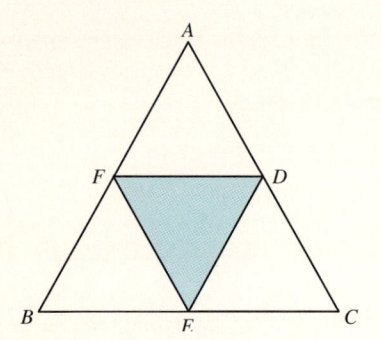

4. Suppose that *x* and *y* are any two positive real numbers. Show that there will always exist positive real numbers *u* and *v* so that $\sqrt{x+y} = \sqrt{u} + \sqrt{v}$. Are there any values of *x* and *y* for which *u* will equal *x* and *v* will equal *y*? [*Hint:* Consider the expression $(\sqrt{x+y} - \sqrt{y})^2$.]

1.7 RATIONAL EXPONENTS

STUDENT GUIDEPOSTS

1. Definition of a Rational Exponent
2. Conversion Between Radicals and Rational Exponents
3. Simplifying Rational-Exponent Expressions
4. Simplifying Radicals Using Rational Exponents

1 DEFINITION OF A RATIONAL EXPONENT

In Section 1.2 we reviewed the properties of integer exponents. Fractional or rational exponents are defined in such a way that the same properties are true. For example, suppose we consider $a^{1/2}$. Since the product rule for exponents should hold, it should be true that

$$a^{1/2} \cdot a^{1/2} = a^{1/2+1/2} = a^1 = a.$$

Then, we can define $a^{1/2}$ to be $\sqrt{a}$. Similarly,

$$a^{1/3} \cdot a^{1/3} \cdot a^{1/3} = a^{1/3+1/3+1/3} = a^1 = a,$$

so $a^{1/3}$ should equal $\sqrt[3]{a}$. Generalizing, we have the following.

1.7 RATIONAL EXPONENTS

Definition of $a^{1/n}$

Let a be a nonnegative real number with n a natural number greater than 1. Then

$$a^{1/n} = \sqrt[n]{a}.$$

Notice in this definition that a is assumed to be nonnegative. This prevents having to consider the special cases when n is even and when n is odd. To define $a^{m/n}$, we again consider the rules of exponents, this time the power rule.

$$a^{m/n} = a^{m \cdot 1/n} = (a^m)^{1/n} \quad \text{and} \quad a^{m/n} = a^{1/n \cdot m} = (a^{1/n})^m$$

Definition of $a^{m/n}$

Let a be a nonnegative real number with m and n natural numbers, n greater than 1. Then

$$a^{m/n} = \sqrt[n]{a^m} = (\sqrt[n]{a})^m.$$

❷ CONVERSION BETWEEN RADICALS AND RATIONAL EXPONENTS

In view of the above definitions, an expression having rational exponents can always be changed into a radical expression, and conversely.

EXAMPLE 1 Changing to Radical Notation

Change from exponential form to radical form and simplify.

(a) $x^{3/5} = \sqrt[5]{x^3}$ or $(\sqrt[5]{x})^3$

(b) $9^{-1/2} = \dfrac{1}{9^{1/2}} = \dfrac{1}{\sqrt{9}} = \dfrac{1}{3}$ Use definition of negative exponent

(c) $27^{2/3} = (\sqrt[3]{27})^2 = (3)^2 = 9$

We could also write $27^{2/3} = \sqrt[3]{27^2} = \sqrt[3]{729} = \sqrt[3]{9^3} = 9$. Notice, however, that it is easier to take the root first then the power, to avoid finding a root of such a large number.

PRACTICE EXERCISE 1

Change from exponential form to radical form and simplify.

(a) $a^{1/4}$

(b) $8^{-1/3}$

(c) $36^{-3/2}$

Answers: (a) $\sqrt[4]{a}$ (b) $\frac{1}{2}$ (c) $\frac{1}{216}$

EXAMPLE 2 Changing to Exponential Notation

Change from radical form to exponential form.

(a) $\sqrt[3]{x^2} = x^{2/3}$

(b) $\dfrac{1}{\sqrt{7}} = \dfrac{1}{7^{1/2}} = 7^{-1/2}$

(c) $\sqrt[4]{a^2} = a^{2/4} = a^{1/2}$ (which is $\sqrt{a}$)

Notice that we could have simplified $\sqrt[4]{a^2}$ using the index rule for radicals, but it is easier to use rational exponents and simply reduce the fraction.

PRACTICE EXERCISE 2

Change from radical form to exponential form.

(a) $\sqrt[5]{a^3}$

(b) $\dfrac{1}{\sqrt[3]{2}}$

(c) $\sqrt[8]{y^6}$

Answers: (a) $a^{3/5}$ (b) $2^{-1/3}$ (c) $y^{3/4}$ (or $\sqrt[4]{y^3}$ in radical form)

3 SIMPLIFYING RATIONAL-EXPONENT EXPRESSIONS

Note When converting from rational exponent form to radical form, and conversely, simply remember that the denominator on the fractional exponent becomes the index on the radical, and the numerator on the fraction becomes the power. This is easy to remember if you keep in mind the simple example $a^{1/2} = \sqrt{a}$ (the index is 2).

Expressions with rational exponents can be simplified in the same way that we simplified expressions with integer exponents.

EXAMPLE 3 SIMPLIFYING RATIONAL EXPONENTS

Simplify and write the result using radicals.

(a) $a^{1/2} \cdot a^{1/3} = a^{1/2+1/3}$ Product rule
$= a^{3/6+2/6}$
$= a^{5/6} = \sqrt[6]{a^5}$

(b) $\dfrac{y^{5/9}}{y^{1/3}} = y^{5/9-1/3}$ Quotient rule
$= y^{5/9-3/9}$
$= y^{2/9} = \sqrt[9]{y^2}$

(c) $(a^{2/3})^{3/4} = a^{2/3 \cdot 3/4}$ Power rule
$= a^{2/4}$
$= a^{1/2} = \sqrt{a}$

(d) $(x^{2/3} + x^{1/3})x^{3/5}$
$= x^{2/3}x^{3/5} + x^{1/3}x^{3/5}$ Distributive law
$= x^{2/3+3/5} + x^{1/3+3/5}$ Product rule
$= x^{19/15} + x^{14/15}$
$= \sqrt[15]{x^{19}} + \sqrt[15]{x^{14}}$
$= x\sqrt[15]{x^4} + \sqrt[15]{x^{14}}$

PRACTICE EXERCISE 3

Simplify and write the result using radicals.

(a) $x^{1/2} \cdot x^{1/6}$

(b) $\dfrac{w^{1/2}}{w^{2/3}}$

(c) $(m^{3/2})^{-2/3}$

(d) $(y^{1/2} - y^{3/2})^2$

Answers: (a) $\sqrt[3]{x^2}$ (b) $\dfrac{1}{\sqrt[6]{w}}$ (c) $\dfrac{1}{m}$ (d) $y - 2y^2 + y^3$

4 SIMPLIFYING RADICALS USING RATIONAL EXPONENTS

Remember the product rule for radicals from Section 1.6.
$$\sqrt{a}\sqrt{b} = \sqrt{ab}$$

As we noted then, this property is easy to verify by changing from radical notation to exponential notation.
$$\sqrt{a}\sqrt{b} = a^{1/2} \cdot b^{1/2} = (ab)^{1/2} \quad a^n b^n = (ab)^n$$
$$= \sqrt{ab}$$

You will be asked to verify the quotient rule and index rule in the exercises using the same technique. In fact, many radical expressions can sometimes be simplified more easily by converting to rational exponents, using the properties of exponents, and then changing back to radicals. This is especially true for expressions that involve the index rule.

EXAMPLE 4 Simplifying Radicals Using Exponents

Use rational exponents to simplify.

(a) $(\sqrt[3]{-8x^9y^6})^2 = (-8x^9y^6)^{2/3}$ $\qquad (\sqrt[3]{a})^2 = a^{2/3}$
$\qquad\qquad\qquad\;\; = [(-2)^3]^{2/3}(x^9)^{2/3}(y^6)^{2/3}$ $\quad -8 = (-2)^3$
$\qquad\qquad\qquad\;\; = (-2)^2 x^6 y^4 = 4x^6 y^4$

(b) $\sqrt[4]{\dfrac{16x^9}{81y^8}} = \left(\dfrac{2^4 x^9}{3^4 y^8}\right)^{1/4}$ $\qquad \sqrt[4]{a} = a^{1/4}$

$\qquad\quad = \dfrac{(2^4)^{1/4}(x^9)^{1/4}}{(3^4)^{1/4}(y^8)^{1/4}}$

$\qquad\quad = \dfrac{2x^{9/4}}{3y^2} = \dfrac{2x^2 x^{1/4}}{3y^2}$ $\quad x^{9/4} = x^{8/4+1/4} = x^{8/4}x^{1/4} = x^2 x^{1/4}$

$\qquad\quad = \dfrac{2x^2\sqrt[4]{x}}{3y^2}$

(c) $\sqrt[3]{\sqrt{5}} = (\sqrt{5})^{1/3} = (5^{1/2})^{1/3} = 5^{1/6} = \sqrt[6]{5}$

(d) $\sqrt{x}\sqrt[5]{x^2} = x^{1/2}x^{2/5} = x^{1/2+2/5} = x^{9/10} = \sqrt[10]{x^9}$

PRACTICE EXERCISE 4

Use rational exponents to simplify.

(a) $\sqrt[4]{16a^8 b^{24}}$

(b) $\left(\dfrac{243a^4 b^{-10}}{64a^{-6} b^5}\right)^{-1/5}$

(c) $\sqrt[4]{\sqrt[3]{a^2}}$

(d) $\sqrt[3]{\sqrt[4]{x+y}}$

Answers: (a) $2a^2 b^6$ (b) $\dfrac{2b^3\sqrt[5]{2}}{3a^2}$
(c) $\sqrt[6]{a}$ (d) $\sqrt[12]{x+y}$

Notice in Example 4(c) and (d) that rational exponents make it much easier to simplify expressions involving either a radical radicand or a product of radicals with different indexes.

If an approximation is needed for a radical that has an index other than 2, the $\boxed{y^x}$ key on a calculator can be used along with knowledge of rational exponents. For example, to approximate $\sqrt[3]{15}$, use the following steps.

ALG: 15 $\boxed{y^x}$ 3 $\boxed{1/x}$ $\boxed{=}$ $\boxed{2.466212074}$
RPN: 15 $\boxed{\text{ENTER}}$ 1 $\boxed{\text{ENTER}}$ 3 $\boxed{\div}$ $\boxed{y^x}$ $\boxed{2.466212074}$

EXAMPLE 5 An Agriculture Problem

The length of an edge of a cube with volume V is given by $e = \sqrt[3]{V}$. If a farmer wishes to build a cubical storage bin that will hold 100 ft³ of wheat, to the nearest tenth, what should he use for the length of an edge of the bin?

PRACTICE EXERCISE 5

Find the approximate length of an edge of a cube with volume 36.8 in³.

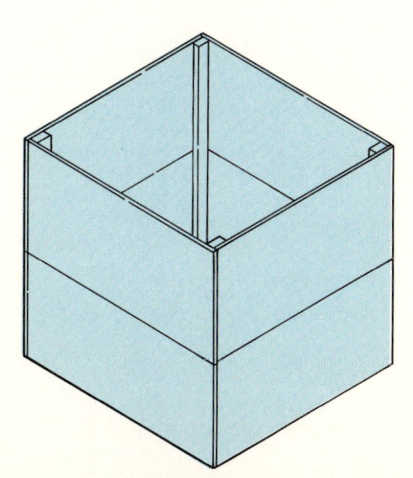

Figure 1.7

Substitute 100 for V in the formula
$$e = \sqrt[3]{V}$$
and use a calculator to find *e*. Remember that $\sqrt[3]{V}$ is the same as $V^{1/3}$.
Use these steps.

ALG: 100 [y^x] 3 [1/x] [=] [4.641588834]
RPN: 100 [ENTER] 1 [ENTER] 3 [÷] [y^x] [4.641588833]

Thus, the edge of the bin must be about 4.6 ft.

Answer: 3.3 in

1.7 EXERCISES A

In Exercises 1–6, change each expression from exponential form to radical form and simplify.

1. $7^{2/5}$
2. $y^{3/4}$
3. $16^{1/4}$
4. $4^{-3/2}$
5. $32^{-2/5}$
6. $(ab^6)^{1/2}$

In Exercises 7–12, change each expression from radical form to exponential form and simplify.

7. $\sqrt[5]{3^2}$
8. $\dfrac{1}{\sqrt{x}}$
9. $\sqrt[6]{a^2}$
10. $\sqrt[3]{\sqrt{5}}$
11. $\sqrt{a}\sqrt[4]{a}$
12. $\sqrt[3]{27^{-1}}$

In Exercises 13–21, simplify each expression and write the result using radicals.

13. $w^{1/3} \cdot w^{1/6}$
14. $\dfrac{a^{3/4}}{a^{1/4}}$
15. $2x^{1/3}x^{2/3}x^{1/5}$
16. $(m^{3/4})^{2/3}$
17. $(y^{1/4} + y^{3/4})y^{7/4}$
18. $(a^{1/3} + a^{2/3})^2$
19. $\dfrac{x^{1/4}y^{2/3}}{x^{1/8}y^{4/3}}$
20. $\left(\dfrac{x^6}{8y^3}\right)^{-1/3}$
21. $(16m^4n^{-2})^{-1/4}$

In Exercises 22–33, simplify each radical expression by converting to rational exponents. Give the result in radical notation.

22. $\sqrt{125x^2}$
23. $\sqrt[6]{64a^{18}b^{24}}$
24. $\sqrt{75x^5y^7}$
25. $\sqrt{\dfrac{25a^3b^4}{4a^{-3}b^2}}$
26. $\sqrt[4]{32(x+y)^5}$
27. $\sqrt[4]{(16x^{12}y^{-8})^{-3}}$
28. $\sqrt[3]{\sqrt{10}}$
29. $\sqrt[4]{\sqrt[3]{x^5}}$
30. $\sqrt{m}\sqrt[3]{m^2}$
31. $\sqrt[7]{y^2\sqrt{y^3}}$
32. $\sqrt[5]{\sqrt[3]{a+b}}$
33. $\dfrac{\sqrt[4]{a}\sqrt[3]{a^2}}{\sqrt{a}}$

In Exercises 34–37, use a calculator to approximate each root correct to two decimal places.

34. $\sqrt[4]{41}$ **35.** $\sqrt[3]{7^2}$ **36.** $18.5^{3/5}$ **37.** $(\sqrt[7]{37})^2$

38. ECONOMICS A manufacturing index I, related to the number of units in stock u, in thousands of units, is approximated by $I = 0.712(1 + u)^{1/3}$. Find I, to the nearest hundredth, when $u = 0.521$.

39. INVESTMENT The annual rate of return on an investment of P dollars that is worth A dollars after t years is given by

$$r = \left(\frac{A}{P}\right)^{1/t} - 1.$$

Use this formula to find the approximate rate of return on a rare book purchased for $2500 and sold 5 years later for $7200.

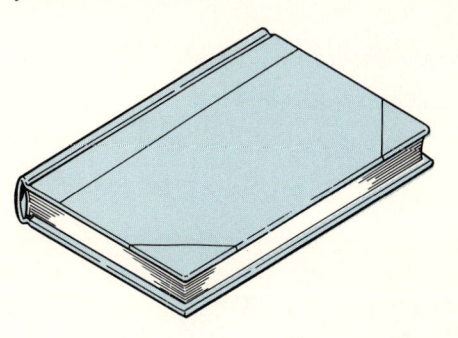

40. CRIMINOLOGY The speed at which a car was traveling at the time of an accident can be estimated from the length of the skid mark l as follows. A test car (the car involved if possible) is driven under the same conditions at a test speed S, and skidded to a stop. If L is the length of the test skid,

$$s = S\left(\frac{l}{L}\right)^{1/2}$$

After stealing a car, Fast Eddie crashed the car into a light pole while making his getaway. A policeman determined that $S = 40$ mph, $l = 185$ ft, and $L = 62$ ft. What was the approximate speed of the stolen car at the time of the accident?

41. Use rational exponents to prove the quotient rule for radicals:

$$\sqrt{\frac{a}{b}} = \frac{\sqrt{a}}{\sqrt{b}}$$

42. What is wrong with the following?
$-5 = (-5)^1 = (-5)^{2/2} = [(-5)^2]^{1/2} = [25]^{1/2} = 5$

43. Give a value of x which shows that $(x^2)^{1/2}$ *does not necessarily* equal x.

FOR REVIEW

*Simplify the radicals in Exercises 44–47. For these exercises **do not** assume that x and y are nonnegative. That is, use absolute value bars when appropriate.*

44. $\sqrt{x^6}$
45. $\sqrt[3]{x^3y^3}$
46. $\sqrt[4]{x^4y^8}$
47. $\sqrt[5]{32x^{20}}$

48. **PHYSICS** Use the formula $T = 2\pi\sqrt{\frac{L}{32}}$ to find the time in seconds it takes for a pendulum of length 30.5 feet to make one complete swing.

ANSWERS: 1. $\sqrt[5]{7^2}$ or $\sqrt[5]{49}$ 2. $\sqrt[4]{y^3}$ 3. 2 4. $\frac{1}{8}$ 5. $\frac{1}{4}$ 6. $b^3\sqrt{a}$ 7. $3^{2/5}$ 8. $x^{-1/2}$ 9. $a^{1/3}$ 10. $5^{1/6}$ 11. $a^{3/4}$
12. $\frac{1}{3}$ 13. $\sqrt{w}$ 14. $\sqrt{a}$ 15. $2x\sqrt[5]{x}$ 16. $\sqrt{m}$ 17. $y^2 + y^2\sqrt{y}$ 18. $\sqrt[3]{a^2} + 2a + a\sqrt[3]{a}$ 19. $\frac{\sqrt[8]{x}}{\sqrt[3]{y^2}}$ or $\frac{\sqrt[8]{x}\sqrt[3]{y}}{y}$ 20. $\frac{2y}{x^2}$
21. $\frac{\sqrt{n}}{2m}$ 22. $5x\sqrt{5}$ 23. $2a^3b^4$ 24. $5x^2y^3\sqrt[4]{3xy}$ 25. $\frac{5a^3b}{2}$ 26. $2(x+y)\sqrt[4]{2(x+y)}$ 27. $\frac{y^6}{8x^9}$ 28. $\sqrt[6]{10}$ 29. $\sqrt[12]{x^5}$
30. $m\sqrt[6]{m}$ 31. $\sqrt[6]{y}$ 32. $\sqrt[15]{a+b}$ 33. $\sqrt[12]{a^5}$ 34. 2.53 35. 3.66 36. 5.76 37. 2.81 38. 0.82 39. 23.6%
40. 69 mph 41. $\sqrt{\frac{a}{b}} = \left(\frac{a}{b}\right)^{1/2} = \frac{a^{1/2}}{b^{1/2}} = \frac{\sqrt{a}}{\sqrt{b}}$ 42. $[(-5)^2]^{1/2}$ is the same as $\sqrt{(-5)^2} = \sqrt{25}$ which is the principal root, 5, not -5. 43. Any negative value for x will show this. 44. $|x^3|$ 45. xy 46. $|xy^2|$ 47. $2x^4$ 48. approximately 6.1 sec

1.7 EXERCISES B

In Exercises 1–6, change each expression from exponential form to radical form and simplify.

1. $13^{2/7}$
2. $a^{4/5}$
3. $81^{1/4}$
4. $9^{-3/2}$
5. $-243^{-3/5}$
6. $(x^2y^8)^{1/2}$

In Exercises 7–12, change each expression from radical form to exponential form and simplify.

7. $\sqrt[7]{5^3}$
8. $\frac{1}{\sqrt[3]{y}}$
9. $\sqrt[8]{a^2}$
10. $\sqrt[4]{\sqrt{3}}$
11. $\sqrt{w}\sqrt[5]{w}$
12. $\sqrt[4]{16^{-1}}$

In Exercises 13–21, simplify each expression and write the result using radicals.

13. $u^{3/4} \cdot u^{1/2}$
14. $\frac{m^{3/5}}{m^{1/5}}$
15. $3a^{1/4}a^{1/2}a^{3/4}$
16. $(y^{3/7})^{2/3}$
17. $(y^{1/3} - y^{5/3})y^{2/3}$
18. $(w^{1/2} + w^{1/4})^2$
19. $\frac{a^{1/3}b^{5/6}}{a^{7/3}b^{1/6}}$
20. $\left(\frac{x^8}{16y^4}\right)^{-1/4}$
21. $(32u^5v^{-10})^{-1/5}$

In Exercises 22–33, simplify each radical expression by converting to rational exponents. Give the result in radical notation.

22. $\sqrt[3]{125x^4}$
23. $\sqrt[3]{-27x^6y^9}$
24. $\sqrt[3]{24a^{10}b^{11}}$
25. $\sqrt[3]{\frac{8x^{-3}y^7}{27x^3y^{-2}}}$
26. $\sqrt{5a^2 + 20a + 20}$
27. $\sqrt{(a^3b^2)(a^2b^3)^3}$
28. $\sqrt[4]{\sqrt[5]{3}}$
29. $\sqrt[3]{\sqrt[4]{x^7}}$
30. $\sqrt{u}\sqrt[4]{u^3}$
31. $\sqrt[6]{y^3\sqrt{y}}$
32. $\sqrt[5]{\sqrt{a-b}}$
33. $\frac{\sqrt[5]{m}\sqrt[3]{m^2}}{\sqrt{m}}$

In Exercises 34–37, use a calculator to approximate each root correct to two decimal places.

34. $\sqrt[5]{29}$ 35. $\sqrt[4]{5^3}$ 36. $23.2^{4/3}$ 37. $(\sqrt[6]{25})^5$

38. **ECONOMICS** A manufacturing index I, related to the number of units in stock u, in thousands of units, is approximated by $I = 0.853(1 + u)^{1/3}$. Find I, to the nearest hundredth, when $u = 0.635$.

39. **TRANSPORTATION** A city traffic engineer uses the formula $H = 0.00025d^{9/4}$ to determine the height of signs to be painted on a street relative to the distance d at which they are viewed from a vehicle. If Mill Street has a blind corner at a distance of 150 feet from an intersection, how high (to the nearest foot) should the letters in the word *STOP* be painted at the intersection to give a driver the best chance of reading them?

40. **NAVIGATION** Use the formula $d = 1.4h^{1/2}$ to find the viewing distance on the horizon d (in miles) from a height h (in feet) to determine whether a ship 5 miles from a submarine can be viewed from the submarine periscope raised to a height of 8 ft above the surface of the ocean.

41. Use rational exponents to prove the index rule for radicals:
$$\sqrt[m]{\sqrt[n]{a}} = \sqrt[mn]{a}$$

42. What is wrong with the following?
$$3 = 3^1 = 3^{3/3} = (3^3)^{1/3} = 27^{1/3} = \frac{1}{3}(27) = 9$$

43. Give an example to show that $(a - b)^{1/2}$ is *not* the same as $a^{1/2} - b^{1/2}$.

FOR REVIEW

*Simplify the radicals in Exercises 44–47. For these exercises **do not** assume that x and y are nonnegative. That is, use absolute value bars when necessary.*

44. $\sqrt[4]{y^4}$ 45. $\sqrt[5]{x^5 y^{10}}$ 46. $\sqrt{x^2 y}$ 47. $\sqrt[6]{64 x^6}$

48. **PHYSICS** Use the formula $T = 2\pi\sqrt{\frac{L}{32}}$ to find the time in seconds it takes for a pendulum of length 41.5 feet to make one complete swing.

1.7 EXERCISES C

Simplify the following expressions.

1. $\dfrac{(x^2 + 1)^{1/2}(2) - (2x)\left(\dfrac{1}{2}\right)(x^2 + 1)^{-1/2}(2x)}{[(x^2 + 1)^{1/2}]^2}$

2. $\dfrac{(2x^2 - 1)^{1/3}(4) - (4x + 3)\left(\dfrac{1}{3}\right)(2x^2 - 1)^{-2/3}(4x)}{[(2x^2 - 1)^{1/3}]^2}$

HEALTH One formula for approximating the surface area, in square feet, of a human body is
$$A = (0.12)\sqrt[5]{w^2}\sqrt[10]{h^7}$$
where w is the weight of the person in pounds and h is the height of the person in inches. Use this formula, which can help in determining body fat, in Exercises 3–4.

3. Estimate the surface area of a man 6 ft tall who weighs 180 pounds. Give the answer correct to the nearest tenth.
[Answer: 19.1 ft²]

4. What is the effect on the surface area of the body of the man in Exercise 3 if he gains 10 pounds?

5. Show that $\sqrt{x^{\sqrt{x}}} = (\sqrt{x})^{\sqrt{x}}$.

6. Show that $[(\sqrt{a})^{\sqrt{a}}]^{\sqrt{a}} = a^{a/2}$.

CHAPTER 1 EXTENSION PROJECT

Have you ever wondered how your calculator arrives at the approximate square root of a number? How were square roots found before the existence of calculators? Actually a lengthy algorithm (procedure) was used that took many steps and arithmetic calculations. You might be interested in researching this algorithm in an old edition of an algebra book. Your calculator also uses an algorithm to find roots, but an algorithm different from the one taught in old algebra books.

A very simple algorithm you could use involves finding repeated approximations of a root. For example, to find the square root of 50, you might begin by noting that

$$7 < \sqrt{50} < 8 \quad \text{since} \quad 7^2 = 49 < 50 < 64 = 8^2.$$

You could then square 7.1, 7.2, 7.3, . . . , 7.9 until discovering that

$$7.0^2 = 49.00 < 50.00 < 50.41 = 7.1^2.$$

Now you know that

$$7.0 < \sqrt{50} < 7.1.$$

Subdividing again, square 7.01, 7.02, 7.03, . . . , 7.09 until you find that

$$7.07^2 = 49.9849 < 50.0000 < 50.1264 = 7.08^2.$$

Now you know that

$$7.07 < \sqrt{50} < 7.08.$$

Continuing in this manner you could approximate $\sqrt{50}$ to any desired degree of accuracy.

In another quicker technique for finding $\sqrt{x}$ for any positive real number x, you make an estimate for $\sqrt{x}$, call it a_1, and use it in the formula $a_2 = \frac{1}{2}\left(a_1 + \frac{x}{a_1}\right)$. Continue this process letting

$$a_{n+1} = \frac{1}{2}\left(a_n + \frac{x}{a_n}\right),$$

for $n = 1, 2, 3, \ldots$. You will discover that the numbers $a_1, a_2, a_3, \ldots$ quickly approach the value of $\sqrt{x}$ given by your calculator.

Suppose we try to approximate $\sqrt{50}$ this way. As a first estimate, we might try $a_1 = 7$ since $7^2 = 49$. Then

$$a_2 = \frac{1}{2}\left(a_1 + \frac{x}{a_1}\right) = \frac{1}{2}\left(7 + \frac{50}{7}\right) = 7.071428571.$$

Now use $a_2 = 7.071428571$ to find a_3.

$$a_3 = \frac{1}{2}\left(a_2 + \frac{50}{a_2}\right) = 7.071067821$$

Use a_3 to find $a_4 = 7.071067812$. Notice that a_4 does not differ much from a_3. In fact, if you evaluate $\sqrt{50}$ on your calculator, you should get a value very close to a_4.

THINGS TO DO:

1. Use the methods illustrated to approximate $\sqrt{5}$, $\sqrt{25}$, and $\sqrt{\pi}$.
2. Try to figure out why the second algorithm works.
3. Write a BASIC computer program that uses the algorithm and prints the values $a_1, a_2, a_3, \ldots, a_{10}$. Notice how these values quickly approach your calculator value of the square root of a number.
4. What happens if you vary the first estimate, a_1? For example try $a_1 = 3$ as an estimate for $\sqrt{50}$ and see how many steps you have to use to get the value your calculator gives. [It should happen at about a_6.] What conclusions might you draw?
5. Consult a history of mathematics book and research the evolution of the radical, $\sqrt{}$.
6. Can you discover an algorithm for approximating $\sqrt[3]{x}$ that is similar to the one given for $\sqrt{x}$?

CHAPTER 1 REVIEW

KEY TERMS

1.1
1. A **set** is a collection of objects called **elements** of the set.
2. **Natural (counting) numbers:**
   $\{1, 2, 3, \ldots\}$
3. **Whole numbers:**
   $\{0, 1, 2, 3, \ldots\}$
4. **Integers:**
   $\{\ldots, -3, -2, 1, 0, 1, 2, 3, \ldots\}$
5. **Rational numbers:**
   $\left\{\frac{a}{b}\right\}$ such that a and b are integers, $b \neq 0$

CHAPTER 1 REVIEW

6. **Irrational numbers:**
   {x such that x is not rational}

7. **Real numbers:**
   {x such that x is rational or x is irrational}

8. A **number line** is used to picture real numbers. The point on a number line corresponding to 0 is the **origin.**

9. Two real numbers a and b are **equal** if they correspond to the same point on a number line.

10. If a and b are real numbers, a **is less than** b ($a < b$) or b **is greater than** a ($b > a$) if $b - a$ is positive.

11. The **absolute value** of a real number a, $|a|$, is the distance from a to 0 on a number line. More formally,
$$|a| = \begin{cases} a \text{ if } a \geq 0 \\ -a \text{ if } a < 0. \end{cases}$$

12. A **theorem** is a property that can be proved using axioms and definitions.

1.2

1. An expression such as a^n is in **exponential notation** where a is the **base** and n is the **exponent.**

2. A number is written in **scientific notation** when it is expressed in the form $a \times 10^n$ where $1 \leq a < 10$ and n is an integer.

3. A number in scientific notation, $a \times 10^n$, where a has been rounded to k decimal places, has $k + 1$ **significant digits.**

1.3

1. A **variable** is a letter used to represent an arbitrary element (usually a number) of a set containing more than one element.

2. A **constant** is a symbol used to represent a specific number.

3. An **algebraic expression** includes sums, differences, products, and quotients of numbers and variables raised to various powers.

4. A **term** of an algebraic expression includes only the product of a number, called the **numerical coefficient,** and variables raised to powers.

5. A **polynomial** is an algebraic expression with terms that are products of numbers and variables raised to whole-number exponents.

6. A **monomial** is a polynomial with one term, a **binomial** is a polynomial with two terms, and a **trinomial** is a polynomial with three terms.

7. The **degree of a term of a polynomial** is the sum of the powers on the variables in the term. The **degree of a polynomial** is the same as the degree of the term of highest degree.

1.4

To factor a polynomial, write the polynomial as a product.

1.5

1. A quotient of algebraic expressions is called an **algebraic fraction.**

2. A **rational expression** is an algebraic fraction formed by taking the quotient of two polynomials.

3. Two fractions are **equivalent** if one can be obtained from the other by multiplying or dividing the numerator and the denominator by the same nonzero expression.

4. A rational expression has been **reduced to lowest terms** when 1 (or -1) is the only polynomial that will divide both the numerator and the denominator.

5. A **complex fraction** is an algebraic fraction that contains at least one fraction in its numerator or denominator.

1.6

1. If $b^2 = a$, b is a **square root** of a. If $b^3 = a$, b is a **cube root** of a. If $b^k = a$, b is a **kth root** of a.

2. If k is even and a is nonnegative, $\sqrt[k]{a}$ denotes the nonnegative kth root of a, called the **principal kth root of a.**

3. The symbol $\sqrt{}$ is called a **radical,** and for kth roots when $k > 2$, the **index** k is used on the radical, $\sqrt[k]{}$.

4. The expression written under a radical is called the **radicand.**

5. Two radical expressions (or simply two radicals) are **like radicals** if they have the same radicand and the same index.

6. Removing all radicals from a denominator (numerator) is called **rationalizing the denominator (numerator).**

7. The expressions $c\sqrt{x} + d\sqrt{y}$ and $c\sqrt{x} - d\sqrt{y}$ are **conjugates** of each other.

1.7

1. If a is a nonnegative real number with n a natural number greater than 1, then $a^{1/n} = \sqrt[n]{a}$.

2. If a is a nonnegative real number with m and n natural numbers, n greater than 1, then $a^{m/n} = \sqrt[n]{a^m} = (\sqrt[n]{a})^m$.

78 CHAPTER 1 REVIEW OF FUNDAMENTAL CONCEPTS

REVIEW EXERCISES

Part I

1.1 *Answer* true *or* false *in Exercises 1–4. If the statement is false, explain why.*

1. $\sqrt{5}$ is a rational number.

2. Some integers are natural numbers.

3. If a is a rational number, then $\sqrt{a}$ is rational.

4. For all positive real numbers x and y, $\sqrt{x+y}$ is positive.

State the property illustrated in Exercises 5–6.

5. $3x + (y + 2) = (3x + y) + 2$

6. If $a < 5$ and $5 < b$, then $a < b$.

Simplify each expression in Exercises 7–9.

7. $-[a - (b + a)]$

8. $|-6| - |-13|$

9. $|-x|$ if $x < 0$

Find the distance between the given numbers in Exercises 10–11.

10. $-\dfrac{1}{5}$ and $\dfrac{3}{10}$

11. $\sqrt{7}$ and $-3\sqrt{7}$

Perform the indicated operations in Exercises 12–17.

12. $(-8) + (-4)$

13. $(-5) - (-4)$

14. $(-8) \div 0$

15. $0 \div (-8)$

16. $\left(-\dfrac{3}{4}\right) \cdot \left(\dfrac{1}{9}\right)$

17. $\left(-\dfrac{2}{3}\right) \div \left(-\dfrac{4}{3}\right)$

1.2 *In Exercises 18–21 simplify each expression and write it without negative exponents.*

18. $5x^3 \cdot x^{-5}$

19. $3y^{-1}$

20. $\left(\dfrac{8^0 x^{-1}}{y^3}\right)^{-2}$

21. $\left(\dfrac{5x^2 y}{x^{-1} y^2}\right)^{-3}$

In Exercises 22–23 use your calculator to perform the indicated operations. Round to three significant digits and give answers in scientific notation.

22. $\dfrac{(0.000\ 571)^2}{992{,}000}$

23. $\dfrac{(0.000\ 021\ 6)(8{,}360{,}000)}{(0.000\ 000\ 011\ 5)}$

24. **PHYSICS** If the speed of light is 3.00×10^5 km/sec, calculate the value of a light year in kilometers. (A light year is the distance light travels in a year.)

1.3 *Evaluate each expression in Exercises 25–30 when $a = -2$, $b = 3$, and $c = -1$.*

25. $(-2ab)^{-1}$

26. $(3a)^2$

27. $3a^2$

28. $(a + b + c)^2$

29. $a^2 + bc$

30. $(c - b)^3$

31. Add $13x^2 y^2 - 7xy + 8$ and $-2x^2 y^2 + 3xy + 1$

32. Subtract $3x^2 + 2y^2 - 5x + 4y$ from $2x^2 - 5y^2 - 6x + 2y$

33. $(-4a^2 b^2 - 3ab + 5) - (4a^2 b^2 - 4ab - 6)$

34. $(8u^3 v^2 - 2u^2 v^2) + (-4u^2 v^3 + 2u^2 v^2) - (u^2 v^2 - 2)$

Multiply in Exercises 35–40. Use special products when applicable.

35. $5x^3 y^2 (-2xy + 7x^2 y)$

36. $(6u - v)(4u + v)$

37. $(7x - 2y)^2$

38. $(5x + 7y)^2$

39. $(3a + 8b)(3a - 8b)$

40. $(2a^2 + b)(5a^2 - 2b)$

Divide in Exercises 41–42.

41. $\dfrac{14a^3b^4 - 28a^2b^3 + 49ab^5}{-7a^2b^3}$

42. $x - 3y \overline{)x^3 - 4x^2y + 2xy^2 - 3y^3}$

43. **ENGINEERING** The height in feet of a rocket t seconds after firing is $-16t^2 + 240t$. Find the height after 3.25 seconds.

1.4 *Factor each polynomial in Exercises 44–55. Check all answers by multiplying.*

44. $10x^2y^3 - 15xy^2$

45. $2x^2 - 4xy - 3x + 6y$

46. $u^2 - 7uv + 6v^2$

47. $u^2 + 13u + 42$

48. $9a^2 - 16b^2$

49. $3a^2 + ab - 14b^2$

50. $54x^3 + 2y^6$

51. $25x^2 + 30xy + 9y^2$

52. $8u^2 - 2uv - 15v^2$

53. $125x^3 - 64y^3$

54. $x^2 + 4y^2$

55. $16x^2 - 56xy + 49y^2$

1.5 *Reduce to lowest terms in Exercises 56–58.*

56. $\dfrac{36a^2b^2c^2}{75ab^4c^2}$

57. $\dfrac{x - 5}{5 - x}$

58. $\dfrac{x^2 + 5x - 24}{x^2 - 10x + 21}$

Perform the indicated operations in Exercises 59–62.

59. $\dfrac{a^2 - 5a + 6}{a^2 - 6a + 9} \cdot \dfrac{a^2 + 4a - 21}{a^2 + 2a - 35}$

60. $\dfrac{a^3 - b^3}{a^2 + 4a - 5} \div \dfrac{a^2 + ab + b^2}{a^2 + 10a + 25}$

61. $\dfrac{5 - x}{x^2 - 12x + 35} + \dfrac{x}{x^2 - 14x + 49}$

62. $\dfrac{3}{y^2 + y} - \dfrac{2}{y + 1}$

Find the LCD of the fractions given in Exercises 63–64.

63. $\dfrac{8ab}{a^3 - b^3}$ and $\dfrac{7}{(a - b)^2}$

64. $\dfrac{a + 1}{a^2 - 7a + b}$ and $\dfrac{a + 5}{a^2 - 12a + 36}$

Simplify each complex fraction in Exercises 65–66.

65. $\dfrac{\dfrac{a}{b} - 1}{\dfrac{b}{a} - 1}$

66. $\dfrac{a - 7 + \dfrac{10}{a}}{a - 10 + \dfrac{25}{a}}$

67. **CALCULUS** Simplify the expression that comes from taking a derivative in calculus.

$$\dfrac{4x^3(4x) - (2x^2 - 1)12x^2}{16x^6}$$

1.6 *Simplify each radical expression in Exercises 68–82. Rationalize denominators when necessary.*

68. $\sqrt[3]{-27}$

69. $\sqrt[5]{-1}$

70. $\sqrt[6]{-64}$

71. $\sqrt{9a^4}$

72. $\sqrt[3]{8x^6y^9}$

73. $\sqrt[4]{81a^8b^{20}}$

74. $\sqrt{a^2 - 14a + 49}$

75. $3\sqrt{4x}\sqrt{12xy^2}$

76. $\dfrac{\sqrt{20a^3b^7}}{\sqrt{5ab}}$

77. $\dfrac{\sqrt[4]{48a^5b^{10}}}{\sqrt[4]{243ab^5}}$

78. $4\sqrt{125} + 5\sqrt{80}$

79. $-7\sqrt[3]{54} + 10\sqrt[3]{128}$

80. $4\sqrt[3]{3xy^3} - 6y\sqrt[3]{81x}$

81. $\dfrac{5 - \sqrt{175}}{10}$

82. $\dfrac{\sqrt{8} - \sqrt{3}}{\sqrt{2} + \sqrt{3}}$

Simplify each expression in Exercises 83–88 and write the answer in radical notation.

83. $a^{7/5}$

84. $(-8)^{2/3}$

85. $(a^2b^4)^{1/2}$

86. $(y^{3/4})^{1/3}$

87. $\left(\dfrac{8a^5b^7}{a^{-4}b^{-3}}\right)^{1/3}$

88. $\left(\dfrac{27^{2/3}x^{-5/3}y^2}{x^{-10/3}y^{-2}}\right)^{-3}$

Part II

In Exercises 89–91, find the values of the variable or variables that must be excluded.

89. $\dfrac{x+7}{x^2-3x+2}$

90. $\dfrac{x^2+5x-7}{5}$

91. $\dfrac{4a^2+5b}{(a^2-4)(b+7)}$

Are the fractions given in Exercises 92–94 equivalent? Explain.

92. $\dfrac{x+y}{x^2-y^2},\ \dfrac{1}{x-y}$

93. $\dfrac{a^2+1}{a^2},\ \dfrac{a^2+2}{a^2+1}$

94. $\dfrac{x-2}{x^2+x-6},\ \dfrac{1}{x+3}$

Write in scientific notation in Exercises 95–97.

95. 5,860,000

96. 0.000 000 005 86

97. 47.2

Write as a decimal in Exercises 98–100.

98. 5.86×10^{-3}

99. 5.86×10^4

100. 3.81×10^{-5}

101. **BUSINESS** All employees of Mutter Manufacturing received a 12% raise. Find a polynomial expression for the new salary if s is the old salary. What was Pat Wood's new salary if she made $24,500 before the raise?

Evaluate each expression in Exercises 102–107 when $a = -3$, $b = -1$, and $c = 5$.

102. $-(a-c)$

103. $3a+b-c$

104. $|2a-(-b)+c|$

105. a^2

106. $-3a^2$

107. $-a^2$

Perform the indicated operations in Exercises 108–113.

108. $\dfrac{4x^2+4x+1}{16x+8} \cdot \dfrac{x^2-8x+12}{2x^2-11x-6}$

109. $\dfrac{x+2}{x-3} + \dfrac{2x-1}{x-3}$

110. $\dfrac{2xy}{x^2-y^2} + \dfrac{y}{y-x} + 3$

111. $\dfrac{2x^2+x-15}{x^2-x-12} \div \dfrac{2x^2+3x-20}{16x-x^3}$

112. $(2u^2-3uv+v^2)(5u+8v)$

113. $(3x-2y-5)^2$

Simplify the radicals in Exercises 114–119.

114. $\dfrac{\sqrt{x}+\sqrt{y}}{2\sqrt{x}-\sqrt{y}}$

115. $5\sqrt{40a} + \dfrac{8\sqrt{2a}}{\sqrt{5}}$

116. $\sqrt{\dfrac{125a^5b^{-2}}{4a^{-2}b^{-6}}}$

117. $\dfrac{\sqrt[3]{-81x^3y^{-3}}}{\sqrt[3]{3x^{-3}y^3}}$

118. $\sqrt[4]{m}\sqrt[3]{m^2}$

119. $\sqrt[5]{\sqrt{a^3}}$

Factor the polynomials in Exercises 120–125.

120. $10a^2 + 3ab - 18b^2$ **121.** $5x^3y^3 - 40z^3$ **122.** $ab + xb - ay - xy$

123. $2u^4 - u^2v^2 - v^4$ **124.** $x^2 - y^2 + 10y - 25$ **125.** $9x^{2n} - y^{2n}$

Use a calculator to approximate each expression in Exercises 126–129, correct to the nearest hundredth.

126. $13^{1/3}$ **127.** $\sqrt[4]{30}$ **128.** $\sqrt[3]{(15)^2}$ **129.** $\sqrt[5]{\dfrac{4}{3}}$

130. The radius r (in yards) of an oil spill on water is related to the number of gallons in the spill by the formula $r = 7.75\sqrt{g}$. Find the radius of a spill consisting of 150 gallons of oil, correct to the nearest tenth.

ANSWERS: 1. false 2. true 3. false 4. true 5. associative property of addition 6. transitive property of < 7. b 8. -7 9. $-x$ 10. $\frac{1}{2}$ 11. $4\sqrt{7}$ 12. -12 13. -1 14. undefined 15. 0 16. $-\frac{1}{12}$ 17. $\frac{1}{2}$ 18. $\frac{5}{x^2}$ 19. $\frac{3}{y}$ 20. x^2y^6 21. $\frac{y^3}{125x^9}$ 22. 3.29×10^{-13} 23. 1.57×10^{10} 24. 9.46×10^{12} km 25. $\frac{1}{12}$ 26. 36 27. 12 28. 0 29. 1 30. -64 31. $11x^2y^2 - 4xy + 9$ 32. $-x^2 - 7y^2 - x - 2y$ 33. $-8a^2b^2 + ab + 11$ 34. $8u^3v^2 - 4u^2v^3 - u^2v^2 + 2$ 35. $-10x^4y^3 + 35x^5y^3$ 36. $24u^2 + 2uv - v^2$ 37. $49x^2 - 28xy + 4y^2$ 38. $25x^2 + 70xy + 49y^2$ 39. $9a^2 - 64b^2$ 40. $10a^4 + a^2b - 2b^2$ 41. $-2ab + 4 - \frac{7b^2}{a}$ 42. $x^2 - xy - y^2 - \frac{6y^3}{x-3y}$ 43. 611 ft 44. $5xy^2(2xy - 3)$ 45. $(2x-3)(x-2y)$ 46. $(u-v)(u-6v)$ 47. $(u+6)(u+7)$ 48. $(3a-4b)(3a+4b)$ 49. $(3a+7b)(a-2b)$ 50. $2(3x+y^2)(9x^2-3xy^2+y^4)$ 51. $(5x+3y)^2$ 52. $(2u-3v)(4u+5v)$ 53. $(5x-4y)(25x^2+20xy+16y^2)$ 54. cannot be factored 55. $(4x-7y)^2$ 56. $\frac{12a}{25b^2}$ 57. -1 58. $\frac{x+8}{x-7}$ 59. $\frac{a-2}{a-5}$ 60. $\frac{(a-b)(a+5)}{a-1}$ 61. $\frac{7}{(x-7)^2}$ 62. $\frac{3-2y}{y(y+1)}$ 63. $(a-b)^2(a^2+ab+b^2)$ 64. $(a-1)(a-6)^2$ 65. $-\frac{a}{b}$ 66. $\frac{a-2}{a-5}$ 67. $\frac{-2x^2+3}{4x^4}$ 68. -3 69. -1 70. not a real number 71. $3a^2$ 72. $2x^2y^3$ 73. $3a^2b^5$ 74. $a-7$ 75. $12xy\sqrt{3}$ 76. $2ab^3$ 77. $\frac{2ab\sqrt[4]{b}}{3}$ 78. $40\sqrt{5}$ 79. $19\sqrt[3]{2}$ 80. $-14y\sqrt[3]{3x}$ 81. $\frac{1-\sqrt{7}}{2}$ 82. $3\sqrt{6} - 7$ 83. $a\sqrt[5]{a^2}$ 84. 4 85. ab^2 86. $\sqrt[4]{y}$ 87. $2a^3b^3\sqrt[3]{b}$ 88. $\frac{1}{729x^5y^{12}}$ 89. $x=1, x=2$ 90. none 91. $a=2, a=-2, b=-7$ 92. yes (if $x \neq -y$) 93. no 94. yes (if $x \neq 2$) 95. 5.86×10^6 96. 5.86×10^{-9} 97. 4.72×10^1 98. 0.00586 99. 58,600 100. 0.0000381 101. 1.12s; \$27,440 102. 8 103. -15 104. 2 105. 9 106. -27 107. -9 108. $\frac{x-2}{8}$ 109. $\frac{3x+1}{x-3}$ 110. $\frac{3x+4y}{x+y}$ 111. $-x$ 112. $10u^3 + u^2v - 19uv^2 + 8v^3$ 113. $9x^2 - 12xy - 30x + 4y^2 + 20y + 25$ 114. $\frac{2x+3\sqrt{xy}+y}{4x-y}$ 115. $\frac{58\sqrt{10a}}{5}$ 116. $\frac{5a^3b^2\sqrt{5a}}{2}$ 117. $\frac{-3x^2}{y^2}$ 118. $\sqrt[12]{m^{11}}$ 119. $\sqrt[10]{a^3}$ 120. $(5a-6b)(2a+3b)$ 121. $5(xy-2z)(x^2y^2+2xyz+4z^2)$ 122. $(a+x)(b-y)$ 123. $(u+v)(u-v)(2u^2+v^2)$ 124. $(x+y-5)(x-y+5)$ 125. $(3x^n+y^n)(3x^n-y^n)$ 126. 2.35 127. 2.34 128. 6.08 129. 1.06 130. 94.9 yd

CHAPTER 1 TEST

Answer true *or* false.

1. Every rational number is an integer.

2. If a and b are real numbers, then $a + b = b + a$ by the associative property of addition.

Simplify and write without negative exponents.

3. $-(-(-8))$

4. $|3 \cdot 0 + 5(-4)|$

5. $(2a^{-2}b^3)^{-4}$

6. $\left(\dfrac{2x^2y}{x^{-1}y^2}\right)^{-3}$

7. The national debt of a country is $86,500,000,000 and the population is 3,920,000. What is the amount of debt per person? Give your answer in scientific notation correct to three significant digits.

Perform the indicated operations.

8. $(2a^2b^2 + ab - 3) - (a^2b^2 - 3ab + 1) - (3a^2b^2 - ab + 6)$

9. $(2x^2 - 3y)^2$

10. $(2x + y)(x - 3y)$

11. $a - 2b \overline{)a^3 + 2a^2b - 6ab^2 - 4b^3}$

Factor.

12. $4a^2 - 4ab - 3b^2$

13. $u^2 - 4v^2$

14. $2x^3 + 250$

15. $a^2 - 4b^2 + 2a + 4b$

16. In a business operation, the manufacturing cost in dollars is given by $M = 2u^2 + u - 6$ and the wholesale cost in dollars is given by $W = 3u^2 + u + 1$. The total cost is given by $C = M + W$. Find C when $u = 5$.

CHAPTER 1 TEST
CONTINUED

Perform the indicated operations.

17. $\dfrac{x^2 - 4}{x^2 - x} \cdot \dfrac{x^2 + 5x - 6}{x^2 + 5x + 6} \div \dfrac{x^2 + 3x - 10}{x^2 + 3x}$

17. _____

18. $\dfrac{x}{x^2 - y^2} + \dfrac{y}{x^2 - 2xy + y^2}$

18. _____

19. $\dfrac{2a + b}{a^2 - ab} - \dfrac{3b}{a^2 - b^2}$

19. _____

20. Simplify. $\dfrac{a - \dfrac{a}{b}}{b - \dfrac{b}{a}}$

20. _____

Simplify. Assume all variables are positive, rationalize denominators, and write all answers in radical notation.

21. $\sqrt[3]{2ab^2}\sqrt[3]{16a^2b^5}$

21. _____

22. $\left(\dfrac{9x^{-1}y}{x^2y^3}\right)^{-1/2}$

22. _____

23. $\dfrac{\sqrt[3]{2xy}}{\sqrt[3]{y^2}}$

23. _____

24. $\dfrac{\sqrt{a} + 1}{\sqrt{a} - 1}$

24. _____

25. $\sqrt[4]{m^3}\sqrt[3]{m}$

25. _____

26. $\sqrt{\sqrt[4]{(a + b)^7}}$

26. _____

27. A manufacturing index I is related to the number of units in stock u, in thousands of units, by the formula $I = 0.435(1 + u)^{1/4}$. Find I correct to three decimal places when $u = 0.448$.

27. _____

28. Use the formula $T = 2\pi\sqrt{\dfrac{L}{32}}$ to find the time (in seconds) required for a pendulum 1.75 ft long to make one complete swing. Give answer correct to the nearest tenth.

28. _____

CHAPTER 2

Equations, Inequalities, and Problem Solving

AN APPLIED PROBLEM IN ECOLOGY

A park ranger wants to estimate the number of trout in a lake. He catches 200 fish, tags their fins, and returns them to the lake. After a period of time allowing the tagged fish to mix with the total population of fish, he catches another sample of 150 fish and discovers that 3 of these are tagged. Use a proportion to estimate the number of trout in the lake.

Analysis

To solve this problem, we will use a proportion, an equation that states that two ratios (fractions) are equal. A proportion is solved using the cross-product equation. In this case, the ratio of the total number of trout tagged to the total number of trout in the lake must equal the ratio of the number of tagged trout in the second sample to the number of trout in the second sample.

Tabulation

200 = total number of tagged trout in the lake

3 = number of tagged trout in the second sample

150 = number of trout in the second sample

Let x = the number of trout in the lake.

Translation

The following proportion completely describes the problem.

$$\frac{200}{x} = \frac{3}{150}$$

Solving equations is an important part of algebra. Equations, such as the proportion at the bottom of this page (see Example 8 in Section 2.2), often arise as mathematical models of applied problems. In this chapter we continue our review of basic topics starting with solution techniques for equations. Next we consider a strategy for *problem solving*, an approach that will be used repeatedly throughout the rest of the text, and consider a variety of applications involving simple everyday problems as well as more complex problems from business and economics, science, music, and other fields. We conclude the chapter by studying linear and absolute value inequalities and interval notation.

2.1 LINEAR AND ABSOLUTE VALUE EQUATIONS

STUDENT GUIDEPOSTS

1. Equations
2. Linear Equations
3. Radical Equations
4. Fractional Equations
5. Absolute Value Equations
6. Literal Equations

❶ EQUATIONS

An **equation** is a statement that two quantities are equal. Some equations are true, some are false, and for some, such as

$$x + 3 = 5,$$

the truth value depends on the value of the variable x. If the variable can be replaced by a number which makes the equation true, that number is a **solution** or **root** of the equation. When we **solve** an equation, we find all solutions or roots to the equation.

Equations such as $x + 3 = 5$ which are true for certain replacements of the variable (when $x = 2$) and false for other replacements are called **conditional equations.** An equation of the type

$$x + 2 = 2 + x$$

which is true for all replacements of the variable (every number is a solution) is called an **identity.** Finally, an equation such as

$$x + 2 = x - 2,$$

which has no solutions, is called a **contradiction.**

Two equations which have exactly the same solutions are **equivalent.** Notice that

$$x + 3 = 5 \quad \text{and} \quad x = 2$$

are equivalent. To solve an equation, we change it into an equivalent equation which can be solved by direct inspection. This is accomplished by changing the given equation, through a succession of steps, into an equivalent equation with the variable isolated on one side.

❷ LINEAR EQUATIONS

In this section we concentrate on **linear equations** in one variable. Since the variable in a linear equation is raised to the first power only, they are often called **first-degree equations.** Every linear equation in one variable x can be written as an equivalent equation in the form

$$ax + b = 0,$$

where a and b are known real numbers and $a \neq 0$.

We review the basic equation-solving rules in the following theorem.

Equation-Solving Rules

An equivalent equation will be obtained when

1. The same expression is added to or subtracted from both sides of an equation. That is,

    if $a = b$ then $a + c = b + c$ and $a - c = b - c$.

2. Each side of an equation is multiplied by or divided by the same non-zero expression. That is,

    if $a = b$ then $ac = bc$ and $\dfrac{a}{c} = \dfrac{b}{c}$ $(c \neq 0)$.

The proof of this theorem follows by applying the axioms of equality for real numbers given in Chapter 1.

EXAMPLE 1 Solving Linear Equations

Solve.

(a) $3(x - 4) + 5x = 4$

$$3x - 12 + 5x = 4 \quad \text{Clear parentheses first}$$
$$8x - 12 = 4 \quad \text{Collect like terms}$$
$$8x - 12 + 12 = 4 + 12 \quad \text{Add 12 to both sides}$$
$$8x = 16$$
$$\dfrac{8x}{8} = \dfrac{16}{8} \quad \text{Divide both sides by 8}$$
$$x = 2$$

Check: $3(2 - 4) + 5(2) \stackrel{?}{=} 4$
$3(-2) + 10 \stackrel{?}{=} 4$
$-6 + 10 \stackrel{?}{=} 4$
$4 = 4$

The solution is 2.

(b) $1.2y - 3.6 = 2.4$

When an equation involves decimals or fractions, solving might be easier if they are cleared before proceeding.

$$10(1.2y - 3.6) = 2.4(10) \quad \text{Multiply both sides by 10 to clear decimals}$$
$$12y - 36 = 24 \quad \text{Clear parentheses}$$
$$12y - 36 + 36 = 24 + 36 \quad \text{Add 36 to both sides}$$
$$12y = 60$$
$$\dfrac{12y}{12} = \dfrac{60}{12} \quad \text{Divide both sides by 12}$$
$$y = 5$$

Check: $1.2(5) - 3.6 \stackrel{?}{=} 2.4$
$6.0 - 3.6 \stackrel{?}{=} 2.4$
$2.4 = 2.4$

The solution is 5.

PRACTICE EXERCISE 1

Solve.

(a) $6x + 1 - 4x = 3(4 - x) - 11 + x$

(b) $1.5w + 2.5 = 0.5w - 7.5$

(c) $\dfrac{5}{2}z + \dfrac{3}{2} = \dfrac{3}{2}z - (3-z)$

$2\left[\dfrac{5}{2}z + \dfrac{3}{2}\right] = 2\left[\dfrac{3}{2}z - (3-z)\right]$ Multiply both sides by 2, the LCD of the fractions

$5z + 3 = 3z - 2(3-z)$ Clear brackets

$5z + 3 = 3z - 6 + 2z$ Clear parentheses

$5z + 3 = 5z - 6$ Collect like terms

$5z - 5z + 3 = 5z - 5z - 6$ Subtract $5z$ from both sides

$3 = -6$

Whenever a contradiction is obtained, such as $3 = -6$, we know that the original equation is also a contradiction. Thus, there is no solution to the equation.

(d) $a^2 - 5 = (a+2)(a-3) + 8$

$a^2 - 5 = a^2 - a - 6 + 8$ Clear parentheses by multiplying the binomials

$a^2 - 5 = a^2 - a + 2$ Collect like terms

$-5 = -a + 2$ Subtract a^2 from both sides

$-7 = -a$ Subtract 2 from both sides

$7 = a$ Multiply both sides by -1

The solution is 7, which does check.

(c) $\dfrac{5}{3}z + 1 = \dfrac{2}{3}z + 1 + z$

(d) $(x-1)^2 + 3x = x^2 - 7$

Answers: (a) 0   (b) -10
(c) every real number (identity)
(d) -8

Note In Example 1(c) we saw that when a contradiction occurs the equation has no solution. In a similar manner, if at any step an identity occurs (like in Practice Exercise 1(c), the original equation is also an identity and every real number is a solution. Also, on first glance an equation might not appear to be linear. Such was the case in Example 1(d) and in Practice Exercise 1(d).

3 RADICAL EQUATIONS

An equation is called a **radical equation** if the variable appears in a radicand. Some radical equations can be transformed into linear equations. The first step involves using the following theorem.

Power Theorem

If a and b are two algebraic expressions, and if $a = b$, then $a^n = b^n$, for any natural number n.

When the power theorem is used on an equation involving a variable, the resulting equation may have more solutions than the original. For example, the equation

$$x = 4$$

has only one solution, 4, but

$$x^2 = 16$$

has two solutions, 4 and -4. Thus, when the power theorem is used, all possible solutions *must* be checked in the original equation. Solutions which do not check are called **extraneous roots** and must be discarded.

88 CHAPTER 2 EQUATIONS, INEQUALITIES, AND PROBLEM SOLVING

> **To Solve an Equation Involving Radicals**
> 1. Isolate a radical on one side of the equation.
> 2. Raise both sides to a power equal to the index on that radical.
> 3. Solve the equation. If a radical remains, isolate it and raise to the power again.
> 4. Check all possible solutions in the *original* equation.

EXAMPLE 2 A RADICAL EQUATION (SQUARING ONCE)

Solve. $\sqrt{x^2 + 5} - x + 5 = 0$

$\sqrt{x^2 + 5} = x - 5$ Isolate the radical
$(\sqrt{x^2 + 5})^2 = (x - 5)^2$ Square both sides
$x^2 + 5 = x^2 - 10x + 25$ $(x - 5)^2$ is *not* $x^2 - 25$
$5 = -10x + 25$
$-20 = -10x$
$2 = x$

Check: $\sqrt{2^2 + 5} - 2 + 5 \stackrel{?}{=} 0$
$\sqrt{9} - 2 + 5 \stackrel{?}{=} 0$
$3 - 2 + 5 \neq 0$

The equation has no solution.

PRACTICE EXERCISE 2

Solve.

$3\sqrt{2y - 5} - \sqrt{y + 23} = 0$

[*Hint:* Isolating one radical isolates both.]

Answer: 4

CAUTION

A common mistake with some radical equations is to square term by term rather than to isolate a radical and square both sides. For example, to solve

$$\sqrt{y + 3} - \sqrt{y - 2} = 1$$

it is *incorrect* to square term by term and obtain

$$(y + 3) - (y - 2) = 1.$$

It is easy to see that this technique does not work by considering a numerical example.

$$\sqrt{9} - \sqrt{4} = 3 - 2 = 1$$

However, $(\sqrt{9})^2 - (\sqrt{4})^2 = 9 - 4 = 5$, not 1^2 or 1.

The correct way to solve this type of equation is shown in Example 3.

EXAMPLE 3 A RADICAL EQUATION (SQUARING TWICE)

Solve. $\sqrt{y + 3} - \sqrt{y - 2} = 1$

$\sqrt{y + 3} = 1 + \sqrt{y - 2}$ Isolate $\sqrt{y + 3}$
$(\sqrt{y + 3})^2 = (1 + \sqrt{y - 2})^2$ Use power theorem
$y + 3 = 1 + 2\sqrt{y - 2} + y - 2$ Don't forget the middle term, $2\sqrt{y - 2}$, when squaring
$3 = 2\sqrt{y - 2} - 1$

PRACTICE EXERCISE 3

Solve. $\sqrt{w + 2} + \sqrt{w + 6} = 4$

$$4 = 2\sqrt{y-2}$$ Isolate remaining radical
$$2 = \sqrt{y-2}$$ Divide by 2 to simplify
$$4 = y - 2$$ Square again
$$6 = y$$

Check: $\sqrt{6+3} - \sqrt{6-2} \stackrel{?}{=} 1$
$\sqrt{9} - \sqrt{4} \stackrel{?}{=} 1$
$3 - 2 = 1$

The solution is 6.

Answer: $\frac{1}{4}$

Note Try to solve the equation $\sqrt{x+1} + \sqrt{x+7} = -1$. If you do, you will discover that it has no solution since the root you obtain, $\frac{21}{4}$, will not check. Remember that the radical is used to represent the principal (nonnegative) root. Can you see that this equation can have no solution without going through all the steps of trying to solve it only to discover that you come up with an extraneous root? Since $\sqrt{x+1}$ and $\sqrt{x+7}$ must both be nonnegative, their sum must be nonnegative, and can never equal -1.

4 FRACTIONAL EQUATIONS

A **fractional equation** is an equation that contains one or more algebraic fractions. We eliminate or clear the fractions by multiplying both sides of the equation by the LCD of all fractions. The resulting equation may be a linear equation that can be solved using previous techniques.

To Solve a Fractional Equation

1. Find the LCD of all fractions in the equation.
2. Multiply both sides of the equation by the LCD to clear fractions. Make sure that *all* terms are multiplied.
3. Solve this equation.
4. Check your solution in the original equation to be sure that it does not make a denominator zero. If this occurs, the solution must be discarded.

EXAMPLE 4 A FRACTIONAL EQUATION

Solve. $\dfrac{x+2}{x-1} + \dfrac{3}{x} = 1$

PRACTICE EXERCISE 4

Solve. $\dfrac{y-2}{y+1} = \dfrac{y}{y-3}$

The LCD of the fraction is $x(x-1)$ so we multiply both sides by this expression.

$$x(x-1)\left[\frac{x+2}{x-1} + \frac{3}{x}\right] = x(x-1)[1]$$

$$x(x-1)\frac{x+2}{x-1} + x(x-1)\frac{3}{x} = x(x-1) \quad \text{Clear brackets}$$

$$x(x+2) + 3(x-1) = x(x-1) \quad \text{Divide common factors}$$

$$x^2 + 2x + 3x - 3 = x^2 - x \quad \text{Clear parentheses}$$

$$6x = 3 \quad \text{Collect terms and isolate the variable term}$$

$$x = \frac{1}{2}$$

Check: $\dfrac{\frac{1}{2}+2}{\frac{1}{2}-1} + \dfrac{3}{\frac{1}{2}} \stackrel{?}{=} 1$

$\dfrac{\frac{5}{2}}{-\frac{1}{2}} + \dfrac{3}{\frac{1}{2}} \stackrel{?}{=} 1$

$-5 + 6 \stackrel{?}{=} 1$

$1 = 1$

The solution is $\dfrac{1}{2}$.

Answer: 1

CAUTION

A common mistake is to confuse addition or subtraction of algebraic fractions with solving a fractional equation. For example, compare the following two problems.

$$\text{Solve } \frac{x+2}{x-1} + \frac{3}{x} = 1. \qquad \text{Add } \frac{x+2}{x-1} + \frac{3}{x} + 1.$$

In each case, the LCD of all fractions must be found; however, it is used in two very different ways. In the first case, since we have an *equation*, we can multiply both sides by the LCD, $x(x-1)$, to find the numerical solution as in Example 4. In the second case, we rewrite each of the three terms with the denominator the LCD and then add numerators.

$$\frac{x+2}{x-1} + \frac{3}{x} + 1 = \frac{x(x+2)}{x(x-1)} + \frac{3(x-1)}{x(x-1)} + \frac{x(x-1)}{x(x-1)}$$

$$= \frac{x(x+2) + 3(x-1) + x(x-1)}{x(x-1)}$$

$$= \frac{x^2 + 2x + 3x - 3 + x^2 - x}{x(x-1)}$$

$$= \frac{2x^2 + 4x - 3}{x(x-1)}$$

EXAMPLE 5 A Fractional Equation (No Solution)

Solve. $\dfrac{y^2+9}{y^2-9} - \dfrac{y}{y-3} = \dfrac{3}{y+3}$

Multiply both sides by the LCD of all fractions, $(y-3)(y+3)$.

$(y-3)(y+3)\left[\dfrac{y^2+9}{(y-3)(y+3)} - \dfrac{y}{y-3}\right] = (y-3)(y+3)\left[\dfrac{3}{y+3}\right]$

PRACTICE EXERCISE 5

Solve.

$\dfrac{x}{x+4} = \dfrac{4}{x-4} + \dfrac{x^2+16}{x^2-16}$

Clear brackets and divide common factors.

$$y^2 + 9 - y(y + 3) = 3(y - 3)$$
$$y^2 + 9 - y^2 - 3y = 3y - 9 \quad \text{\color{blue}Clear parentheses}$$
$$-6y = -18$$
$$y = 3$$

Check: $\dfrac{3^2 + 9}{3^2 - 9} - \dfrac{3}{3 - 3} \stackrel{?}{=} \dfrac{3}{3 + 3}$

$\dfrac{18}{0} - \dfrac{3}{0} \stackrel{?}{=} \dfrac{3}{6}$

Since division by zero is undefined, we must discard 3. There is no solution.

Answer: no solution

Remember that to solve a fractional equation, we multiply both sides of the equation by the LCD, which usually includes a variable. Since the LCD may be zero for particular values of the variable, we may be multiplying both sides by an expression that equals zero. As a result, we may obtain an equation that is not equivalent to the original. It is a good idea to determine the values for the variable that must be excluded in the fractions before starting to solve the equation. For example, in Example 5, both 3 and -3 must be excluded. Thus, when we obtain 3 as a "solution" we know immediately that it must be discarded.

5 ABSOLUTE VALUE EQUATIONS

Equations that have at least one variable within absolute value bars, such as

$$|x| = 4, \quad |3x - 1| = 5, \quad \text{and} \quad |1 - 2x| = 0,$$

are **absolute value equations.** To solve such an equation, for example $|x| = 4$, we solve two related equations since

$$|x| = \begin{cases} x \text{ if } x \geq 0 \\ -x \text{ if } x < 0. \end{cases}$$

The two equations to solve are

$$x = 4 \quad \text{and} \quad -x = 4.$$

The second equation becomes $x = -4$, giving us the two equations

$$x = 4 \quad \text{and} \quad x = -4,$$

which also gives the two solutions 4 and -4.

EXAMPLE 6 Solving Absolute Value Equations	Practice Exercise 6

Solve.

(a) $|3x - 1| = 5$

Since 5 and -5 are the only two numbers whose absolute value is 5, $3x - 1$ must be equal to one of these two numbers. We need to solve the following two equations.

$$3x - 1 = 5 \quad \text{and} \quad 3x - 1 = -5$$
$$3x = 6 \qquad\qquad\qquad 3x = -4$$
$$x = 2 \qquad\qquad\qquad x = -\dfrac{4}{3}$$

Solve.

(a) $|1 - 2y| = 3$

The solutions are 2 and $-\dfrac{4}{3}$. Check.

(b) $|1 + 2x| = |x|$

The only way that the absolute values of two expressions can be equal is if the two expressions are either equal or opposite in sign. Thus, we need to solve

$1 + 2x = x$ or $1 + 2x = -x$.
$\quad\quad 1 = -x \quad\quad\quad\quad 1 = -3x$
$\quad -1 = x \quad\quad\quad -\dfrac{1}{3} = x$

The solutions are -1 and $-\dfrac{1}{3}$. Check.

(c) $|8x + 2| = -3$

Be careful with this type of equation. Since $|8x + 2|$ must be nonnegative, that is, $|8x + 2| \geq 0$ for every choice of x, there are no values of x for which $|8x + 2|$ can equal -3. Thus, the equation has no solution.

(d) $|1 - 2x| = 0$

The only number whose absolute value is 0 is 0 itself. Thus we have only one equation to solve,

$$1 - 2x = 0.$$
$$-2x = -1$$
$$x = \dfrac{1}{2}$$

The only solution is $\dfrac{1}{2}$.

(b) $|a| = |3a + 4|$

(c) $|5 - 4y| = -10$

(d) $|7a + 21| = 0$

Answers: (a) 2, -1 (b) -2, -1
(c) no solution (d) -3

⚠ CAUTION

Be careful when solving absolute value equations and watch for the special cases:

|EXPRESSION| = negative number   has   **no solution.**
|EXPRESSION| = zero   has   **one solution.**
|EXPRESSION| = positive number   has   **two solutions.**

Also, don't just ignore the absolute value bars and solve the one equation $3x - 1 = 5$, for example, when solving $|3x - 1| = 5$.

---

6 LITERAL EQUATIONS

A **literal equation** includes two or more variables. Formulas such as $P = 2l + 2w$ and $d = rt$ are examples of literal equations. Remember that the letters in a literal equation represent numbers. All the rules for solving equations with numerals also apply to literal equations. In the following examples, we solve a similar numerical equation along with each literal equation in order to clarify the solution procedures.

EXAMPLE 7 Solving Literal Equations

Solve $\sqrt{xy + z} = z$ for x.

$\sqrt{xy + z} = z$
$(\sqrt{xy + z})^2 = z^2$
$xy + z = z^2$
$xy = z^2 - z$
$\dfrac{xy}{y} = \dfrac{z(z-1)}{y}$
$x = \dfrac{z(z-1)}{y}$

Similar numerical example

$\sqrt{3x + 7} = 5$
$(\sqrt{3x + 7})^2 = 5^2$
$3x + 7 = 25$
$3x = 18$
$\dfrac{3x}{3} = \dfrac{18}{3}$
$x = 6$

PRACTICE EXERCISE 7

Solve $a^2 = \dfrac{3}{b + c}$ for c.

Answer: $c = \dfrac{3 - a^2 b}{a^2}$

EXAMPLE 8 A Physics Problem

The resistance R of an electrical circuit is given by

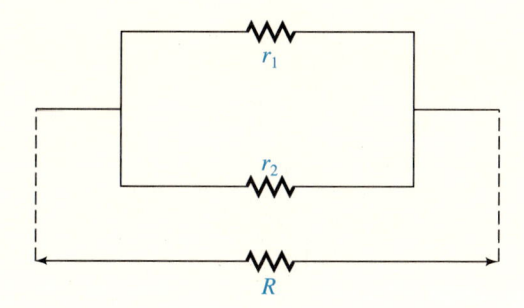

$$\dfrac{1}{R} = \dfrac{1}{r_1} + \dfrac{1}{r_2}$$

Figure 2.1

where r_1 and r_2 are resisters in parallel as shown in Figure 2.1. Solve this equation for r_1.

Similar numerical example

$\dfrac{1}{R} = \dfrac{1}{r_1} + \dfrac{1}{r_2}$

$Rr_1 r_2 \left(\dfrac{1}{R}\right) = \left(\dfrac{1}{r_1} + \dfrac{1}{r_2}\right) Rr_1 r_2$

$r_1 r_2 = Rr_2 + Rr_1$

$r_1 r_2 - Rr_1 = Rr_2$

$r_1(r_2 - R) = Rr_2$

$\dfrac{r_1(r_2 - R)}{r_2 - R} = \dfrac{Rr_2}{r_2 - R}$

$r_1 = \dfrac{Rr_2}{r_2 - R}$

$\dfrac{1}{5} = \dfrac{1}{r_1} + \dfrac{1}{3}$

$15 r_1 \left(\dfrac{1}{5}\right) = \left(\dfrac{1}{r_1} + \dfrac{1}{3}\right) 15 r_1$

$3 r_1 = 15 + 5 r_1$

$3 r_1 - 5 r_1 = 15$

$(3 - 5) r_1 = 15$ This step is usually omitted

$\dfrac{-2 r_1}{-2} = \dfrac{15}{-2}$

$r_1 = -\dfrac{15}{2}$

PRACTICE EXERCISE 8

The formula for the area of a trapezoid is

$$A = \dfrac{1}{2}(b_1 + b_2)h$$

where b_1 and b_2 are the lengths of its bases, and h is its height. Solve this equation for b_2.

Answer: $b_2 = \dfrac{2A - b_1 h}{h}$

2.1 EXERCISES A

Solve each equation in Exercises 1–27. Be sure to check for extraneous roots in radical and fractional equations.

1. $3y + 5 = 7 - y$

2. $1.8x - 4.1 = 6.5 + 0.2x$

3. $\dfrac{z}{2} + \dfrac{3}{8} - \dfrac{z}{4} = \dfrac{5}{4}$

4. $4(y + 2) = 7(3 - y)$

5. $5 + 7x = 7x + 4$

6. $5 + 7z = 7z + 5$

7. $3y \quad (y - 2) - 8$

8. $2(3x - 2) - 2(4x + 3) = 4$

9. $z^2 + 2 = (z - 1)(z + 1) + z$

10. $2[z - (1 - 3z)] = 2(3 + 2z)$

11. $\sqrt{2z + 5} - 3\sqrt{z - 1} = 0$

12. $\sqrt{x^2 + 8} - x - 4 = 0$

13. $\sqrt{z - 4} + \sqrt{z - 8} = -2$

14. $\sqrt{2x + 10} - \sqrt{2x - 5} = 3$

15. $\dfrac{z - 2}{z + 1} = \dfrac{z}{z - 3}$

16. $\dfrac{2x}{x - 3} - \dfrac{3}{x + 4} = 2$

17. $\dfrac{3}{z + 5} - \dfrac{2}{z - 2} = \dfrac{4}{z^2 + 3z - 10}$

18. $\dfrac{1}{z^2 - z - 2} - \dfrac{3}{z^2 - 2z - 3} = \dfrac{1}{z^2 - 5z + 6}$

19. $\sqrt[3]{a + 1} = \sqrt[3]{2a + 7}$

20. $\sqrt[3]{3a + 2} + 4 = 6$

21. $\sqrt[4]{3x} - \sqrt[4]{4 - x} = 0$

22. $|x + 1| = 3$

23. $|x - 1| = -2$

24. $|y + 2| = 0$

25. $|z| = z + 1$

26. $|y - 1| = y - 1$

27. $|7 - 10x| = 7$

In Exercises 28–33 solve the literal equation for the indicated variable. Assume that all expressions are defined so that division by zero and negative numbers under radicals are avoided.

28. $3a + 2b = 5c$ for a

29. $3uvw = p$ for v

30. $\sqrt{\dfrac{x}{y}} = z$ for y

31. $xy = w - y$ for y

32. $\sqrt{a + b} + \sqrt{b} = \sqrt{c}$ for a

33. $1 = \dfrac{1}{x} + \dfrac{1}{y}$ for x

In Exercises 34–37 solve the formula for the indicated variable.

34. GEOMETRY Solve for h in the formula for the area of a trapezoid.
$$A = \frac{1}{2}(b_1 + b_2)h$$

35. BANKING Solve for r in the simple interest formula.
$$A = P(1 + rt)$$

36. PHYSICS Solve for f in the focal length formula.
$$\frac{1}{f} = \frac{1}{d_1} + \frac{1}{d_2}$$

37. GEOMETRY Solve for h in the formula for the surface area of a cylinder.
$$A = 2\pi rh + 2\pi r^2$$

In Exercises 38–41, either solve the given equation or perform the indicated operation.

38. $\dfrac{3a + 5}{6} = \dfrac{4 + 3a}{5}$

39. $\dfrac{3a + 5}{6} + \dfrac{4 + 3a}{5}$

40. $\dfrac{2}{x + 1} - \dfrac{3}{x} = 0$

41. $\dfrac{2}{x + 1} - \dfrac{3}{x}$

42. Is the equation $a = 9$ equivalent to the equation $\sqrt{a} + 3 = 0$?

43. Is the equation $x = 5$ equivalent to the equation
$$\frac{1}{x - 5} + \frac{1}{x} = \frac{5}{x^2 - 5x}?$$

FOR REVIEW

In Exercises 44–47 simplify each expression and write it without negative exponents.

44. $(b^{-4})^{-3}$ **45.** $(3y)^{-1}$ **46.** $3y^{-1}$ **47.** $-3y^{-1}$

Perform the indicated operations in Exercises 48–49.

48. $(6u - v)(4u + v)$

49. $\dfrac{2x^2 + x - 15}{x^2 - x - 12} \div \dfrac{2x^2 + 3x - 20}{16x - x^3}$

Simplify each expression in Exercises 50–52. Rationalize the denominator, and write the answer in radical notation.

50. $\left(\dfrac{125a^5b^{-2}}{4a^{-2}b^{-6}}\right)^{1/2}$

51. $\sqrt[4]{w}\sqrt[5]{w^8}$

52. $\sqrt{\sqrt[3]{x^2}}$

ANSWERS: 1. $\frac{1}{2}$ 2. 6.625 3. $\frac{7}{2}$ 4. $\frac{13}{11}$ 5. no solution 6. every real number 7. 3 8. -7 9. 3 10. 2 11. 2 12. -1 13. no solution 14. 3 15. 1 16. -11 17. 20 18. $\frac{2}{3}$ 19. -6 20. 2 21. 1 22. 2, -4 23. no solution 24. -2 25. $-\frac{1}{2}$ 26. all numbers $y \geq 1$ 27. 0, $\frac{7}{5}$ 28. $a = \frac{5c - 2b}{3}$ 29. $v = \frac{P}{3uw}$ 30. $y = \frac{x}{z^2}$ 31. $y = \frac{w}{x + 1}$ 32. $a = c - 2\sqrt{cb}$ 33. $x = \frac{y}{y - 1}$ 34. $h = \frac{2A}{b_1 + b_2}$ 35. $r = \frac{A - P}{Pt}$ 36. $f = \frac{d_1 d_2}{d_1 + d_2}$ 37. $h = \frac{A - 2\pi r^2}{2\pi r}$ 38. $\frac{1}{3}$ 39. $\frac{33a + 49}{30}$ 40. -3 41. $\frac{-x - 3}{x(x + 1)}$ 42. no 43. no 44. b^{12} 45. $\frac{1}{3y}$ 46. $\frac{3}{y}$ 47. $-\frac{3}{y}$ 48. $24u^2 + 2uv - v^2$ 49. $-x$ 50. $\frac{5a^3b^2\sqrt{5a}}{2}$ 51. $w\sqrt[20]{w^{17}}$ 52. $\sqrt[3]{x}$

2.1 EXERCISES B

Solve each equation in Exercises 1–27. Be sure to check for extraneous roots in radical and fractional equations.

1. $4z - 3 = 3 - 2z$
2. $2.5y - 6.1 = 7.4 + 1.3y$
3. $\dfrac{x}{5} + \dfrac{1}{10} - \dfrac{x}{2} = \dfrac{1}{20}$
4. $5(z - 1) = 8(z + 2)$
5. $2(y - 1) = 4 + 2y$
6. $3(x - 5) = 3(1 + x) - 18$
7. $5z - (1 - z) = 11$
8. $4(2y - 1) - (3y + 2) = 0$
9. $(3y + 2)(y - 1) = (y - 4)(3y + 5)$
10. $5[2a - (3 - a)] = 5(a + 3)$
11. $\sqrt{5a - 1} - 2\sqrt{a + 1} = 0$
12. $\sqrt{9y^2 + 5} + 3y - 1 = 0$
13. $\sqrt{a + 5} + \sqrt{a - 16} = -1$
14. $\sqrt{9 - y} - \sqrt{18 - y} = -1$
15. $\dfrac{a + 1}{a + 2} = \dfrac{a + 2}{a - 1}$
16. $\dfrac{2y}{2y - 1} - \dfrac{6}{y + 5} = 1$
17. $\dfrac{5}{a - 3} - \dfrac{3}{a + 3} = \dfrac{7}{a^2 - 9}$
18. $\dfrac{2}{a^2 + a - 6} = \dfrac{2}{a^2 - 3a + 2} - \dfrac{1}{a^2 + 2a - 3}$
19. $\sqrt[3]{w - 1} = \sqrt[3]{3w + 7}$
20. $\sqrt[4]{a - 5} + 1 = 3$
21. $\sqrt[5]{2x} - \sqrt[5]{6 - x} = 0$
22. $|5 - 2z| = 13$
23. $|w + 3| = -6$
24. $|a - 5| = 0$
25. $|1 - 3x| = 2 - x$
26. $|z + 1| = z + 1$
27. $|5 - 8y| = 5$

In Exercises 28–33 solve the literal equation for the indicated variable. Assume that all expressions are defined so that division by zero and negative numbers under radicals are avoided.

28. $3a - 2b = 5c$ for b
29. $3uvw = p$ for w
30. $\sqrt{\dfrac{x}{y}} = z$ for x
31. $xy = w + y$ for y
32. $\sqrt{a + b} - \sqrt{a} = \sqrt{c}$ for b
33. $1 = \dfrac{1}{x} + \dfrac{1}{y}$ for y

In Exercises 34–37 solve the formula for the indicated variable.

34. **GEOMETRY** Solve for a in the surface area formula for a rectangular solid.
$$S = 2ab + 2ac + 2bc$$

35. **METEOROLOGY** Solve for F in the temperature conversion formula.
$$C = \dfrac{5}{9}(F - 32)$$

36. **PHYSICS** Solve for r_2 in the formula for resistance in an electrical circuit.
$$\dfrac{1}{R} = \dfrac{1}{r_1} + \dfrac{1}{r_2}$$

37. **PHYSICS** Solve for g in the formula for the height of a falling object.
$$h = \dfrac{1}{2}gt^2 + vt$$

In Exercises 38–41, either solve the given equation or perform the indicated operation.

38. $\dfrac{5a - 2}{3} = \dfrac{2 + 7a}{4}$
39. $\dfrac{5a - 2}{3} - \dfrac{2 + 7a}{4}$

40. $\dfrac{5}{y+2} + \dfrac{2}{y} = 0$

41. $\dfrac{5}{y+2} + \dfrac{2}{y}$

42. Is the equation $y = 4$ equivalent to the equation
$$\sqrt{y} + 2 = 0?$$

43. Is the equation $a = 3$ equivalent to the equation
$$\dfrac{1}{a-3} + \dfrac{1}{a} = \dfrac{3}{a(a-3)}?$$

FOR REVIEW

In Exercises 44–47 simplify each expression and write it without negative exponents.

44. $(x^{-5})^3$ **45.** $(2w)^{-3}$ **46.** $2w^{-3}$ **47.** $-2w^{-3}$

Perform the indicated operations in Exercises 48–49.

48. $(-4uv + 5)(6u^2v^2 + 7uv - 3)$

49. $\dfrac{5-x}{x^2 - 12x + 35} + \dfrac{x}{x^2 - 14x + 49}$

Simplify each expression in Exercises 50–52. Rationalize the denominator, and write the answer in radical notation.

50. $\left(\dfrac{27x^4y^{-1}}{8xy^5}\right)^{1/3}$ **51.** $\sqrt[5]{y}\sqrt[3]{y^8}$ **52.** $\sqrt[3]{\sqrt{a^5}}$

2.1 EXERCISES C

In Exercises 1–4 solve the equation for the unknown first without making any calculations. Then use a calculator to simplify the answer and round it to three decimal places. Store all results in the calculator's memory and do not do any rounding until the final step.

1. $9.328(y + 12.465) = 2.003(4.135 - y)$
[Answer: -9.531]

2. $z^2 + 6.557 = (z - 4.055)^2$

3. $2\sqrt{x^2 - 4.25} - 2x + 2.8 = 0$
[Answer: 2.218]

4. $\dfrac{a}{1.035} - \dfrac{a}{0.541} = 1000$

5. Find a value of m so that the equations are equivalent.
$$3x - m = x + 8 \text{ and } 7(x - 1) = 23 + x$$
[Answer: $m = 2$]

6. Find a value of n so that the equation has only the solution -5.
$$\dfrac{10}{x} + \dfrac{15}{x} = n - 3$$
[Answer: $n = -2$]

7. Find a value of p so that the equation has only the solution -5.
$$\sqrt{x + 9} = p + 1$$

8. Find a value of m and b so that the equation $mx + b = 0$ has the solution $\tfrac{2}{5}$. Are these the only values for m and b?

2.2 PROBLEM SOLVING AND APPLICATIONS OF LINEAR EQUATIONS

STUDENT GUIDEPOSTS

1. Strategy for Problem Solving
2. Percent Applications
3. Geometry Applications
4. Work Applications
5. Rate-Motion Applications
6. Proportion Applications

1 STRATEGY FOR PROBLEM SOLVING

To solve an *applied problem,* a problem that is normally stated in words, we must understand what is being asked, and then translate the *words* of the problem into a mathematical model, usually an equation or formula. Finally, we solve the equation or substitute known values into the formula. Unfortunately, there is no precise rule that we can use to solve all applied problems. However, we can describe a general strategy. "How do we *attack* such a problem?" Every applied problem can be approached, or attacked, using the following:

A STRATEGY FOR PROBLEM SOLVING

A – **A**nalyze and familiarize: Try to understand the problem and what is being asked. Read it several times to determine what type of problem you have and whether some formula (perhaps from geometry, such as $A = lw$, or relating to motion, such as $d = rt$, etc.) might be applicable.

T – **T**abulate and sketch: Tabulate or list the known facts presented in the problem. Also list what you wish to find, that is, what is unknown. Use a variable to describe this unknown. A sketch might help picture what is stated, and a chart or table summarizing the information can also be useful.

T – **T**ranslate and solve: Translate the "words" and symbols into an equation, perhaps using the formulas identified as appropriate. The equation should then be solved by the techniques we have already learned.

A – **A**pproximate and estimate: Ask yourself what type of answer is reasonable and appropriate. Does your answer fit this description? For example, if a problem involved looking for the value of a new car, and you obtained an answer of $25, you would see that something was wrong, and you should check your work to try and discover the problem.

C – **C**heck and verify: If your answer seems reasonable, check to see if it fits the original problem, that it satisfies the words of the problem (it doesn't just solve the equation which, of course, is something you came up with and may not be correct). Also, make sure that you have answered the question actually posed in the problem.

K – The **K**ey to problem solving is: *Try something.* What you try might not be correct, but at least you learn something from your mistakes and efforts. Don't be discouraged; if you need to, follow the steps again.

2.2 PROBLEM SOLVING AND APPLICATIONS OF LINEAR EQUATIONS

In a given problem, depending on the difficulty, you may use only some of these steps, and you may do many of them mentally. For example, as you analyze a problem, you probably won't write out your analysis. However, you should begin to write as you tabulate the known facts and identify the unknown quantity. Sometimes we will solve a problem in the text that uses all of these steps, and sometimes not. Keep in mind that some of the problems we solve are not particularly relevant, but contrived to give you practice at problem solving before we approach more "real-world applications" that sometimes are a bit more challenging. However, with practice, you will become more and more proficient at problem solving when you use these steps to **ATTACK** such problems.

To be successful at translating words to symbols and ultimately to an equation, we begin by reviewing several common terms and their translations.

Terms	Symbol
and, sum, sum of, added to, increased by	+
minus, less, subtracted from, less than, diminished by, difference between, difference, decreased by	−
times, product, product of, multiplied by, of	·
divided by, quotient of, ratio	÷
equals, is equal to, is as much as, is, is the same as	=

Unknown quantities in a problem can be represented by a variable. Several typical statements and their translations are given below.

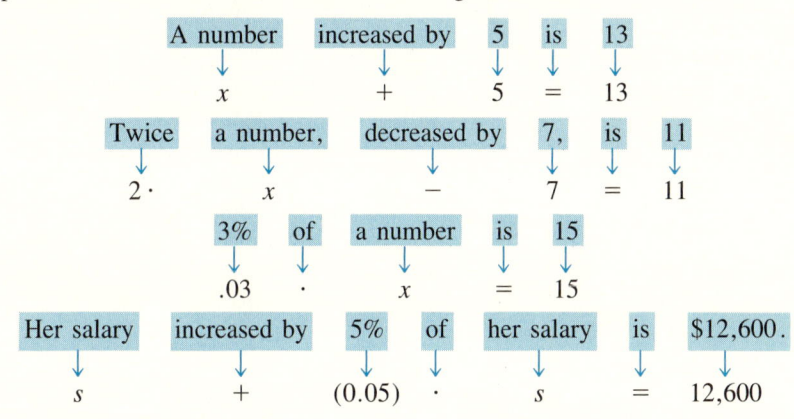

A number increased by 5 is 13
$x + 5 = 13$

Twice a number, decreased by 7, is 11
$2 \cdot x - 7 = 11$

3% of a number is 15
$.03 \cdot x = 15$

Her salary increased by 5% of her salary is $12,600.
$s + (0.05) \cdot s = 12{,}600$

This sentence would probably come from a problem such as:

Mary is presently earning a salary of $12,600. If she received a 5% raise last year, what was her former salary?

To illustrate the strategy for problem solving we start with a basic example involving numbers.

EXAMPLE 1 NUMBER PROBLEM

The sum of three consecutive even integers is 72. Find the integers.

Analysis: We need to find three consecutive even integers. Remember that consecutive even integers are even integers that follow each other in the regular counting order.

Tabulation: Let x = the first even integer,

$x + 2$ = next consecutive even integer,

$x + 4$ = third consecutive even integer.

PRACTICE EXERCISE 1

The ratio of 3 more than a number and 5 less than a number is $-13/3$. Find the number.

Translation: $x + (x + 2) + (x + 4) = 72$ Their sum is 72
$$3x + 6 = 72$$
$$3x = 66$$
$$x = 22$$
$$x + 2 = 24$$
$$x + 4 = 26$$

Approximation: For three consecutive even integers to add to 72, they must all be close to one-third of 72, which is 24. The answers we obtained fill this description.

Check: 22, 24, and 26 are indeed consecutive even integers and $22 + 24 + 26 = 72$.

Thus, the integers are 22, 24, and 26.

Answer: 7/2

Note In Example 1 we showed every detail for solving this problem. Of course, as you progress, you will not write out every detail; some steps will be performed mentally. Probably you will write only the information in the *Tabulation*, *Translation*, and *Check*.

2 PERCENT APPLICATIONS

A wide variety of problems involve the notion of percent.

EXAMPLE 2   A COMMISSION PROBLEM

What were the total sales on which Bart received a commission of $692 if the commission rate was 8%?

Analysis: Remember that the commission received by a salesperson is a percent of the total sales, and given by:

commission = (commission rate)(total sales).

In this problem we need to find Bart's total sales.

Tabulation: Let x = Bart's total sales,
   0.08 = Bart's commission rate.

Translation: commission = (commission rate)(total sales)
   $692 = (0.08)\ x$

$$\frac{692}{0.08} = x$$

$$\$8650 = x$$

Approximation: Since Bart's commission was about $700, and 8% is about 10%, we would expect his sales to be somewhere around $7000. Since $8650 is somewhat close to this figure, we conclude that the answer is at least reasonable.

Check: 8% of $8650 is $(0.08)(8650) = 692$.

Thus, Bart had total sales amounting to $8650.

PRACTICE EXERCISE 2

Nancy received a commission of $9327.50 selling automobiles. If her sales for January amounted to $143,500.00, what is her commission rate?

Answer: 6.5%

2.2 PROBLEM SOLVING AND APPLICATIONS OF LINEAR EQUATIONS

EXAMPLE 3 A Salary Problem

After she received a 7% raise, Leslie's new salary is $22,470. What was her former salary?

Analysis:

$$\text{new salary} = \text{former salary} + \text{raise}$$
$$\text{raise} = (0.07)(\text{former salary})$$

Remember that you take a percent of the former salary (not the present salary) to determine a raise.

Tabulation: Let x = Leslie's former salary,
$0.07x$ = Leslie's raise in salary,
$x + 0.07x$ = Leslie's present salary.

Translation: $x + 0.07x = 22{,}470$
$1.07x = 22{,}470$ Distributive law
$x = \dfrac{22{,}470}{1.07}$
$x = 21{,}000$

Approximation: With a present salary of $22,470, after a raise of 7%, we would estimate that the former salary was a bit less than $22,470; so $21,000 seems reasonable.

Check: $21{,}000 + (0.07)(21{,}000) =$
$21{,}000 + 1470 = 22{,}470$

Thus, Leslie's former salary was $21,000.

Practice Exercise 3

The sales-tax rate in Livermore is 5.5%. Roy Troutman bought some computer software for $400.90, including sales tax. What was the price of the software before tax?

Answer: $380.00

③ GEOMETRY APPLICATIONS

Many applications involve the perimeter, area, or volume of some type of geometric figure. As you begin to analyze such a problem, identify a formula that will assist you, and be sure to make a sketch when you *tabulate* the information given and requested.

EXAMPLE 4 A Geometry Problem

The second angle of a triangle is six times as large as the first angle. If the third angle is 45° more than twice the first, find the measure of each angle.

Analysis: We know that the sum of the measures of the angles of a triangle is 180°. We must find the measures of the three angles.

Tabulation: Let x = measure of first angle,
$6x$ = measure of second,
$2x + 45$ = measure of third angle.

Make a sketch of the triangle as in Figure 2.2.

Practice Exercise 4

The perimeter of a rectangular garden is 104 yd. If the length of the garden is 2 yd less than five times the width, find the dimensions of the garden.

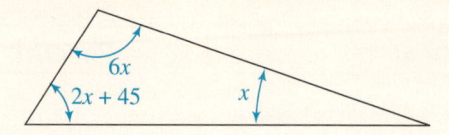

Figure 2.2

Translation: $x + 6x + (2x + 45) = 180$ The sum of the angles is 180°
$$9x + 45 = 180$$
$$9x = 135$$
$$x = 15$$
$$6x = 6(15) = 90$$
$$2x + 45 = 2(15) + 45 = 75$$

Approximation: Notice that 15°, 90°, and 75° are reasonable for the measures of the angles of a triangle. Had we obtained an angle with measure 200°, for example, we would know that something is wrong.

Check: $15° + 90° + 75° = 180°$, $6(15°) = 90°$, and $75° = 2(15°) + 45°$.

Thus, the measures are 15°, 90°, and 75°.

Answer: 43 yd by 9 yd

④ WORK APPLICATIONS

Some problems translate into fractional equations. Consider the following example. Jim can do a certain job in 3 hours and Dave can do the same job in 7 hours. How long would it take them to do the job if they worked together?

This type of problem is usually called a *work problem,* and three important principles must be kept in mind.

1. The time it takes to do a job when two individuals work together must be less than the time it takes for the faster worker to complete the job by himself. The time together is *not* the average of the two times.

2. If a job can be done in t hours, in 1 hour $\frac{1}{t}$ of the job can be completed. For example, since Jim can do the job in 3 hours, in 1 hour he could do $\frac{1}{3}$ of the job.

3. The work done by Jim in 1 hour added to the work done by Dave in 1 hour equals the work done by them together in 1 hour. Thus, if t is the time it takes to complete the job working together,

   (amount Jim does in 1 hour) + (amount Dave does in 1 hour)
   = (amount done together in 1 hour)

   which translates into

   $$\frac{1}{3} + \frac{1}{7} = \frac{1}{t}.$$

Solving by multiplying both sides by the LCD, $21t$, we obtain

$$21t\left(\frac{1}{3} + \frac{1}{7}\right) = 21t\left(\frac{1}{t}\right)$$
$$7t + 3t = 21$$
$$10t = 21$$
$$t = \frac{21}{10}.$$

Working together, it would take Jim and Dave 21/10 hours (2 hours, 6 minutes) to complete the job.

EXAMPLE 5 A BUSINESS (WORK) PROBLEM

C&C Painters have the painting contract for all the new homes in a subdivision. When Claude and Clyde work together, it takes 5 days to paint one house. If Claude paints one house by himself, it takes 7 days to complete the job. How long would it take Clyde to paint one house if he worked alone?

Analysis: Recognize this as a work problem and recall the basic format for such a problem.

Tabulation: We have the following:

5 = no. days to paint one house together,

$\frac{1}{5}$ = amount painted together in 1 day,

7 = no. days for Claude to paint one house,

$\frac{1}{7}$ = amount painted by Claude in 1 day.

Let x = no. days for Clyde to paint one house,

$\frac{1}{x}$ = amount painted by Clyde in 1 day.

Translation:

(amount by Claude) + (amount by Clyde) = (amount together)

$$\frac{1}{7} + \frac{1}{x} = \frac{1}{5}$$

Multiply both sides of the equation by $35x$, the LCD.

$$35x\left(\frac{1}{7} + \frac{1}{x}\right) = 35x\left(\frac{1}{5}\right)$$
$$5x + 35 = 7x$$
$$35 = 2x$$
$$\frac{35}{2} = x$$

Approximation: Since Claude can do the whole job in 7 days, and with help from Clyde the time required goes down only 2 days, we would expect that Clyde is much slower than Claude; so $\frac{35}{2}$ days, or $17\frac{1}{2}$ days seems reasonable for an answer.

Thus, it would take Clyde $17\frac{1}{2}$ days to paint one house by himself. Check this.

PRACTICE EXERCISE 5

When Mary and Jane work together, it takes them 15 min to sort the mail. If Mary can do the job in 21 min by herself, how long will it take Jane to sort the mail if she works alone?

Answer: $52\frac{1}{2}$ min

5 RATE-MOTION APPLICATIONS

The distance d that an object travels in a given time t at a constant or uniform rate r is given by the following formula.

$$\text{distance} = (\text{rate})(\text{time})$$
$$d = rt$$

If we know two of the variables in this formula, we can calculate the third. For example, if you travel 165 miles at an average rate of 55 mph, it will take

$$t = \frac{d}{r} = \frac{165}{55} = 3 \text{ hours}$$

to make the trip. Problems that use this formula have come to be called *rate-motion problems*. Consider the following examples.

EXAMPLE 6   A Transportation Problem

Two cars leave the same point traveling in opposite directions. The second car travels 15 km/hr faster than the first, and after 3 hours they are 465 km apart. How fast is each car traveling?

Analysis: Use the formula $d = rt$.

Tabulation: Let d_1 = distance first car travels in km,
  r = rate of first car in km/hr,
  d_2 = distance second car travels in km,
  $r + 15$ = rate of second car in km/hr.

Then since $d = rt$, or $d = tr$, we have:

$$d_1 = 3r$$
$$d_2 = 3(r + 15)$$

Make a sketch showing this information as in Figure 2.3.

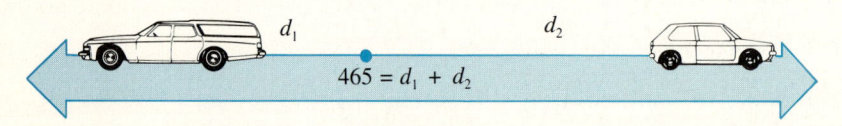

Figure 2.3

Translation:

$$d_1 + d_2 = 465$$
$$3r + 3(r + 15) = 465$$
$$3r + 3r + 45 = 465 \quad \text{Distributive law}$$
$$6r = 420$$
$$r = 70$$
$$r + 15 = 70 + 15 = 85$$

Approximation: Are 70 km/hr and 85 km/hr reasonable rates for an automobile? Do they differ by 15 km/hr?

Thus, the first car travels 70 km/hr and the second 85 km/hr. Check these.

Practice Exercise 6

Two boats leave an island at 8:00 A.M. sailing due north. If one is going 5 mph faster than the other, how far apart will they be at 3:00 P.M. the same day?

Answer: 35 mi

The following motion problem results in a fractional equation. We use the principle that if a boat travels downstream, its speed relative to the bank of the stream is its speed in still water *increased by* (plus) the speed of the stream. Likewise, when the boat travels upstream, its speed relative to the bank is its speed in still water *decreased by* (minus) the speed of the stream.

This time we show the solution as you normally will without writing the specific problem-solving steps. After reading the problem, analyze it mentally, note the tabulation of facts, and ask yourself if the answer is reasonable.

EXAMPLE 7 A NAVIGATION PROBLEM

The speed of a stream is 4 mph. A boat travels 48 miles upstream in the same time it takes to travel 72 miles downstream. What is the speed of the boat in still water?

Let x = the speed of the boat in still water,

$x + 4$ = the speed of the boat when traveling downstream,

$x - 4$ = the speed of the boat when traveling upstream,

$\dfrac{72}{x + 4}$ = the time of travel downstream $\left(t = \dfrac{d}{r}\right)$,

$\dfrac{48}{x - 4}$ = the time of travel upstream.

Sometimes it helps to organize the information in a table.

	Distance	Speed	Time
Downstream	72	$x + 4$	$\dfrac{72}{x + 4}$
Upstream	48	$x - 4$	$\dfrac{48}{x - 4}$

$$\frac{72}{x + 4} = \frac{48}{x - 4} \quad \text{Times are equal}$$

$$(x - 4)(x + 4)\frac{72}{x + 4} = \frac{48}{x - 4}(x - 4)(x + 4)$$

$$72(x - 4) = 48(x + 4)$$

$$72x - 288 = 48x + 192$$

$$24x = 480$$

$$x = 20$$

The speed of the boat in still water is 20 mph.

PRACTICE EXERCISE 7

An airplane can fly 2400 miles with the wind in the same time that it could fly 1800 miles against the wind. If the speed of the plane in still air is 350 mph, what is the speed of the wind?

Answer: 50 mph

6 PROPORTION APPLICATIONS

The **ratio** of one number a to another number b is the fraction $\frac{a}{b}$ sometimes written $a:b$. A **proportion** is an equation that states that two ratios are equal. In the proportion

$$\frac{a}{b} = \frac{c}{d}$$

the first term, a, and the fourth term, d, are called the **extremes** of the proportion. The second term, b, and the third term, c, are the **means**. If the second and third terms are equal, as in

$$\frac{a}{b} = \frac{b}{c},$$

b is the **mean proportional** between a and c. When one or more terms of a proportion involve a variable, a quick way to solve the equation is to form the **cross-product** equation. For example, the cross-product equation for

$$\frac{a}{b} = \frac{c}{d} \quad \text{is} \quad ad = bc.$$

CAUTION

Never attempt to find the cross-product equation when the original equation is not a proportion. For example, the cross-product technique would not apply to a fractional equation such as

$$\frac{2}{y} + \frac{3}{y-1} = 8.$$

EXAMPLE 8 A PROPORTION PROBLEM IN GEOMETRY

If a man who is 6 ft tall stands in the sunlight and casts a shadow 14 ft long, while a pole casts a shadow 35 ft long, how tall is the pole?

Make a sketch, as in Figure 2.4.

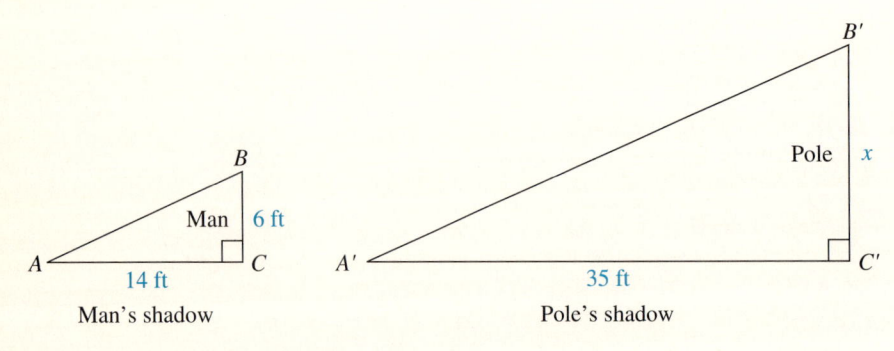

Figure 2.4 Similar Triangles

We assume that $\angle ACB = \angle A'C'B' = 90°$ (a right angle). Since both measurements are taken at the same time, the sun's rays make the same angle with level ground, that is, $\angle BAC = \angle B'A'C'$. Thus, $\triangle ABC$ is similar to $\triangle A'B'C'$ (the corresponding angles are equal) making the corresponding sides proportional. Thus,

$$\frac{AC}{A'C'} = \frac{BC}{B'C'}.$$

$$\frac{14}{35} = \frac{6}{x} \qquad \text{Substituting for } AC, BC, A'C'$$

$$14x = 35 \cdot 6 \qquad \text{Cross-product equation}$$

$$x = \frac{35 \cdot 6}{14} = 15$$

The pole is 15 ft tall.

PRACTICE EXERCISE 8

In a sample of 100 replacement disk drives for a computer, 2 were discovered to be defective. How many defective drives would be expected in a shipment of 900?

Answer: 18 disk drives

CAUTION

Do not take shortcuts and try to do too many parts of an applied problem in your head. Remember, the *analysis* and *approximation* steps are usually mental, but *tabulation,* including making a sketch when appropriate, should be written in complete detail as shown in Example 8. In this example our analysis followed the tabulation and sketch and was provided for your understanding, but normally would not be written out. Following this method will save time in the long run and provide organization for your work.

We conclude this section by solving the application given in the introduction to Chapter 2.

EXAMPLE 9 AN ECOLOGY PROBLEM

A park ranger wants to estimate the number of trout in a lake. He catches 200 fish, tags their fins, and returns them to the lake. After a period of time allowing the tagged fish to mix with the total population of fish, he catches another sample of 150 fish and discovers that 3 of these are tagged. Use a proportion to estimate the number of trout in the lake.

Analysis: The key to solving this problem is the proportion:

$$\frac{\text{total number tagged trout}}{\text{total number trout in lake}} = \frac{\text{number tagged trout in sample}}{\text{number trout in sample}}$$

Tabulation: 200 = no. tagged trout in the lake,

3 = no. tagged trout in second sample,

150 = no. trout in the second sample.

Let x = no. trout in the lake.

Translation: Substitute into the proportion above.

$$\frac{200}{x} = \frac{3}{150}$$

$(200)(150) = 3x$ Cross-product equation

$10{,}000 = x$

Thus, there are approximately 10,000 trout in the lake.

PRACTICE EXERCISE 9

Suppose the ranger in Example 9 followed the same tagging procedure, but in a second sample of 100 fish he found that 8 were tagged. Estimate the number of fish in the lake under these circumstances.

Answer: 2500 trout

2.2 EXERCISES A

Solve.

1. If five times a number is decreased by 7, the result is twice the number, increased by 10. Find the number.

2. The sum of 3 positive, consecutive odd integers is 29 less than four times the first. Find the integers.

3. **EDUCATION** If Jorge Ortiz scored 61, 89, and 86 on three tests, what must he score on a fourth test to have an average score of 80?

4. **CONSTRUCTION** A board is 23 feet long. It is to be cut into 3 pieces in such a way that the second piece is twice as long as the first and the third is 3 feet more than the second. Find the length of each piece.

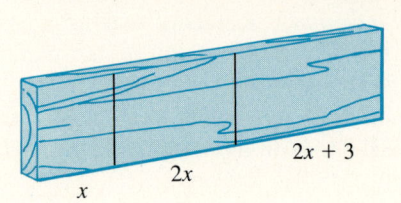

5. **CONSUMER** If the tax rate in West, Texas, is 4%, what is the tax on a purchase of $52.50?

6. **BUSINESS** If Maria received a 12% raise and is now making $25,760 a year, what was her salary before the raise?

7. **GEOMETRY** The first angle of a triangle is 5° more than three times the second and the third angle is 10° less than the second. Find the measure of each.

8. **AGRICULTURE** The perimeter of a rectangular wheat field is 1760 m. The length is 40 m more than the width. Find the dimensions of the field.

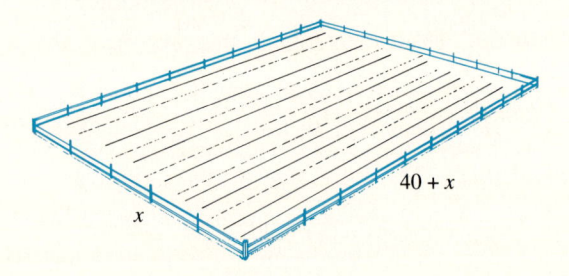

9. **CONSTRUCTION** Jeff can frame a shed in 8 days and Barry can do the same job in 9 days. To the nearest tenth of a day, how long would it take them to frame the shed if they worked together?

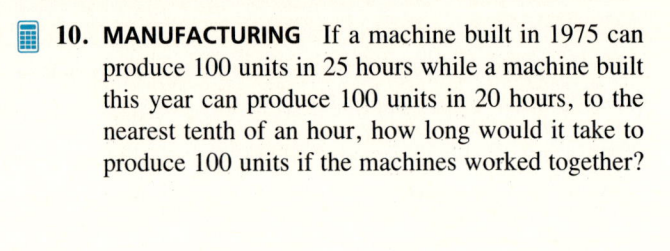

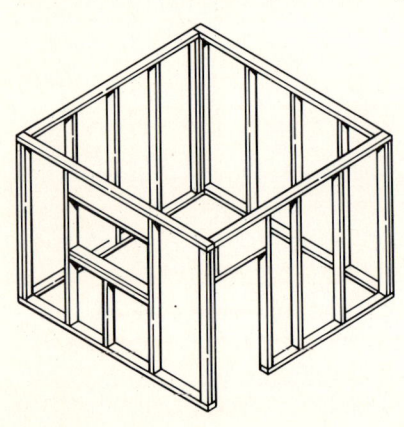

10. **MANUFACTURING** If a machine built in 1975 can produce 100 units in 25 hours while a machine built this year can produce 100 units in 20 hours, to the nearest tenth of an hour, how long would it take to produce 100 units if the machines worked together?

11. **SPORTS** If Mike Nesbitt, head trainer at NAU, can tape the ankles of a team in 30 min, and with the help of a student trainer the job takes 20 min, how long would it take the student to do the job by himself?

12. **RECREATION** A small swimming pool can be filled by an inlet pipe in 2 hr. It can be emptied in 10 hr through a drain pipe. If the two pipes were opened at the same time, how long would it take to fill an empty pool?

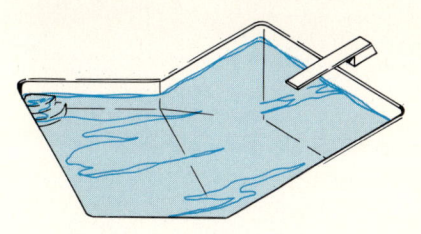

13. **AERONAUTICS** Two airplanes leave an airport at the same time, one flying east and the other west. One plane is traveling 50 mph faster than the other. After 3 hr they are 3030 mi apart. How fast is each plane flying?

14. **AERONAUTICS** A plane flies 480 mi with the wind and 330 mi against the wind in the same time. If the speed of the plane in still air is 135 mph, what is the speed of the wind?

15. **COMMUNICATIONS** Two printing presses print 31,500 fliers. If their rates of production differ by 30 fliers per minute, what is the rate of each if they both work 210 minutes to complete the job?

16. **CHEMISTRY** A chemist wishes to make a mixture of concentrated hydrochloric acid and distilled water so that the ratio of acid to water is 7:3. If he starts with 21 L of acid, how much water should be mixed with the acid?

17. **GEOMETRY** A boy 5 ft tall casts a shadow 12 ft long at the same time that a tower casts a shadow 252 ft long. How tall is the tower?

18. **GEOMETRY** If a tangent PA and a secant PB are drawn to a circle from an external point, the length of the tangent is the mean proportional between the length of the secant and its external segment PC. Find the length of the tangent if the secant is 49 cm and its external segment is 4 cm.

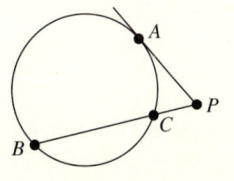

110 CHAPTER 2 EQUATIONS, INEQUALITIES, AND PROBLEM SOLVING

FOR REVIEW

Solve each equation in Exercises 19–22.

19. $\sqrt{4a^2 + 8} - 2a - 4 = 0$

20. $\dfrac{7}{x-6} + \dfrac{5}{x-8} = \dfrac{2}{x^2 - 14x + 48}$

21. $|6 - 5z| = 1$

22. $S = \dfrac{a_1}{1-r}$ for r

Perform the operations and rationalize the denominator in Exercises 23–25. These procedures will be used in the next section.

23. $(4 + 2\sqrt{2})(-3 - 5\sqrt{2})$

24. $(2 - 3\sqrt{2})(2 - 3\sqrt{2})$

25. $\dfrac{1}{5 + 4\sqrt{2}}$

ANSWERS: 1. $\frac{17}{3}$ 2. 35, 37, 39 3. 84 4. 4 ft, 8 ft, 11 ft 5. $2.10 6. $23,000 7. 116°, 37°, 27° 8. 420 m by 460 m 9. 4.2 days 10. 11.1 hr 11. 60 min 12. 2.5 hr 13. 480 mph, 530 mph 14. 25 mph 15. 90 fliers per minute, 60 fliers per minute 16. 9 L 17. 105 ft 18. 14 cm 19. $-\frac{1}{2}$ 20. $\frac{22}{3}$ 21. $1, \frac{7}{5}$ 22. $r = \frac{S-a_1}{S}$ 23. $-32 - 26\sqrt{2}$ 24. $22 - 12\sqrt{2}$ 25. $\frac{4\sqrt{2}-5}{7}$

2.2 EXERCISES B

Solve.

1. If twice the sum of a number and 5 is increased by 2, the result is 26 times the number. Find the number.

2. The sum of 4 consecutive integers is 8 more than 3 times the largest. Find the integers.

3. **SPORTS** Three of the four starting linemen on the Kansas State football team weigh 256 lb, 240 lb, and 242 lb. If the average weight of the starting line is 249.5 lb, how much does the fourth lineman weigh?

4. **AGE** If Sam is twice as old as Harry and the sum of their ages is the same as five times Harry's age, decreased by 28, how old is each?

5. **SPORTS** A quarterback completed 27 passes in 45 attempts. What was his completion percentage?

6. **RETAILING** A retailer bought a stereo for $285 and put it on sale at a 70% markup rate. What was the price of the stereo?

7. **GEOMETRY** If two angles are **supplementary** (their measures total 180°), and the second is 30° less than twice the first, find the measure of each angle.

8. **GEOMETRY** An **isosceles triangle** has two equal sides called *legs* and a third side called the *base*. If each leg of an isosceles triangle is 2 m less than three times the base, and the perimeter is 31 m, find the length of each leg and the length of the base.

9. **BUSINESS** If Bill and Morris can write a computer program in 18 min working together, and Bill needs 42 min working alone, how long would it take Morris to write the program by himself?

10. **AGRICULTURE** A 4-inch pipe can fill an irrigation reservoir in 25 days, a 6-inch pipe can fill it in 10 days, and a 9-inch pipe takes 5 days. How long would it take if all three pipes were turned on together (to the nearest tenth of a day)?

11. **CONSTRUCTION** If Sue can paint a house in 12 hr, Larry can paint the same house in 16 hr, and Eve needs 20 hr to paint the house, how long would it take if they all worked together (to the nearest tenth of an hour)?

12. **AGRICULTURE** A grain conveyor can fill a silo in 2 days. If it takes 3 days to empty the silo into railroad cars, how long would it take to fill an empty silo if both doors were opened simultaneously?

13. **TRANSPORTATION** Two families agree to meet in Boise for a reunion. They both leave at 8:00 A.M., one travels at a rate 10 mph faster than the other, and they meet in Boise at 12:00 noon. If the total distance they traveled was 480 mi, how fast was each traveling?

14. **AERONAUTICS** A plane flies 1160 km with the wind and 840 km against the wind in the same length of time. If the speed of the wind is 40 km/hr, what is the speed of the plane in calm air?

15. **MANUFACTURING** A machine made in 1975 has a production rate of 20 units per hour less than a machine made in 1990. If they are turned on at the same time and produce 1120 units in 8 hours, what is the production rate of each?

16. **CONSTRUCTION** If 150 ft of electrical wire weighs 45 lb, what will 270 ft of the same wire weigh?

17. **GEOMETRY** Given two similar triangles, the sides of one are 6 in, 7 in, and 12 in. If the shortest side of the other is 18 in, find the other two sides.

18. **EDUCATION** In a class of 40 students, the ratio of boys to girls is 5:3. After the first test, a certain number of boys drop the class making the ratio of boys to girls 7:5. How many boys dropped?

FOR REVIEW

Solve each equation in Exercises 19–22.

19. $\sqrt[3]{a+3} - \sqrt[3]{2a+7} = 0$

20. $\dfrac{2a}{2a-1} - \dfrac{6}{a+5} = 1$

21. $|3 - 8y| = -1$

22. $3x + ay = mx$ for x

Perform the operations and rationalize the denominator in Exercises 23–25. These procedures will be used in the next section.

23. $(3 - \sqrt{5}) - (-2 - 4\sqrt{5})$

24. $(1 + 2\sqrt{3})^2$

25. $\dfrac{2 + \sqrt{5}}{1 - 3\sqrt{5}}$

2.2 EXERCISES C

Solve.

1. **METEOROLOGY** The temperature F on the Fahrenheit scale and the temperature C on the Celsius scale are given by the formula $C = \frac{5}{9}(F - 32)$. What temperature is the same on both scales?

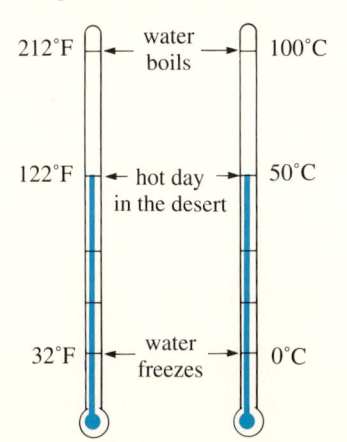

2. **RECREATION** To balance a teeter-totter, the weight w_1 of the first child times her distance d_1 from the fulcrum (the central point of the board) must equal the weight w_2 of the second child times his distance d_2 from the fulcrum. If one child weighing 85 lb is 6 ft from the fulcrum, how far from the fulcrum must the second child be in order to balance the teeter-totter, if he weighs 51 lb?

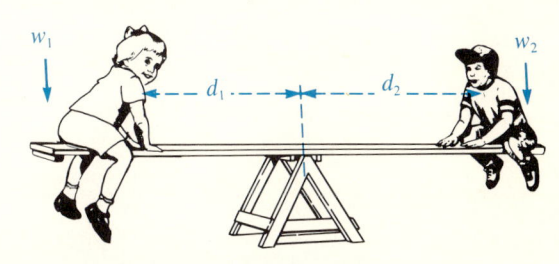

3. **ECONOMICS** On Monday an investor bought 100 shares of stock. On Tuesday, the value of the shares went up 6%, and on Wednesday, the value fell 5%. How much did the investor pay for the 100 shares on Monday if he sold them on Wednesday for $1460.15?

4. **DEMOGRAPHY** The population of a town increased 4% during one year, then decreased 7% the next. What was the original population if it was 2418 at the end of these two years?

5. Find the mean proportional between 25 and 9.

6. The fourth term of a proportion is called the **fourth proportional** to the first, second, and third terms. Find the fourth proportional to 20, −12, and −5.

7. What number must be added to each of 20, 8, 32, and 14 to give four numbers that are in proportion in that order?

8. **MUSIC** The notes in a major chord in music have frequencies in the ratio of 4 to 5 to 6. If the first note of a chord has a frequency of 264 hertz, what are the frequencies of the other two notes?

9. **SURVEYING** A pole is standing in a lake. The ratios of the length of the pole above the water to the length in the water to the length in the sand at the bottom of the lake are 3 to 4 to 2. If 40 ft of the pole are in the water, how much of the pole is out of the water and how much is in the sand?

10. **INVESTMENT** Jay Arnote invested $12,000 in an account that pays 12% simple interest. How much additional money must be invested in an account that pays 15% simple interest so that the average return on the two investments amounts to 13%?

11. **AGRICULTURE** An irrigation canal has a cross section in the shape of an isosceles trapezoid. If the depth of the water is 3 ft, the bottom of the canal is 5 ft wide, and the cross-sectional area is 18 ft^2, what is the width of the canal at the water level?

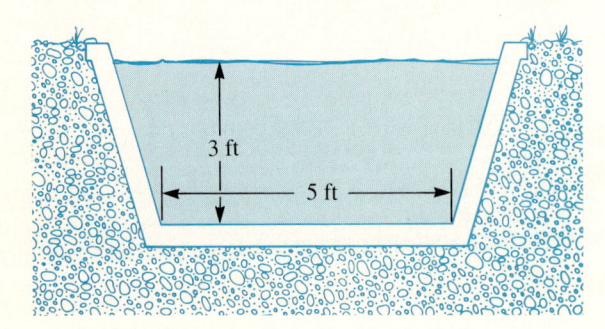

12. **MANUFACTURING** A refinery operates two smoke stacks. The first pollutes the air 2.5 times as fast as the second. If both stacks are operating together, they yield a certain amount of pollutants in 12.5 hours. How long would it take each to emit the same amount of pollutants if operated by itself?

13. **NAVIGATION** A small boat traveling at 20 knots (a knot is 1 nautical mile per hour) is 15 nautical miles from an island when a Coast Guard cutter initiates pursuit on the same course traveling 30 knots. How long will it take for the cutter to overtake the boat?

14. **GEOMETRY** The length of the altitude drawn to the hypotenuse of a right triangle is the mean proportional between the lengths of the segments of the hypotenuse determined by it. Find the length of the altitude if the segments are 9 inches and 16 inches in length.

ANSWERS: 1. −40° 2. 10 ft 3. $1450 4. 2500 5. ±15 6. 3 7. 4 8. 330 hertz, 396 hertz 9. 30 ft, 20 ft 10. $6000 11. 7 ft 12. 17.5 hr, 43.75 hr 13. 1.5 hr 14. 12 in

2.3 COMPLEX NUMBERS

STUDENT GUIDEPOSTS

1. Imaginary Numbers
2. Complex Numbers
3. Operations on Complex Numbers
4. Equations With Complex-Number Coefficients

1 IMAGINARY NUMBERS

Up to this point we have worked with the system of real numbers. This system has several shortcomings, however. For example, we saw that the square root of a negative number is not a real number. As a result, certain basic equations that require taking such a root could not be solved if all we had available were real numbers. To provide solutions for these equations and square roots for negative numbers we introduce a new kind of number called an *imaginary number*. Suppose we agree to follow rules of radicals similar to those for real numbers; then all we will need is one new numeral, i.

Definition of Imaginary Number

Define the number $i = \sqrt{-1}$, that is, $i^2 = -1$. An **imaginary number** is a number of the form bi, where b is a real number.

Consider the number $\sqrt{-4}$, for example. We can simplify and recognize $\sqrt{-4}$ as an imaginary number as follows:

$$\sqrt{-4} = \sqrt{4(-1)} = \sqrt{4}\sqrt{-1} = 2i.$$

EXAMPLE 1 SIMPLIFYING IMAGINARY NUMBERS

Simplify.

(a) $\sqrt{-25} = \sqrt{25(-1)} = \sqrt{25}\sqrt{-1} = 5i$

(b) $\sqrt{-3}\sqrt{-7} = \sqrt{3(-1)}\sqrt{7(-1)}$
$= \sqrt{3}\sqrt{-1}\sqrt{7}\sqrt{-1}$
$= \sqrt{3} \cdot i \cdot \sqrt{7} \cdot i$
$= \sqrt{3}\sqrt{7}i^2$
$= \sqrt{21}(-1)$ $i^2 = -1$
$= -\sqrt{21}$

(c) $\dfrac{\sqrt{-30}}{\sqrt{-5}} = \dfrac{\sqrt{30(-1)}}{\sqrt{5(-1)}} = \dfrac{\sqrt{30}\sqrt{-1}}{\sqrt{5}\sqrt{-1}} = \dfrac{\sqrt{30}i}{\sqrt{5}i} = \sqrt{\dfrac{30}{5}} = \sqrt{6}$

(d) $\sqrt{-36} + \sqrt{-100} = \sqrt{36}i + \sqrt{100}i = 6i + 10i = 16i$

PRACTICE EXERCISE 1

Simplify.

(a) $\sqrt{-121}$

(b) $\sqrt{-2}\sqrt{-6}$

(c) $\dfrac{\sqrt{-50}}{\sqrt{-10}}$

(d) $\sqrt{-25} - \sqrt{-49}$

Answers: (a) $11i$ (b) $-2\sqrt{3}$ (c) $\sqrt{5}$ (d) $-2i$

Note An important principle is illustrated in Example 1(b). Had we written $\sqrt{-3}\sqrt{-7} = \sqrt{(-3)(-7)} = \sqrt{21}$, we would not obtain the same result. It is necessary to express the square root of a negative number in the form $i\sqrt{b}$ *before* making simplifications. Recall that the rule $\sqrt{ab} = \sqrt{a}\sqrt{b}$ was given only for $a \geq 0$ and $b \geq 0$, so $\sqrt{(-3)(-7)} = \sqrt{-3}\sqrt{-7}$ is incorrect. However, in the special case where $a \geq 0$ and $b = -1$, we can use a similar rule to express square roots of negative numbers in terms of the imaginary number i.

Powers of the imaginary number i are interesting. Consider the following powers:

$i^1 = i$

$i^2 = -1$

$i^3 = (i^2)(i) = (-1)(i) = -i$

$i^4 = (i^2)(i^2) = (-1)(-1) = 1$

$i^5 = (i^4)(i) = (1)(i) = i$

$i^6 = (i^4)(i^2) = (1)(-1) = -1$

$i^7 = (i^4)(i^3) = (1)(-i) = -i$

$i^8 = (i^4)(i^4) = (1)(1) = 1$

Notice that these powers cycle through the values i, -1, $-i$, and 1. Any power of i can be simplified by identifying a factor that is a power of i^4, which is 1, times a remaining factor of i, -1, $-i$, or 1.

EXAMPLE 2 Powers of i

Simplify.

(a) $i^{41} = (i^4)^{10}(i) = (1)^{10}i = (1)i = i$

(b) $i^{106} = (i^4)^{26}(i^2) = (1)^{26}(i^2) = (1)(-1) = -1$

PRACTICE EXERCISE 2

Simplify.

(a) i^{63}

(b) i^{128}

Answers: (a) $-i$ (b) 1

❷ COMPLEX NUMBERS

Numbers formed by adding a real number to an imaginary number give us a powerful number system with many applications in mathematics. One important use as we shall see involves providing solutions to basic equations that are not solvable otherwise.

Definition of Complex Number

A **complex number** is a number of the form $a + bi$, where a and b are real numbers and $i^2 = -1$. The number a is the **real part** and b is the **imaginary part**. If $b = 0$, the complex number is simply the real number a, while if $a = 0$, it is the **imaginary number** bi.

In view of this definition, we see that the complex numbers include both the real numbers and the imaginary numbers. For example, the real number 7 could be represented as $7 + 0i$, and the imaginary number $-2i$ could be written as $0 - 2i$.

To investigate some of the properties of complex numbers we first need to decide when two complex numbers are *equal*.

Equality of Complex Numbers

Two complex numbers $a + bi$ and $c + di$ are **equal**,

$$a + bi = c + di,$$

if and only if their real parts are equal ($a = c$) and their imaginary parts are equal ($b = d$).

EXAMPLE 3 EQUALITY OF COMPLEX NUMBERS

(a) If $a + bi = 3 + \sqrt{2}i$, then $a = 3$ and $b = \sqrt{2}$.
(b) If $c + di = 5$, then $c = 5$ and $d = 0$ since $5 = 5 + 0i$.
(c) If $u + vi = -\sqrt{3}i$, then $u = 0$ and $v = -\sqrt{3}$ since $-\sqrt{3}i = 0 - \sqrt{3}i$.
(d) If $2x + yi = 3x - 2 + 7i$, then

$$2x = 3x - 2 \quad \text{and} \quad y = 7$$
$$-x = -2$$
$$x = 2.$$

PRACTICE EXERCISE 3

(a) If $x + yi = 3 - 7i$, find x and y.
(b) If $p + qi = -3$, find p and q.
(c) If $m + ni = \sqrt{5}i$, find m and n.
(d) If $3x - yi = x + 1 + 2yi$, find x and y.

Answers: (a) $x = 3$, $y = -7$ (b) $p = -3$, $q = 0$ (c) $m = 0$, $n = \sqrt{5}$ (d) $x = 1/2$, $y = 0$

CAUTION

When a radical appears as a coefficient of i as in $\sqrt{2}i$ and $-\sqrt{3}i$ in Example 3, notice that i is *not* under the radical. Since this is a common mistake, we often write $i\sqrt{2}$ and $-i\sqrt{3}$ to avoid the problem.

3 OPERATIONS ON COMPLEX NUMBERS

We now consider the four basic operations on complex numbers, addition, subtraction, multiplication, and division.

Addition and Subtraction of Complex Numbers

Let $a + bi$ and $c + di$ be two complex numbers. Their **sum** is

$$(a + bi) + (c + di) = (a + c) + (b + d)i,$$

and their **difference** is

$$(a + bi) - (c + di) = (a - c) + (b - d)i.$$

Notice that for addition (or subtraction), we add (or subtract) the real parts and then add (or subtract) the imaginary parts.

EXAMPLE 4 ADDING AND SUBTRACTING COMPLEX NUMBERS

Perform the indicated operations.

(a) $(4 + 6i) + (1 - 5i) = (4 + 1) + (6 - 5)i = 5 + i$

(b) $(2 - i\sqrt{5}) - (-7 + 3i\sqrt{5}) = (2 + 7) + (-\sqrt{5} - 3\sqrt{5})i$
$$= 9 - 4i\sqrt{5}$$

PRACTICE EXERCISE 4

Perform the indicated operations.

(a) $(3 - 8i) + (-2 + i)$
(b) $(3 + i\sqrt{2}) - (1 - i\sqrt{2})$

Answers: (a) $1 - 7i$ (b) $2 + 2i\sqrt{2}$

Note Adding and subtracting complex numbers is exactly the same as adding and subtracting binomials such as $a + bx$ and $c + dx$ by collecting like terms. In fact, it is easier to add and subtract this way than by memorizing the definitions.

Multiplication of Complex Numbers

Let $a + bi$ and $c + di$ be two complex numbers. Their **product** is

$$(a + bi)(c + di) = (ac - bd) + (ad + bc)i.$$

Although the multiplication of complex numbers may seem strange at first, it becomes clearer when we multiply as if finding the product of binomials using the FOIL method.

$$
\begin{aligned}
(a + bi)(c + di) &= \overset{F}{ac} + \overset{O}{adi} + \overset{I}{bci} + \overset{L}{bdi^2} \\
&= ac + bd(-1) + adi + bci \quad i^2 = -1 \\
&= (ac - bd) + (ad + bc)i
\end{aligned}
$$

In fact, it might be best to use the FOIL method for finding products in specific cases, keeping in mind that $i^2 = -1$.

EXAMPLE 5 MULTIPLYING COMPLEX NUMBERS

Find the products.

(a) $(3 + 6i)(2 - i) = \overset{F}{6} - \overset{O}{3i} + \overset{I}{12i} - \overset{L}{6i^2}$

$\qquad = 6 - 6i^2 - 3i + 12i$

$\qquad = (6 + 6) + (-3 + 12)i$

$\qquad = 12 + 9i$

(b) $(4 + i\sqrt{3})(4 - i\sqrt{3}) = 16 - 4i\sqrt{3} + 4i\sqrt{3} - 3i^2$

$\qquad = 16 - 3i^2$

$\qquad = 16 + 3 = 19$

PRACTICE EXERCISE 5

Find the products.

(a) $(2 - 5i)(1 + 3i)$

(b) $(4 - 3i)(4 + 3i)$

Answers: (a) $17 + i$ (b) 25

We mentioned earlier that complex numbers provide solutions to other types of equations. For example, consider the equation $x^2 - 2x + 2 = 0$. We can show that $1 - i$ is a solution by substituting and using the definitions and rules for the order of operations.

$$
\begin{aligned}
(1 - i)^2 - 2(1 - i) + 2 &= (1 - 2i + i^2) - 2(1 - i) + 2 & \text{Square first} \\
&= (1 - 2i - 1) - 2 + 2i + 2 & i^2 = -1, \text{ and} \\
&= -2i - 2 + 2i + 2 & \text{distribute} \\
&= 0
\end{aligned}
$$

In the next section we will see how this solution was obtained.

A special product was illustrated in Example 5(b). The numbers $4 + i\sqrt{3}$ and $4 - i\sqrt{3}$ have the property that their product is a real number, 19. Complex numbers like these are called *conjugates*.

Definition of Conjugates

Two complex numbers of the form $a + bi$ and $a - bi$ are **conjugates** of each other.

2.3 COMPLEX NUMBERS

When two conjugates are multiplied, $(a + bi)(a - bi)$, the product is always a nonnegative real number. You will be asked to prove this in the exercises. One application of conjugates is in division of complex numbers.

> **Division of Complex Numbers**
>
> Let $a + bi$ and $c + di$ be two complex numbers. The **quotient** of $a + bi$ and $c + di$ can be simplified by multiplying the numerator and denominator by the conjugate of the denominator. That is,
>
> $$\frac{a + bi}{c + di} = \frac{(a + bi)(c - di)}{(c + di)(c - di)} = \frac{(ac + bd) + (bc - ad)i}{c^2 + d^2}.$$

EXAMPLE 6 DIVIDING COMPLEX NUMBERS

Divide $2 + 5i$ by $1 - 3i$.

$$\frac{2 + 5i}{1 - 3i} = \frac{(2 + 5i)(1 + 3i)}{(1 - 3i)(1 + 3i)} \quad \text{$1 + 3i$ is the conjugate of $1 - 3i$}$$

$$= \frac{2 + 6i + 5i + 15i^2}{1 + 3i - 3i - 9i^2}$$

$$= \frac{2 + 11i - 15}{1 + 9} \quad i^2 = -1$$

$$= \frac{-13 + 11i}{10} = -\frac{13}{10} + \frac{11}{10}i$$

PRACTICE EXERCISE 6

Divide $3 + 7i$ by $2 - 5i$.

Answer: $-1 + i$

Finding the reciprocal of a complex number is the same as dividing the complex number 1 ($1 = 1 + 0i$) by the given complex number.

EXAMPLE 7 FINDING A RECIPROCAL

Find the reciprocal of $5 + 4i$ and express it in the form $a + bi$.

The reciprocal of $5 + 4i$ is $\frac{1}{5 + 4i}$. Thus, we must divide the complex number $1 = 1 + 0i$ by the complex number $5 + 4i$. In effect, since $i = \sqrt{-1}$, we are rationalizing the denominator.

$$\frac{1}{5 + 4i} = \frac{1(5 - 4i)}{(5 + 4i)(5 - 4i)}$$

$$= \frac{5 - 4i}{25 + 16}$$

$$= \frac{5 - 4i}{41} = \frac{5}{41} - \frac{4}{41}i$$

PRACTICE EXERCISE 7

Find the reciprocal of $3 - 2i$ and express it in the form $a + bi$.

Answer: $\frac{3}{13} + \frac{2}{13}i$

4 EQUATIONS WITH COMPLEX-NUMBER COEFFICIENTS

A linear equation that has complex numbers for its coefficients is solved in exactly the same way that we solved linear equations with real-number coefficients.

EXAMPLE 8 An Equation with Complex Coefficients

Solve for x.

$$2 + 3i + 4ix = 2i - (1 + i)x$$
$$2 + 3i + 4ix + (1 + i)x = 2i \quad \text{Add } (1 + i)x$$
$$2 + 3i + (1 + 5i)x = 2i \quad \text{Distributive law}$$
$$(1 + 5i)x = 2i - (2 + 3i) \quad \text{Subtract } 2 + 3i$$
$$(1 + 5i)x = -2 - i \quad \text{Simplify}$$
$$x = \frac{-2 - i}{1 + 5i} \quad \text{Divide by } 1 + 5i$$
$$x = \frac{(-2 - i)(1 - 5i)}{(1 + 5i)(1 - 5i)} \quad \text{Multiply numerator and denominator by } 1 - 5i, \text{ the conjugate of } 1 + 5i$$
$$x = \frac{-2 + 10i - i + 5i^2}{1 - 25i^2} \quad \text{Multiply}$$
$$x = \frac{-2 + 9i + 5(-1)}{1 - 25(-1)} \quad i^2 = -1$$
$$x = \frac{-7 + 9i}{26} = -\frac{7}{26} + \frac{9}{26}i$$

Practice Exercise 8

Solve for x.

$$(2 - i)x + 3i = 4ix + 5$$

Answer: $\frac{25}{29} + \frac{19}{29}i$

2.3 EXERCISES A

In Exercises 1–9, simplify and express each result as a real number or in terms of i.

1. $\sqrt{-16}$
2. $\sqrt{-24}$
3. $\sqrt{-3}\sqrt{-11}$
4. $-\sqrt{-8}\sqrt{-2}$
5. $\dfrac{\sqrt{-45}}{\sqrt{-9}}$
6. $\dfrac{\sqrt{-64}}{\sqrt{-8}}$
7. $\dfrac{-\sqrt{-121}}{11}$
8. $\sqrt{-9 - 36}$
9. $\sqrt{-9} - \sqrt{-36}$

Use the definition of equality of complex numbers in Exercises 10–11 to find x and y.

10. $5x + (3y + 2)i = -i$
11. $x + yi = \sqrt{9} - \sqrt{-9}$

Perform the indicated operations in Exercises 12–23.

12. $(4 - 3i) + (-2 - 5i)$
13. $(-1 - 2i) - (5 - 4i)$
14. $(3 - 2i) + 8i$
15. $(1 - 3i)(3 + 5i)$
16. $(4 - i)(-2 + 5i)$
17. $2i(3 - 7i)$
18. $3(5 - 12i)$
19. $(4 - 3i)^2$
20. $(-1 + i)^3$
21. i^{15}
22. i^{37}
23. i^{74}

Give the conjugate of each complex number in Exercises 24–27.

24. $3 - 10i$ **25.** $15i$ **26.** -3 **27.** $4 - i\sqrt{7}$

Find the quotients in Exercises 28–30.

28. $\dfrac{4 - 5i}{1 + i}$ **29.** $\dfrac{-2 + 3i}{4 + 3i}$ **30.** $\dfrac{1 + i}{(1 - i)^2}$

Find the reciprocal of each number in Exercises 31–33 and express it in the form a + bi.

31. $2 - i$ **32.** $4i$ **33.** $\dfrac{3 - i}{1 + i}$

Solve each equation in Exercises 34–35.

34. $3ix + 2 = 5 - 2ix$ **35.** $(1 + i)x - 2i = 3ix + 5$

36. The **absolute value of a complex number** is defined by $|a + bi| = \sqrt{a^2 + b^2}$. Use this definition to evaluate the following.
  (a) $|3 + 4i|$ **(b)** $|1 - i|$ **(c)** $|-9i|$

FOR REVIEW

Solve.

37. EDUCATION If Henry's average score on four quizzes is 76 and he knows that three of his scores are 82, 63, and 92, what score did he make on the other quiz?

38. RATE-MOTION Mary Kim drives 12 mph faster than her mother. If Mary Kim travels 310 mi in the same time that her mother travels 250 mi, find the speed of each.

39. RECREATION If the speed of a stream is 8 km/hr and a boat can travel 105 km downstream in the same time that it can travel 49 km upstream, what is the speed of the boat in still water?

40. $6(x + 5) - 2(4x - 1) = 0$

The following exercises will help you prepare for the next section. Factor and show that each expression in Exercises 41–42 is a perfect square.

41. $x^2 - 12x + 36$

42. $x^2 + \dfrac{2}{3}x + \dfrac{1}{9}$

43. Evaluate $\sqrt{b^2 - 4ac}$ when $a = -1$, $b = 5$, and $c = -4$.

ANSWERS: 1. $4i$ 2. $2i\sqrt{6}$ 3. $-\sqrt{33}$ 4. 4 5. $\sqrt{5}$ 6. $2\sqrt{2}$ 7. $-i$ 8. $3i\sqrt{5}$ 9. $-3i$ 10. $x = 0, y = -1$ 11. $x = 3, y = -3$ 12. $2 - 8i$ 13. $-6 + 2i$ 14. $3 + 6i$ 15. $18 - 4i$ 16. $-3 + 22i$ 17. $14 + 6i$ 18. $15 - 36i$ 19. $7 - 24i$ 20. $2 + 2i$ 21. $-i$ 22. i 23. -1 24. $3 + 10i$ 25. $-15i$ 26. -3 27. $4 + i\sqrt{7}$ 28. $-\tfrac{1}{2} - \tfrac{9}{2}i$ 29. $\tfrac{1}{25} + \tfrac{18}{25}i$ 30. $-\tfrac{1}{2} + \tfrac{1}{2}i$ 31. $\tfrac{2}{5} + \tfrac{1}{5}i$ 32. $-\tfrac{1}{4}i$ 33. $\tfrac{1}{5} + \tfrac{2}{5}i$ 34. $-\tfrac{3}{5}i$ 35. $\tfrac{1}{5} + \tfrac{12}{5}i$ 36. (a) 5 (b) $\sqrt{2}$ (c) 9 37. 67 38. 62 mph, 50 mph 39. 22 km/hr 40. 16 41. $(x-6)^2$ 42. $\left(x + \tfrac{1}{3}\right)^2$ 43. 3

2.3 EXERCISES B

In Exercises 1–9, simplify and express each result as a real number or in terms of i.

1. $\sqrt{-8}$

2. $\sqrt{-20}$

3. $\sqrt{-7}\sqrt{-5}$

4. $-\sqrt{-27}\sqrt{-3}$

5. $\dfrac{\sqrt{-35}}{\sqrt{-7}}$

6. $\dfrac{\sqrt{-49}}{\sqrt{-7}}$

7. $\dfrac{-\sqrt{-81}}{9}$

8. $\sqrt{-4 \cdot -25}$

9. $\sqrt{-4} \cdot \sqrt{-25}$

Use the definition of equality of complex numbers in Exercises 10–11 to find x and y.

10. $(2x + 1) + 4yi = 3$

11. $x + yi = \sqrt{4} + \sqrt{-4}$

Perform the indicated operations in Exercises 12–23.

12. $(3 + 2i) + (-1 + 5i)$

13. $(-1 - i) - (8 - 3i)$

14. $(2 - 5i) + 7i$

15. $(1 + 2i)(3 - 4i)$

16. $(2 - i)(-3 + 4i)$

17. $3i(2 - 4i)$

18. $5(2 - 3i)$

19. $(-3 - 4i)^2$

20. $(-3 - 4i)^3$

21. i^{20}

22. i^{39}

23. i^{93}

Give the conjugate of each complex number in Exercises 24–27.

24. $5 - 7i$

25. $-8i$

26. 7

27. $-3 + i\sqrt{5}$

Find the quotients in Exercises 28–30.

28. $\dfrac{2 + 3i}{1 - i}$

29. $\dfrac{-3 + 5i}{2 + 3i}$

30. $\dfrac{2 - 3i}{(1 + i)(3 - 2i)}$

Find the reciprocal of each number in Exercises 31–33 and express it in the form a + bi.

31. $2 + i$

32. $-3i$

33. $\dfrac{2 + i}{3 - i}$

Solve each equation in Exercises 34–35.

34. $5ix + 6 = -2ix + 1$

35. $(2 + i)x - 7 = -3ix + 12$

36. The **absolute value of a complex number** is defined by $|a + bi| = \sqrt{a^2 + b^2}$. Use this definition to evaluate the following.
   (a) $|2 - 3i|$
   (b) $|1 + i|$
   (c) $|-5i|$

FOR REVIEW

Solve.

37. ECONOMICS What amount of money invested at 4 percent simple interest will increase to $1248 by the end of one year?

38. TRAVEL If 3/4 inch on a map represents 20 miles, how many miles are represented by 12 inches?

39. What number when subtracted from 12, 22, 3, and 4 will make the resulting numbers, in that order, proportional?

40. $\sqrt{y - 5} - \sqrt{y + 3} = -2$

The following exercises will help you prepare for the next section. Factor and show that each expression in Exercises 41–42 is a perfect square.

41. $x^2 + 14x + 49$

42. $x^2 - x + \dfrac{1}{4}$

43. Evaluate $\sqrt{b^2 - 4ac}$ when $a = 2$, $b = -3$, and $c = -1$.

2.3 EXERCISES C

If $x = a + bi$ is a complex number, we often denote the conjugate of x, $a - bi$, by $\bar{x}$. Prove each statement in Exercises 1–9 is true for any complex numbers $x = a + bi$ and $y = c + di$.

1. $\overline{x + y} = \bar{x} + \bar{y}$

2. $\overline{xy} = \bar{x}\bar{y}$

3. $|x|^2 = x\bar{x}$

4. If x is a real number, then $\bar{x} = x$. [*Hint:* Let $x = a + 0i$.]

5. $x + \bar{x}$ is a real number.

6. $x - \bar{x}$ is an imaginary number.

7. $\dfrac{1}{2}(x + \bar{x}) = a$

8. $\dfrac{1}{2}i(\bar{x} - x) = b$

9. $\dfrac{1}{x} = \dfrac{a}{a^2 + b^2} - \dfrac{b}{a^2 + b^2}i$

10. Let $x = a + bi$, and solve the equation $x^2 = -4$ by finding the values for a and b. [Answer: $\pm 2i$]

2.4 QUADRATIC EQUATIONS

STUDENT GUIDEPOSTS

1. Solving by Factoring
2. Solving by Taking Roots
3. Completing the Square
4. The Quadratic Formula
5. The Discriminant
6. Properties of Solutions
7. Equations from Solutions
8. Approximating Solutions

1 SOLVING BY FACTORING

In Chapter 1 we learned an important property of the real number system stating that if $a = 0$ or $b = 0$, then $ab = 0$. The converse of this property, stated in the next theorem, is also true.

Zero-Product Rule

If a and b represent two real numbers, and $ab = 0$, then $a = 0$ or $b = 0$.

Proof Suppose that $ab = 0$. Then either $a = 0$ or $a \neq 0$. If $a = 0$, then the theorem is proved. Thus, we may assume that $a \neq 0$, in which case $1/a$ is defined. Multiplying both sides of $ab = 0$ by $1/a$ gives $b = 0$. Hence the theorem is true.

The zero-product rule can be used to solve equations that, although not linear, can be reduced to two linear equations. For example, if

$$(x - 2)(x + 3) = 0$$

then each factor in the product can be set equal to zero, and we solve the resulting equations.

$$x - 2 = 0 \quad \text{or} \quad x + 3 = 0$$
$$x = 2 \qquad\qquad x = -3$$

Thus, the solutions are 2 and -3. Had we started with the equation

$$x^2 + x - 6 = 0,$$

we could have factored the left side, obtaining

$$(x - 2)(x + 3) = 0,$$

and by following the procedure just outlined, we would have solved a *quadratic equation*.

Quadratic Equations

Any equation that can be written in the form

$$ax^2 + bx + c = 0$$

where a, b, and c are constant real numbers with $a \neq 0$, is called a **quadratic equation** or a **second-degree equation.**

Notice that if $a = 0$, the x^2 term would be missing and we would have the linear equation $bx + c = 0$. We call $ax^2 + bx + c = 0$ the **general form** of a quadratic equation; usually it is helpful to write a quadratic equation in this form before trying to solve it. The following examples illustrate the **factoring method** of solving a quadratic equation.

EXAMPLE 1 SOLVING BY FACTORING

Solve.

(a) $3x^2 - 3x = 18$

$\begin{aligned} 3x^2 - 3x - 18 &= 0 & &\text{Write in general form} \\ 3(x^2 - x - 6) &= 0 & &\text{Factor out common factor 3} \\ x^2 - x - 6 &= 0 & &\text{Divide both sides by 3} \\ (x - 3)(x + 2) &= 0 & &\text{Factor} \\ x - 3 = 0 \quad \text{or} \quad x + 2 &= 0 & &\text{Zero-product rule} \\ x = 3 \qquad\qquad x &= -2 \end{aligned}$

The solutions are 3 and -2. Check by substituting in the original equation.

PRACTICE EXERCISE 1

Solve.

(a) $2y^2 - 8y = 10$

(b) $x^2 + 6x = 0$ No constant term

$x(x + 6) = 0$ Factor

$x = 0$ or $x + 6 = 0$ Zero-product rule

$x = -6$

The solutions are 0 and -6. Check.

(c) $-4x^2 + 2x = \dfrac{1}{4}$

$-4x^2 + 2x - \dfrac{1}{4} = 0$ Write in general form

$16x^2 - 8x + 1 = 0$ Multiply both sides by -4 to clear the fraction and make the coefficient of x^2 positive

$(4x - 1)(4x - 1) = 0$

$4x - 1 = 0$ or $4x - 1 = 0$

$4x = 1$ $4x = 1$

$x = \dfrac{1}{4}$ $x = \dfrac{1}{4}$

There is only one solution, $\dfrac{1}{4}$. Check.

(b) $z^2 - 7z = 0$

(c) $x + \dfrac{1}{4} = -x^2$

Answers: (a) $5, -1$ (b) $0, 7$ (c) $-1/2$

///////////// **CAUTION** /////////////

In Example 1(b), do not divide both sides of the equation by x, or the solution $x = 0$ will be lost. Whenever a quadratic equation has no constant term ($c = 0$), one solution is always 0.

/////////

When the two linear equations obtained using the zero-product rule have the same solution, as in Example 1(c), the solution is sometimes called a **double root** or a **root of multiplicity two.**

2 SOLVING BY TAKING ROOTS

If $b = 0$ in a quadratic equation, the equation has no linear term and takes the form

$$ax^2 + c = 0.$$

If we subtract c from both sides, then divide both sides by a (remembering $a \neq 0$), we obtain

$$x^2 = -\dfrac{c}{a}.$$

Suppose we let $d = -\dfrac{c}{a}$. Then the given equation becomes

$$x^2 = d.$$

By taking the square root of both sides and recalling that $\sqrt{x^2} = |x|$, we have

$$|x| = \sqrt{d}.$$

Since $|x| = x$ if $x \geq 0$ and $|x| = -x$ if $x < 0$, this equation can be expressed as two equations

$$x = \sqrt{d} \text{ and } x = -\sqrt{d},$$

often written more compactly as

$$x = \pm\sqrt{d}.$$

Thus, the two solutions to the original equation are $\sqrt{d}$ and $-\sqrt{d}$. If $d \geq 0$, then the solutions are real numbers. When $d < 0$, the solutions are imaginary numbers. Notice that whenever c and a have opposite signs, the solutions are real. The method shown above and illustrated in the next example is called the **method of taking roots.**

EXAMPLE 2   THE METHOD OF TAKING ROOTS

Solve.

(a) $4y^2 - 5 = 0$

$$4y^2 = 5$$
$$y^2 = \frac{5}{4}$$
$$y = \pm\sqrt{\frac{5}{4}}$$
$$y = \pm\frac{\sqrt{5}}{2}$$

Thus, the solutions are $\sqrt{5}/2$ and $-\sqrt{5}/2$. Notice that in this case $4y^2 - 5$ cannot be factored in the usual manner using integer coefficients.

(b) $5x^2 + 25 = 0$

$$5x^2 = -25$$
$$x^2 = -5$$
$$x = \pm\sqrt{-5}$$
$$x = \pm\sqrt{5(-1)}$$
$$x = \pm\sqrt{5}\sqrt{-1} = \pm i\sqrt{5}$$

Thus, the solutions are $i\sqrt{5}$ and $-i\sqrt{5}$.

PRACTICE EXERCISE 2

Solve.

(a) $3x^2 - 24 = 0$

(b) $2y^2 + 18 = 0$

Answers: (a) $\pm 2\sqrt{2}$ (b) $\pm 3i$

---

3 COMPLETING THE SQUARE

Factoring and taking roots are the easiest and best methods for solving many quadratic equations. However, since some quadratic equations cannot be solved using these techniques, alternative methods must be found.

Consider the equation

$$(x + 2)^2 = 3.$$

If we extend the method of taking roots and take the square root of both sides,

$$x + 2 = \pm\sqrt{3}$$
$$x = -2 \pm \sqrt{3}.$$

This gives us the two solutions $-2 + \sqrt{3}$ and $-2 - \sqrt{3}$. Suppose now that we write out the squared term on the left side of the original equation, then put the result in the general form of a quadratic equation.

$$x^2 + 4x + 4 = 3$$
$$x^2 + 4x + 1 = 0 \quad \text{General form}$$

If we had started with this quadratic equation, by what process could we work backwards to obtain $(x + 2)^2 = 3$, and then the two solutions? The process, called **completing the square,** is illustrated at the top of the next page.

$$x^2 + 4x + 1 = 0$$
$$x^2 + 4x = -1 \quad \text{Isolate the constant term on the right side, leaving space}$$
$$x^2 + 4x + (2)^2 = -1 + (2)^2 \quad \text{Add (1/2 the coefficient of } x)^2$$
$$x^2 + 4x + 4 = 3 \quad \text{Simplify}$$
$$(x + 2)^2 = 3 \quad \text{Left side is a perfect square}$$

More generally, consider the equation
$$(x + d)^2 = x^2 + 2dx + d^2.$$

In the expression $x^2 + 2dx + d^2$, the coefficient of x is $2d$. If we have only $x^2 + 2dx$, we can complete the square by taking one half the coefficient of x, $1/2 \cdot (2d) = d$, squaring it, and adding it to $x^2 + 2dx$. Consider the following examples.

To complete square on	Add $\left(\frac{1}{2} \text{ the coefficient of } x\right)^2$		Obtaining the perfect square
$x^2 + 6x$	9	$\left[9 = \left(\frac{1}{2} \cdot 6\right)^2\right]$	$x^2 + 6x + 9 = (x + 3)^2$
$x^2 - 10x$	25	$\left[25 = \left(\frac{1}{2} \cdot (-10)\right)^2\right]$	$x^2 - 10x + 25 = (x - 5)^2$
$x^2 - x$	$\frac{1}{4}$	$\left[\frac{1}{4} = \left(\frac{1}{2} \cdot (-1)\right)^2\right]$	$x^2 - x + \frac{1}{4} = \left(x - \frac{1}{2}\right)^2$

EXAMPLE 3 Solving by Completing the Square

Solve $x^2 + 2x - 15 = 0$ by completing the square.

We observe that $x^2 + 2x - 15 = (x - 3)(x + 5)$. Using the method of factoring, we have $x - 3 = 0$ or $x + 5 = 0$, giving $x = 3$ and $x = -5$ as solutions. Compare this with the following.

$$x^2 + 2x = 15 \quad \text{Isolate the constant on the right side}$$
$$x^2 + 2x + 1 = 15 + 1 \quad \text{Add 1 to both sides } (1 = [1/2 \text{ coefficient of } x]^2)$$
$$(x + 1)^2 = 16$$
$$x + 1 = \pm\sqrt{16} \quad \text{Take square root of both sides}$$
$$x + 1 = \pm 4$$
$$x = -1 \pm 4$$

$x = -1 + 4$ or $x = -1 - 4$
$x = 3$ $\qquad x = -5$

The solutions are 3 and -5, the same two solutions that were obtained using the factoring method.

PRACTICE EXERCISE 3

Solve $x^2 - 5x - 24 = 0$ by completing the square. Check your work by factoring.

Answers: 8, -3

If the coefficient of x^2 in a quadratic equation is not 1, we *cannot* complete the square simply by taking one half the coefficient of x and squaring it. For example, $(2x - 3)^2 = 4x^2 - 12x + 9$. However, if we are given $4x^2 - 12x$ and attempt to complete the square by adding $(1/2 \cdot 12)^2$, we obtain 36, not 9. In such instances, we must first divide through by the coefficient of x^2, as illustrated in the next example.

126 CHAPTER 2 EQUATIONS, INEQUALITIES, AND PROBLEM SOLVING

EXAMPLE 4 COMPLETING THE SQUARE WHEN $a \neq 1$

Solve $2x^2 + 2x - 3 = 0$ by completing the square.

$2x^2 + 2x = 3$ Isolate the constant

$x^2 + x = \dfrac{3}{2}$ Divide by 2 to make the coefficient of x^2 equal to 1

$x^2 + x + \dfrac{1}{4} = \dfrac{3}{2} + \dfrac{1}{4}$ $\left(\dfrac{1}{2} \cdot 1\right)^2 = \dfrac{1}{4}$

$\left(x + \dfrac{1}{2}\right)^2 = \dfrac{7}{4}$

$x + \dfrac{1}{2} = \pm\sqrt{\dfrac{7}{4}}$

$x + \dfrac{1}{2} = \pm\dfrac{\sqrt{7}}{\sqrt{4}}$ $\sqrt{\dfrac{a}{b}} = \dfrac{\sqrt{a}}{\sqrt{b}}$

$x + \dfrac{1}{2} = \dfrac{\pm\sqrt{7}}{2}$

$x = -\dfrac{1}{2} \pm \dfrac{\sqrt{7}}{2} = \dfrac{-1 \pm \sqrt{7}}{2}$

The solutions are $\dfrac{-1 + \sqrt{7}}{2}$ and $\dfrac{-1 - \sqrt{7}}{2}$.

PRACTICE EXERCISE 4

Solve by completing the square.
$3y^2 + y - 1 = 0$

Answer: $\dfrac{-1 \pm \sqrt{13}}{6}$

---

❹ THE QUADRATIC FORMULA

Rather than continuing to solve particular quadratic equations by completing the square, we now use this technique to solve a general quadratic equation and in the process derive the quadratic formula. Recall the general form of a quadratic equation.

$ax^2 + bx + c = 0$ $a \neq 0$

$ax^2 + bx = -c$ Isolate the constant

$\dfrac{\cancel{a}x^2}{\cancel{a}} + \dfrac{b}{a}x = -\dfrac{c}{a}$ Divide by a to make the coefficient of x^2 equal to 1

$x^2 + \dfrac{b}{a}x + \left(\dfrac{b}{2a}\right)^2 = -\dfrac{c}{a} + \left(\dfrac{b}{2a}\right)^2$ Add $\left(\dfrac{1}{2} \cdot \dfrac{b}{a}\right)^2 = \left(\dfrac{b}{2a}\right)^2$

$\left(x + \dfrac{b}{2a}\right)^2 = -\dfrac{c}{a} + \dfrac{b^2}{4a^2}$

$\left(x + \dfrac{b}{2a}\right)^2 = -\dfrac{4ac}{4a^2} + \dfrac{b^2}{4a^2}$ $4a^2$ is LCD

$\left(x + \dfrac{b}{2a}\right)^2 = \dfrac{b^2 - 4ac}{4a^2}$ Subtract

$x + \dfrac{b}{2a} = \pm\sqrt{\dfrac{b^2 - 4ac}{4a^2}}$

$x + \dfrac{b}{2a} = \dfrac{\pm\sqrt{b^2 - 4ac}}{\sqrt{4a^2}}$ $\sqrt{\dfrac{a}{b}} = \dfrac{\sqrt{a}}{\sqrt{b}}$

$x + \dfrac{b}{2a} = \dfrac{\pm\sqrt{b^2 - 4ac}}{2a}$

$$x = -\frac{b}{2a} \pm \frac{\sqrt{b^2 - 4ac}}{2a} \qquad \text{Subtract } \frac{b}{2a}$$

$$x = \frac{-b \pm \sqrt{b^2 - 4ac}}{2a} \qquad \text{Note that 2a is the denominator of the entire expression}$$

This formula is called the **quadratic formula** and it must be memorized. To use it to solve a quadratic equation, identify the constants a, b, and c, substitute them into the quadratic formula, and simplify the numerical expression.

EXAMPLE 5 USING THE QUADRATIC FORMULA	**PRACTICE EXERCISE 5**

Solve using the quadratic formula.

(a) $x^2 - 6x + 8 = 0$

The equation is in the simplest general form and therefore $a = 1$, $b = -6$ (not 6), and $c = 8$.

$$x = \frac{-b \pm \sqrt{b^2 - 4ac}}{2a}$$

$$= \frac{-(-6) \pm \sqrt{(-6)^2 - 4(1)(8)}}{2(1)}$$

$$= \frac{6 \pm \sqrt{36 - 32}}{2}$$

$$= \frac{6 \pm \sqrt{4}}{2} = \frac{6 \pm 2}{2}$$

$$x = \frac{6 + 2}{2} = \frac{8}{2} = 4 \quad \text{or} \quad x = \frac{6 - 2}{2} = \frac{4}{2} = 2.$$

The solutions are 4 and 2. Notice that it would have been easier to solve this equation by factoring.

(b) $x^2 - 2x = -\frac{1}{2}$

Write the equation in general form and clear the fraction.

$$x^2 - 2x + \frac{1}{2} = 0$$

$$2x^2 - 4x + 1 = 0$$

Thus, $a = 2$, $b = -4$, and $c = 1$.

$$x = \frac{-b \pm \sqrt{b^2 - 4ac}}{2a}$$

$$= \frac{-(-4) \pm \sqrt{(-4)^2 - 4(2)(1)}}{2(2)}$$

$$= \frac{4 \pm \sqrt{16 - 8}}{4} \qquad \text{Multiply 4(2)(1) before subtracting}$$

$$= \frac{4 \pm \sqrt{8}}{4}$$

$$= \frac{4 \pm 2\sqrt{2}}{4} = \frac{2(2 \pm \sqrt{2})}{2 \cdot 2} = \frac{2 \pm \sqrt{2}}{2}$$

The solutions are $\frac{2 + \sqrt{2}}{2}$ and $\frac{2 - \sqrt{2}}{2}$.

Solve using the quadratic formula.

(a) $x^2 - 9x + 18 = 0$

(b) $x^2 - 1 = \frac{5}{3}x$

(c) $x^2 + x + 1 = 0$

We have $a = 1$, $b = 1$, and $c = 1$.

$$x = \frac{-b \pm \sqrt{b^2 - 4ac}}{2a}$$

$$= \frac{-(1) \pm \sqrt{(1)^2 - 4(1)(1)}}{2(1)}$$

$$= \frac{-1 \pm \sqrt{1 - 4}}{2}$$

$$= \frac{-1 \pm \sqrt{-3}}{2}$$

$$= \frac{-1 \pm i\sqrt{3}}{2}$$

This time the solutions are complex numbers, $\frac{-1 + i\sqrt{3}}{2}$ and $\frac{-1 - i\sqrt{3}}{2}$.

(c) $2x^2 + x + 1 = 0$

Answers: (a) 3, 6 (b) $\frac{5 \pm \sqrt{61}}{6}$
(c) $\frac{-1 \pm i\sqrt{7}}{4}$

CAUTION

In Example 5(b), students often make the mistake of canceling the 4's in $\frac{4 \pm 2\sqrt{2}}{4}$ obtaining an incorrect answer, $1 \pm 2\sqrt{2}$. Also, remember that both terms in the numerator must be placed over the denominator. DO NOT write $2 \pm \frac{\sqrt{2}}{2}$ for the answer $\frac{2 \pm \sqrt{2}}{2}$.

Factoring and taking roots are the easiest methods for solving quadratic equations. If these techniques are not appropriate, it is usually better to go directly to the quadratic formula rather than to complete the square.

5 THE DISCRIMINANT

With complex numbers available, every quadratic equation can be solved. Sometimes the solutions will be real, and sometimes they will be complex (nonreal). To determine the nature of the solutions to a given quadratic equation, suppose we look again at the quadratic formula

$$x = -b \pm \sqrt{b^2 - 4ac} \; 2ad.$$

The number under the radical, $b^2 - 4ac$, called the **discriminant**, can be used to determine the types of solutions to a quadratic equation without actually solving it.

Types of Solutions to a Quadratic Equation

The quadratic equation $ax^2 + bx + c = 0$ has

1. two real-number solutions if the discriminant is positive;
2. exactly one real-number solution (sometimes considered as two real but equal solutions) if the discriminant is zero;
3. two complex-number (not real) solutions if the discriminant is negative.

2.4 QUADRATIC EQUATIONS 129

EXAMPLE 6 Using the Discriminant

Without solving, use the discriminant to determine the nature of the solutions.

(a) $2x^2 - 3x - 7 = 0$

$b^2 - 4ac = (-3)^2 - 4(2)(-7)$ $a = 2, b = -3, c = -7$

$= 9 + 56 = 65 > 0$

There are two real solutions.

(b) $2x^2 - 3x + 7 = 0$

$b^2 - 4ac = (-3)^2 - 4(2)(7)$ $a = 2, b = -3, c = 7$

$= 9 - 56 = -47 < 0$

There are two complex (nonreal) solutions.

(c) $4x^2 - 12x + 9 = 0$

$b^2 - 4ac = (-12)^2 - 4(4)(9)$ $a = 4, b = -12, c = 9$

$= 144 - 144 = 0$

There is one real solution.

PRACTICE EXERCISE 6

Without solving, use the discriminant to determine the nature of the solutions.

(a) $5x^2 - 2x + 3 = 0$

(b) $3x^2 - 8x + 1 = 0$

(c) $9x^2 + 12x + 4 = 0$

Answers: (a) two complex (nonreal) (b) two real (c) one real

Some equations, although not quadratic themselves, can be factored and solved using the zero-product rule. A more detailed discussion of these types of equations is given in Chapter 4, but we now consider a special case.

EXAMPLE 7 A More Complex Equation

Solve. $x^3 - 8 = 0$

Factoring using the difference-of-cubes formula we obtain

$(x - 2)(x^2 + 2x + 4) = 0.$

$x - 2 = 0$ or $x^2 + 2x + 4 = 0$ Zero-product rule

$x = 2$

$x = \dfrac{-2 \pm \sqrt{4 - 4(1)(4)}}{2(1)}$

$= \dfrac{-2 \pm \sqrt{-12}}{2}$

$= \dfrac{-2 \pm 2i\sqrt{3}}{2}$

$= -1 \pm i\sqrt{3}$

The three numbers 2, $-1 + i\sqrt{3}$, and $-1 - i\sqrt{3}$ are all cube roots of 8. Prior to this section, we were only aware of the real-number cube root, 2.

PRACTICE EXERCISE 7

Solve. $x^3 + 27 = 0$

Answer: -3 and $\dfrac{3 \pm 3i\sqrt{3}}{2}$

---

⑥ PROPERTIES OF SOLUTIONS

An interesting property of the solutions to a quadratic equation involves their sum and product. Suppose that

$x_1 = \dfrac{-b + \sqrt{b^2 - 4ac}}{2a}$ and $x_2 = \dfrac{-b - \sqrt{b^2 - 4ac}}{2a}$

are the two solutions to the quadratic equation $ax^2 + bx + c = 0$. Then

$$\begin{aligned} x_1 + x_2 &= \frac{-b + \sqrt{b^2 - 4ac}}{2a} + \frac{-b - \sqrt{b^2 - 4ac}}{2a} \\ &= \frac{-b + \sqrt{b^2 - 4ac} - b - \sqrt{b^2 - 4ac}}{2a} \\ &= \frac{-2b}{2a} = -\frac{b}{a}. \end{aligned}$$

We have just proved the first part of the following theorem. The proof of the second part is left as an exercise.

> **Sum and Product of Solutions**
>
> If x_1 and x_2 are solutions to $ax^2 + bx + c = 0$, then
>
> $$x_1 + x_2 = -\frac{b}{a} \quad \text{and} \quad x_1 x_2 = \frac{c}{a}.$$

EXAMPLE 8 Sum and Product of Solutions

Find the sum and product of the solutions to $3x^2 - 7x + 8 = 0$.
 Since $a = 3$, $b = -7$, and $c = 8$, the sum of the solutions is $-\frac{b}{a} = -\frac{-7}{3} = \frac{7}{3}$, and their product is $\frac{c}{a} = \frac{8}{3}$.

Practice Exercise 8

Find the sum and product of the solutions to $2x^2 - 3x + 6 = 0$.

Answer: sum: 3/2; product: 3

7 EQUATIONS FROM SOLUTIONS

When a quadratic equation is solved by the factoring method, the equation often takes the form

$$(x - x_1)(x - x_2) = 0,$$

which has solutions x_1 and x_2. By reversing this procedure we can find a quadratic equation when its solutions x_1 and x_2 are given.

> **Finding a Quadratic Equation with Given Solutions**
>
> If x_1 and x_2 are two real numbers or two conjugate complex numbers, the equation
>
> $$(x - x_1)(x - x_2) = 0$$
>
> is a quadratic equation with real coefficients having solutions x_1 and x_2.

EXAMPLE 9 Finding an Equation with Given Solutions

Find a quadratic equation that has the given solutions.
(a) $x_1 = -3$ and $x_2 = 7$

$$\begin{aligned} (x - x_1)(x - x_2) &= 0 \\ (x - (-3))(x - 7) &= 0 \quad \text{Substitute} \\ (x + 3)(x - 7) &= 0 \\ x^2 - 4x - 21 &= 0 \end{aligned}$$

Practice Exercise 9

Find a quadratic equation that has the given solutions.
(a) $x_1 = -4$ and $x_2 = 1$

(b) $x_1 = 1 + 3i$ and $x_2 = 1 - 3i$

$$(x - x_1)(x - x_2) = 0$$
$$[x - (1 + 3i)][x - (1 - 3i)] = 0$$
$$x^2 - (1 - 3i)x - (1 + 3i)x + (1 + 3i)(1 - 3i) = 0$$
$$x^2 - 2x + 10 = 0$$

(b) $x_1 = 2 - i$ and $x_2 = 2 + i$

Answers: (a) $x^2 + 3x - 4 = 0$
(b) $x^2 - 4x + 5 = 0$

8 APPROXIMATING SOLUTIONS

Often an applied problem will translate into a quadratic equation with decimal coefficients. A calculator is helpful for approximating the solutions. To do this, compute $\sqrt{b^2 - 4ac}$ and store this value in memory. Add this value to $-b$ and divide by $2a$ to obtain the first solution. Then subtract this value from $-b$ and divide by $2a$ to obtain the second solution.

EXAMPLE 10 APPROXIMATING SOLUTIONS

Solve. $3.5x^2 - 81.6x + 2.8 = 0$

The solutions are

$$x = \frac{81.6 \pm \sqrt{(-81.6)^2 - 4(3.5)(2.8)}}{2(3.5)}$$

1. Sequence of steps to calculate $\sqrt{b^2 - 4ac}$:

ALG: 4 × 3.5 × 2.8 = +/− + 81.6 x^2 = √ STO

RPN: 81.6 x^2 ENTER 4 ENTER 3.5 × 2.8 × − √ STO

At this point the display will show 81.359449, and this value has also been stored in the memory. Do not clear the display since it is used in the next step.

2. Sequence of steps to calculate the first solution:

ALG: + 81.6 = ÷ 2 ÷ 3.5 =

RPN: 81.6 + 2 ÷ 3.5 ÷

The display now shows the first solution, 23.279921.

3. Sequence of steps to calculate the second solution:

ALG: 81.6 − RCL = ÷ 2 ÷ 3.5 =

RPN: 81.6 ENTER RCL − 2 ÷ 3.5 ÷

The display now shows the second solution, 0.0343644.

PRACTICE EXERCISE 10

Solve.

$0.3x^2 - 25.2x + 1.4 = 0$

Answer: 83.944408 and 0.0555924

2.4 EXERCISES A

In Exercises 1–6 solve each equation by the factoring method.

1. $x^2 - 3x - 10 = 0$

2. $2y^2 - 10y = 12$

3. $y^2 + y = -\frac{1}{4}$

4. $2x^2 + 3x = 0$

5. $5y(y + 2) = 7y$

6. $4(x - 1)(x + 1) = -3$

In Exercises 7–9 solve each equation by the square root method.

7. $x^2 - 36 = 0$

8. $3y^2 + 75 = 0$

9. $(y + 1)^2 = 9$

What must be added to complete the square in each expression in Exercises 10–12?

10. $x^2 - 8x +$ ____

11. $x^2 - x +$ ____

12. $y^2 + \frac{1}{3}y +$ ____

In Exercises 13–14 solve each equation by completing the square.

13. $x^2 - 5x - 24 = 0$

14. $3y^2 - 5y + 1 = 0$

In Exercises 15–18 solve by using the quadratic formula.

15. $x^2 - 5x - 24 = 0$

16. $3y^2 - 5y + 1 = 0$

17. $\frac{1}{2}x^2 - \frac{3}{4}x = 1$

18. $(y + 1)(y - 1) + 2y = 0$

Solve each equation in Exercises 19–27. (First try factoring or taking roots, and if those fail, use the quadratic formula.)

19. $x^2 - 25 = 0$

20. $3(y^2 + y) = -10y - 4$

21. $z^2 = 3(z + 1)$

22. $y^2 - 4y = -1$

23. $3x^2 = x + 1$

24. $y^2 = 3y$

25. $3z^2 + 12 = 0$

26. $x^2 - x = -1$

27. $z(2z - 1) = -7$

Solve for x in Exercises 28–30. Assume a and b are positive constants.

28. $x^2 - 36a^2 = 0$

29. $x^2 - (a + b)^2 = 0$

30. $x^2 - ax - 2a^2 = 0$

Solve each equation in Exercises 31–33. (See Example 7.)

31. $y^3 - 1 = 0$

32. $x^3 + 4x = 0$

33. $z^4 - 1 = 0$

In Exercises 34–36, use the discriminant to determine the nature of the solutions to each equation (two real, one real, or two complex).

34. $3x^2 = 5x - 8$

35. $x^2 - 10x + 25 = 0$

36. $-5x^2 + 2x + 1 = 0$

Without solving, find the sum and product of the solutions to the equations in Exercises 37–38.

37. $x^2 - 10x = -25$

38. $\frac{1}{2}x^2 - \frac{1}{3} = x$

The converse of the theorem on the sum and product of the roots of a quadratic equation is also true and can be used to determine quickly whether two numbers are solutions to a given quadratic equation. Without solving, determine if x_1 and x_2 are solutions to each equation in Exercises 39–40.

39. $x^2 - 3x + 2 = 0$; $x_1 = 2$, $x_2 = 1$

40. $4x^2 + 11x - 3 = 0$; $x_1 = 4$, $x_2 = -\frac{1}{4}$

In Exercises 41–43, find a quadratic equation that has the given solutions.

41. $-2, 3$

42. $5i, -5i$

43. $5 - i, 5 + i$

FOR REVIEW

Perform the indicated operations in Exercises 44–47.

44. $(2 + 5i) - (3 - 8i)$

45. $\dfrac{1}{2 - 3i}$

46. $(4 + 3i)(2 - i)$

47. $(5 - 3i)^2$

ANSWERS: 1. $5, -2$ 2. $6, -1$ 3. $-\frac{1}{2}$ 4. $0, -\frac{3}{2}$ 5. $0, -\frac{3}{5}$ 6. $\frac{1}{2}, -\frac{1}{2}$ 7. ± 6 8. $\pm 5i$ 9. $-4, 2$ 10. 16 11. $\frac{1}{4}$ 12. $\frac{1}{36}$ 13. $8, -3$ 14. $\frac{5 \pm \sqrt{13}}{6}$ 15. $8, -3$ 16. $\frac{5 \pm \sqrt{13}}{6}$ 17. $\frac{3 \pm \sqrt{41}}{4}$ 18. $-1 \pm \sqrt{2}$ 19. ± 5 20. $-\frac{1}{3}, -4$ 21. $\frac{3 \pm \sqrt{21}}{2}$ 22. $2 \pm \sqrt{3}$ 23. $\frac{1 \pm \sqrt{13}}{6}$ 24. $0, 3$ 25. $\pm 2i$ 26. $\frac{1 \pm i\sqrt{3}}{2}$ 27. $\frac{1 \pm i\sqrt{55}}{4}$ 28. $\pm 6a$ 29. $\pm(a + b)$ 30. $2a, -a$ 31. $1, \frac{-1 \pm i\sqrt{3}}{2}$ 32. $0, 2i, -2i$ 33. $\pm 1, \pm i$ 34. two complex (nonreal) 35. one real 36. two real 37. $x_1 + x_2 = 10$; $x_1 x_2 = 25$ 38. $x_1 + x_2 = 2$; $x_1 x_2 = -\frac{2}{3}$ 39. Yes, 2 and 1 are solutions. 40. No, 4 and $-\frac{1}{4}$ are not solutions. 41. $x^2 - x - 6 = 0$ 42. $x^2 + 25 = 0$ 43. $x^2 - 10x + 26 = 0$ 44. $-1 + 13i$ 45. $\frac{2}{13} + \frac{3}{13}i$ 46. $11 + 2i$ 47. $16 - 30i$

2.4 EXERCISES B

In Exercises 1–6 solve each equation by the factoring method.

1. $y^2 + 3y - 18 = 0$

2. $3x^2 + 19x - 14 = 0$

3. $z^2 = \dfrac{11}{3}z + \dfrac{4}{3}$

4. $4y^2 - 7y = 0$

5. $5z(z + 2) = 5(z + 1)^2$

6. $4(2x - 3) = (2x - 3)(x + 1)$

In Exercises 7–9 solve each equation by the square root method.

7. $y^2 - 100 = 0$

8. $5x^2 + 80 = 0$

9. $(z - 3)^2 = 25$

What must be added to complete the square in each expression in Exercises 10–12?

10. $x^2 - 4x +$ ____

11. $z^2 + z +$ ____

12. $z^2 - \dfrac{1}{4}z +$ ____

In Exercises 13–14 solve each equation by completing the square.

13. $x^2 + 3x + 1 = 0$

14. $2z^2 + 2z - 1 = 0$

134 CHAPTER 2 EQUATIONS, INEQUALITIES, AND PROBLEM SOLVING

In Exercises 15–18 solve by using the quadratic formula.

15. $y^2 + 4y - 21 = 0$

16. $2x^2 - 3x - 4 = 0$

17. $x^2 + 5x = 21\left(\dfrac{1}{2} + \dfrac{x}{2}\right)$

18. $(2z - 1)(z - 2) = 5 - 3z$

Solve each equation in Exercises 19–27. (First try factoring or taking roots, and if those fail, use the quadratic formula.)

19. $\dfrac{1}{3}x^2 - 12 = 0$

20. $4(y^2 + 2y) = 3(1 - y)$

21. $x^2 = 2(x + 3)$

22. $2x^2 + 9x = 5$

23. $5y^2 - 14y = 3$

24. $2z^2 = -z$

25. $\dfrac{1}{2}z^2 + 8 = 0$

26. $3y^2 + 2y + 1 = 0$

27. $x(2x + 1) = -2$

Solve for x in Exercises 28–30. Assume a and b are positive constants.

28. $x^2 - 100b^2 = 0$

29. $x^2 - (2a + b)^2 = 0$

30. $x^2 + 2bx - 3b^2 = 0$

Solve each equation in Exercises 31–33. (See Example 7.)

31. $x^3 + 8 = 0$

32. $z^3 + 1 = 0$

33. $y^3 - 4y^2 + 5y = 0$

In Exercises 34–36, use the discriminant to determine the nature of the solutions to each equation (two real, one real, or two complex).

34. $5x^2 + x - 7 = 0$

35. $4x^2 + 25 = 20x$

36. $7x^2 + 3x + 4 = 0$

Without solving, find the sum and product of the solutions to the equations in Exercises 37–38.

37. $-5x^2 + 1 = -2x$

38. $3x^2 + x = 6$

The converse of the theorem on the sum and product of the roots of a quadratic equation is also true and can be used to determine quickly whether two numbers are solutions to a given quadratic equation. Without solving, determine if x_1 and x_2 are solutions to each equation in Exercises 39–40.

39. $x^2 - 4x - 12 = 0$; $x_1 = -2$, $x_2 = 6$

40. $3x^2 + 8x - 35 = 0$; $x_1 = \dfrac{4}{3}$, $x_2 = -4$

In Exercises 41–43, find a quadratic equation that has the given solutions.

41. $-4, 2$

42. $2i, -2i$

43. $3 + i, 3 - i$

FOR REVIEW

Perform the indicated operations in Exercises 44–47.

44. $(5 - 9i) - (-2 + i)$

45. $\dfrac{2 - i}{2 + i}$

46. $(3 + 4i)(3 - 2i)$

47. $(5 + 3i)^2$

2.4 EXERCISES C

In Exercises 1–4 use a calculator to find approximate solutions to the quadratic equations. Give answers correct to the nearest hundredth.

1. $x^2 - 2.1x - 4.3 = 0$
   [Answer: 3.37, −1.27]

2. $y^2 - 5.3y - 1.6 = 0$

3. $2.15x^2 + 3.2x - 4.65 = 0$
   [Answer: $0.90, -2.39$]

4. $3.25y^2 - 2.17y - 5.07 = 0$

Solve each quadratic equation. Leave answer in exact form, that is, do not use a calculator.

5. $2x^2 + \sqrt{3}x - \sqrt{5} = 0$
   $\left[\text{Answer: } \frac{-\sqrt{3} \pm \sqrt{3 + 8\sqrt{5}}}{4}\right]$

6. $\sqrt{2}x^2 - \pi x - \sqrt{2} = 0$

7. Prove that if x_1 and x_2 are solutions to the quadratic equation $ax^2 + bx + c = 0$, then $x_1 x_2 = \frac{c}{a}$.

8. Prove that a quadratic equation $ax^2 + bx + c = 0$, where a, b, and c are integers, has a single rational-number solution if the discriminant is 0.

9. The equation $mx^2 - 7x - 2m = 0$ has one solution 4. First find the value of m, and then find the other solution.
   $\left[\text{Answer: } m = 2; -\frac{1}{2}\right]$

10. The equation $x^2 - mx + 8 = 0$ has one solution $2 + 2i$. First find the value of m, and then find the other solution.

11. If $a \neq 0$ and $c \neq 0$, prove that the solutions to $ax^2 + bx + c = 0$ and $cx^2 + bx + a = 0$ are reciprocals of each other.

12. Find the values of m and n in the equation $2x^2 + mx + n = 0$ if the sum of the solutions to the equation is $-\frac{5}{2}$ and the product of the solutions is $-\frac{3}{2}$.

2.5 EQUATIONS THAT RESULT IN QUADRATIC EQUATIONS

STUDENT GUIDEPOSTS

1 Equations Quadratic in Form
2 Fractional Equations
3 Radical Equations

1 EQUATIONS QUADRATIC IN FORM

Often, an equation is not quadratic in a particular variable, but rather is quadratic in an expression containing the variable. Equations such as this are called **quadratic in form.** For example,

$$x^4 - 5x^2 + 4 = 0$$

is not a quadratic equation. However, if we substitute u for x^2, $x^4 - 5x^2 + 4 = 0$ becomes

$$u^2 - 5u + 4 = 0 \quad u^2 = (x^2)^2 = x^4$$

which is quadratic in the variable u. The original equation is quadratic in form, that is, quadratic in the expression x^2. Similarly,

$$(z - 4)^2 - 5(z - 4) + 6 = 0$$

is quadratic in $u = z - 4$ since it becomes

$$u^2 - 5u + 6 = 0.$$

EXAMPLE 1 An Equation Quadratic in Form

Solve. $x^4 - 5x^2 + 4 = 0$

$$u^2 - 5u + 4 = 0 \quad \text{Let } u = x^2. \text{ Then } u^2 = x^4$$
$$(u - 4)(u - 1) = 0$$
$$u - 4 = 0 \quad \text{or} \quad u - 1 = 0$$
$$u = 4 \qquad\qquad u = 1$$

Since $u = x^2$, $\quad x^2 = 4 \qquad\qquad x^2 = 1$
$$x = \pm\sqrt{4} \qquad\qquad x = \pm\sqrt{1}$$
$$x = \pm 2 \qquad\qquad x = \pm 1$$

Check: $x^4 - 5x^2 + 4 = 0 \qquad x^4 - 5x^2 + 4 = 0$
$\quad\quad\;\; (2)^4 - 5(2)^2 + 4 \stackrel{?}{=} 0 \qquad (1)^4 - 5(1)^2 + 4 \stackrel{?}{=} 0$
$\quad\quad\quad\;\; 16 - 20 + 4 \stackrel{?}{=} 0 \qquad\qquad 1 - 5 + 4 \stackrel{?}{=} 0$
$\quad\quad\quad\quad\quad\quad\quad 0 = 0 \qquad\qquad\qquad\quad\; 0 = 0$

2 (and clearly, -2) checks.   1 (and also -1) checks.

The solutions are 2, -2, 1, -1.

PRACTICE EXERCISE 1

Solve.

$$(z - 4)^2 - 5(z - 4) + 6 = 0$$

Answer: 6 and 7

///////////// **CAUTION** /////////////

Do not give 4 and 1 as the solutions to the original equation in Example 1. Remember that 4 and 1 are solutions to the equation in the variable u and are not solutions relative to the variable x. In other words, do not forget to substitute and find the solutions to the original equation when using the u-substitution method.

EXAMPLE 2 An Equation Quadratic in Form

Solve. $u^{2/3} + 2u^{1/3} = 15$

$$u^{2/3} + 2u^{1/3} - 15 = 0$$
$$w^2 + 2w - 15 = 0 \quad \text{Let } w = u^{1/3}. \text{ Then } w^2 = u^{2/3}$$
$$(w + 5)(w - 3) = 0$$
$$w + 5 = 0 \quad \text{or} \quad w - 3 = 0$$
$$w = -5 \qquad\qquad w = 3$$

Since $w = u^{1/3}$,

$$u^{1/3} = -5 \qquad\qquad u^{1/3} = 3$$
$$(u^{1/3})^3 = (-5)^3 \qquad (u^{1/3})^3 = 3^3$$
$$u = -125 \qquad\qquad u = 27$$

The solutions are 27 and -125. Check these.

PRACTICE EXERCISE 2

Solve.

$$y^{2/3} - 5y^{1/3} - 6 = 0$$

Answer: -1 and 216

2 FRACTIONAL EQUATIONS

In Section 2.1 we solved fractional equations that resulted in linear equations when the fractions were cleared. Recall that to clear the fractions we multiply both sides of the equation by the LCD of all fractions. Clearing some fractional equations results in quadratic equations as shown in the next two examples. Remember that any potential solutions to a fractional equation *must always* be checked in the original equation to be sure that they do not make one of the denominators equal to zero.

EXAMPLE 3 A Fractional Equation

Solve. $\dfrac{y}{y+2} = \dfrac{2}{y-1}$

The LCD $= (y+2)(y-1)$. Multiply both sides by $(y+2)(y-1)$.

$$(y+2)(y-1)\dfrac{y}{(y+2)} = \dfrac{2}{(y-1)}(y+2)(y-1)$$

$$(y-1)y = 2(y+2)$$
$$y^2 - y = 2y + 4$$
$$y^2 - 3y - 4 = 0$$
$$(y+1)(y-4) = 0$$
$$y + 1 = 0 \quad \text{or} \quad y - 4 = 0$$
$$y = -1 \qquad\qquad y = 4$$

The solutions are -1 and 4. Check these to verify that neither makes one of the original denominators equal to zero.

Practice Exercise 3

Solve. $\dfrac{x}{3} = 1 + \dfrac{6}{x}$

Answer: -3 and 6

CAUTION

Remember to notice the values for the variable that must be excluded when solving a fractional equation. If you obtain one of these values as a potential solution, it must be discarded.

EXAMPLE 4 A Fractional Equation

Solve. $\dfrac{x}{x-2} + 1 = \dfrac{x^2+4}{x^2-4}$

The LCD $= (x-2)(x+2)$.

$$(x-2)(x+2)\left[\dfrac{x}{x-2} + 1\right] = \left[\dfrac{x^2+4}{(x-2)(x+2)}\right](x-2)(x+2)$$

$$(x+2)x + (x-2)(x+2) = x^2 + 4$$
$$x^2 + 2x + x^2 - 4 = x^2 + 4$$
$$x^2 + 2x - 8 = 0$$
$$(x+4)(x-2) = 0$$
$$x + 4 = 0 \quad \text{or} \quad x - 2 = 0$$
$$x = -4 \qquad\qquad x = 2$$

Practice Exercise 4

$\dfrac{3}{x+1} = 1 - \dfrac{5}{x-1}$

[*Hint:* The only values to exclude are 1 and -1.]

Check: $\dfrac{-4}{-4-2} + 1 \stackrel{?}{=} \dfrac{(-4)^2 + 4}{(-4)^2 - 4}$ $\dfrac{2}{2-2} + 1 \stackrel{?}{=} \dfrac{2^2 + 4}{2^2 - 4}$

$\dfrac{4}{6} + 1 \stackrel{?}{=} \dfrac{20}{12}$ $\dfrac{2}{0} + 1 \stackrel{?}{=} \dfrac{8}{0}$

$\dfrac{5}{3} = \dfrac{5}{3}$ $\dfrac{2}{0}$ and $\dfrac{8}{0}$ are not defined.

The only solution is -4.

Answer: $4 \pm \sqrt{19}$

❸ RADICAL EQUATIONS

Some radical equations become quadratic equations when the radicals are removed by raising both sides to a power. Always check your results since use of the power theorem may introduce extraneous roots.

EXAMPLE 5 A Radical Equation

Solve. $\sqrt{2y + 11} - y - 4 = 0$

$\sqrt{2y + 11} = y + 4$ Isolate the radical
$(\sqrt{2y + 11})^2 = (y + 4)^2$ Square both sides. Don't forget the middle term 8y
$2y + 11 = y^2 + 8y + 16$
$y^2 + 6y + 5 = 0$
$(y + 1)(y + 5) = 0$
$y + 1 = 0$ or $y + 5 = 0$
$y = -1$ $y = -5$

Check: $\sqrt{2(-1) + 11} - (-1) - 4 \stackrel{?}{=} 0$ $\sqrt{2(-5) + 11} - (-5) - 4 \stackrel{?}{=} 0$
$\sqrt{-2 + 11} + 1 - 4 \stackrel{?}{=} 0$ $\sqrt{-10 + 11} + 5 - 4 \stackrel{?}{=} 0$
$\sqrt{9} + 1 - 4 \stackrel{?}{=} 0$ $\sqrt{1} + 5 - 4 \stackrel{?}{=} 0$
$3 + 1 - 4 \stackrel{?}{=} 0$ $1 + 5 - 4 \stackrel{?}{=} 0$
$0 = 0$ $2 \neq 0$

The only solution is -1.

Practice Exercise 5

Solve.

$\sqrt{a^2 - 7a + 15} - \sqrt{4a - 13} = 0$

Answer: 7 and 4

EXAMPLE 6 A Radical Equation

Solve. $\sqrt{3x + 1} - \sqrt{x + 9} = 2$

$\sqrt{3x + 1} = 2 + \sqrt{x + 9}$ Isolate one radical on the left side
$(\sqrt{3x + 1})^2 = (2 + \sqrt{x + 9})^2$ Square both sides
$3x + 1 = 4 + 4\sqrt{x + 9} + x + 9$ $(a + b)^2 = a^2 + 2ab + b^2$, and $4\sqrt{x + 9}$ is the $2ab$
$2x - 12 = 4\sqrt{x + 9}$ Isolate the radical
$x - 6 = 2\sqrt{x + 9}$ Simplify
$(x - 6)^2 = (2\sqrt{x + 9})^2$ Square both sides
$x^2 - 12x + 36 = 4(x + 9)$ Don't forget to square the 2
$x^2 - 12x + 36 = 4x + 36$

Practice Exercise 6

Solve.

$\sqrt{w - 1} + \sqrt{3w + 3} = 4$

$x^2 - 16x = 0$
$x(x - 16) = 0$
$x = 0 \quad \text{or} \quad x - 16 = 0$
$\qquad\qquad\qquad x = 16$

Check: $\sqrt{3(0) + 1} - \sqrt{0 + 9} \stackrel{?}{=} 2 \qquad\qquad \sqrt{3(16) + 1} - \sqrt{16 + 9} \stackrel{?}{=} 2$
$\qquad\qquad\quad \sqrt{1} - \sqrt{9} \stackrel{?}{=} 2 \qquad\qquad\qquad\quad \sqrt{49} - \sqrt{25} \stackrel{?}{=} 2$
$\qquad\qquad\qquad\quad 1 - 3 \neq 2 \qquad\qquad\qquad\qquad\qquad 7 - 5 = 2$

The only solution is 16.

Answer: 2

CAUTION

Is $(2 + \sqrt{x + 9})^2$ the same as $4 + (x + 9)$? The answer is no. Remember that $(a + b)^2 = a^2 + \mathbf{2ab} + b^2$. Don't forget the middle term, which is $4\sqrt{x + 9}$ in this case. And again, check all possible solutions in the *original* equation.

2.5 EXERCISES A

Solve each equation in Exercises 1–22.

1. $x^4 - 10x^2 + 9 = 0$
2. $(y + 2)^2 - 13(y + 2) + 42 = 0$
3. $z^{2/3} - 5z^{1/3} + 6 = 0$

4. $(y^2 + 4y)^2 - (y^2 + 4y) - 20 = 0$
5. $z^{-2} + 2z^{-1} - 15 = 0$
6. $2\left(\dfrac{2x - 1}{x}\right)^2 + 7\left(\dfrac{2x - 1}{x}\right) - 4 = 0$

7. $y^{1/2} + 6y^{-1/2} = 5$
8. $\sqrt[6]{a} + \sqrt[3]{a} - 6 = 0$
9. $1 - \dfrac{2}{x} = \dfrac{3}{x^2}$

10. $\dfrac{1}{y} = \dfrac{-6}{y^2 + 5}$
11. $3y - \dfrac{2}{y + 1} = \dfrac{4}{y + 1}$
12. $\dfrac{3z}{2z + 1} = \dfrac{2}{4z^2 - 1} + \dfrac{z}{2z - 1}$

13. $\dfrac{x}{x^2 - x - 2} - \dfrac{2}{x^2 - 5x + 6} = \dfrac{-3}{x^2 - 2x - 3}$
14. $\sqrt{y + 4} + 8 = y$

15. $2\sqrt{3y - 2} + \sqrt{3y^2 + 2y} = 0$
16. $\sqrt{x^2 + 6x} + \sqrt{2x + 21} = 0$

17. $\sqrt{3a + 1} - 3 = \sqrt{a - 4}$
18. $\sqrt[3]{x - 1} = 2$

19. $(2x + 11)^{1/2} - x - 4 = 0$
20. $(3x + 3)^{1/2} = 4 - (x - 1)^{1/2}$

140   CHAPTER 2   EQUATIONS, INEQUALITIES, AND PROBLEM SOLVING

FOR REVIEW

Solve each equation in Exercises 21–22.

21. $x^2 + 5x = 21\left(\dfrac{1}{2} + \dfrac{x}{2}\right)$

22. $(y + 2)(y - 2) + 3y = -4$

ANSWERS: 1. $\pm 1, \pm 3$ 2. $4, 5$ 3. $8, 27$ 4. $-5, 1, -2$ 5. $\frac{1}{3}, -\frac{1}{5}$ 6. $\frac{2}{3}, \frac{1}{6}$ 7. $4, 9$ 8. 64 9. $3, -1$
10. $-5, -1$ 11. $1, -2$ 12. $\dfrac{1 \pm \sqrt{3}}{2}$ 13. $-2, 4$ 14. 12 15. no solution   16. no solution   17. $5, 8$ 18. 9 19. -1
20. 2 21. $7, -\frac{3}{2}$ 22. $0, -3$

2.5 EXERCISES B

Solve each equation in Exercises 1–20.

1. $y^4 - 29y^2 + 100 = 0$

2. $(x + 4)^2 - 5(x + 4) + 6 = 0$

3. $y^{2/3} - 5y^{1/3} + 6 = 0$

4. $(x^2 - 2x)^2 - 2(x^2 - 2x) - 3 = 0$

5. $y^{-2} + y^{-1} - 2 = 0$

6. $3\left(\dfrac{y + 2}{y}\right)^2 + \left(\dfrac{y + 2}{y}\right) - 2 = 0$

7. $3x^{1/2} + 12x^{-1/2} = 13$

8. $4\sqrt[4]{y} + \sqrt{y} - 5 = 0$

9. $z + \dfrac{5}{2} = \dfrac{12}{2z}$

10. $\dfrac{5}{y^2 - 14} = \dfrac{1}{y}$

11. $y - \dfrac{3}{y + 2} = \dfrac{-2}{y + 2}$

12. $\dfrac{y + 2}{y - 1} = 1 - \dfrac{y^2 - 7}{y^2 - 1}$

13. $\dfrac{1}{y^2 + 2y - 3} = \dfrac{2}{y^2 - 3y + 2} - \dfrac{y}{y^2 + y - 6}$

14. $6 + \sqrt{3y + 1} = 2y$

15. $\sqrt{x^2 - 7x + 15} - \sqrt{4x - 13} = 0$

16. $\sqrt{10y - 3} - \sqrt{3y^2 - 7y + 7} = 0$

17. $\sqrt{2y + 5} + \sqrt{y + 2} = 5$

18. $\sqrt[4]{y^2 - 9} = 2$

19. $(y + 4)^{1/2} - y + 8 = 0$

20. $(2y + 5)^{1/2} - (y + 2)^{1/2} = 1$

FOR REVIEW

Solve each equation in Exercises 21–22.

21. $\dfrac{1}{5}x^2 - 20 = 0$

22. $y(5y - 1) = 1$

2.5 EXERCISES C

Solve each equation in Exercises 1–6.

1. $\sqrt{x + \sqrt{x - 2}} = 2$
   [Answer: 3]

2. $\sqrt{y + \sqrt{y + 1}} = 1$

3. $\sqrt{x + 8} - \sqrt{3x + 1} = \sqrt{2x - 1}$
   [Answer: 1]

4. $\sqrt{x + 9} + \sqrt{x + 4} = \sqrt{x + 25}$

5. $\sqrt[4]{x+1} = \sqrt[8]{x+1}$
[Answer: 0, −1]

6. $\sqrt{x-1} - \sqrt[4]{x-1} = 6$

In Exercises 7–10 use a calculator to approximate the solutions to the equations. Give answers correct to the nearest hundredth.

7. $x^4 - 6x^2 + 1 = 0$

8. $\dfrac{x}{1.5} + \dfrac{1}{x} = 3.2$
[Answer: 4.46, 0.34]

9. $y + \sqrt{y} - 2.45 = 0$

10. $\sqrt{2y + 5.5} + \sqrt{y} = 4.58$
[Answer: 2.12]

2.6 PROBLEM SOLVING AND APPLICATIONS OF QUADRATIC EQUATIONS

STUDENT GUIDEPOSTS

1. A Review of Problem Solving
2. Geometry Problems
3. Consumer Problems
4. Work Problems
5. Rate-Motion Problems

1 A REVIEW OF PROBLEM SOLVING

In this section we consider a variety of applications that result in a quadratic equation. It is a good idea to return to Section 2.2 and review the problem-solving technique presented there. Recall that we **ATTACK** a problem by **A**nalyzing it, **T**abulating information given and needed (perhaps using a sketch or a table), **T**ranslating the information into an equation to solve, **A**pproximating or estimating to see if our answer seems reasonable, and **C**hecking the answer in the words of the problem. Remember, the **K**ey to becoming a good problem solver is *to try* and not to be discouraged if your first effort fails.

We begin this section with a simple number problem, and use it to review the method of problem solving.

EXAMPLE 1 A NUMBER PROBLEM

One-fourth the product of two consecutive positive even integers is 56. Find the integers.

Analysis: Two consecutive even integers can be represented by x and $x + 2$. Be sure to keep in mind that you only want *positive* integers.

Tabulation:

Let $x =$ first positive even integer,

$x + 2 =$ next positive even integer,

$x(x + 2) =$ the product of the two integers,

$\dfrac{1}{4}x(x+2) = \dfrac{x(x+2)}{4} =$ one-fourth their product.

PRACTICE EXERCISE 1

If the square of Mary's age is decreased by 250, the result is 10 more than seven times her age. How old is Mary?

142 CHAPTER 2 EQUATIONS, INEQUALITIES, AND PROBLEM SOLVING

Translation: The equation to solve is:

$$\frac{x(x+2)}{4} = 56 \quad \text{The product is 56}$$

$$x(x+2) = 224 \quad \text{Clear the fraction}$$

$$x^2 + 2x = 224 \quad \text{Distributive law}$$

$$x^2 + 2x - 224 = 0 \quad \text{Write in general form}$$

$$(x-14)(x+16) = 0 \quad \text{Factor}$$

$x - 14 = 0$ or $x + 16 = 0$ Zero-product rule

$x = 14$ $x = -16$

$x + 2 = 14 + 2 = 16$ $x + 2 = -16 + 2 = -14$

Approximation: We ended up with two solutions: 14 and 16, and -16 and -14. But since the integers must be positive, we discard the second pair. Notice that 14 and 16 are even, positive, and consecutive so they are reasonable answers to the problem.

Check: $\frac{1}{4}(16)(14) = (4)(14) = 56$, so these integers do check.

Thus, the desired numbers are 14 and 16.

Answer: Mary is 20 years old.

Note The details shown in the solution of Example 1 were extensive and designed to outline and review the problem-solving technique. Some of the steps that were written out, of course, would normally be performed mentally.

❷ GEOMETRY PROBLEMS

Example 2 uses the Pythagorean theorem which states that the sum of the squares of the lengths of the legs in a right triangle is equal to the square of the length of the hypotenuse. This geometry theorem is often useful for problem solving.

| **EXAMPLE 2**   **A GEOMETRY PROBLEM** | **PRACTICE EXERCISE 2** |

A ladder 13 ft long leans against a wall. If the base of the ladder is 5 ft from the wall, how far would the lower end have to be pulled out so the upper end slides down the same amount?

Analysis: As we begin to analyze this problem, it is clear that a sketch of the initial position of the ladder, as shown in Figure 2.5, will be necessary.

Find the length of the base and the height of a triangular garden if the base is 4 m longer than the height, and the area is 48 m².

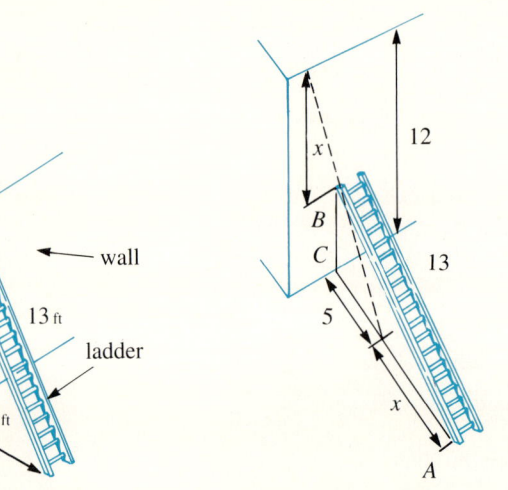

Figure 2.5

Figure 2.6

Tabulation: It will be helpful to find the value of y using the Pythagorean theorem.

$$5^2 + y^2 = 13^2$$
$$25 + y^2 = 169$$
$$y^2 = 144$$
$$y = \pm\sqrt{144} = \pm 12$$

Since y cannot be negative, we discard -12. Thus, the ladder originally reached 12 ft up the wall. Now we sketch what happens when the lower end is pulled away from the wall, as shown in Figure 2.6, with $x =$ the distance it is pulled out (also the distance it slides down). The dashed line shows the initial position of the ladder. We can apply the Pythagorean theorem again, this time to triangle ABC.

Translation:
$$(x + 5)^2 + (12 - x)^2 = 13^2$$
$$x^2 + 10x + 25 + 144 - 24x + x^2 = 169$$
$$2x^2 - 14x + 169 = 169$$
$$2x^2 - 14x = 0$$
$$2x(x - 7) = 0$$
$$2x = 0 \quad \text{or} \quad x - 7 = 0$$
$$x = 0 \qquad\qquad x = 7$$

Approximation: Although 0 ft is really an answer to the problem (if we don't pull the ladder out, the top won't slide down!), we would assume that 0 is not a reasonable answer, and discard it.

Thus, if the ladder is pulled out 7 ft, the top slides down 7 ft. Check this in the words of the problem.

Answer: base: 12 m, height: 8 m

3 CONSUMER PROBLEMS

Many applications involve problems that might confront us on a daily basis. These problems often pertain to money.

EXAMPLE 3   AN INVESTMENT PROBLEM

The amount of money A that will result if a principal P is invested at r percent interest compounded annually for two years is given by the formula $A = P(1 + r)^2$. Use this formula to find the rate of interest if \$5000 grows to \$6272 in two years.

Let $r =$ the rate of interest. Substituting 6272 for A and 5000 for P in the interest formula gives the equation

$$6272 = 5000(1 + r)^2.$$
$$1.2544 = (1 + r)^2 \quad \text{Divide both sides by 5000}$$
$$\pm\sqrt{1.2544} = 1 + r$$
$$\pm 1.12 = 1 + r$$
$$-1 \pm 1.12 = r$$
$$-2.12, 0.12 = r$$

But since r cannot be negative, we discard -2.12. Thus $r = 0.12 = 12\%$.

PRACTICE EXERCISE 3

Use the formula in Example 3 to find the interest rate if \$1500 grows to \$1798.54 in two years when the interest is compounded annually.

Answer: 9.5%

In Example 3 we did not specifically write out all the problem-solving steps. The work there appears in the form that you should begin to approach.

EXAMPLE 4 A CONSUMER PROBLEM

A group of men plan to share equally in the $210 cost of painting a building. At the last minute, two men drop out. This raises the share of each remaining man $28. How many men were in the group at the outset?

Analysis: In a cost problem like this, the fundamental principle is:

$$\text{total cost} = (\text{no. of men})(\text{cost/man})$$

Tabulation: Let n = no. men at outset,

c = cost/man at outset.

Then we know that $210 = nc$. Also,

$n - 2$ = no. men after 2 drop out,

$c + 28$ = cost/man after 2 drop out.

Thus, $210 = (n - 2)(c + 28)$.

Since we want to know the number of men, we need an equation in n. Suppose we solve each equation above for c.

$$c = \frac{210}{n} \quad \text{and} \quad c + 28 = \frac{210}{n-2}$$

$$c = \frac{210}{n-2} - 28$$

Translation: Setting the two expressions for c equal, we obtain the desired equation in n.

$$\frac{210}{n} = \frac{210}{n-2} - 28$$

Multiply both sides by the LCD = $n(n - 2)$.

$$n(n-2)\frac{210}{n} = \left[\frac{210}{n-2} - 28\right]n(n-2)$$

$$210(n - 2) = 210n - 28n(n - 2)$$

$$210n - 420 = 210n - 28n^2 + 56n$$

$$28n^2 - 56n - 420 = 0$$

$$n^2 - 2n - 15 = 0 \quad \text{Divide both sides by 28}$$

$$(n + 3)(n - 5) = 0$$

$n + 3 = 0 \quad \text{or} \quad n - 5 = 0$

$n = \cancel{-3} \qquad\qquad n = 5$

Since the number of men cannot be negative, discard -3. There were 5 men in the group, a number that is reasonable and does check.

PRACTICE EXERCISE 4

Jeff bought a number of lots for a total price of $45,000. He sold all but four of them and received a total of $45,000. If the selling price of each was $4000 more than what Jeff paid for it, how many lots did he purchase originally?

Answer: 9 lots

4 WORK PROBLEMS

Recall from Section 2.2 that in a work problem, when A and B work together to complete a job, the equation used to solve the problem is:

(amount done by A in 1 unit of time) +
(amount done by B in 1 unit of time) =
(amount done together in 1 unit of time).

EXAMPLE 5 A WORK PROBLEM

When each works alone, Sam can do a job in 3 hours less time than Walt. When they work together, it takes them 2 hours to complete the job. How long does it take each to do the job by himself?

Let t = the number of hours it takes Walt to do the job,

$t - 3$ = the number of hours it takes Sam to do the job,

2 = the number of hours it takes to do the job working together,

$\dfrac{1}{t}$ = the amount done by Walt in 1 hour,

$\dfrac{1}{t-3}$ = the amount done by Sam in 1 hour,

$\dfrac{1}{2}$ = the amount done in 1 hour when working together.

We have the following equation.

$$\frac{1}{t} + \frac{1}{t-3} = \frac{1}{2}$$

$$2t(t-3)\left[\frac{1}{t} + \frac{1}{t-3}\right] = 2t(t-3)\frac{1}{2} \quad \text{The LCD} = 2t(t-3)$$

$$2\cancel{t}(t-3)\frac{1}{\cancel{t}} + 2t\cancel{(t-3)}\frac{1}{\cancel{(t-3)}} = \cancel{2}t(t-3)\frac{1}{\cancel{2}}$$

$$2t - 6 + 2t = t^2 - 3t$$

$$0 = t^2 - 7t + 6$$

$$0 = (t-6)(t-1)$$

$$t - 6 = 0 \quad \text{or} \quad t - 1 = 0$$

$$t = 6 \quad\quad\quad t = 1$$

But since $t - 3$ would be negative if $t = 1$, we discard 1 as a possible solution. Thus, Walt can do the job in 6 hours, and Sam can do it in 3 hours.

PRACTICE EXERCISE 5

Andy and his brother Zach can split and stack a load of wood in 4 hours. When each works alone, Zach takes 2 hours longer than Andy. To the nearest tenth of an hour, how long does it take Zach to split and stack the wood by himself?

Answer: 9.1 hr

5 RATE-MOTION PROBLEMS

Recall from Section 2.2 that any time we are dealing with distance, rate, and time, the formula $d = rt$ might be used. The next example combines this formula with the Pythagorean theorem.

EXAMPLE 6 A RECREATION (RATE-MOTION) PROBLEM

Two hikers leave the same campground at the same time, one hiking north and the other hiking east. If one is hiking 1 mph faster than the other, and if after 3 hours they are 15 miles apart, at what rate is each hiking?

Let x = rate one is traveling,
$x + 1$ = rate the other is traveling.
Since they both hike for 3 hours,
$3x$ = distance one travels,
$3(x + 1)$ = distance the other travels.

PRACTICE EXERCISE 6

A trucker transported a heavy load of steel girders from Los Angeles to Albuquerque, a distance of 825 miles, and returned to LA empty. His rate going was 11 mph slower than his rate returning. If the total time of travel was 27.5 hours, what was his rate of speed with the full load? [*Hint:* Find an expression for both times, going and returning, add them, and set the result equal to 27.5.]

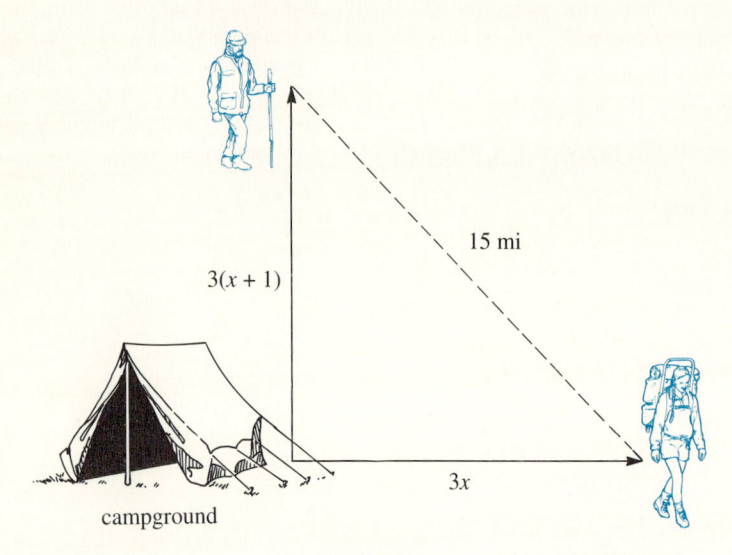

Figure 2.7

Make a sketch describing the problem as in Figure 2.7. Since the triangle is a right triangle, we use the Pythagorean theorem to write the equation in x.

$$(3x)^2 + [3(x + 1)]^2 = 15^2$$
$$9x^2 + 9x^2 + 18x + 9 = 225$$
$$18x^2 + 18x - 216 = 0$$
$$x^2 + x - 12 = 0$$
$$(x - 3)(x + 4) = 0$$
$$x - 3 = 0 \quad \text{or} \quad x + 4 = 0$$
$$x = 3 \qquad\qquad x = -4$$

We can discard -4 as a solution since a rate cannot be negative, giving 3 mph and 4 mph as the desired rates.

Answer: 55 mph

Other motion problems are about finding the height of an object above the ground at a given time after it has been dropped or thrown upward or downward. Let h be the height, measured in feet, of an object above the ground. Let t be the time in seconds following a starting time which is time $t = 0$. Let v_0 be the **initial velocity** of the object, measured in feet per second, where v_0 is positive if the object is thrown upward and negative if thrown downward at time $t = 0$. And let

h_0 be the **initial height** of the object above ground level at time $t = 0$. Then the equation

$$h = -16t^2 + v_0 t + h_0$$

gives the height in terms of the time $t \geq 0$. Notice that if the constants v_0 and h_0 are known, and if a particular height h is given, we have a quadratic equation in the variable t.

EXAMPLE 7   A PHYSICS (RATE-MOTION) PROBLEM

A rock falls from a cliff 420 feet high into a river at the base of the cliff (see Figure 2.8). How long will it take for the rock to hit the water (to the nearest tenth of a second)?

Figure 2.8

Since the initial height is $h_0 = 420$, the initial velocity is $v_0 = 0$ (because the rock falls, and is not thrown, downward), and $h = 0$ when the rock hits the water, we need to solve

$$0 = -16t^2 + 420.$$
$$16t^2 = 420$$
$$t^2 = 26.25$$
$$t = \pm\sqrt{26.25} \approx \pm 5.1$$

Since t must be positive, we discard -5.1. Thus, the rock will hit the water in approximately 5.1 seconds.

PRACTICE EXERCISE 7

While climbing vertically at 440 ft/sec, the hatch on a fighter jet tore loose at an altitude of 10,000 feet and fell into the ocean. Approximately how long did it take for the hatch to hit the water? [*Hint:* $v_0 = 440$ and $h_0 = 10,000$.]

Answer: 42.3 sec

Note   Often the mathematical equation used as a model for a physical problem will have more solutions than are actually applicable. For instance, although $t = -5.1$ is a mathematical solution to the equation in Example 7, it is not a solution to the applied problem since negative time has no meaning in this case. Always check your work and make sure that your answers are meaningful.

2.6 EXERCISES A

Solve each applied problem in Exercises 1–16.

1. **NUMBER** One-third the product of two consecutive positive odd integers is 65. Find the integers.

2. **NUMBER** The principal square root of 4 more than a number is the same as the number less 8. Find the number.

3. **GEOMETRY** If the length of a rectangle is 11 inches more than the width, and the area is 80 in^2, find its dimensions.

4. **MANUFACTURING** A box with no top and a square base is to be made from a square piece of cardboard by cutting out squares with 4-in sides from each corner and folding up the sides. If the box must hold 400 in^3, what must be the length of the sides of the original piece of cardboard?

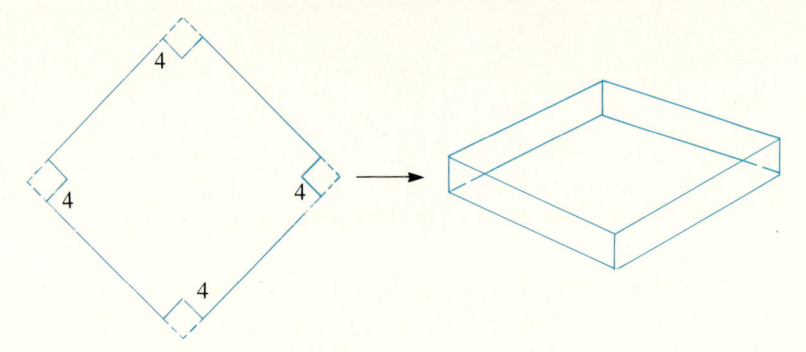

5. **AGRICULTURE** A rectangular garden with dimensions 18 yd by 24 yd is to have a sidewalk of uniform width placed completely around the inside of its border in such a way that the area of the remaining garden is 216 yd^2. How wide is the sidewalk?

6. **ECONOMICS** Use the formula $A = P(1 + r)^2$ to find the rate of interest necessary for $10,000 to grow to $12,321 in two years, if the interest is compounded annually.

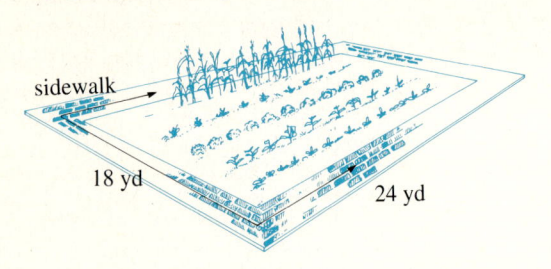

7. RETAILING A store owner earns $2000.00 a week on the sale of a particular type of toy. By reducing the price $2.00 per toy, he can generate more business and sell 50 more toys per week while still earning $2000.00. At what price did he sell each toy originally?

8. MANUFACTURING A manufacturer has discovered that the total monthly cost c of operating her plant is given by $c = 10x^2 - 100x - 2000$, where x is the number of items made per month. How many items were produced during a month when the total costs amounted to $10,000?

9. BUSINESS A wholesaler sells sweaters to a booster club at a rate of $25 per sweater for all orders of 250 or less. If more than 250 are ordered (up to 300), the price per sweater is reduced at a rate of $0.05 times the total number ordered. If the club is billed $3080 for an order, how many sweaters were purchased?

10. CONSTRUCTION It takes Jim 9 hours longer to build a cabinet than his father. If they work together, they can build it in 20 hours. How long would it take each to build the cabinet if he worked alone?

11. COMMUNICATIONS Two printers can print mailing labels for an alumni group in 10 hours. If the newer printer can do the job in 5 hours less time than the older one, how many hours would it take the older printer to print the labels when the newer one was sent in for repair? (Give answer to the nearest tenth.)

12. TRANSPORTATION Two trucks leave the same city at the same time traveling at right angles to each other. If the first travels 10 mph faster than the second, and if after two hours they are 100 miles apart, find the speed of each truck.

13. SPORTS On a recruiting trip, basketball coach Gary Heintz drove 400 miles at a certain rate in a certain time. If he had gone 10 mph faster, the time of the trip would have been shortened 2 hours. What was his speed?

14. PHYSICS A child drops a coin from the top of the Empire State Building at a height of 1472 ft. To the nearest tenth of a second, how long will it take for the coin to reach the ground?

15. PHYSICS A man fires a gun from ground level, with a muzzle velocity of 2400 ft/sec, at a helicopter 3000 feet directly above him. How long (to the nearest hundredth of a second) will it take for the bullet to hit the helicopter?

16. PHYSICS A man drops an object from a balloon at an elevation of 256 ft. If the balloon is rising vertically at a velocity of 16 ft/sec, how long will it take for the object to hit the ground? (Give answer to the nearest hundredth of a second.)

FOR REVIEW

Solve each equation in Exercises 17–18.

17. $\left(\dfrac{x-1}{x+2}\right)^2 + \dfrac{x-1}{x+2} = 6$

18. $\sqrt{x^2+1} + \sqrt{x+1} = 0$

19. Prove that the sum of the two solutions to the equation $x^2 + bx + c = 0$ is $-b$.

20. Find the value of m so that the equation $3x^2 + 3x + m = 0$ has only one real (rational) solution.

21. Determine whether $5 + i$ and $5 - i$ are solutions to the equation $2x^2 - 20x - 52 = 0$.

22. Find a quadratic equation that has -3 and $\tfrac{1}{2}$ as solutions.

ANSWERS: 1. 13, 15   2. 12   3. 5 in by 16 in   4. 18 in   5. 3 yd   6. 11%   7. $10.00   8. 40 items   9. 280 sweaters   10. Jim: 45 hr; father: 36 hr   11. 22.8 hr   12. 30 mph, 40 mph   13. 40 mph   14. 9.6 sec   15. 1.26 sec   16. 4.53 sec   17. $-5, -\tfrac{5}{4}$ 18. no solution   19. Solve the equation using the quadratic formula and add the two solutions to get $-b$.   20. $m = \tfrac{3}{4}$ 21. No, they are not solutions.   22. $2x^2 + 5x - 3 = 0$

2.6 EXERCISES B

Solve each applied problem in Exercises 1–16.

1. **NUMBER** The square of 2 more than a number is the same as 4 less than 10 times the number. Find all numbers that satisfy this condition.

2. **NUMBER** A number decreased by 2 is the same as the principal square root of 4 more than the number. Find the number.

3. **GEOMETRY** If the sides of a square are lengthened by 5 yd, the area of the square will be 256 yd². Find the length of a side.

4. **GEOMETRY** A piece of wire 100 inches long is cut into two pieces. If each piece is bent into the shape of a square and the combined area of the two squares is 337 in², how long is each piece of wire?

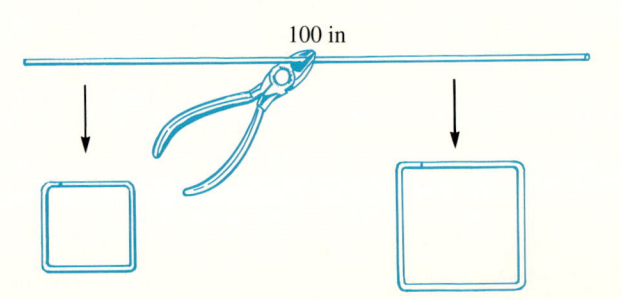

2.6 PROBLEM SOLVING AND APPLICATIONS OF QUADRATIC EQUATIONS

5. **AGRICULTURE** A farmer wishes to enclose a rectangular pasture that borders a river. The pasture is to have a length parallel to the river equal to twice the width. If no fencing is needed along the side that borders the river, and the enclosed area is to be 39,200 yd², how many yards of fence are needed?

6. **ECONOMICS** Use the formula $A = P(1 + r)^2$ to find the rate of interest if $8000 grows to $10,305.80 in 2 years.

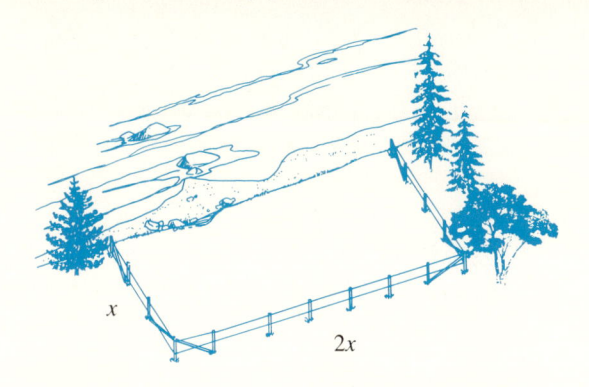

7. **CONSUMER** A group of women plan to share the $180 cost of having a party. At the last minute, 3 women must be out of town, and this raises the share of each remaining woman by $10. How many women were in the group at the outset?

8. **MANUFACTURING** The weekly cost of producing x items is given by the equation $c = 2x^2 + 50x - 30$. What number of items was produced in a week when the cost of production was $270?

9. **ECONOMICS** The overhead cost for selling a product is $165 per day. In order to sell x items a day, the price per item must be adjusted to $26 - x$ dollars. To break even on a given day, the store owner must sell enough items to pay his overhead costs. What are the break-even points in this situation?

10. **RECREATION** When each valve is turned on separately, a larger pipe can fill a swimming pool in 3 hours less time than a smaller pipe. When they are turned on simultaneously, it takes 2 hours to fill it. How long does it take each pipe separately?

11. **CONSTRUCTION** It takes John 8 days longer to build a redwood deck than his father. If together they can build the deck in fourteen days, how long would it take John's father to build it by himself? (Give answer to the nearest tenth.)

12. **NAVIGATION** Two boats leave an island at 12:00 noon, one sailing due south and the other due west. If both are traveling at 30 knots, to the nearest tenth, how many nautical miles apart will they be at 2:00 P.M. that afternoon?

13. **BUSINESS** On a business trip, Marguerite drove from Cleveland to Boston, a distance of 660 miles. She drove 6 mph faster on the return trip, cutting one hour off the trip. At what speed did she drive from Boston to Cleveland?

14. **PHYSICS** A rocket is fired upward from ground level with an initial velocity of 352 ft/sec. How many seconds will pass before it returns to the ground?

15. **RATE-MOTION** A child standing on a bridge 200 ft above the level of a river throws a rock downward with an initial velocity of 20 ft/sec. How long will it take for the rock to hit the water? (Give answer to the nearest hundredth of a second.)

16. **RATE-MOTION** To the nearest hundredth of a second, how long would it take for the rock to hit the water if the child on the bridge in Exercise 15 threw the rock upward with initial velocity 20 ft/sec? (Do the two answers to Exercises 15 and 16 seem reasonable?)

FOR REVIEW

Solve each equation in Exercises 17–18.

17. $x - \dfrac{20}{2x - 5} = \dfrac{5}{2x - 5}$

18. $\left(\dfrac{x^2 - 4}{x}\right) + 15\left(\dfrac{x}{x^2 - 4}\right) - 8 = 0$

152 CHAPTER 2 EQUATIONS, INEQUALITIES, AND PROBLEM SOLVING

19. Prove that the product of the two solutions to the equation $x^2 + bx + c = 0$ is c.

20. Find the value of m so that the equation $2x^2 - 5x + m = 0$ has only one real (rational) solution.

21. Determine whether $\frac{-1+i\sqrt{3}}{2}$ and $\frac{-1-i\sqrt{3}}{2}$ are solutions to the equation $x^2 + x + 1 = 0$.

22. Find a quadratic equation that has 4 and $-\frac{1}{3}$ as solutions.

2.6 EXERCISES C

Solve each applied problem in Exercises 1–14.

1. **AGRICULTURE** Grain from a conveyor is falling in a pile with the shape of a right circular cone. If the diameter of the base of the cone is always three times the height of the pile, what is the diameter of the base, to the nearest tenth, when the volume of the grain is 200 ft³? [*Hint:* the volume of a cone is $V = \frac{1}{3}\pi r^2 h$.]

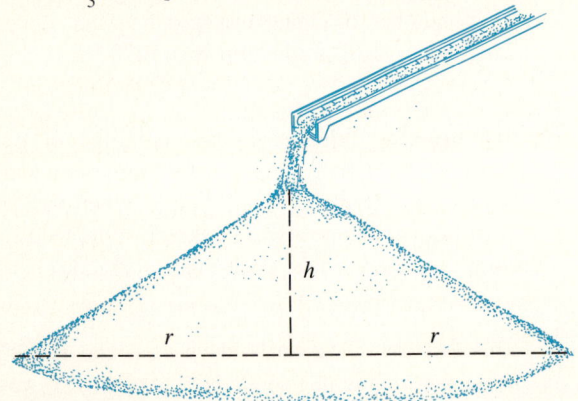

2. **GEOMETRY** A cylindrical container 4 inches high holds 30 in³ of water. If the volume of a cylinder is given by $V = \pi r^2 h$, what is the radius of the cylinder, to the nearest hundredth?

3. **ARCHITECTURE** A window in a building is to be constructed in the shape of an equilateral triangle with height 12 ft. To the nearest tenth of a foot, how long are the sides of the triangle?

4. **PHYSICS** The pressure, in pounds per square foot, of a wind blowing V miles per hour can be approximated by the equation

$$P = \frac{3V^2}{1000}.$$

If the pressure of the wind against the side of a building is 11.25 lb/ft², what is the approximate velocity of the wind?

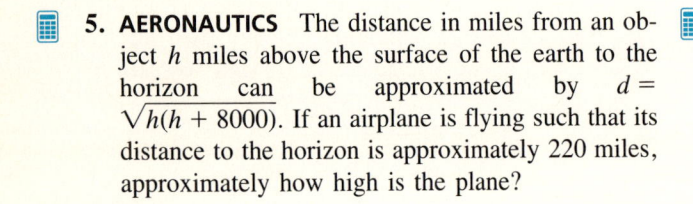

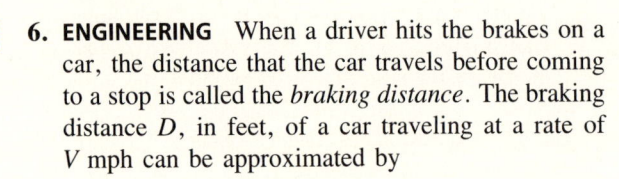

5. **AERONAUTICS** The distance in miles from an object h miles above the surface of the earth to the horizon can be approximated by $d = \sqrt{h(h + 8000)}$. If an airplane is flying such that its distance to the horizon is approximately 220 miles, approximately how high is the plane?

6. **ENGINEERING** When a driver hits the brakes on a car, the distance that the car travels before coming to a stop is called the *braking distance*. The braking distance D, in feet, of a car traveling at a rate of V mph can be approximated by

$$D = \frac{V(V + 20)}{20}.$$

If a child runs into the street 100 feet in front of your car and you react instantaneously and hit the brakes, what is the maximum speed you can be traveling and still avoid hitting the child?

2.6 PROBLEM SOLVING AND APPLICATIONS OF QUADRATIC EQUATIONS

7. GEOMETRY It can be shown that a polygon with m sides has

$$\frac{m^2 - 3m}{2}$$

diagonals. If a polygon has 27 diagonals, how many sides does it have?

8. METEOROLOGY A weather balloon is inflated with 14,137.2 in³ of helium. It is then discovered that an additional 10,000 in³ of helium will be needed to lift the necessary equipment. To the nearest hundredth, how much does this change the diameter of the balloon?

9. PHYSICS The Gateway Arch in St. Louis is 640 feet tall. If an object is dropped from the top of the arch, how long does it take to reach the ground? (Give answer correct to the nearest tenth of a foot.)

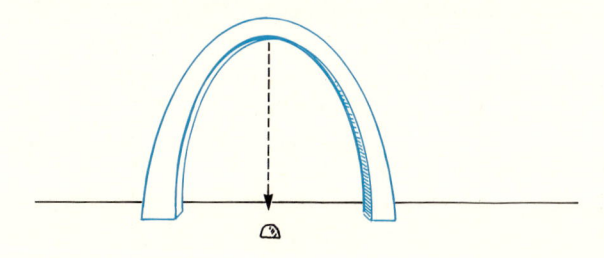

10. PHYSICS A rock is dropped into a vertical mine shaft and two seconds later the sound of it hitting the bottom is heard. Use the fact that the speed of sound is about 1100 ft/sec to determine the depth of the shaft, to the nearest foot. [*Hint:* Let t_1 be the time for the rock to hit the bottom and t_2 the time for the sound to reach the top. Solve the equation $t_1 + t_2 = 2$.]

11. GEOMETRY A circle with circumference 4π cm has an equilateral triangle inscribed in it. Find the length of the sides of the triangle, and then find the area of the triangle. Leave answers in exact form.

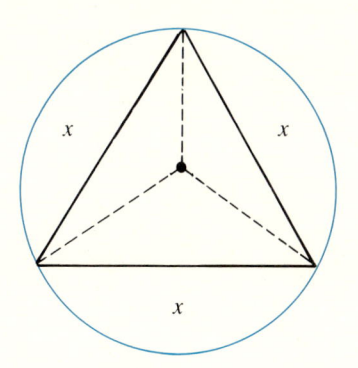

12. A power line is to be run from a power plant on one side of a river to a substation 10 miles downstream and on the other side of the river. It costs $5000 per mile to lay the line on land, and $8000 per mile to lay the line under water. Determine how the line should be run, that is, find the point P (in the figure) if the total funds available for the project amount to $56,000.

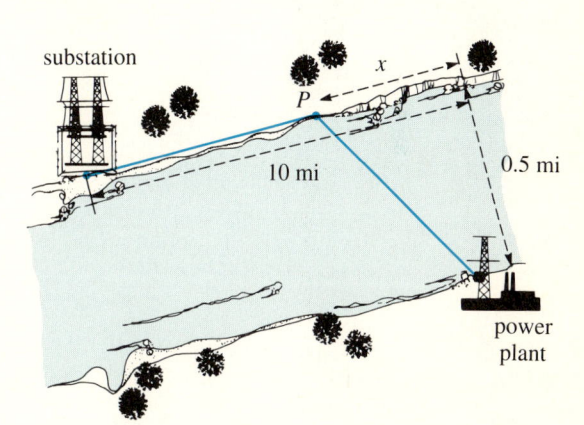

ANSWERS: 1. 13.2 ft 2. 1.55 in 3. 13.9 ft 4. 61.2 mph 5. approximately 6 mi 6. approximately 36 mph
7. 9 sides 8. increases the diameter 5.86 in 9. 6.3 sec 10. 61 ft 11. $2\sqrt{3}$ cm, $3\sqrt{3}$ cm² 12. P should be located at the point where x is about 1.82 mi.

2.7 LINEAR AND ABSOLUTE VALUE INEQUALITIES

STUDENT GUIDEPOSTS

1. Linear Inequalities
2. Graphing Inequalities
3. Compound Inequalities
4. Absolute Value Inequalities

1 LINEAR INEQUALITIES

In Chapter 1 we introduced the inequality symbols $<$, $>$, $\leq$, and $\geq$. Remember that if a and b are real numbers, $a < b$ means that $b - a$ is a positive number. Statements such as

$$x + 2 > 7 \quad 4x \leq 9 \quad 2x + 1 \geq 3x - 5 \quad 2(x + 1) < 3(x + 2) + 5,$$

in which x represents a real number, are called **linear inequalities.** We now consider several properties of inequalities which will be used to solve linear inequalities in a manner similar to solving linear equations.

If x is replaced with a real number which makes an inequality true, that number is a **solution** to the inequality. To solve an inequality is to find all of its solutions. As was the case with solving equations, we transform inequalities into **equivalent inequalities** which have the same solutions, ending with an inequality having the variable isolated on one side. The next theorem summarizes four properties used in the solution process.

Properties of Inequalities

Let a, b, and c be real numbers.

1. If $a < b$, then $a + c < b + c$.
2. If $a < b$, then $a - c < b - c$.
3. If $a < b$ and $c > 0$, then $ac < bc$.
4. If $a < b$ and $c < 0$, then $ac > bc$.

Proof of 1: Since $a < b$, by definition $b - a$ is positive. Since $(b + c) - (a + c) = b - a$, $(b + c) - (a + c)$ is positive. Thus, $a + c < b + c$.

Proof of 2: Since $a < b$, using part 1, $a + (-c) < b + (-c)$, or, equivalently, $a - c < b - c$.

Proof of 3: Since $a < b$, $b - a$ is positive. With c also positive, the product $(b - a)c$ is positive. Then $bc - ac$ is positive so that $ac < bc$.

Proof of 4: Since $a < b$, $b - a$ is positive. With c negative, we know that $-c$ is positive. Then the product $(b - a)(-c)$ is positive. Thus $ac - bc$ is positive so that $bc < ac$, or, equivalently, $ac > bc$.

Note Similar results to those in the preceding theorem are also true for the inequalities $>$, $\leq$, and $\geq$. The only difference between these properties and the corresponding ones for equations occurs when an inequality is multiplied (or divided) on both sides by a negative number. In this case *always remember to reverse the sense of the inequality.*

2.7 LINEAR AND ABSOLUTE VALUE INEQUALITIES

EXAMPLE 1 A LINEAR INEQUALITY

Solve. $3x - 1 < 5x - 7$

$3x - 5x - 1 < 5x - 5x - 7$ Subtract 5x from both sides

$-2x - 1 < -7$

$-2x - 1 + 1 < -7 + 1$ Add 1 to both sides

$-2x < -6$

$\left(-\frac{1}{2}\right)(-2x) > \left(-\frac{1}{2}\right)(-6)$ Multiply both sides by $-1/2$ and reverse the inequality

$x > 3$

Thus, the solutions are all real numbers greater than 3. Usually we do not state the solution in a sentence, and simply give the solution as $x > 3$.

PRACTICE EXERCISE 1

Solve. $2y + 3 \le 4y - 9$

Answer: $y \ge 6$

Inequalities with parentheses are treated in the same way as equations with parentheses.

EXAMPLE 2 A LINEAR INEQUALITY WITH PARENTHESES

Solve. $y - 3(2 + y) \ge 2(3y - 2) + 2$

$y - 6 - 3y \ge 6y - 4 + 2$ Clear parentheses

$-2y - 6 \ge 6y - 2$ Collect like terms

$-8y - 6 \ge -2$ Subtract 6y from both sides

$-8y \ge 4$ Add 6 to both sides

$y \le -\frac{1}{2}$ Divide both sides by -8 and reverse the inequality

PRACTICE EXERCISE 2

Solve.

$x - (1 - 3x) < 3(2x + 5) - 8$

Answer: $x > -4$

As is the case with equations, some inequalities are **identities** that have every real number as a solution, and others are **contradictions** that have no solution.

EXAMPLE 3 SPECIAL LINEAR INEQUALITIES

Solve.

(a) $3(x + 2) < 5 + 3x$

$3x + 6 < 5 + 3x$ Clear parentheses

$6 < 5$ Subtract 3x from both sides

Since $6 < 5$ is a false inequality equivalent to the original inequality, the original inequality is also false regardless of the choice of replacement for the variable x. When this occurs, the inequality has no solution and is a contradiction.

PRACTICE EXERCISE 3

(a) $2(y - 3) \ge y - (1 - y)$

(b) $2(x + 5) + x \geq 3(x + 1)$

$\qquad 2x + 10 + x \geq 3x + 3$

$\qquad\quad 3x + 10 \geq 3x + 3$

$\qquad\qquad\quad 10 \geq 3 \quad$ Subtract $3x$ from both sides

Since $10 \geq 3$ is a true inequality equivalent to the original inequality, the original inequality is also true regardless of the choice of replacement for the variable x. When this happens, the inequality has every real number for a solution and is an identity.

(b) $5y - (4 - 2y) < 3(2y - 1) + y$

Answers: (a) no solution (contradiction) (b) every real number (identity)

Some applications translate into linear inequalities as shown in the next example.

EXAMPLE 4 A Business Problem

In the manufacture and sale of record albums, the revenue made on the sale of x albums is $\$2.20x$, and the cost of producing x albums is $\$1.30x + \$4500$. In order to make a profit, the revenue received must be greater than the costs of production. For what values of x will a profit be returned?

Analysis: The words "greater than" seem to indicate an inequality. We want revenue > costs.

Tabulation: Let x = no. albums sold.

$\qquad 2.20x$ = revenue received

$1.30x + 4500$ = cost of production

Translation: We must solve the following inequality:

$$2.20x > 1.30x + 4500$$
$$0.9x > 4500$$
$$x > 5000$$

Thus, a profit will be made when more than 5000 albums are sold. This answer seems reasonable, and does check.

Practice Exercise 4

To get an A in biology, Jay must have 90% of the total points in the class. The grade is determined by three 100-point tests and a 150-point final exam. If he has 85, 96, and 92 on the tests, what range of scores on the final will give him an A? Let s = Jay's score on the final.

Answer: Jay's score s must be greater than or equal to 132 to get an A; that is, $s \geq 132$.

2 GRAPHING INEQUALITIES

To graph an equation, we plot the point or points on a number line which correspond to solutions of the equation. For example, to graph $2x - 1 = 3$, we first solve the equation, obtaining $x = 2$, and then graph the solution as in Figure 2.9.

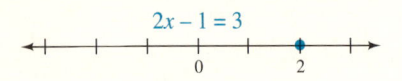

$$2x - 1 = 3$$
$$\longleftarrow \underset{0}{\;|\;} \underset{}{\;|\;} \underset{}{\;|\;} \underset{2}{\;\bullet\;} \underset{}{\;|\;} \longrightarrow$$

Figure 2.9

Graphing inequalities is somewhat more interesting. For example, to graph $3(x - 2) > 2x - 4$, we solve the inequality, obtaining $x > 2$, and graph the result as shown in Figure 2.10. The small circle at the left end of the colored arrow indicates that the number 2 is not included while all points to the right of

2 are included. Figure 2.11 is the graph of $x < 0$ or $0 > x$. The graph of $x \geq -2$ or $-2 \leq x$ is given in Figure 2.12. The circle at -2 is solid to indicate that the number -2 is included in the graph. Finally, the graph of $x \leq 0$ is given in Figure 2.13.

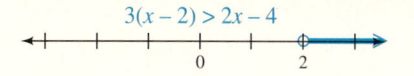

Figure 2.10 $3(x-2) > 2x - 4$

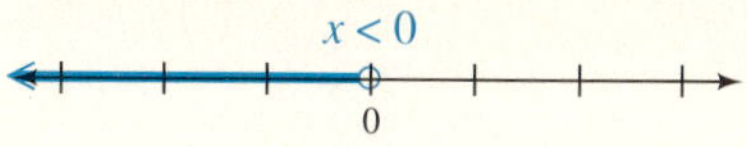

Figure 2.11 $x < 0$

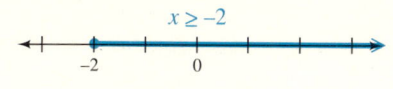

Figure 2.12 $x \geq -2$

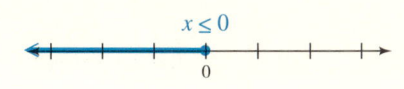

Figure 2.13 $x \leq 0$

3 COMPOUND INEQUALITIES

Sometimes two inequalities are combined or joined together to form **compound inequalities** by using the connectives *and* or *or*. For example, the compound inequality

$$x > -3 \text{ and } x < 1$$

could be given by

$$-3 < x \quad \text{and} \quad x < 1$$

or by the single chain of inequalities

$$-3 < x < 1.$$

The numbers that satisfy this chain of inequalities are between -3 and 1, and the graph is shown in Figure 2.14(a). Figure 2.14(b) shows an alternative method of graphing this chain where the open circles have been replaced with parentheses. This notation results from expressing a compound inequality in *interval notation*.

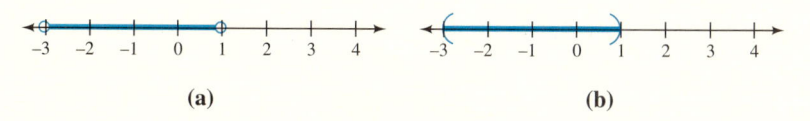

Figure 2.14

Definition of Open Interval

If a and b are real numbers with $a < b$, the symbol (a, b), defined by

$$(a, b) = \{x \text{ such that } a < x < b\},$$

is an **open interval** with **endpoints** a and b.

Notice that the endpoints -3 and 1, of the open interval $(-3, 1)$ graphed in Figure 2.14(b), are not part of the interval. If an endpoint is to be included in an interval, brackets are used instead of parentheses to form a *closed interval*.

158 CHAPTER 2 EQUATIONS, INEQUALITIES, AND PROBLEM SOLVING

Definition of Closed and Half-Open Intervals

If a and b are real numbers with $a < b$, the symbol $[a, b]$, defined by

$$[a, b] = \{x \text{ such that } a \leq x \leq b\},$$

is a **closed interval**. Similarly, **half-open intervals** are defined by

$$(a, b] = \{x \text{ such that } a < x \leq b\}$$

and $\quad [a, b) = \{x \text{ such that } a \leq x < b\}.$

The closed interval $[-2, 3]$ is shown in Figure 2.15 and the half-open interval $[-1, 2)$ is shown in Figure 2.16. Notice again that brackets and parentheses are used in the graphs to replace solid circles and open circles, respectively, at the endpoints.

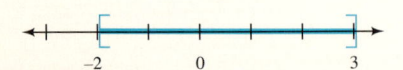

Figure 2.15 Closed Interval $[-2, 3]$

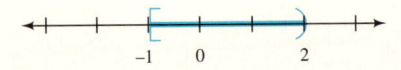

Figure 2.16 Half-Open Interval $[-1, 2)$

It is sometimes convenient to consider another type of interval, an *infinite interval,* that can be used to describe inequalities such as

$$x < 2, \quad x > -1, \quad x \leq 3, \quad x \geq -4.$$

To do this we use the symbols ∞ and $-\infty$, read "infinity" and "minus infinity." These symbols do not represent real numbers; rather, they are used to show that the real numbers extend infinitely far, both positively and negatively.

Definition of Infinite Intervals

Let a be a real number. The four **infinite intervals** are denoted and defined by the following:

$$(-\infty, a) = \{x \text{ such that } x < a\}$$
$$(-\infty, a] = \{x \text{ such that } x \leq a\}$$
$$(a, \infty) = \{x \text{ such that } x > a\}$$
$$[a, \infty) = \{x \text{ such that } x \geq a\}$$

Notice that a parenthesis is used next to ∞ and $-\infty$, not a bracket, in all four infinite intervals. In this notation, the four inequalities above can be represented by the infinite intervals $(-\infty, 2)$, $(-1, \infty)$, $(-\infty, 3]$, and $[-4, \infty)$, respectively. The entire set of real numbers could also be written as $(-\infty, \infty)$.

EXAMPLE 5 GRAPHING A COMPOUND INEQUALITY *(and)*	PRACTICE EXERCISE 5

Solve and graph. $\quad -1 < 2x + 1 \leq 3$

$-1 - 1 < 2x + 1 - 1 \leq 3 - 1 \quad$ Subtract 1 throughout each portion

$\qquad -2 < 2x \leq 2$

$\dfrac{1}{2}(-2) < \dfrac{1}{2}(2x) \leq \dfrac{1}{2}(2) \qquad$ Multiply throughout by 1/2

$\qquad -1 < x \leq 1$

Solve and graph.

$1 \geq 1 - 3x > -5$

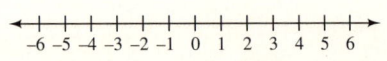

In interval notation, the solution is $(-1, 1]$, and its graph is given in Figure 2.17.

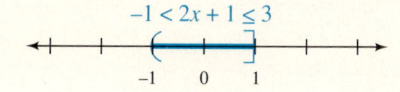

$$-1 < 2x + 1 \leq 3$$

Figure 2.17

Answer: Graph $0 \leq x < 2$; or $[0, 2)$

The second type of compound inequality uses the connective *or*. For example,

$$x < -1 \quad or \quad x > 2.$$

Do *not* form a single chain when the word *or* is used. Notice that $2 < x < -1$ is a meaningless expression (2 is not less than -1). The graph of an *or* combination is usually two segments of a number line. For example, Figure 2.18 shows the graph of $x < -1$ or $x > 2$, which in interval notation is $(-\infty, -1)$ or $(2, \infty)$.

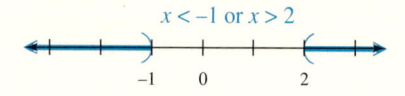

$x < -1$ or $x > 2$

Figure 2.18

$x < -2$ or $x \geq 1$

Figure 2.19

Similarly, $x < -2$ or $x \geq 1$, in interval notation $(-\infty, -2)$ or $[1, \infty)$, is graphed in Figure 2.19.

EXAMPLE 6 GRAPHING A COMPOUND INEQUALITY (*or*)

Solve and graph.

$$3 - 4x < -1 \quad or \quad 3 - 4x \geq 9$$
$$-4x < -4 \quad or \quad -4x \geq 6 \qquad \text{Subtract 3}$$
$$x > 1 \quad or \quad x \leq -\frac{6}{4} \qquad \text{Reverse}$$
$$x \leq -\frac{3}{2}$$

The solution is $x > 1$ or $x \leq -\frac{3}{2}$, graphed in Figure 2.20. Using intervals, the solution is $(1, \infty)$ or $\left(-\infty, -\frac{3}{2}\right]$.

$3 - 4x < -1$ or $3 - 4x \geq 9$

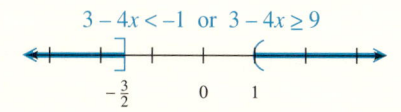

Figure 2.20

PRACTICE EXERCISE 6

Solve and graph.

$$2x + 3 \leq -3 \quad or \quad 2x + 3 > 1$$

Answer: Graph $x \leq -3$ or $x > -1$; $(-\infty, -3]$ or $(-1, \infty)$.

CAUTION

An *and* compound, such as $-4 < x$ *and* $x \leq 3$, can often be written as a single **chain of inequalities,** $-4 < x \leq 3$, and generally has as its graph a single (noninfinite) interval on a number line. An *or* compound, such as $x < -1$ *or* $x > 2$, *cannot* be written as a chain of inequalities; the word "or" must remain in the answer. [*Never write* $2 < x < -1$ for this *or* compound since this would imply that $2 < -1$, which is false.] The graph of an *or* compound generally has as its graph two (infinite) intervals on a number line.

4 ABSOLUTE VALUE INEQUALITIES

Inequalities that have at least one variable within absolute value bars, such as

$$|x| < 5 \quad \text{and} \quad |2 - 3x| \geq 11,$$

are **absolute value inequalities.** When we solved absolute value equations in Section 2.1, recall that we translated the absolute equation into two equations that did not involve absolute value. An absolute value inequality can be translated into two inequalities connected with the words *and* or *or*, that is, into a compound inequality. Translation depends on the following theorem.

Theorem on Absolute Value Inequalities

Let a be a real number, $a > 0$.

1. $|x| < a$ is equivalent to $-a < x < a$.
2. $|x| > a$ is equivalent to $x < -a$ or $x > a$.

Intuitively, since $|x|$ can be interpreted geometrically as the distance from x to 0 on a number line, if $|x| < a$, then x must be *inside* the interval $(-a, a)$ which means $-a < x < a$. Also, if $|x| > a$, then x must be *outside* the interval $(-a, a)$ which means $x < -a$ or $x > a$. Similar results hold for $|x| \leq a$ and $|x| \geq a$.

EXAMPLE 7 SOLVING ABSOLUTE VALUE INEQUALITIES

Solve the following absolute value inequalities.

(a) $|x| < 5$

Translated to the equivalent compound inequality, the solution is $-5 < x < 5$, or in interval notation, $(-5, 5)$.

(b) $|x| \geq 5$

Translating, we obtain the solution $x \leq -5$ or $x \geq 5$; equivalently, $(-\infty, -5]$ or $[5, \infty)$.

(c) $|x| < 0$

Since $|x|$ is never negative, this inequality has no solution.

PRACTICE EXERCISE 7

Solve the following absolute value inequalities.

(a) $|y| \leq 4$

(b) $|y| > 4$

(c) $|y| > 0$

(d) $|y| \geq 0$

(e) $|y| > -4$

(d) $|x| \leq 0$

The only solution is $x = 0$ since $|x|$ cannot be negative.

(e) $|x| < -5$

Since $|x| \geq 0$ for every real number x, there is no solution.

Answers: (a) $-4 \leq y \leq 4$; $[-4, 4]$ (b) $y < -4$ or $y > 4$; $(-\infty, -4)$ or $(4, \infty)$ (c) all reals except 0; $(-\infty, 0)$ or $(0, \infty)$ (d) all reals; $(-\infty, \infty)$ (e) all reals; $(-\infty, \infty)$

Note When solving absolute value inequalities, be sure to think about what you are solving. For example, since the absolute value of any real number is always nonnegative, the inequalities in Example 7 (c) and (e) have no solution.

EXAMPLE 8 MORE COMPLEX INEQUALITIES

Solve.

(a) $|2 - 3x| < 11$

$-11 < 2 - 3x < 11$ Equivalent compound inequality

$-13 < -3x < 9$ Subtract 2 throughout

$\dfrac{13}{3} > x > -3$ Divide through by -3 and reverse the inequalities

The solution is $-3 < x < \frac{13}{3}$ or $\left(-3, \frac{13}{3}\right)$.

(b) $|2x - 1| \geq 5$

$2x - 1 \leq -5$ or $2x - 1 \geq 5$ Equivalent compound inequality

$2x \leq -4$ or $2x \geq 6$

$x \leq -2$ or $x \geq 3$

The solution is $x \leq -2$ or $x \geq 3$, or $(-\infty, -2]$ or $[3, \infty)$. Do not write $3 \leq x \leq -2$ for the answer!

(c) $|5x + 10| > 0$

Since $|5x + 10| \geq 0$ for all values of x, the only real number x that will *not* be a solution is when $5x + 10 = 0$, that is when $x = -2$. Thus, the solution is all real numbers except -2, or in interval notation, $(-\infty, -2)$ or $(-2, \infty)$.

PRACTICE EXERCISE 8

Solve.

(a) $|2x + 1| \leq 5$

(b) $|3 - x| > 1$

(c) $|5x + 10| < -2$

Answers: (a) $-3 \leq x \leq 2$; $[-3, 2]$ (b) $x < 2$ or $x > 4$; $(-\infty, 2)$ or $(4, \infty)$ (c) no solution (the absolute value of $5x + 10$ is always ≥ 0)

2.7 EXERCISES A

Solve and graph each inequality in Exercises 1–12. Give answers using both inequalities and intervals.

1. $4x - 2 \leq 6$

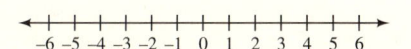

2. $2y + 5 < 4y - 9$

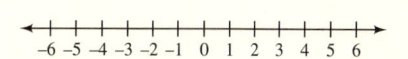

3. $(z - 1)^2 < z(z + 2)$

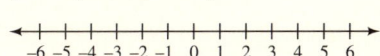

4. $(4y + 1)(y - 3) \geq (2y + 1)(2y - 1)$

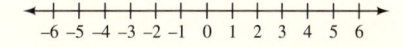

5. $\dfrac{3}{2y+4} > 0$

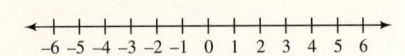

   [*Hint:* The denominator must be positive.]

6. $(z+4)^{-1} > 0$

7. $\dfrac{y+1}{3} \le \dfrac{y+1}{2}$

8. $4 \le 2x - 6 < 10$

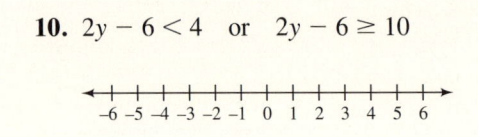

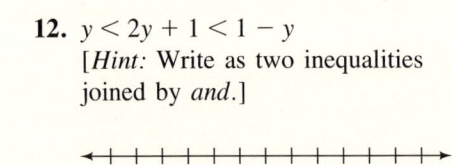

9. $0 \le \dfrac{z+1}{2} < 1$

10. $2y - 6 < 4$ or $2y - 6 \ge 10$

11. $2z + 1 > z$ or $z + 3 < 0$

12. $y < 2y + 1 < 1 - y$

    [*Hint:* Write as two inequalities joined by *and*.]

In Exercises 13–15, find the values of x that make the given expression positive.

13. $3(x+5)$
14. $-3(x+5)$
15. $(x+5)^2$

In Exercises 16–18, find the values of x that make the given expression negative.

16. $3(x+5)$
17. $-3(x+5)$
18. $-(x+5)^2$

Solve each applied problem in Exercises 19–21.

19. **SCIENCE** Fahrenheit (F) and Celsius (C) temperatures are related by $C = \tfrac{5}{9}(F - 32)$. If the Celsius temperature is between $-15°$ C and $30°$ C, inclusive, what is the range of Fahrenheit temperatures?

20. **BUSINESS** To make a profit on the sale of video recorders, the total income on sales S, must exceed the total costs involved C. If x is the number of recorders sold, $S = 200x$, and $C = 150x + 300$, what is the smallest number of recorders that must be sold to make a profit?

21. **ENGINEERING** The force F in pounds it takes to stretch a spring x inches beyond its normal length is given by $F = 5.5x$. What are the corresponding values for x if F is between 2.75 pounds and 6.60 pounds, inclusive?

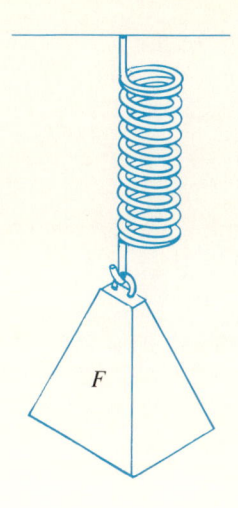

Solve each inequality in Exercises 22–29. Give answers using both inequalities and intervals.

22. $|x| < 7$

23. $|z| \leq 0$

24. $|y| > 0$

25. $|x + 2| > 3$

26. $|1 - 5z| < 6$

27. $\left|\dfrac{y + 7}{3}\right| > 3$

28. $|2x - 7| < -1$

29. $|2z - 7| \leq 0$

FOR REVIEW

Solve.

30. **CONSUMER** A man pays $300 for a number of shares of stock. If each share had cost $3 less, he could have bought 5 more shares for the same $300. How many shares did he buy?

31. **RATE-MOTION** A child fires a toy rocket from the ground into the air with an initial velocity of 128 ft/sec.
    (a) How many seconds later will the rocket return to the ground?
    (b) After how many seconds will the rocket reach a height of 240 ft? Why are there two answers?
    (c) After how many seconds will the rocket reach a height of 256 ft? Why is there only one answer?
    (d) After how many seconds will the rocket reach a height of 300 ft? Explain.

164 CHAPTER 2 EQUATIONS, INEQUALITIES, AND PROBLEM SOLVING

32. $(a^2 - 2)^2 - 6(a^2 - 2) - 7 = 0$ **33.** $y^4 - 2y^2 - 63 = 0$

ANSWERS:

1. $x \leq 2$; $(-\infty, 2]$
2. $y > 7$; $(7, \infty)$
3. $z > \frac{1}{4}$; $\left(\frac{1}{4}, \infty\right)$
4. $y \leq -\frac{2}{11}$; $\left(-\infty, -\frac{2}{11}\right]$
5. $y > -2$; $(-2, \infty)$
6. $z > -4$; $(-4, \infty)$
7. $y \geq -1$; $[-1, \infty)$
8. $5 \leq x < 8$; $[5, 8)$
9. $-1 \leq z < 1$; $[-1, 1)$
10. $y < 5$ or $y \geq 8$; $(-\infty, 5)$ or $[8, \infty)$
11. $z < -3$ or $z > -1$; $(-\infty, -3)$ or $(-1, \infty)$
12. $-1 < y < 0$; $(-1, 0)$

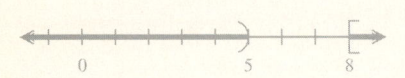

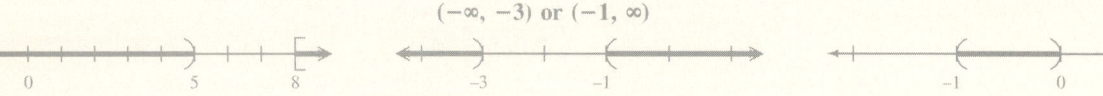

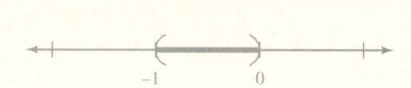

13. $x > -5$ 14. $x < -5$ 15. all reals except -5; that is, $x < -5$ or $x > -5$ 16. $x < -5$ 17. $x > -5$ 18. all reals except -5; that is, $x < -5$ or $x > -5$ 19. $5° \leq F \leq 86°$ 20. $x > 6$; that is, at least 7 records must be sold to realize a profit   21. 0.5 in $\leq x \leq 1.2$ in   22. $-7 < x < 7$; $(-7, 7)$ 23. $z = 0$ 24. all reals except 0; that is, $y < 0$ or $y > 0$; $(-\infty, 0)$ or $(0, \infty)$ 25. $x < -5$ or $x > 1$; $(-\infty, -5)$ or $(1, \infty)$ 26. $-1 < z < 7/5$; $(-1, 7/5)$ 27. $y < -16$ or $y > 2$; $(-\infty, -16)$ or $(2, \infty)$ 28. no solution   29. $z = 7/2$ 30. 20 shares   31. (a) 8 sec (b) at 3 seconds on the way up and at 5 seconds on the way down (c) 4 sec; this is the maximum height of the rocket. (d) The rocket will not reach a height of 300 ft (the maximum height is 256 ft). When solving the equation for t with $h = 300$, the solutions are complex, not real numbers.   32. $\pm 1, \pm 3$ 33. $\pm 3, \pm i\sqrt{7}$

2.7 EXERCISES B

Solve and graph each inequality in Exercises 1–12. Give answers using both inequalities and intervals.

1. $3x - 1 \geq 5$ **2.** $5(y + 2) - 3 > 3(y - 1)$

3. $(x + 2)^2 \geq x(x - 2)$ **4.** $(z - 1)(z + 2) > (z + 1)^2$

5. $\dfrac{3}{2y + 4} < 0$ **6.** $(x + 4)^{-1} < 0$

7. $\dfrac{2(z + 2)}{5} < \dfrac{3(z + 2)}{7}$ **8.** $3 \leq 5y - 2 \leq 18$

9. $0 \leq \dfrac{3(x + 1)}{6} < 2$ **10.** $5z - 2 \leq 3$ or $5z - 2 > 18$

11. $3 - x > x + 3$ or $1 - x \leq x - 3$ **12.** $z - 2 < 2z + 1 \leq 2 + z$

In Exercises 13–15, find the values of x that make the given expression positive.

13. $2(x - 3)$ **14.** $-2(x - 3)$ **15.** $(x - 3)^2$

In Exercises 16–18, find the values of x that make the given expression negative.

16. $2(x - 3)$ **17.** $-2(x - 3)$ **18.** $-(x - 3)^2$

Solve each applied problem in Exercises 19–21.

19. **EDUCATION** To receive a grade of C in a course, a student must have an average mark between 70% and 80%, inclusive. If Jonas has 68%, 83%, and 79% on the first three tests this semester, what range of scores on the fourth test would give him a C?

20. **EDUCATION** Burford is in the same class as Jonas (see Exercise 19). Burford has 45%, 32% and 58% on the first three tests. What range of scores on the fourth test would give him a C?

21. **ENGINEERING** The power measured in watts W on an electrical circuit is related to the pressure in volts E and the current in amperes I by the equation $W = EI$. If the power demands on a 110-volt circuit in a garage vary between 220 watts and 2310 watts, what is the range of current in the circuit?

Solve each inequality in Exercises 22–29. Give answers using both inequalities and intervals.

22. $|y| > 7$
23. $|x| < 0$
24. $|z| \geq 0$
25. $|y - 2| \leq 5$
26. $|5 - z| \geq 2$
27. $\left|\dfrac{x + 5}{2}\right| \leq 10$
28. $|2(z + 3) - 1| < 9$
29. $|2y - 7| \geq 1$

FOR REVIEW

Solve.

30. **PHYSICS** If a car is traveling v mph, the shortest distance d required to bring the car to a full stop under ideal conditions (including the reaction time of the driver) can be approximated by $d = 0.05v^2 + 1.05v$. If a dog runs into the street 60 ft in front of your car while you are traveling 25 mph, can you stop in time to avoid hitting the dog?

31. **CONSTRUCTION** Fred can paint a house in 1 day less than it takes Charlie. If together they can paint it in 3 days, to the nearest tenth of a day, how long would it take Fred to do the job by himself?

32. $15y^{-2} + 2y^{-1} = 1$

33. $a^{2/3} - 12 = a^{1/3}$

2.7 EXERCISES C

Use the discriminant to find all values of m in Exercises 1–4 that make the quadratic equation satisfy the given condition.

1. $4x^2 - 4x + m = 0$; two complex (nonreal) solutions
   [Answer: $m > 1$]

2. $4x^2 - 4x + m = 0$; two real solutions

3. $mx^2 - x + 2 = 0$; two real solutions
   [Answer: $m < \frac{1}{8}$]

4. $mx^2 - x + 2 = 0$; two complex (nonreal) solutions

Solve each inequality in Exercises 5–7. Give the answer using intervals.

5. $\dfrac{1}{x^2 + 1} > 0$

6. $\dfrac{1}{(x - 1)^2} > 0$
   [Answer: $(-\infty, 1)$ or $(1, \infty)$]

7. $\dfrac{1}{|x + 3|} > 0$

8. Prove that if a and b are real numbers satisfying $0 < a < b$, then $a^2 < b^2$. Is this still true if the condition $0 < a$ is deleted?

9. Prove the transitive law which states that if a, b, and c are real numbers with $a > b$ and $b > c$, then $a > c$.

CHAPTER 2 EXTENSION PROJECT

Ratios and proportions have many applications that occur in unusual and interesting ways. Consider the rectangle below with width 1 unit and length g units in which a square has been constructed.

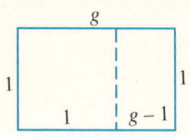

Suppose the smaller remaining rectangle with length 1 unit and width $g - 1$ units is similar to (has the same shape as) the original rectangle. As with similar triangles, rectangles are similar if their sides are proportional. Thus, we would like g to satisfy the proportion

$$\frac{g}{1} = \frac{1}{g-1}.$$

If you solve this proportion for g, $g = \frac{1+\sqrt{5}}{2} \approx 1.618034$. Rectangles like the one above, whose sides are in the ratio $g:1$, or approximately $1.62:1$ are called **golden rectangles** and g is called the **golden ratio.** The early Greeks thought that golden rectangles were aesthetically the most pleasing to the eye. Before the top of the Parthenon (pictured above) was destroyed, the front of the building could be placed into a golden rectangle.

The golden ratio is related to other interesting results in mathematics. One of the finest medieval mathematicians, Leonardo of Pisa (1170–1250?), better known as Fibonacci, wrote numerous books in mathematics, but is perhaps best known for the sequence of numbers

1, 1, 2, 3, 5, 8, 13, 21, 34, 55, 89, 144, . . .

called the **Fibonacci sequence.** After the first two terms, 1 and 1, each remaining term is obtained by adding the two preceding terms. Fibonacci is said to have discovered this pattern by considering a reproduction problem involving rabbits. Notice the ratios of consecutive terms in the sequence, for example,

$$\frac{55}{34} = 1.6176471, \frac{89}{55} = 1.6181818, \frac{144}{89} = 1.6179775.$$

As we go out further in the sequence, these ratios approach the golden ratio, certainly an unexpected result!

Numerous other occurrences of the golden ratio can be found in nature. Growth patterns of the southwestern century plant, certain properties of the honeycomb, and the spiral of a nautilus shell, the home of a snail-like animal, are a few examples.

THINGS TO DO:

1. Solve the proportion $\frac{g}{1} = \frac{1}{g-1}$ to obtain the exact value of the golden ratio.
2. Consider a standard 3×5 inch index card. How close does the card come to being a golden rectangle?
3. Does a normal 8.5×11 inch sheet of notebook paper have the shape of a golden rectangle?
4. Can you find other objects, for example a book, calculator, etc. that have the shape of a golden rectangle?
5. Write out the next ten Fibonacci numbers, consider the ratio of consecutive terms, and verify that the ratios do indeed get closer to the golden ratio.
6. Consult a book on the history of mathematics or on interesting mathematical diversions and investigate other relationships between the golden ratio and Fibonacci numbers, including their occurrences in nature.

CHAPTER 2 REVIEW

KEY TERMS

2.1
1. An **equation** is a statement that two quantities are equal.
2. A number is a **solution** or **root** of an equation if the equation is true when the number replaces the variable.
3. A **conditional equation** is an equation that is false for some replacements of the variable and true for others.
4. An **identity** is an equation that is true for all replacements of the variable.

5. A **contradiction** is an equation that is false for all replacements of the variable.

6. Two equations are **equivalent** if they have exactly the same solution(s).

7. A **linear** or **first-degree equation** is an equation that can be written as an equivalent equation in the form $ax + b = 0$.

8. A **radical equation** is an equation that has at least one variable in a radicand.

9. An **extraneous root** of a radical equation is a "solution" that does not check in the original equation.

10. A **fractional equation** is an equation that contains at least one algebraic fraction.

11. An equation that has at least one variable within absolute value bars is an **absolute value equation**.

12. A **literal equation** is an equation that has two or more variables.

2.2
1. The **ratio** of one number a to another number b is the fraction $\frac{a}{b}$.

2. A **proportion** is an equation stating that two ratios are equal.

3. In the proportion $\frac{a}{b} = \frac{c}{d}$, a and d are the **extremes** and b and c the **means**. If $b = c$, b (or c) is the **mean proportional** between a and d. The equation $ad = bc$ is the **cross-product equation**.

2.3
1. An **imaginary number** is a number of the form bi, where b is a real number and $i = \sqrt{-1}$.

2. A number of the form $a + bi$, where a and b are real numbers and $i^2 = -1$, is a **complex number** with **real part** a and **imaginary part** b.

3. Two complex numbers are **equal** if their real parts are equal and their imaginary parts are equal.

4. If $a + bi$ and $c + di$ are two complex numbers, their **sum** is
$$(a + bi) + (c + di) = (a + c) + (b + d)i,$$
their **difference** is
$$(a + bi) - (c + di) = (a - c) + (b - d)i,$$
their **product** is
$$(a + bi)(c + di) = (ac - bd) + (ad + bc)i,$$
and their **quotient** is
$$\frac{a + bi}{c + di} = \frac{(ac + bd) + (bc - ad)i}{c^2 + d^2}.$$

5. The **absolute value** of complex number $a + bi$ is $|a + bi| = \sqrt{a^2 + b^2}$.

2.4
1. Any equation that can be written in the form $ax^2 + bx + c = 0$, where a, b, and c are constant real numbers with $a \neq 0$, is a **quadratic** or **second-degree equation**.

2. If a quadratic equation has only one real root, the root is a **double root** or a **root of multiplicity two**.

3. If $ax^2 + bx + c = 0$, the number $b^2 - 4ac$ is the **discriminant** of the equation.

2.5 An equation that is quadratic in some expression other than a single variable is **quadratic in form**.

2.7
1. An inequality that can be written in one of the forms
$$ax + b > 0, \quad ax + b < 0,$$
$$ax + b \leq 0, \quad \text{or } ax + b \geq 0$$
is a **linear inequality**.

2. A **solution** to an inequality is any number that makes the inequality true when substituted for the variable.

3. Two inequalities are **equivalent inequalities** if they have exactly the same solutions.

4. A **compound inequality** is an inequality formed by joining two inequalities using either the word *and* or the word *or*.

5. Let a and b be real numbers with $a < b$. An **open interval** is
$$(a, b) = \{x \text{ such that } a < x < b\},$$
and a **closed interval** is
$$[a, b] = \{x \text{ such that } a \leq x \leq b\}.$$
The intervals
$$(a, b] = \{x \text{ such that } a < x \leq b\}$$
and $[a, b) = \{x \text{ such that } a \leq x < b\}$
are **half-open intervals**. The four **infinite intervals** are:
$$(-\infty, a) = \{x \text{ such that } x < a\}$$
$$(-\infty, a] = \{x \text{ such that } x \leq a\}$$
$$(a, \infty) = \{x \text{ such that } x > a\}$$
$$[a, \infty) = \{x \text{ such that } x \geq a\}$$

6. An **absolute value inequality** is an inequality that has at least one variable within the absolute value bars.

REVIEW EXERCISES

Part I

2.1 Solve each equation in Exercises 1–13.

1. $6z - 5 = 2z + 8$
2. $\dfrac{2y}{3} - \dfrac{5}{6} = -\dfrac{3y}{2} - \dfrac{1}{3}$
3. $6.5 + 2.5z = 3z + \dfrac{3}{2}$
4. $3x - 2 = x + (2x + 1)$
5. $\left(\dfrac{z}{2} - 1\right) - \left(z - \dfrac{1}{2}\right) = 1$
6. $4(x - 2) + 3(3 - x) = 3x - 5$
7. $|2x + 3| = 3$
8. $|3 - 4x| = 0$
9. $|3 - 4x| = -1$
10. $\dfrac{z - 2}{z - 5} = \dfrac{z + 1}{z + 3}$
11. $\dfrac{3}{x + 6} - \dfrac{4}{x - 2} = \dfrac{2}{x^2 + 4x - 12}$
12. $2\sqrt[3]{a - 6} - 3 = -2$
13. $\sqrt{z - 5} - \sqrt{z + 3} = -2$

2.2 Solve each applied problem in Exercises 14–21.

14. **NUMBER** The sum of three positive, consecutive, even integers is 40 more than twice the largest. Find the integers.

15. **AVERAGE** The average weight of Jane, Bill, Henry, and Sue is 78 kg. If Sue weighs 65 kg, Bill weighs 84 kg, and Henry weighs 88 kg, how much does Jane weigh?

16. **AGE** Joe is 11 years older than Mary. Twice Joe's age plus three times Mary's age is 172. How old is each?

17. **ECONOMICS** If after receiving an 8% raise Gloria now makes $30,240 a year, what was her former salary?

18. **GEOMETRY** The perimeter of a rectangle is 70 m. If the length is 1 m less than twice the width, find the dimensions of the rectangle.

19. **GEOMETRY** The first angle of a triangle is 15° more than the third, and the second is 5° less than six times the first. Find the measure of each angle.

20. **WORK** Working together, Beth and Peggy can paint a garage in 6 days. Working alone, Peggy can do the job in 10 days. How many days would it take Beth to do the job alone?

21. **PROPORTION** A wire 64 ft long is cut into two pieces that have the ratio 11:5. How long is each piece?

2.3

22. Find the quotient.
$$\dfrac{3 - 5i}{-2 + i}$$

23. Find the reciprocal of $4 - 3i$.

Perform the indicated operations in Exercises 24–28.

24. $(2 - 3i) + (5 - i)$
25. $[-(\sqrt{4} - \sqrt{-9})]^{31}$
26. $|-4 + 3i|$
27. $(2 - i)(3 + 2i)$
28. $(1 - 4i)^2$

2.4 Solve each equation in Exercises 29–34.

29. $x^2 - 2x - 48 = 0$
30. $y(y + 1) = 6(3 - y)$
31. $3x^2 - x - 1 = 0$
32. $2y(y - 2) = -1$
33. $3x^2 + 15 = 0$
34. $2x^2 - x + 1 = 0$

Use the discriminant to determine the nature of the solutions to each equation in Exercises 35–37.

35. $2x^2 - 3 = 0$
36. $4x^2 = 4x - 1$
37. $3x^2 + 3x + 1 = 0$

2.5 Solve each equation in Exercises 38–41.

38. $y^4 - 2y^2 - 63 = 0$
39. $\left(\dfrac{y - 3}{y + 1}\right)^2 - 6\left(\dfrac{y - 3}{y + 1}\right) + 8 = 0$

40. $2x + \dfrac{x}{x+7} = \dfrac{8}{x+7}$

41. $\sqrt{3y+1} - \sqrt{y+4} = 1$

2.6 *Solve each applied problem in Exercises 42–45.*

42. RATE-MOTION Two cars leave the same city traveling at right angles to each other. If the first travels 15 mph faster than the second, and if after two hours they are 150 miles apart, find the speed of each.

43. RECREATION A group of men plan to share equally the $132 cost of a fishing trip. At the last minute, two men decide not to go, and this raises the share of each remaining man by $11. How many men were planning to go initially?

44. WORK Jan can do a job in 12 minutes less time than her mother, and together they can do the job in 8 minutes. How long would it take each to do the job alone?

45. GEOMETRY The Eiffel Tower in Paris is 984 ft tall. If an observer at the top of the tower on a clear day views the horizon, how far away does she see? Use 4000 mi for the radius of the earth, and consider the figure below.

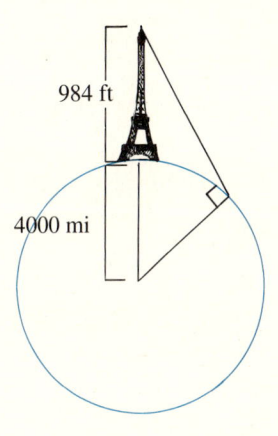

2.7 *Solve and graph each inequality in Exercises 46–53. Give answers using both inequalities and intervals.*

46. $5(x+2) - 3x \leq x + 7$

47. $3(x-1) + x < 4(x+1)$

48. $-3 < 2x - 3 < 3$

49. $2x - 3 \leq -3$ or $2x - 3 \geq 3$

50. $|2 - x| < 1$

51. $|4x - 3| \geq 1$

52. $|2 - x| \leq 0$

53. $|2 - x| < -1$

Part II

Solve each equation in Exercises 54–67.

54. $(3y - 2) - (y + 1) = 0$

55. $5\sqrt{x-1} - 3\sqrt{2x+5} = 0$

56. $\sqrt{z^2 + 7} - z - 1 = 0$

57. $\dfrac{x-3}{x-1} - \dfrac{5}{x} = 1$

58. $\dfrac{z}{z+2} - \dfrac{2}{z-2} = \dfrac{z^2 + 4}{z^2 - 4}$

59. $5x^2 + 2x + 1 = 0$

60. $x^{2/3} + 2x^{1/3} + 1 = 0$

61. $\dfrac{3y}{y+1} = \dfrac{2}{y^2-1} + \dfrac{y}{y-1}$

62. $\sqrt{x^2+2} - \sqrt{3x+6} = 0$

63. $|x+5| = |5-x|$

64. $\dfrac{1}{x} - \dfrac{1}{y} = \dfrac{1}{z}$ for y

65. $ab = ac + d$ for a

66. $2x^2 - 8 = 0$

67. $2x^2 + 8 = 0$

Solve and graph each inequality in Exercises 68–71. Give answers using both inequalities and intervals.

68. $3(x+1) - 2x \geq 2(x+1)$

69. $2 + 3x < -1$ or $2 + 3x \geq 5$

70. $|4 - x| < 1$

71. $|4 - x| \geq 1$

Perform the indicated operations in Exercises 72–74.

72. $(8 - 6i) - (-3 + i)$

73. $2i - \sqrt{-25}$

74. $\dfrac{1}{3-i}$

75. Without solving, find the sum and product of the solutions to the equation $5x^2 - 10x + 25 = 0$.

Solve each applied problem in Exercises 76–81.

76. WORK John can do a job in 24 minutes and Sean can do the same job in 18 minutes. How long would it take if they worked together? Give answer correct to the nearest tenth of a minute.

77. RATE-MOTION Mark and Pam live 159 miles apart. If Mark averages 50 mph driving while Pam averages 56 mph, and they both leave at 8:00 A.M. and drive toward each other, at what time will they meet?

78. AERONAUTICS A plane flies 700 miles with the wind and 500 miles against the wind in the same length of time. If the speed of the wind is 20 mph, what is the speed of the plane in calm air?

79. PROPORTION A tree casts a shadow 32 ft long at the same time that a 100-ft tower casts a shadow 64 ft long. How tall is the tree?

80. PHYSICS Fernando throws a baseball into the air with an initial velocity of 64 ft/sec. If the ball is released from a height of 6 ft, at what time will the ball be at a height of 54 ft?

81. ECOLOGY A state game officer wants to estimate the number of antelope in a preserve. He catches 100 antelope, tags each one's ear, and returns them to the preserve. After allowing his sample to mix thoroughly with the rest of the herd, he catches another sample of 50 antelope and discovers that 5 of these are tagged. Estimate the number of antelope in the game preserve.

82. Use a calculator to solve $7.3x^2 - 3.2x - 8.4 = 0$. Give the answer rounded to the nearest hundredth.

83. Solve for x.
$$\sqrt{\frac{x+y}{x}} = z$$

84. Solve for y.
$$2x^2 - 3y + 4z^2 = 9$$

85. Find a quadratic equation that has $1 - 2i$ and $1 + 2i$ for its solutions.

86. Solve $x^2 - 4x + 1 = 0$ by completing the square.

ANSWERS: 1. $\frac{13}{4}$ 2. $\frac{3}{13}$ 3. 10 4. no solution (contradiction) 5. -3 6. 3 7. $-3, 0$ 8. $\frac{3}{4}$ 9. no solution 10. $\frac{1}{5}$ 11. -32 12. $\frac{49}{8}$ 13. 6 14. 42, 44, 46 15. 75 kg 16. Joe is 41, Mary is 30. 17. \$28,000 18. 12 m, 23 m 19. 25°, 145°, 10° 20. 15 days 21. 44 ft, 20 ft 22. $-\frac{11}{5} + \frac{7}{5}i$ 23. $\frac{4}{25} + \frac{3}{25}i$ 24. $7 - 4i$ 25. $-i$ 26. 5 27. $8 + i$ 28. $-15 - 8i$ 29. 8, -6 30. 2, -9 31. $\frac{1 \pm \sqrt{13}}{6}$ 32. $\frac{2 \pm \sqrt{2}}{2}$ 33. $\pm i\sqrt{5}$ 34. $\frac{1 \pm i\sqrt{7}}{4}$ 35. two real 36. one real 37. two complex (nonreal) 38. $\pm 3, \pm i\sqrt{7}$ 39. $-5, -\frac{7}{3}$ 40. $\frac{1}{2}, -8$ 41. 5 42. 45 mph, 60 mph 43. 6 men 44. Jan: 12 min, mother: 24 min 45. 38.6 mi
46. $x \le -3$; $(-\infty, -3]$; 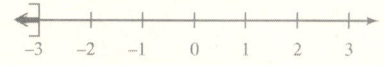
47. every real number; $(-\infty, \infty)$; 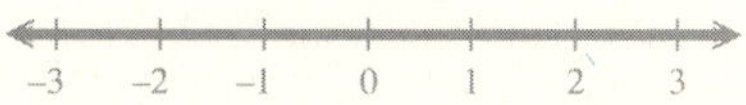
48. $0 < x < 3$; $(0, 3)$;
49. $x \le 0$ or $x \ge 3$; $(-\infty, 0]$ or $[3, \infty)$;
50. $1 < x < 3$; $(1, 3)$;
51. $x \le 1/2$ or $x \ge 1$; $(-\infty, 1/2]$ or $[1, \infty)$;
52. $x = 2$;
53. no solution; no graph 54. $\frac{3}{2}$ 55. 10 56. 3 57. $\frac{5}{7}$ 58. no solution 59. $\frac{-1 \pm 2i}{5}$ 60. -1 61. $1 \pm \sqrt{2}$ 62. 4, -1 63. 0 64. $y = \frac{xz}{z-x}$ 65. $a = \frac{d}{b-c}$ 66. ± 2 67. $\pm 2i$
68. $x \le 1$; $(-\infty, 1]$; 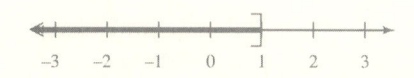
69. $x < -1$ or $x \ge 1$; $(-\infty, -1)$ or $[1, \infty)$;
70. $3 < x < 5$; $(3, 5)$;
71. $x \le 3$ or $x \ge 5$; $(-\infty, 3]$ or $[5, \infty)$;
72. $11 - 7i$ 73. $-3i$ 74. $\frac{3}{10} + \frac{1}{10}i$ 75. $x_1 + x_2 = 2$; $x_1 x_2 = 5$ 76. 10.3 min 77. 9:30 A.M. 78. 120 mph 79. 50 ft 80. 1 sec and 3 sec 81. 1000 antelope 82. 1.31, -0.88 83. $x = \frac{y}{z^2 - 1}$ 84. $y = \frac{2x^2 + 4z^2 - 9}{3}$ 85. $x^2 - 2x + 5 = 0$ 86. $2 \pm \sqrt{3}$

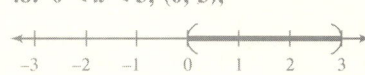

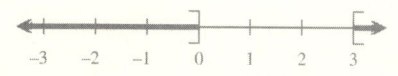

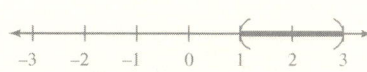

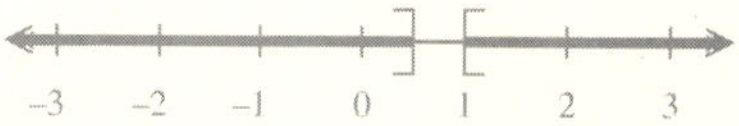

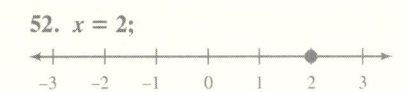

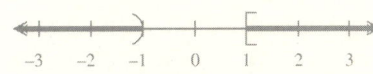

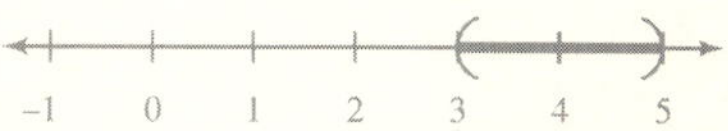

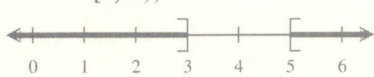

CHAPTER 2 TEST

Solve.

1. $\sqrt{3x+2} - \sqrt{3x-1} = 1$

2. $|3x+1| = 4$

3. $az = ay + w$ for a

4. Jan can wash the dishes in 30 minutes, and Wilma can wash them in 20 minutes. How long would it take them to wash the dishes if they worked together?

5. A plane flies 680 km with the wind and 520 km against the wind in the same amount of time. If the wind speed is 40 km/hr, what is the speed of the plane in still air?

6. A rope 104 ft long is cut into two pieces that have the ratio 9:4. How long is each piece?

7. $x^2 + 6x - 1 = 0$

Solve.

8. $15x^{-2} + 2x^{-1} - 1 = 0$

9. $\dfrac{1}{x+1} + \dfrac{1}{12} = \dfrac{1}{x}$

10. Tony buys some stock for $200.00. If each share had cost $2.00 less, he could have bought 5 more shares for the same $200.00. How many shares of stock did he buy?

172

CHAPTER 2 TEST
CONTINUED

11. Two trains leave the same town traveling at right angles. One travels 10 mph faster than the other. In 1 hr they are 50 miles apart. Find the speed of each train.

11. _____

12. $2x^2 - 3x + 2 = 0$

12. _____

13. Divide $\frac{4-i}{2+3i}$ and express the quotient in the form $a + bi$.

13. _____

14. Use the discriminant to determine the nature of the solutions to $x^2 - 2x + 5 = 0$ (two real, one real, or two complex).

14. _____

15. Graph the inequality.
$$3x - 3(x - 1) \geq 5 - x$$

15.

$\longleftarrow\!+\!+\!+\!+\!+\!+\!+\!+\!+\!+\!+\!\longrightarrow$
−6 −5 −4 −3 −2 −1 0 1 2 3 4 5 6

16. Solve, give the answer using interval(s), and graph the solution.
$$2x - 4 < 6 \quad \text{or} \quad 2x - 4 \geq 12$$

16. _____

$\longleftarrow\!+\!+\!+\!+\!+\!+\!+\!+\!+\!+\!\longrightarrow$
−1 0 1 2 3 4 5 6 7 8 9

Solve. Give answers using interval(s).

17. $|x + 4| < 11$

17. _____

18. $|3x - 1| > 10$

18. _____

19. $|x + 1| = |x - 1|$

19. _____

20. Use the formula $h = -16t^2 + v_0 t + h_0$ to find the time that it will take for an object dropped from a height of 500 ft to reach the ground. Give answer to the nearest tenth of a second.

20. _____

CHAPTER 3

Relations, Functions, and Graphs

AN APPLIED PROBLEM IN BUSINESS

A car agency averages 20 rental customers per day for a model that rents for $32 a day. A survey has shown that the agency could get 2 new customers for each $1 reduction in the daily rental rate. What daily rate would give the company the maximum revenue?

Analysis
The daily revenue is the product of the number of daily rental customers and the daily rental fee. This product gives a quadratic function, and the y-coordinate of the vertex of the graph of this function is the maximum daily revenue.

Tabulation
Let n = number of dollars reduction in the daily rental fee

$2n$ = number of new customers (since there are 2 new customers for each dollar reduction)

$20 + 2n$ = number of customers after reducing the fee

$32 - n$ = new daily rental fee

R = daily revenue

Translation
$$R(n) = (20 + 2n)(32 - n)$$

In this chapter we will introduce the concept of a *relation*. A *function* is a special kind of relation that is basic to all areas of mathematics. In business, science, and engineering, the idea of one variable depending on another, that is, the notion of a function, is used extensively in applications such as the one given at the left (see Example 4 in Section 3.6). We study not only functions and their properties, but also the way to describe functions graphically in a rectangular coordinate system. Linear and quadratic functions are given priority treatment along with mathematical modeling involving these functions; the subject of variation concludes the chapter.

3.1 THE RECTANGULAR COORDINATE SYSTEM

STUDENT GUIDEPOSTS
1. Graphing Functions
2. Plotting Points
3. Graphs
4. Distance Formula
5. Midpoint Formula

① GRAPHING FUNCTIONS

The study of functions is made easier if we know how to picture them in a graph. Thus, to begin our study we discuss graphing equations in two variables. In Chapter 2 we graphed equations and inequalities in one variable on a number line. When a horizontal and vertical number line are superimposed so that the two origins coincide and the lines are perpendicular, the result is a **rectangular coordinate system** or **coordinate plane.** This configuration is also called a **Cartesian coordinate system** after René Descartes, the French mathematician who introduced it.

Figure 3.1 shows the important features of a rectangular coordinate system. The horizontal number line is the **x-axis** and the vertical number line the **y-axis.** The point of intersection of the axes is the **origin;** the system is sometimes called an **x, y-coordinate system.**

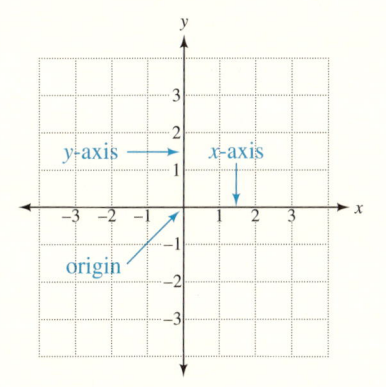

Figure 3.1 Rectangular or Cartesian Coordinate System (Coordinate Plane)

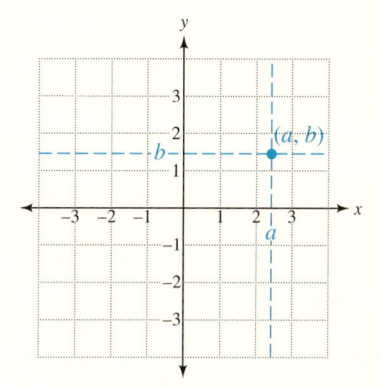

Figure 3.2 Plotting Points

② PLOTTING POINTS

Recall that there is one and only one point on a number line associated with each real number. Similarly, there is a unique pair of numbers associated with each point in the coordinate plane. The pair is written (x, y) where x represents units on the x-axis and y corresponds to units on the y-axis. We call (x, y) an **ordered pair** since the order, x first and y second, must be preserved to properly identify points. The ordered pair (a, b) is plotted in Figure 3.2 at the intersection of a vertical line through a and a horizontal line through b. The number a is called the **x-coordinate** or **abscissa** of the point (a, b), and b is called the **y-coordinate** or **ordinate** of the point.

176 CHAPTER 3 RELATIONS, FUNCTIONS, AND GRAPHS

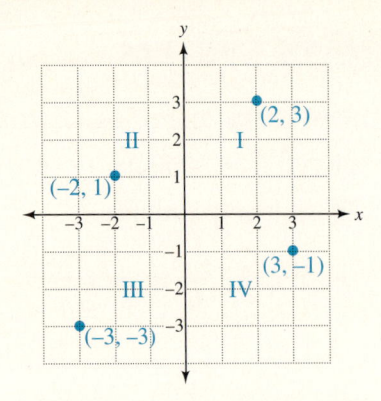

Figure 3.3 Quadrants and Graphs

In Figure 3.3 we have plotted the points (2, 3), (−2, 1), (−3, −3), and (3, −1). Notice also that in this figure the Roman numerals, I, II, III, and IV are included. These designate the four sections in a rectangular coordinate system called **quadrants.** The point (2, 3) is in the first quadrant, (−2, 1) in the second, (−3, −3) in the third, and (3, −1) in the fourth. The signs of the x-coordinate and y-coordinate in each of the quadrants are as follows.

$$\text{I: } (+, +), \quad \text{II: } (-, +), \quad \text{III: } (-, -), \quad \text{IV: } (+, -)$$

❸ GRAPHS

The points plotted in Figure 3.3 are the **graph** of the set {(2, 3), (−2, 1), (−3, −3), (3, −1)}. That is, the complete graph of this set is the four points in Figure 3.3. Any set that consists of a finite number of ordered pairs will have a graph similar to this one.

A more common type of graph in mathematics describes a set that contains infinitely many elements. For example, to graph the set of solutions of $y = -2x + 3$ we need to show all the ordered pairs for which the equation is true. In Figure 3.4 we plotted a few of the solutions, (−1, 5), (0, 3), (1, 1), (2, −1), (3, −3), and (4, −5). We saw that all the points lie in a straight line and, knowing that the graph is a continuous, connected curve, we constructed the line. The graph consists of the infinitely many points on the line.

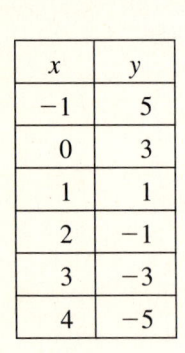

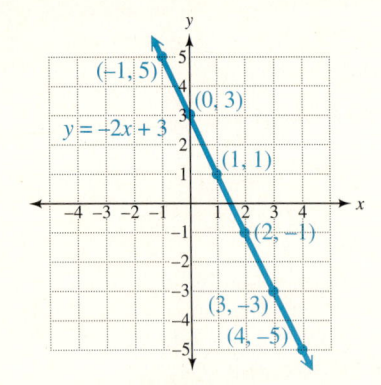

x	y
−1	5
0	3
1	1
2	−1
3	−3
4	−5

Figure 3.4

Notice that a table containing the points plotted is given along with the graph in Figure 3.4. One procedure for graphing an equation is to select values of one of the variables (usually x), calculate values of the other variable, and store the results in such a table. We plot enough points to determine the shape of the curve. As we continue the study of graphing we will discover methods that will allow us to decrease the number of points we use.

3.1 THE RECTANGULAR COORDINATE SYSTEM   177

EXAMPLE 1 Graphing an Infinite Solution Set

Graph $y = x^2 - 3$.

Select values of x which give y-values that can be plotted in a coordinate system of reasonable size. For example, for

$$x = -3 \quad y = (-3)^2 - 3 = 6,$$
$$x = -2 \quad y = (-2)^2 - 3 = 1.$$

Other points are calculated for the table and plotted in Figure 3.5. Enough points are included to allow us to connect them with the indicated curve. The graph of $y = x^2 - 3$ is all the points on the curve.

x	y
-3	6
-2	1
-1	-2
0	-3
1	-2
2	1
3	6

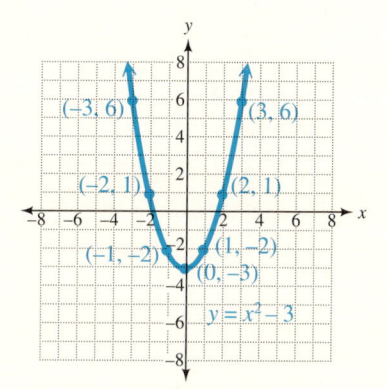

Figure 3.5

Practice Exercise 1

Graph $y = -x^2 + 3$.

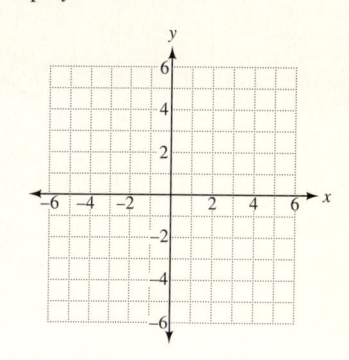

Answer: like Figure 3.5 except opening down

④ DISTANCE FORMULA

Two formulas that are used throughout this chapter deal with finding the distance between two points and finding the midpoint of the line segment joining two points.

Distance Formula

The distance between two points with coordinates (x_1, y_1) and (x_2, y_2) is given by the **distance formula**

$$d = \sqrt{(x_1 - x_2)^2 + (y_1 - y_2)^2}.$$

To derive this formula, consider Figure 3.6. Apply the Pythagorean theorem to the right triangle with vertices P, Q, and R (the sum of the squares of the legs of a right triangle is equal to the square of the hypotenuse).

$$d^2 = (x_1 - x_2)^2 + (y_1 - y_2)^2$$
$$d = \sqrt{(x_1 - x_2)^2 + (y_1 - y_2)^2}$$

Observe that since both $(x_1 - x_2)$ and $(y_1 - y_2)$ are squared, if P is (x_2, y_2) and Q is (x_1, y_1), the formula yields the same result for the distance. Also, if the points are placed in different quadrants, the same formula applies.

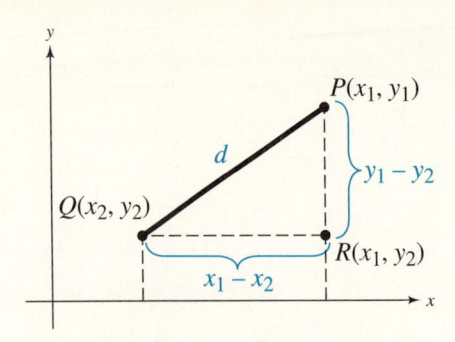

Figure 3.6 Distance Formula

EXAMPLE 2 USING THE DISTANCE FORMULA

Find the lengths of the sides of the triangle ABC with vertices $A(9, 4)$, $B(-1, 2)$, and $C(0, -3)$. Is ABC a right triangle?

First we draw ABC as in Figure 3.7.

PRACTICE EXERCISE 2

Determine if the triangle with vertices $D(2, 1)$, $E(-3, 2)$, and $F(-1, -4)$ is a right triangle.

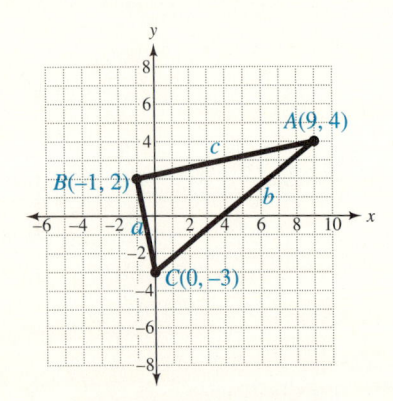

Figure 3.7 Right Triangle

$$a = \sqrt{(0-(-1))^2 + (-3-2)^2} = \sqrt{1^2 + (-5)^2} = \sqrt{26}$$
$$b = \sqrt{(9-0)^2 + (4-(-3))^2} = \sqrt{9^2 + 7^2} = \sqrt{130}$$
$$c = \sqrt{(-1-9)^2 + (2-4)^2} = \sqrt{(-10)^2 + (-2)^2} = \sqrt{104} = 2\sqrt{26}.$$

If ABC is a right triangle, $\sqrt{130}$ must be the length of the hypotenuse (it is the longest side), which together with the legs, must satisfy the conditions of the Pythagorean theorem.

$$a^2 + c^2 = (\sqrt{26})^2 + (\sqrt{104})^2 = 26 + 104 = 130 = (\sqrt{130})^2 = b^2$$

Thus, the triangle is a right triangle.

Answer: no

---

5 MIDPOINT FORMULA

The midpoint formula can be derived by considering Figure 3.8. Let $(\bar{x}, \bar{y})$ (read x-bar, y-bar) be the midpoint of the line segment joining $P(x_1, y_1)$ and $Q(x_2, y_2)$. From geometry we know that the point $(\bar{x}, 0)$ is the midpoint between $(x_2, 0)$ and $(x_1, 0)$. Thus, the distance d_1 on the x-axis between $(x_2, 0)$ and $(\bar{x}, 0)$ is the same as the distance d_2 between $(\bar{x}, 0)$ and $(x_1, 0)$.

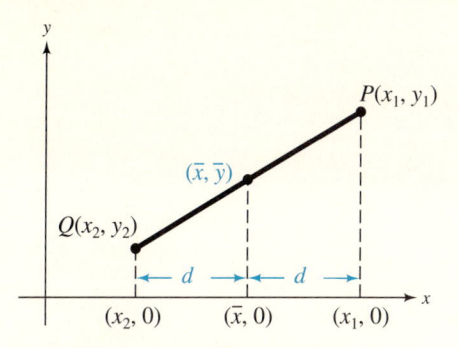

Figure 3.8 Midpoint Formula

$$d_1 = d_2$$
$$|\bar{x} - x_2| = |x_1 - \bar{x}|$$
$$\bar{x} - x_2 = x_1 - \bar{x} \quad \text{Since } \bar{x} > x_2 \text{ and } x_1 > \bar{x}$$
$$2\bar{x} = x_1 + x_2$$
$$\bar{x} = \frac{x_1 + x_2}{2}$$

Similarly, by considering distance on the y-axis, we can show $\bar{y} = \frac{1}{2}(y_1 + y_2)$. This gives the following theorem.

Midpoint Formula

The coordinates $(\bar{x}, \bar{y})$ of the midpoint of the line segment joining (x_1, y_1) and (x_2, y_2) are given by the **midpoint formula**

$$(\bar{x}, \bar{y}) = \left(\frac{x_1 + x_2}{2}, \frac{y_1 + y_2}{2}\right).$$

Notice that the midpoint is found by finding the average of the x-coordinates and the average of the y-coordinates.

EXAMPLE 3 USING THE MIDPOINT FORMULA

Find the midpoint between the points $(-5, 2)$ and $(7, 3)$.

$$(\bar{x}, \bar{y}) = \left(\frac{x_1 + x_2}{2}, \frac{y_1 + y_2}{2}\right)$$

$$= \left(\frac{-5 + 7}{2}, \frac{2 + 3}{2}\right) = \left(\frac{2}{2}, \frac{5}{2}\right) = \left(1, \frac{5}{2}\right)$$

PRACTICE EXERCISE 3

Find a and b such that $(-3, 5)$ is the midpoint between $(a, 6)$ and $(-1, b)$.

Answer: $a = -5$, $b = 4$

3.1 EXERCISES A

Graph the set of ordered pairs in Exercises 1–2.

1. $\{(2, 5), (-1, 3)\}$

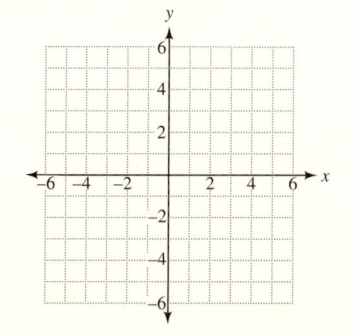

2. $\{(-3, 4), (-3, 2), (-3, 0), (-3, -2), (-3, -3)\}$

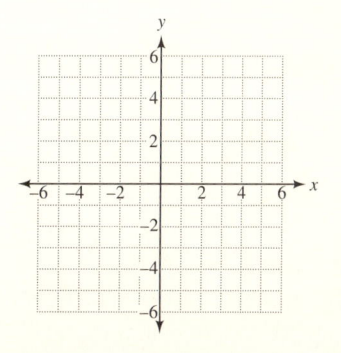

180 CHAPTER 3 RELATIONS, FUNCTIONS, AND GRAPHS

In Exercises 3–6, give the quadrant in which the point is located.

3. $(-2, 1)$ **4.** $(\sqrt{2}, -\sqrt{2})$ **5.** $\left(-\dfrac{1}{2}, -\dfrac{1}{4}\right)$ **6.** $(88, 90)$

Graph the solutions to each equation (graph the equation) in Exercises 7–12.

7. $y - x = 0$ **8.** $y = -\dfrac{1}{2}x + 4$ 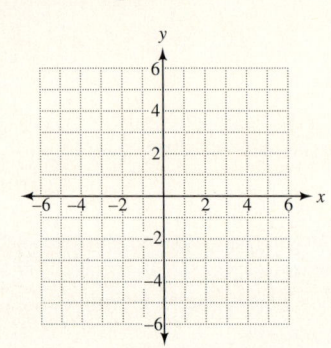   **9.** $2x = -5$

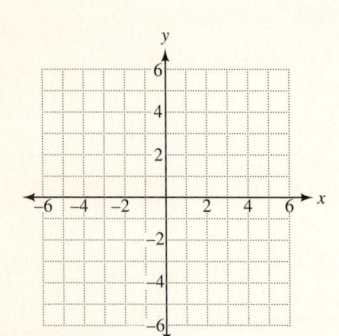

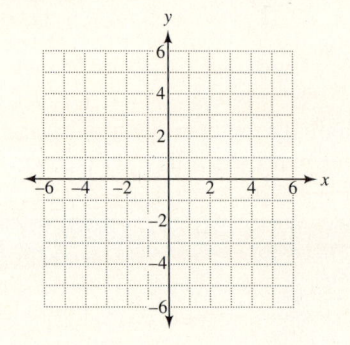

10. $2y = 7$ 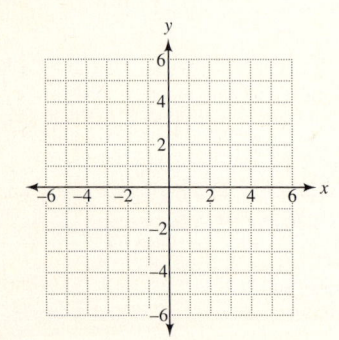   **11.** $y + x^2 = 0$ 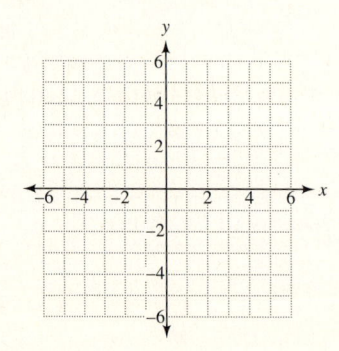   **12.** $y^2 = x + 2$

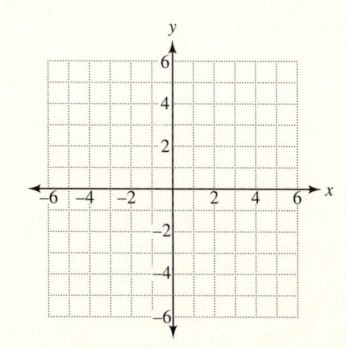

In Exercises 13–18 find the distance between the points and the midpoint of the line segment joining them.

13. $(4, 3)$ and $(1, 7)$ **14.** $(7, -1)$ and $(5, 8)$ **15.** $(-6, 2)$ and $(-6, -5)$

16. $(a, 3)$ and $(2a, 4)$ **17.** $(2\sqrt{x}, 3\sqrt{y})$ and $(-\sqrt{x}, 2\sqrt{y})$ **18.** $(1.634, -2.147)$ and $(7.812, 4.115)$

19. If $(3, 2)$ is the midpoint of the line segment joining $(a, 8)$ and $(7, b)$, find a and b.

20. Find all numbers a such that the distance from $(a, -2)$ to $(8, 1)$ is equal to 5.

21. Use the distance formula and the Pythagorean theorem to determine if triangle ABC with vertices $A(5, 6)$, $B(-2, -1)$, and $C(2, -3)$ is a right triangle.

22. Determine if $(6, -1)$, $(1, -6)$, $(-7, 2)$, and $(-2, 7)$ are the vertices of a rectangle. [*Hint:* What can be said about the lengths of the diagonals of a rectangle?]

23. AGRICULTURE A large ranch is marked with electronic devices at the intersection of each mile line to aid in the location of herds of cattle. If one herd is at the point (4, 7) and another at (9, 3), find the distance between them, to the nearest tenth of a mile.

24. ECOLOGY If rainfall was 2.5 inches in January, 3.7 inches in February, and 6.3 inches in March, use J, F, and M for the months to list ordered pairs that represent this data.

FOR REVIEW

Solve each inequality in Exercises 25–27. Give answers using both inequalities and intervals.

25. $x^2 + 5x + 6 > 0$

26. $\dfrac{x-5}{2x+3} \leq 0$

27. $|4x - 3| \geq 1$

28. Consider the equation $x + 2y - 6 = 0$. **(a)** Assume that $x = 0$ and solve for y. **(b)** Assume that $y = 0$ and solve for x. **(c)** Solve the equation for y in terms of x.

ANSWERS: 1. [graph with points (2, 5) and (−1, 3)] 2. [graph with points (−3, 4), (−3, 2), (−3, 0), (−3, −2), (−3, −3)] 3. II 4. IV 5. III 6. I

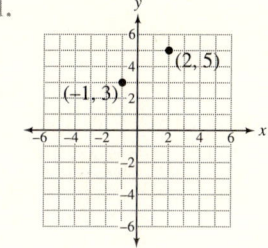

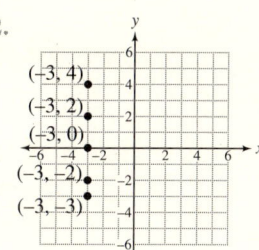

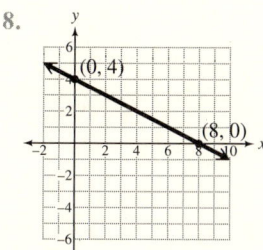

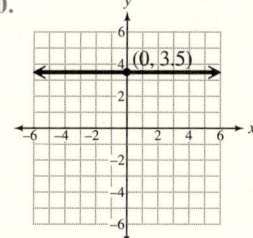

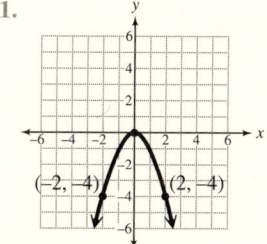

7. [graph through (4, 4)] 8. [graph through (0, 4) and (8, 0)] 9. [graph through $\left(-\tfrac{5}{2}, 0\right)$]

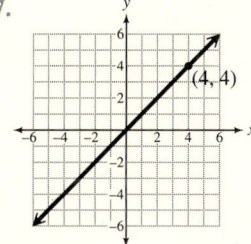

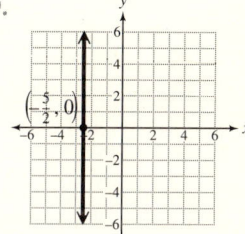

10. [graph through (0, 3.5)] 11. [graph through (−2, −4) and (2, −4)] 12. [graph through (2, 2) and (2, −2)]

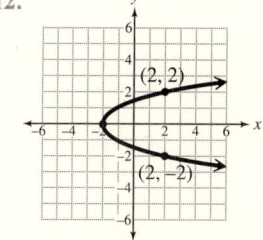

13. 5; $\left(\tfrac{5}{2}, 5\right)$ 14. $\sqrt{85}$; $\left(6, \tfrac{7}{2}\right)$ 15. 7; $\left(-6, -\tfrac{3}{2}\right)$ 16. $\sqrt{a^2 + 1}$; $\left(\tfrac{3a}{2}, \tfrac{7}{2}\right)$ 17. $\sqrt{9x + y}$; $\left(\tfrac{\sqrt{x}}{2}, \tfrac{5\sqrt{y}}{2}\right)$ 18. 8.797; (4.723, 0.984) 19. $a = -1$, $b = -4$ 20. 4, 12 21. sides: $7\sqrt{2}$, $2\sqrt{5}$, $3\sqrt{10}$; not a right triangle 22. yes 23. 6.4 mi 24. (J, 2.5), (F, 3.7), (M, 6.3) 25. $x < -3$ or $x > -2$; $(-\infty, -3)$ or $(-2, \infty)$ 26. $-\tfrac{3}{2} < x \leq 5$; $\left(-\tfrac{3}{2}, 5\right]$ 27. $x \leq \tfrac{1}{2}$ or $x \geq 1$; $\left(-\infty, \tfrac{1}{2}\right]$ or $[1, \infty)$ 28. (a) $y = 3$ (b) $x = 6$ (c) $y = -\tfrac{1}{2}x + 3$

3.1 EXERCISES B

Graph the set of ordered pairs in Exercises 1–2.

1. $\{(-2, 1), (3, -4), (2, 2)\}$
2. $\{(-1, -3), (-4, -4), (1, 5), (2, 0)\}$

In Exercises 3–6 give the quadrant in which the point is located.

3. $(-1, -5)$
4. $(1.2, 3.4)$
5. $(100, -2)$
6. $(-8, 7.5)$

Graph the solutions to each equation (graph the equation) in Exercises 7–12.

7. $y + x = 0$
8. $y = 1.2x - 2$
9. $3x - 7 = 0$
10. $6 - 4y = 0$
11. $y = x^2 - 1$
12. $y = \sqrt{x + 2}$

In Exercises 13–18 find the distance between the points and the midpoint of the line segment joining them.

13. $(5, 8)$ and $(-3, -2)$
14. $(-3, 6)$ and $(2, 7)$
15. $(4, -8)$ and $(6, -8)$
16. $(5, b)$ and $(-1, 3b)$
17. $(-3\sqrt{x}, 4\sqrt{y})$ and $(5\sqrt{x}, 6\sqrt{y})$
18. $(-3.667, -8.925)$ and $(6.421, -0.844)$

19. If $\left(\frac{7}{2}, -7\right)$ is the midpoint of the line segment joining (a, b) and $(3, -5)$, determine a and b.

20. Find all points $(1, b)$ that are at a distance of 10 units from $(7, 6)$.

21. Use the distance formula and the Pythagorean theorem to determine if triangle ABC with vertices $A(5, 9)$, $B(0, 1)$, and $C(4, -1)$ is a right triangle.

22. Determine if $(-3, 5)$, $(5, -3)$, $(2, -6)$, and $(-6, 2)$ are the vertices of a rectangle. [*Hint:* What can be said about the lengths of the diagonals of a rectangle?]

23. **AGRICULTURE** A large ranch is marked with electronic devices at the intersection of each mile line to aid in the location of herds of cattle. A herd at $(4, 7)$ is to be combined with one at $(9, 3)$ at a point midway between them. To what point should the ranch foreman fly in his helicopter to assist in combining the herds?

24. **ECOLOGY** In 1970 the number of deer in a forest management area was 2800. By 1975 the number had dropped to 2100, but was back up to 2600 in 1980 and 3100 in 1985. Present this data as a set of ordered pairs.

FOR REVIEW

Solve each inequality in Exercises 25–27.

25. $4x^2 - 20x + 25 \geq 0$
26. $\dfrac{x + 4}{3x - 5} > 0$
27. $|2 - 5x| < 6$

28. Consider the equation $2x - y + 8 = 0$.
    (a) Assume that $x = 0$ and solve for y.
    (b) Assume that $y = 0$ and solve for x.
    (c) Solve the equation for y in terms of x.

3.1 EXERCISES C

Let (x_1, y_1) and (x_2, y_2) be the coordinates of two points on the same horizontal line.

1. Use the distance formula to find a simplified formula for the distance between the two points.

2. Use the midpoint formula to find a simplified formula for the midpoint of the line segment joining the points.

Let (x_1, y_1) and (x_2, y_2) be the coordinates of two points on the same vertical line.

3. Use the distance formula to find a simplified formula for the distance between the two points.

4. Use the midpoint formula to find a simplified formula for the midpoint of the line segment joining the points.

3.2 LINEAR EQUATIONS

STUDENT GUIDEPOSTS

1. General Form
2. Intercepts
3. Slope
4. Point-Slope Form
5. Slope-Intercept Form
6. Parallel and Perpendicular Lines

1 GENERAL FORM

In Section 3.1 we graphed equations by plotting points. If we can recognize certain types of equations and their graphs, we can reduce the work needed in point plotting. A **linear equation** is an equation in two variables x and y that has as its graph a straight line. Such equations can be written in the **general form**

$$ax + by + c = 0$$

where a, b, and c are real-number constants, a and b not both zero. Since the variables occur raised to the first power only, linear equations are also called **first-degree equations.** As examples,

$$3x + 2y = 5, \quad -4x + 7y = 0, \quad 5y = 6, \quad x - 2 = 0$$

are linear equations while

$$x^2 + y^2 = 6, \quad 3xy = 5, \quad \frac{1}{x} + y = 5, \quad y = x^3$$

are *not* linear equations. Their graphs are curves that are not straight lines.

2 INTERCEPTS

Now that we know that the graph of a linear equation is a straight line, we need to plot only two points. For most equations the **intercepts,** the points where the graph crosses the axes, can be used. The **x-intercept** or intersection with the x-axis has the form $(a, 0)$ and can be found for an equation by letting $y = 0$ and solving for x. Likewise, the **y-intercept,** $(0, b)$, is found by letting $x = 0$ and solving for y.

For example, to graph $x - 2y - 4 = 0$, let $x = 0$ to find the y-intercept $(0, -2)$, and let $y = 0$ to find the x-intercept $(4, 0)$. The graph is shown in Figure 3.9.

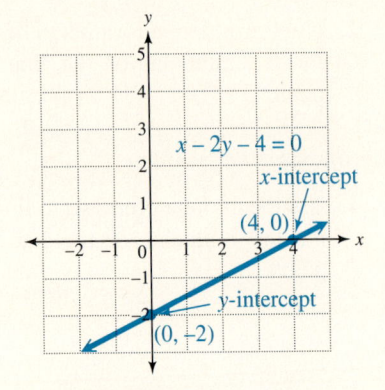

x	y
0	-2
4	0

Figure 3.9 Using Intercepts

For the equation $2x + 3y = 0$, the one point $(0, 0)$, is both x-intercept and y-intercept. The line goes through the origin and we need one other point to graph the equation. If $x = 3$, then $y = -2$. The graph is shown in Figure 3.10.

x	y
0	0
3	-2

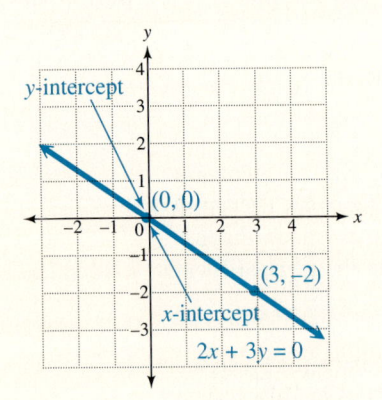

Figure 3.10 Line Through Origin

Thus, an equation of the form $ax + by = 0$ has as its graph a line through the origin.

$$ax + by = 0 \quad \text{Line through the origin}$$

Two other special cases should be considered. It can be shown that an equation of the form $x = a$ has a vertical line as its graph.

$$x = a \quad \text{Vertical line}$$

Also, $y = b$ has a horizontal line as its graph.

$$y = b \quad \text{Horizontal line}$$

For example, $2x - 4 = 0$ can be written as $x = 2$ and has the graph shown in Figure 3.11. The graph of $y = -\frac{5}{2}$ is shown in Figure 3.12.

x	y
2	0
2	-3
2	4

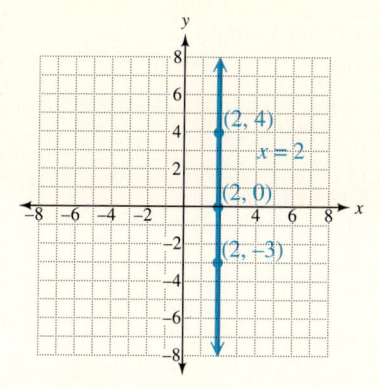

Figure 3.11 Vertical Line

x	y
0	-5/2
-3	-5/2
4	-5/2

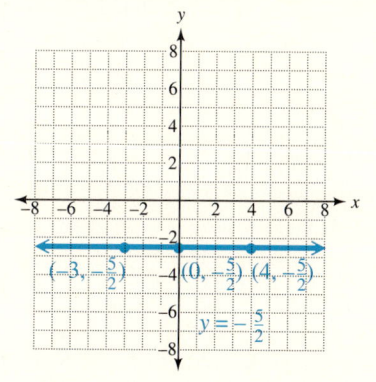

Figure 3.12 Horizontal Line

3 SLOPE

Notice that the graph in Figure 3.9 slopes upward as x increases, while in Figure 3.10 the graph slopes downward for increasing x. Figure 3.11 shows a vertical line, and Figure 3.12 a horizontal line. We can describe these differences better by defining *slope* of a line. Consider Figure 3.13 to help understand the definition.

Slope of a Line

Let (x_1, y_1) and (x_2, y_2) be two distinct points on a nonvertical line. The **slope** m of the line is given by

$$m = \frac{y_2 - y_1}{x_2 - x_1}.$$

The slope of a vertical line is undefined.

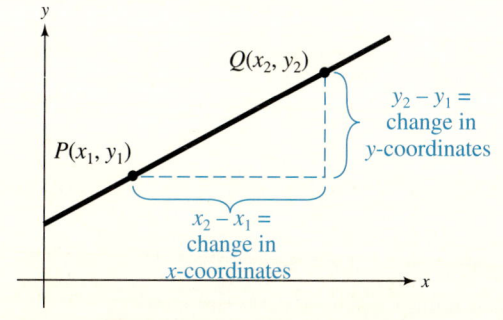

Figure 3.13 Slope of a Line

Note Since $y_2 - y_1$ is the change in y and $x_2 - x_1$ is the change in x, slope is a measure of the steepness of a line. Also, since

$$\frac{y_2 - y_1}{x_2 - x_1} = \frac{y_1 - y_2}{x_1 - x_2},$$

the slope does not depend on which point is called $P(x_1, y_1)$ and which one is called $Q(x_2, y_2)$.

EXAMPLE 1 FINDING SLOPE

Find the slope of the line passing through the given points.

(a) $(6, 3)$ and $(-2, 1)$

Let $P(x_1, y_1)$ be identified with $(6, 3)$ and $Q(x_2, y_2)$ with $(-2, 1)$.

$$m = \frac{y_2 - y_1}{x_2 - x_1}$$

$$= \frac{1 - 3}{-2 - 6} = \frac{-2}{-8} = \frac{1}{4}$$

(b) $(-2, 3)$ and $(1, -4)$

Let $P(x_1, y_1)$ be $(-2, 3)$ and $Q(x_2, y_2)$ be $(1, -4)$.

$$m = \frac{y_2 - y_1}{x_2 - x_1}$$

$$= \frac{-4 - 3}{1 - (-2)} = \frac{-7}{1 + 2} = -\frac{7}{3}$$

(c) $(4, 3)$ and $(-2, 3)$

Let $(x_1, y_1) = (4, 3)$ and $(x_2, y_2) = (-2, 3)$.

$$m = \frac{y_2 - y_1}{x_2 - x_1}$$

$$= \frac{3 - 3}{-2 - 4} = \frac{0}{-6} = 0$$

(d) $(-2, 0)$ and $(-2, -3)$

Let $(x_1, y_1) = (-2, 0)$ and $(x_2, y_2) = (-2, -3)$.

$$m = \frac{y_2 - y_1}{x_2 - x_1}$$

$$= \frac{-3 - 0}{-2 - (-2)} = \frac{-3}{0}$$

Since $\frac{-3}{0}$ is not defined, the slope is undefined and the line is a vertical line.

PRACTICE EXERCISE 1

Find the slope of the line passing through the given points.

(a) $(4, -7)$ and $(6, 3)$

(b) $(-2, 1)$ and $(3, -9)$

(c) $(1, 7)$ and $(-3, 7)$

(d) $(-4, 9)$ and $(-4, 6)$

Answers: (a) 5   (b) -2 (c) 0   (d) undefined

The lines through the points in Example 1 are graphed in Figure 3.14. These summarize the four types of possible outcomes in determining slope.

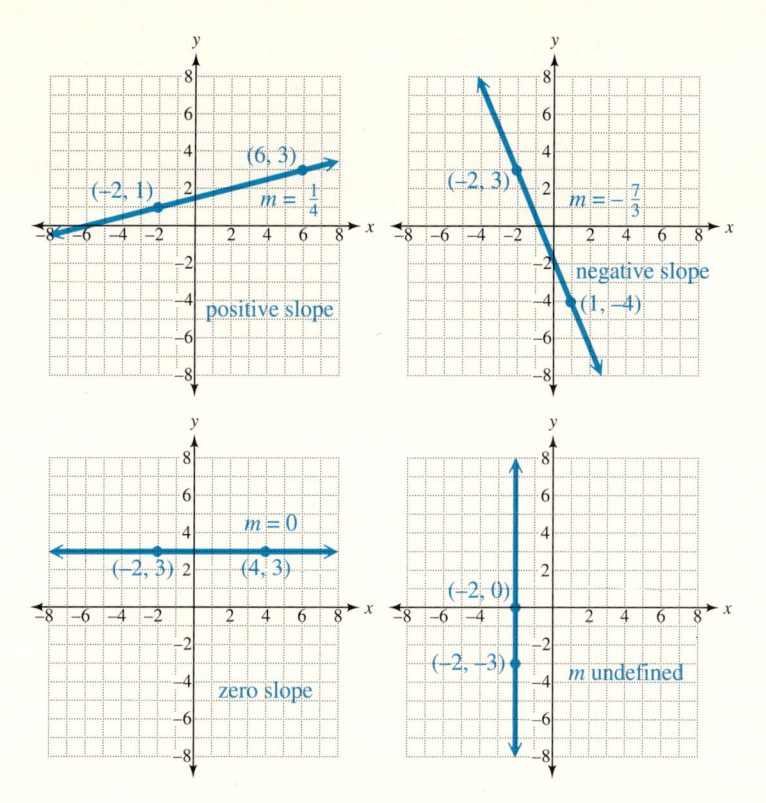

Figure 3.14 Types of Slopes

④ POINT SLOPE FORM

From the notion of slope, other forms of the equation of a line can be developed. Let $P(x_1, y_1)$ be a fixed point on the line with slope m. (See Figure 3.15.) If (x, y) is any other point on the line with $x \neq x_1$, then

$$m = \frac{y - y_1}{x - x_1}$$

or
$$y - y_1 = m(x - x_1).$$

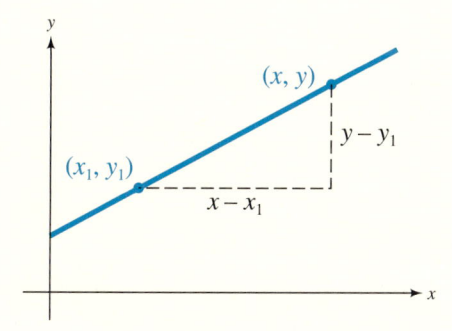

Figure 3.15 Point-Slope Form

Point-Slope Form

An equation for the line with slope m passing through the point (x_1, y_1) is

$$y - y_1 = m(x - x_1).$$

EXAMPLE 2   USING THE POINT-SLOPE FORM

Find the general form of the equation of a line passing through the points (3, −4) and (6, 1).

First determine the slope of the line.

$$m = \frac{1-(-4)}{6-3} = \frac{5}{3} \qquad (3, -4) = (x_1, y_1) \text{ and } (6, 1) = (x_2, y_2)$$

Now use the point-slope form and change to the general form.

$$y - y_1 = m(x - x_1) \qquad \text{Point-slope form}$$

$$y - (-4) = \frac{5}{3}(x - 3) \qquad m = \frac{5}{3} \text{ and } (x_1, y_1) = (3, -4)$$

$$y + 4 = \frac{5}{3}(x - 3)$$

$$3(y + 4) = 5(x - 3) \qquad \text{Multiply by 3}$$

$$3y + 12 = 5x - 15$$

$$-5x + 3y + 27 = 0$$

$$5x - 3y - 27 = 0 \qquad \text{General form}$$

PRACTICE EXERCISE 2

Find the general form of the equation of the line through (−5, 6) and (−1, −4).

Answer: $5x + 2y + 13 = 0$

5 SLOPE-INTERCEPT FORM

By substituting the y-intercept (0, b) for (x_1, y_1) in the point-slope form, we obtain

$$y - b = m(x - 0)$$
$$y = mx + b.$$

> **Slope-Intercept Form**
>
> An equation for the line with slope m and y-intercept (0, b) is
>
> $$y = mx + b.$$

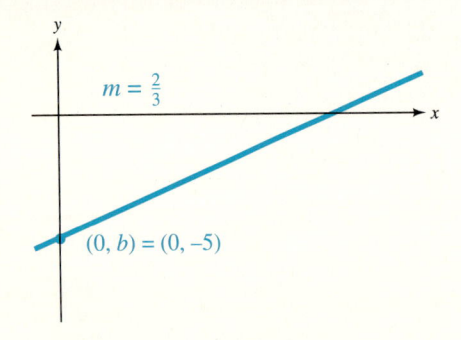

Figure 3.16 Slope-Intercept Form

Thus, if m and (0, b) are given, we can find an equation for the line. For example, if $m = \frac{2}{3}$ and the y-intercept is (0, −5) as in Figure 3.16, then

$$y = mx + b$$
$$y = \frac{2}{3}x - 5.$$

To put the equation in general form,

$$3y = 2x - 15 \quad \text{Multiply by 3}$$
$$-2x + 3y + 15 = 0$$
$$2x - 3y - 15 = 0. \quad \text{General form}$$

This shows that we can obtain the general form from the slope-intercept form.

Conversely, if we are given the general form, $ax + by + c = 0$, we can find the slope and y-intercept by writing the equation in slope-intercept form.

EXAMPLE 3 FINDING SLOPE AND y-INTERCEPT	PRACTICE EXERCISE 3

Find the slope and y-intercept of the line $3x + 7y - 14 = 0$.

Since the equation is in general form, we first convert it to the slope-intercept form, and then read m and b directly.

$$3x + 7y - 14 = 0$$
$$7y = -3x + 14$$
$$y = -\frac{3}{7}x + \frac{14}{7}$$
$$y = -\frac{3}{7}x + 2$$

Thus, $m = -\frac{3}{7}$ (not $-\frac{3}{7}x$) and $(0, b) = (0, 2)$.

Find the slope and y-intercept for $6x - 2y + 22 = 0$.

Answer: $m = 3$; $(0, b) = (0, 11)$

⑥ PARALLEL AND PERPENDICULAR LINES

By writing each of two different lines in slope-intercept form, we can determine if they are *parallel* (they never intersect—see Figure 3.17) or *perpendicular* (they intersect at right angles—see Figure 3.18). The following theorem is needed.

Parallel and Perpendicular Lines

Given two distinct lines with slopes m_1 and m_2:

1. The lines are **parallel** if and only if $m_1 = m_2$ (the slopes are equal).
2. The lines are **perpendicular** if and only if $m_1 = -\frac{1}{m_2}$ or $m_1 m_2 = -1$ (the slopes are negative reciprocals).

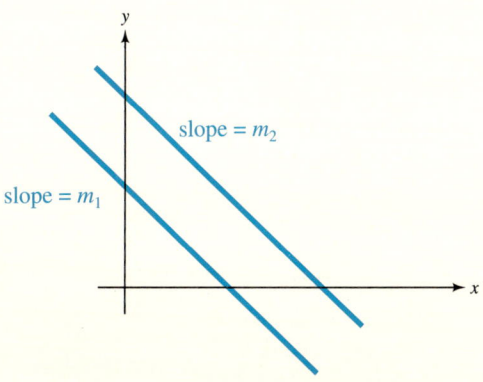

Figure 3.17 Parallel Lines ($m_1 = m_2$)

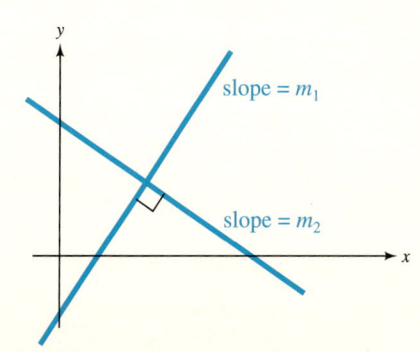

Figure 3.18 Perpendicular Lines ($m_1 m_2 = -1$)

EXAMPLE 4 Relationships Between Two Lines

(a) Decide if the lines $3x - 2y + 5 = 0$ and $10x + 15y = 7$ are parallel or perpendicular.

$$3x - 2y + 5 = 0 \qquad\qquad 10x + 15y = 7$$
$$-2y = -3x - 5 \qquad\qquad 15y = -10x + 7$$
$$y = \frac{3}{2}x + \frac{5}{2} \qquad\qquad y = -\frac{10}{15}x + \frac{7}{15}$$
$$\text{slope} = \frac{3}{2} \qquad\qquad y = -\frac{2}{3}x + \frac{7}{15}$$
$$\text{slope} = -\frac{2}{3}$$

Since the slopes are $\frac{3}{2}$ and $-\frac{2}{3}$ (negative reciprocals) the lines are perpendicular.

(b) Decide if the lines $6x + 3y + 2 = 0$ and $6y = -12x - 4$ are parallel or perpendicular.

$$6x + 3y + 2 = 0 \qquad\qquad 6y = -12x - 4$$
$$3y = -6x - 2 \qquad\qquad y = -\frac{12}{6}x - \frac{4}{6}$$
$$y = -\frac{6}{3}x - \frac{2}{3} \qquad\qquad y = -2x - \frac{2}{3}$$
$$y = -2x - \frac{2}{3}$$

Since the slopes are the same (both -2), we are tempted to conclude that the lines are parallel. However, we observe that the y-intercepts are also equal (both are $\left(0, -\frac{2}{3}\right)$). Thus, the equations determine the same line.

Practice Exercise 4

Decide if the lines are parallel or perpendicular.

(a) $5x - 2y + 8 = 0$ and $-10x + 4y - 12 = 0$

(b) $-3x - 6y + 7 = 0$ and $10x - 5y - 2 = 0$

Answers: (a) parallel (b) perpendicular

In Example 4(b) there are two equations for the same line. In this case the lines **coincide**.

EXAMPLE 5 Finding the Equation of a Line

Find the general form of the equation of a line with y-intercept $(0, -2)$ perpendicular to a line with slope $-\frac{4}{5}$.

$$y = mx + b$$
$$y = \frac{5}{4}x - 2 \qquad m = -\frac{1}{-\frac{4}{5}} = \frac{5}{4} \text{ and } b = -2$$
$$4y = 5x - 8 \qquad \text{Multiply by 4}$$
$$-5x + 4y + 8 = 0$$
$$5x - 4y - 8 = 0$$

Practice Exercise 5

Find the general form of the equation of a line through $(-1, 6)$ and perpendicular to $x - 5y + 4 = 0$.

Answer: $5x + y - 1 = 0$

EXAMPLE 6 ENGINEERING PROBLEM

In the design of a machine part an engineer needs to find the equation of the perpendicular bisector of the line segment joining the points $(-3, 2)$ and $(3, -6)$.

The perpendicular bisector is the line perpendicular to the segment and passing through its midpoint. First calculate the slope of the line segment.

$$m = \frac{-6 - 2}{3 - (-3)} = \frac{-8}{6} = -\frac{4}{3}$$

The slope of the desired line is the negative reciprocal of $-\frac{4}{3}$.

$$-\frac{1}{m} = -\frac{1}{-\frac{4}{3}} = \frac{3}{4}$$

Now find the midpoint.

$$(\bar{x}, \bar{y}) = \left(\frac{x_1 + x_2}{2}, \frac{y_1 + y_2}{2}\right) = \left(\frac{-3 + 3}{2}, \frac{2 + (-6)}{2}\right) = (0, -2)$$

Use the point-slope form of the equation of a line with $(0, -2)$ as the point and $\frac{3}{4}$ as the slope to find the equation needed in the design.

$$y - (-2) = \frac{3}{4}(x - 0)$$

$$y + 2 = \frac{3}{4}x$$

$4(y + 2) = 3x$ Multiply by 4

$4y + 8 = 3x$

$-3x + 4y + 8 = 0$

$3x - 4y - 8 = 0$ General form

PRACTICE EXERCISE 6

Find the perpendicular bisector of the line segment joining $(4, -7)$ and $(-2, -1)$.

Answer: $x - y - 5 = 0$

3.2 EXERCISES A

In Exercises 1–6 find the intercepts and graph the equation.

1. $3x - 2y - 6 = 0$

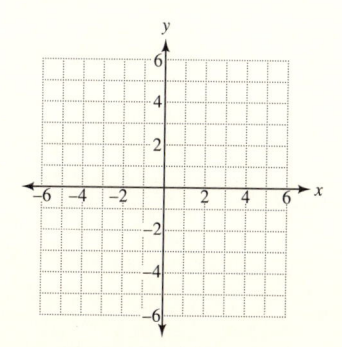

2. $4x + 3y = 12$

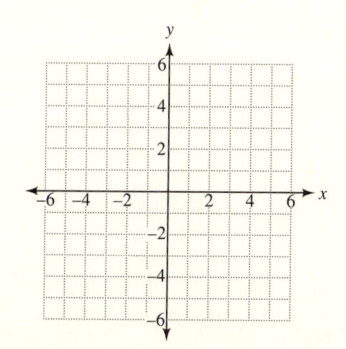

3. $2x - 3 = 0$

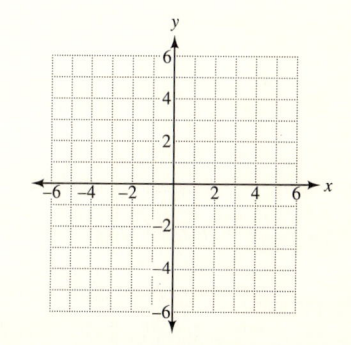

4. $y = 3$

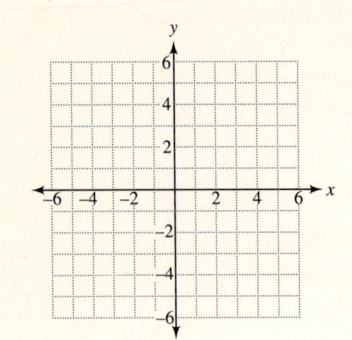

5. $3x - 2y = 0$

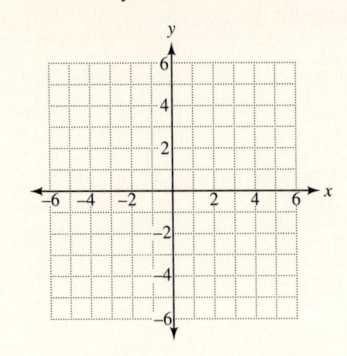

6. $-5x + 2y = 10$

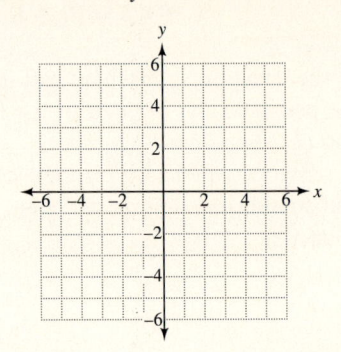

Find the slope of the line through the given points in Exercises 7–9.

7. $(6, -2)$ and $(4, 1)$

8. $(-3, 2)$ and $(-3, -4)$

9. $(-4, 5)$ and $(1, 5)$

In Exercises 10–16 find the general form of the equation of the line satisfying the given conditions.

10. With slope -2 passing through $(-3, 5)$

11. With slope $\frac{4}{3}$ and x-intercept $(3, 0)$

12. Horizontal line through $(6, -5)$

13. Passing through $(-3, 4)$ and $(2, 6)$

14. Passing through $(-3, 7)$ and perpendicular to $5x - 10y + 3 = 0$

15. Passing through $(2, 3)$ and parallel to $2y - 3 = 0$

16. With x-intercept $(-5, 0)$ and parallel to $7x + 4 = 0$

Find the slope and y-intercept of each line in Exercises 17–19.

17. $2x - 7y + 5 = 0$

18. $8x = -3y + 7$

19. $-3y + 6 = 0$

Determine if the pairs of lines in Exercises 20–22 are parallel or perpendicular.

20. $6x - 2y + 7 = 0$ and $x + 3y - 8 = 0$

21. $4x - 2y = -6$ and $10x = 5y + 8$

22. $3y - 8 = 0$ and $4x + 2 = 0$

23. Use the slope to determine if the points $P(4, 3)$, $Q(2, 0)$, and $R(-2, -6)$ are **collinear** (on the same line).

24. Beginning with the point-slope formula, derive the **two-point form** of the equation of a line:

$$y - y_1 = \frac{y_2 - y_1}{x_2 - x_1}(x - x_1).$$

Use the two-point form to find the general equation of a line through $(-3, 6)$ and $(2, -4)$.

25. Find the equation of the perpendicular bisector of the line segment joining $(1, 7)$ and $(-3, -5)$.

26. **CONSUMER** A new house was purchased for $85,000. After 5 years the value of the house was $105,000. Assume that the appreciation in value is given by a linear equation. Find the equation, and find the value of the house 7 years after it was originally purchased.

27. If the house in Exercise 26 is now worth $125,000 how many years ago was it purchased?

28. **ECONOMICS** If $1200 is borrowed at 9% simple interest, the sum to be paid at the end of n years is given by the linear equation

$$S = 1200(1 + 0.09n)$$

(a) Find S when $n = 3$.

(b) Find n when $S = \$1740$.

29. **CHEMISTRY** In a laboratory experiment 3 grams of sulfur were produced in 16 minutes and 8 grams in 36 minutes.
  (a) Use the number of grams (g) for y and time (t) for x to find a linear equation which fits this data.
  (b) Use the equation to find the number of grams produced in 20 minutes.
  (c) Why can you conclude that this equation is not accurate when $t = 0$? This illustrates the fact that empirical equations are often valid only for restricted values of the variable.

FOR REVIEW

In Exercises 30–31 find the distance between the two points and the midpoint of the line segment joining them.

30. $(-1, 2)$ and $(4, 6)$

31. $(4, -5)$ and $(-4, 2)$

194 CHAPTER 3 RELATIONS, FUNCTIONS, AND GRAPHS

In Exercises 32–33 give the values of x for which the given expression is not a real number.

32. $\dfrac{2}{(x+1)(x-5)}$

33. $\dfrac{1}{\sqrt{x+3}}$

ANSWERS:

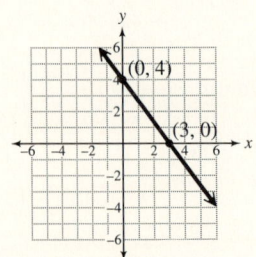

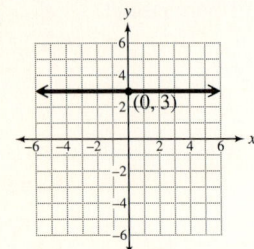

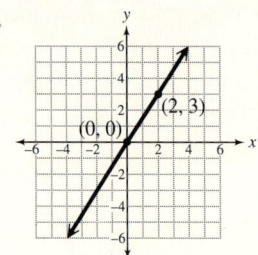

1. [graph: line through $(2,0)$ and $(0,-3)$]
2. [graph: line through $(0,4)$ and $(3,0)$]
3. [graph: vertical line through $(\tfrac{3}{2}, 0)$]
4. [graph: horizontal line through $(0,3)$]
5. [graph: line through $(0,0)$ and $(2,3)$]
6. [graph: line through $(-2,0)$ and $(0,5)$]

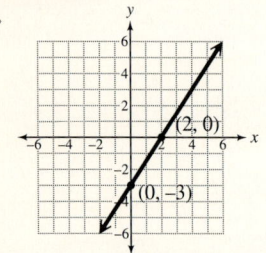

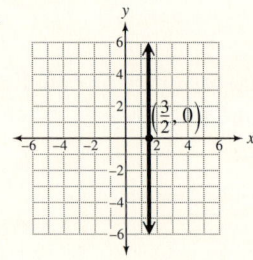

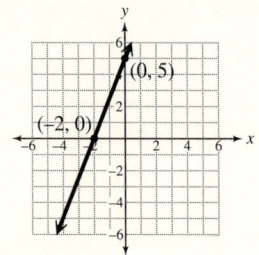

7. $-\tfrac{3}{2}$ 8. undefined 9. 0 10. $2x + y + 1 = 0$ 11. $4x - 3y - 12 = 0$ 12. $y + 5 = 0$ 13. $2x - 5y + 26 = 0$
14. $2x + y - 1 = 0$ 15. $y - 3 = 0$ 16. $x + 5 = 0$ 17. $\tfrac{2}{7}$; $(0, \tfrac{5}{7})$ 18. $-\tfrac{8}{3}$; $(0, \tfrac{7}{3})$ 19. 0; $(0, 2)$ 20. perpendicular
21. parallel 22. perpendicular 23. Yes. The slope of the line through P and Q is the same as the slope of the line through Q and R. 24. let $m = \dfrac{y_2 - y_1}{x_2 - x_1}$; $2x + y = 0$ 25. $x + 3y - 2 = 0$ 26. $y = 4000x + 85{,}000$; \$113,000 27. 10 yr
28. (a) \$1524 (b) 5 yr 29. (a) $g = \tfrac{1}{4}t - 1$ (g is used for y and t for x) (b) 4 gm (c) The grams produced at $t = 0$ should not be -1. 30. $\sqrt{41}$; $(\tfrac{3}{2}, 4)$ 31. $\sqrt{113}$; $(0, -\tfrac{3}{2})$ 32. $x = -1, 5$ 33. $x \leq -3$

3.2 EXERCISES B

In Exercises 1–6 find the intercepts and graph the equation.

1. $2x + 3y = 6$ **2.** $2x - 5y = 10$ **3.** $3x + 9 = 0$

4. $3y = -12$ **5.** $5x = -10y$ **6.** $4x + 2y = -8$

Find the slope of the line through the given points in Exercises 7–9.

7. $(5, 3)$ and $(-2, -6)$ **8.** $(7, -2)$ and $(-3, -2)$ **9.** $(4, -5)$ and $(4, 8)$

In Exercises 10–16 find the general form of the equation of the line satisfying the given conditions.

10. With slope $\tfrac{1}{3}$ passing through $(2, -1)$

11. With slope $\tfrac{1}{7}$ and y-intercept $(0, 5)$

12. Vertical line through $(6, -5)$

13. Passing through $(3, 0)$ and $(0, -5)$

14. Passing through (2, 1) and parallel to
$4x + 8y - 5 = 0$

15. Passing through (2, 3) and perpendicular to
$2y - 3 = 0$

16. With y-intercept $(0, -5)$ and perpendicular to
$7x + 4 = 0$

Find the slope and y-intercept of each line in Exercises 17–19.

17. $5x - 2y + 8 = 0$
18. $-3x - 2y = 6$
19. $4x + 11 = 0$

Determine if the pair of lines in Exercises 20–22 are parallel or perpendicular.

20. $2x - 5 = 0$ and $3x + 1 = 0$

21. $3x - 8y + 7 = 0$ and $4x + 6y - 15 = 0$

22. $7x - 21y + 8 = 0$ and $12x + 4y - 7 = 0$

23. Use slope to determine if the triangle with vertices $A(-2, 7)$, $B(-3, 3)$, and $C(-6, 8)$ is a right triangle.

24. Show that the **two-intercept form** of the equation of a line,

$$\frac{x}{a} + \frac{y}{b} = 1,$$

where $(a, 0)$ is the x-intercept and $(0, b)$ is the y-intercept, can be written in general form. Use the two-intercept form to find the general form of the equation of the line through $(-2, 0)$ and $(0, 5)$.

25. Find the general form of the equation of the perpendicular bisector of the portion of the line $x + y = 6$ which lies in the first quadrant.

26. **ECONOMICS** A new car was purchased for $14,500, and 3 years later it was worth $9100. Assume that the depreciation in value is given by a linear equation. Find the equation, and find the value of the car at the end of 6 years.

27. How many years will it take for the car in Exercise 26 to have a value of only $100?

28. **ECONOMICS** If $1200 is borrowed at x percent simple interest (in decimal form), the sum to be paid at the end of 2 years is

$$S = 1200(1 + 2x)$$

(a) Find S when $x = 0.12$ (12% interest).
(b) Find the interest rate when $S = \$1428$.

29. **ECOLOGY** An environmental scientist found 30 deer on 800 acres in one area and 50 deer on 1200 acres in another area. (a) Use the number of deer (d) for y and the number of acres (a) for x to find a linear equation relating the number of deer to the number of acres. (b) Find the number of deer to be expected on 960 acres. (c) Use 20 acres to explain why the equation is not accurate for a small number of acres.

FOR REVIEW

In Exercises 30–31 find the distance between the two points and the midpoint of the line segment joining them.

30. $(2, -3)$ and $(5, -1)$
31. $(7, 2)$ and $(-3, 6)$

In Exercises 32–33 give the values of x for which the given expression is not a real number.

32. $\dfrac{x + 1}{(x - 2)(x + 4)}$

33. $\dfrac{x + 2}{\sqrt{5 - x}}$

3.2 EXERCISES C

1. Find the distance between the point (x_0, y_0) and the point (x_0, y_1) where (x_0, y_1) is a point on the line $y = mx + b$.   [Answer: $|mx_0 + b - y_0|$]

2. Use the results of Exercise 1 to find the distance between $(1, 2)$ and $(1, y_1)$ where $(1, y_1)$ is a point on the line $y = 5x - 6$.

3. Find the point (x_0, y_0) which is the intersection of the line $y = x - 1$ and the line through $(2, -3)$ that is perpendicular to $y = x - 1$.

4. Find the distance between $(2, -3)$ and the point of intersection found in Exercise 3. This is the **distance from the point to the line.**

3.3 RELATIONS AND FUNCTIONS

STUDENT GUIDEPOSTS

1. Relation
2. Function
3. Types of Functions
4. Domain
5. Independent and Dependent Variables
6. Ordered-Pair Definitions

1 RELATION

Relationships or correspondences are all around us. For example,

1. To each person there corresponds a height.
2. To each company there corresponds a number of employees.
3. For a given amount of money borrowed there corresponds an amount to be paid back after one year.
4. To each number in {1, 2, 3, 4} there corresponds one or more numbers in the set which are greater than or equal to the number.

Definition of Relation

A correspondence between a first set of objects, the **domain,** and a second set of objects, the **range,** is called a **relation** if to each element of the domain there corresponds *one or more* elements in the range.

2 FUNCTION

The four examples given above are all relations. Notice that in the first three relations only one element in the range is related to each element in the domain. These are examples of *functions*.

Definition of Function

A **function** is a relation with the property that to each element in the domain there corresponds *one and only one* (exactly one) element in the range.

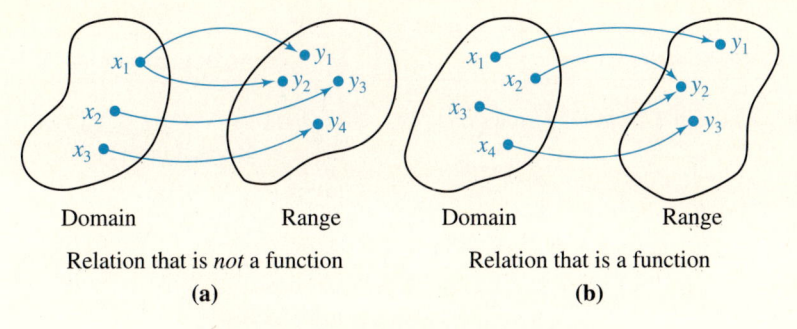

Figure 3.19 Relations and Functions

Figure 3.19 illustrates the difference between a relation that is not a function and a relation that is a function. The relation in Figure 3.19(a) has x_1 related to both y_1 and y_2 and thus is not a function. In Figure 3.19(b) the fact that both x_2 and x_3 are related to y_2 does not keep the relation from being a function.

To each person there corresponds one and only one height, and thus that relation is a function. Notice however, that the relation ≤ (less than or equal to) on the set {1, 2, 3, 4} is not a function since, for example,

$$2 \leq 2, \quad 2 \leq 3, \quad 2 \leq 4.$$

That is, 2 is related to three numbers, 2, 3, and 4.

All the linear equations in Section 3.2, except those representing vertical lines, provide examples of functions. If the equation can be written in the slope-intercept form, $y = mx + b$, then for each x there corresponds one and only one y. Letters such as f and g are frequently used for the names of functions. For example, for $y = 2x - 3$, we write

$$f(x) = 2x - 3$$

where $f(x)$ is read "f of x" or "the value of f at x." The instruction to find y or $f(x)$ when $x = -4$ is written "find $f(-4)$." Thus,

$$f(-4) = 2(-4) - 3 = -8 - 3 = -11.$$

This means that corresponding to the number $x = -4$ is the number $y = -11$ or $f(-4) = -11$.

❸ TYPES OF FUNCTIONS

The general class of functions that have as their graph a straight line and can be written

$$f(x) = mx + b \qquad \text{Linear function}$$

are called **linear functions.** Included in these functions are the **constant functions**

$$f(x) = b \qquad \text{Constant function}$$

and the **identity function**

$$f(x) = x. \qquad \text{Identity function}$$

Another function, studied in detail in Section 3.6, is the **quadratic function.**

$$f(x) = ax^2 + bx + c \qquad \text{Quadratic function}$$

EXAMPLE 1 FUNCTIONAL NOTATION

Find $f(0)$, $f(-3)$, $f(a)$, and $f(x + h)$ for each of the following functions.

(a) $f(x) = -5x + 3$

$f(0) = -5(0) + 3 = 3 \qquad f(a) = -5(a) + 3 = -5a + 3$

$f(-3) = -5(-3) + 3 = 18 \qquad f(x + h) = -5(x + h) + 3 = -5x - 5h + 3$

(b) $f(x) = 6$

$f(0) = 6 \qquad\qquad f(a) = 6$

$f(-3) = 6 \qquad\qquad f(x + h) = 6 \qquad f\text{ (anything)} = 6$

(c) $f(x) = x$

$f(0) = 0 \qquad\qquad f(a) = a$

$f(-3) = -3 \qquad\quad f(x + h) = x + h$

(d) $f(x) = -2x^2 + 5x - 4$

$f(0) = -2(0)^2 + 5(0) - 4 = -4$

$f(-3) = -2(-3)^2 + 5(-3) - 4 = -18 - 15 - 4 = -37$

$f(a) = -2(a)^2 + 5(a) - 4 = -2a^2 + 5a - 4$

$f(x + h) = -2(x + h)^2 + 5(x + h) - 4$

$\qquad\quad = -2(x^2 + 2xh + h^2) + 5x + 5h - 4$

$\qquad\quad = -2x^2 - 4xh - 2h^2 + 5x + 5h - 4$

$\qquad\quad = -2x^2 + (5 - 4h)x - 2h^2 + 5h - 4$

PRACTICE EXERCISE 1

Find $f(2)$, $f(-1)$, $f(a - 1)$, and $f(a + h)$ for each function.

(a) $f(x) = 2x - 4$

(b) $f(x) = -5$

(c) $f(x) = -x$

(d) $f(x) = 3x^2 - 2x + 8$

Answers: (a) 0, -6, $2a - 6$, $2a + 2h - 4$ (b) -5, -5, -5, -5 (c) -2, 1, $-a + 1$, $-a - h$ (d) 16, 13, $3a^2 - 8a + 13$, $3a^2 + 6ah + 3h^2 - 2a - 2h + 8$

④ DOMAIN

To define completely a function we need not only to give the correspondence, but also to specify the domain. When the domain is not stated, we shall assume that it consists of all real numbers for which the function is defined. The linear and quadratic functions are defined for all real numbers. However, if

$$g(x) = \sqrt{x + 2}, \qquad \text{Defined for } x \geq -2$$

then the domain of g is all real numbers x such that $x + 2 \geq 0$ or $x \geq -2$. Notice that the range of g is the set of real numbers $g(x) \geq 0$.

EXAMPLE 2 DETERMINING DOMAIN

Find the set of all real numbers for which the function is defined.

(a) $h(x) = \dfrac{1}{x + 2}$

This function is defined for all real numbers except when $x + 2 = 0$. Thus, the domain is the set of all real numbers except $x = -2$.

(b) $k(x) = \dfrac{x - 3}{\sqrt{x^2 + 1}}$

This function is defined for all real numbers since $x^2 + 1 > 0$ for all x. Thus, the domain is the set of all real numbers.

PRACTICE EXERCISE 2

Find the set of real numbers for which the function is defined.

(a) $g(x) = \dfrac{x + 2}{x - 5}$

(b) $f(x) = \dfrac{x + 1}{\sqrt{1 - x}}$

(c) $f(x) = \dfrac{\sqrt{3-x}}{(x-1)(x+2)}$

There are three conditions: $3 - x \geq 0$ or $3 \geq x$
$x - 1 \neq 0$ or $x \neq 1$
$x + 2 \neq 0$ or $x \neq -2$

Thus, the domain is the set of all real numbers such that $x \leq 3$ and $x \neq 1$ and $x \neq -2$.

(c) $h(x) = \dfrac{\sqrt{2+x}}{(x+6)(x-7)}$

Answers: (a) $x \neq 5$ (b) $x < 1$
(c) $x \geq -2$, $x \neq -6$, and $x \neq 7$

CAUTION

As Example 2 illustrates, we must exclude from a domain numbers that make a denominator zero or give a negative value under an even-indexed radical.

EXAMPLE 3 APPLICATION TO RETAILING

A retailer received a shipment of shirts which cost him $32.00 each. In order to find the total price that his customers will have to pay for a given markup, he wants a relation between total price T and the percent markup x. **(a)** Find the total price function T if the markup is x percent and there is a 5% sales tax on the marked price. **(b)** Find the total price when the markup is 25%.

(a) Let x = percent markup in decimal notation.
The selling price of each shirt is the cost plus markup.

$$32 + 32x = 32(1 + x) \quad \text{Selling price}$$

The total price is $32(1 + x)$ plus 5% of $32(1 + x)$.

$$\begin{aligned}
T(x) &= 32(1 + x) + 0.05(32)(1 + x) \quad \text{Selling price plus tax}\\
&= 32(1 + x)[1 + 0.05]\\
&= 32(1.05)(1 + x)\\
&= 33.6(1 + x) \quad \text{Total price as a function of the markup rate}
\end{aligned}$$

(b) $T(x) = 33.6(1 + x)$

$$\begin{aligned}
T(0.25) &= 33.6(1 + 0.25) \quad 25\% = 0.25\\
&= \$42.00 \quad \text{Price with tax}
\end{aligned}$$

Thus, if each shirt is marked up 25% the total price is $42.00.

PRACTICE EXERCISE 3

A retailer buys dresses for $65.50 each.

(a) Find the total price function T if the markup is x percent and there is a 6% sales tax on the marked price.

(b) Find $T(0.32)$.

Answers: (a) $T(x) = 69.43(1 + x)$
(b) $91.65

EXAMPLE 4 ENGINEERING PROBLEM

Cylindrical storage tanks that are 10 m long and have a radius of r meters are to be painted (including both ends). **(a)** If the paint job costs $0.22 per square meter, express the cost C of painting a tank as a function of the radius r. **(b)** Find the cost of painting a tank with radius 2.6 m.

(a) Let r = radius of the cylindrical tank.

PRACTICE EXERCISE 4

Hollow spheres of radius r feet are to be filled with a liquid costing $0.75 per cubic foot and insulated at a cost of $1.50 per square foot.

(a) Find the cost C of filling and insulating a sphere as a function of r.

Then $2\pi r^2$ = area of both ends of the tank,

$2\pi r(10)$ = area of sides of the tank,

$2\pi r^2 + 20\pi r$ = total surface area of the tank.

The total cost C is the cost per square meter times the surface area.

$$C(r) = 0.22[2\pi r^2 + 20\pi r]$$
$$= 0.22(2\pi)r(r + 10)$$
$$\approx 1.38r(r + 10) \quad \text{Required function}$$

(b) Now find $C(2.6)$.

$$C(2.6) = 1.38(2.6)(2.6 + 10)$$
$$\approx \$45.21$$

The cost of painting a cylinder with radius 2.6 m is approximately $45.21.

(b) Find $C(2.5)$.

Answers: (a) $C(r) = \pi r^2(r + 6)$
(b) $166.90

⑤ INDEPENDENT AND DEPENDENT VARIABLES

If we write the function described by $f(x) = 2x - 1$ as $y = 2x - 1$, we stress that for each x in the domain there is a corresponding y in the range. Thus y depends on x, and we call x the **independent variable** and y the **dependent variable.** The set of all values of the independent variable is the domain of the function, and the set of all values of the dependent variable is the range.

⑥ ORDERED-PAIR DEFINITIONS

Consider again $f(x) = 2x - 1$ or $y = 2x - 1$ and assume that the domain is $\{1, 2, 3\}$. Then

$$f(1) = 1 \text{ or } x = 1 \text{ implies } y = 1,$$
$$f(2) = 3 \text{ or } x = 2 \text{ implies } y = 3,$$

and $\quad f(3) = 5$ or $x = 3$ implies $y = 5$.

The function is then completely described by the set of ordered pairs $\{(1,1), (2, 3), (3, 5)\}$. Since any relation can be described in this manner, we can give the following alternative definition of relation and function.

> **Definition of Relation and Function (Using Ordered Pairs)**
>
> Any set of ordered pairs of numbers is called a **relation.** A **function** is a relation with the property that no two ordered pairs have the same x-coordinate and different y-coordinates.

EXAMPLE 5 DETERMINING FUNCTIONS

State which of the following relations are functions.

(a) $\{(3, 4), (-1, 6), (3, 5)\}$

This relation is not a function because $(3, 4)$ and $(3, 5)$ have the same first coordinate and different second coordinates. Thus, 3 is related to both 4 and 5.

PRACTICE EXERCISE 5

State which relations are functions.

(a) $\{(1, 1), (2, 3), (1, 0)\}$

(b) $\{(0, 0), (1, 1), (-2, 5), (6, 5)\}$
This is a function because no two ordered pairs have the same first and different second coordinates. The fact that -2 and 6 are both related to 5 is allowed in a function.

(b) $\{(1, 4), (5, 4)\}$

Answers: (a) not a function
(b) function

3.3 EXERCISES A

State which of the relations given in Exercises 1–3 are functions if x is the independent variable and y is the dependent variable.

1. $x = 2y + 5$

2. $y = 5x^2$

3. $y = \dfrac{3x^2 - 2}{x + 1}$

In Exercises 4–6 the function is described completely by giving the relation and the domain. Give the range.

4. $f(x) = 5x - 3$; $\{1, 2, 3\}$

5. $g(x) = \sqrt{x + 2}$; $x \geq -2$

6. $h(x) = \dfrac{1}{x - 3}$; all real numbers except 3

Determine all the real numbers for which the functions in Exercises 7–9 are defined.

7. $f(x) = 7x - 8$

8. $g(x) = \sqrt{2 - x}$

9. $k(x) = \dfrac{\sqrt{x + 8}}{(x + 2)(x - 3)}$

Use the most appropriate word, identity, constant, linear, or quadratic, to describe each function in Exercises 10–12.

10. $f(x) = x$

11. $h(x) = 1 - x$

12. $k(x) = 14$

In Exercises 13–15 find $f(0)$, $f(1)$, $f(-3)$, $f(a)$, and $f(a - 1)$.

13. $f(x) = 5x - 3$

14. $f(x) = 7$

15. $f(x) = |x - 5|$

In Exercises 16–17 find $h(a^2)$, $h\left(\dfrac{1}{a}\right)$, and $\dfrac{1}{h(a)}$.

16. $h(x) = 3x - 7$

17. $h(x) = \dfrac{1}{x} + 6$

In Exercises 18–19 find $f(1 + h)$, $\dfrac{f(1 + h) - f(1)}{h}$, $f(x + h)$, and $\dfrac{f(x + h) - f(x)}{h}$.

18. $f(x) = x + 5$

19. $f(x) = x^2 + 2$

State which of the relations in Exercises 20–23 are functions.

20. $\{(1, 2), (-1, 3)\}$

21. $\{(2, 5), (4, 6), (2, 7)\}$

22. $\{(3, -1), (4, -1), (5, -1)\}$

23. $\{(4, 3), (4, 2), (4, 8), (4, 9)\}$

202 CHAPTER 3 RELATIONS, FUNCTIONS, AND GRAPHS

24. RETAILING A retailer wishes to mark up dresses by 28.5%.
  (a) If c is the cost of each dress, give the function S that represents the selling price.
  (b) What is the selling price of a dress that costs $56?

25. RETAILING The Pant Shop usually has a markup of 25% on pants. Those that do not sell are discounted a percent based on the selling price.
  (a) If x represents this percent as a decimal, give a function S for the discount price of pants which cost $40 per pair.
  (b) What would be a function T representing the total price if a 5% sales tax were added?
  (c) Find the percent discount that makes the discount price (excluding sales tax) the same as the original cost.

26. ENGINEERING Spherical tanks of radius r are to be insulated at a cost of $1.25 per square foot.
  (a) Find the total cost T for the insulation job as a function of r.
  (b) Find the total cost when $r = 9.6$ ft.

27. MINING Empirical data has shown that the cost in dollars of extracting x percent of metal from a ton of ore is given by the function $C(x) = 265.9x^2 - 36.8$, where $0.75 \le x \le 0.95$.
  (a) Find the range of this function over the domain $0.75 \le x \le 0.95$.
  (b) Use the number $x = 0.30$ outside the domain to show why the function is restricted to its domain.

FOR REVIEW

28. Find the equation of the line passing through $(-1, 3)$ and $(2, 4)$.

29. Find the equation of the line passing through $(8, 5)$ and perpendicular to $2y - 7 = 0$.

30. Find the intercepts and graph the equation $2x - 5 = 0$.

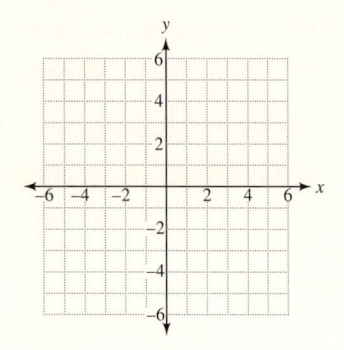

ANSWERS: 1. function 2. function 3. function 4. $\{2, 7, 12\}$ 5. all nonnegative real numbers 6. all real numbers except 0 7. all real numbers 8. $x \le 2$ 9. $x \ge -8$ and $x \ne -2$ and $x \ne 3$ 10. identity 11. linear 12. constant 13. $-3, 2, -18, 5a - 3, 5a - 8$ 14. $7, 7, 7, 7, 7$ 15. $5, 4, 8, |a - 5|, |a - 6|$ 16. $3a^2 - 7, \frac{3}{a} - 7, \frac{1}{3a - 7}$ 17. $\frac{1}{a^2} + 6, a + 6, \frac{a}{1 + 6a}$ 18. $h + 6, 1, x + h + 5, 1$ 19. $h^2 + 2h + 3, h + 2, x^2 + 2xh + h^2 + 2, 2x + h$ 20. function 21. not a function 22. function 23. not a function 24. (a) $S(c) = 1.285c$, (b) $71.96 25. (a) $S(x) = 50(1 - x)$, (b) $T(x) = 52.5(1 - x)$, (c) 20% 26. (a) $T(r) = 5\pi r^2$ (b) $1447.65 27. (a) Range is given by $112.77 \le c(x) \le $203.17 (b) For $x = 0.30$ the cost would be $-$12.87 which is clearly not reasonable. 28. $x - 3y + 10 = 0$ 29. $x - 8 = 0$ 30.

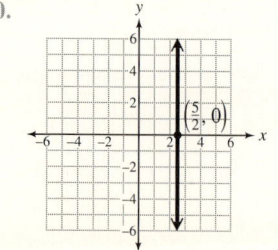

3.3 EXERCISES B

State which of the relations given in Exercises 1–3 are functions if x is the independent variable and y is the dependent variable.

1. $y = 5x^2$
2. $y = \sqrt{x-1}$
3. $x^2 = \dfrac{y-1}{y+1}$

In Exercises 4–6 the function is described completely by giving the relation and the domain. Give the range.

4. $f(x) = x^2 - 1$; $\{-1, 0, 1, 2\}$
5. $g(x) = \dfrac{1}{x}$; $x > 0$
6. $h(x) = \dfrac{-1}{x^2}$; all real numbers except 0

Determine all the real numbers for which the functions in Exercises 7–9 are defined.

7. $f(x) = 3x^2 + 2x + 1$
8. $h(x) = \dfrac{(x+5)(x-5)}{x^2 - 25}$
9. $k(x) = \dfrac{\sqrt{8-x}}{x(x+5)}$

Use the most appropriate word, identity, constant, linear, or quadratic, to describe each function in Exercises 10–12.

10. $g(x) = x^2 + 2$
11. $h(x) = -x$
12. $k(x) = (x+1)(x-1)$

In Exercises 13–15 find $f(0)$, $f(1)$, $f(-3)$, $f(a)$, and $f(a-1)$.

13. $f(x) = x^2 - 10$
14. $f(x) = -8$
15. $f(x) = |x| - 5$

In Exercises 16–17 find $h(a^2)$, $h\left(\dfrac{1}{a}\right)$, and $\dfrac{1}{h(a)}$.

16. $h(x) = 2x^2 - 5$
17. $h(x) = \dfrac{1}{x+6}$

In Exercises 18–19 find $f(1+h)$, $\dfrac{f(1+h) - f(1)}{h}$, $f(x+h)$, and $\dfrac{f(x+h) - f(x)}{h}$.

18. $f(x) = -3x - 4$
19. $f(x) = 3x^2 - 5x - 7$

State which of the relations in Exercises 20–23 are functions.

20. $\{(2, 1), (3, -1)\}$
21. $\{(5, 2), (6, 4), (7, 2)\}$
22. $\{(-1, 3), (-1, 4), (-1, 5)\}$
23. $\{(3, 4), (2, 4), (8, 4), (9, 4)\}$

24. **RETAILING** Lori's Shop has discounted sweaters by 12.5%.
    (a) If s is the selling price of a sweater, find a function P that represents the discounted price.
    (b) What would be the sale price of a sweater that was originally priced at $44.00?

25. **RETAILING** The Appliance Dealer uses a markup of x percent (x in decimal notation) on all products. A sale with markdowns of 50% starts next month.
    (a) Find a function S that represents the sale price of a toaster that originally cost $30.
    (b) What would be a function T representing the total price if a 6% sales tax were added?
    (c) What would have been the markup if the sale price (excluding sales tax) were to be $3.00 below the original cost?

26. **AGRICULTURE** It costs 0.03 cents ($0.0003) per liter to fill a spherical tank of radius r meters with a liquid fertilizer.
    (a) Find the cost C (in dollars) as a function of r.
    (b) Find C for a tank with radius 5.2 m. [*Hint*: 1 m³ = 1000 liters]

27. **ECOLOGY** The cost of cleaning up a small oil spill has been found to be given by the function $C(x) = \dfrac{582.6}{1-x} - 726.4$, $0.5 \leq x \leq 0.9$, where x is the percent of the spill removed.
    (a) Give the range of the function for the domain $0.5 \leq x \leq 0.9$.
    (b) Use a cleanup of 15% to show why the function is restricted to its domain.

FOR REVIEW

28. Find the equation of the line with y-intercept $(0, -3)$ and perpendicular to $4x + 2y - 7 = 0$.

29. Find the equation of the perpendicular bisector of the line segment joining the intercepts of $5x - 2y = 10$.

30. Find the slope of the line through the two points $(-2, 4)$ and $(-8, -3)$.

3.3 EXERCISES C

In Exercises 1–4 find $\frac{f(x + h) - f(x)}{h}$. Simplify your answer as much as possible.

1. $f(x) = \dfrac{1}{x + 1}$

2. $f(x) = \dfrac{x}{x - 1}$

3. $f(x) = \sqrt{x - 3}$

$\left[\text{Answer: } \dfrac{1}{\sqrt{x + h - 3} + \sqrt{x - 3}}\right]$

4. $f(x) = \dfrac{1}{\sqrt{x + 5}}$

$\left[\text{Answer: } -\dfrac{1}{\sqrt{x + h + 5}\sqrt{x + 5}(\sqrt{x + 5} + \sqrt{x + h + 5})}\right]$

3.4 PROPERTIES OF FUNCTIONS AND TRANSFORMATIONS

STUDENT GUIDEPOSTS

1. Vertical Line Test
2. Increasing and Decreasing Functions
3. Shifts in Graphs
4. Even and Odd Functions
5. Tests for Symmetry
6. Noncontinuous Graphs

1 VERTICAL LINE TEST

We now investigate the nature of the graph of a function. By definition a function has the property that for each x in the domain there is one and only one y in the range. Thus, if we draw a vertical line through any point on the x-axis it will cross the graph of a function in at most *one* point. Using this **vertical line test** on the graphs in Figure 3.20, we see that **(a)** and **(b)** are the graphs of functions but **(c)** and **(d)** are not.

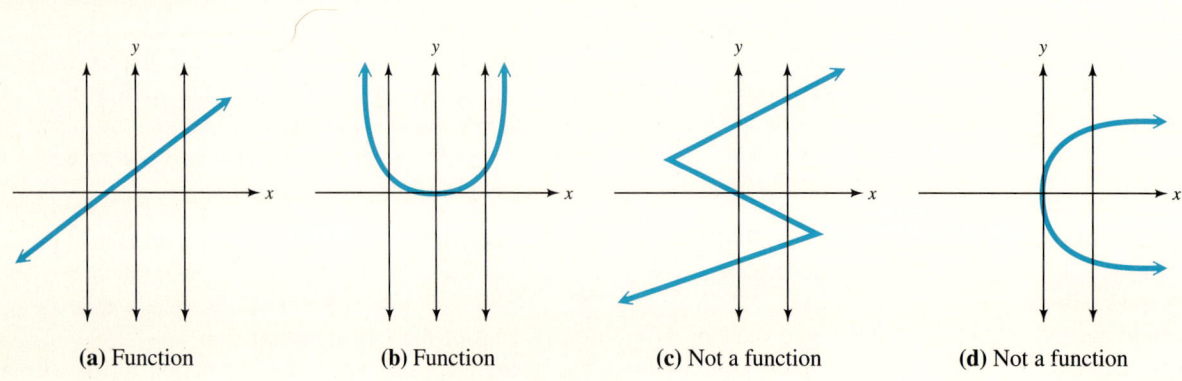

(a) Function **(b)** Function **(c)** Not a function **(d)** Not a function

Figure 3.20 Vertical Line Test

❷ INCREASING AND DECREASING FUNCTIONS

The graph of a function can be classified as either *increasing, decreasing,* or *constant.* The graph of a linear function will be increasing if the slope is positive, decreasing if the slope is negative, and constant if the slope is zero. The graphs of $f(x) = \frac{1}{2}x - 3$, $g(x) = -2x + 4$, and $h(x) = 3$ illustrate these three cases in Figure 3.21.

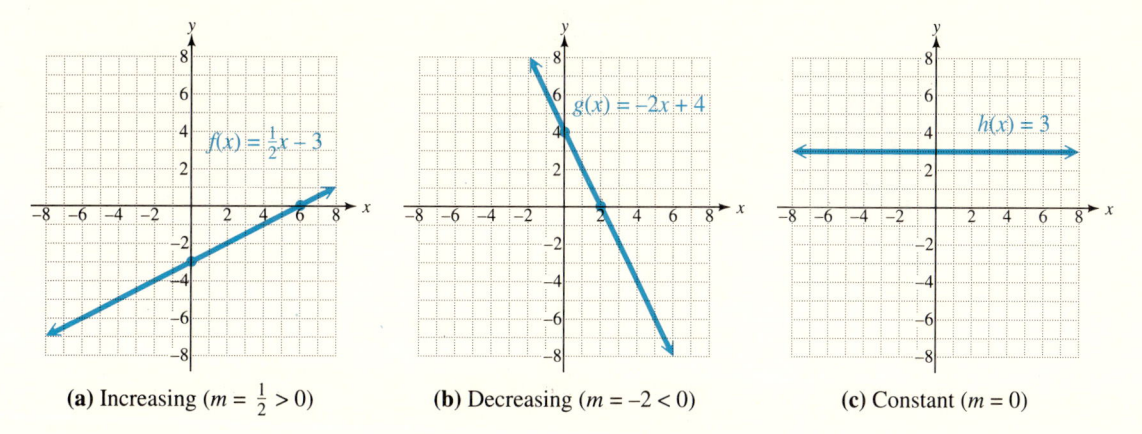

(a) Increasing ($m = \frac{1}{2} > 0$) **(b)** Decreasing ($m = -2 < 0$) **(c)** Constant ($m = 0$)

Figure 3.21 Linear Functions

Increasing, Decreasing, and Constant Functions

If a and b are any numbers in an interval I in the domain of f, then

1. f is **increasing** on I if whenever $a < b$, then $f(a) < f(b)$.
2. f is **decreasing** on I if whenever $a < b$, then $f(b) < f(a)$.
3. f is **constant** on I if $f(a) = f(b)$ for all a and b in I.

Thus, for an increasing function, y increases as x increases, and for a decreasing function, y decreases as x increases.

The linear functions in Figure 3.21 had one of these properties for all values of x. Some functions will increase on one interval and decrease on another. Consider the **absolute value function,**

$$f(x) = |x| = \begin{cases} x, & x \geq 0 \\ -x, & x < 0. \end{cases}$$

To graph this function, graph $f(x) = x$ for all $x \geq 0$ and $f(x) = -x$ for all $x < 0$. The graph is shown in Figure 3.22. Notice that for $x \leq 0$ the function is decreasing while for $x \geq 0$ it is increasing.

x	$f(x)$
-5	5
-1	1
0	0
1	1
5	5

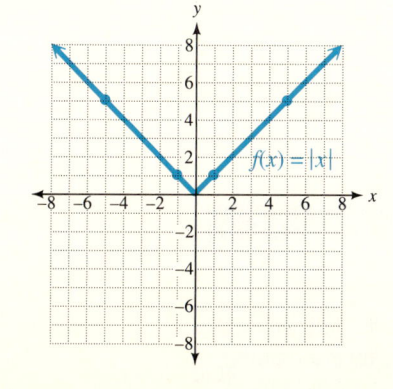

Figure 3.22 Absolute Value Function

(b) $h(x) = |x - 2| + 1$

This is a function of the form $f(x - c_1) + c_2$ where $c_1 = 2$ and $c_2 = 1$. This means that the graph of $f(x)$ is shifted two units to the right and one unit upward as shown in Figure 3.26(b).

(b) $h(x) = -|x + 2| - 1$

Answers: (a) up 2; right 1
(b) down 1; left 2

EXAMPLE 3 GRAPHING OTHER SHIFTS

Using the graph of $f(x) = x^2$ in Figure 3.27(a), find the graph of the given function.

(a) $g(x) = -x^2$

Notice that each ordinate of $g(x) = -x^2$ is the negative of the corresponding ordinate of $f(x) = x^2$. Thus, the graph is the reflection of $f(x)$ across the x-axis. See Figure 3.27(b).

(b) $h(x) = 3x^2$

Each ordinate of $f(x) = x^2$ is multiplied by a factor of 3. Thus, for any x the ordinate of $h(x)$ is 3 times the ordinate of $f(x)$. See Figure 3.27(c).

(c) $k(x) = \frac{1}{2}x^2$

Each ordinate of $f(x) = x^2$ is multiplied by 1/2 or divided by 2. Thus, for any x the graph of $h(x)$ is 1/2 the height of $f(x)$. See Figure 3.27(d).

PRACTICE EXERCISE 3

Graph each function using shifts of $f(x) = x^2$.

(a) $g(x) = 2x^2$

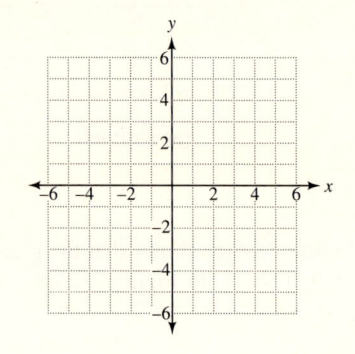

(b) $h(x) = -2x^2$

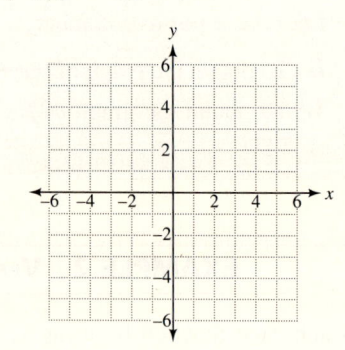

(c) $k(x) = \frac{1}{4}x^2$

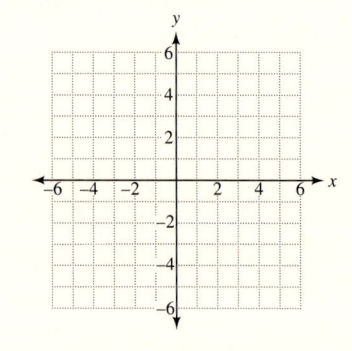

Answers: (a) ordinate times 2
(b) ordinate times -2
(c) ordinate times 1/4

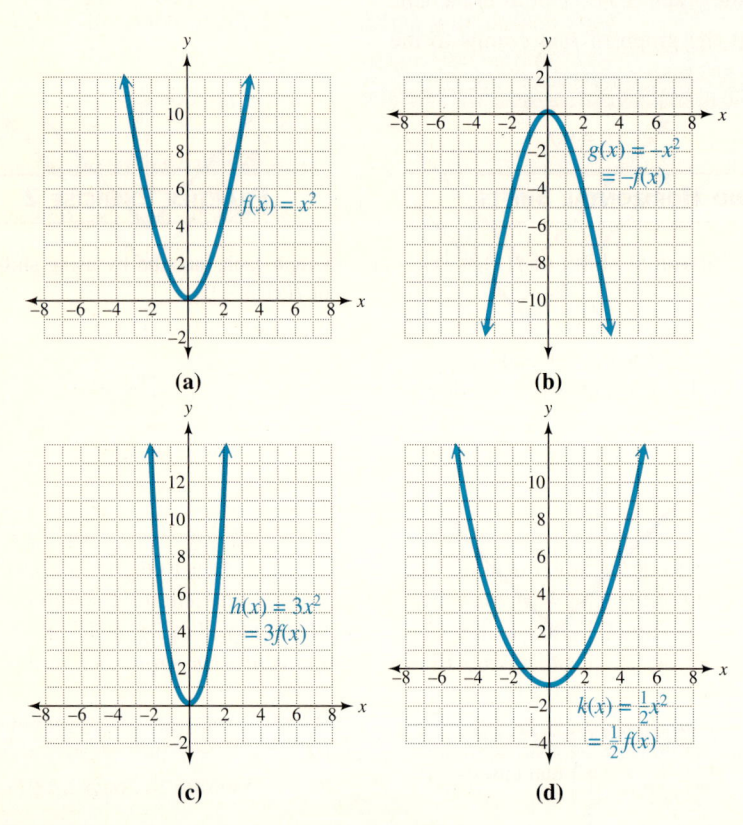

Figure 3.27

❹ EVEN AND ODD FUNCTIONS

Each of the graphs in Figure 3.27 is symmetric with respect to the y-axis. Notice that $f(x) = f(-x)$, that is, the graph has the same ordinate for x and for $-x$.

> **Even and Odd Functions**
>
> If for all x in the domain of f,
>
> 1. $f(x) = f(-x)$ then f is an **even function;**
> 2. $f(x) = -f(-x)$ then f is an **odd function.**

EXAMPLE 4 FINDING EVEN AND ODD FUNCTIONS

Decide if the given function is even or odd.

(a) $f(x) = -x^2 + 3$

We must determine if $f(x) = f(-x)$ or $f(x) = -f(-x)$.

$$f(-x) = -(-x)^2 + 3 = -x^2 + 3 = f(x)$$

Thus f is even. The graph shown in Figure 3.28(a) is symmetric with respect to the y-axis. Thus, for $(x, f(x))$ on the graph $(-x, f(x))$ is also on the graph.

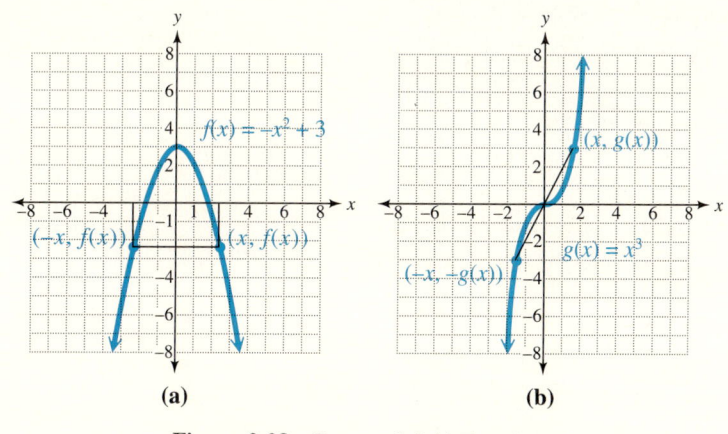

Figure 3.28 Even and Odd Functions

(b) $g(x) = x^3$

$$g(-x) = (-x)^3 = -x^3 = -g(x)$$

Since $g(-x) = -g(x)$ is equivalent to $g(x) = -g(-x)$, the function is odd. Notice that the graph of g in Figure 3.28(b) is symmetric with respect to the origin. That is, for $(x, g(x))$ on the graph $(-x, -g(x))$ is also on the graph.

(c) $h(x) = x^3 + x^2 - 1$

$$h(-x) = (-x)^3 + (-x)^2 - 1 = -x^3 + x^2 - 1$$

Since $h(-x) \neq h(x)$ and $h(-x) \neq -h(x)$, the function is neither even nor odd.

PRACTICE EXERCISE 4

Decide if each function is even or odd.

(a) $f(x) = x^4 + 2x^2$

(b) $g(x) = 2x^3 + 1$

(c) $h(x) = x - 2x^3 + x^5$

Answers: (a) even (b) neither (c) odd

5 TESTS FOR SYMMETRY

For relations in general we test symmetry with respect to the y-axis by replacing x with $-x$. If the equation is unchanged the graph is symmetric with respect to the y-axis. For example, consider the equation of a circle with center at the origin and radius 3, $x^2 + y^2 = 9$. Replace x with $-x$.

$$(-x)^2 + y^2 = 9$$
$$x^2 + y^2 = 9$$

Since $(-x)^2 = x^2$ the graph is symmetric with respect to the y-axis. Replace y with $-y$.

$$x^2 + (-y)^2 = 9$$
$$x^2 + y^2 = 9$$

This means that the graph is symmetric with respect to the x-axis. Observe that a graph that is symmetric with respect to the x-axis cannot be the graph of a function.

The graph of $x^2 + y^2 = 9$ is also symmetric with respect to the origin. Replace x with $-x$ and y with $-y$ to check this.

$$(-x)^2 + (-y)^2 = 9$$
$$x^2 + y^2 = 9$$

Tests for Symmetry

If the equation remains unchanged when:

1. x is replaced by $-x$, there is symmetry with respect to the y-axis.
2. y is replaced by $-y$, there is symmetry with respect to the x-axis.
3. x is replaced by $-x$ and y is replaced by $-y$, there is symmetry with respect to the origin.

EXAMPLE 5 TESTING FOR SYMMETRY

Test the symmetry of each relation.

(a) $y = |x| + 6$

Replace x by $-x$.
$y = 2|-x| + 6 = 2|x| + 6.$ Symmetric with respect to the y-axis

Replace y by $-y$:
$-y = 2|x| + 6$ Not symmetric with respect to x-axis

Replace x by $-x$ and y by $-y$.
$-y = 2|-x| + 6 = 2|x| + 6$ Not symmetric with respect to the origin

(b) $y^2 = x - 5$

Replace x by $-x$.
$y^2 = (-x) - 5 = -x - 5$ Not symmetric with respect to y-axis

Replace y by $-y$.
$(-y)^2 = x - 5$
$y^2 = x - 5$ Symmetric with respect to x-axis

Considering the first two tests together, we can see that the relation is not symmetric with respect to the origin.

PRACTICE EXERCISE 5

Test each relation for symmetry.

(a) $y = 1 - x^2$

(b) $x = 1 - y^2$

(c) $y^4 + x^2 = 5$

Replace x by $-x$ and y by $-y$.

$(-y)^4 + (-x)^2 = 5$

$y^4 + x^2 = 5$ Symmetric with respect to the origin

Notice that the graph is symmetric with respect to the y-axis and x-axis also.

(d) $x^3 + y^3 = xy^2$

Replace x by $-x$ and y by $-y$.

$(-x)^3 + (-y)^3 = xy^2$

$-x^3 - y^3 = -xy^2$ Multiply by -1

$x^3 + y^3 = xy^2$ Symmetric with respect to the origin

Replace x with $-x$.

$(-x)^3 + y^3 = (-x)y^2$

$-x^3 + y^3 = -xy^2$ Not symmetric with respect to the y-axis

Similarly, the graph is not symmetric with respect to the x-axis.

(c) $|x| + |y| = 1$

(d) $x^2y + xy^2 = 5$

Answers: (a) y-axis (b) x-axis (c) x-axis, y-axis, and origin (d) no symmetry

⑥ NONCONTINUOUS GRAPHS

We conclude this section by looking at functions whose graphs are not continuous curves. For example, consider the following function.

$$f(x) = \begin{cases} -2, & x \leq -3 \\ x - 1, & -3 < x < 0 \\ x^2, & x \geq 0 \end{cases}$$

This function is defined by different relations on different intervals of its domain. In Figure 3.29 the solid dot at breaks in the graph shows the value of the function. There is an open dot on the other part of the curve at this value of x.

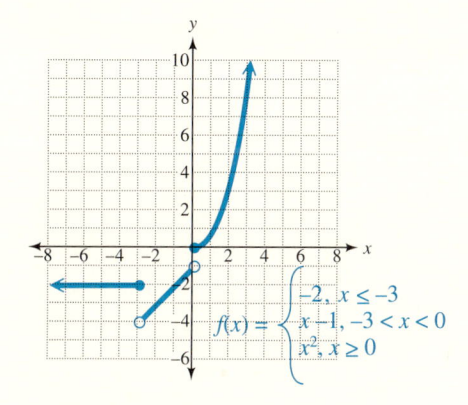

Figure 3.29 Noncontinuous Curve

Figure 3.30 Greatest Integer Function

The graph in Figure 3.30 shows the **greatest integer function.** It is denoted by

$f(x) = [x]$ Greatest integer function

and is defined to be the largest integer n such that $n \leq x$. Thus $f(x) = [x] = -2$ for $-2 \leq x < -1$ and $f(x) = [x] = 1$ for $1 \leq x < 2$. Note that $[\sqrt{2}] = 1$, $[-1.4] = -2$, and $[\pi] = 3$.

3.4 EXERCISES A

Use the vertical line test in Exercises 1–3 to decide if the graph is the graph of a function.

1.

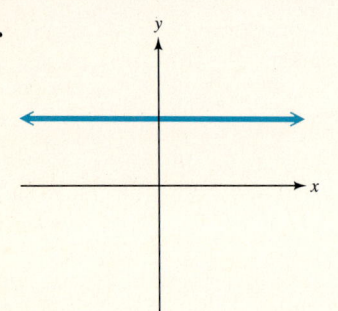

2.

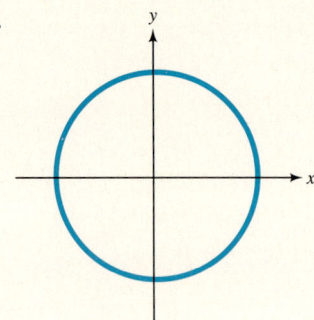

3.

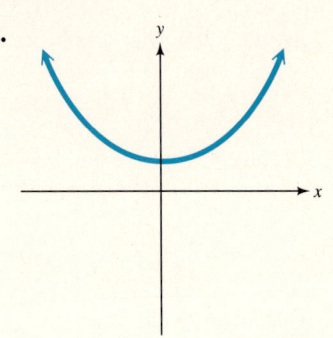

In Exercises 4–7 graph the function and give the values of x for which the function increases and the values for which it decreases.

4. $f(x) = 5x - 3$

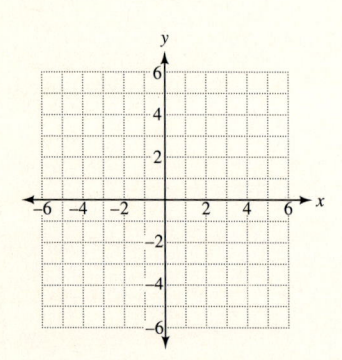

5. $h(x) = -4$

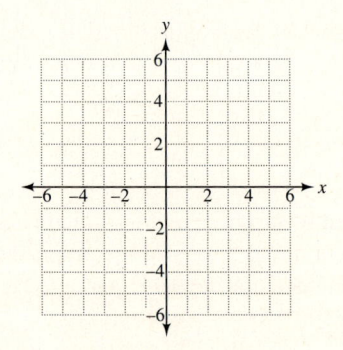

6. $k(x) = |x| - 2$

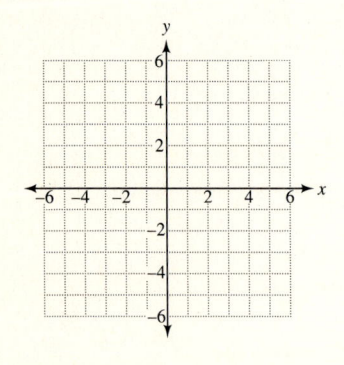

7. $g(x) = -(x - 3)^2$

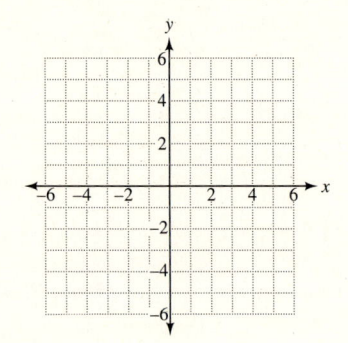

The graph of f is given in the figure. Use this to sketch the graph of each function in Exercises 8–11.

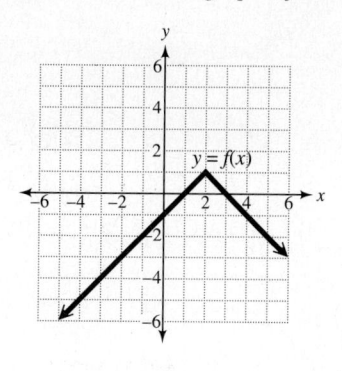

8. $y = f(x) + 3$

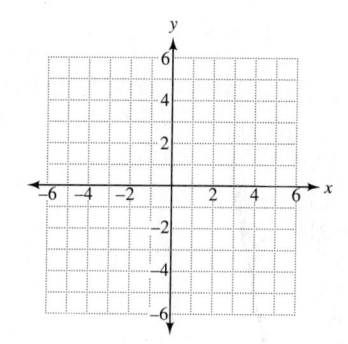

9. $y = f(x + 3)$

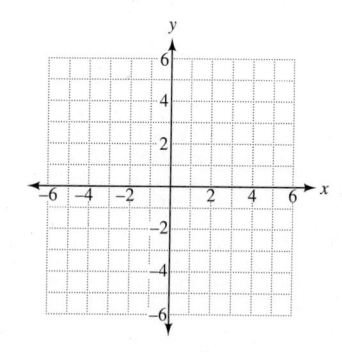

10. $y = -f(x)$

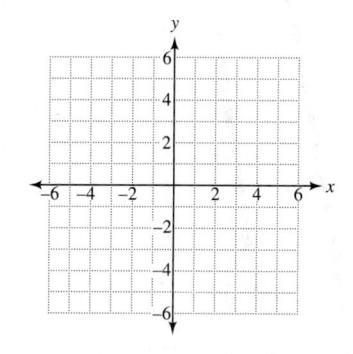

11. $y = f(x + 3) - 2$

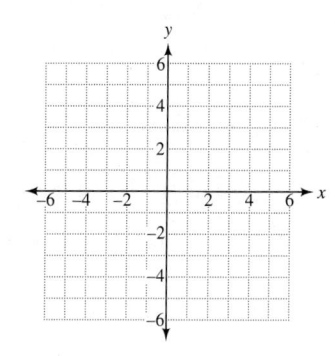

The graph of g is given in the figure. Use this to sketch the graph of each function in Exercises 12–15.

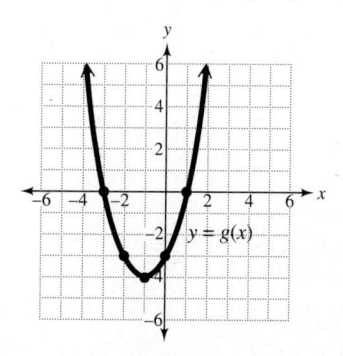

12. $y = g(x) - 1$

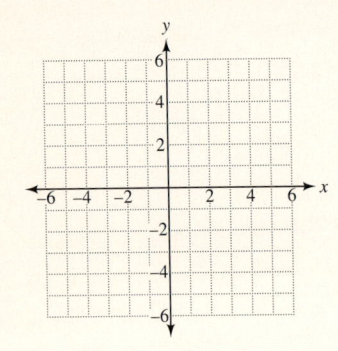

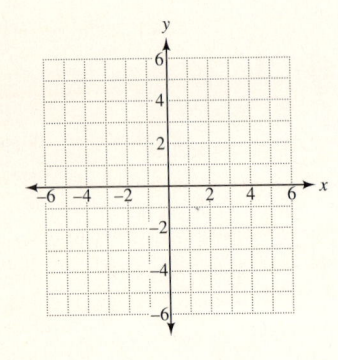

13. $y = g(x - 3)$

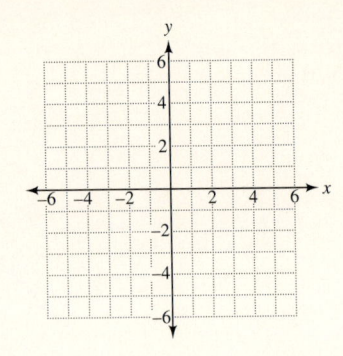

14. $y = \dfrac{1}{2} g(x)$

15. $y = g(x - 2) + 3$

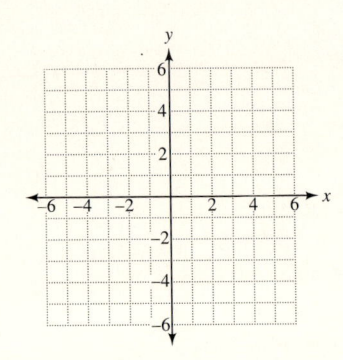

Use the graphs of f, g, and h, given in the figure for Exercises 16–17.

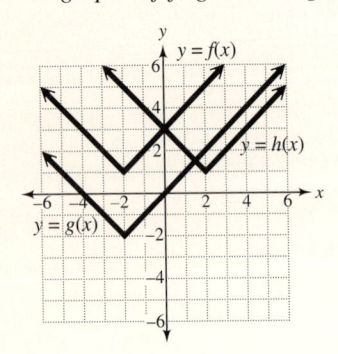

16. Determine $g(x)$ in terms of $f(x)$.

17. Determine $h(x)$ in terms of $g(x)$.

In Exercises 18–20 decide if each function is even or odd.

18. $f(x) = 2x^3 - 3x$

19. $h(x) = 0$

20. $k(x) = 3|x| - 5$

Without graphing, give the symmetry of the graphs in Exercises 21–24.

21. $y^2 + x^2 = 8$

22. $y = 3x^2 + x^4$

23. $y = -3x^2 + 2x$

24. $x^2y + xy^2 = 0$

Graph the functions in Exercises 25–27. The greatest integer function is written [x].

25. $f(x) = \begin{cases} 3, & x \leq 1 \\ -2, & x > 1 \end{cases}$

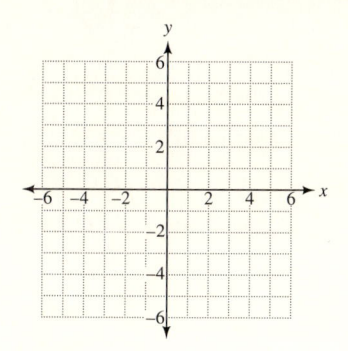

26. $h(x) = \begin{cases} \dfrac{x^2 - 4}{x + 2}, & x \neq -2 \\ -1, & x = -2 \end{cases}$

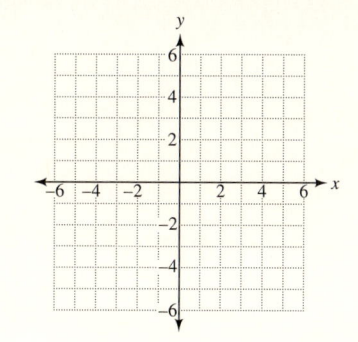

27. $k(x) = [x] + 2$

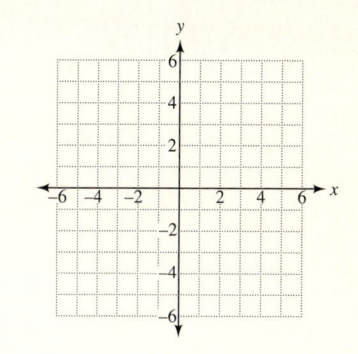

FOR REVIEW

28. Find $f(-2)$, $f(x - 1)$, and $f(x^2)$ for $f(x) = x^2 - 3$.

29. MANAGEMENT All employees of a chemical company were given an 8% increase in salary plus a $1200 increase.
  (a) Find a function S that represents the new salary in terms of the previous salary x.
  (b) If a laboratory technician was making $18,000, what is her new salary?

30. Find the distance between the two points $(-2, 6)$ and $(5, 3)$.

ANSWERS: 1. function 2. not a function 3. function
4. increases for all x 5. constant for all x 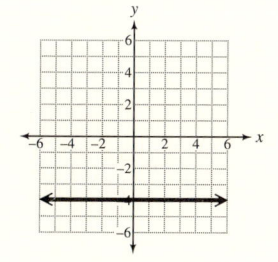   6. decreases for $x \leq 0$ increases for $x \geq 0$ 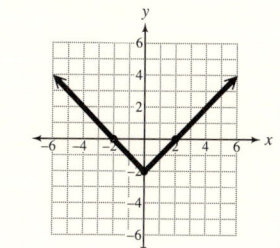   7. increases for $x \leq 3$ decreases for $x \geq 3$

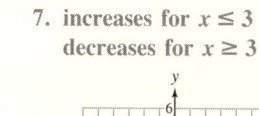

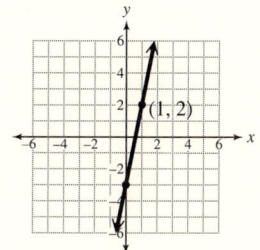

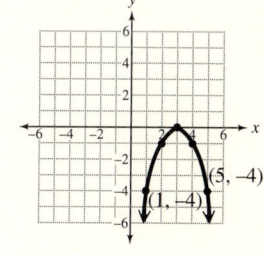

8. 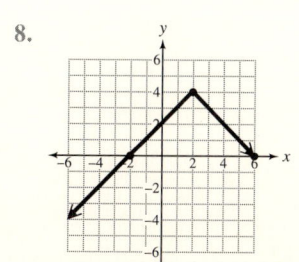   9. 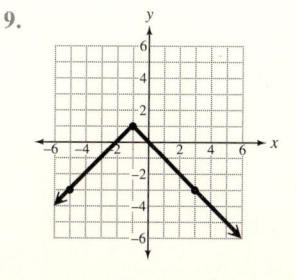   10. 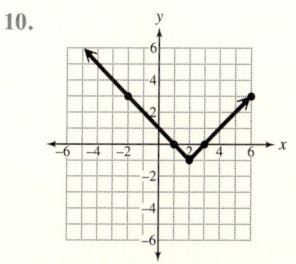   11.

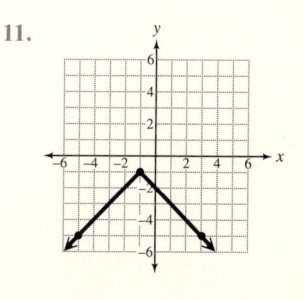

216 CHAPTER 3 RELATIONS, FUNCTIONS, AND GRAPHS

12. 13. 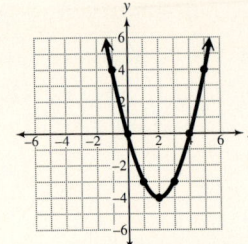 14. 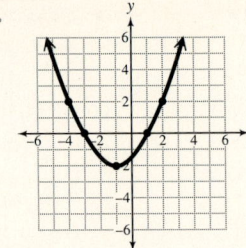 15.

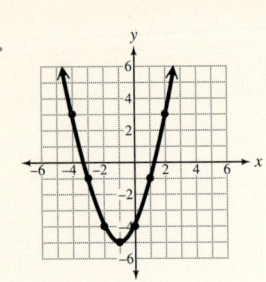

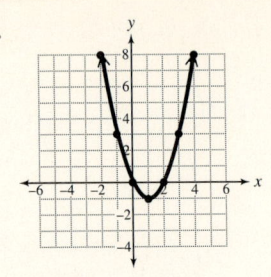

16. $g(x) = f(x) - 3$ 17. $h(x) = g(x - 4) + 3$ 18. odd   19. both even and odd   20. even   21. x-axis, y-axis, origin
22. y-axis   23. none   24. origin
25.   26.

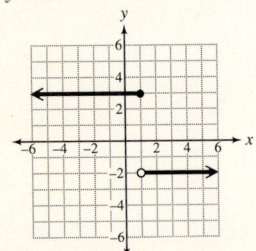

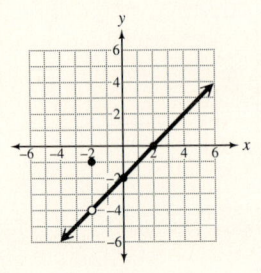

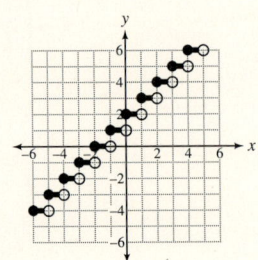

28. $1, x^2 - 2x - 2, x^4 - 3$
29. (a) $S(x) = 1.08x + 1200$
    (b) $20,640   30. $\sqrt{58}$

3.4 EXERCISES B

Use the vertical line test in Exercises 1–3 to decide if the graph is the graph of a function.

1. 2. 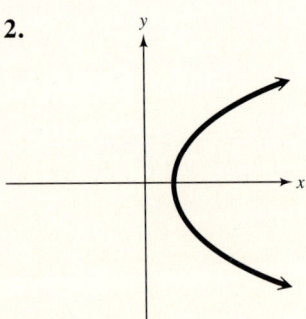 3.

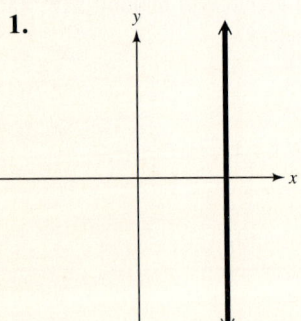

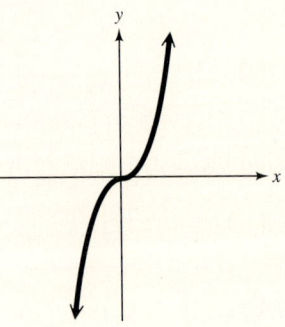

In Exercises 4–7 graph the function and give the values of x for which the function increases and the values for which it decreases.

4. $g(x) = -2x$
5. $h(x) = 4$
6. $f(x) = |x + 1|$
7. $g(x) = x(x - 2)$

The graph of f is given in the figure. Use this to sketch the graph of each function in Exercises 8–11.

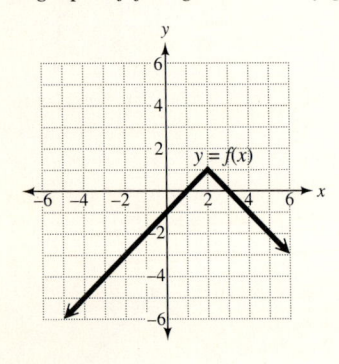

8. $y = f(x) - 2$
9. $y = f(x - 2)$
10. $y = 2f(x)$
11. $y = f(x - 3) + 2$

The graph of g is given in the figure. Use this to sketch the graph of each function in Exercises 12–15.

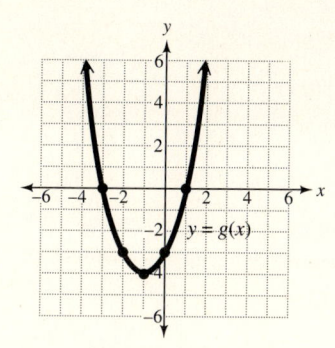

12. $y = g(x) + 4$

13. $y = g(x + 1)$

14. $y = -\dfrac{3}{4} g(x)$

15. $y = g(x + 1) - 1$

Use the graphs of f, g, and h, given in the figure for Exercises 16–17.

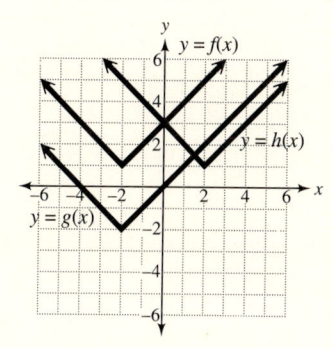

16. Determine $h(x)$ in terms of $f(x)$.

17. Determine $g(x)$ in terms of $h(x)$.

In Exercises 18–20 decide if each function is even or odd.

18. $f(x) = -4x^6 + x^4$

19. $h(x) = 3$

20. $k(x) = 3x^2 - 2x + 1$

Without graphing, give the symmetry of the graphs in Exercises 21–24.

21. $xy = 5$

22. $x^2 + xy^2 = 5$

23. $y = \sqrt{x}$

24. $(x + y)(x - y) = 8$

Graph the functions in Exercises 25–27. The greatest integer function is written [x].

25. $g(x) = \begin{cases} -2, & x < 2 \\ x^2, & -2 \leq x \leq 2 \\ 4, & x > 2 \end{cases}$

26. $h(x) = \begin{cases} \dfrac{x^2 - 1}{x - 1}, & x \neq 1 \\ 3, & x = 1 \end{cases}$

27. $k(x) = [x - 2]$

FOR REVIEW

28. Find $f(-2), f(x - 1),$ and $f(x^2)$ for $f(x) = (x - 3)^2$.

29. **MANUFACTURING** The bottom and sides of shipping boxes are made by cutting squares of length x inches from the corners of a piece of cardboard that is 12 inches by 16 inches, and folding up the sides.
    (a) Find a function $V(x)$ that represents the volume of the box.
    (b) What is the volume when $x = 2$ in?

30. Find the midpoint of the line segment joining the two points $(4, -8)$ and $(2, -2)$.

3.4 EXERCISES C

For each of the functions in Exercises 1–2 find $g(x) = f(x-2) + 2$ and $h(x) = \frac{1}{2}f(2x)$.

1. $f(x) = |x^2 - 4|$

2. $f(x) = [x + 2]$
$\left[\text{Answer: } g(x) = [x] + 2; \; h(x) = \frac{1}{2}[2(x+1)]\right]$

Graph each function in Exercises 3–4 and give the intervals on which it is increasing and the intervals on which it is decreasing.

3. $f(x) = \begin{cases} 3(x-2), & x \geq 1 \\ 2 - x, & x < 1 \end{cases}$

4. $f(x) = \begin{cases} x^2 - 4, & x > 2 \\ 4 - x^2, & -2 \leq x \leq 2 \\ x^2 - 4, & x < -2 \end{cases}$

3.5 COMPOSITE AND INVERSE FUNCTIONS

STUDENT GUIDEPOSTS

1. Algebra of Functions
2. Composite Functions
3. Inverse Functions
4. One-to-One Functions
5. Procedure for Finding Inverses

1 ALGEBRA OF FUNCTIONS

Many applications of functions require that we use the basic algebraic operations to combine functions.

Algebra of Functions

Let f and g be two functions.

$(f + g)(x) = f(x) + g(x)$ Sum of functions
$(f - g)(x) = f(x) - g(x)$ Difference of functions
$(fg)(x) = f(x) \cdot g(x)$ Product of functions
$\left(\dfrac{f}{g}\right)(x) = \dfrac{f(x)}{g(x)}, \; g(x) \neq 0$ Quotient of functions

Suppose we consider $f(x) = -2x + 3$ and $g(x) = -x^2$.

$(f - g)(x) = f(x) - g(x)$ Definition of difference of functions
$\quad\quad\quad\;\; = (-2x + 3) - (-x^2)$ Substitute
$\quad\quad\quad\;\; = x^2 - 2x + 3$

The domain of these combinations of functions is restricted to the points that the domains have in common except for the quotient where we must, as always, exclude division by zero. For example,

$$\left(\frac{g}{f}\right)(x) = \frac{g(x)}{f(x)} = \frac{-x^2}{-2x + 3} = \frac{x^2}{2x - 3}$$

is not defined for $2x - 3 = 0$ or $x = \frac{3}{2}$.

❷ COMPOSITE FUNCTIONS

There is another operation on functions which we illustrate with a basic manufacturing operation. In industry there are machines that receive an input, which we name x, and put out a product, which we name $f(x)$. Figure 3.31 illustrates such a machine which is a model of a function.

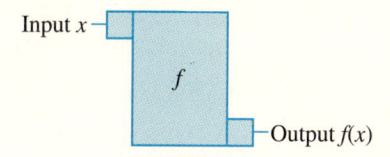

Figure 3.31

Now suppose the output $f(x)$ becomes the input to a second machine g which has an output of $g(f(x))$ as shown in Figure 3.32. If we think of the combination of machines as one machine h, then the output becomes $h(x) = g(f(x))$. For a particular input $x = 3$, we have $h(3) = g(f(3))$.

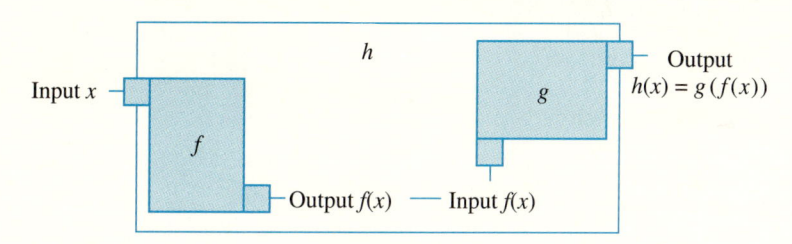

Figure 3.32

If we now consider the functions $f(x) = 3x - 7$ and $g(x) = 5x^2 + 4$ and let $x = 3$, $f(3) = 3(3) - 7 = 2$. Now evaluating g at 2, $g(2) = 5(2)^2 + 4 = 24$. Thus, $g(f(3)) = 24$. This leads to the following definition.

Composite Functions

If the range of f is contained in the domain of g, then the **composite function** $g \circ f$, read "g composed with f," is defined by

$$(g \circ f)(x) = g(f(x)).$$

We generally use the notation $g(f(x))$, read "g of f of x." If $h(x) = (g \circ f)(x) = g(f(x))$, then h is a function that takes x in the domain of f directly to the range of g. This process is illustrated in Figure 3.33.

Domain of f
Domain of h Range of f in domain of g Range of g
Range of h

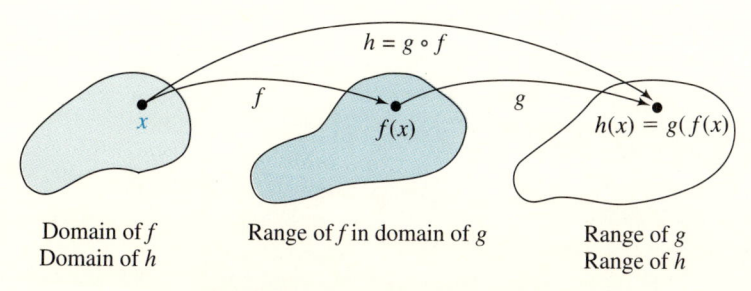

Figure 3.33 Composite Function

EXAMPLE 1 FINDING COMPOSITE FUNCTIONS

Let $f(x) = 2x + 1$ and $g(x) = x^2 - 4$. Find $g(f(x))$ and $f(g(x))$. Are the two functions equal?

$$(g \circ f)(x) = g(f(x)) = g(2x + 1) \qquad f(x) = 2x + 1$$
$$= (2x + 1)^2 - 4 \qquad g(x) = x^2 - 4$$
$$= 4x^2 + 4x - 3$$

$$(f \circ g)(x) = f(g(x)) = f(x^2 - 4) \qquad g(x) = x^2 - 4$$
$$= 2(x^2 - 4) + 1 \qquad f(x) = 2x + 1$$
$$= 2x^2 - 7$$

Thus $g(f(x)) \neq f(g(x))$ or $g \circ f \neq f \circ g$. Hence, composition of functions is not commutative.

PRACTICE EXERCISE 1

Find $g(f(x))$ and $f(g(x))$ if $f(x) = 1 - x$ and $g(x) = -2x^2 + 3$.

Answers: $g(f(x)) = -2x^2 + 4x + 1$ and $f(g(x)) = 2x^2 - 2$

EXAMPLE 2 ENVIRONMENTAL SCIENCE

The radius of an oil spill is increasing with time and given by the function $r(t) = \sqrt{t}$ for $3 \leq t \leq 48$, where t is in hours and r is in miles. The cost C in dollars of cleaning up the spill is given by $C(r) = 2630r^2 + 5820r + 9320$.

(a) Find the cost C as a function of time.

We need to find the composite function $C(r(t))$.

$$C(r(t)) = 2630(\sqrt{t})^2 + 5820(\sqrt{t}) + 9320$$
$$= 2630t + 5820\sqrt{t} + 9320$$

(b) Find the approximate cost if the cleaning process is started when $t = 40$ hr.

$$C(40) = 2630(40) + 5820\sqrt{40} + 9320 \approx \$151,000$$

PRACTICE EXERCISE 2

In Example 2 assume $r(t) = 2t$.

(a) Find C as a function of time.

(b) Determine $C(40)$.

Answers: (a) $C(t) = 10{,}520t^2 + 11{,}640t + 9320$ (b) $\$17{,}306{,}920$

❸ INVERSE FUNCTIONS

Consider the functions $h(x) = x^2$ for $x \geq 0$ and $k(x) = \sqrt{x}$. Suppose we find $h(k(x))$ and $k(h(x))$.

$$h(k(x)) = h(\sqrt{x}) = (\sqrt{x})^2 = x$$
$$k(h(x)) = k(x^2) = \sqrt{x^2} = x \qquad \text{Since } x \geq 0$$

Not only is $h(k(x)) = k(h(x))$, but they are equal to the identity function. The functions h and k are *inverses* of each other.

Inverse Functions

A function g is the **inverse** of f if

$$f(g(x)) = x \quad \text{for all } x \text{ in the domain of } g$$
and
$$g(f(x)) = x \quad \text{for all } x \text{ in the domain of } f.$$

The inverse of f is usually denoted by f^{-1}.

Note Using the f^{-1} notation, we have

$$f(f^{-1}(x)) = x \quad \text{and} \quad f^{-1}(f(x)) = x.$$

4 ONE-TO-ONE FUNCTIONS

Before finding the inverse of a function f, we need to determine which functions have inverses. To show that $h(x) = x^2$ and $k(x) = \sqrt{x}$ were inverses, we had to restrict the domain of h to $x \geq 0$. This restriction makes h a *one-to-one function*.

One-to-One Functions

A function f is a **one-to-one function** if for x_1 and x_2 in the domain of f, $x_1 \neq x_2$ implies $f(x_1) \neq f(x_2)$.

Another way of describing the one-to-one property is to say that for each y in the range of f there is one and only one x in the domain of f. Graphically, if any horizontal line crosses the graph of f in more than one point, f is not one-to-one. This **horizontal line test** in Figure 3.34 shows that $h(x) = x^2$ with domain $x \geq 0$ is one-to-one, while $g(x) = x^2$ with domain all real x is not one-to-one.

(a) One-to-one function

(b) Not a one-to-one function

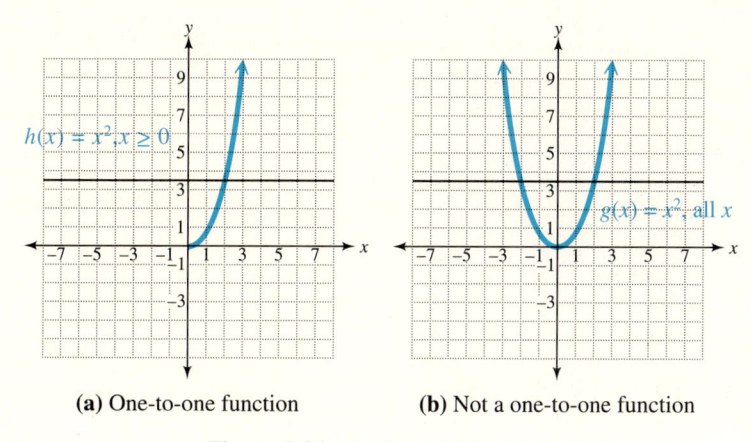

Figure 3.34 Horizontal Line Test

The next theorem helps us decide which functions are one-to-one.

Determining One-to-One Functions

If f is an increasing function for all x in its domain or if f is a decreasing function for all x in its domain, then f is one-to-one.

To see why this is true for increasing functions, consider two different points x_1 and x_2 in the domain of f. If $x_1 < x_2$, then $f(x_1) \neq f(x_2)$ since the increasing property requires $f(x_1) < f(x_2)$. Similarly, the theorem is true for $x_2 < x_1$ and for decreasing functions. This theorem implies that any linear function $f(x) = mx + b$, $m \neq 0$, is a one-to-one function. Only the constant linear functions are not one-to-one.

EXAMPLE 3 FINDING ONE-TO-ONE FUNCTIONS

Decide if the given function is one-to-one.

(a) $f(x) = x^3$, with domain all x
Since the function is increasing for all x (see Figure 3.35), f is one-to-one.

PRACTICE EXERCISE 3

Determine if each function is one-to-one.

(a) $f(x) = 1 - x^2$, all x

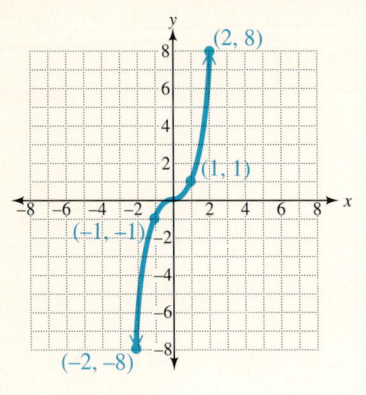

Figure 3.35 $f(x) = x^3$

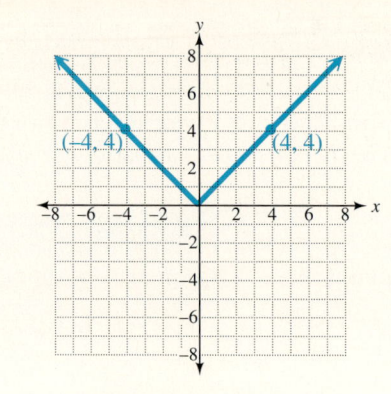

Figure 3.36 $f(x) = |x|$

(b) $f(x) = |x| + 1$, $x > 0$

(b) $f(x) = |x|$, with domain all x
Since $|x| = -x$ if $x < 0$ and $|x| = x$ if $x \geq 0$, f is decreasing for $x \leq 0$ and increasing for $x \geq 0$ (see Figure 3.36). Thus f is not one-to-one.

Answers: (a) no   (b) yes

This leads to the following theorem.

Functions with Inverses

If f is a one-to-one function, then there is a function f^{-1} such that
$$f(f^{-1}(x)) = x = f^{-1}(f(x)).$$

Now that we know which functions have inverses, a technique is needed for finding the inverse of a given function. Consider again the function $h(x) = x^2$ ($x \geq 0$) and $k(x) = \sqrt{x}$. Figure 3.37 shows that h and k are symmetric with respect to the line $y = x$. For example, $(2, 4)$ is on the graph of h and the point $(4, 2)$ is on the graph of k. In general, if (x, y) is on the graph of f, then (y, x) is on the graph of f^{-1}.

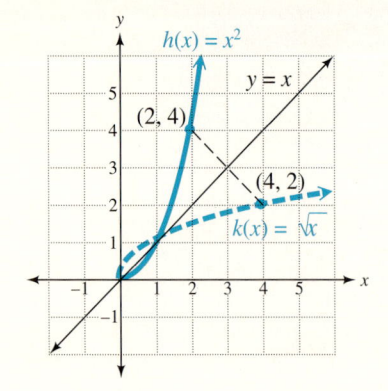

Figure 3.37 Inverse Function

5 PROCEDURE FOR FINDING INVERSES

This discussion suggests the following procedure for finding the inverse of a function f.

Procedure for Finding Inverses

1. Write the function using y for $f(x)$.
2. Interchange x and y in the equation.
3. Obtain the expression for f^{-1} by solving the equation for y.

We can always prove our results by showing $f(f^{-1}(x)) = x = f^{-1}(f(x))$.

EXAMPLE 4 FINDING INVERSES

Find the inverse of the given function if it exists.

(a) $f(x) = 5x - 8$

f^{-1} exists since f is a linear function that is increasing.

$$y = 5x - 8 \quad \text{Write } y \text{ for } f(x)$$
$$x = 5y - 8 \quad \text{Interchange } x \text{ and } y$$
$$x + 8 = 5y$$
$$y = \frac{x + 8}{5} \quad \text{Solve for } y$$
$$f^{-1}(x) = \frac{x + 8}{5} \quad y \text{ is } f^{-1}(x)$$

As a check, determine if $f(f^{-1}(x)) = x$ and $f^{-1}(f(x)) = x$.

$$f(f^{-1}(x)) = f\left(\frac{x+8}{5}\right) = 5\left(\frac{x+8}{5}\right) - 8 = x + 8 - 8 = x$$

$$f^{-1}(f(x)) = f^{-1}(5x - 8) = \frac{(5x - 8) + 8}{5} = \frac{5x - 8 + 8}{5} = x$$

(b) $g(x) = \sqrt{x - 3}, \quad x \geq 3$

Since g is an increasing function where it is defined, it has an inverse.

$$y = \sqrt{x - 3} \quad \text{Write } y \text{ for } g(x)$$
$$x = \sqrt{y - 3} \quad \text{Interchange } x \text{ and } y$$
$$x^2 = y - 3 \quad \text{Square both sides}$$
$$y = x^2 + 3 \quad \text{Solve for } y$$
$$g^{-1}(x) = x^2 + 3 \quad y \text{ is } g^{-1}(x)$$

The check will indicate something about the domain of g^{-1}.

$$g(g^{-1}(x)) = g(x^2 + 3) = \sqrt{x^2 + 3 - 3} = \sqrt{x^2} = x \quad \text{If } x \geq 0$$
$$g^{-1}(g(x)) = g^{-1}(\sqrt{x - 3}) = (\sqrt{x - 3})^2 + 3 = x$$

Notice that for $g(g^{-1}(x))$ to be x, we needed $x \geq 0$. Thus, the domain of g^{-1} is $x \geq 0$. In general, for any function f, the domain of f^{-1} is the range of f and the domain of f is the range of f^{-1}.

(c) $h(x) = x^4 + 2$, all real x

Since $h(x)$ is decreasing for $x \leq 0$ and increasing for $x \geq 0$, it does not have an inverse.

PRACTICE EXERCISE 4

Find the inverse of each function.

(a) $f(x) = -2x + 6$

(b) $g(x) = \sqrt{2 - x}, \, x \leq 2$

(c) $h(x) = (x + 1)^2$, all x

(d) $k(x) = 4 - x^3$
Since k is decreasing for all x, it has an inverse.

$$y = 4 - x^3 \quad \text{Write } y \text{ for } k(x)$$
$$x = 4 - y^3 \quad \text{Interchange } x \text{ and } y$$
$$y^3 = 4 - x$$
$$y = \sqrt[3]{4 - x} \quad \text{Solve for } y$$
$$k^{-1}(x) = \sqrt[3]{4 - x} \quad y \text{ is } k^{-1}(x)$$

Check: $k(k^{-1}(x)) = k(\sqrt[3]{4-x}) = 4 - (\sqrt[3]{4-x})^3 = 4 - (4-x) = x$
$k^{-1}(k(x)) = k^{-1}(4-x^3) = \sqrt[3]{4-(4-x^3)} = \sqrt[3]{x^3} = x$

(d) $k(x) = x^3 - 3$, all x

Answers: (a) $f^{-1}(x) = -\frac{1}{2}x + 3$
(b) $g^{-1}(x) = -x^2 + 2$, $x \geq 0$
(c) no inverse (d) $k^{-1}(x) = \sqrt[3]{x+3}$

If a relation is defined as a set of ordered pairs, the inverse is found by changing each pair (x, y) to (y, x). Thus, every relation has an inverse, but only one-to-one functions have inverses that are functions. Notice that

$$\{(1, 2), (3, 4), (5, 6)\}$$

is a one-to-one function since for each x there is one and only one y *and* for each y there is one and only one x. Thus the inverse relation

$$\{(2, 1), (4, 3), (6, 5)\},$$

found by changing $(1, 2)$ to $(2, 1)$, $(3, 4)$ to $(4, 3)$, and $(5, 6)$ to $(6, 5)$, is also a function. This is what we mean by a function having an inverse. The relation $\{(7, 8), (9, 8)\}$ is a function, but we say that it does not have an inverse since the inverse relation $\{(8, 7), (8, 9)\}$ is not a function.

3.5 EXERCISES A

Find $(f + g)(x)$, $(f - g)(x)$, $(fg)(x)$, *and* $\left(\frac{f}{g}\right)(x)$ *for each pair of functions in Exercises 1–4. For the quotient tell which values of x must be excluded.*

1. $f(x) = 5x$ and $g(x) = 3x - 2$

2. $f(x) = 2x^2$ and $g(x) = -x$

3. $f(x) = -2x + 1$ and $g(x) = x^2 + 2$

4. $f(x) = x^3 + 1$ and $g(x) = \sqrt[3]{x - 1}$

Find $f(g(2))$, $g(f(-1))$, $f(g(x))$, *and* $g(f(x))$ *for each pair of functions in Exercises 5–8.*

5. $f(x) = 5x$ and $g(x) = 3x - 2$

6. $f(x) = 2x^2$ and $g(x) = -x$

7. $f(x) = -2x + 1$ and $g(x) = x^2 + 2$

8. $f(x) = x^3 + 1$ and $g(x) = \sqrt[3]{x - 1}$

In Exercises 9–14 state which functions are one-to-one.

9. $f(x) = -2x + 6$

10. $h(x) = |x + 1|$

11. $k(x) = x^3 + 1$

12. $\{(4, 5), (6, 7)\}$

13. $\{(8, 9), (9, 9)\}$

14. $\{(2, 1), (3, 4), (4, 1)\}$

In Exercises 15–16 decide if the functions are inverses of each other and graph them in the same coordinate system.

15. $f(x) = \sqrt{x-2}$ and $g(x) = x^2 + 2$, $x \geq 0$

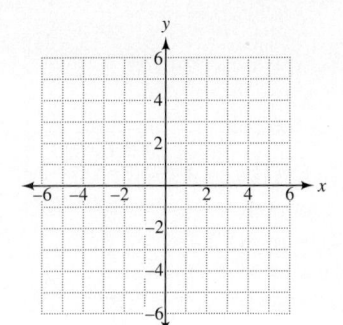

16. $f(x) = \sqrt{x+4}$ and $g(x) = x^2 - 4$, $x \geq 0$

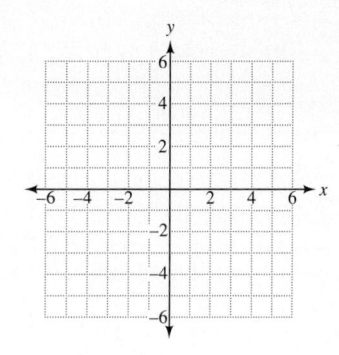

Find the inverse of each function in Exercises 17–22.

17. $f(x) = 2x + 3$

18. $h(x) = \sqrt{x-3}$

19. $k(x) = 2x^3$

20. $g(x) = \sqrt{1-x^2}$, $0 \leq x \leq 1$

21. $\{(8, 8), (9, 9)\}$

22. $\{(6, 4), (7, 8), (8, 3)\}$

23. Find the inverse of $f(x) = mx + b$. Are there any restrictions on m or b?

24. ECOLOGY The radius of a circular grass fire is increasing with time according to the equation $r(t) = 2t^2 - 1$.
   (a) Find the circumference C of the fire as a function of time.
   (b) Find the area burned A as a function of time.

25. ECOLOGY An oil spill is moving down a river which is 22 m wide. The distance in kilometers that it has traveled at time t in hours is given by $d(t) = 3t^2 + 2t + 1$.
   (a) Find the area A, in square meters, covered by the spill as a function of time.
   (b) Find the area after 2 hours.

26. ENGINEERING The cost of cleaning the spill in Exercise 25 is given by the equation $C(d) = 2220d + 9530$, where C is in dollars and $d(t) = 3t^2 + 2t + 1$ is in kilometers.
   (a) Find C as a function of time.
   (b) Find C if the spill was stopped after 2 hours.

In Exercises 27–28 determine what effect the transformation has on the graph of f.

27. $y = f(x - 4) + 2$

28. $y = f(x + 4) - 2$

State whether each function in Exercises 29–30 is even or odd.

29. $f(x) = 4x^3 + 2x$

30. $g(x) = |x^3|$

31. Find the equation of the line passing through $(-6, 2)$ and parallel to the line $3x - 2y + 5 = 0$.

32. Find the equation of the perpendicular bisector of the line segment joining $(4, 6)$ and $(-8, -2)$.

Exercises 33–34 will help you prepare for the next section.

33. What must be added to complete the square in each expression?
(a) $x^2 - 8x$ (b) $x^2 + 3x$

34. Solve. $2x^2 + 9x - 5 = 0$

ANSWERS: 1. $8x - 2$, $2x + 2$, $15x^2 - 10x$, $\frac{5x}{3x-2}$ $\left(x \neq \frac{2}{3}\right)$ 2. $2x^2 - x$, $2x^2 + x$, $-2x^3$, $-2x$ $(x \neq 0)$ 3. $x^2 - 2x + 3$, $-x^2 - 2x - 1$, $-2x^3 + x^2 - 4x + 2$, $\frac{-2x+1}{x^2+2}$ 4. $x^3 + 1 + \sqrt[3]{x-1}$, $x^3 + 1 - \sqrt[3]{x-1}$, $(x^3 + 1)\sqrt[3]{x-1}$, $\frac{x^3+1}{\sqrt[3]{x-1}}$ $(x \neq 1)$
5. 20, -17, $15x - 10$, $15x - 2$ 6. 8, -2, $2x^2$, $-2x^2$ 7. -11, 11, $-2x^2 - 3$, $4x^2 - 4x - 3$ 8. 2, -1, x, x
9. one-to-one 10. not one-to-one 11. one-to-one 12. one-to-one 13. not one-to-one 14. not one-to-one
15. yes 16. yes 17. $f^{-1}(x) = \frac{x-3}{2}$ 18. $h^{-1}(x) = x^2 + 3$, $x \geq 0$
19. $k^{-1}(x) = \sqrt[3]{\frac{x}{2}}$ 20. $g^{-1}(x) = \sqrt{1 - x^2}$, $0 \leq x \leq 1$
21. $\{(8, 8), (9, 9)\}$ 22. $\{(4, 6), (8, 7), (3, 8)\}$
23. $f^{-1}(x) = \frac{x-b}{m}$, $m \neq 0$ 24. (a) $C(t) = 2\pi(2t^2 - 1)$
(b) $A(t) = \pi(2t^2 - 1)^2$ 25. (a) $22{,}000(3t^2 + 2t + 1)$
(b) $374{,}000$ m^2 26. (a) $C(t) = 6660t^2 + 4440t + 11{,}750$
(b) $\$47{,}270$ 27. right 4 units and upward 2 units
28. left 4 units and downward 2 units 29. odd 30. even
31. $3x - 2y + 22 = 0$ 32. $3x + 2y + 2 = 0$ 33. (a) 16
(b) $\frac{9}{4}$ 34. $\frac{1}{2}$, -5

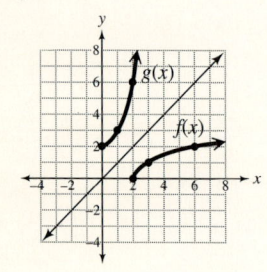

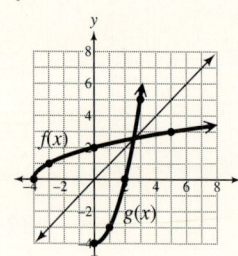

3.5 EXERCISES B

Find $(f + g)(x)$, $(f - g)(x)$, $(fg)(x)$, and $\left(\frac{f}{g}\right)(x)$ for each pair of functions in Exercises 1–4. For the quotient tell which values of x must be excluded.

1. $f(x) = 1 - x$ and $g(x) = 5x + 1$

2. $f(x) = \sqrt{x^2 + 5}$ and $g(x) = x^4$

3. $f(x) = 2x + 3$ and $g(x) = (x - 3)^2$

4. $f(x) = \dfrac{x}{1 - x}$ and $g(x) = \dfrac{x}{x + 1}$

Find $f(g(2))$, $g(f(-1))$, $f(g(x))$, and $g(f(x))$ for each pair of functions in Exercises 5–8.

5. $f(x) = 1 - x$ and $g(x) = 5x + 1$

6. $f(x) = \sqrt{x^2 + 5}$ and $g(x) = x^4$

7. $f(x) = 2x + 3$ and $g(x) = (x - 3)^2$

8. $f(x) = \dfrac{x}{1 - x}$ and $g(x) = \dfrac{x}{x + 1}$

In Exercises 9–14 state which functions are one-to-one.

9. $g(x) = 2 - 3x^2$

10. $h(x) = \dfrac{1}{x}$

11. $k(x) = \sqrt{x - 2}$

12. $\{(6, 1), (1, 6)\}$

13. $\{(4, 3), (5, 1), (6, 2)\}$

14. $\{(5, 7), (4, 6), (3, 6)\}$

In Exercises 15–16 decide if the functions are inverses of each other and graph them in the same coordinate system.

15. $f(x) = -2x + 5$ and $g(x) = \dfrac{5 - x}{2}$

16. $f(x) = \dfrac{1}{2}x^3$ and $g(x) = \sqrt[3]{2x}$

Find the inverse of each function in Exercises 17–22.

17. $g(x) = 1 - x^2, \ x \geq 0$ **18.** $h(x) = \sqrt{x + 2}$ **19.** $k(x) = \sqrt[3]{x - 3}$

20. $g(x) = \sqrt{x^2 - 1}, \ x \geq 1$ **21.** $\{(1, 1), (2, 2), (3, 3)\}$ **22.** $\{(4, 2), (6, 1), (7, 6)\}$

23. Find the inverse of $g(x) = ax^2 + c, \ x \geq 0$. Are there any restrictions on a or c?

24. METEOROLOGY The radius of a weather balloon is increasing with time according to the equation $r(t) = \sqrt{t + 2}$.
  (a) Find the volume V of the balloon as a function of time.
  (b) Find the surface area S of the balloon as a function of time.

25. ECOLOGY An oil spill is moving down a river which is 10 m wide. The distance in kilometers that it has traveled at time t in hours is given by $d(t) = 3t + 2\sqrt{t} + 1$.
  (a) Find the area A, in square meters, covered by the spill as a function of time.
  (b) Find the area after 3 hours.

26. ENGINEERING The cost of cleaning the spill in Exercise 25 is given by the equation $C(d) = 3080d + 4500$, where C is in dollars and $d(t) = 3t + 2\sqrt{t} + 1$ is in kilometers.
  (a) Find C as a function of time.
  (b) Find C if the spill was stopped after 3 hours.

FOR REVIEW

In Exercises 27–28 determine what effect the transformation has on the graph of f.

27. $y = f(x + 2) - 6$ **28.** $y = f(x - 2) + 6$

State whether each function in Exercises 29–30 is even or odd.

29. $f(x) = 4x^2 - x^4 + 2$ **30.** $g(x) = 3x^5 - x$

31. Find the equation of the line passing through $(3, -4)$ and parallel to the line $6x + 2y - 7 = 0$.

32. Find the equation of the perpendicular bisector of the line segment joining $(1, 5)$ and $(4, 7)$.

Exercises 33–34 will help you prepare for the next section.

33. What must be added to complete the square in each expression?
  (a) $x^2 + 10x$ (b) $x^2 - 7x$

34. Solve. $3x^2 - 2x - 4 = 0$

3.5 EXERCISES C

1. Prove that if f and g are both even functions then $f + g$ is even.

2. Prove that if f and g are both odd functions then fg is even.

3. Prove that if f is an even function and g is either even or odd, then $f \circ g$ is even.

4. If $f(x) = mx + b, \ m \neq 0$, find $f^{-1}(x)$.
  $\left[\text{Answer: } f^{-1}(x) = \frac{1}{m}x - \frac{b}{m}\right]$

3.6 QUADRATIC FUNCTIONS

STUDENT GUIDEPOSTS

1. Standard Form
2. Vertex
3. Intercepts
4. Line of Symmetry

In Section 3.4 graphs of functions of the form $f(x) = ax^2 + c$ were used to study various properties of functions. Now we consider this type of function in more detail.

> **Quadratic Functions**
>
> A **quadratic function** is a function of the form
>
> $$f(x) = ax^2 + bx + c$$
>
> where a, b, and c are constant and $a \neq 0$.

In Figure 3.38 the graphs of $f(x) = x^2 + 2x - 3$ and $g(x) = -x^2 + 2x + 3$ are determined by plotting points.

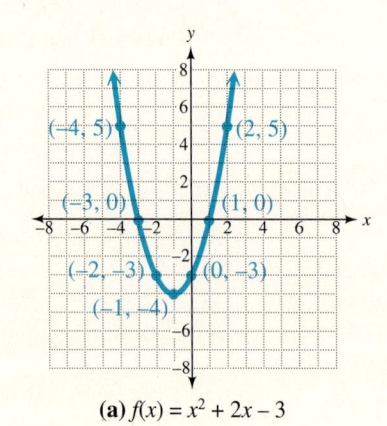

x	$f(x)$
0	-3
1	0
-1	-4
2	5
-2	-3
-3	0
-4	5

(a) $f(x) = x^2 + 2x - 3$

x	$g(x)$
0	3
1	4
-1	0
2	3
-2	-5
3	0
4	-5

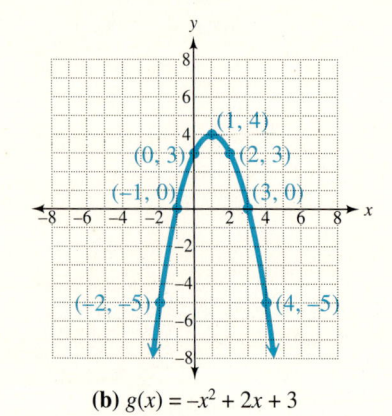

(b) $g(x) = -x^2 + 2x + 3$

Figure 3.38 Parabolas

These graphs are typical of quadratic functions. All are *parabolas* that open up if $a > 0$ as in Figure 3.38(a) and open down if $a < 0$ as in Figure 3.38(b). Notice also that each has a minimum or maximum point called the **vertex**. For example, the vertex in Figure 3.38(a) is $(-1, -4)$.

❶ STANDARD FORM

Graphing quadratic functions is made easier by finding the vertex before plotting points. Suppose we complete the square on the following function.

$$f(x) = 2x^2 + 12x + 13$$
$$= 2(x^2 + 6x \quad) + 13$$
$$= 2(x^2 + 6x + 9) + 13 - 18 \qquad \left(\frac{1}{2} \cdot 6\right)^2 = 9 \text{ and } 9 \cdot 2 = 18$$
$$= 2(x + 3)^2 - 5 \qquad x^2 + 6x + 9 = (x + 3)^2$$
$$= 2[x - (-3)]^2 - 5$$

Thus, the minimum value of f will occur when $[x - (-3)]^2$ is 0, that is, when $x = -3$. Therefore, the vertex is $(h, k) = (-3, -5)$. If we write any quadratic function in **standard form**,

$$f(x) = a(x - h)^2 + k,$$

the vertex can easily be determined. Rather than completing the square on each quadratic function, we can obtain a formula for the vertex (h, k) by considering a general quadratic function.

$$f(x) = ax^2 + bx + c$$
$$= a\left(x^2 + \frac{b}{a}x \quad \right) + c$$
$$= a\left(x^2 + \frac{b}{a}x + \frac{b^2}{4a^2}\right) + c - \frac{b^2}{4a} \qquad \left(\frac{1}{2} \cdot \frac{b}{2}\right)^2 = \frac{b^2}{4a^2} \text{ and } a\left(\frac{b^2}{4a^2}\right) = \frac{b^2}{4a}$$
$$= a\left(x + \frac{b}{2a}\right)^2 + c - \frac{b^2}{4a} \qquad x^2 + \frac{b}{a}x + \frac{b^2}{4a^2} = \left(x + \frac{b}{2a}\right)^2$$
$$= a\left[x - \left(-\frac{b}{2a}\right)\right]^2 + \frac{4ac - b^2}{4a}$$

With $f(x)$ in this form we see that $f\left(-\frac{b}{2a}\right) = \frac{4ac - b^2}{4a}$ and thus the vertex is $\left(-\frac{b}{2a}, f\left(-\frac{b}{2a}\right)\right)$.

❷ VERTEX

This proves the following theorem.

Vertex Formula

If $f(x) = ax^2 + bx + c$ is a quadratic function, then the vertex of the graph is

$$(h, k) = \left(-\frac{b}{2a}, f\left(-\frac{b}{2a}\right)\right).$$
---

230 CHAPTER 3 RELATIONS, FUNCTIONS, AND GRAPHS

EXAMPLE 1 FINDING THE VERTEX

Find the vertex of $f(x) = 3x^2 - 2x - 2$.
   First find h.

$$h = -\frac{b}{2a} = -\frac{(-2)}{2(3)} = \frac{1}{3}$$

Now find $f(h)$.

$$f\left(\frac{1}{3}\right) = 3\left(\frac{1}{3}\right)^2 - 2\left(\frac{1}{3}\right) - 2$$

$$= \frac{1}{3} - \frac{2}{3} - \frac{6}{3} = -\frac{7}{3}$$

The vertex is $\left(\frac{1}{3}, -\frac{7}{3}\right)$.

PRACTICE EXERCISE 1

Find the vertex of $g(x) = -2x^2 + 4x - 7$.

Answer: $(1, -5)$

---

3 INTERCEPTS

In addition to using the vertex when graphing, we can sometimes find the **x-intercepts.** These are the points where $f(x) = 0$, and can be found by solving the quadratic equation

$$ax^2 + bx + c = 0.$$

From Section 2.4,

$$x = \frac{-b \pm \sqrt{b^2 - 4ac}}{2a}.$$

Three cases occur as shown in Figure 3.39.

$b^2 - 4ac > 0$
Two x-intercepts

$b^2 - 4ac = 0$
One x-intercept

$b^2 - 4ac < 0$
No x-intercepts

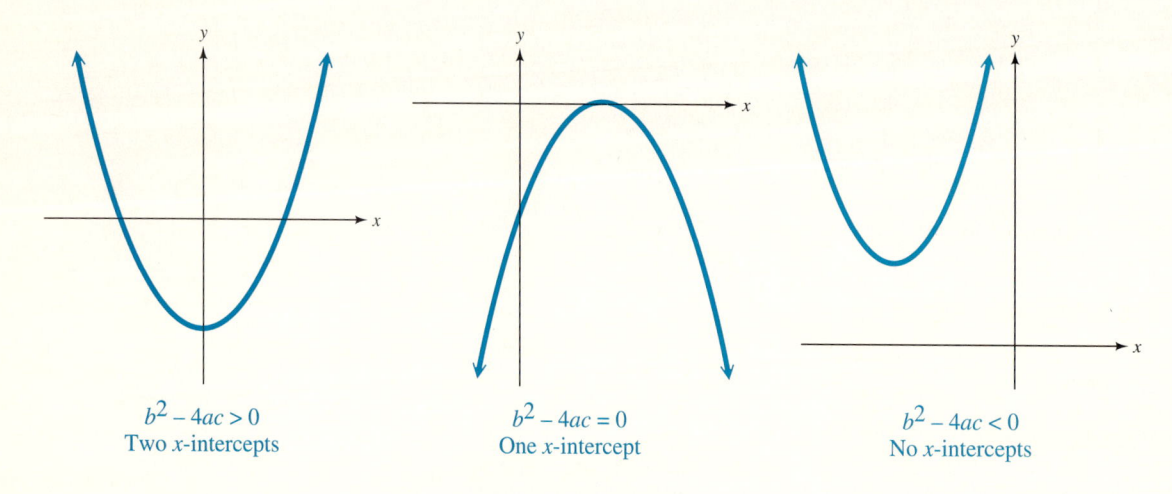

Figure 3.39 Intercepts

4 LINE OF SYMMETRY

The vertical line through the vertex has equation $x = -\frac{b}{2a}$ and is the **line of symmetry** of a parabola. Using this symmetry property, the vertex, the x-intercepts, and a few additional points we can now sketch the graph of a quadratic function.

EXAMPLE 2 GRAPHING A QUADRATIC FUNCTION

Find the vertex, find any x-intercepts, and graph the quadratic function.

(a) $f(x) = x^2 - 6x + 5$

Note that $a = 1 > 0$, so the graph opens upward. Now we find the vertex (h, k).

$$h = -\frac{b}{2a} = -\frac{-6}{2(1)} = 3$$

$$k = f\left(-\frac{b}{2a}\right) = f(3) = 3^2 - 6(3) + 5 = -4$$

The vertex is $(3, -4)$. We solve the equation $x^2 - 6x + 5 = 0$ to find the x-intercepts.

$$x^2 - 6x + 5 = 0$$
$$(x - 1)(x - 5) = 0$$
$$x - 1 = 0 \quad \text{or} \quad x - 5 = 0$$
$$x = 1 \quad\quad\quad\quad x = 5$$

The x-intercepts are $(1, 0)$ and $(5, 0)$. Now choosing an $x < 1$ and an $x > 5$, we plot two additional points $(0, 5)$ and $(6, 5)$. Notice that the graph shown in Figure 3.40 is symmetric with respect to the line of symmetry $x = 3$.

x	$f(x)$
3	-4
1	0
5	0
0	5
6	5

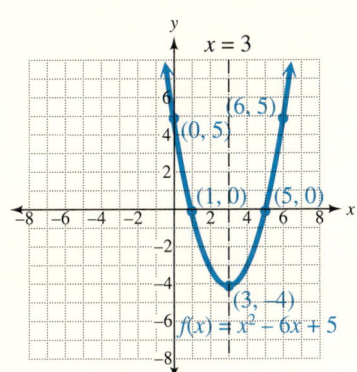

Figure 3.40

(b) $g(x) = -3x^2 + 9x - 4$

The graph opens downward since $a = -3 < 0$. Now we find (h, k).

$$h = -\frac{b}{2a} = -\frac{9}{2(-3)} = \frac{3}{2}$$

$$k = f\left(-\frac{b}{2a}\right) = f\left(\frac{3}{2}\right) = (-3)\left(\frac{3}{2}\right)^2 + 9\left(\frac{3}{2}\right) - 4 = \frac{11}{4}$$

The vertex is $\left(\frac{3}{2}, \frac{11}{4}\right)$. Now we use the quadratic formula to solve $-3x^2 + 9x - 4 = 0$ and find the x-intercepts.

$$x = \frac{-b \pm \sqrt{b^2 - 4ac}}{2a}$$

$$= \frac{-9 \pm \sqrt{(9)^2 - 4(-3)(-4)}}{2(-3)}$$

$$= \frac{-9 \pm \sqrt{81 - 48}}{-6} = \frac{-9 \pm \sqrt{33}}{-6} = \frac{9 \pm \sqrt{33}}{6}$$

PRACTICE EXERCISE 2

Tell whether the graph opens up or down, find the vertex, find any x-intercepts, and graph the quadratic function.

(a) $f(x) = x^2 - 4x + 3$

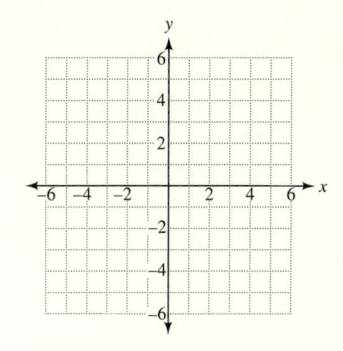

(b) $g(x) = -2x^2 - 6x + 1$

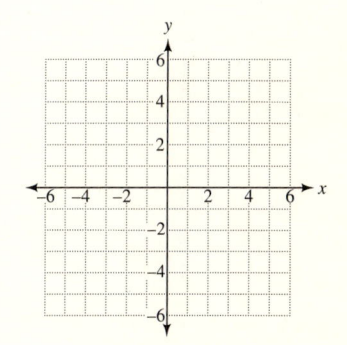

We use a calculator to find $x \approx 0.54$ and $x \approx 2.46$. Thus, the x-intercepts are (0.54, 0) and (2.46, 0). We plot the additional points and graph the function in Figure 3.41. Notice that the line of symmetry is $x = \frac{3}{2}$. Symmetry should be used when we are finding the additional points to plot.

x	$g(x)$
$\frac{3}{2}$	$\frac{11}{4}$
1	2
2	2
0.54	0
2.64	0
0	−4
3	−4

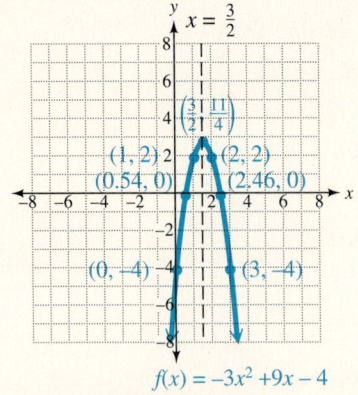

Figure 3.41

Answers: (a) opens up; vertex (2, −1); intercepts (1, 0), (3, 0); (b) opens down; vertex $\left(-\frac{3}{2}, \frac{11}{2}\right)$; intercepts (0.16, 0), (−3.16, 0)

The function $f(x) = \frac{1}{2}x^2 + 5x + 13$ illustrates the case when there are no x-intercepts. Notice that

$$b^2 - 4ac = (5)^2 - 4\left(\frac{1}{2}\right)(13) = 25 - 26 = -1 < 0.$$

Since the discriminant is negative there are no x-intercepts. The graph opens upward, and the vertex is (−5, 0.5). Use symmetry with respect to $x = -5$ and plot enough points to sketch the graph in Figure 3.42.

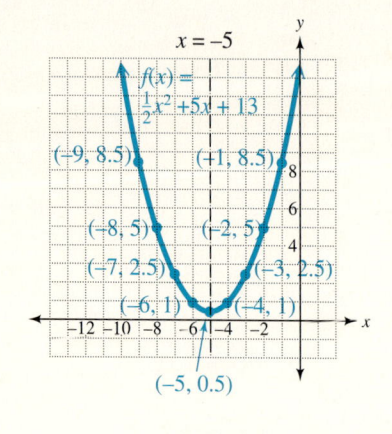

x	$f(x)$
−5	0.5
−4	1
−6	1
−3	2.5
−7	2.5
−2	5
−8	5
−1	8.5
−9	8.5

Figure 3.42

Many applications of quadratic functions are solved by finding the maximum value (if $a < 0$) or the minimum value (if $a > 0$) using the vertex.

EXAMPLE 3 APPLICATION TO AGRICULTURE

A rancher needs to build a holding pen for cattle along an existing fence. She has available 600 m of fencing to build the 3 sides of the rectangular pen shown in Figure 3.43. Find the dimensions of the pen that give the maximum area and also find the maximum area.

PRACTICE EXERCISE 3

If a ball is thrown upward, the height in feet after t seconds is $h(t) = -16t^2 + 64t$. Find its maximum height.

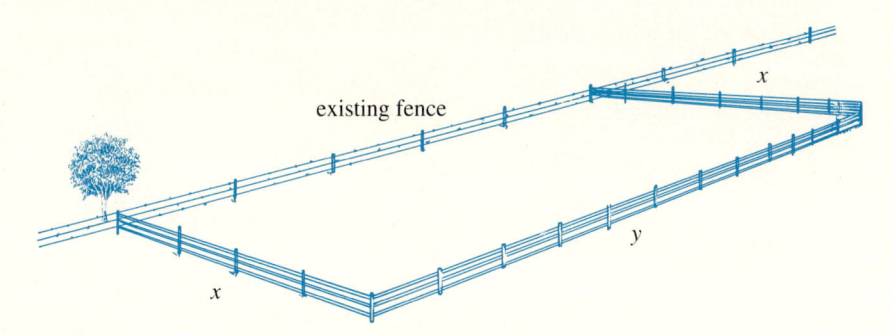

Figure 3.43

Let x = length of sides perpendicular to the existing fence,
y = length of side parallel to the existing fence.

Since 600 m of fencing is available,

$2x + y = 600$ Length of the sides is 600 m
$y = 600 - 2x$.   Write y in terms of x

The quantity to be maximized is the area A of the pen.

$A = xy$
$A(x) = x(600 - 2x)$ $y = 600 - 2x$ and A is now a function of x alone
$\quad = 600x - 2x^2$
$\quad = -2x^2 + 600x$ A is a quadratic function

Notice that in the function, $a = -2 < 0$, and thus the parabola opens downward and has a maximum. We can find the value of x that gives the maximum by finding h in the vertex (h, k).

$$h = -\frac{b}{2a} = -\frac{600}{2(-2)} = 150$$

Hence, each side perpendicular to the existing fence should be 150 m to give the maximum area. Now find y.

$y = 600 - 2x = 600 - 2(150) = 300$

The side parallel to the existing fence is 300 m. The maximum area can be found by finding k in (h, k) or by finding the product of x and y.

$A = xy = (150)(300) = 45{,}000 \text{ m}^2$

The maximum area is 45,000 m².

Answer: 64 ft

The following application was first presented in the introduction to this chapter.

EXAMPLE 4 APPLICATION TO BUSINESS

A car rental agency averages 20 customers per day for a model which rents for $32 per day. A survey shows that the agency could get 2 new customers for each $1 reduction in the daily rental rate. What daily rate would give the company the maximum revenue?

Let n = the number of dollars reduction in the daily rental fee. Since there are 2 new customers for each dollar reduction, $2n$ is the number of new customers. Then

$20 + 2n$ = number of customers after decreasing the daily fee by n dollars,

$32 - n$ = new daily rental fee.

The product $(20 + 2n)(32 - n)$ is the daily revenue that is to be maximized.

$$R(n) = (20 + 2n)(32 - n) = 640 + 44n - 2n^2$$
$$= -2n^2 + 44n + 640$$

Since $a = -2 < 0$, the quadratic function has a maximum. Find h in (h, k).

$$h = -\frac{b}{2a} = -\frac{44}{2(-2)} = 11$$

Thus, the daily rental fee should be reduced by $11 to give the maximum revenue. This would make the new fee $21.

PRACTICE EXERCISE 4

If the car rental agency in Example 4 only got one new customer per $1 reduction in daily rate, what daily rate would give maximum revenue?

Answer: $26

3.6 EXERCISES A

Without graphing, find the vertex and any x-intercepts of the quadratic functions given in Exercises 1–6.

1. $f(x) = x^2 - 5x + 6$

2. $h(x) = x^2 + 2x + 2$

3. $k(x) = x^2 + 8x$

4. $m(x) = 2x^2 - 4$

5. $f(x) = 2x^2 + 8x - 5$

6. $h(x) = \frac{1}{3}x^2 - 2x + 5$

In Exercises 7–12 find the vertex, find any x-intercepts, and graph the function.

7. $g(x) = -x^2 - 2x + 3$

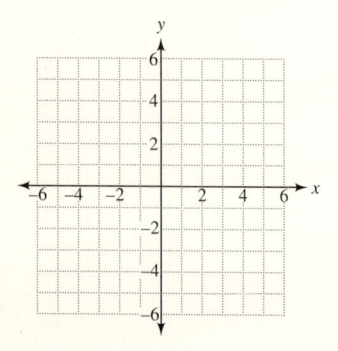

8. $k(x) = -x^2 + 5x - 2$

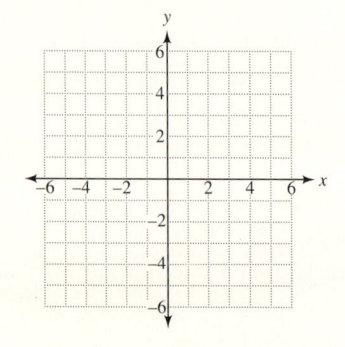

9. $f(x) = x^2 + 4x + 5$

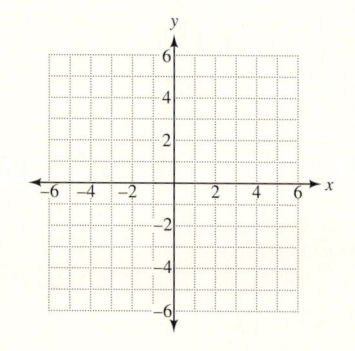

10. $g(x) = \dfrac{1}{2}x^2 - 7x + 20$

11. $k(x) = -4x^2 + 8x - 4$

12. $m(x) = 2(x + 1)^2 + 2$

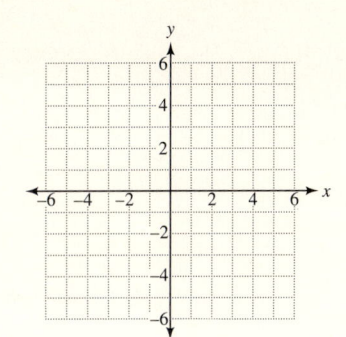

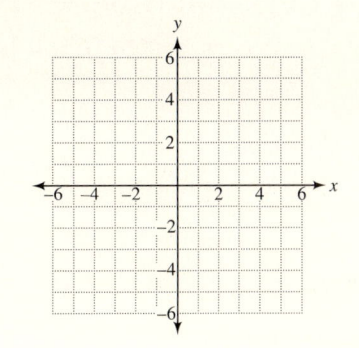

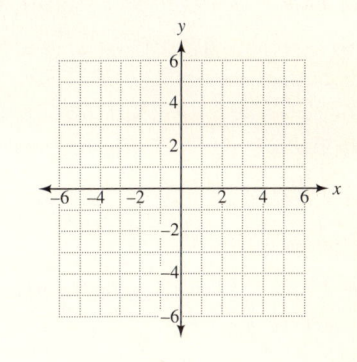

13. AGRICULTURE A rancher needs to build a pen with 1000 yd of fencing. What are the dimensions of the largest rectangular pen that he can enclose? What is the maximum area?

14. ENGINEERING The towers for a suspension bridge are 100 ft apart. If a coordinate system is set up as indicated in the figure, the cable between the towers follows the curve $f(x) = 0.01x^2 + 20$. At what height on each tower is the cable connected?

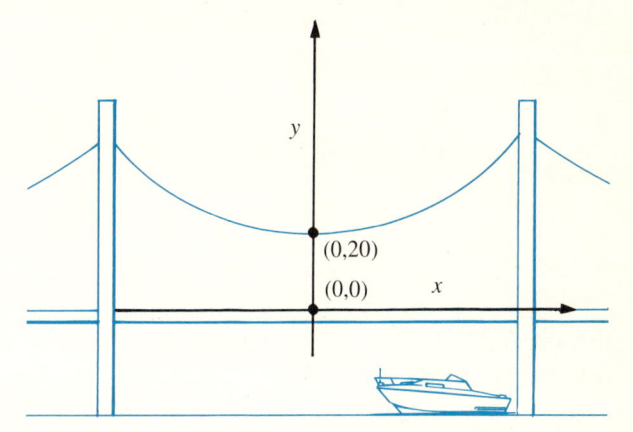

15. PHYSICS The height in feet of a small rocket t seconds after it is fired is given by the function $h(t) = -16t^2 + 256t$.
 (a) At what time will the rocket reach its maximum height?
 (b) What is the maximum height?
 (c) At what time will it reach the ground again?

16. MANUFACTURING The daily profit in a manufacturing plant is given by $P(x) = -x^2 + 120x - 1000$, where x is the number of machine parts produced. What is the number of parts that will give the maximum profit? What is the maximum profit?

17. MANUFACTURING The cost in a manufacturing operation is given by a quadratic function. The minimum cost occurs at the point (30, 200). If the curve also goes through the point (0, 2000), find the quadratic function.

18. BUSINESS Electronic components are sold by a wholesaler for $40 each on orders of 200 or less. For larger orders the price of each component over 200 is reduced at a rate of $0.20 times the number ordered over 200. What size order would give the wholesaler the largest revenue?

236 CHAPTER 3 RELATIONS, FUNCTIONS, AND GRAPHS

FOR REVIEW

In Exercises 19–20 find $f(g(0))$, $g(f(-2))$, $f(g(x))$, and $g(f(x))$.

19. $f(x) = -3x + 4$ and $g(x) = 2x - 5$

20. $f(x) = |x + 5|$ and $g(x) = x^2$

In Exercises 21–23 give the inverse of each function.

21. $f(x) = 7x + 8$

22. $g(x) = \sqrt{x + 4}$

23. $h(x) = \sqrt[3]{x + 1}$

In Exercises 24–25 graph the function and give the values of x for which the function increases and the values for which it decreases.

24. $g(x) = -5x - 3$

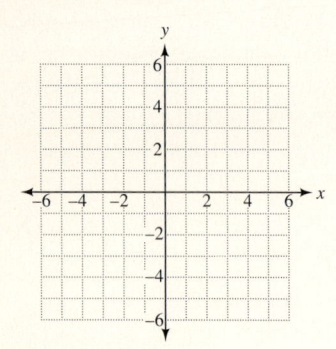

25. $f(x) = |x| + 3$

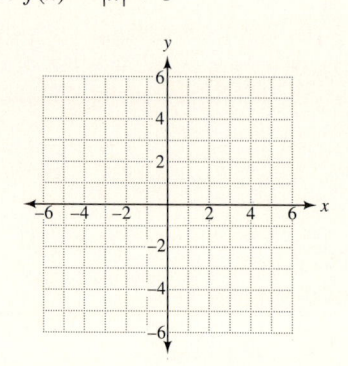

ANSWERS: 1. $\left(\frac{5}{2}, -\frac{1}{4}\right)$; $(2, 0)$ and $(3, 0)$ 2. $(-1, 1)$; no x-intercepts 3. $(-4, -16)$; $(0, 0)$ and $(-8, 0)$ 4. $(0, -4)$; $(\sqrt{2}, 0)$ and $(-\sqrt{2}, 0)$ 5. $(-2, -13)$; $\left(\frac{-4 + \sqrt{26}}{2}, 0\right)$ and $\left(\frac{-4 - \sqrt{26}}{2}, 0\right)$ 6. $(3, 2)$; no x-intercepts

7. 8. 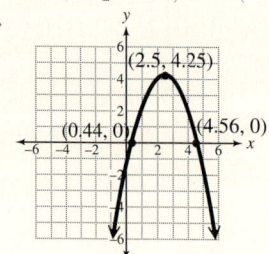 9. 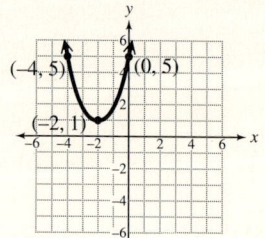 10.

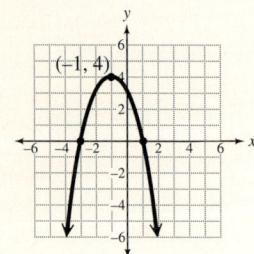

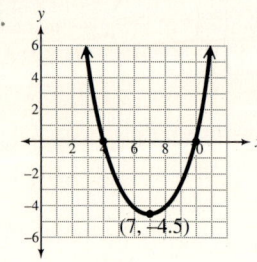

11. 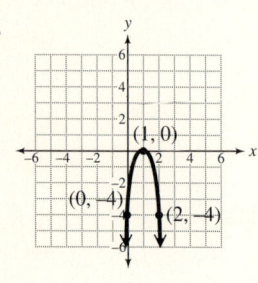 12. 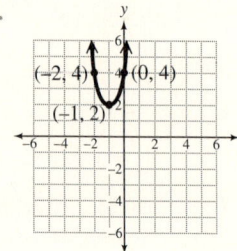 13. 250 yd by 250 yd; 62,500 yd² 14. 45 ft
15. (a) 8 sec (b) 1024 ft (c) 16 sec 16. 60; $2600
17. $C(x) = 2x^2 - 120x + 2000$ 18. 300 units
19. 19, 15, $-6x + 19$, $-6x + 3$ 20. 5, 9, $x^2 + 5$, $(x + 5)^2$
21. $f^{-1}(x) = \frac{x - 8}{7}$ 22. $g^{-1}(x) = x^2 - 4$, $x \geq 0$
23. $h^{-1}(x) = x^3 - 1$
24. decreases for all x 25. decreases for $x \leq 0$
increases for $x \geq 0$

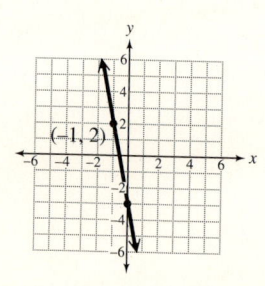

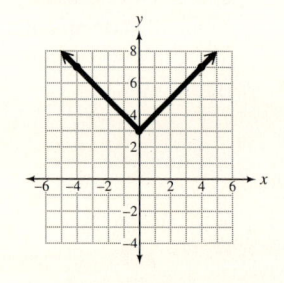

3.6 EXERCISES B

Without graphing, find the vertex and any x-intercepts of the quadratic functions in Exercises 1–6.

1. $g(x) = -x^2 - 6x - 8$
2. $h(x) = x^2 - 2x + 1$
3. $k(x) = -x^2 - 8x$
4. $m(x) = -3x^2 + 6$
5. $g(x) = -3x^2 + 4x + 3$
6. $k(x) = -4x^2 + 12x + 5$

In Exercises 7–12 find the vertex, find any x-intercepts, and graph the function.

7. $h(x) = x^2 - 7x + 6$
8. $k(x) = -x^2 - 4x + 2$
9. $f(x) = x^2 - 3x - 5$
10. $h(x) = -\frac{2}{3}x^2 - \frac{4}{3}x + 4$
11. $k(x) = \frac{1}{2}x^2 + 4x + 8$
12. $m(x) = -(x + 2)^2 - 2$

13. **AGRICULTURE** A rectangular cattle pen is to be built along a straight river bank. If no fence is needed along the river and there are 240 ft of fencing available, what are the dimensions of the largest rectangular pen possible? What is the maximum area?

14. **ENGINEERING** If the cable in the figure must be connected 30 ft up each tower and the cable follows the curve $f(x) = 0.01x^2 + 20$, how far apart must the towers be placed?

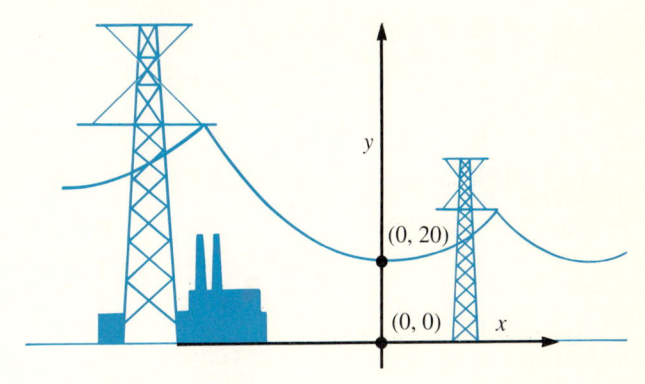

15. **MINING** Coal is dropped from a conveyor belt to the bed on a truck. The height in feet of coal above the bed of the truck is described by $h(x) = -16t^2 + 16$ where t is the number of seconds after the coal leaves the belt. If the velocity of the coal is given by $v(t) = -32t$ in ft/sec, what is the velocity with which the first coal hits the bed of the truck?

16. **MANUFACTURING** The daily manufacturing cost of a small plant is given by $C(x) = 2x^2 - 180x + 5200$ where x is the number of pairs of shoes produced. What is the number of pairs of shoes which will minimize the cost? What is the minimum cost?

17. **RETAILING** The profit in a retail operation is given by a quadratic function. The maximum profit occurs at the point (100, 400). If the curve also goes through the point (80, 200), find the quadratic function.

18. **BUSINESS** Shirts are sold by a wholesaler for $30 each on orders of 100 or less. For larger orders the price of each shirt over 100 is reduced at a rate of $0.15 times the number ordered over 100. What size order would give the wholesaler the largest revenue?

FOR REVIEW

In Exercises 19–20 find $f(g(0))$, $g(f(-1))$, $f(g(x))$, and $g(f(x))$.

19. $f(x) = 5x - 1$ and $g(x) = 1 - 2x$
20. $f(x) = |x - 2|$ and $g(x) = 1 - x^2$

In Exercises 21–23 give the inverse of each function.

21. $f(x) = -4x + 1$
22. $g(x) = \sqrt{x - 6}$
23. $h(x) = \sqrt[3]{x} + 1$

In Exercises 24–25 graph the function and give the values of x for which the function increases and the values for which it decreases.

24. $g(x) = 4 - x^2$
25. $f(x) = |x + 3|$

3.6 EXERCISES C

In Exercises 1–3 find the vertex of each quadratic function by writing it in the form $f(x) = a(x - h)^2 + k$.

1. $f(x) = x^2 + c$ **2.** $f(x) = x^2 + bx$ **3.** $f(x) = x^2 + bx + c$

4. The gasoline mileage m for an economy car is given by

$$m = -0.03s^2 + 2.7s \quad (25 < s < 65)$$

where s is the speed of the car in miles per hour. What is the most economical speed on the highway?

3.7 MATHEMATICAL MODELING

STUDENT GUIDEPOSTS

1 Examples of Models   **3** Nature of Models
2 Linear Models   **4** Quadratic Models

① EXAMPLES OF MODELS

In Example 4 of Section 3.6 we solved the business problem that was stated in the chapter introduction. The daily revenue function

$$R(n) = (20 + 2n)(32 - n)$$

was maximized to determine the best daily rental fee. In Section 3.3 (Example 4) we found that the approximate cost of painting a storage tank of radius r was

$$C(r) \approx 1.38r(r + 10).$$

In fact, since Section 2.2 where we solved many problems including those involving rate of motion, percent increase, and work rate, we have been working with mathematical models.

When we translate a problem described in words into mathematical symbols we are creating a mathematical model. Many models, such as

$$d = rt \quad \text{Distance} = \text{rate} \cdot \text{time}$$

come from our knowledge of physical laws while

$$S(x) = 1.07x \quad \text{Salary after a 7\% raise}$$

is derived from knowledge of percent increase. Still other models are derived from empirical data. We considered such a problem in Section 3.3 (Exercises A, #27). There a cost function

$$C(x) = 265.9x^2 - 36.8, \text{ where } 0.75 \leq x \leq 0.95,$$

was given as derived from empirical data.

In all mathematical models we have to be careful to consider the domain. For example, the cost model above was shown to be of no value for $x = 0.30$. We must also realize that most models only approximate the system they represent. For example, the height of a projectile at time t,

$$h(t) = \frac{1}{2}gt^2 + v_0t + h_0,$$

is derived by neglecting friction forces. Thus, calculations of height will be somewhat inaccurate, but the model is certainly accurate enough for many applications.

2 LINEAR MODELS

To illustrate some new techniques, we will start with a linear model, one which results in a linear function. Suppose we drop a stone from the top of a cliff and measure its velocity at two-second intervals to obtain the following data.

Time (sec)	Velocity (ft/sec)
0	0
2	55
4	140
6	180
8	265
10	310

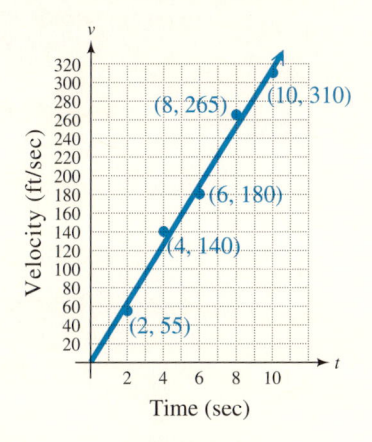

Figure 3.44

In Figure 3.44 we have plotted the points (t, v) and have drawn what we consider to be the best approximation of a curve to represent this data. In this case the curve is a straight line, and we can obtain an equation by selecting any two points on the line, calculating the slope, and then using the slope-intercept form since the line has y-intercept $(0, 0)$. The line we have drawn also appears to contain the point $(10, 315)$.

$$v(t) = mt + b$$
$$= \frac{315 - 0}{10 - 0} t + 0$$
$$= 31.5t$$

Thus, $v(t) = 31.5t$ is the linear model that we derived from the empirical data. In this case we have a check on our work since a derivation from the laws of motion gives $v(t) = 32t$ which is close enough to give us confidence in our work.

3 NATURE OF MODELS

We must understand that there are always inaccuracies associated with measurements and within the mathematical models. For example, frictional forces and strong winds can even change the velocity of a falling rock. As a matter of fact a strong updraft near the bottom of the cliff might slow the rock considerably. Thus, our model is best used within the domain of the collected data, $0 \leq t \leq 10$. The laws of motion might suggest that the model is good for $t > 10$, but the empirical data would not give the point $(12, 0)$ if the rock hit the ground at $t = 11$! The process of calculating value beyond collected data is called **extrapolation.** Values determined can be informative but must be used with care.

In the preceding discussion and in the following examples we are relying on the problem-solving strategy developed in Chapter 2. We try to analyze, tabulate and sketch, translate and solve, approximate, and check. Notice that the analysis in the problem above led us to important information about the domain of the function.

EXAMPLE 1 DEVELOPING A LINEAR MODEL

The following data represent the approximate average monthly salary of the employees of the Wesco Corporation.

Year	Average Monthly Salary
1970	$1550
1974	1700
1977	1900
1980	2450
1983	2500
1986	2800
1990	3250

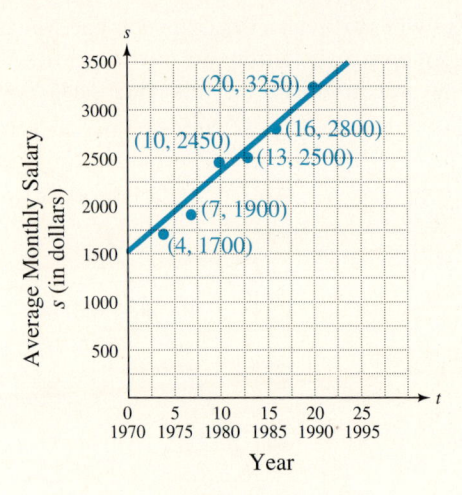

Figure 3.45

The data is graphed in Figure 3.45 and an approximate curve (straight line) is drawn through the data. We have designated 1970 as $t = 0$ in order to work with smaller numbers in the model.

(a) Find an approximate linear model (equation) to represent this data.

We can choose any two points that appear to be on the line, such as (6, 2000) and (19, 3100), calculate the slope and then use the point-slope form.

$$m = \frac{3100 - 2000}{19 - 6} \approx 84.6$$

Use the point (6, 2000) in the point-slope equation.

$$s - 2000 = 84.6(t - 6)$$
$$s(t) = 84.6t - 84.6(6) + 2000$$
$$\approx 84.6t + 1492$$

(b) Use this linear model to find the approximate average salary in 1979.

For 1979 use $t = 9$.

$$s(t) = 84.6t + 1492$$
$$s(9) = 84.6(9) + 1492 \approx 2253$$

The average annual salary in 1979 was approximately $2253.

(c) What would you estimate the average annual salary to be in 1995?

Since 1995 is five years beyond the last data point, we are extrapolating in this case but should get a good estimate. For 1995, $t = 25$.

$$s(t) = 84.6t + 1492$$
$$s(25) = 84.6(25) + 1492 \approx 3610$$

Employees can expect about $3610 in 1995.

PRACTICE EXERCISE 1

Refer to the data given in Example 1.

(a) Choose (0, 1500) and (25, 3600) from the graph in Figure 3.45 to determine a linear model for the data.

(b) Use the model from (a) to calculate $s(9)$.

(c) Use the model found in (a) to estimate $s(t)$ for 1965.

Answers: (a) $s(t) = 84t + 1500$
(b) $2256 (c) $1080

4 QUADRATIC MODELS

The first two models mentioned in this section used linear functions. In the next example we use a quadratic model to estimate the number of customers in a department store at any hour of the day.

EXAMPLE 2 Using a Quadratic Model

When a department store opens, the number of customers in the store increases from zero to a maximum number then down to zero again at closing time. Empirical data has shown that the number N of customers in the store is given as a function of time t by $N(t) = -15t^2 + 180t$, where $t = 0$ corresponds to the time the store opens at 10:00 A.M.

(a) At what time will the store have the maximum number of customers?

The graph of $N(t)$ is a parabola opening downward. The first coordinate of the vertex will give the time requested. For $N(t) = -15t^2 + 180t$, $a = -15$ and $b = 180$, the t coordinate of the vertex is

$$-\frac{b}{2a} = -\frac{180}{2(-15)} = 6,$$

and $t = 6$ means six hours after 10:00 A.M. or 4:00 P.M. The maximum number customers are in the store at 4:00 P.M.

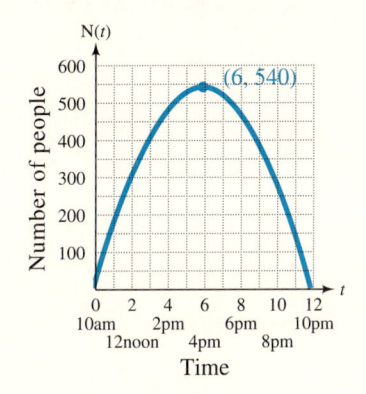

Figure 3.46

(b) What is the maximum number of customers?

The maximum number is the second coordinate of the vertex or $N(6)$.

$$N(t) = -15t^2 + 180t$$
$$N(6) = -15(6)^2 + 180(6) = 540$$

Thus the maximum number of customers is 540 at 4:00 P.M.

(c) At what time does the store close?

Since $N = 0$ at closing time,

$$-15t^2 + 180t = 0.$$
$$-15t(t - 12) = 0$$
$$t = 0 \quad \text{or} \quad t - 12 = 0$$

Since $t = 0$ corresponds to opening time, use $t = 12$ for closing. Thus closing is 12 hours after 10:00 A.M. which is 10:00 P.M.

PRACTICE EXERCISE 2

Use all the conditions in Example 2 with $N(t) = -12t^2 + 120t$ to answer the following questions.

(a) At what time will the store have the maximum number of customers?

(b) What is the maximum number of customers?

(c) At what time does the store close?

Answers: (a) 3:00 P.M. (b) 300 (c) 8:00 P.M.

3.7 EXERCISES A

Solve.

1. A car starts accelerating at $t = 0$ sec and the following measurements are taken on its velocity.

Time (sec)	Velocity (ft/sec)
0	0
2	25
4	50
6	90
8	130
10	155

   (a) Graph this data on the given grid and develop a linear model.
   (b) Use the model to find the velocity at $t = 5$ sec.
   (c) Do you think that this model is accurate for $t = 20$ sec?

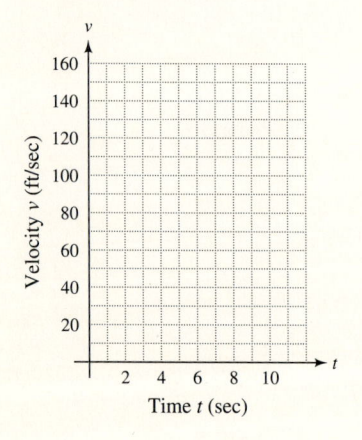

2. Starting August 15 ($t = 0$) the Rocky Mountain Lodge experienced the following decrease in rooms occupied by summer tourists.

Date	Rooms Occupied
August 15	68
August 23	60
August 27	50
September 6	38
September 10	30
September 14	27

   (a) Graph this data and develop a linear model.
   (b) If the lodge will close two days after the number of rooms occupied drops to 10, use the linear model to predict the closing date.
   (c) Would this model be of any value for December 15 after the ski season has started?

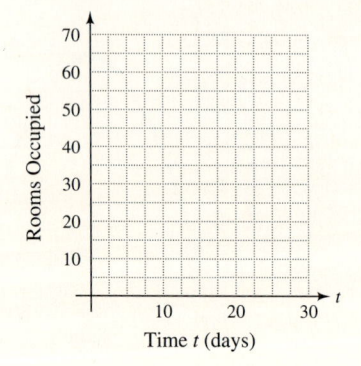

3. An appliance manufacturer has found that it costs $220 to produce 8 small appliances and $360 to produce 18 appliances.
   (a) Use these two data points, (8, 220) and (18, 360), to develop a linear model for this operation.
   (b) Use the model to find the cost of producing 12 appliances.
   (c) What are the fixed costs for this operation (the cost when no units are made)?

4. In a forest preserve a ranger counted 310 trees on 20 acres and 640 trees on 50 acres.
   (a) Use this data to develop a linear model.
   (b) Use the model to estimate the number of trees on 70 acres.
   (c) Would this model be useful for calculating the number of trees on 1 or 2 acres?

3.7 MATHEMATICAL MODELING 243

5. The fixed cost in a manufacturing operation is $1500. The cost per unit produced is $12. Revenue is produced by selling each unit for $22.
   (a) Determine a function C representing the total cost of producing x units and use it to find $C(27)$.
   (b) Find the revenue function R and calculate $R(27)$.
   (c) Give the profit function P if $P(x) = R(x) - C(x)$. Find $P(27)$.
   (d) What is the break-even point, the value of x for which the profit is zero? For what values of x is a profit made ($P(x) > 0$)? For what values of x is there a loss?

6. If n is the number of items sold in a wholesale operation, then empirical data has shown that $C(n) = n^2 - 70n + 900$ is the cost function and $R(n) = 2n^2 - 80n - 300$ is the revenue function.
   (a) Find the profit function $P(n) = R(n) - C(n)$.
   (b) What is the break-even point ($P(n) = 0$)?

7. The height of a projectile at time t is given by

$$h(t) = \frac{1}{2}gt^2 + v_0 t + h_0$$

where $g = -32$ ft/sec², v_0 is the initial velocity, and h_0 is the initial height.
   (a) If a baseball is hit upward from ground level with an initial velocity of 240 ft/sec, find the quadratic model.
   (b) What is the domain of this function?
   (c) What is the range of this function?

8. From the laws of motion we know that $d = rt$, distance is equal to the rate times the time.
   (a) If $r(x) = 3x - 6$ and $t(x) = 4x - 20$, where $0 \le x \le 10$, find the model for $d(x)$.
   (b) For what values of x is the function 0?
   (c) What is the range of $d(x)$?

FOR REVIEW

Find the vertex and x-intercepts for each quadratic function in Exercises 9–10.

9. $f(x) = 2x^2 + 4x - 48$

10. $g(x) = -x^2 + 16x - 63$

ANSWERS:
1. (a) $v(t) = 15t$ (b) 75 ft/sec (c) No. The model predicts 300 ft/sec which is probably beyond the car's capability. 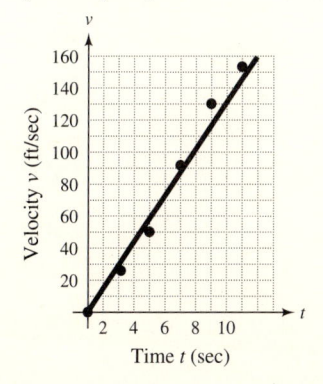
2. (a) $r(t) = -1.5t + 70$ (b) September 26 (c) No. Ski season will bring in new tourists. 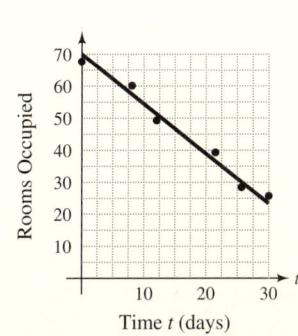
3. (a) $C(x) = 14x + 108$ (b) $276 (c) $108
4. (a) $t(a) = 11a + 90$ (b) 860 trees (c) No. It gives 90 trees on 0 acres.
5. (a) $C(x) = 12x + 1500$; $1824
   (b) $R(x) = 22x$; $594 (c) $P(x) = 10x - 1500$; $-$1230 (d) 150 units; $x > 150$; $0 \le x < 150$
6. $P(n) = n^2 - 10n - 1200$ (b) 40 items
7. (a) $h(t) = -16t^2 + 240t$ (b) $0 \le t \le 15$ (c) $0 \le h(t) \le 900$
8. (a) $d(x) = 12x^2 - 84x + 120$ (b) 2, 5 (c) $-27 \le d(x) \le 480$
9. $(-1, -50)$; $(4, 0)$, $(-6, 0)$
10. $(8, 1)$; $(7, 0)$, $(9, 0)$

3.7 EXERCISES B

Solve.

1. A truck starts decelerating at $t = 0$ sec and the following measurements are taken on its velocity.

Time (sec)	Velocity (ft/sec)
0	92
2	68
4	50
6	38
8	22
10	0

   (a) Graph this data and develop a linear model.
   (b) Use the model to find the velocity at $t = 1.5$ sec.
   (c) What is the domain of this function? Is the model of value outside the domain?

2. Starting November 1 ($t = 1$) the Rocky Mountain Lodge experienced the following increase in number of rooms occupied by skiers.

Date	Rooms Occupied
November 1	48
November 8	80
November 25	84
December 10	98
December 20	115
December 30	136

   (a) Graph this data and develop a linear model.
   (b) Which of the data points plotted is off from the linear model the most?
   (c) Would this model be of any value for January 15 after the Christmas break is over?

3. A car dealer makes a profit of $2500 when 10 cars are sold. If 22 cars are sold the profit is $6700.
   (a) Use these two data points to develop a linear model for this operation.
   (b) What is the profit if no cars are sold?
   (c) What is the minimum number of cars that must be sold to make a profit?

4. A ranger counted 15 antelope on 13 acres and 45 antelope on 33 acres.
   (a) Use this data to develop a linear model.
   (b) Use the model to estimate the number of antelope on 200 acres.
   (c) Would this model be useful for calculating the number of antelope on 1 or 2 acres?

5. The fixed cost in a retail operation is $990. The cost to sell one suit is $10 and the selling price is $120.
   (a) Give a function C representing the total cost of selling n suits.
   (b) What is the revenue function R?
   (c) Give the profit function P.
   (d) What is the break-even point?

6. Empirical data has shown that if n is the number of units produced then $C(n) = n^2 - 20n + 1200$ is the cost function and $R(n) = 2n^2 - 90n - 600$ is the revenue function.
   (a) Find the profit function $P(n) = R(n) - C(n)$.
   (b) What is the break-even point ($P(n) = 0$)?

7. The height of a projectile at time t is given by
   $$h(t) = \frac{1}{2}gt^2 + v_0 t + h_0$$
   where $g = -32$ ft/sec^2, v_0 is the initial velocity, and h_0 is the initial height.
   (a) If a ball is thrown upward from a height of 96 ft with an initial velocity of 80 ft/sec, find the quadratic model.
   (b) What is the domain of this function?
   (c) What is the range of this function?

8. In an electrical circuit, $E = IR$.
   (a) If $I(t) = 2t - 8$ and $R(t) = 5t + 20$, where $0 \leq t \leq 10$, find the model for $E(t)$.
   (b) For what values of t is the function 0?
   (c) What is the range of $E(t)$?

FOR REVIEW

Find the vertex and x-intercepts for each quadratic function in Exercises 9–10.

9. $f(x) = 6x^2 - 13x - 8$

10. $g(x) = -2x^2 + 24x - 40$

3.7 EXERCISES C

To find the equation $y = mx + b$ of the best straight line through three points (x_1, y_1), (x_2, y_2), (x_3, y_3), use the formulas

$$m = \frac{3(x_1y_1 + x_2y_2 + x_3y_3) - (x_1 + x_2 + x_3)(y_1 + y_2 + y_3)}{3(x_1^2 + x_2^2 + x_3^2) - (x_1 + x_2 + x_3)^2}$$

and

$$b = \frac{1}{3}[(y_1 + y_2 + y_3) - m(x_1 + x_2 + x_3)].$$

In Exercises 1–2 find the equation of the best straight line through the given points.

1. (2, 1), (5, 6), (−4, −4)

2. (−3, 4), (0, 0), (5, −4)
   $\left[\text{Answer: } y = -\frac{48}{49}x + \frac{32}{49}\right]$

3. For the points in Exercise 2 and the function $f(x) = -\frac{48}{49}x + \frac{32}{49}$, give the value of
   $$[f(x_1) - y_1]^2 + [f(x_2) - y_2]^2 + [f(x_3) - y_3]^2$$
   correct to two decimal places.

4. Use the points in Exercise 2 to show that for $g(x) = -x + \frac{3}{5}$, the value of
   $$[g(x_1) - y_1]^2 + [g(x_2) - y_2]^2 + [g(x_3) - y_3]^2$$
   is greater than the corresponding sum in Exercise 3.

3.8 VARIATION

STUDENT GUIDEPOSTS

1 Direct Variation
2 Inverse Variation
3 Joint Variation

1 DIRECT VARIATION

Expressions such as "the cost of production *varies* with the number of units produced" and "the volume of a gas *varies* with the temperature" are frequently used. *Variation* can be defined precisely with one variable represented as a function of one or more other variables. For example, if a car is moving at a constant rate of 55 miles per hour, the distance traveled is a function of the time.

$$d = 55t$$

Expressed in terms of variation we say "the distance *varies directly* as the time." This is an example of the first type of variation called *direct variation*.

> **Direct Variation**
>
> If two variables x and y are so related that
>
> $$y = cx$$
>
> where c is a constant, we say that y **varies directly as** x, y **is proportional to** x, or simply y **varies as** x. We call c the **constant of variation** or **constant of proportionality**.

Notice from the definition that the subject of variation is closely related to proportion studied in Section 2.2. In fact, variation is another way of looking at proportion problems.

EXAMPLE 1 Using Direct Variation

Find the equation of variation.

(a) y varies directly as x, and when $x = \frac{3}{2}$ then $y = 10$.

First we give the equation described by y *varies directly as* x.

$$y = cx$$

Find the value of the constant by using the information $y = 10$ when $x = \frac{3}{2}$.

$$10 = c \cdot \frac{3}{2}$$

$$\frac{20}{3} = c$$

Thus, the equation of variation is $y = \frac{20}{3}x$.

(b) The volume V of a gas under constant pressure varies directly as the temperature T. The volume is 220 cm³ when the temperature is 320°K (degrees Kelvin). The equation of variation is

$$V = cT.$$

Using $V = 220$ cm³ when $T = 320°K$, we have

$$220 = c(320).$$

$$c = \frac{220}{320} = \frac{11}{16}$$

Thus
$$V = \frac{11}{16}T.$$

PRACTICE EXERCISE 1

Find the equation of variation.

(a) u varies directly as v, and when $v = 1.4$ then $u = 28$.

(b) At a constant rate of speed, distance d varies directly as time t, and $d = 55$ when $t = 0.11$.

Answers: (a) $u = 20v$ (b) $d = 500t$

Notice in Example 1(b) that after it has been determined that $V = \frac{11}{16}T$, we can find the volume for any temperature. For example, if $T = 480°K$, then

$$V = \frac{11}{16}(480) = 330 \text{ cm}^3.$$

Also, if $V = 77$ cm³, we can find T.

$$77 = \frac{11}{16}T$$

$$T = \frac{16}{11}(77) = 112°K$$

In the direct variation relation, $y = cx$, the variable y is a linear function of x. We can also consider y varying as powers of x. The following table gives several possibilities.

Variation Statement	Equation
y varies as the square of x	$y = cx^2$
u varies as the square root of v	$u = c\sqrt{v}$
the volume V of a sphere varies as the cube of the radius r	$V = cr^3 \quad \left(c = \frac{4}{3}\pi\right)$

EXAMPLE 2 APPLICATION TO PHYSICS

The period P of a simple pendulum varies directly as the square root of the length l. If the period is 3.2 sec when the length is 12.5 cm, find the period when $l = 20.2$ cm.

The equation of variation is

$$P = c\sqrt{l}.$$

To find c, use $P = 3.2$ sec when $l = 12.5$ cm.

$$3.2 = c\sqrt{12.5}$$

$$c = \frac{3.2}{\sqrt{12.5}} \approx 0.91$$

Thus, $\qquad P = 0.91\sqrt{l}.$

When $l = 20.2$,

$$P = (0.91)\sqrt{20.2} \approx 4.1 \text{ sec}$$

PRACTICE EXERCISE 2

The surface area S of a sphere varies directly as the square of the radius r, and $S = 32.5$ m^2 when $r = 2.6$ m. Find S when $r = 17.6$.

Answer: $S \approx 1490$ m^2

2 INVERSE VARIATION

Consider now a second kind of variation called *inverse variation*.

Inverse Variation

If two variables x and y are so related that

$$y = \frac{c}{x}$$

where c is a constant, then we say that y **varies inversely as** x or y is **inversely proportional to** x. Again, c is called the **constant of variation.**

In direct variation, y increases as x increases when $c > 0$. In contrast, for inverse variation y decreases as x increases ($c > 0$). For example, over a fixed distance of 100 miles the time t varies inversely as the rate r of travel.

$$t = \frac{100}{r}$$

Thus, as the rate r increases, the time t to travel the 100 mi decreases.

EXAMPLE 3 APPLICATION TO PHYSICS

The intensity of illumination I of a light source varies inversely as the square of the distance d from the source. When $d = 20$ ft, then $I = 5000$ candlepower. Find I when d is 60 ft.

First write the equation of variation and then solve for c.

$$I = \frac{c}{d^2}$$

$$5000 = \frac{c}{(20)^2} \qquad I = 5000 \text{ candlepower when } d = 20 \text{ ft}$$

$$c = 2{,}000{,}000$$

PRACTICE EXERCISE 3

At constant temperature the volume V of a gas varies inversely as the pressure P, and $V = 0.44$ ft^3 when $P = 5.26$ lb/ft^2. Find P when $V = 2.81$ ft^3.

$$I = \frac{2,000,000}{d^2} \qquad \text{Equation of variation}$$

$$I = \frac{2,000,000}{(60)^2} \qquad \text{Let } d = 60 \text{ ft}$$

$$\approx 556 \text{ candlepower}$$

Answer: 0.82 lb/ft^3

3 JOINT VARIATION

A variable z can vary with two or more variables. This is called *joint variation*.

> **Joint Variation**
>
> If three variables x, y, and z are so related that
>
> $$z = cxy$$
>
> where c is a constant, we say that z varies directly as x and directly as y or that z **varies jointly** as x and y.

Other variation equations can be found by combining the various types of variation. The following table gives several examples.

Variation Statement	Equation
y varies jointly with x and z and inversely as w	$y = \dfrac{cxz}{w}$
the volume V of a gas varies directly as the temperature T and inversely as the pressure P	$V = \dfrac{cT}{P}$
the resistance R of a wire varies directly as the length l and inversely as the square of the diameter d	$R = \dfrac{cl}{d^2}$

EXAMPLE 4 AN APPLICATION TO ENGINEERING

The resistance R of a wire at a constant temperature varies directly as the length l and inversely as the square of the diameter d. A section of wire having a diameter of 0.01 inches and a length of 1 foot has a resistance of 8.2 ohms. Find the resistance of a 1-mile-long wire with diameter 0.05 inches.

The equation is $R = cl/d^2$. Substitute the first set of values.

$$8.2 = \frac{c \cdot 1}{(0.01)^2}$$

$$(0.01)^2(8.2) = c$$

$$0.00082 = c$$

Thus, the variation equation is $R = 0.00082l/d^2$. Substitute $d = 0.05$ and $l = 5280$ ft (1 mi = 5280 ft).

$$R = \frac{(0.00082)(5280)}{(0.05)^2} \approx 1732$$

The resistance is approximately 1732 ohms.

PRACTICE EXERCISE 4

The volume V of a gas varies directly as the temperature T and inversely as the pressure. When $T = 480°$ and $P = 16 \text{ lb/ft}^2$, then $V = 22 \text{ ft}^3$. Find T when $P = 24 \text{ lb/ft}^2$ and $V = 11 \text{ ft}^3$.

Answer: 360°

3.8 EXERCISES A

Find the equation of variation in Exercises 1–5.

1. y varies directly as x, and when $x = 16$ then $y = 8$.

2. z varies inversely as w, and $z = 12$ when $w = 5$.

3. z varies jointly as x and y. When $x = 2$ and $y = 7$, then $z = 42$.

4. a is directly proportional to the square of b and inversely proportional to the square root of d. When $b = 3$ and $d = 25$, then $a = 405$.

5. w varies jointly as the square of x and the cube root of y and inversely as the square root of z. If $x = 40$, $y = 125$, and $z = 400$, then $w = 330$.

6. **CONSUMER** The simple interest I paid on a loan for one year varies directly as the amount borrowed A. If $250 interest was paid on a loan of $2000, what interest must be paid on a loan of $9250?

7. **MANUFACTURING** The daily cost C of production varies directly as the number n of units produced, when $3000 \leq n \leq 7000$. If the cost is $25,200 when 4100 units are produced, what will be the cost if production is increased to 6400?

8. **MANUFACTURING** In a manufacturing operation the productivity P varies directly as the square of the efficiency e and inversely as the square root of the downtime d. If $P = 7.6$ when $e = 7$ and $d = 16$, find P for $e = 8$ and $d = 9$.

9. **PHYSICS** The volume V of a gas varies directly as the temperature T and inversely as the pressure P. If $V = 16$ in^3 when $T = 480°$K and $P = 45$ lb/in^3, find V when $T = 1200°$K and $P = 15$ lb/in^2.

10. **SCIENCE** The weight w of a body above the surface of the earth is inversely proportional to the square of its distance d from the center of the earth. What is the effect on the weight when the distance is doubled?

11. **PHYSICS** The period of vibration P of a pendulum varies directly as the square root of the length l. If the period of vibration is 1.56 sec when the length is 9.25 cm, what is the period when $l = 16.8$ cm?

12. **PHYSICS** The intensity I of light varies inversely as the square of the distance d from the source. If the intensity must be doubled, what must be done to the distance?

13. **MANUFACTURING** The time T necessary to make an enlargement of a photo negative varies directly as the area A of the enlargement. If it takes 25 seconds to make a 5-by-7 enlargement, find the time it takes for an 11-by-14 enlargement.

14. **ENGINEERING** The speed s of a car when the brakes are applied can be estimated by measuring the length l of the skid marks. This is possible since s is directly proportional to the square root of l. If a car traveling 45 mph makes a skid mark of 65 feet, estimate the speed of a car that leaves a mark 120 ft long.

15. ENGINEERING The maximum safe load m of a horizontal beam varies jointly with the width w and the square of the thickness t and inversely as the length l. If a second beam has width $W = 5w$, thickness $T = \frac{1}{2}t$, and length $L = 3l$, what is the relationship between the safe load M on the second beam and the load on the first?

FOR REVIEW

In Exercises 16–17 find the mathematical model relating f and x.

16. $f(x)$ varies directly with x, and when $x = 6.5$ then $f(x) = 13.0$.

17. $f(x)$ is inversely proportional to x and $f(x) = 0.25$ when $x = 8$.

18. BUSINESS The cost function C varies directly as the number n of units produced and when $n = 25$ then $C(n) = \$100$. The revenue function R varies directly as the square of n, and when $n = 25$ then $R(n) = \$125$.
(a) Give a linear model for C.

(b) Develop a quadratic model for R.

(c) If the profit $P(n) = R(n) - C(n)$, find the break-even point, where $P(n) = 0$.

ANSWERS: 1. $y = \frac{1}{2}x$ 2. $z = \frac{60}{w}$ 3. $z = 3xy$ 4. $a = \frac{225b^2}{\sqrt{d}}$ 5. $w = \frac{0.825x^2\sqrt[3]{y}}{\sqrt{z}}$ 6. \$1156.25 7. \$39,337 8. 13.2 9. 120 in³ 10. weight is divided by 4 11. 2.10 sec 12. distance must be multiplied by $\frac{1}{\sqrt{2}}$ 13. 110 sec 14. 61 mph 15. $M = \frac{5}{12}m$ 16. $f(x) = 2x$ 17. $f(x) = \frac{2}{x}$ 18. (a) $C(n) = 4n$ (b) $R(n) = 0.2n^2$ (c) 0 units or 20 units

3.8 EXERCISES B

Find the equation of variation in Exercises 1–5.

1. p varies directly as q, and when $q = 40$ then $p = 400$.

2. u varies inversely as v, and $u = 0.025$ when $v = 0.4$.

3. m varies jointly as p and q. When $p = 0.25$ and $q = 9.5$ then $m = 8.55$.

4. P is directly proportional to the cube root of R and inversely proportional to the fourth power of T. When $R = 0.027$ and $T = 0.1$, then $P = 3300$.

5. L varies directly as the square root of M and inversely as the square of N and the cube of K. If $M = 0.09$, $N = 0.2$, and $K = 2$, then $L = 1.5$.

6. CONSUMER The yearly interest I received on savings varies directly as the interest rate r. If \$468 in interest was received when the rate was 9%, what interest would have been received if the rate were 7%?

7. BUSINESS The daily profit P varies directly as the number of units sold n, if $200 \leq n \leq 1000$. If the profit is \$6520 when 580 units are sold, what is the profit when only 340 units are sold?

8. MANAGEMENT In a wholesale operation the movement factor M is directly proportional to the square root of the number n of workers on the job and inversely proportional to the square of the break time t. If $M = 5.8$ when $n = 9$ and $t = 2.2$, find M when $n = 16$ and $t = 3.1$.

9. **PHYSICS** The volume V of a gas varies directly as the temperature T and inversely as the pressure P. If $V = 16$ in^3 when $T = 480°$K and $P = 45$ lb/in^2, find the pressure P when $V = 6$ in^3 and $T = 360°$K.

10. **ENGINEERING** The horsepower P required to move a ship varies directly as the cube of the speed s. What is the effect on P when the speed is cut in half?

11. **SCIENCE** The gravitational attraction A between two masses varies inversely as the square of the distance d between them. The force of attraction is 17.6 lb when the masses are 12.4 ft apart. What is the attraction when the masses are 34.6 ft apart?

12. **PHYSICS** The force F of attraction between two charged particles varies jointly as the charges q_1 and q_2 and inversely as the square of the distance d between them. What happens to the force when q_1 is doubled, q_2 is divided by 3, and d is doubled?

13. **PHYSICS** The weight w of a liquid varies directly as its volume V. If the weight of the liquid in a cylinder of radius 2 cm and height 12 cm is 142 g, find the weight in a sphere of radius 5 cm.

14. **ENGINEERING** The speed s of a car when the brakes are applied can be estimated by measuring the length l of the skid marks. This is possible since s is directly proportional to the square root of l. If a car traveling 45 mph makes a skid mark of 65 feet, estimate the length of a skid mark of a car traveling 90 mph when the brakes are applied.

15. **ENGINEERING** The total pressure P of the wind on a flat surface varies jointly as the area A of the surface and the square of the velocity of the wind v. If the area of the surface is doubled, what must happen to the speed of the wind to keep P the same?

FOR REVIEW

In Exercises 16–17 find the mathematical model relating f to x.

16. $f(x)$ varies directly as x, and when $x = 52$ then $f(x) = 13$.

17. $f(x)$ is inversely proportional to x and $f(x) = 84$ when $x = 0.02$.

18. **BUSINESS** The cost function C varies directly as the square of the number n of units sold, and when $n = 60$ then $C(n) = \$240$. The revenue function R also is directly proportional to the square of n, and $R(60) = 360$.
    (a) Give a quadratic model for C.

    (b) Develop a quadratic model for R.

    (c) Find the break-even point.

3.8 EXERCISES C

1. If w varies directly as the square of u, give the equation of variation for u in terms of w.

2. If a varies inversely as the cube root of b, give the equation of variation of b in terms of a.

3. If r varies jointly as the cube of s and the square root of t and inversely as the square of u, give the equation of variation of u in terms of the other variables.

4. If w varies directly as the square root of x and inversely as the square of y and the cube root of z, find the equation of variation of x in terms of the other variables.

CHAPTER 3 EXTENSION PROJECT

In Example 1 of Section 3.7 we plotted data on average monthly salaries of the Wesco Corporation (see Figure 3.45) to obtain what is called a **scatter diagram**. We then drew a straight line that seemed to fit the data the best. By reading two points from the graph we found the linear model, $s(t) = 84.6t + 1492$. However, in Practice Exercise 1, by choosing two different points, we determined the linear model $s(t) = 84t + 1500$. We knew at the time that we were working with approximations and did not worry about the small differences.

Now we would like to find out which of these models is the better representation of the data and show a procedure for determining the best of all models. To find the better of these two we calculate for each point the difference between the given second coordinate of a point and the second coordinate determined by the linear model. If we square each of these numbers to make them positive, we can add them to get a measure of the distance that the set of points is from the graph. We do this in the following tables using d for the difference between second coordinates.

Since the first linear model has the smaller sum of squares, we assume that it is the better of the two. This method uses the **least-squares criterion** for determining the best line through data.

It can be shown by minimizing the sum of the d^2 that the best-fitting line through a set of n points (x, y) has the **regression equation** $y = mx + b$ where

$$m = \frac{n(\Sigma xy) - (\Sigma x)(\Sigma y)}{n(\Sigma x^2) - (\Sigma x)^2} \quad \text{and} \quad b = \frac{1}{n}(\Sigma y - m\Sigma x).$$

The symbol Σ means summation. Thus, Σxy means to find the products of the xy and find the sum. Notice the difference between Σx^2 which means to find the x^2 and then add while $(\Sigma x)^2$ means find the sum of the x and then square. For calculations of this nature a computer is most helpful. We give below a computer program and a run on the data on average salaries.

The best-fitting line has equation $y = 87.6x + 1431$ or, using the s and t variables, $s(t) = 87.6t + 1431$.

$s(t) = 84.6t + 1492$

Point	$s(t)$	d	d^2
(0, 1550)	1492	58	3364
(4, 1700)	1830	−130	16900
(7, 1900)	2084	−184	33856
(10, 2450)	2338	112	12544
(13, 2500)	2592	92	8464
(16, 2800)	2846	−46	2116
(20, 3250)	3184	66	4356
			81,600

$s(t) = 84t + 1500$

Point	$s(t)$	d	d^2
(0, 1550)	1500	50	2500
(4, 1700)	1836	−136	18496
(7, 1900)	2088	−188	35344
(10, 2450)	2340	110	12100
(13, 2500)	2592	92	8464
(16, 2800)	2844	−44	1936
(20, 3250)	3180	70	4900
			83,740

```
10 INPUT N
20 XY=0
30 X=0
40 Y=0
50 XS=0
60 FOR I = 1 TO N
70 INPUT X(I),Y(I)
80 NEXT I
90 FOR I = 1 TO N
100 XY = XY + X(I)*Y(I)
110 X = X + X(I)
120 Y = Y + Y(I)
130 XS = XS + X(I)^2
140 NEXT I
150 M = (N*XY−X*Y)/(N*XS−X^2)
160 B = (Y−M*X)/N
170 PRINT "y = mx + b = (";M;")x + (";B;")"

RUN
? 7
? 0,1550
? 4,1700
? 7,1900
? 10,2450
? 13,2500
? 16,2800
? 20,3250
y = mx + b = ( 87.58621 )x + ( 1431.281 )
```

THINGS TO DO:

1. Collect and plot a set of data, draw a line and approximate its equation, and then use a computer or calculator to determine the regression equation.
2. Calculate Σd^2 for $s(t) = 87.6t + 1431$ and compare the results to those of the other two lines.
3. Study the literature to learn more about the least-squares criterion and regression equations.
4. Investigate the variety of applications of this type of mathematics.

CHAPTER 3 REVIEW

KEY TERMS

3.1
1. A **rectangular** or **Cartesian coordinate system** is formed when a horizontal and vertical number line are superimposed so that the two origins coincide and the lines are perpendicular.
2. In a rectangular coordinate system the horizontal number line is the **x-axis**, the vertical number line the **y-axis**, and the point of intersection is the **origin**.
3. We call (x, y) an **ordered pair** since the order, x first and y second, must be preserved to properly identify points.
4. In the ordered pair (a, b) a is the **x-coordinate** or **abscissa** and b is the **y-coordinate** or **ordinate**.
5. **Quadrants** are the four sections of a rectangular coordinate system.
6. We **graph** a set when we plot the points in a rectangular coordinate system.
7. The distance between two points with coordinates (x_1, y_1) and (x_2, y_2) is given by the **distance formula**
$$d = \sqrt{(x_1 - x_2)^2 + (y_1 - y_2)^2}.$$
8. The midpoint of the line segment joining (x_1, y_1) and (x_2, y_2) is given by the **midpoint formula**
$$(\bar{x}, \bar{y}) = \left(\frac{x_1 + x_2}{2}, \frac{y_1 + y_2}{2}\right).$$

3.2
1. A **linear** or **first-degree equation** is an equation that has as its graph a straight line.
2. The **general form** of a linear equation is
$$ax + by + c = 0.$$
3. The **x-intercept** is the point where a line crosses the x-axis, and the **y-intercept** is the point where a line crosses the y-axis.
4. The **slope** of a line through (x_1, y_1) and (x_2, y_2) is given by
$$m = \frac{y_2 - y_1}{x_2 - x_1}.$$
5. The **point-slope form** of the equation of a line with slope m and passing through the point (x_1, y_1) is
$$y - y_1 = m(x - x_1).$$
6. The **slope-intercept form** of the equation of a line with slope m and y-intercept $(0, b)$ is
$$y = mx + b.$$
7. Two distinct lines with slopes m_1 and m_2 are **parallel** if and only if $m_1 = m_2$.
8. Two distinct lines with slopes m_1 and m_2 are **perpendicular** if and only if $m_1 m_2 = -1$.

3.3
1. A correspondence between a first set of objects, the **domain**, and a second set of objects, the **range**, is a **relation** if to each element of the domain there corresponds one or more elements in the range. Equivalently, a **relation** is a set of ordered pairs.
2. A **function** is a relation with the property that to each element in the domain there corresponds one and only one (exactly one) element in the range. In terms of ordered pairs, a **function** is a relation with the property that no two ordered pairs have the same x-coordinates and different y-coordinates.
3. A **linear function** is of the form
$$f(x) = mx + b.$$
4. A **constant function** is of the form
$$f(x) = b.$$
5. The **identity function** is
$$f(x) = x.$$
6. A **quadratic function** can be written
$$f(x) = ax^2 + bx + c. \quad (a \neq 0)$$

3.4
1. The **vertical line test** is used to determine if a graph is the graph of a function.
2. A function is **increasing** on I if whenever $a < b$ then $f(a) < f(b)$.
3. A function is **decreasing** on I if whenever $a < b$ then $f(a) > f(b)$.
4. A function is **constant** on I if $f(a) = f(b)$ for all a and b in I.
5. The **absolute value function** is
$$f(x) = \begin{cases} x, & x \geq 0 \\ -x, & x < 0 \end{cases}.$$
6. A **transformation** is a shift in the graph of a function.
7. If $f(x) = f(-x)$ then f is an **even function**.
8. If $f(x) = -f(-x)$ then f is an **odd function**.
9. If an equation is unchanged when x is replaced by $-x$, then the graph is **symmetric with respect to the y-axis**.
10. If an equation is unchanged when y is replaced by $-y$, then the graph is **symmetric with respect to the x-axis**.
11. If an equation is unchanged when x is replaced by $-x$ and y is replaced by $-y$, then the graph is **symmetric with respect to the origin**.
12. The **greatest integer function** is written
$$f(x) = [x].$$

3.5
1. The **composite function** $g \circ f$, read "g composed with f," is defined by
$$(g \circ f)(x) = g(f(x)).$$
2. A function g is the **inverse** of f if
$f(g(x)) = x$ for all x in the domain of g
and $g(f(x)) = x$ for all x in the domain of f.

The inverse of f is denoted f^{-1}.
3. A function is a **one-to-one function** if for x_1 and x_2 in the domain of f, $x_1 \neq x_2$ implies $f(x_1) \neq f(x_2)$.

3.6
1. A quadratic function is in **standard form** if it is written
$$f(x) = a(x - h)^2 + k.$$
2. The **vertex** of the quadratic function $f(x) = ax^2 + bx + c$ is
$$(h, k) = \left(-\frac{b}{2a}, f\left(-\frac{b}{2a}\right)\right).$$
3. The **x-intercepts** of a quadratic function are found by solving the quadratic equation $ax^2 + bx + c = 0$.
4. The **line of symmetry** of $f(x) = ax^2 + bx + c$ is
$$x = -\frac{b}{2a}.$$

3.7
1. When we translate a problem described in words into mathematical symbols we are creating a **mathematical model**.
2. A **linear model** is a model which results in a linear function.
3. A **quadratic model** is a model which results in a quadratic function.

3.8
1. If $y = cx$ then y **varies directly as** x or y is **proportional to** x. The constant c is the **constant of variation** or **constant of proportionality**.
2. If $y = \frac{c}{x}$ then y **varies inversely as** x or y is **inversely proportional to** x.
3. If $z = cxy$ then z **varies jointly as** x **and** y.

REVIEW EXERCISES

Part I

3.1
1. Graph the solutions to the equation $3x - 2y = 6$.
2. Find the distance between the points $(-2, 6)$ and $(7, 3)$.

3. If (7, 2) is the midpoint of the line segment joining (a, b) and (8, 5), find (a, b).

4. Find all points $(a, 2)$ which are 5 units from $(2, -1)$.

5. Find the slope of the line through the points $(-2, 8)$ and $(-3, -1)$.

3.2 *Find the general form of the equation of a line in Exercises 6–7.*

6. Passing through $(-3, -7)$ and $(-1, 2)$

7. With x-intercept $(-2, 0)$ and parallel to $4x - 3y + 6 = 0$

8. Find the slope and y-intercept of $7x - 2y = 4$.

State whether the lines are parallel or perpendicular in Exercises 9–10.

9. $8x + 4y - 7 = 0$ and $-12x - 6y + 5 = 0$

10. $x + 3y - 2 = 0$ and $6x - 2y + 7 = 0$

11. Is the triangle with vertices $A(5, 6)$, $B(-2, 1)$, and $C(3, -6)$ a right triangle?

12. Give the general form of the perpendicular bisector of the line segment joining $(-2, -4)$ and $(4, 8)$.

3.3 *In Exercises 13–14 state whether the relation is a function if x is the independent variable and y the dependent variable.*

13. $y = 2x^2 + 3$

14. $x = 2y^2 + 3$

Give the real numbers for which each function is defined in Exercises 15–16.

15. $f(x) = \dfrac{1}{\sqrt{x+1}}$

16. $g(x) = \dfrac{x+5}{x^2 - 9}$

Use the most appropriate word, identity, constant, linear, *or* quadratic *for each function in Exercises 17–18.*

17. $g(x) = 1 - x$

18. $h(x) = 16$

19. If $f(x) = 2x^2 + 3x$, find $f(0)$, $f(-2)$, $f\left(\dfrac{1}{2}\right)$, $f(a)$, and $f(a + 1)$.

20. If $g(x) = 3x - 5$, find $g(a^3)$, $g\left(\dfrac{1}{a}\right)$, and $g(\sqrt{a})$.

3.4 *Use the vertical line test in Exercises 21–22 to determine if the graph is the graph of a function.*

21.

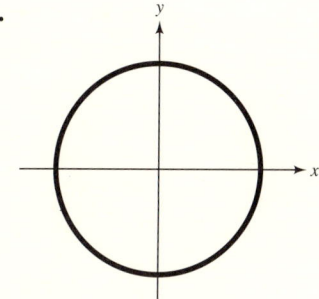

22.

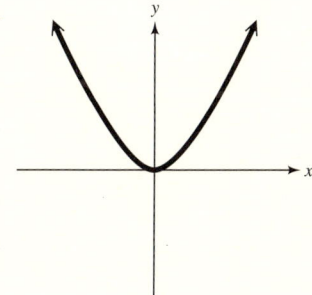

Give the values of x for which the function increases and the values for which it decreases in Exercises 23–24.

23. $g(x) = -\dfrac{1}{3}x - 2$

24. $k(x) = \dfrac{4}{x}$

The graph of f is given in the figure. Use this to sketch the graphs of each function in Exercises 25–28.

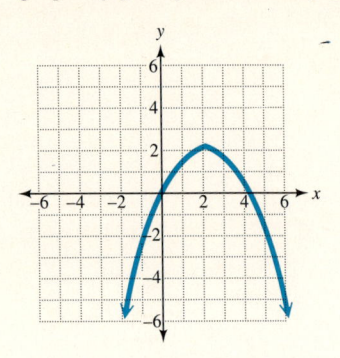

25. $y = -f(x)$

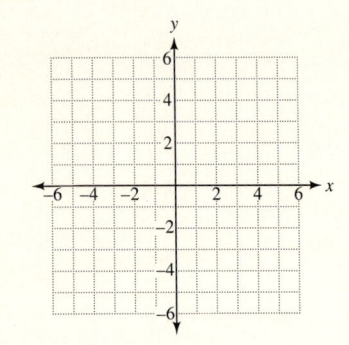

26. $y = 2f(x)$

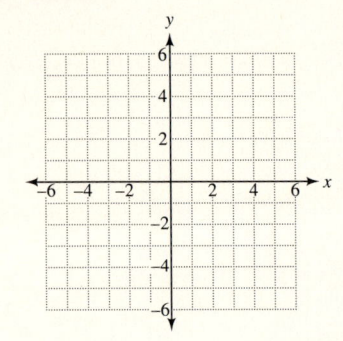

27. $y = f(x + 1) - 2$

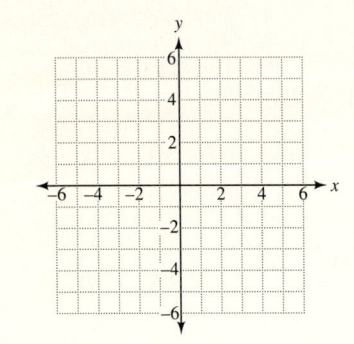

28. $y = f(x - 1) + 2$

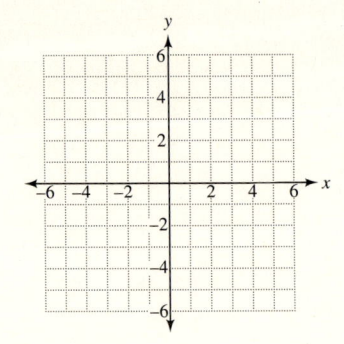

In Exercises 29–31 decide if the function is even or odd.

29. $f(x) = 2x^2 - 3$

30. $g(x) = 4x^3 + 2x$

31. $h(x) = |x| + x$

In Exercises 32–34 determine the symmetry of the graphs without graphing the relation.

32. $3x^2 + 2y^2 = 5$

33. $8xy = 1$

34. $y^2 + 2x = 5$

Graph each function in Exercises 35–36.

35. $f(x) = \begin{cases} -x, & x \leq -2 \\ -2, & -2 < x \leq 1 \\ 2x - 1, & x > 1 \end{cases}$

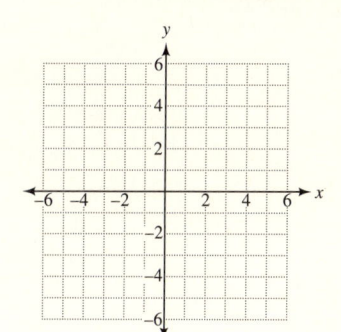

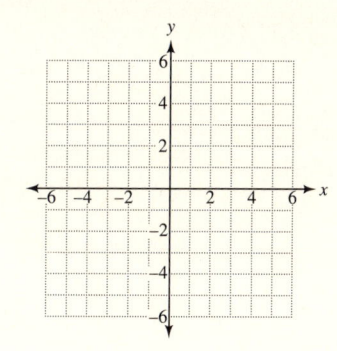

36. $g(x) = \dfrac{1}{2}[x]$

3.5 *Find $f(g(3))$, $g(f(3))$, $f(g(x))$, and $g(f(x))$ for each pair of functions in Exercises 37–38.*

37. $f(x) = 3x^2 + 2$ and $g(x) = x - 5$

38. $f(x) = |x - 2|$ and $g(x) = 1 - x^2$

Determine if the functions in Exercises 39–40 are one-to-one.

39. $f(x) = \sqrt{x - 4}$

40. $f(x) = |x - 4|$

Find the inverse of each function in Exercises 41–44.

41. $f(x) = \dfrac{1}{2}x + 3$

42. $g(x) = \sqrt{x + 5}$

43. $h(x) = \dfrac{1}{2}x^2 - 2, \; x \geq 0$

44. $k(x) = \dfrac{1}{2}x^3$

3.6 *In Exercises 45–46 find the vertex and any x-intercepts of the quadratic function.*

45. $f(x) = 2x^2 + 9x - 5$

46. $g(x) = -3x^2 - 2x + 8$

In Exercises 47–48 find the vertex and any x-intercepts, and graph the function.

47. $f(x) = -2x^2 + 8x - 5$

48. $g(x) = \dfrac{1}{2}x^2 + 3x - 2$

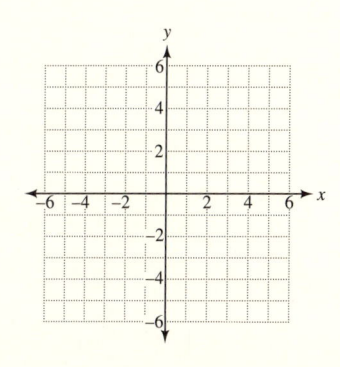

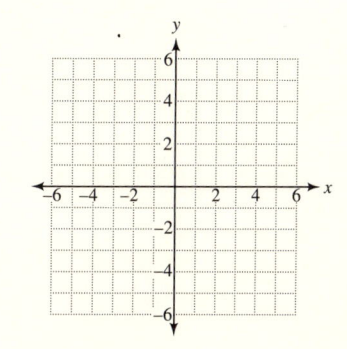

49. BUSINESS Two outdoor storage areas are to be built along an existing fence as indicated in the figure. If 120 ft of fencing is to be used, what are the dimensions x and y that will enclose the maximum area?

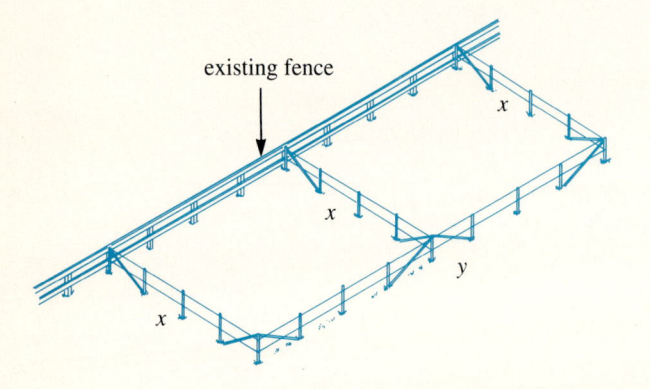

50. BUSINESS The daily cost of a warehouse operation is given by $C(x) = 2x^2 - 80x + 2200$ where x is the number of trucks loaded. What number of trucks gives the minimum cost? What is the minimum cost?

3.7

51. RETAILING A retailer has a markup of 30% and then a markdown of x percent. If sales tax is 5%, find a model T that represents the total price of a coat that originally cost $120. Use this model to find the total price of the coat if the markdown rate is 40%.

52. MANUFACTURING The profit in a small manufacturing operation is given by a quadratic function. The maximum profit occurs at the point (150, 850). If the curve also goes through the point (130, 50), find the quadratic model.

53. MINING Empirical data gives the cost C in dollars of extracting x percent of a metal from a ton of ore as the quadratic model $C(x) = 923.6x^2 - 1756x + 46.8$. Find the percent that minimizes the cost.

54. BUSINESS Spherical storage tanks of radius r are to be painted at a cost of $3.20 per square meter. Find the model for the cost C as a function of r. Find the cost when $r = 3.2$ m.

3.8 *Find the equation of variation in Exercises 55–56.*

55. u varies directly as the square of v and inversely as w. When $v = 2$ and $w = 5$ then $u = 20$.

56. z varies jointly as x and the square root of y and inversely as the cube of r. When $x = 6$, $y = 9$, and $r = 3$, then $z = 8$.

57. RETAILING The revenue R received from sales varies directly as the number n of units sold. If $R = \$2500$ on sales of 320 units, what would be the revenue on 920 units?

58. ENGINEERING The time T it takes for an elevator to lift a weight w varies jointly as the weight and the distance d to be lifted and inversely as the power p of the motor. If its takes 24 seconds for a 3-horsepower motor to lift 500 lb a distance of 60 ft, what power would be needed to lift 700 lb a distance of 100 ft in 35 sec?

Part II

In Exercises 59–64 find the vertex and any x-intercepts of each quadratic function.

59. $f(x) = x^2 - 7x + 6$

60. $f(x) = -x^2 - 12x - 11$

61. $g(x) = -2x^2 + 3x + 20$

62. $g(x) = 3x^2 + 4x - 4$

63. $h(x) = 4x^2 - 4x + 1$

64. $h(x) = -4x^2 + 2x - 2$

In Exercises 65–68 find the general form of the equation of the line satisfying the given conditions.

65. With slope -8 and x-intercept $(-4, 0)$

66. Passing through $(4, -2)$ and parallel to a line with slope $\frac{2}{3}$

67. With x-intercept $(3, 0)$ and perpendicular to $2x - 5y = 4$

68. With y-intercept $(0, -8)$ and parallel to $3x - 9y = 5$

69. CONSUMER If x dollars are borrowed for 3 years at 11% simple interest and S is the amount to be paid back, find a linear model that relates S to x. What amount must be paid back when $2200 is borrowed?

70. ECOLOGY The radius r of an oil spill is increasing with time t in hours according to the function $r(t) = 2t + 1$. If the cost in dollars of cleaning the spill is $C(r) = 30r^2 + 8000$, find C as a function of time. Determine the cost when $t = 3$ hr.

71. AERONAUTICS The height in feet of a small rocket, t seconds after firing, is given by the function $h(t) = -16t^2 + 96t$. What is the maximum height the rocket will reach, and at what time will it hit the ground?

72. PHYSICS The period p of a pendulum varies directly as the square root of the length l. How would the length have to be changed to make the period three times as long?

73. Plot the points $(0, 10)$, $(2, 25)$, $(4, 55)$, $(6, 75)$, and $(8, 85)$. Construct a line through the data, and use the points on your graph at $x = 0$ and $x = 8$ to determine a linear model for this data.

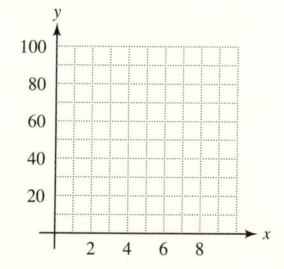

260 CHAPTER 3 RELATIONS, FUNCTIONS, AND GRAPHS

ANSWERS:
1. 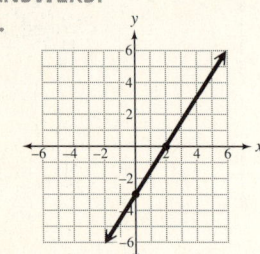 [graph]   2. $3\sqrt{10}$ 3. $(6, -1)$ 4. $(6, 2), (-2, 2)$ 5. 9   6. $9x - 2y + 13 = 0$ 7. $4x - 3y + 8 = 0$
8. $\frac{7}{2}$; $(0, -2)$ 9. parallel   10. perpendicular   11. yes   12. $x + 2y - 5 = 0$
13. function   14. not a function   15. $x > -1$ 16. $x \neq \pm 3$ 17. linear   18. constant
19. 0, 2, 2, $2a^2 + 3a$, $2a^2 + 7a + 5$ 20. $3a^3 - 5$, $\frac{3 - 5a}{a}$, $3\sqrt{a} - 5$ 21. not a function
22. function   23. decreases for all x 24. decreases for all x where it is defined

25. 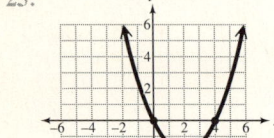 [graph]   26. 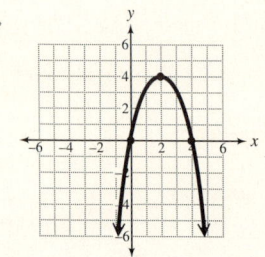 [graph]   27. 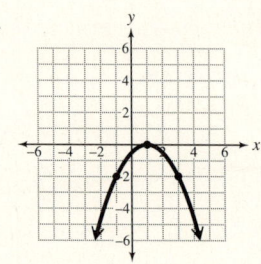 [graph]   28. 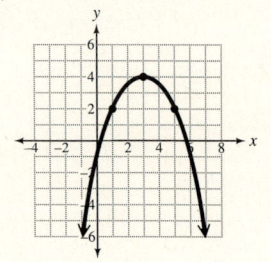 [graph]

29. even   30. odd   31. neither   32. x-axis, y-axis, origin   33. origin   34. x-axis
35. 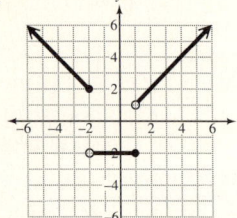 [graph]   36. 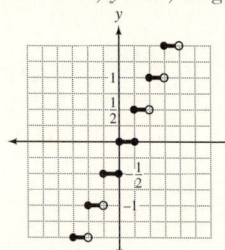 [graph]   37. 14, 24, $3x^2 - 30x + 77$, $3x^2 - 3$ 38. 10, 0, $x^2 + 1$, $-x^2 + 4x - 3$ 39. one-to-one   40. not one-to-one
41. $f^{-1}(x) = 2x - 6$ 42. $g^{-1}(x) = x^2 - 5$, $x \geq 0$
43. $h^{-1}(x) = \sqrt{2x + 4}$ 44. $k^{-1}(x) = \sqrt[3]{2x}$ 45. $\left(-\frac{9}{4}, -\frac{121}{8}\right)$; $\left(\frac{1}{2}, 0\right)$, $(-5, 0)$ 46. $\left(-\frac{1}{3}, \frac{25}{3}\right)$; $\left(\frac{4}{3}, 0\right)$, $(-2, 0)$

47. 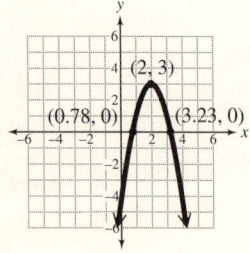 [graph with points (2, 3), (0.78, 0), (3.23, 0)]   48. 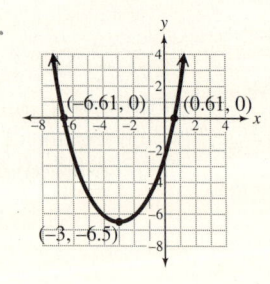 [graph with points (-6.61, 0), (0.61, 0), (-3, -6.5)]   49. $x = 20$ ft, $y = 60$ ft   50. 20; $1400   51. $T(x) = 163.8(1 - x)$; $98.28   52. $f(x) = -2x^2 + 600x - 44{,}150$
53. 95.1%   54. $C(r) = 12.8\pi r^2$, $411.77   55. $u = \frac{25v^2}{w}$
56. $z = \frac{12x\sqrt{y}}{r^3}$ 57. $7187.50   58. 4.8 horsepower
59. $\left(\frac{7}{2}, -\frac{25}{4}\right)$; $(1, 0), (6, 0)$ 60. $(-6, 25)$; $(-1, 0)$, $(-11, 0)$ 61. $\left(\frac{3}{4}, \frac{169}{8}\right)$; $(4, 0)$, $\left(-\frac{5}{2}, 0\right)$ 62. $\left(-\frac{2}{3}, -\frac{16}{3}\right)$; $(-2, 0)$, $\left(\frac{2}{3}, 0\right)$ 63. $\left(\frac{1}{2}, 0\right)$; $\left(\frac{1}{2}, 0\right)$ 64. $\left(\frac{1}{4}, -\frac{7}{4}\right)$; none
65. $8x + y + 32 = 0$ 66. $2x - 3y - 14 = 0$
67. $5x + 2y - 15 = 0$ 68. $x - 3y - 24 = 0$
69. $S = 1.33x$; $2926   70. $C(t) = 120t^2 + 120t + 8030$; $9470   71. 144 ft; 6 sec   72. l would have to be 9 times as long.
73. $y = 10x + 10$

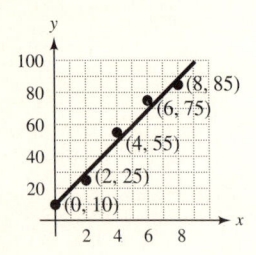

[graph with points (0, 10), (2, 25), (4, 55), (6, 75), (8, 85)]

CHAPTER 3 TEST

1. Find the distance between $(-2, 3)$ and $(1, 7)$.

2. Find the midpoint of the line segment joining $(-2, 7)$ and $(1, 3)$.

3. Find the general form of the equation of the line passing through $(-2, 7)$ and $(1, 3)$.

4. Find the slope and y-intercept of the line with equation $x - 2y - 4 = 0$.

5. Find the general form of the equation of the line passing through $(1, -2)$ parallel to $x + 2y - 3 = 0$.

6. A new house was purchased for $80,000. After 4 years the value of the house was $96,000. Assume that the appreciation in value is given by a linear equation. Find the equation and use it to find the value of the house 10 years after it was purchased.

7. Find the domain of $f(x) = \frac{1}{3x - 1}$.

8. Find the range of the function described by $f(x) = 2x + 5$ with domain $\{1, 2, 3\}$.

9. Use the vertical line test to determine if the graph is the graph of a function.

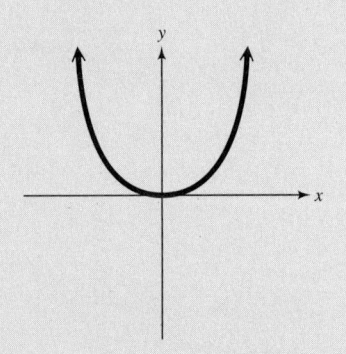

10. Give the values of x for which the function $f(x) = 4x + 7$ is increasing.

11. If the graph of $y = h(x)$ is known, describe how to obtain the graph of $y = h(x - 1) + 2$ using the graph of $y = h(x)$.

12. Determine if $f(x) = x^3 - 4x$ is even, odd, or neither even nor odd.

13. Without graphing, determine the symmetry of the graph of $y = 3x^2 - 1$. [Consider the x-axis, y-axis, and origin.]

CHAPTER 3 TEST
CONTINUED

14. Graph. $f(x) = \begin{cases} 2 & \text{if } x \geq 1 \\ x - 1 & \text{if } x < 1 \end{cases}$

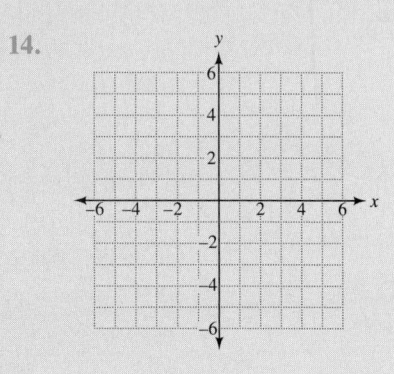

14. _____

15. Find $g[f(x)]$ if $f(x) = 3x - 5$ and $g(x) = 2x^2 - 1$.

15. _____

16. Determine if $f(x) = 5x + 6$ is one-to-one.

16. _____

17. Find the inverse of $f(x) = x^2 + 1$ for $x \geq 0$.

17. _____

18. Use the vertex and the x-intercepts to graph $f(x) = x^2 + 4x$.

18. _____

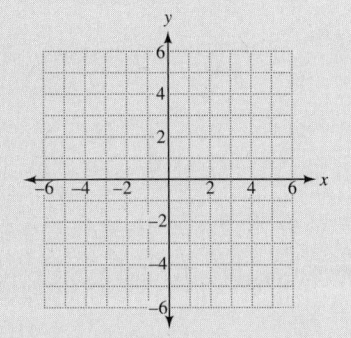

19. What is the minimum product of two numbers whose difference is 24? What are the two numbers?

19. _____

20. The fixed costs in a manufacturing operation is $1200. The cost of producing each unit is $20, and revenue is generated by selling each unit for $30.

    (a) Give a linear model for the cost C as a function of the number n of units produced.

    20.(a) _____

    (b) Use Profit = Revenue − Cost $[P(n) = R(n) − C(n)]$ to find a linear model for P.

    (b) _____

    (c) Find the break-even point $[P(n) = 0]$.

    (c) _____

21. The volume V of a given mass of gas varies directly as the temperature T and inversely as the pressure P. If $V = 231$ in^3 when $T = 420°$ and $P = 20$ lb/in^2, what is the volume when $T = 300°$ and $P = 15$ lb/in^2?

21. _____

CHAPTER 4

Polynomial and Rational Functions

AN APPLIED PROBLEM IN ENGINEERING

An engineer has contracted to design the tank on a petroleum delivery truck. The tank must be in the shape of a cylinder with a hemisphere attached at each end. A municipal code specifies that the total length of the tank can be no more than 20 ft. The contractor wants the tank to be exactly 20 ft long with a storage capacity of 162π ft^3 (approximately 509 ft^3). What should be the radius of the cylinder?

Analysis

This problem involves the volume of a tank that is made up of a cylinder and two hemispheres. The volume of a cylinder is $V = \pi r^2 h$, where r is the radius and h is the height. The volume of a sphere is $V = \frac{4}{3}\pi r^3$, where r is its radius. Since a hemisphere is half of a sphere, its volume is $\frac{2}{3}\pi r^3$. The volume of the tank can be found using a combination of these formulas since the volume of the tank is the sum of the volume of the cylinder and the volumes of the two hemispheres.

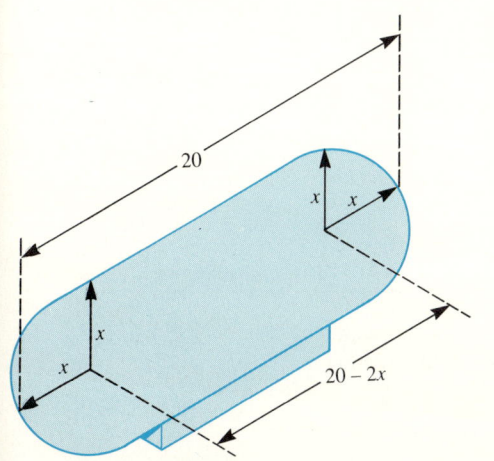

Tabulation

Let x = radius of the cylinder. A sketch of the tank is helpful.

$\pi x^2 (20 - 2x)$ = volume of cylinder

$\frac{2}{3}\pi x^3 + \frac{2}{3}\pi x^3 = \frac{4}{3}\pi x^3$ = volume of the two hemispheres

162π = volume of the tank

Translation

We need to solve:

$$162\pi = \pi x^2(20 - 2x) + \frac{4}{3}\pi x^3$$

In Chapter 1 we reviewed the operations on polynomials. A polynomial in one variable defines a function with domain the set of all real numbers. These *polynomial functions* play a major role in mathematics. Once we have developed the theory of polynomial functions we will have the tools for solving the applied problem at the left (see Example 4 in Section 4.4) and for graphing polynomial functions. We conclude the chapter with a study of *rational functions,* functions formed by taking quotients of polynomials, and solve quadratic and rational inequalities.

263

4.1 POLYNOMIALS AND SYNTHETIC DIVISION

STUDENT GUIDEPOSTS

1. Polynomials
2. Polynomial Equations
3. Zeros of a Polynomial
4. Synthetic Division

1 POLYNOMIALS

In Chapter 2 we reviewed the technique for solving equations that can be transformed into a linear or first-degree equation of the form

$$ax + b = 0 \quad (a \neq 0).$$

We also learned how to solve quadratic or second-degree equations of the form

$$ax^2 + bx + c = 0 \quad (a \neq 0).$$

One of the primary objectives of the present chapter is to learn how to solve equations of degree higher than the second. Formulas exist for solving *third-degree* or *cubic equations* such as

$$ax^3 + bx^2 + cx + d = 0 \quad (a \neq 0)$$

and *fourth-degree* or *quartic equations* such as

$$ax^4 + bx^3 + cx^2 + dx + e = 0 \quad (a \neq 0).$$

However, they are seldom used since they are extremely complicated. No formulas for solving equations of degree higher than the fourth exist. Consequently, when solving general equations of degree higher than the second, we search for solutions based on various theorems. Often solutions are only approximated using numerical techniques.

> **Polynomials with Real Coefficients**
>
> An expression of the form
>
> $$a_n x^n + a_{n-1} x^{n-1} + a_{n-2} x^{n-2} + \cdots + a_2 x^2 + a_1 x + a_0$$
>
> where $a_n, a_{n-1}, a_{n-2}, \ldots, a_2, a_1, a_0$ are constant real numbers and n is a nonnegative integer, is called a **polynomial with real coefficients in the variable x**. Each of the expressions $a_n x^n, a_{n-1} x^{n-1}, \ldots, a_1 x$, and a_0 is a **term** of the polynomial. The numbers
>
> $$a_n, \; a_{n-1}, \; \ldots, \; a_1, \; a_0$$
>
> are **coefficients**, and $a_n \neq 0$ is the **leading coefficient**. If all coefficients are zero, the polynomial is a **zero polynomial**. The polynomial $a_n x^n + a_{n-1} x^{n-1} + \cdots + a_1 x + a_0$ $(a_n \neq 0)$ has **degree n** or is **of nth degree**, and the zero polynomial does not have a degree.

EXAMPLE 1 PROPERTIES OF POLYNOMIALS

Give the degree of each polynomial and the leading coefficient.

(a) $2x^4 - 3x^3 + x^2 - x + 7$

The polynomial has degree 4, and the leading coefficient is 2, the coefficient of x^4.

PRACTICE EXERCISE 1

Give the degree of each polynomial and the leading coefficient.

(a) $7 - x^2$

(b) 0

(b) 15

The constant 15 can be thought of as $15x^0$ since $x^0 = 1$. Thus, 15 has degree 0 and leading coefficient 15.

Answers: (a) degree 2; leading coefficient -1 (b) no degree; no leading coefficient

② POLYNOMIAL EQUATIONS

We now formally define a *polynomial equation*.

> **Polynomial Equations**
>
> An equation of the form
>
> $$a_n x^n + a_{n-1} x^{n-1} + \cdots + a_2 x^2 + a_1 x + a_0 = 0 \quad (a_n \neq 0)$$
>
> is a **polynomial equation**. A **solution** to a polynomial equation is a number (either real or complex) which, when substituted for the variable, makes the equation true.

If we use functional notation with polynomials we can shorten our work considerably. For example, if we let

$$P(x) = a_n x^n + a_{n-1} x^{n-1} + \cdots + a_2 x^2 + a_1 x + a_0,$$

the phrase "evaluate the polynomial $P(x)$ when x is equal to 2" can be shortened to "find $P(2)$." The corresponding polynomial equation can be represented by $P(x) = 0$.

EXAMPLE 2 SOLUTIONS TO POLYNOMIAL EQUATIONS

Decide whether each number is a solution to the polynomial equation

$$P(x) = x^4 + 5x^2 - 36 = 0.$$

(a) -2

Find $P(-2)$.

$$P(-2) = (-2)^4 + 5(-2)^2 - 36$$
$$= 16 + 5(4) - 36$$
$$= 16 + 20 - 36 = 0$$

Thus, -2 is a solution.

(b) 3

$$P(3) = (3)^4 + 5(3)^2 - 36$$
$$= 81 + 5(9) - 36$$
$$= 81 + 45 - 36 = 90$$

Thus, 3 is not a solution.

(c) $-3i$

$$P(-3i) = (-3i)^4 + 5(-3i)^2 - 36$$
$$= 81i^4 + 5(9)i^2 - 36$$
$$= 81(1) + 5(9)(-1) - 36$$
$$= 81 - 45 - 36 = 0$$

Thus, $-3i$ is a solution.

PRACTICE EXERCISE 2

Decide whether each number is a solution to the polynomial equation $P(x) = x^3 - 3x^2 + 4x - 2 = 0.$

(a) 1

(b) -2

(c) $1 + i$

Answers: (a) yes (b) no (c) yes

3 ZEROS OF A POLYNOMIAL

> **Zeros of a Polynomial**
>
> If $P(x)$ is a polynomial and r is a number (real or complex) such that $P(r) = 0$, then r is a **zero** of the polynomial $P(x)$ and a **solution** or **root** of the polynomial equation $P(x) = 0$.

The letter P used to represent a polynomial has no special importance; other letters can also be used, just as letters other than x can be used for the variable. Thus, we might see polynomials such as $Q(y)$, $F(t)$, or $f(x)$.

EXAMPLE 3 ZEROS OF A POLYNOMIAL

Show that -3, i, and $-i$ are all zeros of the polynomial $Q(t) = t^3 + 3t^2 + t + 3$.

$$Q(-3) = (-3)^3 + 3(-3)^2 + (-3) + 3 = -27 + 27 - 3 + 3 = 0$$
$$Q(i) = (i)^3 + 3(i)^2 + (i) + 3 = -i - 3 + i + 3 = 0$$
$$Q(-i) = (-i)^3 + 3(-i)^2 + (-i) + 3 = i - 3 - i + 3 = 0$$

Thus, -3, i, and $-i$ are zeros of $Q(t)$, and solutions to the polynomial equation $Q(t) = 0$.

PRACTICE EXERCISE 3

Show that 2, $2 - i$, and $2 + i$ are all zeros of the polynomial $F(y) = y^3 - 6y^2 + 13y - 10$.

Answer: $F(2)$, $F(2 - i)$, and $F(2 + i)$ are all 0.

4 SYNTHETIC DIVISION

Solutions to polynomial equations may be real or complex numbers as shown in Example 2. At this point we can see whether a given number is a solution to a polynomial equation; however, we do not yet have a method for finding possible solutions. Although not apparent now, the ability to divide quickly a polynomial $P(x)$ by a binomial of the form $x - r$ is an extremely helpful tool for solving equations. The method of long division, reviewed in Chapter 1, can be shortened with a process called **synthetic division.** Suppose we divide $P(x) = 2x^3 - 3x^2 + 3x - 4$ by $x - 2$.

$$\begin{array}{r} 2x^2 + x + 5 \\ x - 2 \overline{) 2x^3 - 3x^2 + 3x - 4} \\ \underline{2x^3 - 4x^2} \\ x^2 + 3x \\ \underline{x^2 - 2x} \\ 5x - 4 \\ \underline{5x - 10} \\ 6 \end{array}$$

The divisor → $x - 2$; ← The quotient $Q(x)$; ← The dividend $P(x)$; ← The remainder R

We can shorten our work by writing only the coefficients.

$$\begin{array}{r} 2 + 1 + 5 \\ 1 - 2 \overline{) 2 - 3 + 3 - 4} \\ \underline{2 - 4} \\ 1 + 3 \\ \underline{1 - 2} \\ 5 - 4 \\ \underline{5 - 10} \\ 6 \end{array}$$

$2 + 1 + 5$ corresponds to $2x^2 + x + 5$

If $P(x)$ is of degree n, then $Q(x)$ is of degree $n - 1$. In this case, since $2x^3 - 3x^2 + 3x - 4$ is of degree 3, $Q(x)$ is of degree 2.

The first term in each line (in color above) is unnecessary since it is subtracted in the next step. Hence, we can abbreviate by eliminating these numbers.

$$
\begin{array}{r}
2 + 1 + 5 \\
1 - 2 \overline{)2 - 3 + 3 - 4} \\
-4 \\
\overline{1 + 3 } \\
-2 \\
\overline{5 - 4} \\
-10 \\
\overline{6}
\end{array}
$$

The remainder R is always a constant, for if R involves the variable x to some power, the process of division is continued until x does not appear.

The 1 in the divisor $1 - 2$ is unnecessary since we are no longer writing the first number in other lines. Also, "bringing down" the next term merely wastes space since all subtractions can be performed in one line, as shown in color below.

$$
\begin{array}{r}
2 + 1 + 5 \\
-2\overline{)2 - 3 + 3 - 4} \\
-4 - 2 - 10 \\
\overline{1 + 5 + 6}
\end{array}
$$

The initial 2 in the quotient $2 + 1 + 5$ is always the same as the initial term in the dividend $2 - 3 + 3 - 4$, and the remaining terms in the quotient $(1 + 5)$ are duplicated in front of the remainder (which is 6) in the bottom line $1 + 5 + 6$. Thus, we could abbreviate further by writing the quotient only in the bottom line by first "bringing down" the 2.

$$
\begin{array}{r}
-2\overline{)2 - 3 + 3 - 4} \\
-4 - 2 - 10 \\
\overline{2 + 1 + 5 + 6}
\end{array}
$$

Finally, since the division lines are unnecessary, we omit them. Also, since most of us can add more quickly than we subtract, by changing the sign of the divisor (from -2 to $+2$) we can obtain the coefficients in the quotient by addition instead of subtraction.

$$
\begin{array}{r|r}
+2 & 2 - 3 + 3 - 4 \\
& +4 + 2 + 10 \\
\hline
& 2 + 1 + 5 + 6
\end{array}
$$

The first 2 is brought down below the line and multiplied by the divisor $+2$, and the product, 4, is placed above the line. Add -3 to 4, obtaining 1, multiply 1 by the divisor $+2$, and place the product, 2, above the line. Add 3 to 2, obtaining 5, multiply 5 by the divisor $+2$, and place the product, 10, above the line. The final addition of -4 and 10 gives 6, the last entry below the line, which is the remainder while $2 + 1 + 5$ corresponds to the quotient $2x^2 + x + 5$.

CAUTION

When dividing a polynomial $P(x)$ by a binomial $x - r$ using synthetic division, always arrange $P(x)$ in descending powers of x, and insert a zero for all terms with zero coefficients.

EXAMPLE 4 Using Synthetic Division

Use synthetic division to find each quotient and remainder.

(a) $(x^4 - 2x^2 + x + 3) \div (x - 1)$

The divisor $x - r$ is $x - 1$ so that $r = 1$. Be sure to supply the zero in the position of the missing x^3-term. The polynomial is $x^4 + 0x^3 - 2x^2 + x + 3$.

$$\underline{1|}\ \ \begin{array}{cccccc} 1 & +0 & -2 & +1 & +3 \\ & +1 & +1 & -1 & +0 \\ \hline 1 & +1 & -1 & +0 & +3 \end{array}$$

The quotient is $x^3 + x^2 - x$ and the remainder is 3.

(b) $(3x^5 - 2x^3 + 7 - x + x^2) \div (x + 2)$

The divisor $x - r$ is $x + 2$ so we have $x + 2 = x - (-2)$ making $r = -2$. Be sure to supply the coefficient zero and rearrange the terms in descending order. The polynomial is $3x^5 + 0x^4 - 2x^3 + x^2 - x + 7$.

$$\underline{-2|}\ \ \begin{array}{cccccc} 3 & +0 & -2 & +1 & -1 & +7 \\ & -6 & +12 & -20 & +38 & -74 \\ \hline 3 & -6 & +10 & -19 & +37 & -67 \end{array}$$

The quotient is $3x^4 - 6x^3 + 10x^2 - 19x + 37$, and the remainder is -67.

Practice Exercise 4

Use synthetic division to find each quotient and remainder.

(a) $(x^4 - 3x^3 + 2x - 5) \div (x - 2)$

(b) $(2x^2 - x^4 + x^5 - 1 - x) \div (x + 1)$

Answers: (a) $x^3 - x^2 - 2x - 2$, remainder: -9 (b) $x^4 - 2x^3 + 2x^2 - 1$, remainder: 0

With practice, the process of synthetic division becomes mechanical, and can be performed in a short time. With the aid of a calculator, we can increase our speed even more, especially when the coefficients of the polynomial are large in absolute value or are decimals. This is illustrated in the next example.

EXAMPLE 5 Synthetic Division on a Calculator

Divide $6x^4 - 48x^3 + 210x^2 - 457x - 5043$ by $x + 3$ using synthetic division and a calculator.

Practice Exercise 5

Divide $5x^5 - 40x^3 + 410x^2 - 3350x - 5280$ by $x - 5$.

$$\underline{-3|}\ \ \begin{array}{ccccc} 6 & -48 & 210 & -457 & -5043 \\ & -18 & 198 & -1224 & 5043 \\ \hline 6 & -66 & 408 & -1681 & 0 \end{array}$$

FIRST:
- ALG: 3 [+/−] [STO] [×] 6 [+] 48 [+/−] [=]
- RPN: 3 [CHS] [STO] [ENTER] 6 [×] 48 [CHS] [+]

SECOND:
- ALG: [DISPLAY] [×] [RCL] [+] 210 [=]
- RPN: [DISPLAY] [RCL] [×] 210 [+]

THIRD:
- ALG: [DISPLAY] [×] [RCL] [+] 457 [+/−] [=]
- RPN: [DISPLAY] [RCL] [×] 457 [CHS] [+]

FOURTH:
- ALG: [DISPLAY] [×] [RCL] [+] 5043 [+/−] [=]
- RPN: [DISPLAY] [RCL] [×] 5043 [CHS] [+]

The numbers -18, 198, -1224, 5043 shown in color are not written when the calculator steps are performed. They are provided here simply for verifying the steps used. Also, the symbol DISPLAY does not represent a calculator entry. Rather it is the number shown in the display at the end of each step. Notice how the sequence

ALG: $\quad$ × $\;$ RCL $\;$ + $\;$ Number $\;$ +/− $\;$ =

RPN: $\quad$ RCL $\;$ × $\;$ Number $\;$ CHS $\;$ +

occurs repeatedly where +/− or CHS is used only in the case of a negative coefficient.

Answer: $5x^4 + 25x^3 + 85x^2 + 835x + 825$, remainder: -1155

Note Synthetic division of a polynomial $P(x)$ by a binomial was developed only for binomials of the form $x - r$. If we wish to divide by a binomial of the form $ax - r$, $a \neq 1$, we must first change the form of $P(x)$ and $ax - r$. For example, since

$$\frac{9x^3 + 18x^2 - 3x + 6}{3x - 1} = \frac{\left(\frac{1}{3}\right)(9x^3 + 18x^2 - 3x + 6)}{\left(\frac{1}{3}\right)(3x - 1)} = \frac{3x^3 + 6x^2 - x + 2}{x - \frac{1}{3}},$$

we can transform the given problem into the correct form for synthetic division.

4.1 EXERCISES A

For each polynomial in Exercises 1–4 give (a) its degree, (b) the leading coefficient, (c) $P(2)$, and (d) $P(-1)$.

1. $P(x) = x^3 - 3x^2 + x - 5$

2. $P(x) = -2x + \sqrt{2}$

3. $P(y) = 2y^2 - y^4 - 1$

4. $P(t) = 8$

5. Find whether 1, -1, and 2 are zeros of the polynomial $f(x) = 3x^4 - 2x^2 - 1$.

6. Find whether 0, 1, and -1 are solutions to the polynomial equation $2x^4 - 5x^3 + 3x = 0$.

7. Decide whether -1, $\frac{3 + i\sqrt{3}}{2}$, and $\frac{3 - i\sqrt{3}}{2}$ are solutions to the polynomial equation $x^3 - 3x^2 + 3x = 0$.

8. **RATE-MOTION** The height in feet of an object which is propelled upward with initial velocity 240 ft/sec from an initial height of 1600 ft can be expressed as a polynomial in the variable t (time) by $h(t) = -16t^2 + 240t + 1600$. Show that 20 is a zero of $h(t)$ and interpret this result.

9. **MEDICINE** The polynomial $M(t) = -0.05t^3 + 2t + 2$ gives the concentration of a particular medicine in parts per million after a time t in hours. Find the concentration after
   (a) 1 hr   (b) 3 hr   (c) 6 hr.

270 CHAPTER 4 POLYNOMIAL AND RATIONAL FUNCTIONS

Find each quotient and remainder in Exercises 10–15 using synthetic division.

10. $(3y^3 - 2y^2 + y - 5) \div (y - 2)$

11. $(3z^4 - 2z^2 + z - 3) \div (z - 3)$

12. $(2x^5 - 3x^4 - 6x^3 + x + 8) \div (x - 3)$

13. $(y^3 - y + y^4 - 3) \div (y + 1)$

14. $(6x^3 - x + 4) \div \left(x - \dfrac{1}{2}\right)$

15. $(7x^5 - 25x^4 + 80x^3 - 170x^2 + 305x + 2600) \div (x + 2)$

16. Is it possible for a polynomial to have no zeros?

17. Show that the polynomial $P(x) = -x^7 + x^6 - x^5 + x^4 - x^3 + x^2 - x + 1$ cannot have a zero that is a negative real number.

18. Find a value of m so that $x^3 + x^2 - x + m$ has a remainder of 0 when divided by $x - 1$.

19. Let $P(x) = x^3 - 2x^2 + 3x + 1$. **(a)** Find $P(1)$. **(b)** Find the remainder when $P(x)$ is divided by $x - 1$. **(c)** Compare the results of parts **(a)** and **(b)**.

20. Find a formula for calculating the degree of the product of two polynomials with real coefficients that have degree m and n, respectively.

FOR REVIEW

21. Find the midpoint of the line segment joining $(6, -1)$ and $(3, 5)$.

22. Find the slope of the line passing through the points $(4, -3)$ and $(7, -3)$.

23. Find the general form of the equation of the line passing through $(6, 1)$ and perpendicular to $3x + 6y - 2 = 0$.

24. What is the conjugate of each complex number? **(a)** $3 - 2i$ **(b)** $5 + i\sqrt{7}$

ANSWERS: 1. (a) 3 (b) 1 (c) −7 (d) −10 2. (a) 1 (b) −2 (c) $-4 + \sqrt{2}$ (d) $2 + \sqrt{2}$ 3. (a) 4 (b) −1 (c) −9 (d) 0 4. (a) 0 (b) 8 (c) 8 (d) 8 5. $f(1) = 0, f(-1) = 0, f(2) = 39$; 1 and −1 are zeros 6. 0 and 1 are solutions. 7. $\dfrac{3 + i\sqrt{3}}{2}$ and $\dfrac{3 - i\sqrt{3}}{2}$ are solutions 8. Since $h(20) = 0$, the height of the object after 20 seconds is zero which means that it takes 20 seconds for the object to hit the ground. 9. (a) 3.95 parts per million (b) 6.65 parts per million (c) 3.2 parts per million 10. $Q: 3y^2 + 4y + 9$; $R: 13$ 11. $Q: 3z^3 + 9z^2 + 25z + 76$; $R: 225$ 12. $Q: 2x^4 + 3x^3 + 3x^2 + 9x + 28$; $R: 92$ 13. $Q: y^3 - 1$; $R: -2$ 14. $Q: 6x^2 + 3x + \tfrac{1}{2}$; $R: \tfrac{17}{4}$ 15. $Q: 7x^4 - 39x^3 + 158x^2 - 486x + 1277$; $R: 46$ 16. Yes; a constant polynomial a_0 ($a_0 \neq 0$) has no zeros. 17. Since all odd powers of x have a negative coefficient, each of these terms would be positive for a negative value of x. Since all even-powered terms would also be positive in this case, the polynomial will always be positive, never zero, for a negative value of x. 18. $m = -1$ 19. (a) 3 (b) 3 (c) The remainder is equal to $P(1)$. 20. The degree of the product would be $m + n$, the sum of the degrees. 21. $\left(\tfrac{9}{2}, 2\right)$ 22. 0 23. $2x - y - 11 = 0$ 24. (a) $3 + 2i$ (b) $5 - i\sqrt{7}$

4.1 EXERCISES B

For each polynomial in Exercises 1–4 give (a) its degree, (b) the leading coefficient, (c) P(2), and (d) P(−1).

1. $P(z) = -z^4 - 3z^2 + 5$

2. $P(u) = \sqrt{5} - u^3$

3. $P(x) = 3x^3 - x^5 + 2$

4. $P(t) = 0$

5. Find whether 1, 2, and 3 are zeros of the polynomial $P(x) = x^3 - 2x^2 - 2x - 3$.

6. Find whether 3, $1 + i$, and $1 - i$ are zeros of the polynomial $Q(x) = x^3 - 5x^2 + 8x - 6$.

7. Decide whether 1, $2i$, and $-2i$ are solutions to the polynomial equation $x^3 - x^2 + 4x - 4 = 0$.

8. **BUSINESS** The revenue in dollars brought in on the sale of x items is given by $R(x) = 200x$. The cost in dollars of producing x of these items is given by $C(x) = 1000 + 100x$. The zeros of the polynomial $P(x) = R(x) - C(x)$ are called **break-even points**. Find $P(x)$ and use it to determine whether 10 is a break-even point.

9. **MEDICINE** The polynomial $M(t) = -0.05t^3 + 2t + 2$ gives the concentration of a particular medicine in parts per million after a time t in hours. Find the concentration after **(a)** 2 hr, **(b)** 4 hr, **(c)** 5 hr.

Find each quotient and remainder in Exercises 10–15 using synthetic division.

10. $(x^3 - x^2 + 2x - 3) \div (x - 1)$

11. $(2x^3 - 3x + 5) \div (x - 2)$

12. $(t^4 - 2t^3 + t^2 - t + 5) \div (t + 2)$

13. $(z^2 - z + 2 + z^4) \div (z - 2)$

14. $(6t^3 - t^2 + 2) \div \left(t + \dfrac{1}{2}\right)$

15. $(5x^4 - 30x^3 + 180x^2 - 640x + 800) \div (x - 3)$

16. Is it possible for a polynomial to have every real number as a zero?

17. Show that the polynomial $P(x) = x^7 + x^6 + x^5 + x^4 + x^3 + x^2 + x + 1$ cannot have a zero that is a positive real number.

18. Find a value of m so that $x^3 - 2x^2 + mx + 1$ has a remainder of 5 when divided by $x + 2$.

19. Let $P(x) = 2x^4 - 4x^2 + x - 2$. **(a)** Find $P(-2)$. **(b)** Find the remainder when $P(x)$ is divided by $x + 2$. **(c)** Compare the results of parts **(a)** and **(b)**.

20. Find a formula for calculating the degree of the sum of two polynomials with real coefficients that have degree m and n, respectively.

FOR REVIEW

21. Find the midpoint of the line segment joining $(2, -3)$ and $(4, 2)$.

22. Find the slope of the line passing through the points $(-1, 5)$ and $(2, -3)$.

23. Find the general form of the equation of the line passing through $(2, -1)$ and perpendicular to $2x - y + 3 = 0$.

24. What is the conjugate of each complex number? **(a)** $4 + 3i$ **(b)** $3 - i\sqrt{2}$

4.1 EXERCISES C

Use a calculator and synthetic division to find each quotient and remainder in Exercises 1–2.

1. $(6.5x^3 - 2.77x^2 + 11.413x - 456.329) \div (x + 3.6)$
   [Answer: Q: $6.5x^2 - 26.17x + 105.625$; R: -836.579]

2. $(2.1x^3 - 3.45x^2 + 6.865x + 7.5312) \div (x - 1.2)$

Find each quotient and remainder using synthetic division.

3. $(t^3 + 3t^2 + t + 3) \div (t + i)$
   [Answer: Q: $t^2 + (3 - i)t - 3i$; R: 0]

4. $(2y^3 + 3y^2 + 7y + 3) \div (2y + 1)$

5. Recall that a function P is even whenever $P(-x) = P(x)$, for all x. Can you find conditions that will make a polynomial function $P(x)$ of a real variable x even?

6. Recall that a function P is odd whenever $P(-x) = -P(x)$, for all x. Can you find conditions that will make a polynomial function $P(x)$ of a real variable x odd?

7. Prior to making a dive, a diver stands at the end of a diving board 15 ft in length. The deflection y (in feet) of the board x ft from the platform is given by the polynomial $y = P(x) = 0.0003x^2(45 - x)$.
   (a) Give the amount of deflection at the very tip of the board. [Answer: 2.03 ft]
   (b) Show that the deflection is 1 foot for some value of x between 9.7 ft and 9.8 ft.

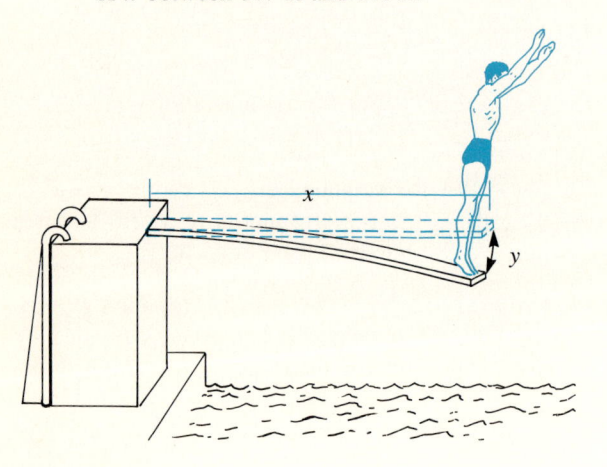

4.2 THE REMAINDER AND FACTOR THEOREMS

STUDENT GUIDEPOSTS

1. The Division Algorithm
2. The Remainder Theorem
3. The Factor Theorem

❶ THE DIVISION ALGORITHM

A familiar property from arithmetic, called the **division algorithm,** states that if p, d, q, and r are whole numbers with $d \neq 0$, such that $p \div d = q$ with a remainder of r, then

$$p = qd + r.$$

For example, if we divide 13 by 4, the quotient is 3 with a remainder of 1, and
$$13 = 3 \cdot 4 + 1.$$
A similar result holds for polynomials.

> **The Division Algorithm**
>
> Let $P(x)$ and $D(x)$ be polynomials with $D(x)$ not a constant polynomial. Then there exist unique polynomials $Q(x)$ and $R(x)$ such that
> $$P(x) = Q(x)D(x) + R(x)$$
> is true for all real numbers x. Also either $R(x)$ is the zero polynomial or else the degree of $R(x)$ is less than the degree of $D(x)$.

EXAMPLE 1 THE DIVISION ALGORITHM

Let $P(x) = 3x^3 - 4x^2 + x - 7$ and $D(x) = x - 2$. Find $Q(x)$ and $R(x)$ so that $P(x) = Q(x)D(x) + R(x)$.

Suppose we divide $P(x)$ by $D(x)$ using synthetic division.

$$\underline{2|}\;\; 3 - 4 + 1 - 7$$
$$\phantom{\underline{2|}\;\;}\;\; \underline{ 6 + 4 + 10}$$
$$\phantom{\underline{2|}\;\;}\;\; 3 + 2 + 5 + 3$$

Then, the quotient is $Q(x) = 3x^2 + 2x + 5$, and the remainder is $R(x) = 3$. Then
$$P(x) = Q(x)D(x) + R(x)$$
$$= (3x^2 + 2x + 5)(x - 2) + 3.$$

PRACTICE EXERCISE 1

Let $P(x) = 2x^4 - 3x^2 + x - 1$ and $D(x) = x + 1$. Find $Q(x)$ and $R(x)$ so that $P(x) = Q(x)D(x) + R(x)$.

Answer: If $Q(x)$ is $2x^3 - 2x^2 - x + 2$ and $R(x)$ is -3, then $P(x) = Q(x)D(x) + R(x)$.

❷ THE REMAINDER THEOREM

When dividing a polynomial $P(x)$ by a binomial $x - r$, the remainder $R(x)$ must be the zero polynomial 0, or must have degree 0 since its degree must be less than 1, the degree of $x - r$. As a result, we write R for $R(x)$ in many such cases emphasizing that R is a constant ($R(x)$ is a constant polynomial). Suppose that $P(x)$ is a polynomial of degree n, and we divide $P(x)$ by $x - r$. Then $Q(x)$ is of degree $n - 1$, R is a constant, and
$$P(x) = Q(x)(x - r) + R.$$
With $P(x)$ expressed in this form, suppose we find $P(r)$. Substituting,
$$P(r) = Q(r)(r - r) + R$$
$$= Q(r) \cdot 0 + R$$
$$= R.$$
Thus, the remainder when $P(x)$ is divided by $x - r$ is $P(r)$. We have just proved the following theorem, which was demonstrated in Exercises A-19 and B-19 in Section 4.1.

> **The Remainder Theorem**
>
> If $P(x)$ is a polynomial, r is a complex number, and $P(x)$ is divided by $x - r$, then the remainder obtained is equal to $P(r)$.

EXAMPLE 2 THE REMAINDER THEOREM

Use the remainder theorem to find the value of the polynomial $P(x) = x^4 + 3x^3 + x^2 + x - 6$ for the given number.

(a) $r = 2$

We divide $P(x)$ by $(x - 2)$ using synthetic division.

$$\begin{array}{r|rrrrr} 2 & 1 + 3 + & 1 + & 1 - & 6 \\ & 2 + 10 + 22 + 46 \\ \hline & 1 + 5 + 11 + 23 + \mathbf{40} \end{array} \leftarrow \text{Remainder}$$

Thus, by the remainder theorem, $P(2) = 40$. We can check this result by direct substitution.

$$P(2) = 2^4 + 3(2)^3 + (2)^2 + (2) - 6$$
$$= 16 + 24 + 4 + 2 - 6 = 40$$

(b) $r = -3$

Divide $P(x)$ by $x + 3$ using synthetic division.

$$\begin{array}{r|rrrrr} -3 & 1 + 3 + 1 + 1 - 6 \\ & -3 + 0 - 3 + 6 \\ \hline & 1 + 0 + 1 - 2 + \mathbf{0} \end{array} \leftarrow \text{Remainder}$$

Thus, by the remainder theorem, $P(-3) = 0$. This means that -3 is a zero of $P(x)$, and since

$$P(x) = Q(x)(x - r) + R,$$

we have
$$P(x) = (x^3 + x - 2)(x + 3) + 0$$
$$= (x^3 + x - 2)(x + 3).$$

PRACTICE EXERCISE 2

Use the remainder theorem to find the value of the polynomial $P(x) = x^3 - 3x^2 + x + 2$ for the given number.

(a) $r = -3$

(b) $r = 2$

Answers: (a) $P(-3) = -55$
(b) $P(2) = 0$

Notice how synthetic division can be used to evaluate a polynomial quickly with the aid of the remainder theorem.

③ THE FACTOR THEOREM

In Example 2(b) we discovered that

$$P(x) = x^4 + 3x^3 + x^2 + x - 6 = (x^3 + x - 2)(x + 3) + 0$$
$$= (x^3 + x - 2)(x + 3).$$

When the remainder is 0, as in this case, the polynomial $P(x)$ has $x + 3$ as a factor. This was a special instance of the next important theorem.

> **The Factor Theorem**
>
> Let $P(x)$ be a polynomial, and let r be a complex number. Then r is a zero of $P(x)$ if and only if $x - r$ is a factor of $P(x)$.

Proof Since the factor theorem is an "if and only if" theorem, there are two things to prove. *First,* we assume that r is a zero of $P(x)$ and then show that $x - r$ is a factor. If $P(x)$ is divided by $x - r$, we obtain $P(x) = Q(x)(x - r) + R$. But by the remainder theorem, $R = P(r)$, and since $P(r) = 0$, $P(x) = Q(x)(x - r)$ so that $x - r$ is a factor of $P(x)$. *Second,* suppose that $x - r$ is a factor of $P(x)$. Then $P(x) = Q(x)(x - r)$, and $P(r) = Q(r)(r - r) = Q(r) \cdot 0 = 0$. Thus, r is a zero of $P(x)$.

EXAMPLE 3 THE FACTOR THEOREM

Decide whether $x - 5$ is a factor of
$$P(x) = x^3 - 7x^2 + 13x - 15.$$

Using the factor theorem, we need to find whether $P(5) = 0$. The best way to do this is to use synthetic division.

$$\begin{array}{r|rrrr} 5 & 1 & -7 & +13 & -15 \\ & & +5 & +10 & +15 \\ \hline & 1 & -2 & +3 & 0 \end{array} \leftarrow \text{Remainder} = P(5)$$

Since $P(5) = 0$, $x - 5$ is a factor of $P(x)$. By using synthetic division, not only have we discovered that $P(5) = 0$, but we also know that the quotient when $P(x)$ is divided by $x - 5$ is $Q(x) = x^2 - 2x + 3$. Thus we can represent $P(x)$ in the factored form
$$P(x) = (x^2 - 2x + 3)(x - 5).$$

PRACTICE EXERCISE 3

Decide whether $x + 4$ is a factor of $P(x) = 2x^3 + 8x^2 - x - 4$.

Answer: $x + 4$ is a factor since $P(-4) = 0$.

As soon as we have found one zero of a polynomial, we can use the factor theorem, and look for other zeros that will be zeros of a polynomial of degree 1 less than the original. For example, to solve the polynomial equation
$$x^3 - 7x^2 + 13x - 15 = 0,$$
if we have discovered that 5 is a solution (see Example 3), we can write the equation in the factored form
$$(x^2 - 2x + 3)(x - 5) = 0,$$
and use the zero-product rule to obtain
$$x^2 - 2x + 3 = 0 \quad \text{or} \quad x - 5 = 0.$$

Using the quadratic formula to solve $x^2 - 2x + 3 = 0$, we obtain
$$x = \frac{2 \pm \sqrt{4 - 4(3)}}{2} = \frac{2 \pm \sqrt{-8}}{2} = \frac{2 \pm 2i\sqrt{2}}{2} = 1 \pm i\sqrt{2}.$$

Thus, the solutions to the original equation are 5, $1 + i\sqrt{2}$, and $1 - i\sqrt{2}$.

EXAMPLE 4 USING THE FACTOR THEOREM

Suppose $x - 2$ is a factor of $x^3 + 3x^2 - x + m$. Find the value of m.

Use synthetic division and divide $x^3 + 3x^2 - x + m$ by $x - 2$.

$$\begin{array}{r|rrrr} 2 & 1 & +3 & -1 & +m \\ & & +2 & +10 & +18 \\ \hline & 1 & +5 & +9 & (m + 18) \end{array}$$

Since $x - 2$ is a factor, $m + 18$ must be 0.

Thus,
$$m + 18 = 0$$
$$m = -18$$

PRACTICE EXERCISE 4

Suppose $x + 1$ is a factor of $x^4 - 2x^2 + 3x + m$. Find the value of m.

Answer: $m = 4$

276 CHAPTER 4 POLYNOMIAL AND RATIONAL FUNCTIONS

CAUTION

Remember to watch the sign on r when dividing a polynomial $P(x)$ by $x - r$ using synthetic division. For example, to divide $P(x)$ by $x + 2$ use $r = -2$ since $x - r = x + 2 = x - (-2)$. Also, be sure to write 0's for missing terms in $P(x)$.

4.2 EXERCISES A

In Exercises 1–2 use synthetic division and the remainder theorem to find each value of $P(x)$ where $P(x) = x^3 - 3x^2 + x + 1$.

1. $P(1)$
2. $P(5)$

In Exercises 3–4 use synthetic division and the remainder theorem to find each value of $P(x)$ where $P(x) = x^5 + 3x^3 - x + 1$.

3. $P(-1)$
4. $P(-3)$

In Exercises 5–6 use synthetic division to find the remainder when $P(x) = x^3 - 27$ is divided by each binomial.

5. $x - 3$
6. $x - 4$

In Exercises 7–8 use synthetic division to find the remainder when $P(x) = x^4 + x^2 - 3$ is divided by each binomial.

7. $x + 2$
8. $x - 3$

In Exercises 9–10 use synthetic division and the division algorithm to express $P(x) = x^3 - 6x^2 - 9x + 14$ in the form $P(x) = Q(x)D(x) + R(x)$ for each of the given binomial divisors, $D(x)$.

9. $x - 7$
10. $x + 1$

In Exercises 11–12 use synthetic division and the factor theorem and tell whether $P(x) = 2x^3 - x^2 - 15x + 18$ has the given binomial as a factor.

11. $x + 3$
12. $x - 1$

13. If a polynomial $P(x)$ is divided by $x - r$, what is the value of the remainder?

14. If a polynomial $P(x)$ is divided by $x + r$, what is the value of the remainder?

15. If r is a zero of polynomial $P(x)$, give one factor of $P(x)$.

16. If $-r$ is a zero of polynomial $P(x)$, give one factor of $P(x)$.

17. Show that 3 is a solution to $x^3 - 3x^2 - x + 3 = 0$, and find the remaining solutions.

FOR REVIEW

18. Consider the polynomial $P(x) = x^8 - 3x^4 + x^2 - x^9 + 7$. **(a)** What is the degree of $P(x)$? **(b)** What is the leading coefficient of $P(x)$? **(c)** Find $P(-1)$.

19. BUSINESS The revenue in dollars brought in by selling x items is given by $R(x) = 125x$. The cost in dollars of producing x items is given by $C(x) = 600 + 85x$. If $P(x) = R(x) - C(x)$, find the break-even points by finding the zeros of $P(x)$.

20. Find $P(-x)$ if $P(x) = 15x^4 - 3x^3 + x^2 + 7x - 6$.

ANSWERS: 1. 0   2. 56   3. -2 4. -320 5. 0   6. 37   7. 17   8. 87   9. $P(x) = (x^2 + x - 2)(x - 7) + 0$ 10. $P(x) = (x^2 - 7x - 2)(x + 1) + 16$ 11. yes   12. no   13. $P(r)$ 14. $P(-r)$ 15. $x - r$ 16. $x + r$ 17. Substitution shows that 3 is a solution. The other solutions, 1 and -1, are solutions to $x^2 - 1 = 0$, the polynomial obtained when the original is divided by $x - 3$.   18. (a) 9   (b) -1 (c) 7   19. The only break-even point is 15, $P(15) = 0$.   20. $P(-x) = 15x^4 + 3x^3 + x^2 - 7x - 6$

4.2 EXERCISES B

In Exercises 1–2 use synthetic division and the remainder theorem to find each value of $P(x)$ where $P(x) = x^3 - 3x^2 + x + 1$.

1. $P(-1)$

2. $P(-2)$

In Exercises 3–4 use synthetic division and the remainder theorem to find each value of $P(x)$ where $P(x) = x^5 + 3x^3 - x + 1$.

3. $P(2)$

4. $P(1)$

In Exercises 5–6 use synthetic division to find the remainder when $P(x) = x^3 - 27$ is divided by each binomial.

5. $x + 2$

6. $x + 3$

In Exercises 7–8 use synthetic division to find the remainder when $P(x) = x^4 + x^2 - 3$ is divided by each binomial.

7. $x - 1$

8. $x + 4$

In Exercises 9–10 use synthetic division and the division algorithm to express $P(x) = x^3 - 6x^2 - 9x + 14$ in the form $P(x) = Q(x)D(x) + R(x)$ for each of the given binomial divisors, $D(x)$.

9. $x + 2$

10. $x - 3$

In Exercises 11–12 use synthetic division and the factor theorem and tell whether $P(x) = 2x^3 - x^2 - 15x + 18$ has the given binomial as a factor.

11. $x - 2$

12. $x - \dfrac{3}{2}$

13. If a polynomial $P(x)$ is divided by $x - a$, what is the value of the remainder?

14. If a polynomial $P(x)$ is divided by $x + a$, what is the value of the remainder?

15. If a is a zero of polynomial $P(x)$, give one factor of $P(x)$.

16. If $-a$ is a zero of polynomial $P(x)$, give one factor of $P(x)$.

17. Show that -2 is a solution to $x^3 + 2x^2 + 4x + 8 = 0$, and find the remaining solutions.

FOR REVIEW

18. Consider the polynomial $P(x) = 2x^7 - 3x^5 + x^3 - x^2 + x - 5$. **(a)** What is the degree of $P(x)$? **(b)** What is the leading coefficient of $P(x)$? **(c)** Find $P(-2)$.

19. SPORTS If there are x teams in a conference, and each team plays each other team once during the season, the total number of games played is given by $G(x) = \frac{1}{2}(x^2 - x)$. How many games will be played in a conference with 8 teams?

20. Find $P(-x)$ if $P(x) = -4x^5 + x^3 - x^2 + 3x + 2$.

4.2 EXERCISES C

Use a calculator, synthetic division, and the remainder theorem to find each value in Exercises 1–2.

1. $P(-2.6)$ if $P(x) = 6.5x^5 - 4.1x^4 + 3.25x^3 - 7.51x - 521.4674$
[Answer: -1518.713]

2. $P(5.3)$ if $P(x) = 7.3x^4 - 8.5x^3 + 2.153x^2 - 4.16x - 3219.0574$

3. Prove that $x - r$ is a factor of $x^n - r^n$ for any natural number n.

4. Prove that $x + r$ is a factor of $x^n + r^n$ for every odd natural number n.

5. Prove that $x + r$ is a factor of $x^n - r^n$ for every even natural number n.

6. Prove that $x - r$ is not a factor of $-x^6 - 2x^4 - x^2 - 5$ for any real number r.

7. If $x + 1$ and $x - 1$ are both factors of $x^3 + x^2 - x + m$, find the value of m.
[Answer: $m = -1$]

8. What is the remainder when $3x^{80} + 7x^{60} - 5x^{40} + x^{20} - 4$ is divided by $x - 1$?

4.3 MORE THEOREMS INVOLVING POLYNOMIALS

STUDENT GUIDEPOSTS

❶ The Fundamental Theorem of Algebra
❷ Complex and Irrational Roots
❸ Descartes' Rule of Signs

❶ THE FUNDAMENTAL THEOREM OF ALGEBRA

We know that every linear equation of the form

$$a_1 x + a_0 = 0 \quad (a_1 \neq 0)$$

has one solution. Likewise, every quadratic equation of the form

$$a_2 x^2 + a_1 x + a_0 = 0 \quad (a_2 \neq 0)$$

has at least one solution and at most two solutions that are complex numbers. The next theorem, the proof of which is beyond the level of our work, is a generalization of these observations.

The Fundamental Theorem of Algebra

Every polynomial equation

$$a_n x^n + a_{n-1} x^{n-1} + \cdots + a_2 x^2 + a_1 x + a_0 = 0 \quad (a_n \neq 0)$$

where $n \geq 1$, has at least one root in the set of complex numbers.

Notice that the fundamental theorem of algebra guarantees that a solution exists; however, it does not tell us how to find it. Much of what we do in this and the next section is directed toward finding solutions to polynomial equations. The next theorem follows from the fundamental theorem and the factor theorem.

Theorem on Linear Factors of a Polynomial

Every polynomial of degree n, $n > 0$, can be factored into a product of linear factors.

Proof Let $P(x)$ be a polynomial of degree n, $n > 0$. By the fundamental theorem of algebra, the equation $P(x) = 0$ has a root, call it r_1. By the factor theorem, $x - r_1$ is a factor of $P(x)$ so $P(x)$ can be written as

$$P(x) = (x - r_1) Q_1(x)$$

where $Q_1(x)$ is the quotient obtained when $P(x)$ is divided by $x - r_1$. If the leading coefficient of $P(x)$ is a_n, then the leading coefficient of $Q_1(x)$ must also be a_n, and the degree of $Q_1(x)$ is $n - 1$. If $n - 1$ is greater than 0, then applying the fundamental theorem to $Q_1(x) = 0$, there must be a root, say r_2. By the factor theorem, $Q_1(x) = (x - r_2) Q_2(x)$, where the degree of $Q_2(x)$ is $n - 2$ and its leading coefficient is a_n, making

$$P(x) = (x - r_1)(x - r_2) Q_2(x).$$

Continuing this process, we will finally obtain a quotient $Q_n(x)$ of degree 0 making $Q_n(x) = a_n$. Then $P(x)$ will be expressed as

$$P(x) = a_n (x - r_1)(x - r_2)(x - r_3) \cdots (x - r_n),$$

which completes the proof.

For example, the polynomial $P(x) = x^4 + x^3 - 7x^2 - x + 6$ can be factored as $(x - 1)(x + 1)(x - 2)(x + 3)$ showing that $1, -1, 2,$ and -3 are zeros of $P(x)$. Notice that the degree of $P(x)$ is 4, and $P(x)$ has 4 zeros. The polynomial $T(x) = x^4 + 2x^3 - 3x^2 - 4x + 4$ has factors $(x - 1)(x - 1)(x + 2)(x + 2)$ and zeros 1 and -2. This time the degree is also 4 but $T(x)$ has only two zeros. However, using the idea of **multiple roots,** we could say that both 1 and -2 are **double roots,** or **roots of multiplicity two,** of the equation $T(x) = 0$ since $(x - 1)$ and $(x + 2)$ are double factors of $T(x)$. These observations, together with the preceding theorem, lead to the next important result.

Theorem on Roots of a Polynomial Equation

Every polynomial equation

$$a_n x^n + a_{n-1} x^{n-1} + \cdots + a_2 x^2 + a_1 x + a_0 = 0 \quad (a_n \neq 0)$$

where $n \geq 1$, has at least one and at most n distinct roots which are complex numbers. Alternatively, counting multiplicities, if $P(x)$ is of degree n, the polynomial equation $P(x) = 0$ has *exactly* n roots or solutions.

280 CHAPTER 4 POLYNOMIAL AND RATIONAL FUNCTIONS

Note The terms "roots" and "zeros" of a polynomial are sometimes used interchangeably for simplicity. To be precise, a polynomial has *zeros,* values that make the polynomial 0, and a polynomial equation has *roots* or solutions. However, we often refer to the "roots of a polynomial" when we actually mean the "zeros of a polynomial."

EXAMPLE 1 ZEROS OF A POLYNOMIAL

Discuss the number of zeros for each polynomial.

(a) $P(x) = 12x^7 - 5x^4 + x^3 - 2x^2 + 7x - 1$
Since $P(x)$ has degree 7, $P(x)$ has at least one and at most seven distinct zeros. (The equation $P(x) = 0$ has at least one and at most seven roots.) However, counting multiplicities, should there be any, we could say that $P(x)$ has exactly seven zeros ($P(x) = 0$ has exactly seven roots).

(b) $f(x) = (x - 1)^2(x - 5)^3(x + 3)$
The degree of $f(x)$ is 6, so $f(x)$ has at least one and at most six distinct zeros. Actually, in this case, counting multiplicities, we could say that $f(x)$ has exactly six zeros, namely 1, 1, 5, 5, 5, and -3. We call 1 a zero of multiplicity two, 5 a zero of multiplicity three, and -3 a zero of multiplicity one or simply a single zero.

PRACTICE EXERCISE 1

Discuss the number of zeros for each polynomial.

(a) $F(x) = x^4 - 3x^2 + 2x - 5$

(b) $g(x) = (x - 2)^4(x + 6)$

Answers: (a) $F(x)$ has at least one and at most four zeros. Counting multiplicities, $F(x)$ has exactly four zeros. (b) $g(x)$ has at least one and at most five zeros. Counting multiplicities, $g(x)$ has exactly five zeros, namely, 2, 2, 2, 2, and -6.

It is easy to construct a polynomial equation that has a specified collection of roots by using the factor theorem. This is an extension of our work in Section 2.4 when we determined a quadratic equation with given solutions.

EXAMPLE 2 FINDING EQUATIONS GIVEN THE ROOTS

Find a polynomial equation that has -2 as a double root (a root of multiplicity 2), 4 as a triple root (a root of multiplicity 3), and 1 as a single root (a root of multiplicity 1).

Counting multiplicities, we need a polynomial of degree six ($2 + 3 + 1 = 6$), which has factors $(x - (-2))^2 = (x + 2)^2$, $(x - 4)^3$, and $(x - 1)$. Since

$$(x + 2)^2(x - 4)^3(x - 1) = 0,$$

we can multiply the factors to get

$$x^6 - 9x^5 + 12x^4 + 76x^3 - 144x^2 - 192x + 256 = 0,$$

a polynomial equation having the desired roots.

PRACTICE EXERCISE 2

Find a polynomial $P(x)$ that has 3 as a double zero, -1 as a triple zero, and 5 as a single zero.

Answer:
$P(x) = (x - 3)^2(x + 1)^3(x - 5)$
$= x^6 - 8x^5 + 9x^4 + 40x^3 - 29x^2 - 96x - 45$

---

❷ COMPLEX AND IRRATIONAL ROOTS

The polynomial (quadratic) equation $x^2 - 2x + 5 = 0$ has roots $1 + 2i$ and $1 - 2i$ (found using the quadratic formula). Notice that these solutions are conjugates. This is a result of the following general theorem given without proof.

4.3 MORE THEOREMS INVOLVING POLYNOMIALS

Complex Root Theorem

If the complex number $a + bi$ is a root of the polynomial equation with *real* coefficients,

$$P(x) = a_n x^n + a_{n-1} x^{n-1} + \cdots + a_1 x + a_0 \quad (a_n \neq 0, n \geq 1),$$

then its conjugate $a - bi$ is also a root. That is, complex roots always occur in conjugate pairs.

EXAMPLE 3 POLYNOMIALS WITH COMPLEX ROOTS

(a) Find a polynomial, $P(x)$, of degree three with real coefficients such that the equation $P(x) = 0$ has 3 and $1 + i$ as roots.

Since $1 + i$ is a root, by the above theorem, $1 - i$ is also a root. Thus, $(x - 3)$, $(x - (1 + i))$, and $(x - (1 - i))$ are factors of $P(x)$ by the factor theorem.

$$P(x) = (x - 3)(x - 1 - i)(x - 1 + i)$$
$$= (x - 3)(x^2 - 2x + 2)$$
$$= x^3 - 5x^2 + 8x - 6$$

$$\begin{array}{r} x - 1 - i \\ x - 1 + i \\ \hline x^2 - x - xi \\ -x \quad\quad + 1 + i \\ + xi \quad\quad - i - i^2 \\ \hline x^2 - 2x \quad\quad + 1 \quad + 1 \end{array}$$

(b) Find the polynomial, $P(x)$, of degree four with real coefficients such that the equation $P(x) = 0$ has $2 + i$ as a root of multiplicity two and 7 as a root of multiplicity one.

Since $2 + i$ is a double root, $2 - i$ must also be a double root. If 7 is also a root, the polynomial has five roots, and cannot be of degree four. Thus, no polynomial exists that satisfies the given conditions.

PRACTICE EXERCISE 3

(a) Find a polynomial, $P(x)$, of degree three with real coefficients such that $P(x) = 0$ has 2 and $3 - i$ as roots.

(b) Find a polynomial, $T(x)$, of degree five with real coefficients such that $T(x) = 0$ has $1 - i$ as a double root, and -4 and 3 as single roots.

Answers: (a) $P(x) = x^3 - 8x^2 + 22x - 20$ (b) There is no such polynomial since to have the given roots, the degree would have to be six, not five.

Remember that *the preceding theorem is true only when the coefficients of the polynomial are real numbers*. If an additional restriction of *rational* coefficients is imposed, there is a similar theorem for certain real roots.

Irrational Root Theorem

If the real number $a + b\sqrt{c}$, where a and b are rational but $\sqrt{c}$ is irrational, is a root of the polynomial equation with *rational* coefficients,

$$P(x) = a_n x^n + a_{n-1} x^{n-1} + \cdots + a_1 x + a_0 = 0 \quad (a_n \neq 0, n \geq 1),$$

then $a - b\sqrt{c}$ is also a root.

EXAMPLE 4 POLYNOMIAL WITH IRRATIONAL ROOTS

Find a polynomial, $P(x)$, of degree four with rational coefficients such that $P(x) = 0$ has $1 + \sqrt{2}$ and $i\sqrt{3}$ as roots.

Since $1 + \sqrt{2}$ is a root, $1 - \sqrt{2}$ is a second root. Also, since $i\sqrt{3}$ is a root, $-i\sqrt{3}$ is the fourth root. Thus,

$$P(x) = (x - (1 + \sqrt{2}))(x - (1 - \sqrt{2}))(x - i\sqrt{3})(x + i\sqrt{3})$$
$$= (x^2 - 2x - 1)(x^2 + 3)$$
$$= x^4 - 2x^3 + 2x^2 - 6x - 3.$$

PRACTICE EXERCISE 4

Find a polynomial, $T(x)$, of degree four with rational coefficients such that $T(x) = 0$ has $3 - \sqrt{3}$ and $1 + i$ as roots.

Answer: $T(x) = x^4 - 8x^3 + 20x^2 - 24x + 12$

3 DESCARTES' RULE OF SIGNS

The next theorem, proved in 1636 by the French mathematician René Descartes, gives us information about the number of real roots to a polynomial equation. It depends on the "variation of sign" from term to term in a polynomial written in descending order. For example, consider the polynomial equation

$$P(x) = 2x^5 - x^4 + 3x - 6 = 0.$$

There are three variations in the signs of the terms of $P(x)$, as shown below:

$$P(x) = 2x^5 \overset{1}{-} x^4 \overset{2}{+} 3x \overset{3}{-} 6 = 0.$$

Similarly, there are 4 variations in the signs of $Q(x)$ in the equation

$$Q(x) = 5x^5 \overset{1}{-} 3x^4 \overset{2}{-} 2x^3 \overset{3}{+} x^2 \overset{4}{-} 7x + 10 = 0.$$

Thus, a "variation of signs" occurs whenever successive terms have opposite signs. The theorem also depends on the variation of signs of the equation formed by replacing x with $-x$ throughout. For example,

$$P(-x) = -2x^5 - x^4 - 3x - 6$$

has no variation in signs, while

$$Q(-x) = -5x^5 \overset{1}{-} 3x^4 + 2x^3 + x^2 + 7x + 10$$

has one variation in signs. Descartes' theorem is presented without proof since a proof is beyond the scope of this course.

Descartes' Rule of Signs

Let $P(x)$ be a polynomial with real coefficients having a nonzero constant term.

1. **Positive real solutions:** The number of positive real solutions of the equation $P(x) = 0$ is either equal to the number of variations of signs in $P(x)$ or less than the number of variations by an even number.
2. **Negative real solutions:** The number of negative real solutions of the equation $P(x) = 0$ is either equal to the number of variations of signs in $P(-x)$ or less than the number of variations by an even number.

EXAMPLE 5 Using Descartes' Rule of Signs

What can be said about the number of positive and negative real solutions to the given polynomial equation?

(a) $P(x) = 5x^4 - 6x^3 + x - 9 = 0$

$$P(x) = 5x^4 \overset{1}{-} 6x \overset{2}{+} x \overset{3}{-} 9 \qquad \text{Three variations}$$

$$P(-x) = 5x^4 \overset{1}{+} 6x^3 - x - 9 \qquad \text{One variation}$$

PRACTICE EXERCISE 5

What can be said about the number of positive and negative real solutions to the given polynomial equation? Give the various possibilities in the order: negative, positive, nonreal.

(a) $P(x) = 3x^4 + x^3 - 2x - 5 = 0$

4.3 MORE THEOREMS INVOLVING POLYNOMIALS 283

Thus, the equation has 1 or 3 positive real solutions and 1 negative solution. We can use a table to summarize the different possibilities. Remember that since the degree of $P(x)$ is 4, there must be 4 solutions to the equation.

Total number of solutions	Number of negative solutions	Number of positive solutions	Number of nonreal solutions
4	1	1	2
4	1	3	0

(b) $Q(x) = 4x^6 + 3x^4 - 7x + 8 = 0$

$$Q(x) = 4x^6 + 3x^4 - 7x + 8 \quad \text{Two variations}$$
$$Q(-x) = 4x^6 + 3x^4 + 7x + 8 \quad \text{No variations}$$

Thus, the equation has 2 or 0 positive real solutions and 0 negative solutions.

Total number of solutions	Number of negative solutions	Number of positive solutions	Number of nonreal solutions
6	0	0	6
6	0	2	4

(c) $R(x) = 7x^6 - 2x^4 - 5x^3 + 2x^2 - x = 0$

Since the constant term is zero, Descartes' rule does not apply to $R(x)$. However, notice that we can factor out an x and apply the theorem to the remaining factor. As a result, clearly 0 is one solution to the equation.

$$R(x) = x(7x^5 - 2x^3 - 5x^2 + 2x - 1) = 0.$$

Let
$$S(x) = 7x^5 - 2x^3 - 5x^2 + 2x - 1. \quad \text{Three variations}$$
$$S(-x) = -7x^5 + 2x^3 - 5x^2 - 2x - 1 \quad \text{Two variations}$$

Thus, $S(x) = 0$ has 3 or 1 positive real solutions and 2 or 0 negative real solutions. The possible solutions to $S(x) = 0$ are summarized in the table.

Total number of solutions	Number of negative solutions	Number of positive solutions	Number of nonreal solutions
5	0	1	4
5	0	3	2
5	2	1	2
5	2	3	0

(d) $W(x) = x^6 + 5x^4 + 2x^2 + 8 = 0$

Since there are no variations of sign in either $W(x)$ or $W(-x)$, there are neither positive nor negative real solutions. Thus, the equation must have 6 complex solutions which occur in three conjugate pairs.

(b) $Q(x) = 2x^5 + 3x^3 + x + 1 = 0$

(c) $R(x) = 3x^6 + x^4 - x^3 + x = 0$

(d) $W(x) = x^8 + 3x^6 + x^2 + 2 = 0$

Answers: (a) either 1, 1, 2 or 3, 1, 0   (b) 1, 0, 4   (c) 0 is one root; either 1, 2, 2 or 1, 0, 4   (d) 0, 0, 8 (four pairs of complex conjugates)

4.3 EXERCISES A

1. A polynomial of degree $n \geq 1$ has at least how many distinct zeros?

2. If 5 is a zero of multiplicity 2 of polynomial $P(x)$, give a quadratic factor of $P(x)$.

3. Counting multiplicities, a polynomial of degree 5 has exactly how many real or complex zeros?

4. If $3 + 2i$ is a zero of polynomial $P(x)$, with real coefficients, give another zero of $P(x)$.

5. If $1 - 3\sqrt{2}$ is a zero of polynomial $P(x)$, with rational coefficients, give another zero of $P(x)$.

6. What is the greatest number of distinct solutions the equation $x^5 - 3x^3 + x^2 - 5x + 7 = 0$ can have? The fewest? Exactly how many solutions does it have, counting multiplicities?

In Exercises 7–10 find a polynomial $P(x)$ with rational coefficients of least degree such that the given numbers are roots of $P(x) = 0$.

7. 3 is a double root and $1 + 2i$ is a single root

8. 2 and $3 + \sqrt{2}$ are single roots

9. 0 is a double root and $1 - \sqrt{3}$ is a single root

10. -2, $-\sqrt{2}$, and $-i\sqrt{2}$ are single roots

11. The solutions to $x^3 - 1 = 0$ are called cube roots of 1. How many cube roots of 1 are there? Find them.

12. Show that $1 + \sqrt{2}$ is a zero of $P(x) = x^2 - \sqrt{2}x - 1 - \sqrt{2}$, but $1 - \sqrt{2}$ is not. Does this contradict a theorem given in this section?

In Exercises 13–18 use Descartes' rule of signs to find the possible number of negative, positive, and nonreal solutions to each equation. List the possibilities in order: negative, positive, nonreal. Do not try to find the solutions.

13. $x^3 - x^2 + 2x + 1 = 0$

14. $x^3 + 3x^2 - x - 9 = 0$

15. $-x^4 + x^3 - 2x^2 + x - 5 = 0$

16. $3x^5 - x^4 + x^2 + x + 6 = 0$

17. $2x^6 - 4x^4 + x^2 - 3 = 0$

18. $8x^8 + 6x^6 + 4x^4 + 2x^2 + 1 = 0$

FOR REVIEW

19. Find the remainder when $3x^4 - x^2 + x - 5$ is divided by $x + 1$.

20. Use synthetic division and the remainder theorem to find $P(-2)$ if $P(x) = x^3 - 2x^2 + x + 1$.

21. If -4 is a zero of polynomial $P(x)$, give one factor of $P(x)$.

22. Consider the polynomial $P(x) = 15x^4 - 3x^3 + x^2 - x - 6$. (a) List all the factors of the leading coefficient 15. (b) List all the factors of the constant term -6.

ANSWERS: 1. 1 2. $(x-5)^2$ 3. five 4. $3-2i$ 5. $1+3\sqrt{2}$ 6. five; one; five 7. $P(x) = x^4 - 8x^3 + 26x^2 - 48x + 45$ 8. $P(x) = x^3 - 8x^2 + 19x - 14$ 9. $P(x) = x^4 - 2x^3 - 2x^2$ 10. $P(x) = x^5 + 2x^4 - 4x - 8$ 11. three; $1, \frac{-1 \pm i\sqrt{3}}{2}$ 12. Direct substitution shows that $P(1+\sqrt{2}) = 0$. No; the theorem in this section is true for rational coefficients, not for real (irrational) ones. 13. either 1, 2, 0 or 1, 0, 2 14. either 0, 1, 2 or 2, 1, 0 15. either 0, 0, 4; 0, 2, 2; or 0, 4, 0 16. either 1, 0, 4; 1, 2, 2; 3, 0, 2; or 3, 2, 0 17. either 1, 1, 4; 1, 3, 2; 3, 1, 2; or 3, 3, 0 18. 0, 0, 8 19. -4 20. -17 21. $x+4$ 22. (a) $\pm 1, \pm 3, \pm 5, \pm 15$ (b) $\pm 1, \pm 2, \pm 3, \pm 6$

4.3 EXERCISES B

1. A polynomial of degree $n \geq 1$ has at most how many distinct zeros?

2. If -2 is a zero of multiplicity 3 of polynomial $P(x)$, give a cubic factor of $P(x)$.

3. Counting multiplicities, a polynomial of degree 0 has exactly how many real or complex zeros?

4. If $5 - 6i$ is a zero of polynomial $P(x)$ with real coefficients, give another zero of $P(x)$.

5. If $2 + 3\sqrt{5}$ is a zero of polynomial $P(x)$ with rational coefficients, give another zero of $P(x)$.

6. What is the greatest number of distinct solutions the equation $x^7 + x^4 - 3x^3 + 7x^2 + 1 = 0$ can have? The fewest? Exactly how many solutions does it have, counting multiplicities?

In Exercises 7–10 find a polynomial $P(x)$ with rational coefficients of least degree such that the given numbers are roots of $P(x) = 0$.

7. -2 is a double root and $4 - i$ is a single root

8. 5 and $1 + \sqrt{2}$ are single roots

9. 0 is a triple root and $2 - \sqrt{3}$ is a single root

10. 3, $\sqrt{5}$, and $-i\sqrt{5}$ are single roots

11. The solutions to $x^4 - 1 = 0$ are called fourth roots of 1. How many fourth roots of 1 are there? Find them.

12. Show that $1 + i$ is a zero of $P(x) = x^2 + (1-i)x - 2 - 2i$, but $1 - i$ is not. Does this contradict a theorem given in this section?

In Exercises 13–18 use Descartes' rule of signs to find the possible number of negative, positive, and nonreal solutions to each equation. List the possibilities in order: negative, positive, nonreal. Do not try to find the solutions.

13. $x^3 + 2x^2 - 2x + 5 = 0$

14. $-x^3 - 5x^2 + 2x + 7 = 0$

15. $x^4 - 2x^3 + x^2 - x + 8 = 0$

16. $2x^5 + x^4 - x^2 - x - 1 = 0$

17. $3x^6 + 2x^4 - 5x^2 + 2 = 0$

18. $3x^7 + 2x^5 + x^3 + 4x + 1 = 0$

FOR REVIEW

19. Find the remainder when $-2x^4 + 3x^3 - x + 2$ is divided by $x - 2$.

20. Use synthetic division and the remainder theorem to find $P(3)$ if $P(x) = x^4 - 2x^3 + x - 4$.

21. If -5 is a zero of polynomial $P(x)$, give one factor of $P(x)$.

22. Consider the polynomial $P(x) = 24x^5 - 3x^3 + 2x^2 - 4x + 4$. (a) List all the factors of the leading coefficient 24. (b) List all the factors of the constant term 4.

4.3 EXERCISES C

1. If $P(x) = a_n x^n + a_{n-1} x^{n-1} + \cdots + a_1 x + a_0$ is a polynomial such that $P(x) = 0$ for every real number x, prove that $a_n, a_{n-1}, \ldots, a_1, a_0$ are all zero, that is, prove that $P(x)$ is the zero polynomial. [*Hint:* If $a_n \neq 0$, use the degree of $P(x)$ to show that we have a contradiction. Thus, a_n must be zero. Continue in a similar manner relative to the remaining coefficients.]

2. If $P(x) = a_n x^n + a_{n-1} x^{n-1} + \cdots + a_1 x + a_0$, and $Q(x) = b_n x^n + b_{n-1} x^{n-1} + \cdots + b_1 x + b_0$, and $P(x) = Q(x)$ for every real number x, prove that $a_i = b_i$, for $i = 0, 1, 2, \ldots, n$. [*Hint:* Consider the polynomial $P(x) - Q(x)$ and use Exercise 1.]

3. If a is a positive real number, explain why $x^4 + a = 0$ cannot have a real root. [*Hint:* Consider the number of sign variations for both $f(x) = x^4 + a$ and $f(-x)$.]

4. If a is a positive real number, explain why $x^4 - a = 0$ must have exactly two real roots.

5. If $P(x)$ is a polynomial of odd degree with real coefficients, prove that $P(x)$ has at least one real zero.

6. If $P(x)$ is a polynomial with all coefficients positive real numbers, prove that $P(x)$ cannot have a positive zero.

7. The graph given below is the graph of a polynomial function $y = P(x)$, where $P(x)$ is of degree three. Find $P(x)$.

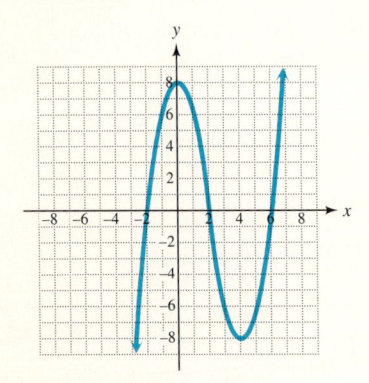

8. **METEOROLOGY** The data in the table below was collected over a twelve-hour period where t is time in hours ($t = 0$ corresponding to midnight) and T is the temperature (in degrees Celsius). Find a third-degree polynomial, $T = P(t)$, that fits this data.

t	0	3	6	9
T	0	0	0	18

[Answer: $T = P(t) = \frac{1}{9} t(t-3)(t-6)$
$= \frac{1}{9} t^3 - t^2 + 2t$]

[Answer: $P(x) = \frac{1}{3}(x+2)(x-2)(x-6)$
$= \frac{1}{3} x^3 - 2x^2 - \frac{4}{3} x + 8$]

4.4 BOUNDS AND THE RATIONAL ROOT THEOREM

STUDENT GUIDEPOSTS

1. Bounds for Real Solutions
2. The Rational Root Theorem

1 BOUNDS FOR REAL SOLUTIONS

The theorems in Section 4.3 gave some information about the nature of the roots to a polynomial equation. Although no specific method for solving general polynomial equations exists, in certain instances we may be able to find some or even all solutions. We begin this section by showing how to locate an interval in which all real roots (if such exist) must be found, and conclude with a theorem that tells how to find all rational-number roots of a polynomial equation.

4.4 BOUNDS AND THE RATIONAL ROOT THEOREM

Consider a polynomial equation $P(x) = 0$. If this equation has no real root greater than a real number a, a is called an **upper bound** for the real roots. Similarly, if $P(x) = 0$ has no real roots less than the real number b, b is a **lower bound** for the real roots. That is, a and b are upper and lower bounds for the real roots of $P(x) = 0$ if

$$b \leq \text{every real root} \leq a.$$

The roots of the polynomial equation $P(x) = (x - 1)(2x - 1)(2x + 1) = 0$ are $-1/2$, $1/2$, and 1. Thus, any number b such that $b \leq -1/2$ is a lower bound for the roots, and any number a such that $a \geq 1$ is an upper bound. Suppose we had started with the factors of $P(x)$ multiplied out,

$$P(x) = 4x^3 - 4x^2 - x + 1 = 0,$$

and did not know the values of the three roots. Let us use synthetic division to divide $P(x)$ by $x + 2$, $x + 1$, $x - 0$, $x - 1$, $x - 2$, and $x - 3$.

```
-2 | 4 -  4 -  1 +  1       -1 | 4 -  4 -  1 +  1        0 | 4 -  4 -  1 +  1
   |   -  8 + 24 - 46          |   -  4 +  8 -  7          |     0 +  0 +  0
     4 - 12 + 23 - 45            4 -  8 +  7 -  6            4 -  4 -  1 +  1

 1 | 4 -  4 -  1 +  1        2 | 4 -  4 -  1 +  1        3 | 4 -  4 -  1 +  1
   |   +  4 +  0 -  1          |   +  8 +  8 + 14          |   + 12 + 24 + 69
     4 +  0 -  1 +  0            4 +  4 +  7 + 15            4 +  8 + 23 + 70
```

Notice that when we divided by -2 and -1, the signs in the row below the line alternated from $+$ to $-$ and back again. When this happens we know that the negative number is a lower bound for the real roots. Also, when we divided by 2 and 3, the signs in the row below the line were all nonnegative. When this happens we know that the positive number is an upper bound for the real roots. We summarize these results in the next theorem, which is given without proof.

Theorem on Bounding Real Roots

Suppose that polynomial $P(x)$ is divided by $x - r$ using synthetic division.

1. If $r < 0$ and the terms in the third row (below the line) alternate from positive to negative (0 can be thought of as either $+0$ or -0), then $P(x) = 0$ has no root less than r, that is, r is a lower bound for the real roots.

2. If $r > 0$ and the terms in the third row are all positive or 0, then $P(x) = 0$ has no root greater than r, that is, r is an upper bound for the real roots.

EXAMPLE 1 FINDING BOUNDS FOR REAL ROOTS

Verify that -2 is a lower bound and 3 is an upper bound for the real roots of $3x^4 - 5x^3 + 19x^2 - 35x - 14 = 0$.

We use synthetic division and divide by -2 and 3.

```
-2 | 3 -  5 + 19 -  35 -  14        3 | 3 -  5 + 19 -  35 -  14
   |   -  6 + 22 -  82 + 234          |   +  9 + 12 + 93 + 174
     3 - 11 + 41 - 117 + 220            3 +  4 + 31 + 58 + 160
     +    -    +    -     +              +    +    +    +    +
```

Since the signs alternate in the first case and are all positive in the second, we know that -2 is a lower bound and 3 is an upper bound for the roots of the equation.

PRACTICE EXERCISE 1

Verify that -5 is a lower bound and 4 is an upper bound for the real roots of $x^4 + x^3 - 16x^2 - 4x + 48 = 0$.

Answer: The signs alternate when dividing by -5 and the signs are all positive when dividing by 4 using synthetic division.

> **CAUTION**
>
> Remember, when using the bounding theorem, that r must be positive for the upper bound test to apply, and r must be negative for the lower bound test to apply.

2 THE RATIONAL ROOT THEOREM

We now consider a method for finding all the rational-number roots to a polynomial equation having integer coefficients.

> **Rational Root Theorem**
>
> Suppose that
>
> $$P(x) = a_n x^n + a_{n-1} x^{n-1} + \cdots + a_2 x^2 + a_1 x + a_0 \quad (a_n \neq 0)$$
>
> is a polynomial for which all coefficients, $a_n, a_{n-1}, \ldots, a_1, a_0$, are integers. If p/q is a rational number reduced to lowest terms, such that p/q is a root of $P(x) = 0$ (that is, $P(p/q) = 0$), then p is a factor of a_0 and q is a factor of a_n.

Proof Suppose that p/q is a root of $P(x)$; then

$$a_n \left(\frac{p}{q}\right)^n + a_{n-1} \left(\frac{p}{q}\right)^{n-1} + \cdots + a_1 \left(\frac{p}{q}\right) + a_0 = 0.$$

Multiply both sides of this equation by q^n.

$$a_n p^n + a_{n-1} p^{n-1} q + \cdots + a_1 p q^{n-1} + a_0 q^n = 0 \quad (1)$$

Solve for $a_n p^n$ and factor out q on the right side.

$$a_n p^n = (-a_{n-1} p^{n-1} - \cdots - a_1 p q^{n-2} - a_0 q^{n-1}) q$$

This shows that q is a factor of $a_n p^n$ since q is a factor of the right side which is equal to $a_n p^n$. But since $\frac{p}{q}$ was reduced to lowest terms, q is not a factor of p, and therefore cannot be a factor of p^n. Thus q is a factor of a_n. In a similar manner, we could solve for $a_0 q^n$ in equation (1).

$$a_0 q^n = (-a_n p^{n-1} - a_{n-1} p^{n-2} q - \cdots - a_1 q^{n-1}) p$$

Then p is a factor of $a_0 q^n$, and since p is not a factor of q, therefore of q^n, p must be a factor of a_0.

Notice that the rational root theorem does not directly give us the roots of a polynomial equation $P(x) = 0$. However, it does allow us to form a limited list of possible rational-number solutions which can then be checked using synthetic division. It is also helpful to keep in mind bounds for the real roots as we test possible rational solutions since we may be able to use this information to pare down our list of possibilities. Also, suppose we discover that p/q is a root, then $x - p/q$ is a factor of $P(x)$ by the factor theorem, and we know that

$$P(x) = \left(x - \frac{p}{q}\right) Q(x),$$

where $Q(x)$ is a polynomial of degree one less than $P(x)$. Additional roots of $P(x) = 0$ must then be roots of $Q(x) = 0$. We then apply the rational root theorem to solve $Q(x) = 0$, and continue the process. Also, to solve $P(x) = 0$, it is wise to consider Descartes' rule of signs first to determine the possible number of positive and negative real solutions. For example, if you determine that an equation has no positive real solutions, then clearly it can have no positive *rational* solutions so it would be a waste of time to try any positive values of p/q.

EXAMPLE 2 USING THE RATIONAL ROOT THEOREM

List the possible rational-number solutions to the equation $P(x) = 2x^4 - 3x^3 + 2x^2 - 6x - 4 = 0$, and determine all solutions.

Since $P(x)$ has three variations of sign, the equation has 3 or 1 positive real solutions. Also, since $P(-x)$ has one variation of sign, there must be 1 negative real solution. Suppose we summarize these results for easy reference.

Total number of solutions	Number of negative solutions	Number of positive solutions	Number of nonreal solutions
4	1	3	0
4	1	1	2

If p/q is a rational-number solution to this equation, then p is a factor of -4 ($a_0 = -4$) and q is a factor of 2 ($a_n = 2$). Thus, the possibilities for p and q are

$$p: 1, -1, 2, -2, 4, -4; \quad q: 1, -1, 2, -2$$

so that the possible rational-number solutions are

$$\frac{p}{q}: 1, -1, 2, -2, 4, -4, \frac{1}{2}, -\frac{1}{2}.$$

If we use synthetic division to find one solution, we will also discover the polynomial of degree 3 that is the other factor of $P(x)$. We can then concentrate on finding its zeros.

Now we begin to try the eight possible solutions.

$$\frac{p}{q} = 1: \quad \underline{1\,|}\ \begin{array}{r} 2 - 3 + 2 - 6 - 4 \\ 2 - 1 + 1 - 5 \\ \hline 2 - 1 + 1 - 5 - 9 \end{array}$$

Thus, $P(p/q) = P(1) = -9 \neq 0$ so that 1 is not a solution.

$$\frac{p}{q} = -1: \quad \underline{-1\,|}\ \begin{array}{r} 2 - 3 + 2 - 6 - 4 \\ -2 + 5 - 7 + 13 \\ \hline 2 - 5 + 7 - 13 + 9 \end{array}$$

Thus, $P(p/q) = P(-1) = 9$ so that -1 is not a solution. Moreover, since the signs in the third row alternate, there is no need to try -2 or -4 since -1 is a lower bound for the roots.

$$\frac{p}{q} = 2: \quad \underline{2\,|}\ \begin{array}{r} 2 - 3 + 2 - 6 - 4 \\ 4 + 2 + 8 + 4 \\ \hline 2 + 1 + 4 + 2 + 0 \end{array}$$

Thus, $P(p/q) = P(2) = 0$ so that 2 is a solution, and by the factor theorem, $x - 2$ is a factor of $P(x)$, with $Q(x) = 2x^3 + x^2 + 4x + 2$ as the other factor. Thus,

$$P(x) = (x - 2)(2x^3 + x^2 + 4x + 2) = 0,$$

which means that we must solve

$$2x^3 + x^2 + 4x + 2 = 0.$$

PRACTICE EXERCISE 2

Find all solutions to the equation $P(x) = 3x^4 + 11x^3 - 7x^2 - 11x + 4 = 0$.
$p: \pm 1, \pm 2, \pm 4$
$q: \pm 1, \pm 3$
$\frac{p}{q}: \pm 1, \pm 2, \pm 4, \pm\frac{1}{3}, \pm\frac{2}{3}, \pm\frac{4}{3}$

Any rational solution p/q of this equation must still be found in the previous list, but now we know that p must be a factor of 2 (a_0 is now 2), and q must be a factor of 2 (a_n is now 2). The only possibilities are

$$\frac{p}{q}: 2, \frac{1}{2}, -\frac{1}{2}$$

since 1, -1, and -2 have already been discarded.

We now try $p/q = 2$. Although 2 was found to be a solution, it must be tried again since it could be a solution to the new polynomial equation $2x^3 + x^2 + 4x + 2 = 0$, that is, it could be a double root of the original equation.

$$\frac{p}{q} = 2: \qquad \underline{2|} \begin{array}{c} 2 + 1 + 4 + 2 \\ 4 + 10 + 28 \\ \hline 2 + 5 + 14 + 30 \end{array}$$

Therefore, 2 is not a double root. Since the signs are all positive, 2 is an upper bound for the real solutions to the equation. The only possibilities remaining are $-1/2$ and $1/2$.

$$\frac{p}{q} = -\frac{1}{2}: \qquad \underline{-\tfrac{1}{2}|} \begin{array}{c} 2 + 1 + 4 + 2 \\ -1 + 0 - 2 \\ \hline 2 + 0 + 4 + 0 \end{array}$$

Thus, $P(p/q) = P(-1/2) = 0$, so that $-1/2$ is a solution, and by the factor theorem, $(x + 1/2)$ is a factor of $2x^3 + x^2 + 4x + 2$ with $2x^2 + 4$ as the other factor. Thus, we have resolved the original equation into

$$(x - 2)\left(x + \frac{1}{2}\right)(2x^2 + 4) = 0$$

and we know that 2 and $-1/2$ are two rational solutions. We now solve the quadratic equation $2x^2 + 4 = 0$.

$$2x^2 + 4 = 0$$
$$x^2 + 2 = 0$$
$$x^2 = -2$$
$$x = \pm\sqrt{-2} = \pm\sqrt{2}i$$

The four solutions to the original equation are 2, $-1/2$, $\sqrt{2}i$ and $-\sqrt{2}i$. Notice that we obtained one negative real solution, one positive real solution, and two complex solutions.

Answer: $-4, -1, 1, \frac{1}{3}$

EXAMPLE 3 Using the Rational Root Theorem

Find all rational solutions to $P(x) = x^3 - 5x^2 + x + 2 = 0$.

Since $P(x)$ has two variations of sign, the equation has 2 or 0 positive real solutions. Also, since $P(-x)$ has one variation of sign, the equation has 1 negative real solution. Thus, the equation has 1 negative real solution and 2 positive real solutions or 1 negative real solution and 2 nonreal solutions.

The possibilities for p are the factors of 2 ($a_0 = 2$), and the possibilities for q are the factors of 1 ($a_n = 1$).

$$p: 1, -1, 2, -2; \qquad q: 1, -1; \qquad \frac{p}{q}: 1, -1, 2, -2$$

PRACTICE EXERCISE 3

Find all solutions to the equation $P(x) = x^3 - 2x^2 + 5x - 10 = 0$.

4.4 BOUNDS AND THE RATIONAL ROOT THEOREM

Whenever the leading coefficient is 1 (as in this case), the only possible values for q are 1 and -1. In these cases, the possibilities for p/q will always be integers (the factors of a_0).

$$\begin{array}{r|rrrr} 2 & 1 & -5 & +1 & +2 \\ & & 2 & -6 & -10 \\ \hline & 1 & -3 & -5 & -8 \end{array} \quad \begin{array}{r|rrrr} 1 & 1 & -5 & +1 & +2 \\ & & 1 & -4 & -3 \\ \hline & 1 & -4 & -3 & -1 \end{array} \quad \begin{array}{r|rrrr} -1 & 1 & -5 & +1 & +2 \\ & & -1 & +6 & -7 \\ \hline & 1 & -6 & +7 & -5 \end{array}$$

Since the signs in the third row alternate when we try -1, there is no need to try -2. The equation has no rational solutions.

Answer: $2, i\sqrt{5}, -i\sqrt{5}$

We conclude this section by solving the applied problem presented in the chapter introduction.

EXAMPLE 4 A PROBLEM IN ENGINEERING

An engineer has contracted to design the tank on a petroleum delivery truck. The tank must be in the shape of a cylinder with a hemisphere attached at each end. A municipal code specifies that the total length of the tank can be no more than 20 ft. The contractor wants the tank to be exactly 20 ft long with a storage capacity of 162π ft^3 (approximately 509 ft^3). What should be the radius of the cylinder?

PRACTICE EXERCISE 4

Assume that the truck in Example 4 has a total length of 30 ft and a storage capacity of 936π ft^3. What is the radius of the cylinder under these conditions?

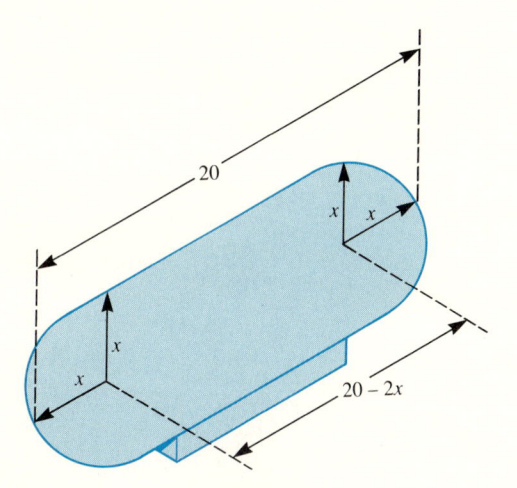

Figure 4.1

Let $x =$ the radius of the cylinder. Consider the sketch of the tank in Figure 4.1. Notice that the radius of the cylinder is the same as the radius of each hemisphere. The volume of the cylinder is $\pi r^2 h = \pi x^2 (20 - 2x)$ and the total volume of the two hemispheres is $\frac{4}{3}\pi r^3 = \frac{4}{3}\pi x^3$. We must solve

$$\pi x^2 (20 - 2x) + \frac{4}{3}\pi x^3 = 162\pi.$$

Dividing through by π and simplifying we have

$$x^3 - 30x^2 + 243 = 0.$$

Since the leading coefficient is 1, the only possible rational solutions must be integers which are divisors of 243. Since $243 = 3^5$, the only possibilities are 3, 9, 27, 81, and 243 (the negative solutions do not need to be considered).

$$\begin{array}{r|rrrr} 3 & 1 - 30 + & 0 + 243 \\ & + 3 - 81 - 243 \\ \hline & 1 - 27 - 81 & 0 \end{array}$$

Therefore, 3 is a solution. The remaining solutions must solve $x^2 - 27x - 81 = 0$, but it is easy to show that one of them is negative, and the other is far too big to satisfy the conditions of the problem. Thus, the radius of the cylinder is 3 ft.

Answer: 6 ft

The theory of equations and the study of methods of solving polynomial equations is an extensive area in mathematics that we have only briefly considered. In the next section we will investigate ways of approximating solutions numerically taking advantage of knowledge about the graphs of polynomial functions.

4.4 EXERCISES A

In Exercises 1–6 use the upper and lower bound test to find the smallest positive integer upper bound and the largest negative integer lower bound for the real solutions to each equation.

1. $4x^4 + 7x^2 - 2 = 0$
2. $2x^3 - 3x^2 + 6x - 9 = 0$
3. $x^4 + x^3 - 7x^2 - 5x + 10 = 0$
4. $9x^4 - 6x^3 + 10x^2 - 30x - 165 = 0$
5. $x^4 - 3x^3 + 9x^2 - 4x = 0$
6. $2x^4 - 15x^3 + 41x^2 - 48x + 20 = 0$

In Exercises 7–14 find all rational solutions (if such exist). If possible, find the other solutions.

7. $x^3 + 2x^2 - x - 2 = 0$
8. $x^3 - x^2 - 14x + 24 = 0$
9. $x^3 - 5x^2 + x + 12 = 0$
10. $x^3 - 6x^2 + 4x - 24 = 0$
11. $3x^3 + x^2 + 12x + 4 = 0$
12. $x^4 - 4x^3 + x^2 + 8x - 6 = 0$
13. $9x^5 + 12x^4 + 10x^3 + x^2 - 2x = 0$
14. $2x^5 - x^4 - x^3 + 4x^2 - 2x = 0$

15. **BUSINESS** A shipping carton is to be made out of cardboard in such a way that the length is three times the width and the height is 2 inches less than the width. Find the dimensions of the carton if its volume is 2400 in^3.

FOR REVIEW

16. Use synthetic division and the remainder theorem to find $P(-3)$ when $P(x) = 4x^4 - x^2 + 3x + 8$.

17. Use synthetic division and the factor theorem to determine whether $x - 5$ is a factor of $P(x) = x^5 - 5x^4 + x^2 - 6x + 5$.

18. What is the greatest number of distinct solutions the equation $x^6 + 3x^4 - x^2 + 5x - 7 = 0$ can have? The fewest? Exactly how many solutions does it have, counting multiplicities?

19. If $5 - 2i$ is a zero of a polynomial $P(x)$ with real coefficients, give another zero of $P(x)$.

20. If $1 + \sqrt{7}$ is a zero of polynomial $P(x)$ with rational coefficients, give another zero of $P(x)$.

21. Find a polynomial $P(x)$ with rational coefficients of least degree such that 2 is a double root, $1 + i$ is a single root, and $1 + \sqrt{2}$ is a single root of $P(x) = 0$.

ANSWERS: 1. upper bound: 1; lower bound: -1 2. upper bound: 2; lower bound: -1 3. upper bound: 3; lower bound: -4 4. upper bound: 3; lower bound: -2 5. upper bound: 3; lower bound: -1 6. upper bound: 8; lower bound: -1 7. $1, -1, -2$ 8. $-4, 2, 3$ 9. 4, $\frac{1 \pm \sqrt{13}}{2}$ 10. 6, $\pm 2i$ 11. $-\frac{1}{3}, \pm 2i$ 12. $1, 3, \pm\sqrt{2}$ 13. $0, \frac{1}{3}, -\frac{2}{3}, \frac{-1 \pm i\sqrt{3}}{2}$ 14. 0 is the only rational root 15. 30 in by 10 in by 8 in 16. $P(-3) = 314$ 17. Since $P(5) = 0$, $x - 5$ is a factor of $P(x)$. 18. six; one; six 19. $5 + 2i$ 20. $1 - \sqrt{7}$ 21. $P(x) = x^6 - 8x^5 + 25x^4 - 38x^3 + 26x^2 - 8$

4.4 EXERCISES B

In Exercises 1–6 use the upper and lower bound test to find the smallest positive integer upper bound and the largest negative integer lower bound for the real solutions to each equation.

1. $9x^3 + 9x^2 - 16x - 16 = 0$

2. $4x^3 + 12x^2 + 3x - 5 = 0$

3. $2x^4 - 3x^3 - 13x^2 + 12x + 20 = 0$

4. $x^4 - 2x^3 + 2x^2 - 10x - 15 = 0$

5. $x^4 + 9x^3 + 24x^2 + 20x = 0$

6. $6x^4 - 7x^3 - 26x^2 + 7x + 20 = 0$

In Exercises 7–14 find all rational solutions (if such exist). If possible, find the other solutions.

7. $x^3 - x^2 - 4x + 4 = 0$

8. $x^3 + x^2 - 14x - 24 = 0$

9. $x^3 + 2x^2 - 6x - 9 = 0$

10. $x^3 - 2x^2 + 9x - 18 = 0$

11. $2x^3 - x^2 + 50x - 25 = 0$

12. $x^4 - x^3 - 5x^2 + 3x + 6 = 0$

13. $8x^5 - 2x^4 + 15x^3 - 4x^2 - 2x = 0$

14. $x^5 + x^4 + 3x^3 + 2x^2 + 2x = 0$

15. AGRICULTURE A cattle range is in the shape of a right triangle with the hypotenuse 4 miles longer than one of the legs. Find the dimensions of the range if its area is 24 mi².

FOR REVIEW

16. Use synthetic division and the remainder theorem to find $P(-2)$ when $P(x) = -x^4 + 3x^3 - 2x + 5$.

17. Use synthetic division and the factor theorem to determine whether $x + 3$ is a factor of $P(x) = x^5 - 4x^4 + 3x^2 - x + 537$.

18. What is the greatest number of distinct solutions the equation $3x^8 - 2x^7 + x^4 - 2x^3 + x - 5 = 0$ can have? The fewest? Exactly how many solutions does it have, counting multiplicities?

19. If $3 + 4i$ is a zero of a polynomial $P(x)$ with real coefficients, give another zero of $P(x)$.

20. If $7 - \sqrt{13}$ is a zero of polynomial $P(x)$ with rational coefficients, give another zero of $P(x)$.

21. Find a polynomial $P(x)$ with rational coefficients of least degree such that 1 is a double root, $2 - i$ is a single root, and $\sqrt{5}$ is a single root of $P(x) = 0$.

4.4 EXERCISES C

1. Explain why $\frac{1}{2}$ cannot be a solution to the equation $x^3 + 3x^2 - x + 7 = 0$.

2. Explain why $\frac{3}{2}$ cannot be a solution to the equation $2x^3 + x^2 - 3x + 1 = 0$.

3. Use the rational root theorem and the equation $x^2 - 2 = 0$ to show that $\sqrt{2}$ is an irrational number.

4. **AGRICULTURE** A corn storage silo is in the shape of a right circular cylinder topped by a hemisphere. If the total height of the silo is 20 ft and it will hold 1377π ft³ (approximately 4326 ft³), what is the radius of the cylinder?

   [Answer: 9 ft]

5. **CONSTRUCTION** The framework for a rectangular shipping crate is to be made from 36 ft of steel rod in such a way that the ends of the crate form a square. Find the dimensions of the crate if its volume is 20 ft³.

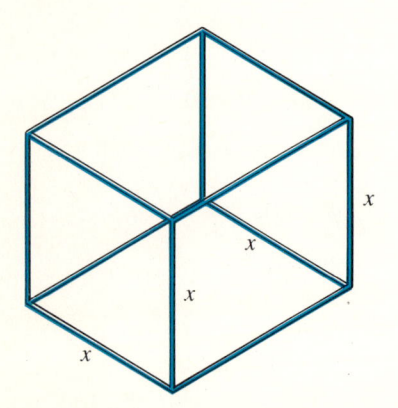

   [Answer: Two boxes exist: one is 5 ft by 2 ft by 2 ft, and the other approximately 1.38 ft by 3.81 ft by 3.81 ft.]

4.5 GRAPHING POLYNOMIAL FUNCTIONS

STUDENT GUIDEPOSTS
1. Nature of Graphs
2. Turning Points
3. Approximating Roots

❶ NATURE OF GRAPHS

In Chapter 3 we discussed several types of functions and their graphs. A linear function is defined by the equation

$$f(x) = a_1 x + a_0 \quad (a_1 \text{ and } a_0 \text{ real numbers}, a_1 \neq 0)$$

and has a straight line for its graph. A quadratic function is given by

$$g(x) = a_2 x^2 + a_1 x + a_0 \quad (a_2, a_1, \text{ and } a_0 \text{ real numbers}, a_2 \neq 0)$$

and has a parabola for its graph. Actually linear and quadratic functions are special cases of *polynomial functions*.

Polynomial Functions

A function defined by

$$P(x) = a_n x^n + a_{n-1} x^{n-1} + \cdots + a_1 x + a_0 \quad (a_n \neq 0)$$

with real coefficients a_i, $i = 1, 2, 3, \ldots, n$, is a **polynomial function (with real coefficients) of degree n,** and has domain the set of all real numbers.

Determining an accurate graph of a polynomial function is usually a topic in more advanced courses. However, we can obtain rough approximations to graphs by making several basic assumptions and using information about the zeros of polynomials learned in previous sections.

Unless a polynomial function is linear, its graph does not contain any straight line segments. Rather, the graph of a polynomial is a "smooth" (no sharp corners) and "continuous" (no breaks nor holes) curve. The graphs in Figure 4.2 are typical graphs of polynomial functions, and the graphs in Figure 4.3 cannot be graphs of polynomial functions.

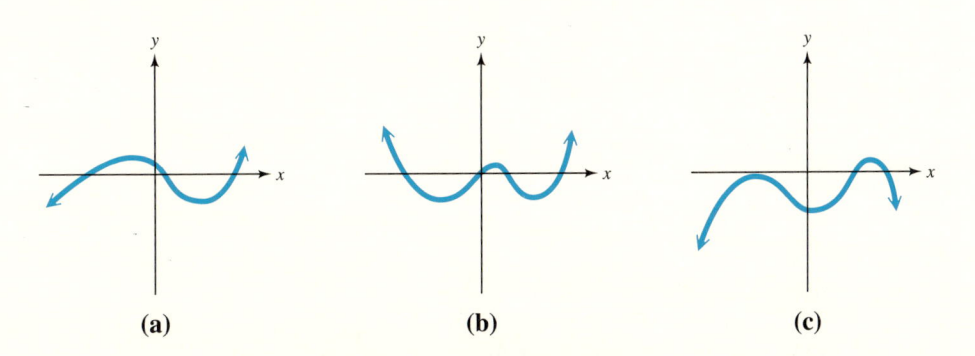

(a)   (b)   (c)

Figure 4.2 Polynomial Graphs

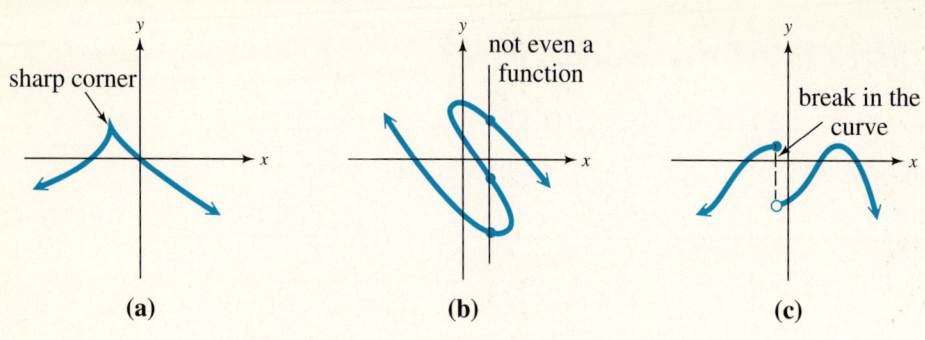

296 CHAPTER 4 POLYNOMIAL AND RATIONAL FUNCTIONS

(a) sharp corner

(b) not even a function

(c) break in the curve

Figure 4.3 Nonpolynomial Graphs

Since a polynomial of degree n can have no more than n real zeros, a polynomial function of degree n can never cross the x-axis more than n times. The graph in Figure 4.2(a) is typical of the graph of a polynomial function of degree 3, and the graphs in Figure 4.2(b) and (c) are typical of polynomials of degree 4. We would know, for example, that the graph in Figure 4.2(b) could not be the graph of a third-degree polynomial since there are four x-intercepts giving four zeros for the polynomial. Multiple zeros are found when the curve is tangent to the x-axis as in the case with the negative zero to the left in Figure 4.2(c). Since this is the graph of a fourth-degree polynomial, and since there are obviously three zeros, one of these must be a double zero in order for the polynomial to have the required four real zeros. A typical third-degree polynomial with one real zero (and two complex) is graphed in Figure 4.4(a), and a typical fourth-degree polynomial with two real zeros (and two complex) is graphed in Figure 4.4(b).

(a)

(b)

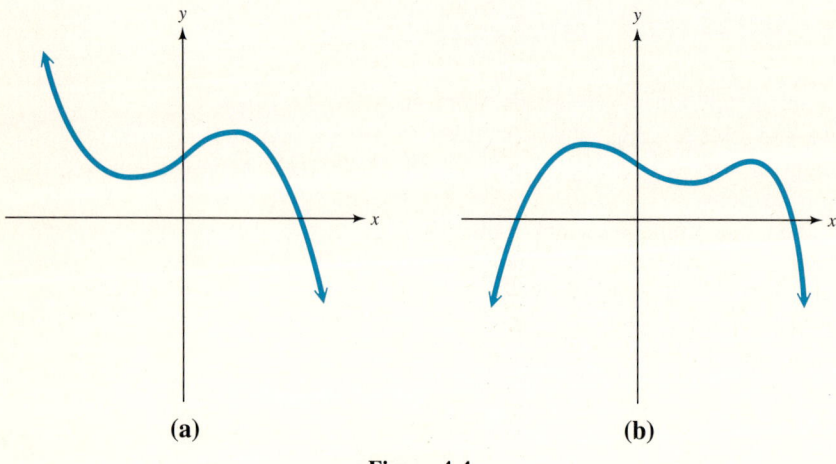

Figure 4.4

When graphing polynomials, it is helpful to know the behavior of the graph for large values of $|x|$. Consider the polynomial

$$P(x) = a_n x^n + a_{n-1} x^{n-1} + \cdots + a_1 x + a_0 \quad (a_n \neq 0).$$

When $|x|$ is very large, the first term of the polynomial, $a_n x^n$, is larger in absolute value than the sum of the remaining terms. As a result, the sign of the first term determines the nature of the graph when $|x|$ is extremely large. The graph of a polynomial will eventually "take off" and increase or decrease, depending on the sign of the leading coefficient, a_n, and whether n is even or odd. The possibilities are summarized in Figure 4.5 in which the arrow indicates the eventual nature of the graph.

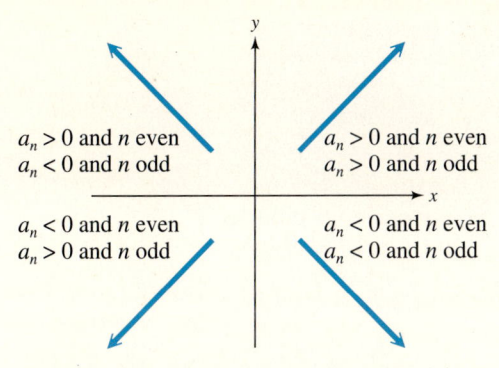

Figure 4.5 Nature of $P(x)$ for Large $|x|$

EXAMPLE 1 NATURE OF A GRAPH FOR LARGE $|x|$

Discuss the eventual nature of the graph for large $|x|$.

(a) $P(x) = x^3 - 3x^2 + x - 1$

Since $n = 3$ (n is odd) and $a_3 = 1 > 0$, the graph of $P(x)$ will eventually go up to the right and down to the left, as shown in Figure 4.6(a).

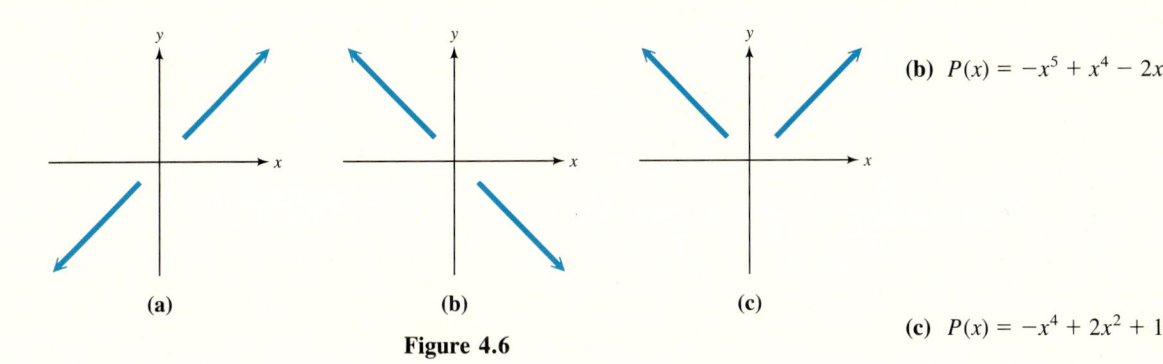

Figure 4.6

(b) $P(x) = -2x^3 + x^2 - 5$

Since $n = 3$ (n is odd) and $a_3 = -2 < 0$, the graph of $P(x)$ will eventually go down to the right and up to the left, as shown in Figure 4.6(b).

(c) $P(x) = 2x^4 - 3x^2 + 7$

Since $n = 4$ (n is even) and $a_4 = 2 > 0$, the graph of $P(x)$ will eventually go up to the right and also up to the left, as in Figure 4.6(c).

PRACTICE EXERCISE 1

Discuss the eventual nature of the graph for large $|x|$.

(a) $P(x) = x^5 + 2x^3 - 3$

(b) $P(x) = -x^5 + x^4 - 2x$

(c) $P(x) = -x^4 + 2x^2 + 1$

Answers: (a) up right; down left (b) down right; up left (c) down right; down left

2 TURNING POINTS

Now that we know how to determine the eventual nature of the graph of a polynomial, both left and right, we turn our attention to the "middle" portion of the graph. Remember that the graph must be a smooth curve. The points on the graph at which the curve "turns" smoothly from upward to downward or downward to upward are the **turning points** of the graph. The next theorem, the proof of which is given in calculus, states an important fact about turning points.

Number of Turning Points

Let $P(x)$ be a polynomial of degree n. The graph of $y = P(x)$ can have at most $n - 1$ turning points.

Thus, for example, a third-degree polynomial can have at most two turning points, although it may have none at all. Two typical third-degree polynomials are sketched in Figure 4.7(a) and (b).

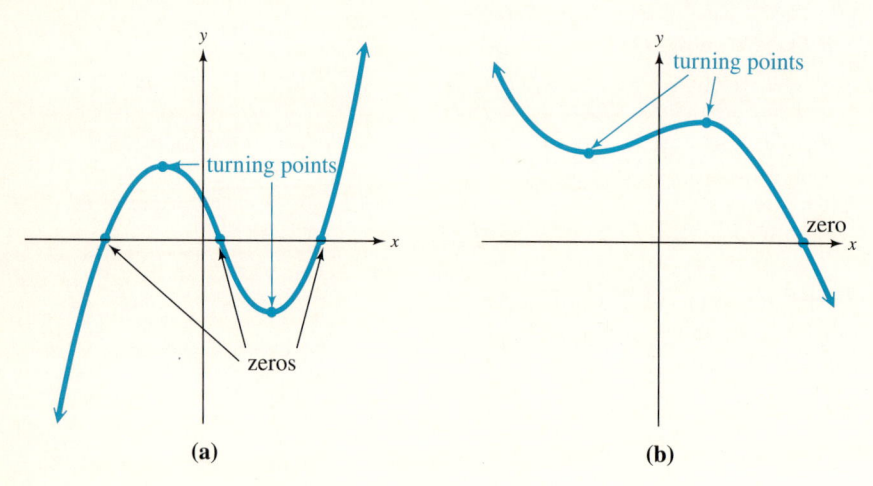

Figure 4.7 Third-Degree Polynomials

For graphing polynomials, it is helpful to use information such as turning points, zeros, eventual appearance for large values of $|x|$, and the fact that all polynomial graphs are "smooth" and "continuous." Other helpful bits of information include the y-intercept and possible symmetries with respect to the y-axis or the origin.

EXAMPLE 2 GRAPHING A THIRD-DEGREE POLYNOMIAL

Graph $y = P(x) = x^3 + 2x^2 - x - 2$.

Since the degree of $P(x)$ is 3, the graph has at most two turning points. Also, since 3 is odd and $a_3 = 1 > 0$, the curve must eventually go up to the right and down to the left. One easy point to determine is the y-intercept, where $x = 0$. Since $P(0) = -2$, $(0, -2)$ is one point on the graph. Other point(s) of interest are the zeros of $P(x)$. Since $P(x)$ has one variation of signs, there will be 1 positive real zero. Also, since $P(-x)$ has two variations of sign, there will either be no negative zeros or 2 negative zeros. We begin by searching for any rational zeros. The possibilities are listed below.

$$p: \pm 1, \pm 2 \qquad q: \pm 1 \qquad \frac{p}{q}: \pm 1, \pm 2$$

$$\underline{1 |} \; \begin{array}{r} 1 + 2 - 1 - 2 \\ 1 + 3 + 2 \\ \hline 1 + 3 + 2 + 0 \end{array}$$

Thus, $P(x) = (x - 1)(x^2 + 3x + 2) = (x - 1)(x + 1)(x + 2)$ and its zeros are 1, -1, and -2. All of this information is helpful in sketching the graph in Figure 4.8.

PRACTICE EXERCISE 2

Graph $y = P(x) = -x^3 - 2x^2 + x + 2$.

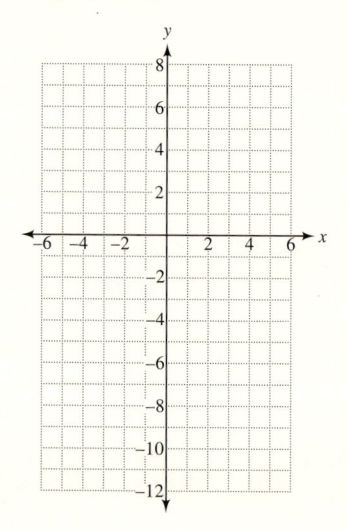

4.5 GRAPHING POLYNOMIAL FUNCTIONS   299

$P(x) = x^3 + 2x^2 - x - 2$

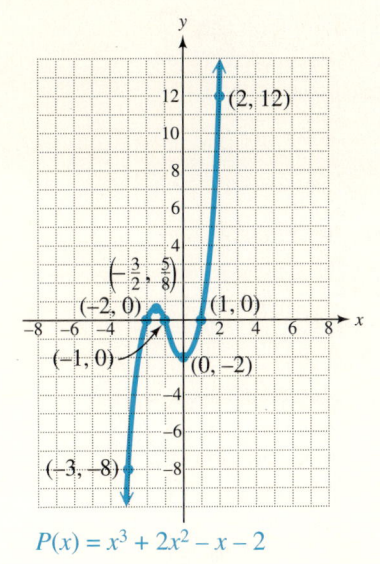

Figure 4.8

Answer: The graph is similar to the one in Figure 4.8 except reflected in the y-axis.

EXAMPLE 3 GRAPHING A FOURTH-DEGREE POLYNOMIAL

Graph $y = P(x) = x^4 - 4x^3 - 3x^2 + 14x - 8$.

Since the degree of $P(x)$ is 4, the graph has at most three turning points. Also, since 4 is even and $a_4 = 1 > 0$, the curve must eventually go up to the right and also up to the left. The y-intercept is $(0, -8)$. Since there are three variations of sign, $P(x)$ must have 1 or 3 positive zeros, and since there is one variation in sign of $P(-x)$, there must be 1 negative zero. Using the rational root theorem, it is not difficult to show that 1 is a zero of multiplicity two, and that -2 and 4 are the other two zeros. Thus, $P(x) = (x - 1)^2(x + 2)(x - 4)$. Using this information and plotting several additional points we can sketch the graph as in Figure 4.9.

PRACTICE EXERCISE 3

Graph $y = P(x) = -x^4 + 4x^3 + 3x^2 - 14x + 8$.

$P(x) = x^4 - 4x^3 - 3x^2 + 14x - 8$

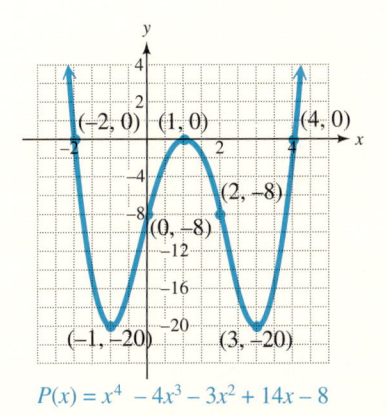

Figure 4.9

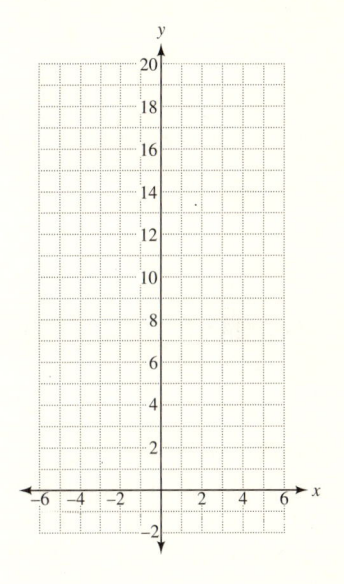

Answer: The graph is similar to the one in Figure 4.9 but reflected in the y-axis.

Note The graph of the polynomial in Example 3 crossed the x-axis at the two zeros of multiplicity 1 [at $(-2, 0)$ and $(4, 0)$] and was tangent to the x-axis at the zero of multiplicity 2 [at $(1, 0)$]. This is true in general. That is, the graph of a polynomial will cross the x-axis at a zero with an odd multiplicity, and it will be tangent to the *x*-axis at a zero with an even multiplicity. This information can be helpful when sketching a graph. Note that, although we know the approximate position of a turning point, we have no way of determining it precisely with methods used in this course.

3 APPROXIMATING ROOTS

When the real roots to a polynomial equation are irrational, the rational root theorem will do us no good. In such cases we can use the graph of the polynomial to help us decide where to look for irrational roots. For example, suppose we were to solve $x^4 - 4 = 0$. We begin by graphing the function $y = P(x) = x^4 - 4$.

The graph will have at most three turning points and four *x*-intercepts since the degree of $P(x)$ is 4. Also, since 4 is even and $a_4 = 1 > 0$, the curve must go up to the right and up to the left. Since both $P(x)$ and $P(-x)$ have one variation of signs, there will be 1 positive zero and 1 negative zero. However, a search for rational zeros yields none so the best we can do is construct a table of values. Notice that since $P(-x) = P(x)$, the graph, given in Figure 4.10, is symmetric with respect to the *y*-axis.

x	$P(x)$
-2	12
-1	-3
0	-4
1	-3
2	12

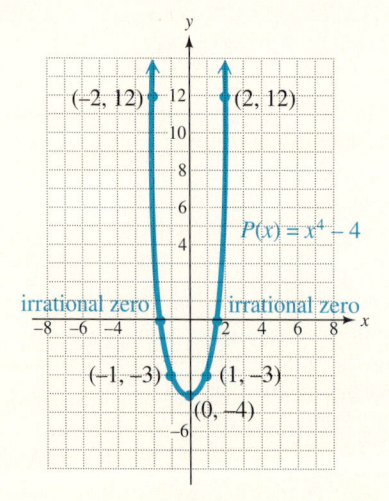

Figure 4.10

Since we know that the polynomial $P(x) = x^4 - 4$ has one positive real zero and one negative real zero, and since the rational root theorem fails to identify any rational roots, we know that the two zeros are irrational. In fact, from the graph in Figure 4.10, we know even more: the positive zero is between 1 and 2, and the negative zero is between -2 and -1. This follows since at 2, $P(2) = 12 > 0$ and at 1, $P(1) = -3 < 0$ so that at some value r between 1 and 2, $P(r) = 0$. This is a special case of the following theorem which is proved in calculus.

The Intermediate Value Theorem

If $P(x)$ is a polynomial, a and b are real numbers with $a < b$, and $P(a) \neq P(b)$, then $P(x)$ takes on every value between $P(a)$ and $P(b)$ in the interval $[a, b]$.

In particular, for $P(x) = x^4 - 4$, since $P(1) = -3$ and $P(2) = 12$, there is at least one real number r in $[1, 2]$ such that $P(r) = 0$. Although the intermediate value theorem guarantees the existence of a zero for $P(x)$ between 1 and 2, it does not indicate how to find it. Since the zero is irrational, often the best we can do is approximate it by rational numbers using a numerical approximation technique. One of the simplest methods, which can be easily programmed on a computer, involves determining successive polynomial values between a known positive value and a known negative value. For example, Figure 4.11 shows the portion of the preceding graph between 1 and 2, enlarged. We divide the segment from 1 to 2 into ten equal parts, with division points at

1, 1.1, 1.2, 1.3, 1.4, 1.5, 1.6, 1.7, 1.8, 1.9, 2.

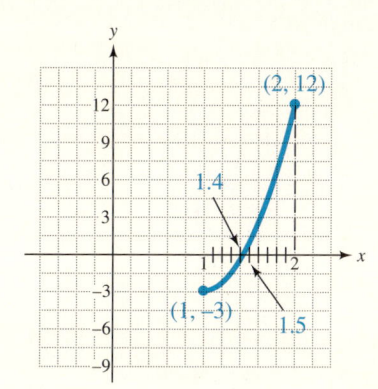

Figure 4.11

The irrational zero that we wish to approximate must be located between two of these values. Using synthetic division or substitution, we calculate the value of the polynomial at the numbers in the above list until the value is less than zero at one value and greater than zero at the next. For example, we find that

$$P(1.4) \approx -0.1584 \quad \text{and} \quad P(1.5) \approx 1.0625.$$

Thus, the desired zero is between 1.4 and 1.5. For greater accuracy, the process can be repeated. Dividing the segment from 1.4 to 1.5 into ten equal parts gives the values

1.40, 1.41, 1.42, 1.43, 1.44, 1.45, 1.46, 1.47, 1.48, 1.49, 1.50.

Again, by synthetic division or direct substitution, we would discover that

$$P(1.41) \approx -0.0475 \quad \text{and} \quad P(1.42) \approx 0.0659.$$

Thus, the desired zero is between 1.41 and 1.42. This procedure can be repeated until we attain any desired degree of accuracy for our approximation.

In this particular example, we could have found the zeros more directly by factoring, since $P(x) = x^4 - 4 = (x^2 - 2)(x^2 + 2)$. The zeros of $P(x)$ are found by setting each factor equal to zero.

$$\begin{array}{ll} x^2 - 2 = 0 \quad \text{or} & x^2 + 2 = 0 \\ x^2 = 2 & x^2 = -2 \\ x = \pm\sqrt{2} & x = \pm\sqrt{2}i \end{array}$$

We have two imaginary zeros and two real zeros. The zero which we were approximating is $\sqrt{2} \approx 1.4142135$.

The method of approximating an irrational zero of a polynomial outlined above is sometimes referred to as *the method of bracketing*. We are finding a smaller and smaller closed interval of the form $[a, b]$ in which the actual zero is located. The term "bracketing" comes from the use of brackets to describe a closed interval. If you have a programmable calculator or access to a personal computer, it is easy to write a simple program for finding values of a given polynomial and use the method of bracketing to close in on an approximation for any root. Another method for approximating zeros is discussed in the *Extension Project* at the end of the chapter.

4.5 EXERCISES A

In Exercises 1–4 match each polynomial with one of the graphs (a), (b), (c), or (d).

1. $P(x) = x^2 - 5x + 4$
2. $P(x) = x^3 - x^2 - 4x + 4$
3. $P(x) = x^3 - 4x^2 - x + 4$
4. $P(x) = x^4 - 5x^2 + 4$

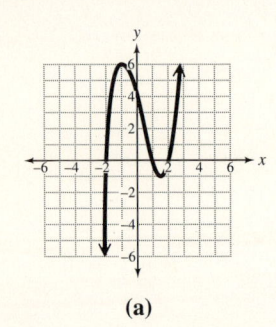

(a) 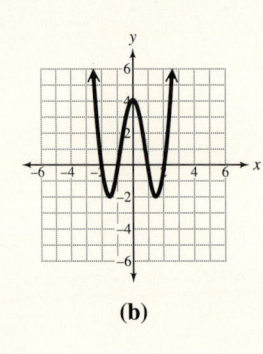   (b) 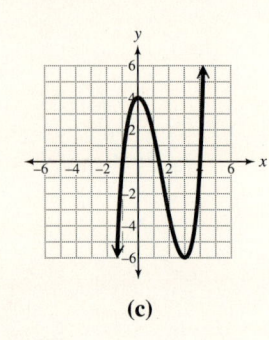   (c) 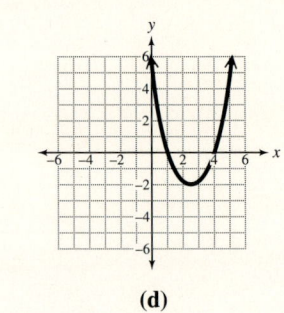   (d)

For each of the polynomial functions in Exercises 5–8 determine (a) the maximum number of real zeros, (b) the maximum number of turning points, (c) the y-intercept, (d) the direction of the graph for large positive values of x, and (e) the direction of the graph for small negative values of x.

5. $P(x) = x^3 + 3x^2 - x + 7$
6. $P(x) = -2x^3 + x^2 - 5x + 12$
7. $P(x) = 5x^4 + x^3 - 3x - 1$
8. $P(x) = -3x^6 + 4x^4 + 2x^2 - x - 5$

In Exercises 9–12 sketch the graph of each polynomial function using the various techniques of this section.

9. $y = P(x) = -x^3 + 4x$ 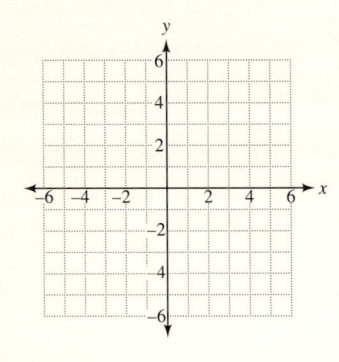
10. $y = P(x) = x^3 - 2x$ 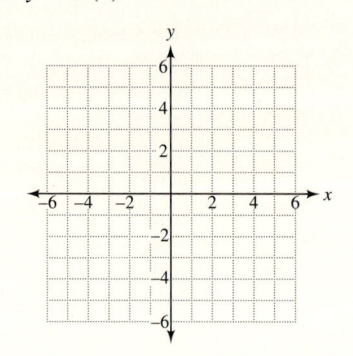
11. $y = P(x) = x^3 + x^2 + x + 1$ 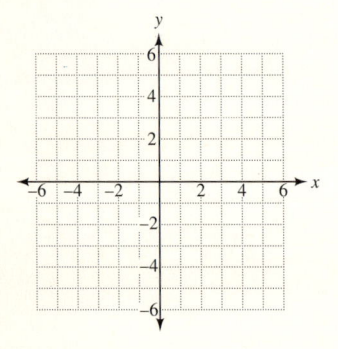
12. $y = P(x) = x^4 - 4x^2$

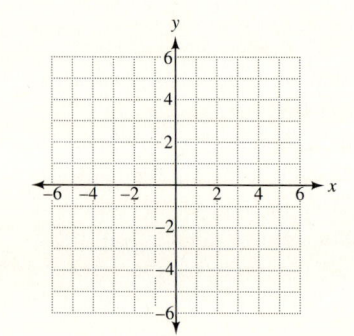

In Exercises 13–14 show that P(x) has a zero between the given values a and b.

13. $P(x) = x^3 + 5x^2 - 2x - 10$;
$a = 1, b = 2$

14. $P(x) = x^4 - 9x^2 + 8$;
$a = -3, b = -2$

In Exercises 15–16 the given polynomial has an irrational zero between a and b. Approximate this zero to the nearest hundredth.

15. $P(x) = x^3 - 2x^2 - 3x + 6$;
$a = 1, b = 2$

16. $P(x) = x^3 + 2x^2 - 5x - 10$;
$a = 2, b = 3$

17. Sketch the graph of $P(x) = x^3$. Use this graph to sketch each of the following graphs and interpret the results.

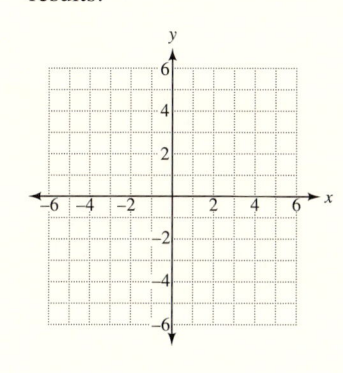

$P(x) = x^3$

(a) $P(x) = x^3 + 2$ **(b)** $P(x) = x^3 - 3$ 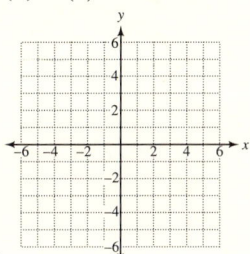   **(c)** $P(x) = (x + 1)^3$ 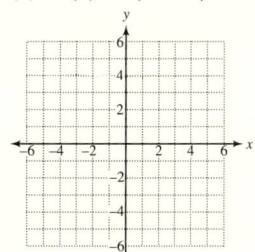   **(d)** $P(x) = (x - 3)^3$

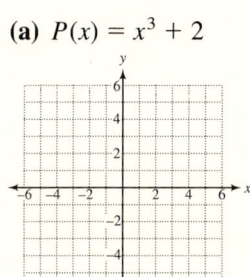

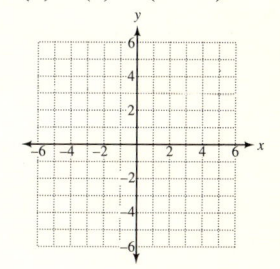

FOR REVIEW

Find all solutions of each polynomial equation in Exercises 18–19.

18. $x^3 - 5x^2 + x - 5 = 0$

19. $2x^4 - 5x^3 - 13x^2 + 25x + 15 = 0$

20. Use synthetic division to find the quotient and remainder when $y^5 - 3y^4 + y^3 - 7y^2 + y - 2$ is divided by $y + 1$.

21. Use a calculator, synthetic division, and the remainder theorem to find $P(4.1)$ if $P(x) = 1.2x^3 - 3.7x^2 - 5.35x + 1.4268$.

22. The solutions to $x^3 + 8 = 0$ are called cube roots of -8. How many cube roots of -8 are there? Find them.

What is the domain of each function in Exercises 23–24? This type of function is considered in the next section.

23. $f(x) = \dfrac{x^2 - 5x - 6}{x + 3}$

24. $g(x) = \dfrac{3x + 1}{x^2 + 5}$

ANSWERS: 1. (d) 2. (a) 3. (c) 4. (b) 5. (a) 3 (b) 2 (c) (0, 7) (d) upward (e) downward 6. (a) 3 (b) 2 (c) (0, 12) (d) downward (e) upward 7. (a) 4 (b) 3 (c) (0, −1) (d) upward (e) upward 8. (a) 6 (b) 5 (c) (0, −5) (d) downward (e) downward

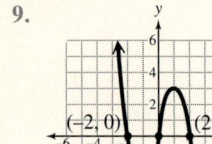

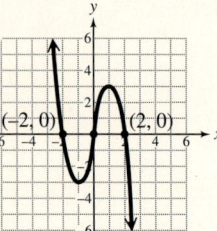

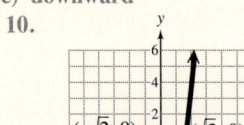

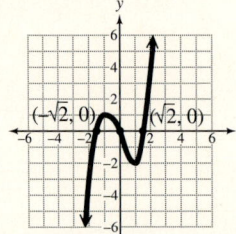

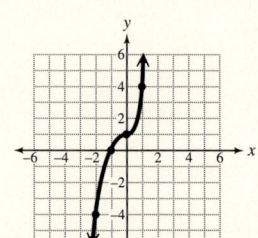

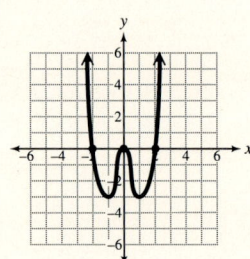

9. 10. 11. 12.

13. $P(1) = -6$ and $P(2) = 14$; so by the intermediate value theorem there is a zero between 1 and 2. 14. $P(-3) = 8$ and $P(-2) = -12$; so by the intermediate value theorem there is a zero between −3 and −2. 15. 1.73 16. 2.24

17.

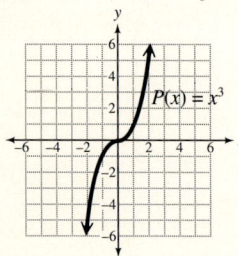

(a) The graph of $P(x) = x^3 + 2$ is the graph of $P(x) = x^3$ moved up 2 units.

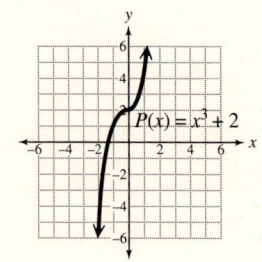

(b) The graph of $P(x) = x^3 - 3$ is the graph of $P(x) = x^3$ moved down 3 units.

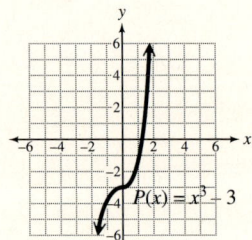

(c) The graph of $P(x) = (x + 1)^3$ is the graph of $P(x) = x^3$ moved left 1 unit.

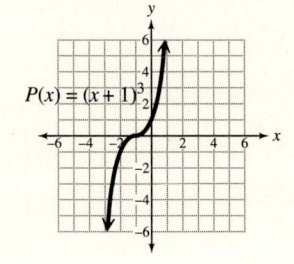

(d) The graph of $P(x) = (x - 3)^3$ is the graph of $P(x) = x^3$ moved right 3 units.

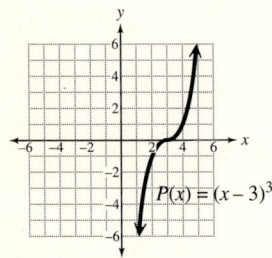

18. 5, i, $-i$ 19. 3, $-\frac{1}{2}$, $\sqrt{5}$, $-\sqrt{5}$ 20. Q: $y^4 - 4y^3 + 5y^2 - 12y + 13$; R: -15 21. $P(4.1) = 0$ 22. 3; -2, $1 \pm i\sqrt{3}$ 23. all real numbers except $x = -3$ 24. all real numbers

4.5 EXERCISES B

In Exercises 1–4 match each polynomial with one of the graphs (a), (b), (c), or (d).

1. $P(x) = x^2 - 2x - 3$
2. $P(x) = -x^3 + x^2 + 4x - 4$
3. $P(x) = x^3 - 2x^2 + x - 2$
4. $P(x) = -x^4 - x^2 + 2$

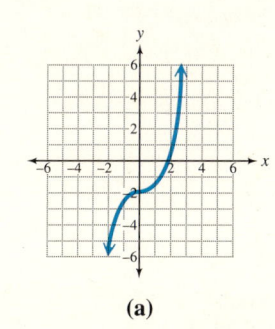

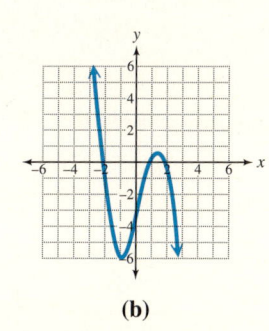

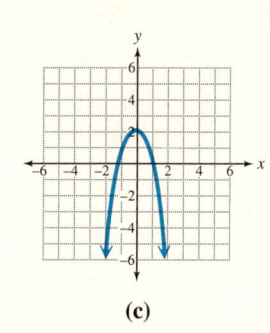

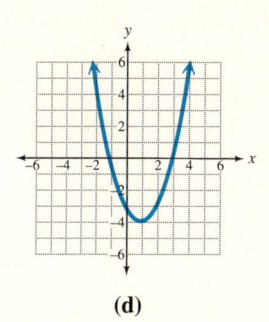

(a) (b) (c) (d)

For each of the polynomial functions in Exercises 5–8 determine (a) the maximum number of real zeros, (b) the maximum number of turning points, (c) the y-intercept, (d) the direction of the graph for large positive values of x, and (e) the direction of the graph for small negative values of x.

5. $P(x) = 2x^5 + x^3 - 3x^2 + x - 2$
6. $P(x) = -5x^5 + 4x^4 - 3x^2 + 2x + 3$
7. $P(x) = -4x^6 + 2x^5 - x^3 + 3x - 6$
8. $P(x) = 7x^8 + 2x^6 - 3x^4 + 4x^2 + 1$

In Exercises 9–12 sketch the graph of each polynomial function using the various techniques of this section.

9. $y = P(x) = x^3 - 9x$
10. $y = P(x) = x^3 - 2x^2 - x + 2$
11. $y = P(x) = -x^3 + 2x^2 - 2x + 4$
12. $y = P(x) = -x^4 + 9x^2$

In Exercises 13–14 show that P(x) has a zero between the given values a and b.

13. $P(x) = x^3 - 4x^2 - 2x + 8$;
    $a = -2,\ b = -1$
14. $P(x) = x^5 - 8x^3 + x^2 - 8$;
    $a = 2,\ b = 3$

In Exercises 15–16 the given polynomial has an irrational zero between a and b. Approximate this zero to the nearest hundredth.

15. $P(x) = x^3 + 4x^2 - 5x - 20$;
    $a = -3,\ b = -2$
16. $P(x) = x^3 + 4x^2 - 10x - 40$;
    $a = 3,\ b = 4$

17. Sketch the graph of $P(x) = x^4$. Use this graph to sketch each of the following graphs and interpret the results.
    (a) $P(x) = x^4 + 2$ (b) $P(x) = x^4 - 3$
    (c) $P(x) = (x + 1)^4$ (d) $P(x) = (x - 3)^4$

FOR REVIEW

Find all solutions of each polynomial equation in Exercises 18–19.

18. $x^3 + x^2 - 16x - 16 = 0$
19. $x^4 - 4x^3 - 20x^2 - 4x - 21 = 0$

20. Use synthetic division to find the quotient and remainder when $y^4 - 2y^3 + 3y + 5$ is divided by $y + 2$.

21. Use a calculator, synthetic division, and the remainder theorem to find $P(-2.1)$ if $P(x) = 1.5x^3 - 3.8x^2 - 4.57x + 2.3842$.

22. The solutions to $x^4 - 16 = 0$ are called fourth roots of 16. How many fourth roots of 16 are there? Find them.

What is the domain of each function in Exercises 23–24? This type of function is considered in the next section.

23. $f(x) = \dfrac{x^2 - x + 5}{x - 1}$

24. $g(x) = \dfrac{2x - 7}{x^2}$

4.5 EXERCISES C

1. Find the value of m so that the graph of $P(x) = x^4 - x^2 + x + m$ passes through the point $(2, -1)$.
[Answer: $m = -15$]

2. Find the value of m so that the graph of $P(x) = x^5 + 4x^4 - x^2 + x + m$ has y-intercept $(0, 5)$.

3. Graph the polynomial function $P(x) = x^3 - 3x$. Approximate the positive irrational zero of $P(x)$, correct to the nearest hundredth.

4. Make a table of values and use it to graph the polynomial function $P(x) = 1.2x^4 - 2.3x^2 + 5.1$.

5. CONSTRUCTION A storage container is to be constructed in the shape of a cube topping a triangular prism as shown below. The length of each edge of the cube is x. Show that the volume of the container is given by the polynomial function $V(x) = x^3 + \tfrac{1}{2}x^2(10 - x)$. Use a calculator to graph the function for $0 \le x \le 5$. If the container is to hold 60 ft³, use your graph to find the value of x, correct to the nearest tenth.

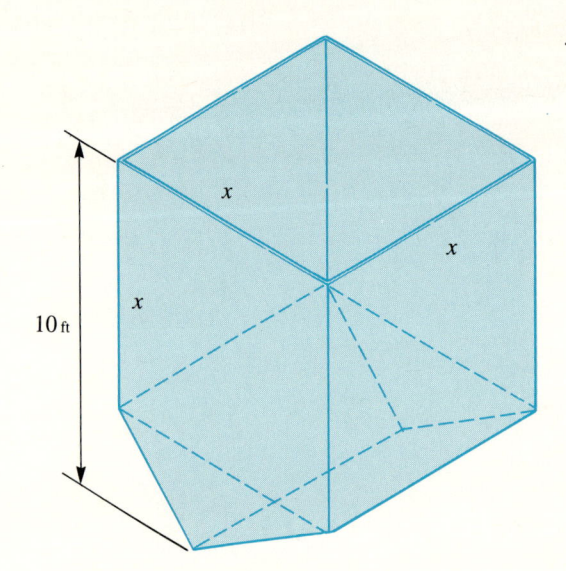

4.6 RATIONAL FUNCTIONS

STUDENT GUIDEPOSTS

1. Rational Functions
2. Asymptotes
3. Graphing Rational Functions
4. Applications of Rational Functions

① RATIONAL FUNCTIONS

In the last section we studied polynomial functions and properties of their graphs. Functions such as

$$f(x) = \frac{x^2 + 2x - 3}{x - 2}, \quad a(x) = \frac{x - 7}{x^2 + x + 3}, \quad b(x) = \frac{2x^3 - 5x^2 + x - 7}{3}$$

which are quotients of polynomials are called *rational functions*.

> **Rational Functions**
>
> A function of the form
>
> $$f(x) = \frac{P(x)}{Q(x)},$$
>
> where $P(x)$ and $Q(x)$ are polynomials and $Q(x)$ is not the zero polynomial, is a **rational function.**

Note that every polynomial function $y = P(x)$ is also a rational function since

$$y = P(x) = \frac{P(x)}{1} = \frac{P(x)}{Q(x)}, \text{ where } Q(x) = 1.$$

In this section we are concerned only with rational functions whose denominators are not constants (otherwise, they could be reduced to polynomial functions). Unlike polynomial functions which are defined for every real number x, rational functions are not defined at any value of x which makes the denominator zero. For example,

$$f(x) = \frac{x^2 + 2x - 3}{x - 2}$$

is not defined when $x = 2$, since

$$f(2) = \frac{(2)^2 + 2(2) - 3}{(2) - 2} = \frac{5}{0},$$

which is undefined. However, $f(x)$ is a number if x is any number other than 2, and thus the domain of $f(x)$ is all real numbers except 2.

② ASYMPTOTES

If the graph of a function gets closer and closer to some line when the values of x get either closer and closer to a given number or larger and larger positively or negatively, the line is an **asymptote** of the graph. Many rational functions have asymptotes. Suppose we consider a very simple rational function, $f(x) = \frac{1}{x}$.

Notice that the domain of this function is all real numbers except 0. If we write y for $f(x)$, the equation can be written as $xy = 1$, and it is easy to see that the graph is symmetric with respect to the origin. (When we replace x with $-x$ and y with $-y$, the same equation will result.) Suppose we use a calculator to find a number of values and put them in a table as shown below. Plotting these points, as shown in Figure 4.12, we discover that they all lie in quadrant I. Notice that as x gets large positively, the corresponding values for $f(x)$ approach 0. Also, as x approaches 0, the values of $f(x)$ get larger and larger. In this case the x-axis and the y-axis are asymptotes of the graph. Using symmetry with respect to the origin, we can obtain the other portion of the graph found in quadrant III.

x	$f(x) = 1/x$
1	1.00
2	0.50
4	0.25
5	0.20
10	0.10
100	0.01
1000	0.001
0.5	2
0.1	10
0.01	100
0.001	1000

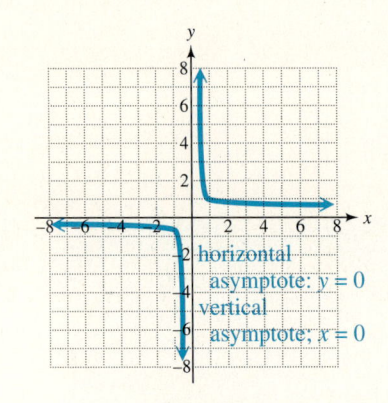

Figure 4.12

Figure 4.13 gives the graphs of two rational functions and shows the three possibilities for asymptotes: *vertical asymptotes* that are equal to or parallel to the y-axis (like $x = 1$ in Figure 4.13(a)), *horizontal asymptotes* that are equal to or parallel to the x-axis (like $y = 0$ in Figure 4.13(a)), and *oblique asymptotes* that are lines not parallel to either of the coordinate axes (like $y = x$ in Figure 4.13(b)).

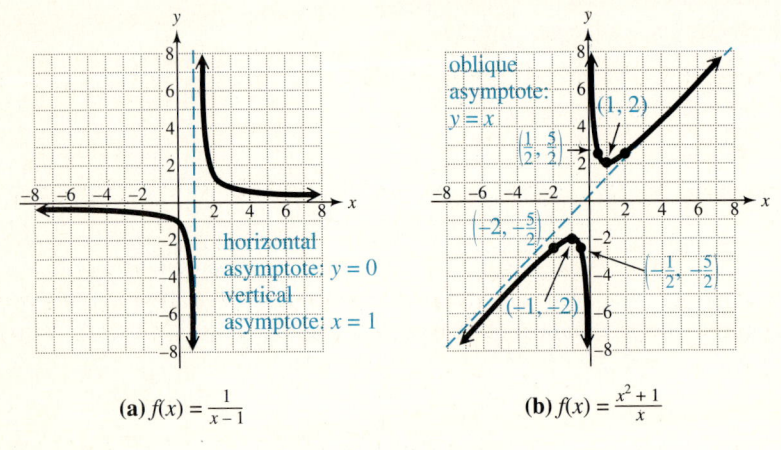

(a) $f(x) = \dfrac{1}{x-1}$

(b) $f(x) = \dfrac{x^2+1}{x}$

Figure 4.13 Asymptotes

To obtain the graph of a rational function, it is clear that we need to find any asymptotes of the graph. We begin with the way to identify vertical asymptotes.

Vertical Asymptotes of a Rational Function

To find the vertical asymptotes of the rational function

$$f(x) = \frac{P(x)}{Q(x)},$$

where $P(x)$ and $Q(x)$ have no common factor, set $Q(x) = 0$ and solve. If r is a real zero of $Q(x)$, then the line $x = r$ is a **vertical asymptote**.

EXAMPLE 1 FINDING VERTICAL ASYMPTOTES

Find the vertical asymptotes of each function.

(a) $f(x) = \dfrac{x^2 + 2x - 3}{x - 2}$

Setting the denominator $x - 2$ equal to zero, we obtain $x = 2$ as a vertical asymptote.

(b) $g(x) = \dfrac{x + 1}{x^2 - 4}$

To solve $x^2 - 4 = 0$, we factor and obtain $x = 2$ and $x = -2$ as the vertical asymptotes.

(c) $h(x) = \dfrac{2x^2}{x^2 + 1}$

Since $x^2 + 1 = 0$ has no real solutions, there are no vertical asymptotes.

PRACTICE EXERCISE 1

Find the vertical asymptotes of each function.

(a) $f(x) = \dfrac{x^2 - 4x + 3}{3x + 6}$

(b) $g(x) = \dfrac{x - 2}{x^2 - 2x - 3}$

(c) $h(x) = \dfrac{x^2}{x^2 + 4}$

Answers: (a) $x = -2$ (b) $x = -1$ and $x = 3$ (c) no vertical asymptotes

---

Horizontal and oblique asymptotes, which describe what happens to the graph when $|x|$ becomes large, can be discovered by considering the degrees of the polynomials in the numerator and denominator. Suppose that the degree of the polynomial in the numerator of a rational function is less than the degree of the polynomial in the denominator, as in the following function, $k(x)$.

$$k(x) = \frac{3x + 5}{x^2 - 2x + 7}$$

Multiply both numerator and denominator by $1/x^2$, or divide each term by x^2.

$$k(x) = \frac{\dfrac{3x}{x^2} + \dfrac{5}{x^2}}{\dfrac{x^2}{x^2} - \dfrac{2x}{x^2} + \dfrac{7}{x^2}} = \frac{\dfrac{3}{x} + \dfrac{5}{x^2}}{1 - \dfrac{2}{x} + \dfrac{7}{x^2}}$$

As x becomes larger, $3/x$ becomes smaller (for example, if $x = 10$, $3/x = 3/10 = 0.3$; if $x = 100$, $3/x = 3/100 = 0.03$; if $x = 1000$, $3/x = 3/1000 = 0.003$). Likewise, when x gets large, $5/x^2$, $2/x$, and $7/x^2$ all approach zero. Thus, the fraction approaches

$$\frac{0 + 0}{1 - 0 + 0} = \frac{0}{1} = 0.$$

310 CHAPTER 4 POLYNOMIAL AND RATIONAL FUNCTIONS

The same is true when x gets larger and larger in the negative direction. Thus, as $|x|$ increases, $k(x)$ approaches 0 so that the x-axis, $y = 0$, is a horizontal asymptote.

If the degree of the numerator is equal to the degree of the denominator, the rational function also has a horizontal asymptote, but this time it is a line parallel to the x-axis. For example, consider

$$s(x) = \frac{2x^2 - x + 3}{5x^2 + 1}.$$

Again, we multiply numerator and denominator by $1/x^2$.

$$s(x) = \frac{2 - \frac{1}{x} + \frac{3}{x^2}}{5 + \frac{1}{x^2}}$$

As $|x|$ increases, $1/x$, $3/x^2$, and $1/x^2$ all approach zero and the fraction approaches

$$\frac{2 - 0 + 0}{5 + 0} = \frac{2}{5}.$$

Thus, as $|x|$ increases, $k(x)$ approaches $2/5$ which means that $y = 2/5$ is a horizontal asymptote.

Horizontal Asymptote of a Rational Function

Let

$$f(x) = \frac{P(x)}{Q(x)} = \frac{a_n x^n + a_{n-1} x^{n-1} + \cdots + a_1 x + a_0}{b_m x^m + b_{m-1} x^{m-1} + \cdots + b_1 x + b_0}, \quad (a_n \neq 0, b_m \neq 0)$$

be a rational function in which $P(x)$ and $Q(x)$ have no common factor.

1. If $n < m$, then the x-axis, $y = 0$, is a horizontal asymptote of the graph of $f(x)$.
2. If $n = m$, then the line $y = a_n/b_m$ is a horizontal asymptote.
3. If $n > m$, there is no horizontal asymptote.

EXAMPLE 2 FINDING HORIZONTAL ASYMPTOTES

Find the horizontal asymptotes of each function.

(a) $f(x) = \dfrac{x^2 + 2x - 3}{x - 2}$

Since $P(x) = x^2 + 2x - 3$ has degree $n = 2$ and $Q(x) = x - 2$ has degree $m = 1$, $n > m$ so there is no horizontal asymptote.

(b) $g(x) = \dfrac{x + 1}{x^2 - 4}$

Since $P(x) = x + 1$ has degree $n = 1$ and $Q(x) = x^2 - 4$ has degree $m = 2$, $n < m$ so that the x-axis, $y = 0$, is a horizontal asymptote.

(c) $h(x) = \dfrac{2x^2}{x^2 + 1}$

PRACTICE EXERCISE 2

Find the horizontal asymptotes of each function.

(a) $f(x) = \dfrac{x^2 - 4x + 3}{3x + 6}$

(b) $g(x) = \dfrac{x - 2}{x^2 - 2x - 3}$

(c) $h(x) = \dfrac{x^2}{x^2 + 4}$

Since $P(x)$ and $Q(x)$ are both of degree 2 ($n = m$), with $a_n = 2$ and $b_m = 1$, the line $y = \frac{a_n}{b_m} = \frac{2}{1} = 2$ is a horizontal asymptote.

Answers: (a) no horizontal asymptotes (b) $y = 0$ (c) $y = 1$

Finally, when the degree of the polynomial in the numerator of a rational function is 1 more than the degree of the polynomial in the denominator, the graph has an oblique asymptote that can be found by dividing the denominator into the numerator.

Oblique Asymptote of a Rational Function

Let
$$f(x) = \frac{P(x)}{Q(x)}$$
be a rational function such that the degree of $P(x)$ is n and the degree of $Q(x)$ is m.

1. If $n = m + 1$ ($m > 0$), and if the quotient of $P(x)$ and $Q(x)$ is $g(x)$ and the remainder is not zero, then the equation
$$y = g(x) = ax + b, \quad a \neq 0$$
is an oblique asymptote of the graph.
2. If $n > m + 1$, there are no oblique asymptotes.

EXAMPLE 3   FINDING OBLIQUE ASYMPTOTES

Find the oblique asymptote of the function
$$f(x) = \frac{x^2 + 2x - 3}{x - 2}.$$
Since the degree of the numerator is 2, which is one more than the degree of the denominator, we divide $x^2 + 2x - 3$ by $x - 2$ and get the quotient $x + 4$ and remainder 5. Thus,
$$f(x) = \frac{x^2 + 2x - 3}{x - 2} = x + 4 + \frac{5}{x - 2}.$$
As $|x|$ gets larger and larger, $5/(x - 2)$ approaches zero so that $f(x)$ approaches $x + 4$. Hence, the line $y = x + 4$ is an oblique asymptote of the graph.

PRACTICE EXERCISE 3

Find the oblique asymptote of the function
$$f(x) = \frac{x^2 - 4x + 3}{3x + 6}.$$

Answer: $y = \frac{1}{3}x - 2$

❸ GRAPHING RATIONAL FUNCTIONS

As we move toward graphing rational functions, another piece of useful information would be where the rational function is zero. A **zero of a rational function** is a value of x that makes the numerator zero while the denominator is nonzero. Of course, the zeros are the x-intercepts of the graph. To graph a rational function $f(x) = \frac{P(x)}{Q(x)}$, find the asymptotes, the x-intercepts, the y-intercept (found by setting $x = 0$), any symmetries, and consider whether $f(x)$ is positive or negative in the intervals with endpoints determined by the zeros of $P(x)$ (the zeros of the rational function) and the zeros of $Q(x)$ (the values that determine vertical asymptotes). Also, it may be necessary to plot several points to obtain a reasonably accurate sketch. We illustrate this technique in the following examples.

EXAMPLE 4   GRAPHING A RATIONAL FUNCTION

Graph $y = f(x) = \dfrac{x^2 + 2x - 3}{x - 2}$.

Vertical asymptotes:	$x = 2$	See Example 1
Horizontal asymptotes:	none	See Example 2
Oblique asymptotes:	$y = x + 4$	See Example 3

We can factor the numerator.

$$\frac{x^2 + 2x - 3}{x - 2} = \frac{(x + 3)(x - 1)}{x - 2} = 0$$

Therefore, -3 and 1 are zeros of $f(x)$.

x-intercepts: $(-3, 0)$ and $(1, 0)$

Substitute 0 for x to find the y-intercept.

$$f(0) = \frac{(0)^2 + 2(0) - 3}{0 - 2} = \frac{-3}{-2} = \frac{3}{2}$$

y-intercept: $(0, 3/2)$

Symmetries: none (The tests for symmetry with respect to the y-axis and origin both fail.)

The intervals determined by the zeros of $f(x)$, -3 and 1, and the vertical asymptote, $x = 2$, are $(-\infty, -3)$, $(-3, 1)$, $(1, 2)$, and $(2, \infty)$. A table summarizes the results.

Interval	$(-\infty, -3)$	$(-3, 1)$	$(1, 2)$	$(2, \infty)$
Test point a	-4	0	$\dfrac{3}{2}$	3
Value of $f(a)$	$f(-4) = -\dfrac{5}{6}$	$f(0) = \dfrac{3}{2}$	$f\left(\dfrac{3}{2}\right) = -\dfrac{9}{2}$	$f(3) = 12$
Sign of $f(x)$	$-$	$+$	$-$	$+$
Location of the graph of $f(x)$	below the x-axis	above the x-axis	below the x-axis	above the x-axis

The graph is shown in Figure 4.14.

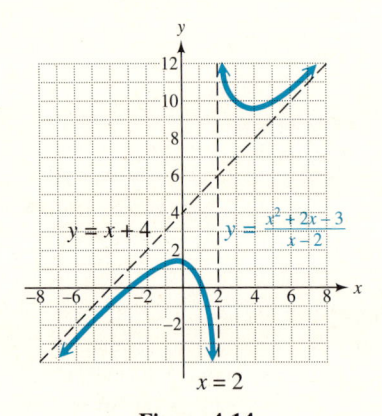

Figure 4.14

PRACTICE EXERCISE 4

Graph $f(x) = \dfrac{x^2 - 4x + 3}{3x + 6}$.

[*Hint:* Take advantage of your work in Practice Exercises 1, 2, and 3.]

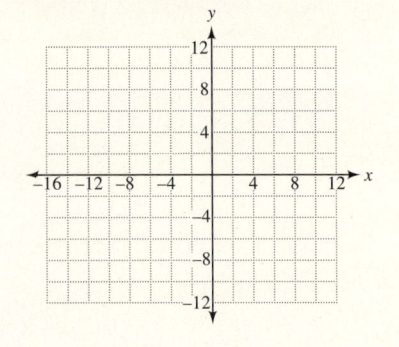

Answer: graph given at the end of this section

EXAMPLE 5 GRAPHING A RATIONAL FUNCTION

Graph $y = h(x) = \dfrac{2x^2}{x^2 + 1}$.

Vertical asymptotes: none
Horizontal asymptotes: $y = 2$
Oblique asymptotes: none
x-intercepts: $(0, 0)$
y-intercept: $(0, 0)$
Symmetries: y-axis

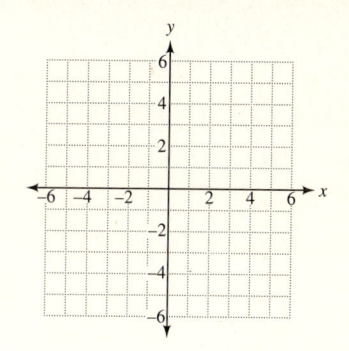

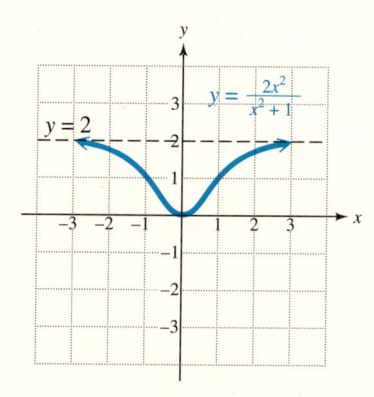

Figure 4.15

This time there are only two intervals to consider, $(-\infty, 0)$ and $(0, \infty)$. Since the graph is symmetric with respect to the y-axis, the sign of $h(x)$ will be the same on both intervals. Using the test point 1 in $(0, \infty)$, we have $h(1) = 1 > 0$, so the sign of $h(x)$ is $+$ on $(0, \infty)$ and also on $(-\infty, 0)$. Thus, the graph of $h(x)$, given in Figure 4.15, is above the x-axis on both intervals.

PRACTICE EXERCISE 5

Graph $g(x) = \dfrac{-2x^2}{x^2 + 4}$.

Answer: graph given at the end of this section

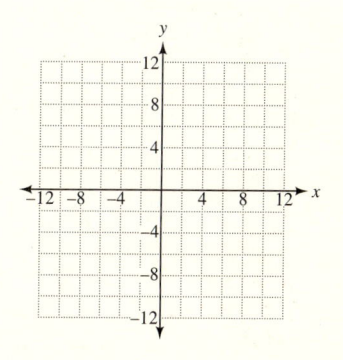

EXAMPLE 6 GRAPHING A RATIONAL FUNCTION

Graph $y = F(x) = \dfrac{x^2 + 1}{x^2 - 1}$.

Vertical asymptotes: $x = 1$ and $x = -1$
Horizontal asymptotes: $y = 1$
Oblique asymptotes: none
x-intercepts: none
y-intercept: $(0, -1)$
Symmetries: y-axis

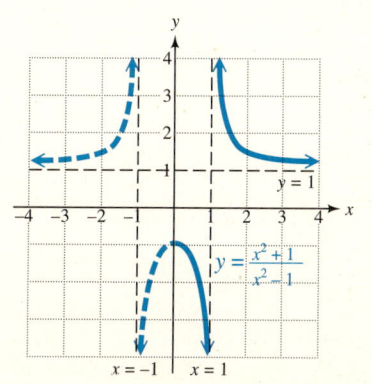

Figure 4.16

PRACTICE EXERCISE 6

Graph $G(x) = \dfrac{2x^2}{x^2 - 4}$.

With no *x*-intercepts, the intervals, determined by the vertical asymptotes $x = 1$ and $x = -1$, are

$$(-\infty, -1), \ (-1, 1), \text{ and } (1, \infty).$$

Using the symmetry with respect to the *y*-axis, the sign of $F(x)$ on $(-\infty, -1)$ will be the same as the sign on $(1, \infty)$.

Interval	$(-1, 1)$	$(1, \infty)$
Test point a	0	2
Value of $F(a)$	$F(0) = -1$	$F(2) = \dfrac{5}{3}$
Sign of $F(x)$	$-$	$+$
Location of the graph of $F(x)$	below the *x*-axis	above the *x*-axis

Thus, the graph of $F(x)$ is above the *x*-axis on $(-\infty, -1)$ and $(1, \infty)$, and below the *x*-axis on $(-1, 1)$. In Figure 4.16, the solid portion of the curve was plotted first; the dashed portion was sketched using symmetry with respect to the *y*-axis.

Answer: graph given at the end of this section

CAUTION

Graphs of rational functions, like polynomial functions, consist of curves that are smooth with no sharp "points" or turns. This fact is illustrated by the graphs in the preceding examples.

❹ APPLICATIONS OF RATIONAL FUNCTIONS

We conclude this section with an application that illustrates how a graph can be helpful to analyze and describe certain information. It also shows that, in a practical situation, we may need to restrict the domain of a function to fit the reality of the conditions implied or stated in the problem.

EXAMPLE 7   A MANUFACTURING PROBLEM

During an eight-hour shift, the number of items produced in a factory is given by $n(t) = t^2 + 16t$ $(0 \leq t \leq 8)$. The total cost in dollars of producing $n(t)$ items is given by $c(t) = 120t + 960$. The average cost of production is a function of t, given by $a(t) = \dfrac{c(t)}{n(t)}$. Graph $a(t)$ concentrating on the portion for $0 < t \leq 8$ and interpret the results.

In order to determine the asymptotes we write $a(t)$ in factored form.

$$y = a(t) = \frac{120t + 960}{t(t + 16)}$$

PRACTICE EXERCISE 7

When a variable resistor with resistance r ohms is placed in parallel in an electrical circuit with a 4-ohm resistor, the total resistance R of the circuit is given by the rational function

$$R = \frac{4r}{4 + r}.$$

Vertical asymptotes: $t = 0$ and $t = -16$
Horizontal asymptotes: $y = 0$
Oblique asymptotes: none
x-intercepts: $(-8, 0)$
y-intercepts: none
Symmetries: none

The graph, given in Figure 4.17, shows the portion of interest ($0 < t \leq 8$) with a solid line.

t	$a(t)$
-8	0
-14	25.7
-2	-25.7
-18	-33.3
-20	-18
-24	-10
2	33.3
4	18
8	10
12	7.1

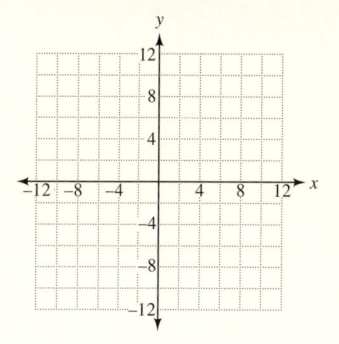

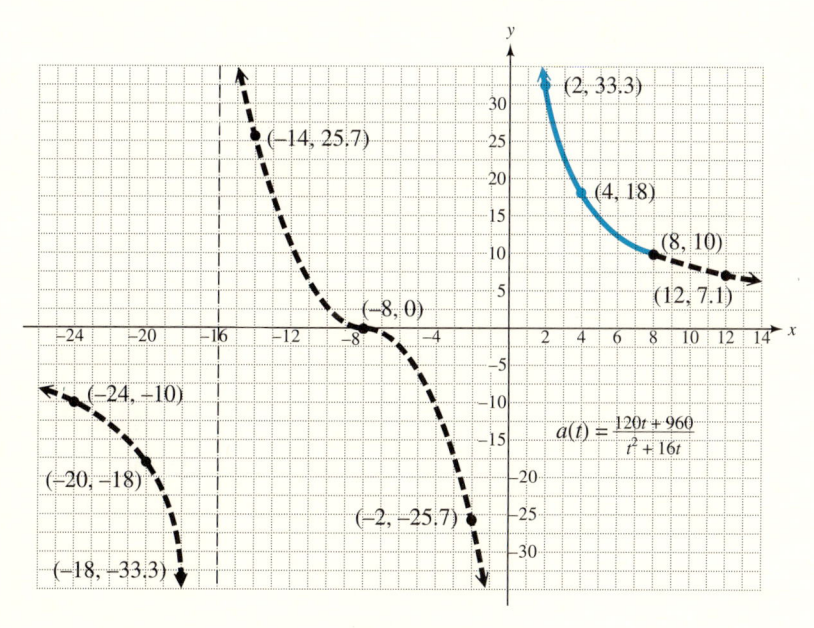

Figure 4.17

The average cost of production decreases to a minimum after 8 hours ($t = 8$), and amounts to $10 per item at that time.

Sketch the graph of this function and use it to determine what happens to the total resistance of the circuit as the variable resistance is allowed to increase without bound.

Answer: The graph and discussion answer are given at the end of this section.

ANSWERS TO PRACTICE EXERCISES:

4. 5. 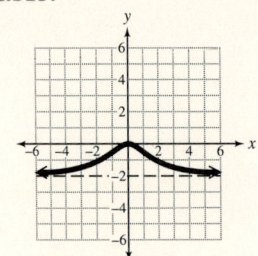 6. 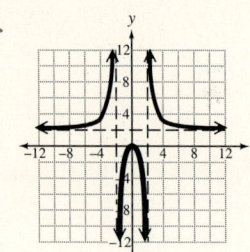 7.

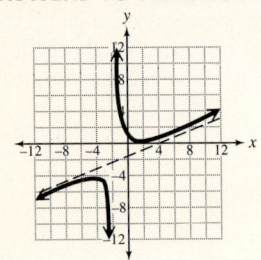

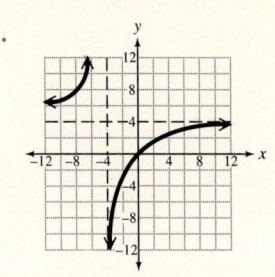

In the graph in number 7, we are interested only in that portion where $r \geq 0$. Only a piece of the second branch of the graph in quadrant II is shown. Notice as r gets larger and larger without bound, the curve approaches the horizontal asymptote $y = 4$, so we see that under these conditions, the total resistance of the circuit is approaching 4 ohms.

4.6 EXERCISES A

In Exercises 1–6 give the (a) vertical asymptotes, (b) horizontal asymptotes, (c) oblique asymptotes (d) x-intercepts, (e) y-intercept, and (f) symmetries of the graph of each rational function.

1. $f(x) = \dfrac{5}{x + 2}$

2. $F(x) = \dfrac{4x}{x - 5}$

3. $f(x) = \dfrac{x + 1}{x^2 + x - 2}$

4. $F(x) = \dfrac{x^2 + x - 6}{x - 5}$

5. $f(x) = \dfrac{5x^4}{x^4 + 1}$

6. $g(x) = \dfrac{x^2 - 9}{x - 1}$

In Exercises 7–12 graph each rational function using information about asymptotes, intercepts, symmetries, and the intervals where the function is above or below the x-axis.

7. $f(x) = \dfrac{2}{x - 2}$

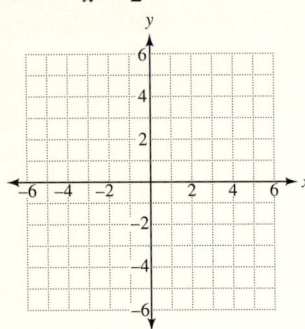

8. $g(x) = \dfrac{x}{x + 1}$

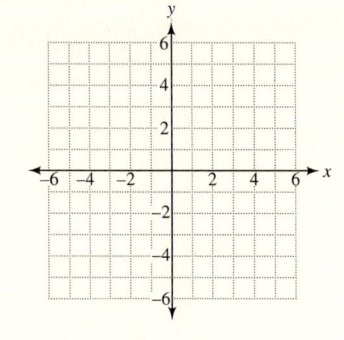

9. $F(x) = \dfrac{1}{x^2 - 4}$

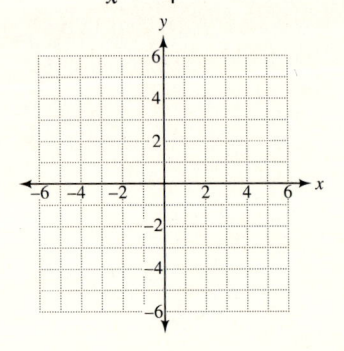

10. $G(x) = \dfrac{x}{x^2 - 4}$

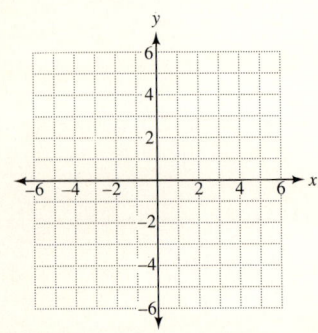

11. $f(x) = \dfrac{2x^2 + 1}{x^2 - 1}$

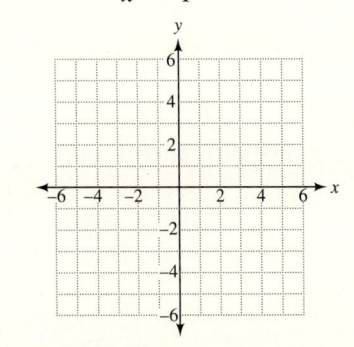

12. $g(x) = \dfrac{5}{x^2 + 5}$

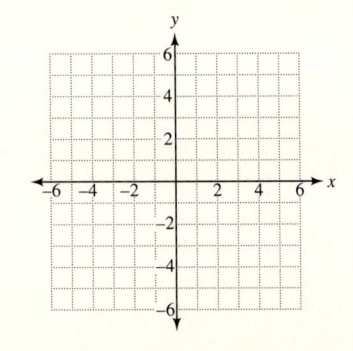

4.6 RATIONAL FUNCTIONS   **317**

13. **MANUFACTURING** During a 12-hour shift, the number of items produced in a factory is given by $n(t) = t^2 + 24t$ $(0 \leq t \leq 12)$. The total cost in dollars of producing $n(t)$ items is given by $c(t) = 144t + 1728$. Graph the rational function $a(t) = c(t)/n(t)$ which expresses the average cost of production for $0 < t \leq 12$ and interpret the results.

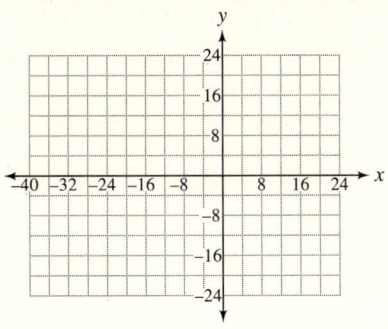

14. Give a rational function whose graph has one vertical asymptote at $x = 1$, one horizontal asymptote at $y = 0$, and a y-intercept $(0, -3)$.

15. Can the graph of a rational function ever cross a horizontal asymptote?

FOR REVIEW

16. Graph the polynomial function $y = P(x) = 2x^2 - x^4$.

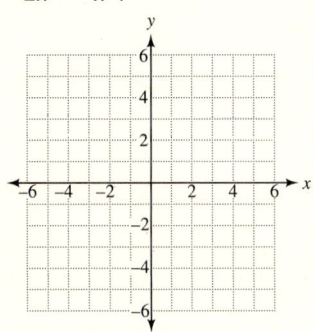

17. Use Descartes' rule of signs to find the possible number of negative, positive, and nonreal solutions to the equation $x^4 + 5x^3 + 2x^2 + x + 2 = 0$.

The following exercises review quadratic equations in preparation for topics in the next section. Solve each equation in Exercises 18–20.

18. $2x^2 + 15x - 8 = 0$

19. $3x^2 + x - 1 = 0$

20. $3x^2 + 9 = 0$

ANSWERS: 1. (a) $x = -2$ (b) $y = 0$ (c) none (d) none (e) $\left(0, \frac{5}{2}\right)$ (f) none 2. (a) $x = 5$ (b) $y = 4$ (c) none (d) $(0, 0)$ (e) $(0, 0)$ (f) none 3. (a) $x = 1$ and $x = -2$ (b) $y = 0$ (c) none (d) $(-1, 0)$ (e) $\left(0, -\frac{1}{2}\right)$ (f) none 4. (a) $x = 5$ (b) none (c) $y = x + 6$ (d) $(2, 0)$ and $(-3, 0)$ (e) $\left(0, \frac{6}{5}\right)$ (f) none 5. (a) none (b) $y = 5$ (c) none (d) $(0, 0)$ (e) $(0, 0)$ (f) y-axis 6. (a) $x = 1$ (b) none (c) $y = x + 1$ (d) $(3, 0)$, $(-3, 0)$ (e) $(0, 9)$ (f) none

7. 8. 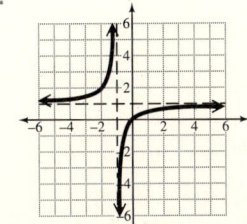 9. 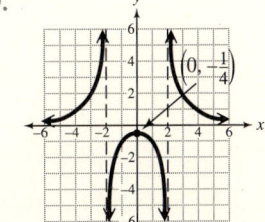 10.

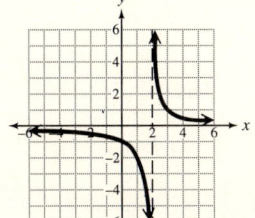

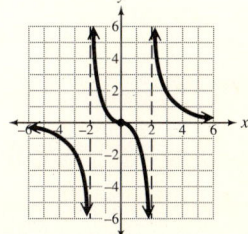

318 CHAPTER 4 POLYNOMIAL AND RATIONAL FUNCTIONS

11.

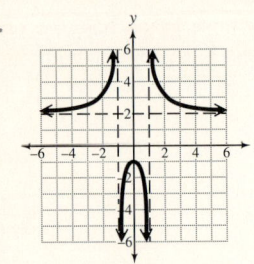

12.

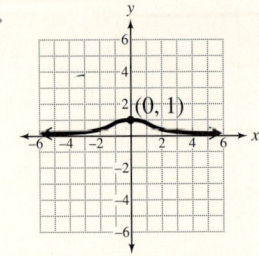

13. The average cost of production decreases to a minimum of $8 per item after 12 hours.

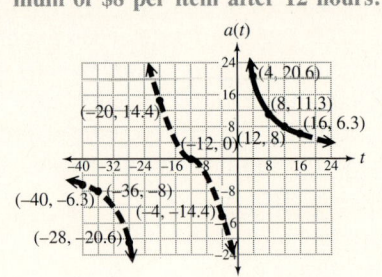

14. $f(x) = \frac{3}{x-1}$ 15. Yes; see Example 7.

16.

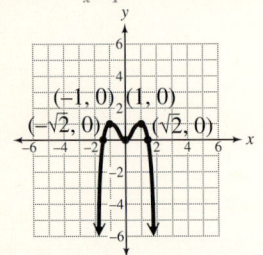

17. either 0, 0, 4; 2, 0, 2; or 4, 0, 0 (in order negative, positive, nonreal)
18. $\frac{1}{2}$, -8 19. $\frac{-1 \pm \sqrt{13}}{6}$ 20. $\pm i\sqrt{3}$

4.6 EXERCISES B

In Exercises 1–6 give the (a) vertical asymptotes, (b) horizontal asymptotes, (c) oblique asymptotes, (d) x-intercepts, (e) y-intercept, and (f) symmetries of the graph of each rational function.

1. $g(x) = \dfrac{-3}{x-4}$

2. $G(x) = \dfrac{-2x}{2x+1}$

3. $g(x) = \dfrac{2(3x-1)}{2x^2+9x-5}$

4. $G(x) = \dfrac{x^2-16}{x+1}$

5. $F(x) = \dfrac{2x^3}{x^3-1}$

6. $g(x) = \dfrac{x^2-4}{x+3}$

In Exercises 7–12 graph each rational function using information about asymptotes, intercepts, symmetries, and the intervals where the function is above or below the x-axis.

7. $f(x) = \dfrac{4}{4-x}$

8. $G(x) = \dfrac{x}{1-x}$

9. $f(x) = \dfrac{-1}{x^2-9}$

10. $g(x) = \dfrac{x}{1-x^2}$

11. $F(x) = \dfrac{x^2+4}{4-x^2}$

12. $f(x) = \dfrac{x^2+x-6}{x+2}$

13. **ELECTRONICS** The power p in watts consumed by an electrical circuit is given by $p = \dfrac{10r}{(1+r)^2}$ where $r \geq 0$ is the resistance in ohms. Sketch the graph of this rational function and estimate the resistance that yields a maximum power.

14. Give a rational function whose graph has one vertical asymptote at $x = 2$, one horizontal asymptote at $y = -1$, and x-intercept $(3, 0)$.

15. Can the graph of a rational function ever cross a vertical asymptote?

FOR REVIEW

16. Graph the polynomial function $y = P(x) = x^3 - x^2 - 16x - 20$.

17. Use the upper and lower bound test to find the smallest positive integer upper bound and the largest negative integer lower bound for the real solutions to the equation $8x^3 + 2x^2 - 15x = 0$.

The following exercises review quadratic equations in preparation for topics in the next section. Solve each equation in Exercises 18–20.

18. $3x^2 - 10x - 8 = 0$ **19.** $5x^2 - 2x - 1 = 0$ **20.** $5x^2 + 10 = 0$

4.6 EXERCISES C

Graph each function in Exercises 1–4.

1. $f(x) = \dfrac{1}{x^2 + 1} - 3$

2. $f(x) = \dfrac{1}{|x - 2|}$

3. $f(x) = \left|\dfrac{1}{x + 2} - 4\right|$

4. $f(x) = \dfrac{x^3 - 2x^2 - 8x}{x^2 - 4x + 3}$

5. Can the graph of a rational function ever cross an oblique asymptote? [*Hint:* Consider the function in Exercise 4.]

6. Give a method for finding the point of intersection of the graph of a rational function and the graph of an asymptote (horizontal or oblique) of that function.

4.7 POLYNOMIAL AND RATIONAL INEQUALITIES

STUDENT GUIDEPOSTS

1. Quadratic Inequalities
2. Polynomial Inequalities
3. Rational Inequalities
4. Applications of Inequalities

1 QUADRATIC INEQUALITIES

In Chapter 2 we solved linear and quadratic equations and linear inequalities. We delayed considering *quadratic inequalities* until now when we had more information about graphing polynomials since these types of inequalities are better understood in this setting.

> **Quadratic Inequalities**
>
> Inequalities of the form
>
> $$ax^2 + bx + c < 0, \qquad ax^2 + bx + c > 0,$$
> $$ax^2 + bx + c \leq 0, \quad \text{and} \quad ax^2 + bx + c \geq 0,$$
>
> where a, b, and c are real numbers, $a \neq 0$, are called **quadratic inequalities**.

Consider the inequality $x^2 - x - 6 \geq 0$. If we set $y = f(x) = x^2 - x - 6$, finding the values of x that make the inequality true, that is finding the **solutions to the inequality,** is equivalent to asking "For what values of x is $y \geq 0$?" Graphically, this is the same as finding the values of x for which the graph of the function is on or above the x-axis. Suppose we graph $f(x) = x^2 - x - 6$. Since $a = 1 > 0$, the parabola opens up. The vertex, $\left(-\dfrac{b}{2a}, f\left(-\dfrac{b}{2a}\right)\right)$ is $\left(\dfrac{1}{2}, -\dfrac{25}{4}\right)$. Since the vertex is below the x-axis and the graph opens up, there will be x-intercepts, $(-2, 0)$ and $(3, 0)$, found by solving the quadratic equation $x^2 - x - 6 = 0$. The graph is given in Figure 4.18.

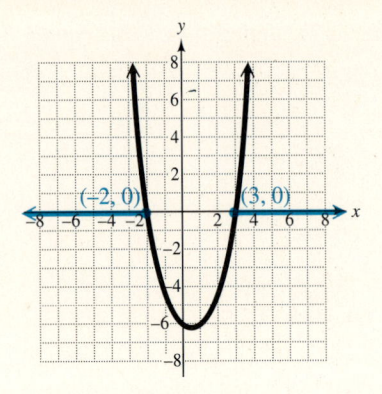

Figure 4.18

Notice that the two intercepts divide the x-axis into three intervals,

$$(-\infty, -2), \quad (-2, 3), \quad \text{and} \quad (3, \infty).$$

Figure 4.19 summarizes information we can see from the graph.

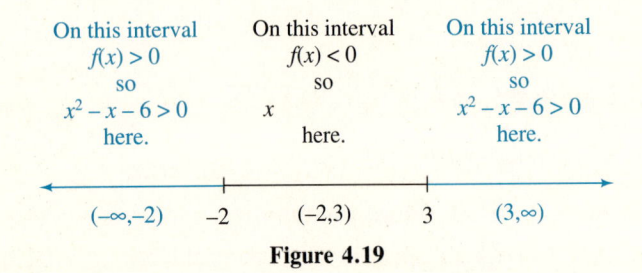

On this interval $f(x) > 0$ so $x^2 - x - 6 > 0$ here.

On this interval $f(x) < 0$ so x here.

On this interval $f(x) > 0$ so $x^2 - x - 6 > 0$ here.

$(-\infty, -2) \quad -2 \quad (-2, 3) \quad 3 \quad (3, \infty)$

Figure 4.19

Since the original inequality was $\geq$, we must include values of x where the expression is 0, that is, we must include -2 and 3 in the solution. Thus, the solution to the inequality is $(-\infty, -2]$ or $[3, \infty)$, or using inequalities, the solution is $x \leq -2$ or $x \geq 3$. Notice that since the solution includes two intervals of the number line, the word *or* is used to join the intervals (or inequalities) when we give the solution. Had we been asked to solve $x^2 - x - 6 > 0$, from the above we see that the solution is

$$(-\infty, -2) \text{ or } (3, \infty) \quad \text{or} \quad x < -2 \text{ or } x > 3.$$

Notice that -2 and 3 are *not* part of the solution since the given inequality is $>$ and does not include $=$ this time. In a similar manner, the solution to $x^2 - x - 6 < 0$, that is, the values of x where the graph is below the x-axis, is

$$(-2, 3) \quad \text{or} \quad -2 < x < 3, \quad \text{(an \textbf{and} compound)}$$

and the solution to $x^2 - x - 6 \leq 0$ is

$$[-2, 3] \quad \text{or} \quad -2 \leq x \leq 3.$$

Do we actually need to graph the quadratic function $f(x) = ax^2 + bx + c$ every time we want to solve the quadratic inequality $ax^2 + bx + c > 0$? The answer is no, since once we have seen what happens, we can shorten our work by finding the values of x for which the expression is 0 (solving the related quadratic equation $ax^2 + bx + c = 0$) and using a test point. Since every quadratic equation with two distinct solutions x_1 and x_2, with $x_1 < x_2$, separates the number line into three intervals,

$$(-\infty, x_1), \quad (x_1, x_2), \quad \text{and} \quad (x_2, \infty)$$

we know the solution will be one of two choices,

$$\text{either} \quad (-\infty, x_1) \text{ or } (x_2, \infty) \quad \text{or else} \quad (x_1, x_2).$$

By choosing one test point r (usually 0 if $x_1 \neq 0$ and $x_2 \neq 0$) and substituting r into the inequality, we will know whether r is a solution or not. If for example, r is a solution, and r is in (x_1, x_2), then (x_1, x_2) is the solution to the inequality. On the other hand, if r is not a solution, and r is in (x_1, x_2), then we know the solution consists of the other two intervals, and is $(-\infty, x_1)$ or (x_2, ∞). Of course, if the given inequality is $\leq$ or $\geq$, the endpoints of the intervals are included in the solution. This solution technique is demonstrated in the next example.

EXAMPLE 1 SOLVING A QUADRATIC INEQUALITY

Solve. $2x^2 + x - 6 < 0$

First solve the quadratic equation.

$$2x^2 + x - 6 = 0$$
$$(2x - 3)(x + 2) = 0 \qquad \text{Factor}$$
$$2x - 3 = 0 \quad \text{or} \quad x + 2 = 0 \qquad \text{Zero-product rule}$$
$$2x = 3 \qquad \qquad x = -2$$
$$x = \frac{3}{2}$$

These solutions separate the number line into the three intervals $(-\infty, -2)$, $(-2, 3/2)$, and $(3/2, \infty)$. Using 0 as a test point, we have

$$2x^2 + x - 6 < 0$$
$$2(0)^2 + (0) - 6 < 0 \qquad \text{Substitute 0 for } x$$
$$-6 < 0,$$

which is true. Since 0 is in $(-2, 3/2)$, the solution to the inequality is $(-2, 3/2)$, or, using inequalities, $-2 < x < 3/2$.

PRACTICE EXERCISE 1

Solve. $2x^2 + x - 6 \geq 0$

Answer: $(-\infty, -2]$ or $[3/2, \infty)$; or, $x \leq -2$ or $x \geq \frac{3}{2}$

When we are solving a quadratic inequality and there is only one solution to the quadratic equation (a root of multiplicity two), a slightly different situation exists. Consider the inequality $x^2 - 6x + 9 > 0$, and graph the function $f(x) = x^2 - 6x + 9$ as shown in Figure 4.20. When we solve $x^2 - 6x + 9 = 0$, factoring, $(x - 3)(x - 3) = 0$, so 3 is a double root. This time the number line is divided into only two intervals, $(-\infty, 3)$ and $(3, \infty)$, and we can see from the graph that $f(x)$ will be above the x-axis, or $x^2 - 6x + 9 > 0$, whenever x is in $(-\infty, 3)$ or $(3, \infty)$. Thus the solution to the inequality is $(-\infty, 3)$ or $(3, \infty)$. Had we been given $x^2 - 6x + 9 \geq 0$, 3 would also be a solution along with $(-\infty, 3)$ or $(3, \infty)$, making the solution all real numbers. On the other hand, $x^2 - 6x + 9 < 0$ has no solution, and $x^2 - 6x + 9 \leq 0$ has only the number 3 for its solution. This illustrates the four different cases when there is only one solution to the related quadratic equation: all real numbers, all real numbers except one, no solution, and just one number. Knowing this, we could use the same test-point technique used in Example 1.

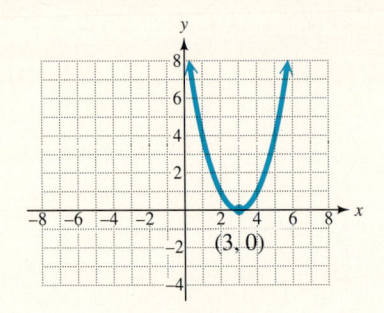

322 CHAPTER 4 POLYNOMIAL AND RATIONAL FUNCTIONS

Figure 4.20

EXAMPLE 2   SOLVING A QUADRATIC INEQUALITY

Solve. $-x^2 + 4x - 4 \leq 0$

First solve

$$-x^2 + 4x - 4 = 0.$$
$$x^2 - 4x + 4 = 0 \quad \text{Multiply by } -1$$
$$(x - 2)(x - 2) = 0 \quad \text{Factor}$$
$$x - 2 = 0 \quad \text{or} \quad x - 2 = 0$$
$$x = 2 \quad\quad\quad\quad x = 2$$

Since there is only one solution, there are only two intervals, $(-\infty, 2)$ and $(2, \infty)$. Use the test point 0.

$$-x^2 + 4x - 4 \leq 0$$
$$-(0)^2 + 4(0) - 4 \leq 0$$
$$-4 \leq 0 \quad \text{This is true}$$

Since the inequality is $\leq$, the intervals $(-\infty, 2)$, $(2, \infty)$, and the value $x = 2$ are all part of the solution making the solution all real numbers.

PRACTICE EXERCISE 2

Solve. $-x^2 + 4x - 4 > 0$

Answer: no solution

    Suppose we try to solve $x^2 + x + 2 \geq 0$. The solutions to the quadratic equation $x^2 + x + 2 = 0$ are complex (nonreal). As a result, there are no real numbers to divide the number line into separate intervals. In this case, as shown in Figure 4.21, the graph of $f(x) = x^2 + x + 2$ does not cross or touch the x-axis. When this happens, the solution to the inequality is one of two: all real numbers or no solution. We can see from the graph that $x^2 + x + 2 \geq 0$ for every real number, but this could also be obtained using the test point 0 since $(0)^2 + (0) + 2 = 2 \geq 0$ is true. Similarly $x^2 + x + 2 > 0$ has every real number for the solution, and both $x^2 + x + 2 < 0$ and $x^2 + x + 2 \leq 0$ have no solution.

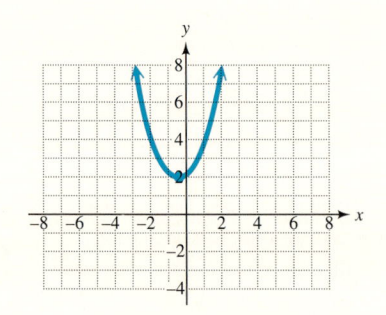

Figure 4.21

When one side of a quadratic inequality is nonzero, first change the form of the inequality, and proceed as before.

EXAMPLE 3 A More Complex Quadratic Inequality

Solve. $(x + 3)(x - 1) \le -2$
First simplify and write in general form.
$$x^2 + 2x - 1 \le 0$$

Solve using the quadratic formula to obtain the solutions $-1 \pm \sqrt{2}$. Approximating $\sqrt{2}$ with 1.4, the solutions are -2.4 and 0.4. Use 0 as the test point noting that 0 is between the two solutions, that is, in $(-1 - \sqrt{2}, -1 + \sqrt{2})$.

$$(0)^2 + 2(0) - 1 \le 0$$
$$-1 \le 0 \quad \text{This is true}$$

Since the inequality is $\le$, the endpoints of the interval must be included so the solution is $[-1 - \sqrt{2}, -1 + \sqrt{2}]$, or, using inequalities, $-1 - \sqrt{2} \le x \le -1 + \sqrt{2}$.

Practice Exercise 3

Solve.
$$x^2 + 2x - 1 > 2(x + 1)$$

Answer: $(-\infty, -\sqrt{3})$ or $(\sqrt{3}, \infty)$; or, $x < -\sqrt{3}$ or $x > \sqrt{3}$

❷ POLYNOMIAL INEQUALITIES

We can extend the method used to solve quadratic inequalities to **polynomial inequalities** such as
$$x^3 - 4x > 0.$$

Factor the left side.
$$x(x - 2)(x + 2) > 0$$

The zeros of the polynomial, -2, 0, and 2, divide the number line into four intervals,
$$(-\infty, -2), \quad (-2, 0), \quad (0, 2), \quad \text{and} \quad (2, \infty).$$

Consider the graph of $f(x) = x^3 - 4x$ in Figure 4.22.

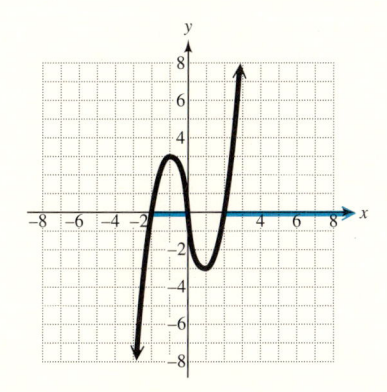

Figure 4.22

Notice that $f(x) > 0$, that is, $x^3 - 4x > 0$, in the two intervals $(-2, 0)$ and $(2, \infty)$. Thus the solution to the inequality is $(-2, 0)$ or $(2, \infty)$. We could use a test point in each interval to arrive at the same conclusion without actually viewing the graph. For example, use test points -3, -1, 1, and 3.

$x = -3$: $x^3 - 4x = (-3)^3 - 4(-3) = -27 + 12 = -15 < 0$
$x = -1$: $x^3 - 4x = (-1)^3 - 4(-1) = -1 + 4 = 3 > 0$
$x = 1$: $x^3 - 4x = (1)^3 - 4(1) = 1 - 4 = -3 < 0$
$x - 3$: $x^3 - 4x = (3)^3 - 4(3) = 27 - 12 = 15 > 0$

Thus, when x is in $(-2, 0)$ or $(2, \infty)$, $x^3 - 4x > 0$. This technique is illustrated in the next example.

EXAMPLE 4 A Polynomial Inequality

Solve. $(x - 1)(x + 3)(x - 5) \leq 0$
The zeros of the polynomial are -3, 1, and 5 which divide the number line into the intervals

$$(-\infty, -3), \quad (-3, 1), \quad (1, 5), \quad \text{and} \quad (5, \infty).$$

Choose a test point in each, say -4, 0, 2, and 6.

$x = -4$: $(x - 1)(x + 3)(x - 5) = (-4 - 1)(-4 + 3)(-4 - 5)$
$= (-5)(-1)(-9) = -45 < 0$
$x = 0$: $(x - 1)(x + 3)(x - 5) = (0 - 1)(0 + 3)(0 - 5)$
$= (-1)(3)(-5) = 15 > 0$
$x = 2$: $(x - 1)(x + 3)(x - 5) = (2 - 1)(2 + 3)(2 - 5)$
$= (1)(5)(-3) = -15 < 0$
$x = 6$: $(x - 1)(x + 3)(x - 5) = (6 - 1)(6 + 3)(6 - 5)$
$= (5)(9)(1) = 45 > 0$

Thus, $(x - 1)(x + 3)(x - 5) < 0$ in $(-\infty, -3)$ and also in $(1, 5)$. Since the inequality is $\leq$, we include the endpoints to obtain the solution $(-\infty, 3]$ or $[1, 5]$, or, using inequalities, $x \leq 3$ or $1 \leq x \leq 5$.

Practice Exercise 4

Solve. $(x + 2)(x + 5)(x - 1) > 0$

Answer: $(-5, -2)$ or $(1, \infty)$; or, $-5 < x < -2$ or $x > 1$

Note Since all we are really interested in is the sign of the polynomial expression at a test point, not the actual value at the test point, you can quickly test a value mentally. For example, to test $x = 0$ in Example 4 above, you know that when x is 0, $(x - 1)$ is negative, $(x + 3)$ is positive, and $(x - 5)$ is negative, so we are multiplying $(-)(+)(-)$. The even number of minus signs makes the product positive. Similarly, to test $x = 2$, you have $(+)(+)(-)$, an odd number of negative numbers, making the product negative.

③ RATIONAL INEQUALITIES

Inequalities involving rational expressions such as

$$\frac{x + 2}{x + 3} \geq 0, \quad \frac{2x + 1}{x - 1} > 3, \quad \text{and} \quad \frac{x^2 + 3x - 18}{x + 2} < 0$$

are called **rational inequalities**. Rational inequalities can be solved using the same test-point technique. However, for these inequalities we find the values of x that make the numerator 0 (zeros of the rational expression) *and* values that make the denominator 0 (values for which the expression is undefined). Suppose we solve $\frac{x+2}{x+3} \geq 0$. It may help initially to look at the graph of $f(x) = \frac{x+2}{x+3}$ with horizontal asymptote $y = 1$ and vertical asymptote $x = -3$, given in Figure 4.23.

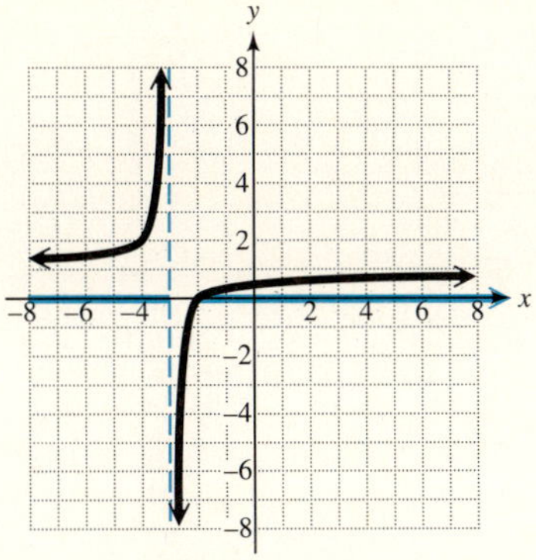

Figure 4.23

From the graph we can see that $f(x) \geq 0$, or $\frac{x+2}{x+3} \geq 0$, when $x < -3$ or $x \geq -2$, or, equivalently, when x is in $(-\infty, -3)$ or $[-2, \infty)$. Notice that -2 and -3 are zeros of the numerator and denominator, respectively. Rather than using the graph we could note that these values divide the number line into the intervals.

$$(-\infty, -3), \quad (-3, -2), \quad \text{and} \quad (-2, \infty).$$

A test point from each interval would show that $\frac{x+2}{x+3} > 0$ in $(-\infty, -3)$ or $(-2, \infty)$. Since the given inequality is $\geq$, we include the value of x, -2, that makes the numerator 0 (but will not include -3 where the denominator is 0) in the solution

$$(-\infty, -3) \text{ or } [2, \infty) \quad \text{or using inequalities} \quad x < -3 \text{ or } x \geq 2.$$

This solution technique is illustrated in the next example. Notice also, similar to quadratic inequalities, we first transform the inequality to a form that has 0 on one side.

EXAMPLE 5 Solving a Rational Inequality

Solve. $\dfrac{2x+1}{x-1} > 3$

We rewrite the inequality so that 0 is on the right side, and simplify the result.

$$\frac{2x+1}{x-1} - 3 > 0 \qquad \text{Subtract 3}$$

$$\frac{2x+1}{x-1} - \frac{3(x-1)}{x-1} > 0 \qquad \text{The LCD} = x - 1$$

$$\frac{2x+1-3x+3}{x-1} > 0 \qquad \text{Watch the signs when subtracting}$$

$$\frac{-x+4}{x-1} > 0$$

The numerator is 0 when $x = 4$, and the denominator is 0 when $x = 1$. Thus, we consider the intervals

$$(-\infty, 1), \quad (1, 4), \quad \text{and} \quad (4, \infty).$$

It may help to use a table to summarize information about test points.

Practice Exercise 5

Solve. $\dfrac{2x+1}{x-1} \leq 3$

Intervals	Test points	Test values	Sign
$(-\infty, 1)$	0	$\dfrac{-0+4}{0-1} = -4$	−
$(1, 4)$	2	$\dfrac{-2+4}{2-1} = 2$	+
$(4, \infty)$	5	$\dfrac{-5+4}{5-1} = -\dfrac{1}{4}$	−

Thus, the solution is $(1, 4)$, or, using inequalities, $1 < x < 4$.

Answer: $(-\infty, 1)$ or $[4, \infty)$; or, $x < 1$ or $x \geq 4$

CAUTION

When solving a rational inequality involving $\geq$ or $\leq$, remember to include values that make the numerator 0; *but never include values that make the denominator 0.*

4 APPLICATIONS OF INEQUALITIES

Some applications translate to inequalities.

EXAMPLE 6 A PHYSICS PROBLEM

A rocket is fired upward from a platform, 10 ft off the ground, with an initial velocity of 128 ft/sec. During what time interval will its height exceed 250 ft?

Recall that the formula for the height h of an object propelled upward from an initial height h_0 with initial velocity v_0 is

$$h = -16t^2 + v_0 t + h_0.$$

In our example, $v_0 = 128$ and $h_0 = 10$. Thus, we are looking for the interval of time for which

$-16t^2 + 128t + 10 > 250.$
$-16t^2 + 128t - 240 > 0$
$-16(t^2 - 8t + 15) > 0$
$t^2 - 8t + 15 < 0$ Divide by -16 and reverse the inequality
$(t - 3)(t - 5) < 0$

The values of t that make the expression 0 are 3 and 5, giving us the intervals

$$(-\infty, 3), \quad (3, 5), \quad \text{and} \quad (5, \infty).$$

Use the test point, 0, in $(-\infty, 3)$.

$(0 - 3)(0 - 5) < 0$
$(-3)(-5) < 0$
$15 < 0$ This is false

PRACTICE EXERCISE 6

During what interval of time will the rocket in Example 7 be at a height less than 250 ft?

Thus the solution is (3, 5), which means that from 3 seconds to 5 seconds, the rocket is higher than 250 ft. Notice that in this problem the interval $(-\infty, 3)$ must actually be restricted to [0, 3) since time t cannot be negative.

Answer: [0, 3) or (5, ∞), that is, the rocket is below 250 ft prior to 3 sec ($0 \le t < 3$) or after 5 sec ($t > 5$).

4.7 EXERCISES A

Solve each inequality in Exercises 1–14. Give answers using both inequalities and intervals.

1. $x^2 + 2x - 15 > 0$

2. $2x^2 + 3x \ge 20$

3. $x^2 - 2x - 2 > 0$

4. $3x^2 - x + 2 > 0$

5. $4x^2 - 20x + 25 \le 0$

6. $\dfrac{x-3}{x+1} \ge 0$

7. $\dfrac{2x-1}{x+3} < 1$

8. $\dfrac{3x+2}{x-3} \ge 2$

9. $(x-2)(x+2)(x-5) \ge 0$

10. $(x-2)(x+2) \ge 3x$

11. $\dfrac{(x-3)(x+6)}{x-1} < 0$

12. $(x^2 - 2x)(x^2 + 8x + 15) \ge 0$

13. $x^3 \ge x$

14. $x + \dfrac{1}{x} < 2$

In Exercises 15–16 determine the interval(s) in which the radical expression defines a real number.

15. $\sqrt{x(x+5)}$

16. $\sqrt{\dfrac{3-x}{x+8}}$

17. Give the interval(s) containing m in which the quadratic equation $x^2 + mx + 4 = 0$ has two complex (nonreal) solutions.

Solve each applied problem in Exercises 18–21.

18. **BUSINESS** The cost of producing t units is $C = t^2 + 6t$, and the revenue generated from sales is $R = 2t^2 + t$. Find the number of units to be sold in order to generate a profit (when $R > C$).

19. **RATE-MOTION** A coin is tossed upward from a balcony 200 ft high with an initial velocity of 48 ft/sec. During what interval of time will the coin be at a height of at least 40 ft?

20. **RATE-MOTION** If a rocket is propelled upward from ground level, its height in meters after t seconds is given by

$$h = -9.8t^2 + 147t.$$

During what interval of time will the rocket be higher than 529.2 m?

21. **PHYSICS** A signal flare fired from the bottom of a gorge is visible to an observer on the rim of the gorge only when the flare is above the level of the rim. If the flare is fired with an initial velocity of 176 ft/sec, and the gorge is 448 ft deep, during what time interval can the flare be seen?

FOR REVIEW

In Exercises 22–23 give the (a) vertical asymptotes, (b) horizontal asymptotes, (c) oblique asymptotes, (d) x-intercepts, (e) y-intercept, and (f) symmetries of the graph of each rational function.

22. $f(x) = \dfrac{-2}{x + 3}$

23. $g(x) = \dfrac{x^2 - 25}{x - 2}$

ANSWERS: 1. $x < -5$ or $x > 3$; $(-\infty, -5)$ or $(3, \infty)$ 2. $x \leq -4$ or $x \geq 5/2$; $(-\infty, -4]$ or $[5/2, \infty)$ 3. $x < 1 - \sqrt{3}$ or $x > 1 + \sqrt{3}$; $(-\infty, 1 - \sqrt{3})$ or $(1 + \sqrt{3}, \infty)$ 4. every real number; $(-\infty, \infty)$ 5. $x = 5/2$ 6. $x < -1$ or $x \geq 3$; $(-\infty, -1)$ or $[3, \infty)$ 7. $-3 < x < 4$; $(-3, 4)$ 8. $x \leq -8$ or $x > 3$; $(-\infty, -8]$ or $(3, \infty)$ 9. $-2 \leq x \leq 2$ or $x \geq 5$; $[-2, 2]$ or $[5, \infty)$ 10. $x \leq -1$ or $x \geq 4$; $(-\infty, -1]$ or $[4, \infty)$ 11. $x < -6$ or $1 < x < 3$; $(-\infty, -6)$ or $(1, 3)$ 12. $x \leq -5$ or $-3 \leq x \leq 0$ or $x \geq 2$; $(-\infty, -5]$ or $[-3, 0]$ or $[2, \infty)$ 13. $-1 \leq x \leq 0$ or $x \geq 1$; $[-1, 0]$ or $[1, \infty)$ 14. $x < 0$; $(-\infty, 0)$ 15. $x \leq -5$ or $x \geq 0$; $(-\infty, -5]$ or $[0, \infty)$ 16. $-8 < x \leq 3$; $(-8, 3]$ 17. $-4 < m < 4$; or $(-4, 4)$ 18. $t > 5$ 19. $0 \text{ sec} \leq t \leq 5 \text{ sec}$ 20. $6 \text{ sec} < t < 9 \text{ sec}$ 21. $4 \text{ sec} < t < 7 \text{ sec}$ 22. (a) $x = -3$ (b) $y = 0$ (c) none (d) none (e) $(0, -2/3)$ (f) none 23. (a) $x = 2$ (b) none (c) $y = x + 2$ (d) $(5, 0)$ and $(-5, 0)$ (e) $(0, 25/2)$ (f) none

4.7 EXERCISES B

Solve each inequality in Exercises 1–14. Give answers using both inequalities and intervals.

1. $x^2 + 2x - 15 \leq 0$

2. $2x^2 + 3x < 20$

3. $x^2 - 2x - 2 \leq 0$

4. $3x^2 - x + 2 \leq 0$

5. $4x^2 - 20x + 25 < 0$

6. $\dfrac{x - 3}{x + 1} < 0$

7. $\dfrac{2x - 1}{x + 3} \geq 1$

8. $\dfrac{3x + 2}{x - 3} < 2$

9. $(x-2)(x+2)(x-5) < 0$

10. $(x-2)(x+2) < 3x$

11. $\dfrac{(x-3)(x+6)}{x-1} \geq 0$

12. $(x^2 - 2x)(x^2 + 8x + 15) < 0$

13. $x^3 < x$

14. $x + \dfrac{4}{x} \geq -4$

In Exercises 15–16 determine the interval(s) in which the radical expression defines a real number.

15. $\sqrt{x^2 - 1}$

16. $\sqrt{2x^2 - 9x - 5}$

17. Give the interval(s) containing m in which the quadratic equation $x^2 + mx + 4 = 0$ has two real solutions.

Solve each applied problem in Exercises 18–21.

18. **BUSINESS** The profit made when t units are sold, $t > 0$, is given by $P = t^2 - 31t + 220$. Find the number of units that would be sold when **(a)** $P = 0$ (the break-even point), **(b)** $P > 0$ (a profit is made), and **(c)** $P < 0$ (a loss is taken).

19. **GEOMETRY** A rectangular enclosure must have an area of at least 900 yd^2. If 200 yd of fencing is to be used, and the width cannot exceed the length, within what limits must the width of the enclosure lie?

20. **BUSINESS** A retailer has found that n games can be sold during the month provided the price of each game is $20 - 0.2n$ dollars. If he purchases each game from a wholesaler for $10, and he wishes to make a profit of at least $120 per month on sales of this game, how many games must he sell each month?

21. **CONSTRUCTION** A rectangular sheet of cardboard is 16 inches long and 12 inches wide. A box is to be made from the sheet by cutting out four square pieces, one in each corner, and folding up the sides. If the area of the base of the box must exceed 32 in^2, what are the possible heights of the box?

FOR REVIEW

In Exercises 22–23 give the (a) vertical asymptotes, (b) horizontal asymptotes, (c) oblique asymptotes, (d) x-intercepts, (e) y-intercept, and (f) symmetries of the graph of each rational function.

22. $f(x) = \dfrac{2x}{x+3}$

23. $g(x) = \dfrac{2x^2}{x^2 + 1}$

4.7 EXERCISES C

Use a calculator to solve each inequality in Exercises 1–2. Give answers rounded to the nearest hundredth.

1. $2.5x^2 - 3.9x - 4.7 > 0$

2. $\dfrac{x - 6.25}{x^2 - 1.5x - 7.3} \leq 0$
   [Answer: $x < -2.05$ or $3.55 < x \leq 6.25$]

Solve each inequality in Exercises 3–5. Give answers using intervals.

3. $\dfrac{x^3 - 2x^2 - 8x}{x^2 - 4x + 3} < 0$
   [*Hint:* You might want to compare this with the graph of the function in Exercise 4, 4.6 EXERCISES C.]

4. $|x^2 - 3x - 1| < 3$
   [*Hint:* Solve $x^2 - 3x - 1 > -3$ and $x^2 - 3x - 1 < 3$.]

5. $|x^2 - 3x - 1| \geq 3$
   [Answer: $(-\infty, -1]$ or $[1, 2]$ or $[4, \infty)$]

CHAPTER 4 EXTENSION PROJECT

In Section 4.5 we discussed one technique for approximating a zero of a polynomial function, the *method of bracketing*. Another method for approximating zeros of many functions involves halving intervals. The nice feature of this method is that it is easy to write a computer program that will give an approximation to a zero without much decision-making on your part. To use the *method of interval halving* to find a zero of polynomial $P(x)$, you need to find two numbers a and b, $a < b$, such that $P(a)$ and $P(b)$ have different signs. By the intermediate value theorem, there must be at least one zero r of $P(x)$ in $[a, b]$. The values of a and b can usually be found quickly by inspection. Refer to Figure 4.24.

1. Choose the number c halfway between a and b, $c = \frac{a+b}{2}$.
2. Calculate $P(c)$. If $P(c) = 0$, we have found a zero of $P(x)$.
3. If $P(c) \neq 0$, then a zero is in either (a, c) or (c, b). If $P(a)P(c) < 0$, then $P(a)$ and $P(c)$ have opposite signs and the zero is in (a, c). If $P(a)P(c) > 0$, then since $P(a)P(b) < 0$ we know that $P(c)P(b) < 0$ and the zero is in (c, b).
4. If the zero is in (a, c), replace c with b and return to Step 1.
5. If the zero is in (c, b), replace a with c and return to Step 1.
6. Repeating this process several times will result in the zero r being in $[a, b]$, with a and b getting closer and closer together.

To prevent an endless loop in the steps above, we can specify an acceptable degree of accuracy by stopping the process when $|b - a|$ is less than some fixed number, say 10^{-4}. The BASIC program below will find the zero of $P(x) = x^3 + 5x^2 - 3x - 15$ which is between 1 and 2. Note that $P(1) = -12$ and $P(2) = 7$. There must be a zero in the interval $(a, b) = (1, 2)$. A copy of the print that results when this program is run is given here.

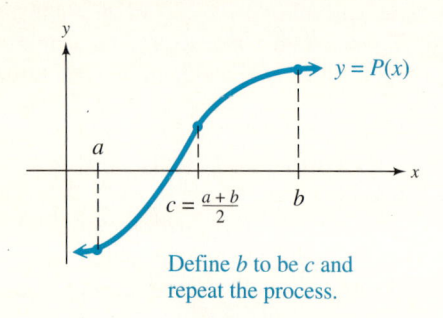

Figure 4.24

Define b to be c and repeat the process.

```
10 INPUT A,B
20 PA=A^3+5*A^2-3*A-15
30 C = (A+B)/2
40 PRINT "A= "A,"B= "B,"C= "C
50 PC=C^3+5*C^2-3*C-15
60 IF PC=0 THEN 140
70 IF PA*PC<0 THEN 110
80 A=C
90 PA=PC
100 GOTO 120
110 B=C
120 IF ABS(B-A)<.0001 THEN 140
130 GOTO 30
140 PRINT"THE APPROXIMATE ZERO: "C
150 END
```

```
RUN
? 1,2
A=  1           B=  2          C=  1.5
A=  1.5         B=  2          C=  1.75
A=  1.5         B=  1.75       C=  1.625
A=  1.625       B=  1.75       C=  1.6875
A=  1.6875      B=  1.75       C=  1.71875
A=  1.71875     B=  1.75       C=  1.734375
A=  1.71875     B=  1.734375   C=  1.726563
A=  1.726563    B=  1.734375   C=  1.730469
A=  1.730469    B=  1.734375   C=  1.732422
A=  1.730469    B=  1.732422   C=  1.731445
A=  1.731445    B=  1.732422   C=  1.731934
A=  1.731934    B=  1.732422   C=  1.732178
A=  1.731934    B=  1.732178   C=  1.732056
A=  1.731934    B=  1.732056   C=  1.731995
THE APPROXIMATE ZERO:   1.731995
```

THINGS TO DO:

1. Find the exact value of the zero approximated in the program.
2. Modify the program to find the zero of $Q(x) = x^3 + 3x^2 - 5x - 15$ in the interval $(2, 3)$. Can you find the exact value? Compare.
3. Figure out why the method of interval halving does not work for the function $f(x) = \frac{1}{x}$ on $(-1, 1)$ even though $f(1) = 1 > -1 = f(-1)$.
4. If you do not have access to a personal computer, use your calculator and follow the steps in the method of interval halving to find the approximate value of the zero of $P(x)$.
5. The *secant method* is another method for approximating zeros. Research this method and compare it with the method of interval halving. Which works better?

CHAPTER 4 REVIEW

KEY TERMS

4.1
1. An expression of the form $a_n x^n + a_{n-1} x^{n-1} + \cdots + a_1 x + a_0$ where a_i is real for $i = 0, 1, 2, \ldots, n$, is a **polynomial with real coefficients in the variable x.** If $a_n \neq 0$ the polynomial has degree n.
2. If $P(x)$ is a polynomial, then $P(x) = 0$ is a **polynomial equation.**
3. If $P(x)$ is a polynomial and r is a number such that $P(r) = 0$, then r is a **zero** of $P(x)$, and a **root** or **solution** to the polynomial equation $P(x) = 0$.

4.2
1. **The Division Algorithm** If $P(x)$ and $D(x)$ are polynomials with $D(x) \neq 0$, then there exist unique polynomials $Q(x)$ and $R(x)$ such that $P(x) = Q(x)D(x) + R(x)$ and $R(x)$ is either the zero polynomial or has degree less than the degree of $D(x)$.
2. **The Remainder Theorem** If $P(x)$ is a polynomial, r is a complex number, and $P(x)$ is divided by $x - r$, then the remainder obtained is equal to $P(r)$.
3. **The Factor Theorem** If $P(x)$ is a polynomial and r is a complex number, then r is a zero of $P(x)$ if and only if $x - r$ is a factor of $P(x)$.

4.3
1. **The Fundamental Theorem of Algebra** Every polynomial equation $P(x) = 0$, where $P(x)$ has degree $n \geq 1$, has at least one root in the set of complex numbers.
2. **Complex Root Theorem** If $a + bi$ is a root of polynomial equation $P(x) = 0$, where $P(x)$ has real coefficients, then $a - bi$ is also a root.
3. **Irrational Root Theorem** If $a + b\sqrt{c}$, where a and b are rational but $\sqrt{c}$ is irrational, is a root of the polynomial equation $P(x) = 0$ with rational coefficients, then $a - b\sqrt{c}$ is also a root.

4.4
1. If $P(x) = 0$ has no real root greater than or equal to real number a, a is an **upper bound** for the real roots. Similarly, if no real root is less than real number b, then b is a **lower bound** for the real roots.
2. **The Rational Root Theorem** If p/q is a rational root of polynomial equation $P(x) = 0$, where the leading coefficient of $P(x)$ is a_n and the constant term is a_0, then p is a factor of a_0 and q is a factor of a_n.

4.5
1. A function of the form $P(x) = a_n x^n + a_{n-1} x^{n-1} + \cdots + a_1 x + a_0$, where the a_i's are real and $a_n \neq 0$, is a **polynomial function of degree n.**
2. The graph of a polynomial function of degree n can have at most $n - 1$ **turning points,** where the curve changes from up to down or down to up.
3. **The Intermediate Value Theorem** If $P(x)$ is a polynomial, a and b are real numbers with $a < b$, and $P(a) \neq P(b)$, then $P(x)$ takes on every value between $P(a)$ and $P(b)$ in the interval $[a, b]$.

4.6
1. A function of the form $f(x) = P(x)/Q(x)$, where $P(x)$ and $Q(x)$ are polynomials, $Q(x)$ not the zero polynomial, is called a **rational function.**
2. An **asymptote** is a line to which the graph of a function gets closer and closer as x increases, decreases, or approaches some value. Asymptotes can be **vertical, horizontal,** or **oblique.**
3. A **zero** of a rational function is a value of x that makes the numerator zero while the denominator is nonzero.

4.7
1. A **quadratic inequality** is an inequality of the form $ax^2 + bx + c > 0$, $ax^2 + bx + c < 0$, $ax^2 + bx + c \geq 0$, or $ax^2 + bx + c \leq 0$, where $a \neq 0$.
2. A **rational inequality** is an inequality of the form $P(x)/Q(x) > 0$, (or < 0, or ≤ 0, or ≥ 0) where $P(x)$ and $Q(x)$ are polynomials.

REVIEW EXERCISES

Part I

4.1
1. Answer the following for the polynomial $P(x) = 3x^4 - 2x^2 + 7x - 8$.
   (a) What is the degree of $P(x)$?
   (b) What is the leading coefficient?
   (c) What is $P(1)$?
   (d) What is $P(-1)$?

2. Determine whether 2, $3i$, and $-3i$ are zeros of $Q(x) = x^3 - 2x^2 + 9x - 18$.

3. Use synthetic division to find the quotient and remainder when $x^4 - 3x^3 + x - 2$ is divided by $x + 3$.

4.2
4. If a polynomial $P(x)$ is divided by $x - 5$, what will be the value of the remainder?

5. What is the name of the theorem used to answer Exercise 4?

6. If $P(x)$ is a polynomial and $P(-3) = 0$, give one factor of $P(x)$.

7. What is the name of the theorem used to answer Exercise 6?

8. **BUSINESS** The revenue in dollars brought in on the sale of x books is given by $R(x) = 10.95x$. The cost in dollars of producing x books is given by $C(x) = 500 + 5.95x$. The zeros of $P(x) = R(x) - C(x)$ are break-even points. Show that 100 is a break-even point.

9. Find a value of m so that $P(x) = x^4 - 2x^3 + mx - 60$ has a remainder of 0 when divided by $x + 3$. What is $P(-3)$?

10. If $P(x) = 2x^4 - 3x^2 + x - 8$, use synthetic division and the remainder theorem to find the following.
    (a) $P(1)$ (b) $P(-1)$ (c) $P(2)$

11. If $P(x) = x^4 - 4x^3 - 3x^2 + 10x + 8$, use synthetic division and the factor theorem to determine if each of the following is a factor of $P(x)$.
    (a) $x - 4$ (b) $x + 1$ (c) $x + 2$

4.3
12. If -4 is a zero of multiplicity three of a polynomial $P(x)$, give a cubic factor of $P(x)$.

13. If $2 - 7i$ is a zero of polynomial $P(x)$, with real coefficients, give another zero of $P(x)$.

14. If $3 + \sqrt{11}$ is a zero of polynomial $P(x)$, with rational coefficients, give another zero of $P(x)$.

15. What is the greatest number of distinct solutions the equation $5x^7 + 3x^5 - 2x^4 + x^2 - 9 = 0$ can have? The least? Exactly how many solutions does it have, counting multiplicities?

16. Give a polynomial $P(x)$ with rational coefficients of least degree such that -2 is a double zero, $1 + \sqrt{2}$ is a single zero, and $1 - i\sqrt{2}$ is a single zero.

17. Use Descartes' rule of signs to find the possible number of negative, positive, and nonreal solutions to the equation $2x^6 - 4x^5 - 2x^3 + 3x^2 + x - 7 = 0$.

4.4
18. Use the upper and lower bound test to find the smallest positive integer upper bound and the largest negative integer lower bound for the real solutions to $8x^4 - 16x^3 - 2x^2 - 16x - 10 = 0$.

19. Suppose a/b is a rational solution to $4x^4 - 3x^3 + 2x^2 - x + 5 = 0$.
    (a) a is a factor of which coefficient?
    (b) b is a factor of which coefficient?

20. Explain why any possible rational solutions of the equation $x^5 - 3x^3 + x^2 + x - 9 = 0$ must be integers.

21. Find all rational solutions of $6x^3 - 11x^2 + 9x - 2 = 0$ (if any exist). If possible, find all solutions.

22. **GEOMETRY** Find the radius of a sphere whose surface area and volume are equal.

4.5 *For each of the polynomial functions in Exercises 23–24, determine (a) the maximum number of real zeros (the maximum number of x-intercepts), (b) the maximum number of turning points, (c) the y-intercept, (d) the direction of the graph for large positive values of x, and (e) the direction of the graph for small negative values of x.*

23. $P(x) = x^3 - 7x^2 + x - 5$

24. $P(x) = -x^4 + 3x^2 - x + 8$

25. Sketch the graph of $y = P(x) = -x^3 - x^2 + 2x$ using information about the x-intercepts, the y-intercept, the turning points, and the eventual direction of the graph.

26. Given that the polynomial $P(x) = x^3 + 5x^2 - 11x - 55$ has an irrational zero between 3 and 4, use the bracketing method to approximate this zero to the nearest hundredth.

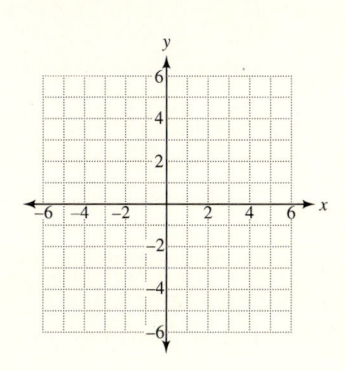

4.6 *In Exercises 27–28 give the (a) vertical asymptotes, (b) horizontal asymptotes, (c) oblique asymptotes, (d) x-intercepts, (e) y-intercept, and (f) symmetries of the graph of each rational function.*

27. $f(x) = \dfrac{10}{x - 9}$

28. $g(x) = \dfrac{6x^2}{9x^2 - 1}$

29. Graph the rational function.
$$y = f(x) = \dfrac{1}{x^2 - 2x + 1}$$

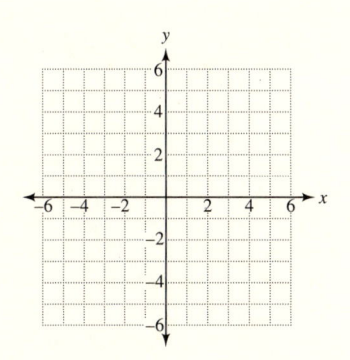

4.7 *Solve each inequality in Exercises 30–33. Give answers using both inequalities and intervals.*

30. $3x^2 + x \leq 10$

31. $x^2 - 3x + 1 \geq 0$

32. $(x - 3)(x^2 - 2x - 35) < 0$

33. $\dfrac{x - 5}{2x - 3} \geq 1$

Part II

34. Which of the following could be the graph of a polynomial function?

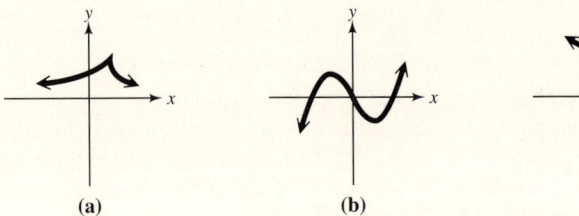

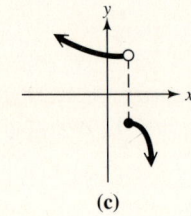

(a) (b) (c)

Answer true *or* false *in Exercises 35–43.*

35. If $P(x)$ is divided by $x - 5$, the remainder is $P(-5)$.

36. The polynomial equation $-3x^4 + 2x^3 - x^2 + x + 5$ has at most four distinct roots.

37. If $P(x)$ is a polynomial and $P(b) = 0$, b is a root of the equation $P(x) = 0$.

38. If $P(5) = 0$, then $x - 5$ is a factor of $P(x)$.

39. If 2, $1 + i$, and $4 + 3i$ are all zeros of polynomial $P(x)$, with real coefficients, then $P(x)$ has degree 3.

40. The graph of $P(x) = 2x^3 + x^2 - x - 12$ will eventually go up to the right and down to the left.

41. If r is a zero of $Q(x)$ in $f(x) = P(x)/Q(x)$, then $x = r$ is a horizontal asymptote.

42. If the degree of $P(x)$ is 5 and the degree of $Q(x)$ is 4, then the graph of $f(x) = P(x)/Q(x)$ has an oblique asymptote.

43. If the degree of $P(x)$ is less than the degree of $Q(x)$ in $f(x) = P(x)/Q(x)$, then the graph has $y = 0$ as a horizontal asymptote.

44. Graph the rational function

$$y = f(x) = \frac{x^3 + 1}{x^2}$$

using information about asymptotes, intercepts, and symmetries.

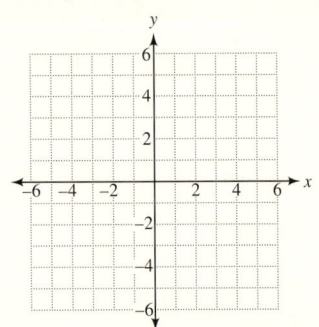

45. Sketch the graph of

$$y = P(x) = 5x^4 - 5$$

using information about the x-intercepts, the y-intercept, the turning points, and the eventual direction of the graph.

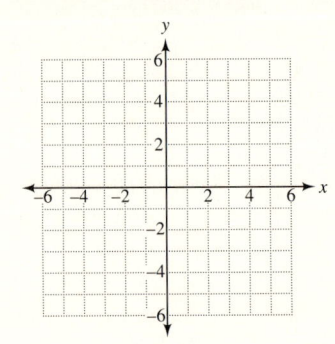

46. Use Descartes' rule of signs to find the possible number of negative, positive, and nonreal solutions to $3x^5 + 7x^3 + 10x + 1 = 0$.

47. Find all rational solutions of $x^3 - 3x^2 - 9x - 5 = 0$ (if they exist). If possible, find all solutions.

48. Use the upper and lower bound test to find the smallest positive integer upper bound and the largest negative integer lower bound for the real solutions to $3x^6 + 13x^5 + 14x^4 + 6x^2 + 26x + 28 = 0$.

49. Prove that a polynomial with positive real coefficients cannot have a positive real zero.

Solve each inequality in Exercises 50–52. Give answers using both inequalities and intervals.

50. $3x^2 - 2x + 1 < 0$ **51.** $3x^2 + x - 10 > 0$ **52.** $\dfrac{x-7}{x+3} \leq 0$

53. SCIENCE The nose section of a missile is in the shape of a right circular cylinder with a hemisphere attached to one end. If the total length of the nose section is 40 inches, and total volume of the section is 5019 in^3, what is the radius of the cylinder (to the nearest tenth of an inch)? [*Hint:* The radius is between 6 inches and 7 inches in length.]

ANSWERS: 1. (a) 4 (b) 3 (c) 0 (d) -14 2. All three are zeros. 3. $x^3 - 6x^2 + 18x - 53, 157$ 4. $P(5)$ 5. remainder theorem 6. $x + 3$ 7. factor theorem 8. Since $P(100) = 0$, 100 is a break-even point. 9. $m = 25$; 0 10. (a) $P(1) = -8$ (b) $P(-1) = -10$ (c) $P(2) = 14$ 11. (a) $x - 4$ is a factor (b) $x + 1$ is a factor (c) $x + 2$ is not a factor 12. $(x + 4)^3$ 13. $2 + 7i$ 14. $3 - \sqrt{11}$ 15. 7; 1; 7 16. $P(x) = x^6 - 6x^4 + 4x^3 + 5x^2 - 28x - 12$ 17. either 1, 3, 2 or 1, 1, 4 18. upper bound: 3; lower bound: -1 19. (a) 5 (b) 4 20. If p/q is a rational solution, then $q = 1$ since q must be a factor of 1. 21. $\frac{1}{3}, \frac{3 \pm i\sqrt{7}}{4}$ 22. $r = 3$ 23. (a) 3 (b) 2 (c) $(0, -5)$ (d) upward (e) downward 24. (a) 4 (b) 3 (c) $(0, 8)$ (d) downward (e) downward 25. 26. 3.32 27. (a) $x = 9$ (b) $y = 0$ (c) none (d) none (e) $(0, -10/9)$ (f) none 28. (a) $x = 1/3$ and $x = -1/3$ (b) $y = 2/3$ (c) none (d) $(0, 0)$ (e) $(0, 0)$ (f) y-axis 29.

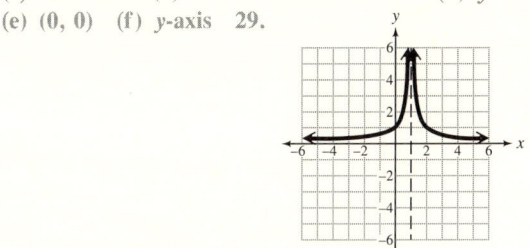

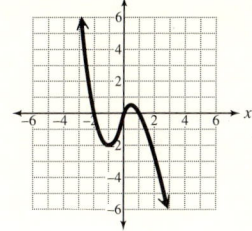

30. $-2 \leq x \leq 5/3$; $[-2, 5/3]$ 31. $x \leq \frac{3-\sqrt{5}}{2}$ or $x \geq \frac{3+\sqrt{5}}{2}$; $\left(-\infty, \frac{3-\sqrt{5}}{2}\right]$ or $\left[\frac{3+\sqrt{5}}{2}, \infty\right)$ 32. $x < -5$ or $3 < x < 7$; $(-\infty, -5)$ or $(3, 7)$ 33. $-2 \leq x < 3/2$; $[-2, 3/2)$ 34. only (b) 35. false 36. true 37. true 38. true 39. false 40. true 41. false 42. true 43. true 44.   45.

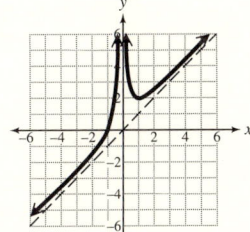

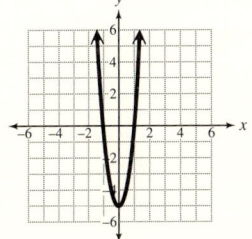

46. 1, 0, 4 47. $-1, -1, 5$ 48. upper bound: 1; lower bound: -5 49. Consider $P(x) = a_n x^n + a_{n-1} x^{n-1} + \cdots + a_0$, with $a_i > 0$ for $i = 0, 1, 2, \ldots, n$. If r is any positive number, then $P(r) > 0$ (not 0) since every term will be positive. 50. no solution 51. $x < -2$ or $x > 5/3$; $(-\infty, -2)$ or $(5/3, \infty)$ 52. $-3 < x \leq 7$; $(-3, 7]$ 53. 6.5 in

CHAPTER 5

Exponential and Logarithmic Functions

AN APPLIED PROBLEM IN AUDIOLOGY

Audiologists generally agree that continuous exposure to a sound level in excess of 90 decibels for more than 5 hours daily may cause long-term hearing problems. If a student listens to a walkman that produces an intensity of 2×10^{-3} watt/m^2 for long periods of time daily, is there a potential danger to her hearing?

Analysis
To calculate a decibel level we use the equation

$$D = 10 \log \frac{S}{S_0}$$

in which D is the loudness of sound in decibels, S is the intensity of a given sound in watt/m^2, and S_0 is the threshold of human hearing, the constant 10^{-12} watt/m^2.

Tabulation
We are given that $S = 2 \times 10^{-3}$, $S_0 = 10^{-12}$, and we need to find D.

Translation
Substitute the known values of S and S_0 into the formula and find the value of D. If D is greater than 90 decibels, we can conclude that there is a potential danger of hearing loss.

Historically, logarithms were developed to help carry out complicated numerical calculations. With the advent of computers and calculators, logarithmic calculations are no longer of much interest. However, logarithmic and exponential equations and functions remain important, with numerous applications in mathematics today. The application given at the left is solved completely in Example 9 in Section 5.6.

A calculator is an invaluable tool for working with logarithms and exponentials. Generally, all computations will be made and stored on a calculator with no rounding until the final answer. Even with this agreement, there can be slight variations resulting from rounding differences among various calculators.

5.1 INTRODUCTION TO LOGARITHMS

STUDENT GUIDEPOSTS

1. Logarithms
2. Logarithmic and Exponential Form
3. Exponential Properties
4. Logarithmic Properties

1 LOGARITHMS

In previous chapters we solved linear, quadratic, and polynomial equations. In these equations the variable does not appear in an exponent. When the variable does occur in an exponent, as in the equation

$$2^x = 8,$$

the equation-solving rules we have already learned are of no help. At present, we would have to discover the solution by inspection. In this case, the problem is not difficult since we would soon recognize that when 2 is cubed, the result is 8. Thus, the solution to the equation is

$$x = 3. \quad 2^3 = 8$$

Alternatively, we might give the solution in words.

x is the exponent on 2 that gives the number 8

The word *logarithm* can be used instead of *exponent*, and the solution is written

x is the logarithm on 2 that gives the number 8.

Since 2 is called the *base* in the expression 2^x, the solution can also be written in the form

x is the logarithm of 8 using 2 as the base

or $\quad x$ is the logarithm to the base 2 of 8.

This final statement is usually symbolized by

$$x = \log_2 8.$$

2 LOGARITHMIC AND EXPONENTIAL FORM

Basically, we have shown that

$$2^x = 8 \quad \text{and} \quad x = \log_2 8$$

are two forms of the same equation, with $2^x = 8$ called the **exponential form** and $x = \log_2 8$ the **logarithmic form.** That is, they are equivalent equations, with the logarithmic form having the desirable quality of being "solved" for the variable x. In this case, one other equivalent equation is

$$x = 3. \quad 3 = \log_2 8$$

For now, however, we cannot always make this additional simplification.

Definition of Logarithm

Let a, x, and y be real numbers, $a > 0$, $a \neq 1$, that satisfy the exponential equation

$$a^x = y.$$

Then x is the **logarithm to the base a** of y, and the equivalent logarithmic equation is

$$x = \log_a y.$$

340 CHAPTER 5 EXPONENTIAL AND LOGARITHMIC FUNCTIONS

Note In the definition of a logarithm, we restrict a to values greater than 0 since if $a < 0$, then a^x would not be defined for numbers such as $x = 1/2$. Also, if $a = 1$ we would have $a^x = a^y$ for all real numbers x and y. Remember that the word *logarithm* means *exponent*.

EXAMPLE 1 IDENTIFYING EQUIVALENT FORMS

Write each equation in an equivalent form.

(a) $9^x = d$

The equivalent logarithmic form is

$$x = \log_9 d.$$

Remember, x is the exponent in 9^x, so x equals the logarithm in the equivalent form.

(b) $w = \log_3 7$

The equivalent exponential form is

$$3^w = 7.$$

Remember, w is the logarithm in the given form so w becomes the exponent in the equivalent form.

(c) $(-3)^y = t$

We do not consider forms such as this since the base for a logarithm must be positive. That is, $\log_{-3} t$ has not been defined.

PRACTICE EXERCISE 1

Write each equation in an equivalent form.

(a) $u^p = 5$

(b) $4 = \log_5 q$

(c) $s = \log_{-5} z$

Answers: (a) $p = \log_u 5$
(b) $5^4 = q$ (c) There is no equivalent form since the base (-5) is negative.

When converting an equation from exponential to logarithmic form, or from logarithmic to exponential form, remember that in both cases the base is written below the level of the logarithm (exponent), as indicated in the following diagram.

$$\overset{\text{exponent}}{a^x = y} \qquad x = \log_a y$$
$$\underset{\text{base}}{}$$

Solutions to simple exponential equations can often be found by direct inspection; solutions to some logarithmic equations are best found by converting to exponential form.

EXAMPLE 2 SOLVING LOGARITHMIC EQUATIONS

Give the numerical value for x.

(a) $x = \log_4 64$

We first convert $x = \log_4 64$ to exponential form.

$$4^x = 64$$

At this point, it might be clear that x must be 3 since $4^3 = 64$. By writing 64 as a power of 4, $64 = 4^3$, we have

$$4^x = 4^3.$$

In this form, it is now obvious that x is 3 since the bases on both sides of the equation are 4, forcing the exponents on both sides to be equal.

PRACTICE EXERCISE 2

Give the numerical value for x.

(a) $x = \log_2 32$

(b) $\log_x \frac{1}{8} = -3$

Convert to exponential form.

$$x^{-3} = \frac{1}{8}$$

$$x^{-3} = 2^{-3} \qquad \tfrac{1}{8} = \tfrac{1}{2^3} = 2^{-3}$$

It is now clear that x is 2 since the exponents on both sides of the equation are -3, forcing the bases on both sides to be equal.

(b) $\log_x \frac{1}{2} = -1$

Answers: (a) $x = 5$ (b) $x = 2$

3 EXPONENTIAL PROPERTIES

Two important properties of exponentials, used in Example 2(a) and (b), are given below. These properties are intuitively clear and will become even more apparent when we consider exponential functions in the next section.

> **Properties of Exponentials**
>
> Let x and y be real numbers, a and b positive real numbers with $a \neq 1$ and $b \neq 1$.
>
> 1. $a^x = a^y$ if and only if $x = y$.
> 2. $a^x = b^x$ if and only if $a = b$.

EXAMPLE 3 SOLVING EQUATIONS

Solve for x.

(a) $3^{x-1} = 27$

Since $27 = 3^3$, $3^{x-1} = 3^3$, so by equating exponents (the bases are equal),

$$x - 1 = 3$$
$$x = 4.$$

(b) $(x-1)^4 + 16$

Since $16 = 2^4$, $(x-1)^4 = 2^4$, so by equating bases (the exponents are equal),

$$x - 1 = 2$$
$$x = 3.$$

(c) $\log_x 36 = 2$

Convert to exponential form.

$$x^2 = 36 = 6^2$$
$$x = 6$$

Notice that $x \neq -6$ since the base for a logarithm must be positive.

(d) $\log_3 x = 1$

$$x = 3^1 = 3 \qquad \text{Exponential form}$$

(e) $\log_3 1 = x$

$$3^x = 1 \qquad \text{Exponential form}$$
$$3^x = 3^0 \qquad 3^0 = 1$$
$$x = 0$$

PRACTICE EXERCISE 3

Solve for x.

(a) $5^{2x} = 25$

(b) $(x+2)^3 = 8$

(c) $\log_x 81 = 4$

(d) $\log_\pi x = 1$

(e) $\log_\pi 1 = x$

Answers: (a) $x = 1$ (b) $x = 0$
(c) $x = 3$ (not -3) (d) $x = \pi$
(e) $x = 0$

4 LOGARITHMIC PROPERTIES

Parts (d) and (e) of Example 3 illustrate the following theorem.

> **Properties of Logarithms**
> For any base a ($a > 0$ and $a \neq 1$),
> 1. $\log_a a = 1$
> 2. $\log_a 1 = 0$.

The proof of this theorem is shown by changing to exponential form since $\log_a a = 1$ is equivalent to $a^1 = a$ and $\log_a 1 = 0$ is equivalent to $a^0 = 1$.

This brief introduction to exponentials and logarithms will help to clarify the formal development of exponential and logarithmic functions given in the next section.

5.1 EXERCISES A

In Exercises 1–4, convert to logarithmic form.

1. $2^3 = 8$
2. $5^v = 9$
3. $a^{-3} = c$
4. $u^{-v} = w$

In Exercises 5–8, convert to exponential form.

5. $\log_3 9 = 2$
6. $\log_a b = 7$
7. $\log_3 \dfrac{1}{27} = b$
8. $\log_a 6 = c$

Solve each equation in Exercises 9–18.

9. $3^y = 243$
10. $b^{-3} = \dfrac{1}{125}$
11. $25^x = 5$
12. $a^{1/2} = 7$
13. $c^{-1/2} = 6$
14. $\log_2 8 = y$
15. $\log_3 x = 4$
16. $\log_a \dfrac{1}{27} = -3$
17. $\log_8 0.125 = w$
18. $2^{x^2} = 16$

19. **ENGINEERING** An electrical engineer uses the formula $D = 10 \log_{10} \dfrac{S}{S_0}$ to calculate the gain on an amplifier. What is the value of D when **(a)** S is 100 and S_0 is 10? **(b)** S is 600 and S_0 is 6?

20. **BUSINESS** A financial advisor uses the formula $A = 100(1 + 0.05)^n$ to calculate the amount of money in an account when interest is compounded semiannually. What is the value of A when n is 2?

5.1 INTRODUCTION TO LOGARITHMS 343

21. **PHYSICS** The atmospheric pressure P on an object, in pounds per square inch, can be approximated by $P = 14.7(2.7)^{-0.2x}$ where x is the height of the object in miles above sea level. What is the pressure on an object 5 miles high? Give answer to the nearest tenth.

FOR REVIEW

22. Given the polynomial function $P(x) = -3x^5 + x^3 + x^2 - 4x - 9$, determine **(a)** the maximum number of real zeros (the maximum number of x-intercepts), **(b)** the maximum number of turning points, **(c)** the y-intercept, **(d)** the direction of the graph for large positive values of x, and **(e)** the direction of the graph for small negative values of x.

23. **ECOLOGY** The annual cost of removing a given percent x of the pollutants from the smoke of a power plant increases tremendously as the percent approaches 100%. The rational function $C(x) = \frac{1000x}{100 - x}$ approximates the cost of this operation for a particular plant in Arizona. Sketch the graph of $C(x)$ for $0 \leq x < 100$, and find the cost of removing 95% of the pollutants.

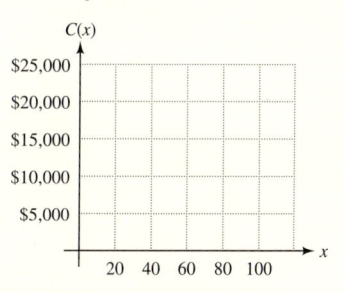

Exercises 24–25, reviewing material from Chapter 3, will help you prepare for the next section. Graph each function.

24. $f(x) = 3x + 1$

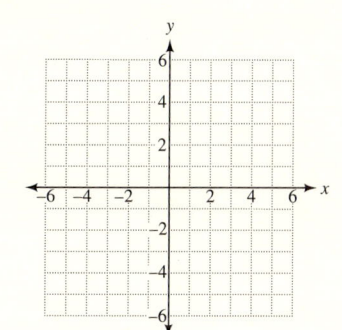

25. $g(x) = x^2 + 2$

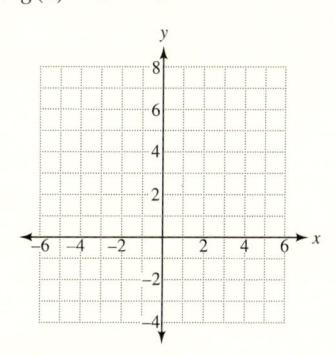

ANSWERS: 1. $\log_2 8 = 3$ 2. $\log_5 9 = v$ 3. $\log_a c = -3$ 4. $\log_u w = -v$ 5. $3^2 = 9$ 6. $a^7 = b$ 7. $3^b = \frac{1}{27}$
8. $a^c = 6$ 9. 5 10. 5 11. $\frac{1}{2}$ 12. 49 13. $\frac{1}{36}$ 14. 3 15. 81 16. 3 17. -1 18. ± 2 19. (a) 10 (b) 20
20. \$110.25 21. 5.4 lb/in² 22. (a) 5 (b) 4 (c) (0, -9) (d) downward (e) upward
23. $C(95) = \$19,000$ 24. 25.

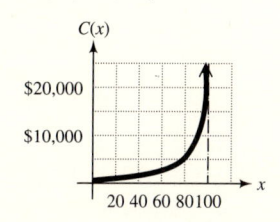

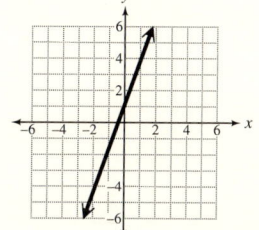

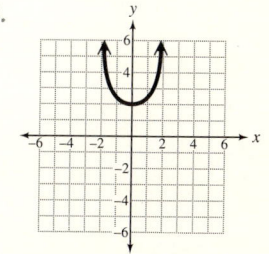

5.1 EXERCISES B

In Exercises 1–4, convert to logarithmic form.

1. $u^5 = 25$ **2.** $a^b = 7$ **3.** $u^v = w$ **4.** $w^{x-1} = z$

In Exercises 5–8, convert to exponential form.

5. $\log_5 \dfrac{1}{25} = -2$ **6.** $\log_a 64 = 3$ **7.** $\log_b 8 = \dfrac{1}{3}$ **8.** $\log_a u = v$

Solve each equation in Exercises 9–18.

9. $2^z = \dfrac{1}{8}$ **10.** $c^3 = 216$

11. $27^x = 3$ **12.** $b^{1/3} = 6$

13. $x^{-1/3} = 2$ **14.** $\log_a 9 = 2$

15. $\log_7 x = -3$ **16.** $\log_x 2 = \dfrac{1}{3}$

17. $\log_4 0.25 = x$ **18.** $3^{x^2} = 81$

19. ENGINEERING An electrical engineer uses the formula $D = 10 \log_{10} \dfrac{S}{S_0}$ to calculate the gain on an amplifier. What is the value of D when **(a)** S is 200 and S_0 is 20? **(b)** S is 200 and S_0 is 2?

20. BUSINESS A financial advisor uses the formula $A = 100(1 + 0.05)^n$ to calculate the amount of money in an account when interest is compounded semiannually. What is the value of A when n is 3?

21. PHYSICS The atmospheric pressure P on an object, in pounds per square inch, can be approximated by $P = 14.7(2.7)^{-0.2x}$ where x is the height of the object in miles above sea level. What is the pressure on an object 5 miles high? Give answer to the nearest tenth.

FOR REVIEW

22. If a polynomial $P(x)$ is divided by $x + 8$, what will be the value of the remainder?

23. Consider the rational function $f(x) = \dfrac{x^2 + 25}{x^2 - 25}$. Give the **(a)** vertical asymptotes, **(b)** horizontal asymptotes, **(c)** oblique asymptotes, **(d)** x-intercepts, **(e)** y-intercept, and **(f)** symmetries of the graph.

Exercises 24–25, reviewing material from Chapter 3, will help you prepare for the next section. Graph each function.

24. $f(x) = 2x - 1$ **25.** $g(x) = x^2 - 3$

5.1 EXERCISES C

Solve each equation in Exercises 1–3.

1. $2^{x^2+1} = 4$ **2.** $4^{2x} = 1$ **3.** $\pi^{2x-1} = 1$
[Answer: 1, −1]

4. **BUSINESS** The trade-in value V at the end of t years of a new car purchased for P dollars is given by $V = 0.815P(0.795)^{t-1}$. If a new Buick is purchased for $19,500, find the trade-in value, to the nearest dollar, after **(a)** 1 year, **(b)** 3 years, **(c)** 10 years.

[Answer: $2016]

5. **LINGUISTICS** One theory used for dating a language assumes that over a long period of time, changes in the language take place at a constant rate. If W_0 is the initial number of base words in a language, the number of base words W remaining in use after a period of years t can be approximated by $W = W_0(0.875)^{t/1000}$. What percent of the base words are lost during the first 1000 years? [Answer: 12.5%]

6. **FINANCE** Accountants use several methods of calculating the depreciation on an item for income-tax purposes. One of these methods assumes that the amount of depreciation that can be taken in a given year is a fixed percent of the present value of the item. If v_0 is the initial value of an item, v is the value of the item after t years, and d is the percent depreciation rate, then $v = v_0(1 - d)^t$. Suppose that after n years, the item has a fixed salvage value of s dollars, and a taxpayer plans to select a depreciation rate making the value of the item after n years s dollars. Show that $d = 1 - \sqrt[n]{s/v_0}$.

5.2 EXPONENTIAL AND LOGARITHMIC FUNCTIONS

STUDENT GUIDEPOSTS

1. Exponential Functions
2. Properties of Exponential Functions
3. Approximating Exponentials
4. Logarithmic Functions
5. Properties of Logarithmic Functions
6. Inverse Function Properties

1 EXPONENTIAL FUNCTIONS

Polynomial functions have constants used as exponents and a variable for a base. In this section we consider *exponential functions* that have variable exponents and a constant for a base.

Exponential Function

Let a be a real number, $a > 0$ and $a \neq 1$. The function

$$f(x) = a^x$$

is an **exponential function** with **base a**.

Evaluating an exponential function when x is an integer n or a rational number p/q poses no problem since a^n and $a^{p/q} = \sqrt[q]{a^p}$ have already been defined. Although we have not defined a^x for irrational x (such as $\sqrt{2}$ or π), we assume that such values could be approximated by using rational number approximations for x. For example, since $\sqrt{2} \approx 1.414$, $a^{\sqrt{2}} \approx a^{1.414} = a^{1414/1000} = \sqrt[1000]{a^{1414}}$. By using more and more accurate approximations of $\sqrt{2}$, we could obtain closer and closer approximations of $a^{\sqrt{2}}$. The idea of defining irrational exponents is treated more thoroughly in calculus, but we will assume that irrational exponents do make sense and that all of the properties of rational exponents, such as the product, quotient, and power rules, also apply to irrational exponents. When we graph an exponential function such as $f(x) = 2^x$, our assumptions about irrational exponents will appear reasonable.

EXAMPLE 1 Graphing an Exponential Function

Graph the function $y = 2^x$.

Selecting values of x and calculating the corresponding values of y, we obtain a table of values. We can use 1.41 as an approximation for $\sqrt{2}$. From the graph in Figure 5.1, we see that 2.7 is an approximation for $2^{\sqrt{2}}$. Similarly, using 3.14 as an approximation for π, we can estimate 2^π to be about 8.8.

x	y
0	1
-1	1/2
1	2
-2	1/4
2	4
-3	1/8
3	8
-4	1/16

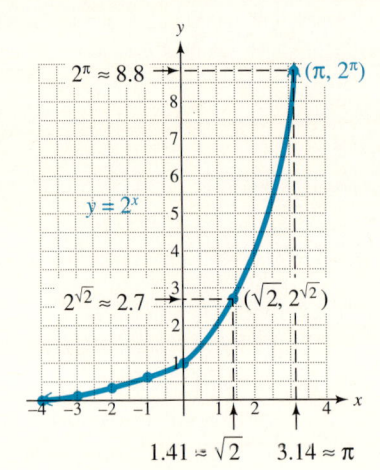

Figure 5.1

Practice Exercise 1

Graph the function $y = 10^x$.

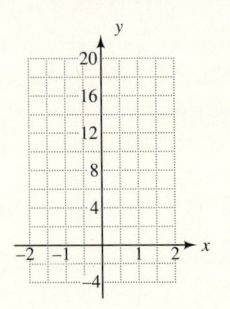

Answer: The graph is part of Figure 5.2 on page 347.

To compare exponential functions, we shall graph several such functions in the same coordinate system. This will help us to generalize about the behavior of these functions.

Graph $y = 2^x$, $y = \left(\frac{1}{2}\right)^x$, $y = 3^x$, $y = \left(\frac{1}{3}\right)^x$, $y = 10^x$, and $y = \left(\frac{1}{10}\right)^x$. First we construct a table of values.

x	0	-1	1	-2	2	-3	3	-4	4
$y = 2^x$	1	$\frac{1}{2}$	2	$\frac{1}{4}$	4	$\frac{1}{8}$	8	$\frac{1}{16}$	16
$y = \left(\frac{1}{2}\right)^x$	1	2	$\frac{1}{2}$	4	$\frac{1}{4}$	8	$\frac{1}{8}$	16	$\frac{1}{16}$
$y = 3^x$	1	$\frac{1}{3}$	3	$\frac{1}{9}$	9	$\frac{1}{27}$	27	$\frac{1}{81}$	81
$y = \left(\frac{1}{3}\right)^x$	1	3	$\frac{1}{3}$	9	$\frac{1}{9}$	27	$\frac{1}{27}$	81	$\frac{1}{81}$
$y = 10^x$	1	$\frac{1}{10}$	10	$\frac{1}{100}$	100	$\frac{1}{1000}$	1000	$\frac{1}{10000}$	10000
$y = \left(\frac{1}{10}\right)^x$	1	10	$\frac{1}{10}$	100	$\frac{1}{100}$	1000	$\frac{1}{1000}$	10000	$\frac{1}{10000}$

The scale for our graphs in Figure 5.2 does not allow us to plot all of the points listed in the table, but the values are included for purposes of comparison.

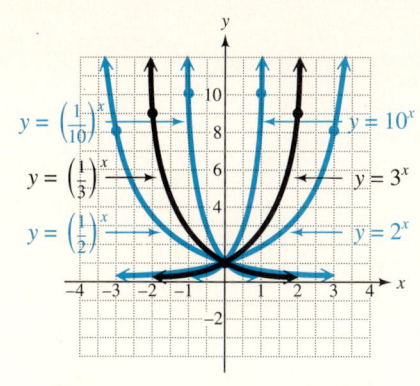

Figure 5.2 Exponential Functions

❷ PROPERTIES OF EXPONENTIAL FUNCTIONS

Several observations can be made in view of the graphs of the exponential functions above.

1. If $a > 1$, $y = a^x$ increases as x increases, and if $a < 1$, $y = a^x$ decreases as x increases. That is, $f(x) = a^x$ is an increasing function when $a > 1$, and a decreasing function when $a < 1$.
2. The graphs of $y = a^x$ and $y = (1/a)^x$ are reflections of each other about the y-axis.
3. The graphs of $y = a^x$ and $y = (1/a)^x$ pass through $(0, 1)$ since $a^0 = 1$ for any $a > 0$.
4. The domain (the possible values of x) of every exponential function is the set of all real numbers.
5. The range (the possible values of y) of every exponential function is the set of all positive real numbers.

❸ APPROXIMATING EXPONENTIALS

A calculator with a $\boxed{y^x}$ key can be extremely helpful for work with exponential functions. For example, suppose we wish to approximate $2^{\sqrt{5}}$. The sequence of keys to use on a calculator is:

ALG: 2 $\boxed{y^x}$ 5 $\boxed{\sqrt{}}$ $\boxed{=}$ RPN: 2 $\boxed{\text{ENTER}}$ 5 $\boxed{\sqrt{}}$ $\boxed{y^x}$.

If this sequence is followed, the display will show 4.7111, correct to four decimal places. Since $\sqrt{5} \approx 2.2$, locate this value for x on the graph of $y = 2^x$ given in Figure 5.2, and notice that the value of y is about 4.7.

EXAMPLE 2 APPROXIMATING EXPONENTIALS	PRACTICE EXERCISE 2
Use a calculator to approximate each number correct to four decimal places. **(a)** $(1.01)^{50}$ ALG: 1.01 $\boxed{y^x}$ 50 $\boxed{=}$ RPN: 1.01 $\boxed{\text{ENTER}}$ 50 $\boxed{y^x}$ Thus, $(1.01)^{50} \approx 1.6446$.	Use a calculator to approximate each number correct to four decimal places. **(a)** $(1.09)^{-40}$ [*Hint:* You will need to use your change sign key when you enter the x value.]

(b) $\pi^{\sqrt{11}}$

ALG: $\boxed{\pi}\ \boxed{y^x}\ 11\ \boxed{\sqrt{}}\ \boxed{=}$ RPN: $\boxed{\pi}\ \boxed{\text{ENTER}}\ 11\ \boxed{\sqrt{}}\ \boxed{y^x}$

Thus, $\pi^{\sqrt{11}} \approx 44.5512$.

(b) $(\sqrt{2})^\pi$

Answers: (a) 0.0318 (b) 2.9707

EXAMPLE 3 GRAPHING A MORE COMPLEX FUNCTION

Sketch the graph of $f(x) = 2^{x^2}$.

A table of values is helpful for graphing functions of this type. Since $f(x) = f(-x)$, the graph is symmetric with respect to the y-axis and is given in Figure 5.3.

x	$f(x)$
0	1
1	2
2	16
3	512

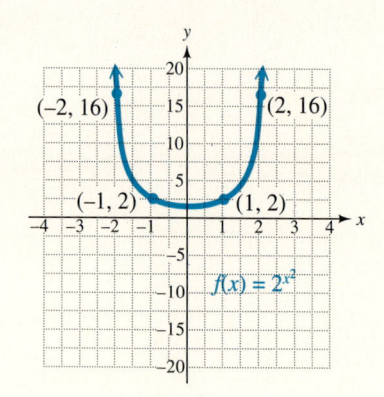

Figure 5.3

PRACTICE EXERCISE 3

Sketch the graph of $f(x) = 2^{x^2} - 4$.

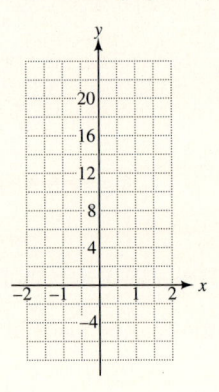

Answer: Slide graph in Figure 5.3 down 4 units.

---

❹ LOGARITHMIC FUNCTIONS

Since each exponential function $y = a^x$ ($a > 0$ and $a \neq 1$) is either increasing or decreasing, it must have an inverse. To find the equation of the inverse, interchange x and y and solve the result for y. Starting with

$$y = a^x$$

we interchange x and y and obtain

$$x = a^y.$$

To solve this equation for y, we convert it to logarithmic form, $y = \log_a x$. Thus, the inverse of the exponential function $y = a^x$ is the **logarithmic function** defined by

$$y = \log_a x.$$

To graph $y = \log_a x$, we need only graph $x = a^y$. Alternatively, since the graphs of inverse functions are symmetric with respect to the line $y = x$, we could graph $y = a^x$ and use it to find the graph of $y = \log_a x$.

EXAMPLE 4 GRAPHING INVERSE FUNCTIONS

Graph $y = 2^x$ and $y = \log_2 x$ ($x = 2^y$) in the same coordinate system.

The tables of values for these functions are on the next page and their graphs are given in Figure 5.4.

PRACTICE EXERCISE 4

Graph $y = 3^x$ and $y = \log_3 x$ in the same coordinate system.

x	$y = 2^x$
0	1
−1	1/2
1	2
−2	1/4
2	4
−3	1/8
3	8

$x = 2^y$	y
1	0
1/2	−1
2	1
1/4	−2
4	2
1/8	−3
8	3

Answer: The graphs are given in Figures 5.2 and 5.5.

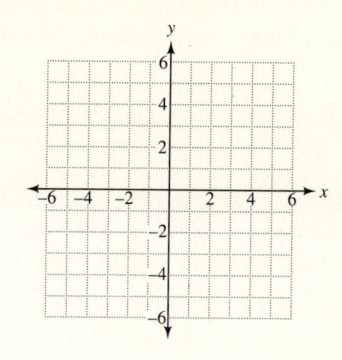

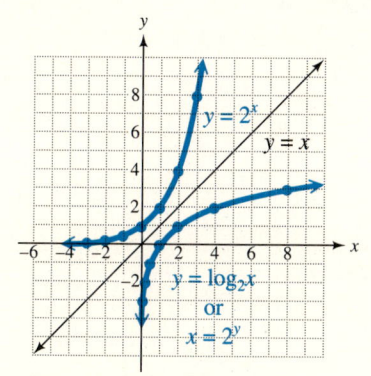

Figure 5.4

The graphs of the three logarithmic functions $y = \log_4 x$, $y = \log_3 x$, and $y = \log_2 x$ are given in Figure 5.5. If we agree to restrict our discussion of logarithmic functions to bases larger than 1, these graphs lead to the observations which follow.

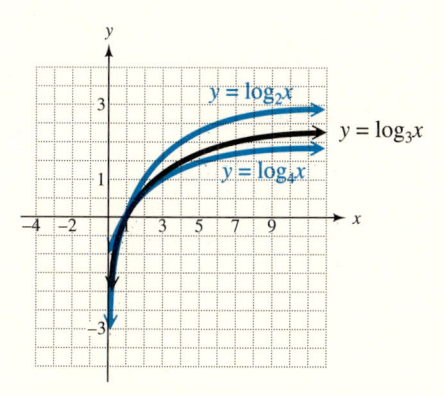

Figure 5.5 Logarithmic Functions

5 PROPERTIES OF LOGARITHMIC FUNCTIONS

1. If $a > 1$, $y = \log_a x$ increases as x increases. That is, $f(x) = \log_a x$ is an increasing function when $a > 1$.
2. The graph of $y = \log_a x$ passes through $(1, 0)$ for every $a > 1$. That is, $\log_a 1 = 0$ for every $a > 1$.
3. The domain (the possible x values) of every logarithmic function is the set of all positive real numbers.
4. The range (the set of all possible values of y) of every logarithmic function is the set of all real numbers.

6 INVERSE FUNCTION PROPERTIES

The fact that logarithmic and exponential functions are inverses gives two special equations. Remember that the inverse of a function f is denoted by f^{-1} and that $f(f^{-1}(x)) = x$ and $f^{-1}(f(x)) = x$. These properties lead to the next theorem.

Inverse Function Properties

For any base a, $a > 0$ and $a \neq 1$,

1. $\log_a a^x = x$, for any real number x.
2. $a^{\log_a x} = x$, for any real number $x > 0$.

Proof We use the fact that $f(x) = a^x$ and $f^{-1}(x) = \log_a x$ are inverses of each other to prove the theorem.

1. $x = f^{-1}(f(x))$
   $= f^{-1}(a^x)$
   $= \log_a a^x$

2. $x = f(f^{-1}(x))$
   $= f(\log_a x)$
   $= a^{\log_a x}$

EXAMPLE 5 Using Inverse Function Properties

Simplify each expression.

(a) $\log_3 3^2 = 2$ since $\log_a a^x = x$ for any x.

(b) $\log_{10} 0.01 = \log_{10} 10^{-2} = -2$

(c) $3^{\log_3 5} = 5$

PRACTICE EXERCISE 5

Simplify each expression.

(a) $\log_\pi \pi^5$

(b) $\log_{100} 10^4$

(c) $4^{\log_4 (5x+2)}$

Answers: (a) 5   (b) 2
(c) $5x + 2$

As with exponential functions, we are often interested in more complex logarithmic functions of the form $f(x) = \log_a p(x)$. Keep in mind that $p(x)$ must be positive for such a function to exist.

EXAMPLE 6 Graphing a More Complex Function

Graph $f(x) = \log_2 x^2$ for $x \neq 0$.

Since $f(-x) = f(x)$, the graph is symmetric with respect to the y-axis. A table of values for $2^y = x^2$ is helpful, and the graph is given in Figure 5.6.

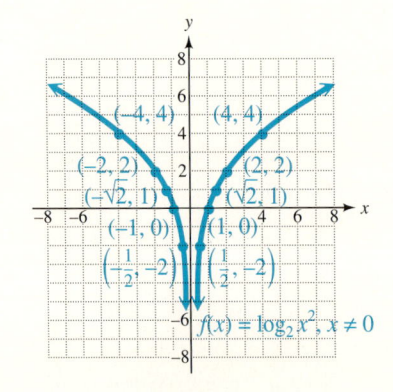

y	x
0	± 1
1	$\pm \sqrt{2} \approx \pm 1.4$
2	± 2
4	± 4
-2	$\pm 1/2$

Figure 5.6

PRACTICE EXERCISE 6

Graph $f(x) = 3 + \log_2 x^2$ for $x \neq 0$.

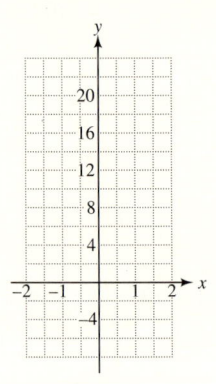

Answer: Slide graph in Figure 5.6 up 3 units.

Many applied problems, such as the one in the next example, involve exponential functions.

EXAMPLE 7   A CHEMISTRY PROBLEM

A radioactive isotope has a half-life of 1200 years. This means that if 1000 grams of the isotope are allowed to decay, the amount A that remains after t years is given by the exponential function $A = (1000)2^{-t/1200}$. Find the amount remaining at the end of 2400 years and sketch the graph of the function for $t \geq 0$.

To find the amount remaining after 2400 years, substitute 2400 for t.

$$A = (1000)2^{-2400/1200} = (1000)2^{-2} = \frac{1000}{4} = 250$$

Thus, 250 grams remain after 2400 years. The graph of the function for $t \geq 0$ is given in Figure 5.7.

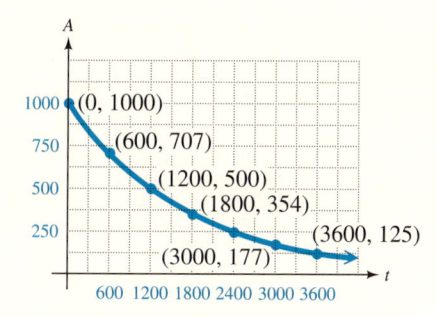

Figure 5.7

PRACTICE EXERCISE 7

The value of a rare coin purchased for $100 triples every year. This means the value after t years is given by $V(t) = (100)3^t$. Graph this function for $t \geq 0$ and find the value of the coin after 5 years.

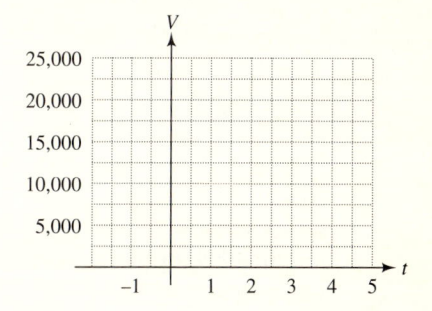

Answer: The graph goes through (3, 2700), (4, 8100), (5, 24,300); $24,300.

5.2 EXERCISES A

In Exercises 1–9 sketch the graph of the function.

1. $f(x) = 5^x$

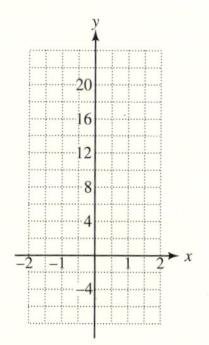

2. $f(x) = \left(\frac{1}{5}\right)^x$

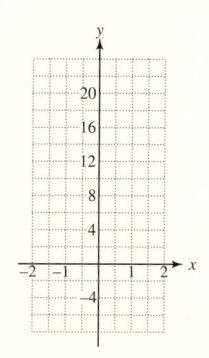

3. $f(x) = -5^x$

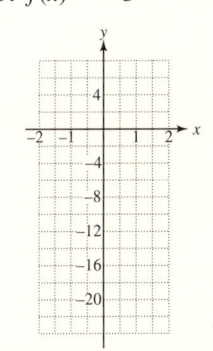

4. $f(x) = 2 + 5^x$

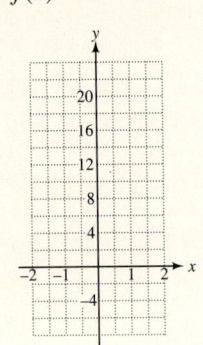

5. $f(x) = 2^{x^2+1}$

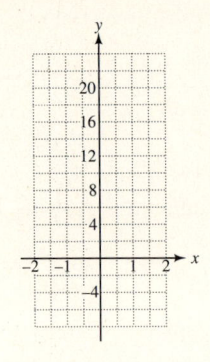

6. $f(x) = 2^x - 2^{-x}$

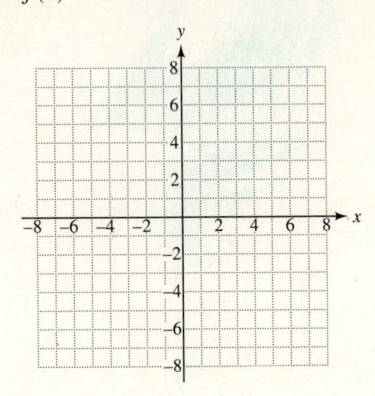

7. $f(x) = \log_8 x$

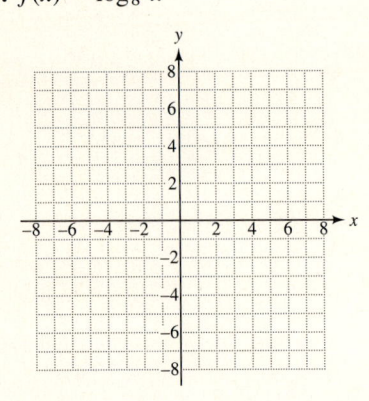

8. $f(x) = \log_8 (-x)$ for $x < 0$

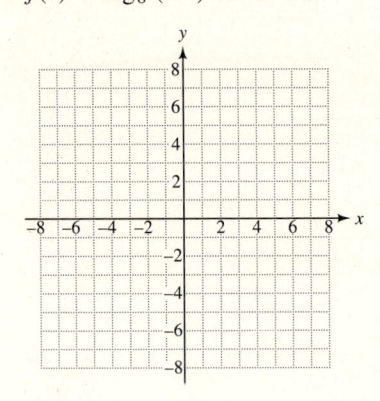

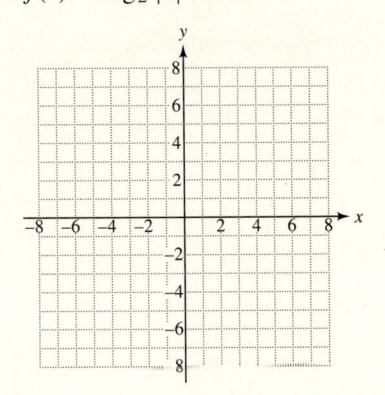

9. $f(x) = \log_2 |x|$ for $x \neq 0$

In Exercises 10–12 simplify each expression.

10. $\log_5 5^3$

11. $\log_{10} 0.001$

12. $e^{\log_e (x+7)}$

In Exercises 13–16 use a calculator to approximate each number correct to four decimal places.

13. $5^{\sqrt{3}}$

14. $-\pi^{\sqrt{3}}$

15. $6^{-\sqrt{2}}$

16. $(1.02)^{60}$

17. The graph of every exponential function passes through which point?

18. What is the domain of every exponential function?

19. What is the domain of every logarithmic function?

20. What is the base of the logarithmic function $y = \log_a x$ if its graph passes through the point $(3, 1)$?

21. Why is 1 omitted as a base for a logarithmic function?

In Exercises 22–23 find the inverse of each function.

22. $f(x) = 2^{x+1} + 1$

23. $f(x) = 2 \log_3 (x - 4)$

24. BIOLOGY The number of bacteria in a culture is given in terms of time t, in hours, by the formula $N = (5000)2^t$.
   (a) How many bacteria are present at the start of the experiment (when $t = 0$)?
   (b) How many bacteria are present in 5 hours?
   (c) How many bacteria are present in 3.25 hours?

25. CHEMISTRY A radioactive isotope has a half-life of 4 years. This means that in 4 years, one-half of any given amount will change to another substance due to radioactive decay. If 50 grams of this isotope are allowed to decay in a reactor, the amount A that will remain at the end of t years is given by $A = 50(1/2)^{t/4}$. How many grams remain at the end of
   (a) 2 years?
   (b) 100 years?
   (c) 5.5 years?

FOR REVIEW

In Exercises 26–28 solve each equation.

26. $\log_a 32 = 5$ **27.** $\log_{10} x = -2$ **28.** $10^{\log_{10}(2x+5)} = 7$

To prepare for the next section, we review the properties of exponents. Answer true *or* false *in Exercises 29–31.*

29. $a^m a^n = a^{m+n}$ **30.** $(a^m)^n = a^{m+n}$ **31.** $\left(\dfrac{a^m a^p}{a^q}\right)^k = a^{mk+pk-qk}$, $a \neq 0$

ANSWERS:

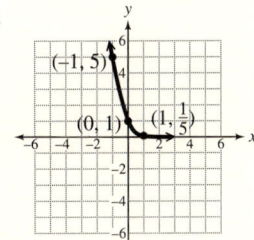

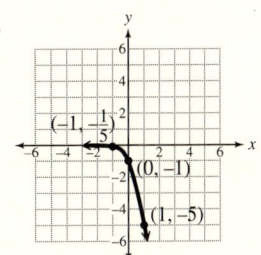

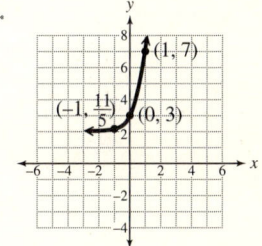

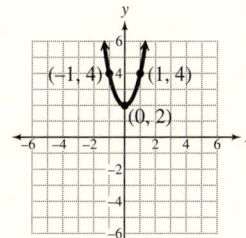

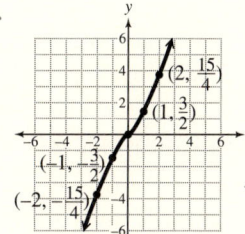

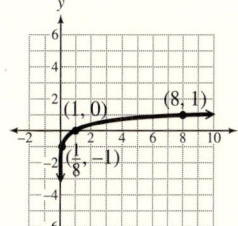

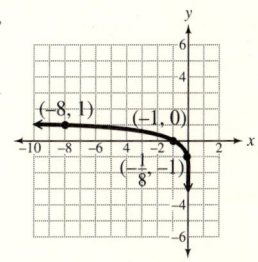

1–9. [graphs]

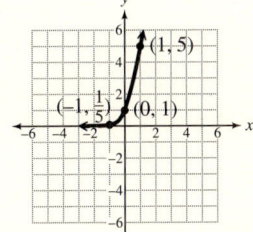

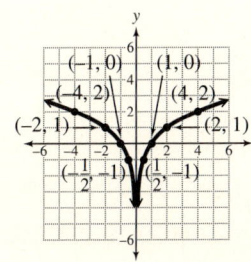

10. 3   11. -3 12. $x + 7$ 13. 16.2425   14. -7.2625 15. 0.0793   16. 3.2810
17. (0, 1)   18. all real numbers   19. all positive numbers   20. $a = 3$ 21. If $a = 1$, $\log_1 x = y$ is equivalent to $1^y = x$, or $x = 1$ for all values of y, which would not be a function.   22. $f^{-1}(x) = -1 + \log_2(x - 1)$; for all $x > 1$ 23. $f^{-1}(x) = 3^{x/2} + 4$; for all real numbers x 24. (a) 5000   (b) 160,000   (c) 47,568   25. (a) 35.36 g   (b) 0.000 001 5 g   (c) 19.28 g   26. 2   27. 0.01   28. 1   29. true   30. false   31. true

5.2 EXERCISES B

In Exercises 1–9 sketch the graph of the function.

1. $f(x) = 8^x$
2. $f(x) = \left(\dfrac{1}{8}\right)^x$
3. $f(x) = -8^x$
4. $f(x) = 3^{x+1}$
5. $f(x) = 3^{x^2+1}$
6. $f(x) = \log_4 x$
7. $f(x) = 1 + \log_4 x$
8. $f(x) = \log_2(x+1)$
9. $f(x) = |\log_2 x|$

In Exercises 10–12 simplify each expression.

10. $\log_{10} 1000$
11. $7^{\log_7 1.5x}$
12. $\pi^{\log_\pi \sqrt{x}}$

In Exercises 13–16 use a calculator to approximate each number correct to four decimal places.

13. $-7^{\sqrt{6}}$
14. $(\sqrt{3})^\pi$
15. $9^{-\pi}$
16. $(1.02)^{-60}$

17. The graph of every logarithmic function passes through which point?

18. What is the range of every exponential function?

19. What is the range of every logarithmic function?

20. What is the base of the exponential function $y = a^x$ if its graph passes through the point (1/2, 8)?

21. Why is every negative number omitted as a base for a logarithmic function?

In Exercises 22–23 find the inverse of each function.

22. $f(x) = 3^{x-1} + 2$
23. $f(x) = 5 \log_2(x+3)$

24. **PHYSICS** The atmospheric pressure P on an object, in pounds per square inch, can be approximated using the formula $P = 14.7(2.7)^{-0.2x}$ where x is the height of the object, in miles, above sea level. What is the pressure on an object **(a)** 1 mile high? **(b)** 5.3 miles high? **(c)** 8.5 miles high?

25. **BUSINESS** The amount of money A in a savings account at the end of t years if $1000 is invested originally at an annual interest rate of 10% compounded quarterly, is given by $A = 1000(1.025)^{4t}$. What amount of money will be in the account at the end of **(a)** 1 year? **(b)** 10 years? **(c)** 3.5 years?

FOR REVIEW

In Exercises 26–28 solve each equation.

26. $\log_5 125 = y$
27. $\log_7 7^{2x-3} = 5$
28. $\log_3 x^2 = 4$

To prepare for the next section, we review the properties of exponents. Answer true *or* false *in Exercises 29–31.*

29. $a^m a^n = a^{mn}$
30. $\dfrac{a^m}{a^n} = a^{m-n}, \quad a \neq 0$
31. $\left(\dfrac{a^m}{a^n a^p}\right)^q = a^{m-n-p+q}$

5.2 EXERCISES C

Use a calculator and graph each function in Exercises 1–2.

1. $f(x) = 1.4 + (2.5)^x$
2. $g(x) = \log_{2.5}(x - 1.4)$

3. What might be concluded from the graphs of f and g in Exercises 1 and 2? Verify your assertion algebraically.

5.3 PROPERTIES OF LOGARITHMS

STUDENT GUIDEPOSTS

1. Basic Rule of Logarithms
2. The Product Rule
3. The Quotient Rule
4. The Power Rule
5. Base Conversion Formula

1 BASIC RULE OF LOGARITHMS

In this section we prove a number of theorems related to logarithmic functions that provide valuable tools for solving logarithmic equations. Recall from Section 5.1 that if $a > 0$ and $a \neq 1$, then

$$a^x = a^y \quad \text{if and only if} \quad x = y.$$

The next theorem is the logarithmic equivalent of this.

Basic Rule of Logarithms

Let a be a base for logarithms, $a > 0$ and $a \neq 1$, with x and y positive real numbers. Then

$$\log_a x = \log_a y \quad \text{if and only if} \quad x = y.$$

Proof We have two things to prove. First, if $x = y$, then clearly $\log_a x = \log_a y$ since $f(x) = \log_a x$ is a function. Conversely, if $\log_a x = \log_a y$, then $a^{\log_a x} = a^{\log_a y}$ by the exponential theorem cited above. Thus, since $x = a^{\log_a x}$ and $y = a^{\log_a y}$, we have $x = y$.

The following example illustrates the usefulness of this theorem for solving certain equations involving logarithms.

EXAMPLE 1 SOLVING A LOGARITHMIC EQUATION

Solve. $\log_3 (4x + 5) = \log_3 x^2$

Using the theorem above with $a = 3$, we have $4x + 5 = x^2$, which is a quadratic equation in x.

$$x^2 - 4x - 5 = 0$$
$$(x + 1)(x - 5) = 0$$
$$x + 1 = 0 \quad \text{or} \quad x - 5 = 0$$
$$x = -1 \qquad\qquad x = 5$$

Check: $x = -1$: $\log_3 (4(-1) + 5) \stackrel{?}{=} \log_3 (-1)^2$
$\log_3 1 = \log_3 1$

$x = 5$: $\log_3 (4(5) + 5) \stackrel{?}{=} \log_3 5^2$
$\log_3 25 = \log_3 25$

Thus, -1 and 5 are both solutions.

PRACTICE EXERCISE 1

Solve.

$$\log_5 (2x + 3) = \log_5 (x + 7)$$

Answer: 4

2 THE PRODUCT RULE

Recall that the word *logarithm* means *exponent*. The next three theorems present laws of logarithms that are based on three familiar laws of exponents. The first is the logarithmic equivalent of the power rule for exponents, $a^m a^n = a^{m+n}$, that is, the exponent (logarithm) on a product is the sum of the exponents (logarithms).

> **The Product Rule**
>
> For any base a ($a > 0$ and $a \neq 1$) and positive real numbers x and y,
>
> $$\log_a xy = \log_a x + \log_a y,$$
>
> that is, **the log of a product is the sum of the logs.**

relates to additive exponents

Proof Let $u = \log_a x$ and $v = \log_a y$. Then $a^u = x$ and $a^v = y$.

$$xy = a^u \cdot a^v$$
$$xy = a^{u+v}$$

From the definition of a logarithm, the last equation $xy = a^{u+v}$ becomes $\log_a xy = u + v$. Since $u = \log_a x$ and $v = \log_a y$, we have

$$\log_a xy = \log_a x + \log_a y.$$

⚠ CAUTION ⚠

$\log_a (x + y) \neq \log_a x + \log_a y$. If these were equal, we would have $\log_a (x + y) = \log_a xy$, so that $x + y = xy$, which is clearly not true.

EXAMPLE 2 USING THE PRODUCT RULE

(a) Express $\log_2 (4 \cdot 8)$ as a sum of logarithms.

$$\log_2 (4 \cdot 8) = \log_2 4 + \log_2 8$$

In this case we can verify that these two values are equal.

$$\log_2 (4 \cdot 8) = \log_2 32 = \log_2 2^5 = 5$$
$$\log_2 4 + \log_2 8 = \log_2 2^2 + \log_2 2^3 = 2 + 3 = 5$$

(b) Express $\log_5 7 + \log_5 w$ as a single logarithm.

$\log_5 7 + \log_5 w = \log_5 7w$ Product rule in reverse

Notice that $\log_5 7 + \log_5 w$ *is not* $\log_5 (7 + w)$.

PRACTICE EXERCISE 2

(a) Express $\log_3 3u$ as a sum of logarithms and simplify.

(b) Express $\log_2 (x + y) + \log_2 (x - y)$ as a single logarithm.

Answers: (a) $1 + \log_3 u$
(b) $\log_2 (x^2 - y^2)$

3 THE QUOTIENT RULE

The next theorem is the logarithmic equivalent of the quotient rule for exponents, $\frac{a^m}{a^n} = a^{m-n}$, that is, the exponent (logarithm) on a quotient is the difference of the exponents (logarithms).

The Quotient Rule

For any base a ($a > 0$ and $a \neq 1$) and positive real numbers x and y,

$$\log_a \frac{x}{y} = \log_a x - \log_a y,$$

that is, **the log of a quotient is the difference of the logs.**

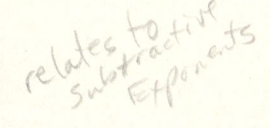

Proof Let $u = \log_a x$ and $v = \log_a y$. Then $a^u = x$ and $a^v = y$.

$$\frac{x}{y} = \frac{a^u}{a^v} \quad y \neq 0 \text{ since the domain of the log function is the set of positive numbers}$$

$$\frac{x}{y} = a^{u-v}$$

$$\log_a \frac{x}{y} = u - v = \log_a x - \log_a y.$$

CAUTION

$\log_a (x - y) \neq \log_a x - \log_a y$. If they were equal, $x - y$ would equal x/y.

EXAMPLE 3 Using the Quotient Rule

(a) Express $\log_2 \frac{32}{8}$ as a difference of logarithms.

$$\log_2 \frac{32}{8} = \log_2 32 - \log_2 8$$

We can verify that these two values are equal.

$$\log_2 \frac{32}{8} = \log_2 4 = \log_2 2^2 = 2$$

$$\log_2 32 - \log_2 8 = \log_2 2^5 - \log_2 2^3 = 5 - 3 = 2$$

(b) Express $\log_b 3 - \log_b w$ as a single logarithm.

$$\log_b 3 - \log_b w = \log_b \frac{3}{w} \quad \text{Quotient rule in reverse}$$

Notice that $\log_b 3 - \log_b w$ **is not** $\log_b (3 - w)$.

Practice Exercise 3

(a) Express $\log_4 \frac{1}{4}$ as a difference of logarithms and simplify.

(b) Express $\log_5 15 - \log_5 3$ as a single logarithm and simplify.

Answers: (a) -1 (b) 1

④ THE POWER RULE

The logarithmic equivalent of the power rule for exponents, $(a^m)^n = a^{mn}$, that is, the exponent (logarithm) on a power is the product of the exponents (logarithms), is the substance of the next theorem.

The Power Rule

For any base a ($a > 0$ and $a \neq 1$), positive real numbers x and y, and real number c,

$$\log_a x^c = c \log_a x,$$

that is, **the log of a number to a power is the power times the log of the number.**

358 CHAPTER 5 EXPONENTIAL AND LOGARITHMIC FUNCTIONS

Proof Begin by letting $u = \log_a x$. Then $a^u = x$, and

$$x^c = (a^u)^c$$
$$x^c = a^{uc}$$
$$\log_a x^c = uc = cu = c \log_a x.$$

EXAMPLE 4 USING THE POWER RULE

Simplify.

(a) $\log_2 4^3$

$$\log_2 4^3 = 3 \log_2 4 \quad \text{Power rule}$$
$$= 3 \log_2 2^2$$
$$= 3(2) = 6 \quad \log_2 2^2 = 2$$

(b) $\log_2 \sqrt[3]{64}$

$$\log_2 \sqrt[3]{64} = \log_2 (64)^{1/3} \quad \sqrt[3]{a} = a^{1/3}$$
$$= \frac{1}{3} \log_2 64 \quad \text{Power rule}$$
$$= \frac{1}{3} \log_2 2^6 = \frac{1}{3}(6) = 2$$

PRACTICE EXERCISE 4

Simplify.

(a) $\log_3 9^2$

(b) $\log_3 \sqrt{27}$

Answers: (a) 4 (b) 3/2

Logarithmic expressions that involve more complicated products, quotients, or powers can be simplified using a combination of rules. Usually the quotient and product rules are used first, followed by the power rule.

EXAMPLE 5 SIMPLIFYING LOGARITHMIC EXPRESSIONS

Use the properties of logarithms to expand the following.

(a) $\log_a \dfrac{x}{yz} = \log_a x - \log_a yz \quad$ Quotient rule

$$= \log_a x - (\log_a y + \log_a z) \quad \text{Product rule, use parentheses to avoid a sign error}$$
$$= \log_a x - \log_a y - \log_a z$$

(b) $\log_a \sqrt{\dfrac{uv}{w^2}} = \log_a \left(\dfrac{uv}{w^2}\right)^{1/2} \quad \sqrt{a} = a^{1/2}$

$$= \frac{1}{2} \log_a \frac{uv}{w^2} \quad \text{Power rule}$$
$$= \frac{1}{2}(\log_a uv - \log_a w^2) \quad \text{Quotient rule}$$
$$= \frac{1}{2}(\log_a u + \log_a v - 2 \log_a w) \quad \text{Product rule, power rule}$$
$$= \frac{1}{2} \log_a u + \frac{1}{2} \log_a v - \log_a w \quad \text{Distribute}$$

PRACTICE EXERCISE 5

Use the properties of logarithms to expand the following.

(a) $\log_a \dfrac{ax}{y}$

(b) $\log_a \dfrac{x\sqrt{y}}{z^3}$

Answers: (a) $1 + \log_a x - \log_a y$
(b) $\log_a x + \frac{1}{2} \log_a y - 3 \log_a z$

A combination of logarithms can often be simplified to a single logarithm by using the rules in reverse. This time we use the power rule first, followed by the product and quotient rules.

EXAMPLE 6 WRITING AS A SINGLE LOGARITHM

Express $\frac{1}{3} \log_a x - 5 \log_a y + \log_a z$ as a single logarithm.

$\frac{1}{3} \log_a x - 5 \log_a y + \log_a z = \log_a x^{1/3} - \log_a y^5 + \log_a z$ Always use power rule first

$\qquad = \log_a \frac{x^{1/3}}{y^5} + \log_a z$ Quotient rule

$\qquad = \log_a \left(\frac{x^{1/3}}{y^5} \cdot z \right)$ Product rule

$\qquad = \log_a \frac{z\sqrt[3]{x}}{y^5}$ Simplify

PRACTICE EXERCISE 6

Express $3 \log_a u - \log_a v - \frac{1}{2} \log_a w$ as a single logarithm.

Answer: $\log_a \frac{u^3}{v\sqrt{w}}$

EXAMPLE 7 APPROXIMATING LOGARITHMS

Suppose that $\log_a 2 = 0.6309$ and $\log_a 5 = 1.4650$. Approximate the numerical value of each of the following.

(a) $\log_a \sqrt{20}$

$\log_a \sqrt{20} = \log_a (2^2 \cdot 5)^{1/2} = \frac{1}{2} \log_a (2^2 \cdot 5)$ Power rule

$\qquad = \frac{1}{2}[\log_a 2^2 + \log_a 5]$ Product rule

$\qquad = \frac{1}{2}[2 \log_a 2 + \log_a 5]$ Power rule

$\qquad = \frac{1}{2}[2(0.6309) + 1.4650]$ Substitute

$\qquad = 1.3634$

(b) $\log_a \frac{5a}{\sqrt[3]{2}}$

$\log_a \frac{5a}{\sqrt[3]{2}} = \log_a 5a - \log_a \sqrt[3]{2}$

$\qquad = \log_a 5 + \log_a a - \frac{1}{3} \log_a 2$

$\qquad = 1.4650 + 1 - \frac{1}{3}(0.6309)$

$\qquad = 2.2547$

PRACTICE EXERCISE 7

Approximate the following, given that $\log_a 2 = 0.3010$ and $\log_a 3 = 0.4771$.

(a) $\log_a 6$

(b) $\log_a \frac{\sqrt{3}}{2a}$

Answers: (a) 0.7781 (b) -1.0625

🛇 BASE CONVERSION FORMULA

The final theorem of this section deals with the relationship between logarithms with different bases. Suppose we consider $\log_a x$ and $\log_b x$, and let

$$u = \log_b x$$
$$b^u = x.$$

Take the logarithm to the base a of both sides.

$$\log_a b^u = \log_a x \qquad \text{If } y = z \text{ then } \log_a y = \log_a z$$
$$u \log_a b = \log_a x \qquad \text{Power rule}$$
$$(\log_b x)(\log_a b) = \log_a x \qquad u = \log_b x$$
$$\log_b x = \frac{\log_a x}{\log_a b} \qquad \text{Not } \log_a \tfrac{x}{b} = \log_a x - \log_a b$$

A special case of this formula is developed by letting $x = a$.

$$\log_b x = \frac{\log_a x}{\log_a b}$$
$$\log_b a = \frac{\log_a a}{\log_a b} \qquad \text{Letting } x = a$$
$$\log_b a = \frac{1}{\log_a b} \qquad \text{Log}_a\, a = 1$$

We have just proved the following.

Base Conversion Formula

If $a > 0$, $b > 0$, $a \neq 1$, $b \neq 1$, and $x > 0$, then

$$\log_b x = \frac{\log_a x}{\log_a b} \quad \text{and} \quad \log_b a = \frac{1}{\log_a b}.$$

Notice that the base conversion formula states that logarithms to the base b are simply multiples of logarithms to the base a where the multiplier is the constant $1/\log_a b$.

EXAMPLE 8 Using the Base Conversion Formula

(a) Find $\log_{16} 8$ by changing to logarithms base 2.

$$\log_{16} 8 = \frac{\log_2 8}{\log_2 16} = \frac{\log_2 2^3}{\log_2 2^4} = \frac{3}{4}$$

Notice that by changing to exponential form, $\log_{16} 8 = 3/4$ becomes $16^{3/4} = 8$, and $(\sqrt[4]{16})^3$ is indeed 8.

(b) Find $\log_9 3$ by changing to logarithms base 3.

$$\log_9 3 = \frac{1}{\log_3 9} = \frac{1}{\log_3 3^2} = \frac{1}{2}$$

Notice that $9^{1/2} = \sqrt{9}$ does indeed equal 3.

PRACTICE EXERCISE 8

(a) Find $\log_{27} 9$ by changing to logarithms base 3.

(b) Find $\log_{16} 4$ by changing to logarithms base 4.

Answers: (a) 2/3   (b) 1/2

5.3 EXERCISES A

Solve each equation in Exercises 1–3.

1. $\log_2 (3x + 1) = \log_2 (5x - 7)$ **2.** $\log_5 x^2 = \log_5 (3x + 4)$ **3.** $\frac{1}{3} \log_2 x^2 = \log_8 2x$

In Exercises 4–12 evaluate each logarithm given that $\log_{10} 2 = 0.3010$, $\log_{10} 3 = 0.4771$, *and* $\log_{10} 5 = 0.6990$.

4. $\log_{10} 500$

5. $\log_{10} 27$

6. $\log_{10} \dfrac{2}{5}$

7. $\log_{10} \dfrac{1}{9}$

8. $\log_{10} \sqrt{6}$

9. $\log_{10} \sqrt[7]{3}$

10. $\log_5 3$

11. $\dfrac{\log_{10} 2^7}{\log_{10} 5}$

12. $\log_{10} 1.8$

In Exercises 13–18 use the properties of logarithms to write each logarithm as a sum, difference, or multiple of logarithms.

13. $\log_a xyz$

14. $\log_a \dfrac{xz^2}{y}$

15. $\log_a \dfrac{y\sqrt{x}}{z^3}$

16. $\log_a z(x+1)^3$

17. $\log_a x^2 \sqrt{\dfrac{y}{z}}$

18. $\log_a \sqrt{x\sqrt{y}}$

In Exercises 19–24 use the properties of logarithms to write each expression as a single logarithm.

19. $\log_a x + \dfrac{1}{2} \log_a y$

20. $3 \log_a x - 2 \log_a xy$

21. $\dfrac{1}{2} \log_a x - 3 \log_a yz + 6 \log_a z$

22. $x \log_a y - 3 \log_a z$

23. $-\log_a (x-1) + \log_a (x^2-1)$

24. $\dfrac{1}{2} \left[\log_a x - \log_a y - \dfrac{1}{2} \log_a z \right]$

In Exercises 25–30 tell whether the equation is true *or* false.

25. $\log_a uv = \log_a u + \log_a v$

26. $\dfrac{\log_a u}{\log_a v} = \log_a \dfrac{u}{v}$

27. $\log_a u - \log_a v = \log_a \dfrac{u}{v}$

28. $\log_a u^v = v \log_a u$

29. $\log_a \sqrt{u} = \sqrt{\log_a u}$

30. $\log_a \dfrac{1}{u} = \log_{1/a} u$

362 CHAPTER 5 EXPONENTIAL AND LOGARITHMIC FUNCTIONS

In Exercises 31–33 use the base conversion formulas to simplify each logarithm relative to the given base.

31. $\log_{32} 8$; base 2

32. $\log_{27} 3$; base 3

33. $\log_9 243$; base 3

34. If $P = \log_2 (m + n)$, calculate the value of P when $m = 8$ and $n = 8$, and show that $\log_2 (m + n) \neq \log_2 m + \log_2 n$.

35. If $R = \log_3 mn$, calculate the value of R when $m = 1$ and $n = 9$, and show that $\log_3 mn \neq (\log_3 m)(\log_3 n)$.

FOR REVIEW

36. Graph the function $f(x) = 2^{x^2} + 2$.

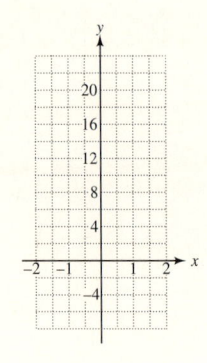

37. Simplify. $\log_\pi \pi^{3x+2}$

38. Evaluate 10^m if $m = \log_{10} 35$.

39. The half-life of an antibiotic in the bloodstream is about 8 hours. This means that when 64 milligrams of the drug are absorbed in your bloodstream, the number of milligrams D remaining after t hours is given by the function $D = (64)2^{-t/8}$. Graph this function for $t \geq 0$ and find the number of milligrams of the antibiotic remaining in the bloodstream after
(a) 0 hours,
(b) 8 hours,
(c) 16 hours, and
(d) 24 hours.

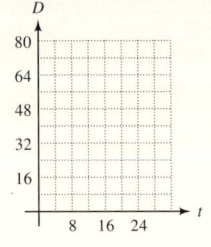

ANSWERS: 1. 4   2. 4, −1   3. 2   4. 2.6990   5. 1.4313   6. −0.3980   7. −0.9542   8. 0.3891   9. 0.0682
10. 0.6825   11. 3.0143   12. 0.2552   13. $\log_a x + \log_a y + \log_a z$ 14. $\log_a x + 2\log_a z - \log_a y$
15. $\log_a y + \frac{1}{2}\log_a x - 3\log_a z$ 16. $\log_a z + 3\log_a (x + 1)$ 17. $2\log_a x + \frac{1}{2}\log_a y - \frac{1}{2}\log_a z$ 18. $\frac{1}{2}\log_a x + \frac{1}{4}\log_a y$
19. $\log_a x\sqrt{y}$ 20. $\log_a \frac{x}{y^2}$ 21. $\log_a \frac{z^3\sqrt{x}}{y^3}$ 22. $\log_a \frac{y^x}{z^3}$ 23. $\log_a (x + 1)$ 24. $\log_a \sqrt{\frac{x}{y\sqrt{z}}}$ 25. true   26. false   27. true
28. true   29. false   30. true   31. $\frac{3}{5}$ 32. $\frac{1}{3}$ 33. $\frac{5}{2}$ 34. $\log_2 (8 + 8) = \log_2 16 = 4$; but $\log_2 8 + \log_2 8 = 3 + 3 = 6$
35. $\log_3 (1 \cdot 9) = \log_3 9 = 2$, but $(\log_3 1)(\log_3 9) = (0)(2) = 0$ 36.   37. $3x + 2$ 38. 35
39. (a) 64 mg
    (b) 32 mg
    (c) 16 mg
    (d) 8 mg

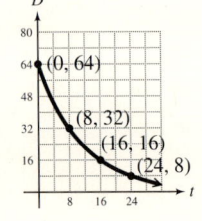

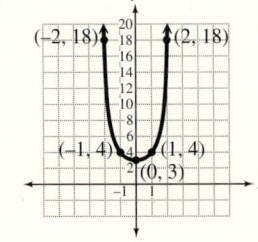

5.3 EXERCISES B

Solve each equation in Exercises 1–3.

1. $\log_5 (1 + x) = \log_5 (1 - x)$
2. $\log_3 (x^2 + 4) = \log_3 4x$
3. $\frac{1}{2} \log_3 2x = \log_9 5x$

In Exercises 4–12 evaluate each logarithm given that $\log_{10} 2 = 0.3010$, $\log_{10} 3 = 0.4771$, *and* $\log_{10} 5 = 0.6990$.

4. $\log_{10} 25$
5. $\log_{10} \frac{5}{2}$
6. $\log_{10} \frac{6}{5}$

7. $\log_{10} \sqrt{5}$
8. $\log_{10} 3^7$
9. $\log_{10} \sqrt{150}$

10. $\log_3 2$
11. $\frac{\log_{10} \sqrt[7]{5}}{\log_{10} \sqrt[3]{2}}$
12. $\log_{10} 7.5$

In Exercises 13–18 use the properties of logarithms to write each logarithm as a sum, difference, or multiple of logarithms.

13. $\log_a x^2 y^3 z^4$
14. $\log_a \left(\frac{xz}{y}\right)^2$
15. $\log_a \frac{(x + y)^2}{z}$

16. $\log_a ax^2$
17. $\log_a \frac{1}{ax^2}$
18. $\log_a \sqrt{\frac{y\sqrt{x}}{z^2}}$

In Exercises 19–24 use the properties of logarithms to write each expression as a single logarithm.

19. $\log_a x - \frac{1}{2} \log_a y$
20. $4 \log_a z + 5 \log_a yx$

21. $\frac{3}{2} \log_a z + \frac{1}{2} \log_a y$
22. $y \log_a z - z \log_a y$

23. $2[\log_a x - \log_a (y + 1) + \log_a z]$
24. $2 \log_a x - y \log_a z + \frac{1}{3} \log_a xy$

In Exercises 25–30, state whether the given equation is true or false.

25. $\log_a uv = (\log_a u)(\log_a v)$
26. $\log_a (u + v) = \log_a uv$

27. $\log_a (u - v) = \log_a \frac{u}{v}$
28. $\log_a au = 1 + \log_a u$

29. $\frac{1}{2} \log_a u = \sqrt{\log_a u}$
30. $\log_a \frac{1}{x} = -\log_a x$

In Exercises 31–33 use the base conversion formulas to simplify each logarithm relative to the given base.

31. $\log_{32} 16$; base 2
32. $\log_{27} 9$; base 3
33. $\log_{25} 5$; base 5

34. If $M = \log_3 \frac{m}{n}$, calculate the value of M when $m = 27$ and $n = 3$, and show that $\log_3 \frac{m}{n} \neq \frac{\log_3 m}{\log_3 n}$.

35. If $T = \log_2 (m - n)$, calculate the value of T when $m = 8$ and $n = 4$, and show that $\log_2 (m - n) \neq \log_2 m - \log_2 n$.

FOR REVIEW

36. Graph the function $f(x) = \log_3 |x|$.

37. Simplify. $10^{\log_{10} 5x^2}$

38. Evaluate $\log_5 m$ if $m = 5^3$.

39. The half-life of radioactive strontium-90, used for fuel in nuclear reactors, is approximately 28 years. This means that if 500 grams of strontium-90 are allowed to decay, the number of grams remaining after t years is given by the exponential function $A = (500)2^{-t/28}$. Graph this function for $t \geq 0$ and find the number of grams of strontium-90 remaining after **(a)** 0 years, **(b)** 28 years, **(c)** 56 years, **(d)** 140 years.

5.3 EXERCISES C

Express each equation in Exercises 1–2 in exponential form, without using logarithms.

1. $\log_a x - \log_a y - z = 0$

2. $\log_a x + \log_a y - z = 0$

Answer true *or* false *in Exercises 3–6. If the statement is false, change the left side of the equation to make it true.*

3. $\log_a \dfrac{3}{x+2} = \log_a 3 - \log_a x + \log_a 2$

4. $\log_a (xa)^x = x \log_a x + x$

5. $(x+1) \log_a (x-1) = \log_a \dfrac{x+1}{x-1}$

6. $3^{\log_3 x^2 + \log_3 y} = x^2 y$

7. If $\log_a u = 5$, show that $\log_a \dfrac{1}{u} = -5$.

8. If $\log_a u = 5$, show that $\log_{1/a} u = -5$.

5.4 COMMON AND NATURAL LOGARITHMS

STUDENT GUIDEPOSTS

1 Common Logarithms
2 Bases Other than 10
3 Antilogarithms
4 Natural Logarithms

❶ COMMON LOGARITHMS

Recall that every positive number, with the exception of the number 1, can be used as a base for logarithms. However, two logarithm bases occur more frequently in applications of science and mathematics than others, base 10 and base e (e is an irrational number, approximately 2.71828, which will be discussed later in this section). Historically, since our number system is based on 10, logarithms to the base 10 were commonly used for computation. As a result, base 10 logarithms are called **common logarithms.** For convenience, we omit the base number 10 in common logarithmic expressions and write, for example,

$$\log 427 \quad \text{instead of} \quad \log_{10} 427.$$

Thus, when a logarithm is written without the base, the base is understood to be 10.

Before common logarithms can be used to solve problems, we must be able to find the approximate value of the logarithm of a given number. Prior to the ready availability of calculators, tables similar to the table of common logarithms in the Appendix were used to approximate logarithms. The Appendix explains how to use the table for those who might not wish to use a calculator. The presentation given in this section and throughout the rest of the chapter will utilize calculators. Keep in mind that finding the value of log 427, for example, means looking for the exponent on 10 which results in 427. Thus, if $a =$ log 427, a is an exponent and in fact, $10^a = 427$. With $10^2 = 100$, $10^3 = 1000$, and $100 < 427 < 1000$, we would suspect that since

$$10^2 = 100 < 427 = 10^a < 1000 = 10^3,$$

a must be a real number satisfying $2 < a < 3$. This number is found in Example 1(a).

EXAMPLE 1 FINDING COMMON LOGARITHMS

Use a calculator to find the following common logarithms.

(a) log 427
The key labeled LOG will give the desired result.
ALG & RPN: 427 LOG → 2.6304279   The display shows log 427

(b) log 0.000427
ALG & RPN: 0.000427 LOG → −3.3695721

(c) log (−427)
ALG: 427 +/− LOG → Error
RPN: 427 CHS LOG → Error 0

In this case the display will show *error* since logarithms of negative numbers are not defined.

(d) log (0.000 000 004 27)
Since most calculator displays allow only eight digits, very small or very large numbers must be entered in scientific notation.
ALG: 4.27 EE 9 +/− LOG → −8.3696 00
RPN: 4.27 EEX 9 CHS LOG → −8.3696 00

The display shows the logarithm −8.3696, correct to four decimal places.

PRACTICE EXERCISE 1

Use a calculator to find the following common logarithms.

(a) log 9254

(b) log 0.00328

(c) log (−5.5)

(d) log 1,298,000,000,000

Answers: (a) 3.9663295
(b) −2.4841262 (c) not defined
(d) 12.1132747

Note There are many variations in the way calculators give logarithms. If your calculator gives more or fewer decimal places of accuracy than demonstrated in Example 1, do not be concerned. Also, as shown in Example 1, numbers greater than 1 have positive logarithms while numbers less than 1 (but greater than zero) have negative logarithms.

❷ BASES OTHER THAN 10

In Section 5.3, we introduced the base conversion formula

$$\log_b x = \frac{\log_a x}{\log_a b},$$

which can be used in conjunction with a calculator to find logarithms to any base b by letting $a = 10$.

EXAMPLE 2 Finding Logarithms to Other Bases

Use a calculator and the base conversion formula to find the following logarithms.

(a) $\log_2 59.4 = \dfrac{\log 59.4}{\log 2}$ Remember that $a = 10$ and $\log_{10} 59.4 = \log 59.4$

Notice that the right side of this equation is a quotient of logs and must be calculated by dividing, not by subtracting.

ALG: 59.4 $\boxed{\text{LOG}}$ $\boxed{\div}$ 2 $\boxed{\text{LOG}}$ $\boxed{=}$ $\rightarrow$ $\boxed{5.8923910}$

RPN: 59.4 $\boxed{\text{LOG}}$ $\boxed{\text{ENTER}}$ 2 $\boxed{\text{LOG}}$ $\boxed{\div}$ $\rightarrow$ $\boxed{5.8923910}$

The final digit in the display may vary slightly from calculator to calculator. However,

$$\log_2 59.4 = 5.8924,$$

correct to four decimal places.

(b) $\log_3 0.00845$

$$\log_3 0.00845 = \dfrac{\log 0.00845}{\log 3}$$

ALG: 0.00845 $\boxed{\text{LOG}}$ $\boxed{\div}$ 3 $\boxed{\text{LOG}}$ $\boxed{=}$ $\rightarrow$ $\boxed{-4.3451078}$

RPN: 0.00845 $\boxed{\text{LOG}}$ $\boxed{\text{ENTER}}$ 3 $\boxed{\text{LOG}}$ $\boxed{\div}$ $\rightarrow$ $\boxed{-4.3451078}$

Thus, correct to four decimal places,

$$\log_3 0.00845 = -4.3451.$$

Practice Exercise 2

Use a calculator and the base conversion formula to find the following logarithms.

(a) $\log_3 26.5$

(b) $\log_{15} 0.00096$

Answers: (a) 2.9830
(b) −2.5659

Always remember that a logarithm is an exponent. In Example 2(a), we found that $\log_2 59.4 \approx 5.8924$. In exponential form, this means that $2^{5.8924} \approx 59.4$. We can use the $\boxed{y^x}$ key on a calculator to verify this result.

ALG: 2 $\boxed{y^x}$ 5.8924 $\boxed{=}$ $\rightarrow$ $\boxed{59.400369}$

RPN: 2 $\boxed{\text{ENTER}}$ 5.8924 $\boxed{y^x}$ $\rightarrow$ $\boxed{59.4003695}$

Similarly, in Example 1(a), we found that $\log 427 \approx 2.6304$. Thus $10^{2.6304} \approx 427$.

ALG: 10 $\boxed{y^x}$ 2.6304 $\boxed{=}$ $\rightarrow$ $\boxed{426.97259}$

RPN: 10 $\boxed{\text{ENTER}}$ 2.6304 $\boxed{y^x}$ $\rightarrow$ $\boxed{426.9725940}$

Notice that, due to round-off errors, we do not obtain 427 in the display, rather an approximation of it.

3 ANTILOGARITHMS

The results demonstrated above can also be found using a different key or set of keys. Some calculators have a $\boxed{10^x}$ key while others use two keys, $\boxed{\text{INV}}$ followed by $\boxed{\text{LOG}}$. Both sequences appear below.

ALG: 2.6304 $\boxed{\text{INV}}$ $\boxed{\text{LOG}}$ $\rightarrow$ $\boxed{426.97259}$

RPN: 2.6304 $\boxed{10^x}$ $\rightarrow$ $\boxed{426.9725940}$

We can use one of these sequences to find a number from its logarithm (the reverse of what we did previously). A number that has a given logarithm is called its **antilogarithm (antilog).** In the equation

$$\log n = x,$$

x is a logarithm of n and n is an antilogarithm of x. In exponential form, $n = 10^x$.

EXAMPLE 3 FINDING ANTILOGARITHMS

Use a calculator to find the following antilogs correct to three significant digits.

(a) $\log x = 0.5623$
We need to find x where $x = 10^{0.5623}$.

ALG: 0.5623 $\boxed{\text{INV}}$ $\boxed{\text{LOG}}$ → $\boxed{3.6500600}$

RPN: 0.5623 $\boxed{10^x}$ → $\boxed{3.6500600}$

Thus, $x = 3.65$, correct to three significant digits.

(b) $\log x = -3.4215$

ALG: 3.4215 $\boxed{+/-}$ $\boxed{\text{INV}}$ $\boxed{\text{LOG}}$ → $\boxed{0.0003789}$

RPN: 3.4215 $\boxed{\text{CHS}}$ $\boxed{10^x}$ → $\boxed{0.0003789}$

Thus, $x = 0.000379$, correct to three significant digits.

PRACTICE EXERCISE 3

Use a calculator to find the following antilogs correct to three significant digits.

(a) $\log x = 2.3597$

(b) $\log x = -0.8665$

Answers: (a) 228   (b) 0.136

---

❹ NATURAL LOGARITHMS

The number e as a base for logarithms arises naturally in many applications, such as population growth, continuously compounding interest, and radioactive decay. As a result, logarithms to the base e are called **natural logarithms.** The full significance of this number cannot be realized without more advanced study. One way to define e is to find the value of

$$\left(1 + \frac{1}{n}\right)^n$$

as n increases through larger and larger values. To evaluate this expression we use the following sequence on a calculator.

ALG: n $\boxed{1/x}$ $\boxed{+}$ 1 $\boxed{=}$ $\boxed{y^x}$ n $\boxed{=}$ → $\boxed{\text{Display}}$

RPN: n $\boxed{1/x}$ $\boxed{\text{ENTER}}$ 1 $\boxed{+}$ $\boxed{\text{ENTER}}$ n $\boxed{y^x}$ → $\boxed{\text{Display}}$

The table below shows this computation for increasing values of n.

n	$\left(1 + \frac{1}{n}\right)^n$
1	2.00000
10	2.59374
100	2.70481
1000	2.71692
10,000	2.71815
100,000	2.71827
1,000,000	2.71828
↓	↓
∞	e

The number e was named for the eighteenth-century mathematician Leonhard Euler. So important is this number that we often call $f(x) = e^x$ ***the*** exponential function, elevating it above all other exponential functions with other bases a. Since logarithms to the base e, natural logarithms, occur so frequently in mathematics, a special notation, ln x, instead of $\log_e x$, is used. Most calculators have a $\boxed{\ln x}$ or $\boxed{\text{LN}}$ key along with the common logarithm key $\boxed{\text{LOG}}$.

EXAMPLE 4 FINDING NATURAL LOGARITHMS

Use a calculator to find the following natural logarithms.

(a) ln 53.6

$$\text{ALG \& RPN:} \quad 53.6 \; \boxed{\ln x} \rightarrow \boxed{3.9815491}$$

Thus, ln 53.6 = 3.9815, correct to four decimal places.

(b) ln 0.000 000 062 5

$$\text{ALG:} \quad 6.25 \; \boxed{\text{EE}} \; 8 \; \boxed{+/-} \; \boxed{\ln x} \rightarrow \boxed{-1.6588 \quad 01}$$

$$\text{RPN:} \quad 6.25 \; \boxed{\text{EEX}} \; 8 \; \boxed{\text{CHS}} \; \boxed{\text{LN}} \rightarrow \boxed{-16.5880993}$$

Thus, ln 0.000 000 062 5 = −16.588, correct to three decimal places.

PRACTICE EXERCISE 4

Use a calculator to find the following natural logarithms.

(a) ln 2395

(b) ln 0.000000554

Answers: (a) 7.7811 (b) −14.4061

Finding antilogarithms involving natural logarithms uses either the $\boxed{e^x}$ key or the $\boxed{\text{INV}}$ and $\boxed{\ln x}$ keys in a manner similar to that for common logarithms.

EXAMPLE 5 FINDING NATURAL ANTILOGARITHMS

Use a calculator to find the following natural antilogs correct to three significant digits.

(a) ln x = 5.9002

We need to find x where ln x = 5.9002.

$$\text{ALG:} \quad 5.9002 \; \boxed{\text{INV}} \; \boxed{\ln x} \rightarrow \boxed{365.11048}$$

$$\text{RPN:} \quad 5.9002 \; \boxed{e^x} \rightarrow \boxed{365.1104827}$$

Thus, x = 365, correct to three significant digits.

(b) ln x = −2.1258

$$\text{ALG:} \quad 2.1258 \; \boxed{+/-} \; \boxed{\text{INV}} \; \boxed{\ln x} \rightarrow \boxed{0.1193375}$$

$$\text{RPN:} \quad 2.1258 \; \boxed{\text{CHS}} \; \boxed{e^x} \rightarrow \boxed{0.1193375}$$

Thus, x = 0.119, correct to three decimal places.

PRACTICE EXERCISE 5

Use a calculator to find the following natural antilogs correct to three significant digits.

(a) ln x = 3.8441

(b) ln x = −0.5459

Answers: (a) 46.7 (b) 0.579

The graphs of the natural logarithm and exponential functions $y = \ln x$ and $y = e^x$ are given in Figure 5.8 with representative values for each in the accompanying tables. Remember, to find e^x when $x = 3$, use your calculator, enter 3, and press either the $\boxed{e^x}$ key or the $\boxed{\text{INV}}$ $\boxed{\ln x}$ keys.

5.4 COMMON AND NATURAL LOGARITHMS

x	$\ln x$
0.5	-0.7
1	0
2	0.7
3	1.1
4	1.4
5	1.6

x	e^x
-3	0.05
-2	0.1
-1	0.4
0	1
1	2.7
2	7.4

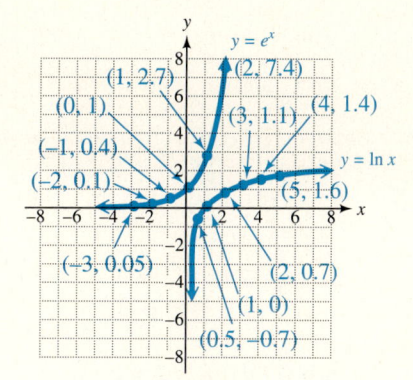

Figure 5.8

Variations of the natural logarithm and exponential functions appear in numerous applications. One variation follows in the final example of this section.

EXAMPLE 6 Graphing an Exponential Function

Graph $f(x) = e^{-x/2}$. Use a calculator to make the following table of values. The graph is given in Figure 5.9.

x	$e^{-x/2}$
0	1.00
-1	1.65
1	0.61
-2	2.72
2	0.37
-3	4.48
3	0.22
-4	7.39
4	0.14

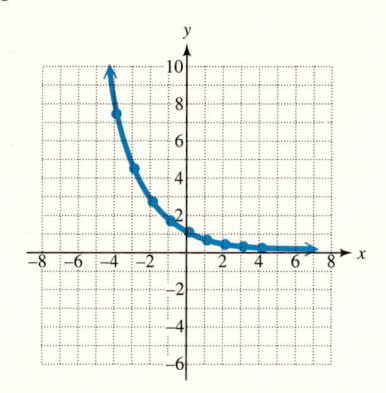

Figure 5.9

PRACTICE EXERCISE 6

Graph $g(x) = -2 \ln x$. Complete the table.

x	$-2 \ln x$
0.1	
0.5	
0.8	
1	
2	
3	
4	
5	
6	

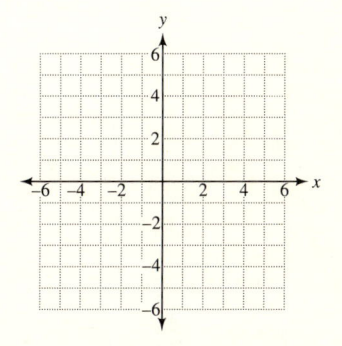

Answer: The graph can be obtained by reflecting the graph in Figure 5.9 in the line $y = x$. (Why?)

5.4 EXERCISES A

In Exercises 1–6 find each common logarithm, correct to four decimal places.

1. log 625

2. log 0.0059

3. log 2.00046

4. log 0.0000669

5. log (5.12×10^{11})

6. log 0.00000000741

Use a calculator and the base conversion formula to find each logarithm, correct to four decimal places, in Exercises 7–9.

7. $\log_4 15.6$

8. $\log_{3.5} 468$

9. $\log_{\sqrt{2}} 348.2$

In Exercises 10–15 find each common antilogarithm, correct to three significant digits.

10. log $x = 1.3565$

11. log $x = -3.2117$

12. log $x = 14.6531$

13. log $x = -0.0035$

14. log $x = -35.2264$

15. log $x = 51.3225$

In Exercises 16–21 find each natural logarithm, correct to four decimal places.

16. ln 846

17. ln 2.003

18. ln 0.000525

19. ln 0.0000387

20. ln (2.66×10^{15})

21. ln π

In Exercises 22–27 find each natural antilogarithm, correct to three significant digits.

22. ln $x = 2.4308$

23. ln $x = -4.1553$

24. ln $x = 21.7439$

25. ln $x = -0.0070$

26. ln $x = -41.4239$

27. ln $x = 17.3591$

In Exercises 28–30 sketch the graph of each function.

28. $y = e^{2x}$

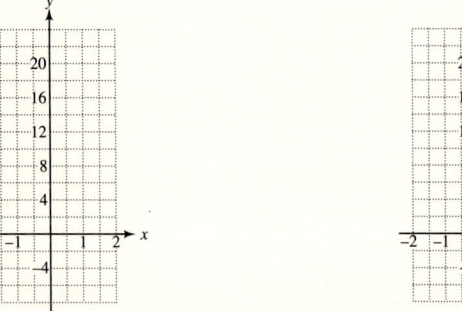

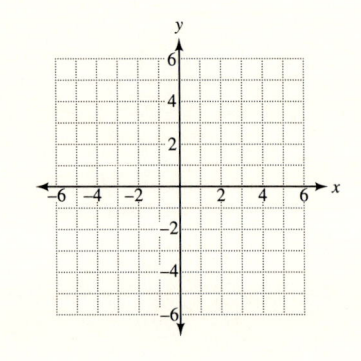

29. $y = e^{-x}$

30. $y = \ln x^2$

31. BUSINESS The demand equation for x units of a product relative to the price per unit, P, is given by
$P = 100 - 0.2e^{0.005x}$.
(a) Find the price if the demand is 100 units.

(b) Find the price if the demand is 1000 units.

32. PHYSICS The atmospheric pressure P, measured in pounds per square inch, at an altitude x, measured in miles, above sea level can be approximated by the formula
$$P = 14.7e^{-0.2x}.$$

(a) What is the atmospheric pressure at sea level?

(b) What is the atmospheric pressure on the outside of a jet airplane cruising at an altitude of 6.25 miles?

33. MEDICINE When a drug is introduced into the human circulatory system, the concentration of the drug decreases as it is eliminated by the liver and kidneys. If an initial dose in the amount of 15 mg is taken orally, the amount A remaining in the system after x hours can be approximated by the formula
$$A = 15e^{-0.3x}.$$

(a) How much of the drug remains in the system after 3 hours?

(b) How much of the drug remains in the system after 1 day?

FOR REVIEW

Use the properties of logarithms in Exercises 34–35 to write each logarithm as a sum, difference, or multiple of logarithms.

34. $\ln \dfrac{e\sqrt{z}}{\sqrt[3]{x}}$

35. $\ln \sqrt{e\sqrt{y}}$

Use the properties of logarithms in Exercises 36–37 to write each expression as a single logarithm.

36. $3 \ln xy + \dfrac{1}{2} \ln z$

37. $3 \ln x - \dfrac{1}{2} \ln y - 5 \ln z$

Solve each equation in Exercises 38–39.

38. $7^x = 7^{2x+3}$

39. $\log_3 x = \log_3 (x^2 - 2)$

ANSWERS: 1. 2.7959 2. −2.2291 3. 0.3011 4. −4.1746 5. 11.7093 6. −8.1302 7. 1.9817 8. 4.9079 9. 16.8875 10. 22.7 11. 0.000 614 12. 4.50×10^{14} 13. 0.992 14. 5.94×10^{-36} 15. 2.10×10^{51} 16. 6.7405 17. 0.6946 18. −7.5521 19. −10.1597 20. 35.5171 21. 1.1447 22. 11.4 23. 0.0157 24. 2.77×10^9 25. 0.993 26. 1.02×10^{-18} 27. 34,600,000
28. 29. 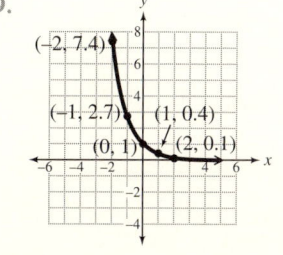 30. 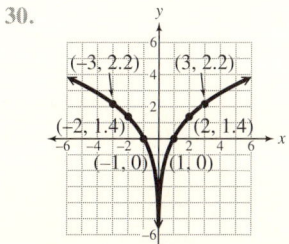 31. (a) $99.67 (b) $70.32 32. (a) 14.7 lb/in² (b) 4.2 lb/in² 33. (a) 6.10 mg (b) 0.01 mg 34. $1 + \dfrac{1}{2} \ln z - \dfrac{1}{3} \ln x$ 35. $\dfrac{1}{2} + \dfrac{1}{4} \ln y$ 36. $\ln x^3 y^3 \sqrt{z}$ 37. $\ln \dfrac{x^3}{z^5 \sqrt{y}}$ 38. −3 39. 2

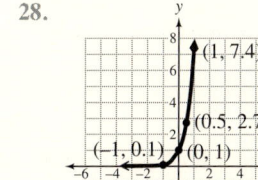

5.4 EXERCISES B

In Exercises 1–6 find each common logarithm, correct to four decimal places.

1. log 725
2. log 0.00311
3. log 694,000
4. log 71.00045
5. log (3.39 × 10⁻⁹)
6. log 229,000,000

Use a calculator and the base conversion formula to find each logarithm, correct to four decimal places, in Exercises 7–9.

7. $\log_{11} 0.566$
8. $\log_{4.6} 2.63$
9. $\log_{\pi} e$

In Exercises 10–15 find each common antilogarithm, correct to three significant digits.

10. log x = 4.9907
11. log x = −2.0075
12. log x = 18.8309
13. log x = −0.0007
14. log x = 42.5761
15. log x = −22.1155

In Exercises 16–21 find each natural logarithm, correct to four decimal places.

16. ln 293
17. ln 1.001
18. ln 605.34
19. ln 215,000,000
20. ln (5.07 × 10⁻¹²)
21. ln √2

In Exercises 22–27 find each natural antilogarithm, correct to three significant digits.

22. ln x = 7.4992
23. ln x = −3.0055
24. ln x = −11.6545
25. ln x = 0.0300
26. ln x = 38.8572
27. ln x = −61.3883

In Exercises 28–30 sketch the graph of each function.

28. $y = e^{x/2}$
29. $y = \ln (x + 1)$
30. $y = \ln |x|$

31. **CHEMISTRY** In chemistry, the **pH (hydrogen potential)** of a substance is defined by

$$pH = -\log [H^+],$$

where [H⁺] is the concentration of hydrogen ions in the solution, measured in moles per liter. The scale of the pH of a solution varies from 0 to 14, with the pH of distilled water equal to 7. If the pH of a substance is less than 7, the substance is called an *acid*, and if the pH of a substance is greater than 7, it is called a *base*.
   (a) Find the pH of a certain brand of coffee whose [H⁺] is 3.89 × 10⁻⁷ moles per liter.
   (b) Find the pH of a sample of seawater whose [H⁺] is 3.15 × 10⁻⁹ moles per liter.
   (c) Find the hydrogen ion concentration of distilled water.
   (d) Find the hydrogen ion concentration of rain water which has a pH of 5.6.

32. **ASTRONOMY** Stars are categorized by brightness on a scale measured in magnitudes, with the faintest star still visible to the naked eye assigned a magnitude of 6. Telescopes are designed to view stars beyond a magnitude of 6. Every telescope has a limiting magnitude denoted by L which depends on the diameter in inches, d, of its lens. The formula which relates these two quantities is

$$L = 8.8 + 2.2 \ln d.$$

   (a) What is the limiting magnitude of a telescope with a 2-inch lens?
   (b) What is the limiting magnitude of a telescope with a 30-inch lens?
   (c) What is the diameter of the lens in a telescope with a limiting magnitude of 10?
   (d) What is the approximate diameter of the lens in a human eye?

33. PSYCHOLOGY In order to determine retention of concepts learned, a group of students was given an exam and retested each month thereafter using an equivalent exam. The average score D was found to satisfy the formula

$$D = 80 - 12 \ln (x + 1),$$

where x is the time in months.
(a) What was the initial average score on the exam?
(b) What was the average score at the end of 1 year?
(c) How many months did it take for the average score to fall below 60?
(d) How many months did it take for the average score to fall below 40?

FOR REVIEW

Use the properties of logarithms in Exercises 34–35 to write each logarithm as a sum, difference, or multiple of logarithms.

34. $\ln \sqrt{\dfrac{x\sqrt{z}}{y^4}}$

35. $\ln \dfrac{ex}{y^2}$

Use the properties of logarithms in Exercises 36–37 to write each expression as a single logarithm.

36. $-\dfrac{1}{2} \ln (x + y) + \dfrac{1}{2} \ln (x^2 - y^2)$

37. $e^{\ln x} - \ln e^x$

Solve each equation in Exercises 38–39.

38. $2^x = 2^{x-1}$

39. $\log_2 x = \log_2 (3x + 8)$

5.4 EXERCISES C

1. Express $2 \ln x + \ln y - 5 = 0$ in exponential form without using logarithms.
[Answer: $x^2 y = e^5$]

2. Solve for x. $\ln (3x - 2) = \ln (2x) + 1$
[*Hint:* Write 1 as $\ln e$.]

Graph the functions in Exercises 3–5.

3. $f(x) = e^{\ln x} + \ln e^x$.
[*Hint:* Simplify first.]

4. $g(x) = \dfrac{e^x + e^{-x}}{2}$

5. $h(x) = \dfrac{e^x - e^{-x}}{2}$

6. Without using a calculator, show that $\dfrac{\log_7 3}{\log_7 9} = \dfrac{1}{2}$.

7. Evaluate $\dfrac{\log_3 5}{\log_2 7}$, correct to four decimal places.
[Answer: 0.5218]

8. Show that $\ln 10$ is equal to $\dfrac{1}{\log e}$.

9. Show that $\log e$ is equal to $\dfrac{1}{\ln 10}$.

5.5 EXPONENTIAL AND LOGARITHMIC EQUATIONS

STUDENT GUIDEPOSTS
1 Solving Exponential Equations
2 Solving Logarithmic Equations

1 SOLVING EXPONENTIAL EQUATIONS

In previous sections we solved several simple equations involving exponentials or logarithms either by inspection or by using a direct application of one of the two theorems

$$a^x = a^y \quad \text{if and only if } x = y$$

$$\text{or} \quad \log_a x = \log_a y \text{ if and only if } x = y.$$

These methods are now expanded to include more complex equations. We begin by considering **exponential equations** in which the variable appears in one or more exponents. In some cases it may be possible to express both sides of an exponential equation as powers using the same base and equate the exponents.

EXAMPLE 1   EXPONENTIAL EQUATION (SAME BASE)

Solve. $3^{5x} = 9^{x-6}$

$$3^{5x} = 9^{x-6}$$
$$3^{5x} = (3^2)^{x-6} \quad 9 = 3^2$$
$$3^{5x} = 3^{2(x-6)} \quad (a^m)^n = a^{m \cdot n}$$
$$5x = 2(x-6) \quad \text{Since both sides are powers of}$$
$$5x = 2x - 12 \quad \text{3, we equate exponents}$$
$$3x = -12$$
$$x = -4$$

Check: $3^{5(-4)} \stackrel{?}{=} 9^{-4-6}$
$$3^{-20} = 9^{-10} = 3^{-20} \quad \text{The solution is } -4$$

PRACTICE EXERCISE 1

Solve. $10^{2x+1} = 100^{x/2}$

Answer: -1

In some equations it may be difficult to express both sides as exponentials with the same base. When this occurs, we take the logarithm of both sides, using a suitable base, and apply the power rule to remove variables from exponents.

EXAMPLE 2   EXPONENTIAL EQUATION (DIFFERENT BASE)

Solve. $3^{2x} = 7^{x+5}$

$$3^{2x} = 7^{x+5}$$
$$\log 3^{2x} = \log 7^{x+5} \quad \text{Take log of both sides}$$
$$2x \log 3 = (x + 5) \log 7 \quad \text{Power rule}$$
$$2x \log 3 = x \log 7 + 5 \log 7 \quad \text{Distributive law}$$
$$2x \log 3 - x \log 7 = 5 \log 7 \quad \text{Write both } x\text{-terms on left}$$
$$(2 \log 3 - \log 7)x = 5 \log 7 \quad \text{Factor out } x$$
$$x = \frac{5 \log 7}{2 \log 3 - \log 7} \quad \text{Divide by } (2 \log 3 - \log 7)$$

Thus, $x \approx 38.7$, correct to three significant digits. Check.

PRACTICE EXERCISE 2

Solve and give the answer correct to three significant digits.

$$5^{2x} = 8$$

[*Hint:* Take the log base 5 of both sides.]

Answer: 0.646

The calculator steps used to find x in Example 2 are:

ALG: 2 [×] 3 [LOG] [=] [−] 7 [LOG] [=] [STO] 5 [×] 7 [LOG] [=] [÷] [RCL] [=] → 38.714652

RPN: 3 [LOG] [ENTER] 2 [×] 7 [LOG] [−] [STO] 7 [LOG] [ENTER] 5 [×] [RCL] [÷] → 38.714652

To Solve an Exponential Equation

1. Try to express each side as a power using the same base, and equate the resulting exponents.
2. If this fails, take the logarithm of each side and use the power rule to eliminate the variable exponents.
3. Solve the resulting equation and check in the original.

Many applied problems result in exponential equations. Recall the "ATTACK" method of problem solving given in Section 2.2.

EXAMPLE 3 A Business Problem

The demand for x units of a product is related to the price per unit of the product, P, by the demand equation, $P = 100 - 0.2e^{0.005x}$. Find the demand if the price is $80.00 per unit.

Analysis: We need to find the value of x for a given value of P when P is substituted into the demand equation.

Tabulation: $P = \$80.00$

Translation: Solve the equation:

$$80.00 = 100 - 0.2e^{0.005x}$$
$$-20 = -0.2e^{0.005x} \quad \text{Subtract 100}$$
$$100 = e^{0.005x} \quad \text{Divide by } -0.2$$
$$\ln 100 = \ln e^{0.005x} \quad \text{Take natural log of both sides}$$
$$\ln 100 = 0.005x \quad \ln e^u = u$$
$$\frac{\ln 100}{0.005} = x \quad \text{Divide by 0.005}$$
$$921.0340372 = x \quad \text{Use a calculator}$$

Approximation: 921 is a reasonable number for the demand of a product. If we had obtained a negative number, for example, we would know that something is wrong.

Check: Substitute 921 into the demand equation.

$$P = 100 - 0.2e^{(0.005)(921)} = 80.00340343$$

Thus, the demand is about 921 units when the price is $80.00.

Key: Remember that the key to problem solving is to try something.

PRACTICE EXERCISE 3

The atmospheric pressure P, in lb/in^2, at an altitude x, in miles, above sea level can be approximated by

$$P = 14.7e^{-0.2x}.$$

At what altitude will the pressure be one-half the pressure at sea level? [*Hint:* The pressure at sea level is found when $x = 0$.]

Answer: 3.47 mi

Note We reviewed all of the problem-solving steps in Example 3. As we stated before, many of these steps are usually done mentally, which will be assumed for most of the remaining examples in this chapter.

376 CHAPTER 5 EXPONENTIAL AND LOGARITHMIC FUNCTIONS

❷ SOLVING LOGARITHMIC EQUATIONS

Equations that contain logarithms of expressions involving a variable are called **logarithmic equations.** Solving such equations requires a thorough knowledge of the basic properties of logarithms.

CAUTION

When solving a logarithmic equation, remember to check possible solutions in the original equation. Since logarithmic functions have domain all positive real numbers, if we obtain a "solution" that requires taking the logarithm of a negative number, that "solution" must be discarded.

EXAMPLE 4   SOLVING A LOGARITHMIC EQUATION

Solve. $\log_2 (x + 1) - \log_2 (x - 6) = 3$

$\log_2 (x + 1) - \log_2 (x - 6) = 3$

$\log_2 \left(\dfrac{x + 1}{x - 6} \right) = 3$ The difference of logs is the log of a quotient

$\left(\dfrac{x + 1}{x - 6} \right) = 2^3$ Convert to exponential form

$\left(\dfrac{x + 1}{x - 6} \right) = 8$

$(x + 1) = 8(x - 6)$ Multiply by the LCD, $(x - 6)$

$x + 1 = 8x - 48$

$-7x = -49$

$x = 7$

Check: $\log_2 (7 + 1) - \log_2 (7 - 6) \stackrel{?}{=} 3$

$\log_2 8 - \log_2 1 \stackrel{?}{=} 3$

$3 - 0 = 3$

The solution is 7.

PRACTICE EXERCISE 4

Solve. $\log (x + 10) - \log (x + 1) = 1$

Answer: 0

EXAMPLE 5   SOLVING A LOGARITHMIC EQUATION

Solve. $\ln (3x - 1) + \ln (x + 1) = \ln (4x + 7)$

$\ln (3x - 1) + \ln (x + 1) = \ln (4x + 7)$

$\ln (3x - 1)(x + 1) = \ln (4x + 7)$ $\ln u + \ln v = \ln uv$

$(3x - 1)(x + 1) = 4x + 7$ $\ln u = \ln v$ means $u = v$

$3x^2 + 2x - 1 = 4x + 7$

$3x^2 - 2x - 8 = 0$

$(3x + 4)(x - 2) = 0$

$3x + 4 = 0$ or $x - 2 = 0$

$x = -\dfrac{4}{3}$ $x = 2$

The only solution is 2 since $-\dfrac{4}{3}$ does not check. (Why?)

PRACTICE EXERCISE 5

Solve. $\ln x + \ln (x + 1) = \ln 2$

Answer: 1 (-2 does not check)

5.5 EXPONENTIAL AND LOGARITHMIC EQUATIONS 377

Notice in Example 4 that we were able to write the equation with a single logarithm on one side, then convert to exponential form. In Example 5, we were able to write the equation with a logarithm to the same base on both sides, then use the basic property of logarithms: $\log_a u = \log_a v$ implies that $u = v$. When an equation involves logarithms using several bases, we begin by converting all of them to the same base. This is illustrated in the next example.

EXAMPLE 6 A Logarithmic Equation

Solve. $\log_2 x + \log_3 x = 1$

$\log_2 x + \log_3 x = 1$ *These expressions cannot be combined since the bases are different*

$\dfrac{\log x}{\log 2} + \dfrac{\log x}{\log 3} = 1$ *Convert to common logarithms*

$\log x \left[\dfrac{1}{\log 2} + \dfrac{1}{\log 3} \right] = 1$ *Factor out log x*

$\log x = \dfrac{1}{\dfrac{1}{\log 2} + \dfrac{1}{\log 3}}$ *Solve for log x*

$\log x = 0.1845757$ *Evaluate using a calculator*

$x \approx 1.53$ *Find the antilog of 0.1845757, correct to three significant digits*

Check this result by substituting and then converting to common logarithms.

Practice Exercise 6

Solve. $\log_5 x - \log_2 x = 3$

Answer: 0.0259

To Solve a Logarithmic Equation

1. Obtain a single logarithmic expression using the same base on one side of the equation, or write each side as a logarithm using the same base.
2. Convert the result to an exponential equation, or use the fact that $\log_a u = \log_a v$ implies $u = v$, and solve.
3. Check all possible solutions in the original equation. Remember, possible solutions that require taking the logarithm of a negative number must be discarded.

5.5 EXERCISES A

Solve each equation in Exercises 1–24. When appropriate, give answer correct to three significant digits.

1. $4^x = 8$
2. $2^{5x} = 8^{x+5}$
3. $2^{x^2} = 16^x$
4. $2^{3x} = 7$
5. $5^{x+1} = 3^{2x}$
6. $(1.08)^x = 100$
7. $8^{\log_2 x} = 125$
8. $2^{x^2} 3^{x^2} = 6^{5x-6}$
9. $7^{2x} 3^x = 10$

10. $\log_2(x+2) = 3$

11. $\log_3 x + \log_3 9 = 5$

12. $\ln(x+1) + \ln(x-1) = \ln(4x+4)$

13. $\log_5(2x-1) - \log_5(x-5) = 1$

14. $\log_5 x + \log_7 x = 2$

15. $\log_4 4x = \log_2(x+1)$

16. $(\log_5 x)^2 = 2\log_5 x$

17. $\ln x^3 = (\ln x)^3$

18. $x^{\ln x} = e^2 x$

19. $\log(\log x) = 1$

20. $\log \sqrt{x} = \sqrt{\log x}$

21. $|\log x| = 2$

22. **BUSINESS** The demand equation for x units of a product relative to the price per unit, P, is given by $P = 100 - 0.2e^{0.005x}$. Find the demand if the price is $60.00.

23. **PHYSICS** The atmospheric pressure P, in lb/in^2, at an altitude x, in miles, above sea level can be approximated by $P = 14.7e^{-0.2x}$. At what altitude will the atmospheric pressure be one-fourth the pressure at sea level?

24. **MEDICINE** When a drug is introduced into the human circulatory system, the concentration of the drug decreases as it is eliminated by the liver and kidneys. If an initial dose in the amount of 15 mg is injected, the amount A remaining in the system after x hours can be approximated by $A = 15e^{-0.3x}$. For the drug to be effective, at least 5 mg must remain in the system. How long will the drug be effective?

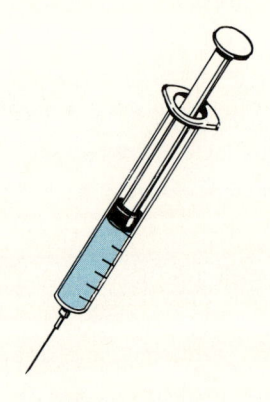

PHYSICS Newton's law of cooling relates the time t, in minutes, that it takes for an object heated to a temperature T_0 to cool to a temperature T when placed in an area of constant temperature T_c. The formula is $T = T_c + (T_0 - T_c)e^{-kt}$, where the constant k depends on the nature of the object considered. Use this information in Exercises 25–26.

25. An iron rod, with constant $k = 0.025$, is heated to 250°F and placed in a room with a constant temperature of 70°F. What will be the temperature of the rod after 10 minutes?

26. An object in a room at a constant temperature of 80°F cools from 400°F to 200°F in 15 minutes. Find the value of the constant k, and use it to determine the temperature of the object after 30 minutes.

27. OCEANOGRAPHY When a beam of light passes through water, the intensity of the light I is diminished by the number of feet x from the surface of the water. The relationship between these two variables is given by $I = I_0 e^{-kx}$, where I_0 is the intensity of the light at the surface of the water, and k is the coefficient of extinction depending on the particular water under consideration. If the coefficient of extinction of Lake Louise is 0.05, at what depth (to the nearest foot) will the intensity of a light be reduced to 50% of that at the surface?

FOR REVIEW

In Exercises 28–29 sketch the graph of each function.

28. $y = \ln(x^2 + 1)$

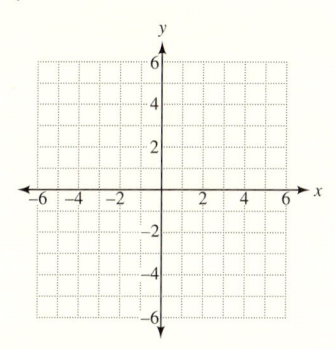

29. $y = e^{-0.5x}$

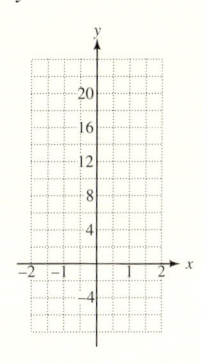

30. CHEMISTRY Use the pH equation, $\text{pH} = -\log[H^+]$, to find the pH of a soft drink with a hydrogen ion concentration of 3.25×10^{-7}.

ANSWERS: 1. $\frac{3}{2}$ 2. $\frac{15}{2}$ 3. 0, 4 4. 0.936 5. 2.74 6. 59.8 7. 5 8. 2, 3 9. 0.461 10. 6 11. 27 12. 5 13. 8 14. 5.82 15. 1 16. 1, 25 17. 1, 5.65, 0.177 18. 7.39, 0.368 19. 10^{10} 20. 1, 10,000 21. 100, 0.01 22. 1060 units 23. 6.93 mi 24. 3.66 hr 25. 210°F 26. $k = 0.0654$; 125°F 27. 14 ft 28. 29. 30. 6.5

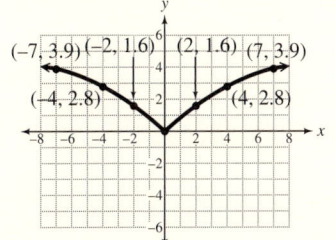

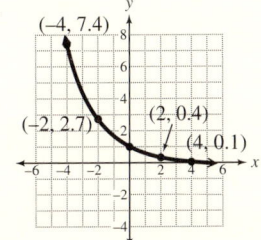

5.5 EXERCISES B

Solve each equation in Exercises 1–24. When appropriate, give answer correct to three significant digits.

1. $9^{2x} = 27$
2. $27^{x-4} = 9^{x+2}$
3. $125^{2x-1} = 25^{x+8}$
4. $3^{x+2} = 4$
5. $7^{x+2} = 4^{3x}$
6. $(1.12)^{x/2} = 1000$
7. $9^{\log_3 x} = 4$
8. $3^{x^2} 5^{x^2} = 15^{4x+5}$
9. $2^{4x} 3^x = 100$
10. $\log_5 (2x + 1) = 2$
11. $\log_2 (x + 3) - \log_2 (x - 6) = \log_2 (x - 5)$
12. $\log_5 (x + 2) + \log_5 (x - 2) = \log_5 (2x - 1)$
13. $\log_3 (2x + 5) - \log_3 (x - 1) = 2$
14. $\log_3 x + \log_9 x = 5$
15. $\log 3^x + \log 4^{2x} = 6$
16. $(\ln x)^2 = 5 \ln x$
17. $(\log x)^2 = \log x^2$
18. $2^{\log x} = 4x$
19. $\ln (\ln x) = 2$
20. $\ln \sqrt[3]{x} = \sqrt[3]{\ln x}$
21. $|\ln x| = 3$

22. **BUSINESS** The demand equation for x units of a product relative to the price per unit, P, is given by $P = 100 - 0.2e^{0.005x}$. Find the demand if the price is $99.00.

23. **PHYSICS** The atmospheric pressure P, in lb/in^2, at an altitude x, in miles, above sea level can be approximated by $P = 14.7e^{-0.2x}$. At what altitude will the atmospheric pressure be one-tenth the pressure at sea level?

24. **MEDICINE** When a drug is introduced into the human circulatory system, the concentration of the drug decreases as it is eliminated by the liver and kidneys. If an initial dose in the amount of 15 mg is injected, the amount A remaining in the system after x hours can be approximated by $A = 15e^{-0.3x}$. What is the half-life of the drug (the time at which one-half the initial dose remains in the system)?

PHYSICS *Newton's law of cooling relates the time t, in minutes, that it takes for an object heated to a temperature T_0 to cool to a temperature T when placed in an area of constant temperature T_c. The formula is $T = T_c + (T_0 - T_c)e^{-kt}$, where the constant k depends on the nature of the object considered. Use this information in Exercises 25–26.*

25. An iron ball, with constant $k = 0.03$, is heated to 320°F and placed in a room with a constant temperature of 75°F. What will be the temperature of the ball after 5 minutes?

26. An object in a room at a constant temperature of 70°F cools from 120°F to 100°F in 30 minutes. Find the value of the constant k, and use it to determine the temperature of the object after 1 hour.

27. **OCEANOGRAPHY** When a beam of light passes through water, the intensity of the light I is diminished by the number of feet x from the surface of the water. The relationship between these two variables is given by $I = I_0 e^{-kt}$, where I_0 is the intensity of the light at the surface of the water, and k is the coefficient of extinction depending on the particular water under consideration. If the coefficient of extinction of Lake Powell is 0.009, at what depth (to the nearest foot) will the intensity of a light be reduced to 10% of that at the surface?

FOR REVIEW

In Exercises 28–29 sketch the graph of each function.

28. $y = e^{-2x}$

29. $y = \ln 2x$

30. ASTRONOMY The formula $L = 8.8 + 2.2 \ln d$ gives the limiting magnitude L of a telescope in terms of its lens diameter d, measured in inches. Find d, to the nearest tenth, when $L = 8$.

5.5 EXERCISES C

Solve each equation in Exercises 1–6. When appropriate, give answers correct to three significant digits.

1. $\dfrac{e^x + e^{-x}}{2} = 1$

[Answer: 0]

2. $\dfrac{e^x - 3e^{-x}}{2} = 1$

3. $\ln |x| = 1$

[Answer: $e, -e$]

4. $|\ln x| = 1$

5. $\dfrac{e^x + e^{-x}}{e^x - e^{-x}} = 2$

[Answer: 0.549]

6. $\dfrac{e^x - e^{-x}}{e^x + e^{-x}} = 1$

7. PHYSICS The formula for Newton's law of cooling is $T = T_c + (T_0 - T_c)e^{-kt}$. Solve for t in terms of T.

[Answer: $t = -\dfrac{1}{k} \ln [(T - T_c)/(T_0 - T_c)]$]

8. CRIMINOLOGY When a coroner arrived at the scene of a murder at 12:00 noon, he discovered that the body temperature of the victim was 80°F. The room in which the victim was found had a temperature of 70°F. Assuming that the constant k is 0.0044, determine the approximate time of death. [*Hint:* The normal body temperature of a living person is 98.6°F.]

9. The height h (in feet) of a certain species of tree can be approximated by the age of the tree t (in years) using the formula $h = 180/(1 + 300e^{-0.25t})$. **(a)** To the nearest foot, what is the height of the tree in 10 years? **(b)** To the nearest year, how many years will it take for the tree to reach a height of 154 ft? **(c)** What happens to h as t becomes larger and larger?

[Answer to **(a)**: 7 ft]

5.6 MORE APPLICATIONS OF EXPONENTIALS AND LOGARITHMS

STUDENT GUIDEPOSTS

1. Compound Interest
2. Amortizing a Loan
3. Population Growth
4. Radioactive Decay
5. Earthquake Magnitudes
6. Intensity of Sound

We have already introduced a variety of applied problems that involve exponentials or logarithms. In this section we concentrate in more detail on six special applications with broad interest and appeal.

1 COMPOUND INTEREST

Suppose that $5000 is deposited in a savings account that pays 12% interest per year. At the end of one year, the amount in the account will be

$$\$5000 + 0.12(\$5000) = (1.12)(\$5000) = \$5600.$$

In order to see a pattern we write $5600 as (1.12)(5000) and note that at the end of the second year, the amount will grow to

$$\$5600 + (0.12)(\$5600) = (1.12)(\$5000) + (0.12)[(1.12)(\$5000)]$$
$$= [1 + 0.12](1.12)(\$5000)$$
$$= (1.12)^2(\$5000) = \$6272.$$

Continuing in this manner, after t years, the value of the account will be

$$\$5000(1.12)^t.$$

In general, if an initial amount P, called the **principal**, is invested for t years at an annual rate of interest r, compounded annually, the amount A will grow to

$$A = P(1 + r)^t.$$

For example, after 30 years, the original $5000 would grow to

$$A = \$5000(1 + 0.12)^{30} = \$5000(1.12)^{30} = \$149{,}799.61.$$

Calculator steps are given below.

ALG: 1.12 $\boxed{y^x}$ 30 $\boxed{=}$ $\boxed{\times}$ 5000 $\boxed{=}$ → $\boxed{149799.61}$

RPN: 1.12 $\boxed{\text{ENTER}}$ 30 $\boxed{y^x}$ 5000 $\boxed{\times}$ → $\boxed{149799.61}$

Remember that calculator values may vary slightly.

Often deposits earn interest by compounding semiannually (twice a year) or quarterly (four times a year). The rate of interest per period is found by dividing the annual interest by the number of compounding periods per year. If $5000 is deposited in an account that pays 12% interest compounded semiannually, the amount in the account at the end of one year (2 compounding periods) is

$$A = \$5000(1 + 0.06)^2 = \$5000(1.06)^2 = \$5618.$$

Notice that by compounding semiannually instead of annually, we increased the value of the account by $18. If compounding were quarterly, the amount at the end of one year would be

$$A = \$5000(1 + 0.03)^4 = \$5000(1.03)^4 = \$5627.55.$$

In general, if a principal P is deposited in an account at an annual rate of interest r, compounded k times a year for t years, the amount in the account is given by the following formula.

5.6 MORE APPLICATIONS OF EXPONENTIALS AND LOGARITHMS

Compound Interest Formula

$$A = P\left(1 + \frac{r}{k}\right)^{kt} \quad (k \text{ compounding periods a year})$$

EXAMPLE 1 A Banking Problem

If $10,000 is deposited in an account in a bank that pays an annual interest rate of 12%, find the amount in the account at the end of 5 years if compounding is quarterly.

Analysis: Since this is a compound interest problem, we recognize that the formula to use is $A = P\left(1 + \frac{r}{k}\right)^{kt}$.

Tabulation: $P = 10,000$
$r = 12\% = 0.12$ Change to decimal
$t = 5$
$k = 4$ Quarterly is 4 times/year

Translation: Substitute these values into the compound interest formula and evaluate.

$$A = P\left(1 + \frac{r}{k}\right)^{kt} = 10,000\left(1 + \frac{0.12}{4}\right)^{(4)(5)}$$
$$= 10,000(1.03)^{20}$$
$$= \$18,061.10$$

Approximation: $18,061.10 is a reasonable amount. The value of the account is $18,061.10. *Check.*

PRACTICE EXERCISE 1

Repeat Example 1 with the interest compounded monthly, and compare the two amounts.

Answer: $18,166.97

A slightly different problem results when A is given and one of the remaining variables must be found, as shown in the next example.

EXAMPLE 2 A Consumer Problem

The Troutmans wish to have $50,000 available for their newborn child's college education. How much should they deposit in a savings account that pays 9% interest, compounded semiannually, to accumulate $50,000 by the end of 18 years?

Analysis: Use the compound interest formula

$$A = P\left(1 + \frac{r}{k}\right)^{kt}.$$

Tabulation: $A = 50,000$, $r = 9\% = 0.09$, $k = 2$, and $t = 18$.

Translation: Substitute these values.

$$50,000 = P\left(1 + \frac{0.09}{2}\right)^{(2)(18)}$$
$$50,000 = P(1.045)^{36}$$
$$\frac{50,000}{(1.045)^{36}} = P$$
$$10251.4087 = P \quad \text{Use a calculator}$$

PRACTICE EXERCISE 2

How much would have to be invested today at 8% interest, compounded quarterly, to have $5000 available at the end of 10 years?

Thus, the Troutmans must deposit $10,251.41 to have the desired $50,000 in 18 years. (Is this reasonable?) *Check*.

Answer: $2264.45

② AMORTIZING A LOAN

When an amount P of money is borrowed from a bank and is to be paid back over t years in equal regular payments of p dollars k times a year, this is called **amortizing** the loan. That is, a loan is **amortized** when a portion of each payment is used to pay the interest and the rest is used to reduce the outstanding principal. Since each payment reduces the principal, that portion of each payment designated for interest will decrease as the payments are made. The formulas for the amount p of each payment and the total interest I charged are given below.

Amortization Formulas

Amount of payment

$$p = \frac{Pr}{k\left[1 - \left(1 + \frac{r}{k}\right)^{-kt}\right]}$$

and

Total interest charged

$$I = kp\left[t - \frac{1 - \left(1 + \frac{r}{k}\right)^{-kt}}{r}\right]$$

EXAMPLE 3 A CONSUMER PROBLEM

In order to purchase a new home a young couple borrows $60,000 at an annual interest rate of 12%. It is to be paid back over a 30-year period in equal monthly payments. What is their monthly payment, and what is the total interest paid over the period of the loan?

To find the monthly payment, substitute 60,000 for P, 12 for k, 0.12 for r, and 30 for t in the first amortization formula.

$$p = \frac{Pr}{k\left[1 - \left(1 + \frac{r}{k}\right)^{-kt}\right]} = \frac{(60{,}000)(0.12)}{12\left[1 - \left(1 + \frac{0.12}{12}\right)^{-12(30)}\right]}$$

$$= 617.17 \quad \text{To nearest cent}$$

The sequence used to calculate p is given below.

ALG: 12 [×] 30 [=] [+/−] [STO] 0.12 [÷] 12 [+] 1 [=] [y^x]
[RCL] [=] [+/−] [+] 1 [=] [×] 12 [=]
[1/x] [×] 60,000 [×] 0.12 [=] → 617.16756

RPN: 12 [ENTER] 30 [×] [CHS] [STO] 0.12 [ENTER] 12 [÷]
1 [+] [RCL] [y^x] [CHS] 1 [+] 12
[×] [1/x] 60,000 [×] 0.12 [×] → 617.16756

Thus, the monthly payment is $617.17. To find the total interest paid, substitute the same values for k, r, and t, along with 617.17 for p in the second amortization formula.

PRACTICE EXERCISE 3

If the couple in Example 3 were able to take a 20-year loan instead of the 30-year loan, how would this change their monthly payment and the total interest paid over the period?

$$I = kp\left[t - \frac{1 - \left(1 + \frac{r}{k}\right)^{-kt}}{r}\right]$$

$$= (12)(617.17)\left[30 - \frac{1 - \left(1 + \frac{0.12}{12}\right)^{-(12)(30)}}{0.12}\right]$$

$$= 162{,}180.96$$

Over 30 years, the couple will pay $162,180.96 in interest.

Answer: $p = \$660.65$, $I = \$98{,}556.15$

③ POPULATION GROWTH

The growth rate of a population of organisms which is not limited by predators, food supply, or living space, can be approximated exponentially. The model that describes the number N of organisms in terms of the initial population I and time t is given by the formula

$$N = Ie^{kt},$$

where k, called the growth constant, is the percent of growth per unit of time. If the population is increasing, k is positive, and if the population is decreasing, k is negative. This model of population growth is called the **Malthusian model** after Thomas Malthus who, in the 18th century, made important observations and predictions about change in populations.

EXAMPLE 4   A DEMOGRAPHY PROBLEM

How long will it take the present population of the United States to double if it continues to grow at an annual rate of 0.9% per year?

If I is the initial population of the United States, we need to find t when $N = 2I$ and $k = 0.009$ in the formula $N = Ie^{kt}$.

$$2I = Ie^{0.009t}$$
$$2 = e^{0.009t} \quad \text{Divide both sides by } I$$
$$\ln 2 = \ln e^{0.009t} \quad \text{Take the natural log}$$
$$\ln 2 = 0.009t \quad \text{of both sides}$$

Thus,
$$t = \frac{\ln 2}{0.009} \approx 77 \text{ years.}$$

PRACTICE EXERCISE 4

The population of a county is 120,000, but decreasing at an annual rate of 2%. How long will it take for the population to be down to 100,000? [*Hint:* k is -0.02.]

Answer: about 9.1 yr

EXAMPLE 5   A DEMOGRAPHY PROBLEM

Suppose the population of West, Texas was 385 people in 1970 and 510 people in 1980. Assuming that the population growth of West can be approximated by the Malthusian model, find the value of k (correct to three decimal places), and use it to determine the number of people who will live there in the year 2010.

When $t = 0$, the initial population I is 385. When $t = 10$, the population is 510. With this information we can find the value of the growth constant k.

PRACTICE EXERCISE 5

The antelope population on a preserve in Utah was 300 in 1980 and 380 in 1990. Predict the antelope population in the year 2000.

$$510 = 385e^{k(10)}$$

$$\left[\frac{510}{385}\right]^{1/10} = e^k$$

Thus,
$$k = \ln\left[\frac{510}{385}\right]^{1/10} \approx 0.028.$$

Using this value of k, we can calculate the population when $t = 40$.

$$N = 385e^{(0.028)(40)} \approx 1180$$

In the year 2010, West will have a population of approximately 1180 people.

Answer: 485 antelope, using $k = 0.024 = 2.4\%$.

Note Remember, to evaluate $385e^{(0.028)(40)} = 385e^{1.12}$ in Example 5 above, the steps on your calculator are:

ALG: 1.12 [INV] [ln x] [×] 385 [=] → [1179.968868]

RPN: 1.12 [e^x] 385 [×] → [1179.968868]

❹ RADIOACTIVE DECAY

Radioactive substances give off radiation (alpha particles, beta particles, and gamma rays) at a rate that depends on the amount of radioactive material remaining in the substance. As the substance gives off radiation, the atomic structure of the substance is changed so that uranium, for example, changes into thorium, radium, and finally lead. Since the rate of change is slowed by the decreasing amount of material present, the time for the material to decay completely cannot be determined theoretically. However, it is possible to find the **half-life** of a substance, the time needed for one-half the substance to decay. The amount A of radioactive material remaining from an initial amount I, at a given time t, can be approximated by

$$A = Ie^{kt},$$

where k is a negative constant determined by the particular nature of the material.

EXAMPLE 6   A CHEMISTRY PROBLEM

Nuclear reactors use radioactive strontium-90 for fuel. If the radioactive decay of strontium-90 can be approximated by $A = Ie^{-0.025t}$, determine its half-life.

We need to solve the equation for t when $A = 0.5I$.

$$0.5I = Ie^{-0.025t}$$
$$0.5 = e^{-0.025t}$$
$$\ln 0.5 = \ln e^{-0.025t}$$
$$\ln 0.5 = -0.025t$$

Thus,
$$t = \frac{\ln 0.5}{-0.025} \approx 27.7 \text{ years}.$$

PRACTICE EXERCISE 6

The half-life of a certain lead isotope is 25 years. How much of a 500-gram sample will remain radioactive after 100 years? [*Hint:* First find k.]

Answer: 30.4 grams, using $k = -0.028$.

Radioactive carbon-14 (C^{14}), an isotope present in the atmosphere, is absorbed by all living organisms through the intake of carbon dioxide. The level of carbon-14 in the organism remains relatively constant while the organism is alive. However, once the organism dies, carbon-14 is not absorbed, and the slow process of decay begins. The decrease in carbon-14 after the death of an organism is used by archaeologists to date very old artifacts by a process called **carbon dating**. The amount A of carbon-14 left after t years is approximated by the formula

$$A = Ie^{-0.000125t},$$

where I is the original amount present.

EXAMPLE 7 AN ARCHAEOLOGY PROBLEM

A tool made of bone was found at an archaeological site near Page, Arizona in 1980. It was determined that approximately 85% of the original carbon-14 still remained. Approximately when was the tool made?

Since 85% of the carbon-14 remained, we need to solve the following equation for t.

$$0.85I = Ie^{-0.000125t}$$
$$0.85 = e^{-0.000125t}$$
$$\ln 0.85 = \ln e^{-0.000125t}$$
$$\ln 0.85 = -0.000125t$$
$$t = \frac{\ln 0.85}{-0.000125} \approx 1300 \text{ years}$$

The tool was made about 1300 years previous to 1980, or in about 680 A.D.

PRACTICE EXERCISE 7

A mummy found recently in an ancient pyramid in Mexico had lost 40% of its carbon-14. Determine how long ago this person died.

Answer: about 4087 yr ago

---

5 EARTHQUAKE MAGNITUDES

For years geologists tried to agree on a scale for measuring the magnitude of an earthquake. Today the scale used most often is called the **Richter scale**, after American seismologist Charles Richter. To establish the scale, a reference earthquake at level zero was defined as any earthquake with a greatest seismic wave of 0.001 millimeter registered on a seismograph placed 100 kilometers from the epicenter of the quake. The **magnitude** of an earthquake is given by

$$M = \log \frac{A}{A_0},$$

where A is the measurement of the seismic wave of the earthquake, and A_0 is the measurement of seismic wave of a level-zero earthquake with the same epicenter. The ratio $\frac{A}{A_0}$ is called the **intensity** of the earthquake.

EXAMPLE 8 A GEOLOGY PROBLEM

The most powerful earthquake ever recorded occurred in 1906 in South America and was $10^{8.9}$ times more powerful than a level-zero earthquake. What was the magnitude of this earthquake on the Richter scale?

Since $A = 10^{8.9} A_0$,

$$M = \log \frac{A}{A_0} = \log \frac{10^{8.9} A_0}{A_0}$$
$$= \log 10^{8.9} = 8.9.$$

The magnitude of the earthquake was 8.9 on the Richter scale.

PRACTICE EXERCISE 8

The 1989 California earthquake during the World Series measured 7.1 on the Richter scale. How many times more intense was this than a reference level-zero earthquake?

Answer: 12,589,254 times more intense

6 INTENSITY OF SOUND

Shortly after Alexander Graham Bell invented the telephone, it became apparent that some means of measuring the intensity of the signal was necessary. The unit of measurement was called a *bel* in his honor. A tenth of a bel, or **decibel**, is approximately the smallest intensity of sound that the human ear can detect. It has been determined that the threshold of human hearing is crossed when a sound wave produces approximately $S_0 = 10^{-12}$ watt/m². This, then, is the zero level for decibels. If S is the number of watt/m² produced by a particular sound wave, the loudness of the sound in decibels is given by

$$D = 10 \log \frac{S}{S_0}.$$

As a frame of reference, a whisper measures about 20 decibels, a gunshot measures about 100 decibels, and sounds measuring over 125 decibels can cause pain.

We conclude this section with the applied problem given at the beginning of the chapter.

EXAMPLE 9 AN AUDIOLOGY PROBLEM

Audiologists generally agree that continuous exposure to a sound level in excess of 90 decibels for more than 5 hours daily may cause long-term hearing problems. If a student listens to a walkman that produces an intensity of 2×10^{-3} watt/m² for long periods of time daily, is there a potential danger to his hearing?

We need to find D when $S = 2 \times 10^{-3}$ watt/m² and $S_0 = 10^{-12}$ watt/m².

$$D = 10 \log \frac{S}{S_0} = 10 \log \frac{2 \times 10^{-3}}{10^{-12}}$$

$$= 10 \log (2 \times 10^9)$$

$$\approx 93 \text{ decibels}$$

Thus, there is a potential danger of hearing loss over a long period of time.

PRACTICE EXERCISE 9

What is the intensity in watt/m² of a gunshot with a decibel reading measured at 100 decibels?

Answer: 10^{-2} watt/m²

Note In several examples we converted an absolute scale of measurement to a more manageable logarithmic scale. Numbers on an absolute scale from $1 = 10^0$ to $10,000,000,000 = 10^{10}$ translate with common logarithms onto a scale from 0 to 10.

5.6 EXERCISES A

BUSINESS *In Exercises 1–4 use the compound interest formula for k compounding periods or one of the two amortization formulas to solve each problem.*

1. Suppose $1000 is deposited in an account that pays an annual rate of 9%. What is the value of the account at the end of 4 years if compounding is (a) annual, (b) semiannual, (c) quarterly, (d) monthly, (e) weekly (use 52 weeks per year), or (f) daily (use 365 days per year)?

2. How long will it take an amount of money to double if it is invested at an interest rate of 12% compounded annually?

3. How long will it take an amount of money to double if it is invested at an interest rate of 12% compounded semiannually?

4. **CONSUMER** Jim Kirk purchased a new truck and camper. He borrowed $18,000 at an annual interest rate of 12%, to be paid back over a 5-year period in equal monthly payments. What is his monthly payment and what is the total interest paid over the period of the loan?

In Exercises 5–8 use the Malthusian model $N = Ie^{kt}$ to solve each problem.

5. **DEMOGRAPHY** The population of a city is growing at an annual rate of 2.1%. How long will it take for the population to double?

6. **DEMOGRAPHY** The population of the United States was 237 million in 1985. If the annual growth rate is approximately 0.9%, what will be the population in the year 2000?

7. **ECOLOGY** An endangered species of African antelope is estimated to number 8000 and is decreasing at a rate of 6.2% per year. In how many years will the population decline to 2000?

8. **DEMOGRAPHY** The population of Alpine, Colorado was 800 in 1960 and 1150 in 1980. Assuming the same growth rate find the value of k (correct to three decimal places) and use it to predict the population of Alpine in the year 2000.

CHEMISTRY *Use the formula for radioactive decay, $A = Ie^{kt}$, to solve Exercises 9–12.*

9. A certain radioactive isotope decays according to the equation $A = Ie^{-0.03t}$ where t is measured in years. Determine the half-life of this isotope.

10. The half-life of radioactive radium is approximately 1650 years. How many years would it take for a given amount of radium to decay to three-fourths that amount?

11. **ARCHAEOLOGY** A wooden death mask was discovered in an ancient burial site in 1987. If it contains 55% of the carbon-14 ($k = -0.000125$) it originally contained, what is the approximate age of the mask?

12. **MEDICINE** A radioactive isotope with a half-life of 3.2 hours is used as a tracer in a medical test. If 30 milligrams are injected into the system at 9:00 A.M., how many milligrams remain in the system at 6:00 P.M. the same day?

GEOLOGY *In Exercises 13–14 use the formula for determining the magnitude of an earthquake on the Richter scale, $M = \log \frac{A}{A_0}$.*

13. A California earthquake in 1983 was 2,511,890 times more powerful than a reference level-zero quake. What was the magnitude of the California earthquake?

14. If an earthquake measured 7.2 on the Richter scale, what was the intensity of the quake?

AUDIOLOGY *In Exercises 15–17 use the sound equation* $D = 10 \log \frac{S}{S_0}$, *where* $S_0 = 10^{-12}$ *watt/m² is the threshold of human hearing.*

15. If the conversation level in a restaurant produces 4.15×10^{-6} watt/m² of power, what is this decibel level?

16. What is the intensity in watt/m² of a noise measured at 60 decibels?

17. The intensity of a sound is 4.6×10^{-7} watt/m². If the intensity is multiplied by 100, will the decibel rate also be multiplied by 100?

FOR REVIEW

Solve each equation in Exercises 18–19.

18. $5^{2x} = 7^{x+1}$

19. $(\log x)^2 = \log x^5$

In Exercises 20–21 find x correct to four decimal places.

20. $x = \log 4{,}820{,}000$

21. $x = \ln(-0.0045)$

In Exercises 22–23 find x correct to three significant digits.

22. $\log x = -9.4994$

23. $\ln x = 13.2317$

24. Use the properties of logarithms to write $\ln \frac{x\sqrt{y}}{z^4}$ as a sum, difference, or multiple of logarithms.

ANSWERS: Many answers are given using an approximate value. 1. (a) $1411.58 (b) $1422.10 (c) $1427.62 (d) $1431.41 (e) $1432.88 (f) $1433.27 2. 6.12 years 3. 5.95 years 4. $400.40; $6024.00 5. 33 years 6. 271 million 7. 22.4 years 8. 1644 9. 23.1 years 10. 685 years 11. 4783 years 12. 4.27 mg 13. 6.4 14. $10^{7.2}$ 15. 66 decibels 16. 10^{-6} watt/m² 17. No; it will increase by about 20 decibels. 18. 1.53 19. 100,000; 1 20. 6.6830 21. cannot find the natural logarithm of a negative number 22. 3.17×10^{-10} 23. 558,000 24. $\ln x + \frac{1}{2} \ln y - 4 \ln z$

5.6 EXERCISES B

BUSINESS *In Exercises 1–4 use the compound interest formula for k compounding periods or one of the two amortization formulas to solve each problem.*

1. Suppose $1000 is deposited in an account that pays an annual rate of 10%. What is the value of the account at the end of 5 years if compounding is (a) annual, (b) semiannual, (c) quarterly, (d) monthly, (e) weekly (use 52 weeks per year), or (f) daily (use 365 days per year)?

2. How long will it take an amount of money to triple if it is invested at an interest rate of 9% compounded annually?

3. How long will it take an amount of money to triple if it is invested at an interest rate of 9% compounded semiannually?

4. **CONSUMER** The Hagoods borrowed $45,000 to purchase a vacation home. If the term of the loan is 20 years and the annual interest rate is 10%, what is their monthly payment, and how much interest will they pay over the 20-year period?

In Exercises 5–8 use the Malthusian model $N = Ie^{kt}$ to solve each problem.

5. **DEMOGRAPHY** How long will it take for the population of a small Caribbean nation to triple if its annual growth rate is 1.8%?

6. **BIOLOGY** The rodent population in a border town is estimated to be 75,000 and growing at an annual rate of 12%. When will the population reach 1 million?

7. **ECONOMICS** When the interstate freeway bypassed a small town in Arizona, the population began to decline from 12,000 residents at an annual rate of 4.1%. How many years will it take for the population to reach 8000?

8. **DEMOGRAPHY** The population of the southwestern United States is expected to double in 30 years. What will be the approximate annual growth rate of this region?

CHEMISTRY *Use the formula for radioactive decay, $A = Ie^{kt}$, to solve Exercises 9–12.*

9. Determine the half-life of carbon-14 if its radioactive decay is given by the formula $A = Ie^{-0.000125t}$.

10. An isotope of thorium has a half-life of approximately 8.5 days. How long will it take 75% of a given amount to decay?

11. **ARCHAEOLOGY** The mummified remains of a woman were discovered in 1960 in a cave in Texas. If approximately 73% of the original carbon-14 ($k = -0.000125$) was still present in the mummy, in approximately what year did she die?

12. **METEOROLOGY** A radioactive isotope is used to power a weather station in a remote location. If the rate of decay reduces the power by 0.025% daily, how many days will the station be functional if it becomes inoperable when the power reaches a level at 40% of the original supply?

GEOLOGY *In Exercises 13–14 use the formula for determining the magnitude of an earthquake on the Richter scale, $M = \log \frac{A}{A_0}$.*

13. The San Francisco earthquake of 1906 was 1.995×10^8 times more powerful than a reference level-zero quake. What was the magnitude of the San Francisco earthquake?

14. The magnitude of an earthquake in Mexico measured 7.8 on the Richter scale. Fourteen years later another quake measured 4.8 on the Richter scale. The first earthquake was how many times more intense than the second?

AUDIOLOGY *In Exercises 15–17 use the sound equation $D = 10 \log \frac{S}{S_0}$, where $S_0 = 10^{-12}$ watt/m² is the threshold of human hearing.*

15. When a jet airplane takes off, the noise produces 0.5 watt/m² of power. What is the decibel level of this noise?

16. The intensity of a sound is 2.5×10^{-3} watt/m². If the intensity is doubled, will the decibel rate also double?

17. A sound measuring 125 decibels can cause pain to the human ear. What is the approximate intensity of such a sound in watt/m²?

FOR REVIEW

Solve each equation in Exercises 18–19.

18. $\log_2 (x + 3) + \log_2 (x - 4) = 3$

19. $|\ln x| = 5$

In Exercises 20–21 find x correct to four decimal places.

20. $x = \log 0.00039$

21. $x = \log(-0.329)$

In Exercises 22–23 find x correct to three significant digits.

22. $\ln x = -4.2215$

23. $\log x = 7.2198$

24. Use the properties of logarithms to write $\ln \dfrac{xy^3}{\sqrt{z}}$ as a sum, difference, or multiple of logarithms.

5.6 EXERCISES C

1. CONSUMER In 1980 the cost to subscribe to a television cable system in a city was $10 per month. In 1990 the same cable subscription was $18 per month. Assume that the cost fits an exponential model $C = C_0 e^{kt}$.
   (a) What is C_0?
   (b) Find the value of k (to three decimal places).
   (c) Use the model to estimate the cost of a cable subscription in the year 2000.
   (d) When will the cost of cable be five times the cost in 1980?

[Answers: (a) 10 (b) 0.059 (c) $32.54 (d) during the year 2007]

2. ECONOMICS The *consumer price index* relates the cost of goods and services in a given year to the base year 1967. Assume that the index fits an exponential model $I = I_0 e^{kt}$, and that $100 worth of goods and services in 1967 cost $190 in 1980.
   (a) What is I_0?
   (b) Find the value of k (to three decimal places).
   (c) Use the model to estimate the cost of the same goods and services in the year 1991.
   (d) When will the same goods and services cost five times the amount in the base year 1967?

CHAPTER 5 EXTENSION PROJECT

Have you ever been confused by interest rates quoted by banks or other savings institutions? Almost daily you can pick up a newspaper and see ads stating that this account pays 8.25% compounded semiannually, that certificate pays 9% compounded quarterly, and so forth. You will also see phrases like "the effective annual rate of earning is 8.3%." What does it all mean? The *effective annual rate* is a percent that gives the actual yearly interest earned. In Section 5.6 we studied the compound interest formula

$$A = P\left(1 + \frac{r}{k}\right)^{kt} \quad (1)$$

and learned that a fixed principal at a fixed rate will yield varying amounts at the end of one year by compounding differently. For example, the table below shows the amount in an account after 1 year if $1000 is deposited at an annual rate of 10% using various compounding periods. The last column gives the effective annual rate (the numbers will vary using different calculators).

COMPOUNDING	NO. PERIODS/YR	AMOUNT	EFFECTIVE ANNUAL RATE
annually	1	$1100.00	10.0000000%
semiannually	2	$1102.50	10.2500000%
quarterly	4	$1103.81	10.3812891%
monthly	12	$1104.71	10.4713067%
daily	365	$1105.16	10.5155781%
hourly	8760	$1105.17	10.5170335%

Are you surprised by these results? After working with compound interest you probably felt that the more periods the better the return of interest, and this is true, of course, but only up to a point. Notice that there is a leveling off. When compounding went from daily to hourly, the interest earned was only 1¢ more, and the change in the effective annual rate was minimal.

Banks are often limited by federal regulations as to the annual rate they can offer. As a result, they "play games" by using different compounding periods to try to gain an advantage over a competitor. But as the table seems to show, there is a limit to what can be done to change the effective annual rate.

Consider the interest formula (1) again, and suppose we let the number of periods k become larger and larger. Rewrite the formula and replace k/r with n.

$$A = P\left(1 + \frac{1}{k/r}\right)^{(k/r)(rt)} = P\left(1 + \frac{1}{n}\right)^{nrt} = P\left[\left(1 + \frac{1}{n}\right)^{n}\right]^{rt}$$

Since r is fixed, as k gets larger, so also does n. But you should recognize $\left(1 + \frac{1}{n}\right)^n$, and recall from Section 5.4 that, as n gets larger and larger, this expression approaches e. Thus we obtain $A = Pe^{rt}$, which is the formula for compound interest when the compounding is **continuous**. Suppose we find A when $P = \$1000$, $r = 10\%$, and $t = 1$, and compare with the results in our table.

$$A = 1000e^{(0.10)(1)} = 1000e^{0.10} = \$1105.17$$

The effective annual rate is 10.5170918%, a rate that cannot be improved upon. Did you have any idea at the start of this discussion that the number e would appear?

THINGS TO DO:

1. Calculate the effective annual rate for continuous compounding at an annual rate of 6%, 8%, and 12%.

2. Add two more lines to the table corresponding to compounding by minute and by second. Are you surprised by the effective annual rates in these cases? Can you explain these results?

3. Try another line in the table for some very large number, say 3×10^9 (about every thousandth of a second). Do you obtain an effective annual rate of 0%? How can this be?

4. If r is an annual rate and R the corresponding effective annual rate with continuous compounding, show that $R = e^r - 1$ and $r = \ln(R + 1)$.

5. What annual rate of interest will have an effective annual rate of 15% when compounding is continuous?

6. Research the number e. Can you find additional interesting properties? If you go on to study calculus, you will discover another very *natural* occurrence of this amazing number.

CHAPTER 5 REVIEW

KEY TERMS

5.1
1. The word *logarithm* means *exponent*.
2. If a, x, and y are real numbers, $a > 0$, $a \neq 1$, such that $a^x = y$, then x is the **logarithm to the base a** of y and $x = \log_a y$.

5.2
1. Let a be a real number, $a > 0$, and $a \neq 1$. Then the function $f(x) = a^x$ is an **exponential function** with **base a**.
2. Let a be a real number, $a > 0$, and $a \neq 1$. Then the function $f(x) = \log_a x$ is a **logarithmic function** with **base a**.

5.4
1. Logarithms to the base 10 are called **common logarithms**.
2. Logarithms to the base e (approximately 2.71828) are called **natural logarithms**.

5.5
1. Equations with the variable appearing in one or more exponents are **exponential equations**.
2. Equations that contain logarithms of expressions in the variable are **logarithmic equations**.

REVIEW EXERCISES

Part I

5.1 1. If a is any base for logarithms, what is the value of (a) $\log_a a$, and (b) $\log_a 1$?

2. Convert $2^x = 5$ to logarithmic form.

3. Convert $\log_3 a = w$ to exponential form.

Solve each equation in Exercises 4–5.

4. $\log_a 64 = 3$

5. $125^x = 5$

6. For any base a and any positive real number x, simplify $a^{\log_a x}$.

7. For any base a and any real number x, simplify $\log_a a^x$.

5.2 *In Exercises 8–9 sketch the graph of each function.*

8. $f(x) = -3^x$

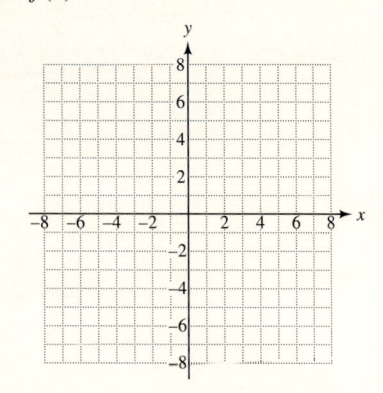

9. $f(x) = \log_5(-x)$ for $x < 0$

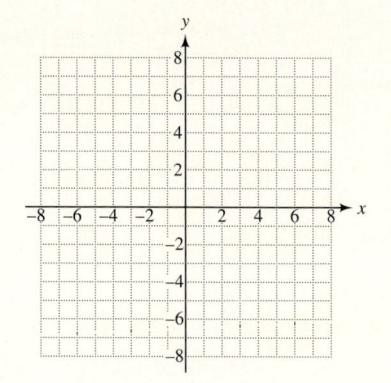

10. Use a calculator to approximate $(\sqrt{5})^\pi$ correct to four decimal places.

5.3 *In Exercises 11–14 evaluate each logarithm given that $\log_a 3 = 0.5283$ and $\log_a 7 = 0.9358$.*

11. $\log_a 21$

12. $\log_a \dfrac{14}{6}$

13. $\log_a \left(\dfrac{7}{3}\right)^{3/2}$

14. $\log_3 7$

15. Use the properties of logarithms to write $\log_a \left(\dfrac{xy}{z}\right)^{2/3}$ as a sum, difference, or multiple of logarithms.

16. Use the properties of logarithms to write $3 \log_a xy + \frac{1}{2} \log_a z$ as a single logarithm.

In Exercises 17–18 state whether the given equation is true *or* false.

17. $\log_a (u + v) = \log_a u + \log_a v$

18. $\log_a \dfrac{u}{a} = \log_a u - 1$

5.4 *Find each logarithm in Exercises 19–20, correct to four decimal places.*

19. $\ln (2.58 \times 10^{-7})$

20. $\log_2 0.0425$

Find each antilogarithm in Exercises 21–22, correct to three significant digits.

21. $\log x = -8.6219$

22. $\ln x = 41.6219$

5.5 *Solve each equation in Exercises 23–28.*

23. $81^{x-2} = 9^{x+5}$

24. $\log_2 (3x + 1) + \log_2 (x - 1) = \log_2 (10x + 14)$

25. $7^{x-5} = 3^{x+7}$

26. $(1.05)^x (1.02)^{2x} = 1000$

27. $(\log_5 x)(\log_2 x) = \log x$

28. $\ln (\ln x) = 0$

5.6 **29. OCEANOGRAPHY** The intensity I of a light is diminished as it is directed through water according to the formula $I = I_0 e^{-kx}$, where I_0 is the intensity of the light at the surface of the water and x is the depth in feet below the surface. If the intensity of a light is reduced by 50% at 20 ft below the surface, find the coefficient of extinction k of the light in water.

30. INVESTMENT If $2000 is deposited in an account that pays an annual interest rate of 14%, what is the value of the account at the end of 2 years if compounding is **(a)** semiannually, and **(b)** monthly?

31. INVESTMENT How long will it take an amount of money to quadruple if it is invested at an interest rate of 10% compounded quarterly?

32. BUSINESS If $30,000 is borrowed at an annual interest rate of 12%, and is to be paid back over a 15-year period in equal monthly payments, what is the amount of each payment? What is the total interest paid over the 15-year period?

33. DEMOGRAPHY Use the Malthusian model $N = Ie^{kt}$, with k computed to three decimal places, to predict the population of Winter, Minnesota in the year 2010 if in 1980 the population was 25,000 and in 1990 the population was 28,000.

34. CHEMISTRY Use the formula for radioactive decay, $A = Ie^{kt}$, to determine the half-life of a radioactive isotope when t is measured in years and $k = -0.0046$.

35. GEOLOGY An earthquake was measured to have an intensity 2.65×10^5 times more powerful than a reference level-zero earthquake. Use the formula $M = \log \frac{A}{A_0}$ to find the magnitude of the earthquake on the Richter scale.

36. AUDIOLOGY A rock band's performance produces an intensity of 5.5×10^{-2} watt/m^2 during a four-hour concert. If the band performs daily, is there a potential danger of hearing loss to the members of the band?

Part II

37. BIOLOGY A bacteria culture of 100 doubles in number each hour. The number N of bacteria present after t hours is given by the function $N = (100)2^t$. Graph this function for $t \geq 0$ and find the number of bacteria present after **(a)** 0 hours, **(b)** 2 hours, **(c)** 4 hours, and **(d)** 6 hours.

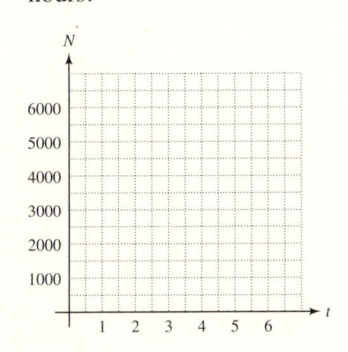

38. Graph $f(x) = e^{-x^2}$

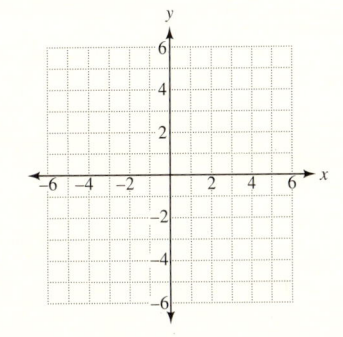

39. Simplify. $\log_\pi \pi^{2x+1}$

40. Use a calculator to approximate $(1.02)^{-50}$ to four decimal places.

41. Use the properties of logarithms to write $2 \log_a (x^2 - y^2) - \log_a (x + y)$ as a single logarithm.

42. Use the properties of logarithms to write $\log_a y^5 \sqrt[3]{\frac{x^2}{z^2}}$ as a sum, difference, or multiple of logarithms.

43. BIOLOGY The number of bacteria N in a culture is given in terms of time t, in hours, by $N = (3000)2^t$. **(a)** How many bacteria are present in 6 hours? **(b)** In how many hours will the bacteria count reach 3,072,000?

44. PHYSICS The atmospheric pressure P in pounds per square inch is approximated by $P = 14.7e^{-0.2x}$, where x is the altitude of the object above sea level in miles. **(a)** What is the pressure on an object 6 miles high? **(b)** A pressure reading of 4.89 lb/in² is taken on the surface of an airplane. How high is the plane?

45. BUSINESS The demand equation for x units of a product relative to the price per unit P is given by $P = 1000 - 0.5e^{0.003x}$. **(a)** Find the price if the demand is 500 units. **(b)** Find the demand if the price is $850.

46. CHEMISTRY The pH of a substance is given by $pH = -\log [H^+]$, where $[H^+]$ is the concentration of hydrogen ions in the solution. **(a)** Find the pH of a mixed drink with hydrogen ion concentration 2.75×10^{-7} moles per liter. **(b)** Find the hydrogen ion concentration of the water in a lake which has a pH of 5.1.

47. ASTRONOMY The limiting magnitude L of a telescope is related to the diameter d of the lens by $L = 8.8 + 2.2 \ln d$. **(a)** What is the limiting magnitude of a telescope with a 3.5-inch lens? **(b)** What is the diameter of the lens of a telescope with a limiting magnitude of 13?

48. MEDICINE The amount of a drug remaining in the circulatory system x hours after a 15-mg dose is taken is approximated by $A = 15e^{-0.25x}$. **(a)** How much of the drug remains in the system after 4 hours? **(b)** For the drug to be effective, at least 10 mg must remain in the system. How long will the dose be effective?

49. PSYCHOLOGY The formula $D = 75 - 15 \ln (x + 1)$ gives the average test score when the test was repeated on monthly intervals where x represents the number of months after testing. **(a)** What was the average score 6 months after the test was first given? **(b)** How many months did it take for the average score to fall below 30?

50. PHYSICS An object placed in a room with a constant temperature of 72°F cools from 130°F to 105°F in 10 minutes. Find the value of the constant k (to three decimal places) in Newton's law of cooling, $T = T_c + (T_0 - T_c)e^{-kt}$, and use it to predict the temperature of the object after 30 minutes.

Find x in Exercises 51–54.

51. $x = \ln 42.5$

52. $\ln x = -0.5138$

53. $x = \log_{3.5} 27.8$

54. $e^{\ln x} = 5$

Solve each equation in Exercises 55–58.

55. $\log_3 (8x + 1) - \log_3 (x - 7) = 3$

56. $\log_2 16^{2x+1} = 8$

57. $\log_a a^{x^2} = x$

58. $e^{-x} = 5$

In Exercises 59–60 state whether the given equation is true *or* false.

59. $\log_a \dfrac{u}{v} = \dfrac{\log_a u}{\log_a v}$

60. $\log_a (u - v) = (\log_a u)(\log_a v)$

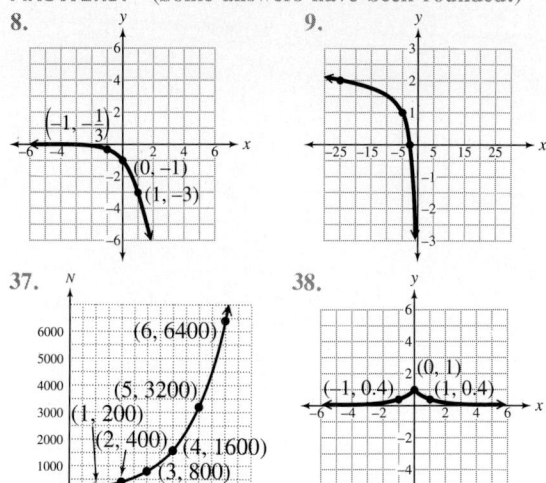

ANSWERS: (Some answers have been rounded.) 1. (a) 1 (b) 0 2. $x = \log_2 5$ 3. $3^w = a$ 4. 4 5. 1/3 6. x 7. x
8. 9. 10. 12.5297 11. 1.4641 12. 0.4075 13. 0.6113 14. 1.7713
15. $\frac{2}{3} \log_a x + \frac{2}{3} \log_a y - \frac{2}{3} \log_a z$ 16. $\log_a x^3 y^3 \sqrt{z}$ 17. false
18. true 19. -15.1703 20. -4.5564 21. 2.39×10^{-9}
22. 1.19×10^{18} 23. 9 24. 5 25. 20.6 26. 78.1 27. 1; 1.62
28. e 29. $k = 0.0347$ 30. (a) $2621.59 (b) $2641.97
31. about 14 yr 32. $360.05; $34,809.03 33. 34,774
34. 150.7 yr 35. 5.4 36. Yes; the decibel level is approximately 107.4, which exceeds 90 decibels by 17.4. 37. (a) 100 (b) 400 (c) 1600 (d) 6400
39. $2x + 1$ 40. 0.3715 41. $\log_a (x - y)^2 (x + y)$
42. $5 \log_a y + \frac{2}{3} \log_a x - \frac{2}{3} \log_a z$ 43. (a) 192,000 (b) 10 hr
44. (a) 4.43 lb/in² (b) 5.5 mi 45. (a) $997.76 (b) 1901
46. (a) 6.6 (b) 7.94×10^{-6} 47. (a) 11.6 (b) 6.7 in
48. (a) 5.5 mg (b) 1.62 hr 49. (a) 45.8 (b) 19 months
50. $k = 0.056$; 82.8°F 51. 3.7495 52. 0.5982 53. 2.6542
54. 5 55. 10 56. 1/2 57. 0, 1 58. $-\ln 5 \approx -1.6094$
59. false 60. false

37. 38.

CHAPTER 5 TEST

Solve.

1. $\log_2 x = 3$

2. $2^a = \frac{1}{4}$

3. Simplify. $\log_6 6^{1.2}$

4. The number of bacteria N in a culture is given in terms of time t, in hours, by the formula $N = (8000)2^t$. How many bacteria are present when $t = 2.5$ hours?

5. Evaluate $\log_a 25$ given that $\log_a 2 = 0.3869$, $\log_a 3 = 0.6131$, and $\log_a 5 = 0.8982$.

6. Use the properties of logarithms to write as a sum, difference, or multiple of logarithms.

$$\log_a \frac{xy^2}{\sqrt[3]{z}}$$

7. Use the properties of logarithms to write as a single logarithm.

$$\frac{1}{2} \log_a x - 2 \log_a z + 3 \log_a y$$

8. Use a calculator and the base conversion formula to find the logarithm correct to four decimal places.

$$\log_5 423$$

9. Graph. $y = e^{3x}$

1. _____

2. _____

3. _____

4. _____

5. _____

6. _____

7. _____

8. _____

9.

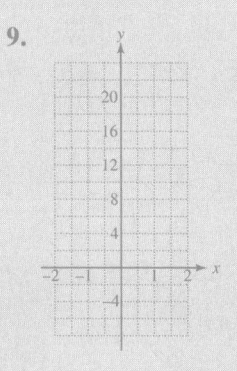

CHAPTER 5 TEST
CONTINUED

Solve. Give answers correct to three significant digits when appropriate.

10. $4^{2x} = 8^{3x-20}$

10. _____

11. $5^x = 3^{2x+1}$

11. _____

12. $\log x + \log (x - 9) = 1$

12. _____

13. $\log_3 x + \log_5 x = 4$

13. _____

Solve.

14. Newton's law of cooling relates the time t, in minutes, that it takes for an object heated to a temperature T_0 to cool to a temperature T when placed in an area of constant temperature T_c. The formula is

$$T = T_c + (T_0 - T_c)e^{-kt}$$

and $k = 0.03$ for an object that is heated to 200°F and placed in a room with a constant temperature of 68°F. What will be the temperature of the object after 10 minutes?

14. _____

15. The compound interest formula is

$$A = P\left(1 + \frac{r}{k}\right)^{kt} \quad (k \text{ compounding periods}).$$

Suppose $5000 is deposited in an account that pays an annual interest rate of 8%. What is the value of the account at the end of 6 years if compounding is quarterly?

15. _____

16. The loudness of a sound in decibels is given by

$$D = 10 \log \frac{S}{S_0}$$

where $S_0 = 10^{-12}$ watt/m^2 and S is the number of watt/m^2 produced by the particular sound wave. What is the loudness of a particular sound (in decibels) that produces an intensity of 2.5×10^{-3} watt/m^2?

16. _____

17. Use the Malthusian model $N = Ie^{kt}$ to find how long it will take for the population of a country to double if its annual growth rate is 1.5%. Round to the nearest year.

17. _____

CHAPTER 6

Systems of Equations and Inequalities

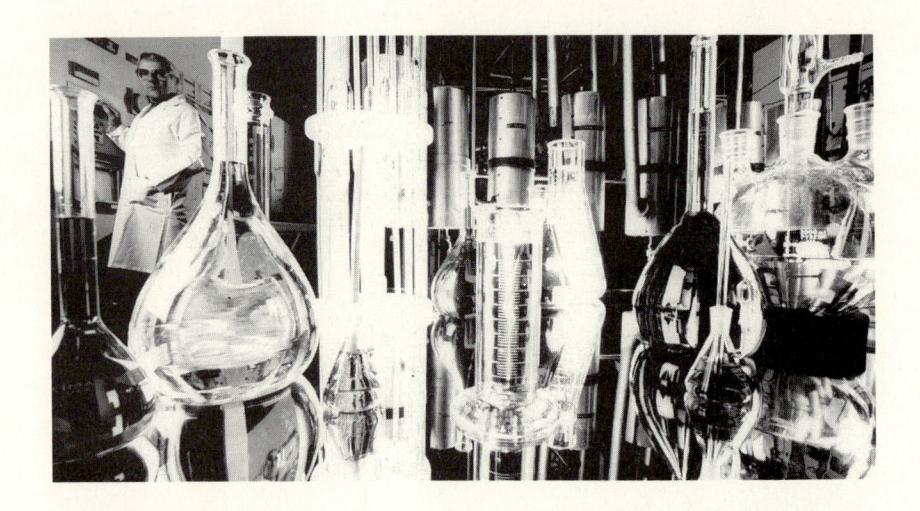

AN APPLIED PROBLEM IN CHEMISTRY

A chemist has two solutions, each containing a certain percentage of acid. If one solution is 5% acid and the other is 15% acid, how many liters of each should be mixed together to obtain 20 L of a solution that is 12% acid?

Analysis

We should recognize that this problem falls into the category of a *combination* or *mixture problem*. Two solutions are being mixed together to form a new solution. In every mixture problem we should be on the lookout for a *quantity equation* and a *value equation*. In this case, the *values* will be the amounts of acid in each of the three different solutions. Keep in mind, for example, that the amount of acid in x liters of a 5% acid solution is $0.05x$. You may want to use a table to summarize the information tabulated below.

Tabulation

Let x = number L of 5% solution used in the mixture,

y = number L of 15% solution used in the mixture.

Then the *quantity equation* is: $\quad x + y = 20$.

$0.05x$ = amount of acid in x L of the 5% solution,

$0.15y$ = amount of acid in y L of the 15% solution,

$(0.12)(20)$ = amount of acid in 20 L of the 12% mixture.

Then the *value equation* is: $\quad 0.05x + 0.15y = (0.12)(20)$.

Translation

We need to solve this system:
$$x + y = 20$$
$$0.05x + 0.15y = (0.12)(20)$$

In Chapter 2 we solved equations and inequalities in one variable. Many applied problems, however, can be solved more easily if we use several equations or inequalities in two or more variables, called a *system of equations or inequalities*. The problem at the left is one example of this (see Example 2 in Section 6.3). Systems of inequalities are especially important in a relatively recent area, *linear programming*, which has numerous applications in business.

6.1 LINEAR SYSTEMS IN TWO VARIABLES

STUDENT GUIDEPOSTS

1. Systems of Equations
2. Types of Systems
3. The Substitution Method
4. The Elimination Method

1 SYSTEMS OF EQUATIONS

In Chapter 3, we defined a linear equation in two variables x and y to be an equation of the form

$$ax + by = c,$$

where a, b, and c are constant real numbers with a and b not both equal to zero. Such equations have infinitely many solutions, ordered pairs of numbers (x, y) that make the equation true. When two linear equations are considered together, the result is a *system of equations*.

> **System of Two Equations in Two Variables**
>
> A **system of two linear equations in two variables,** or simply a **system of equations,** is a pair of linear equations
>
> $$a_1x + b_1y = c_1$$
> $$a_2x + b_2y = c_2,$$
>
> where a_1, b_1, c_1, a_2, b_2, and c_2 are constant real numbers. A **solution to a system of equations** is an ordered pair of numbers that is a solution to *both* equations.

For example, $(2, 1)$ is a solution to both $x - 3y = -1$ and $2x + y = 5$. Thus, $(2, 1)$ is a solution to the system

$$x - 3y = -1$$
$$2x + y = 5.$$

Suppose we graph each equation in this system in the same rectangular coordinate system, as in Figure 6.1. The solution to the system, $(2, 1)$ corresponds to the point of intersection of the two lines. A system can sometimes be solved by this graphing method, but it is clear that identifying the coordinates of the point of intersection of the two lines could be difficult unless they are integer values and we construct very accurate graphs.

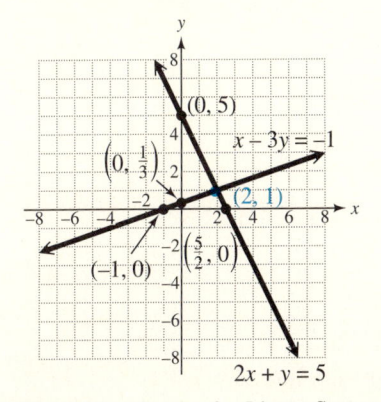

Figure 6.1 Graph of a Linear System

402 CHAPTER 6 SYSTEMS OF EQUATIONS AND INEQUALITIES

② TYPES OF SYSTEMS

Before considering two algebraic techniques for solving a system, we examine the three possible situations that can arise. If two linear equations are graphed in the same rectangular coordinate system, one of the following occurs:

1. The lines coincide, and the system of equations has infinitely many solutions. The system is said to be **dependent**.
2. The lines are parallel, and the system of equations has no solution. The system is said to be **inconsistent**.
3. The lines intersect in exactly one point, and the system of equations has exactly one solution. The system is said to be **consistent** and **independent**.

The following systems of equations and their corresponding graphs in Figure 6.2 illustrate these three cases.

(A) $3x - y = -3$ (C) $3x - y = -3$ (E) $3x - y = -3$
(B) $6x - 2y = -6$ (D) $3x - y = 1$ (F) $3x + 2y = 6$

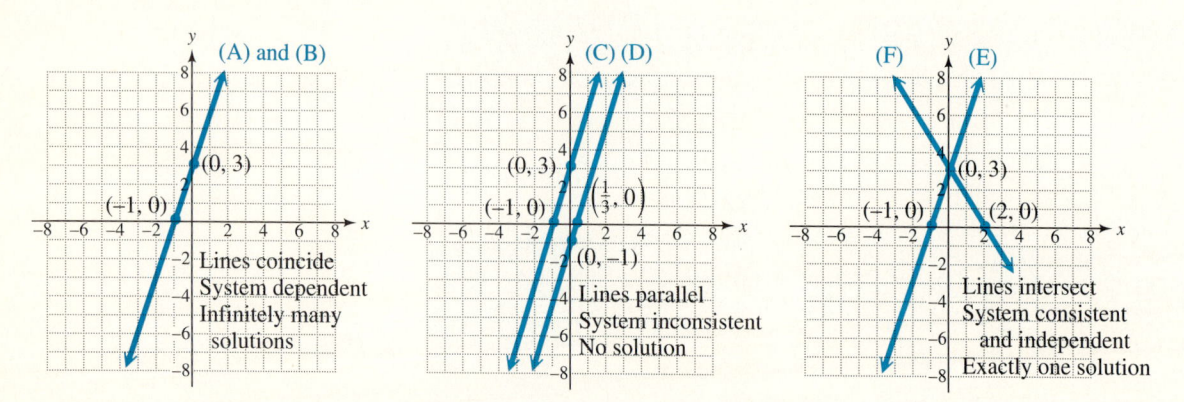

Figure 6.2 Types of Linear Systems

If the equations in a system are written in slope-intercept form, it is easy to tell which of the three cases exists. Recall that parallel lines have equal slopes and that lines with equal slopes and the same y-intercept must coincide. If the slopes are unequal, the lines must intersect.

EXAMPLE 1   FINDING THE NUMBER OF SOLUTIONS

Without solving the system, give the number of solutions.

(a) $2x - y = 1$
       $-4x + 2y = -1$

Write each equation in slope-intercept form.

$$2x - y = 1 \qquad\qquad -4x + 2y = -1$$
$$-y = -2x + 1 \qquad\qquad 2y = 4x - 1$$
$$y = 2x - 1 \qquad\qquad y = 2x - \frac{1}{2}$$

Since both slopes are 2 but the y-intercepts are unequal, the lines are parallel. As a result, the system is inconsistent and has no solution.

PRACTICE EXERCISE 1

Without solving the system, give the number of solutions.

(a) $3x - 2y = -1$
     $6x - 4y = -2$

(b) $\quad 4x - 3y + 2 = 0$
$\quad\quad -8x + 6y - 4 = 0$

Write each equation in slope-intercept form.

$$4x - 3y + 2 = 0 \quad\quad -8x + 6y - 4 = 0$$
$$-3y = -4x - 2 \quad\quad 6y = 8x + 4$$
$$y = \frac{4}{3}x + \frac{2}{3} \quad\quad y = \frac{4}{3}x + \frac{2}{3}$$

Since both slopes are $\frac{4}{3}$ and both y-intercepts are $\left(0, \frac{2}{3}\right)$, the lines coincide. As a result, any solution to either equation is a solution to the system. Since each equation has infinitely many solutions, the system is dependent and has infinitely many solutions.

(c) $\quad 2x - 4 = 0$
$\quad\quad 2x + y = 3$

Since the first equation can be written in the form $x = 2$, its graph is a line parallel to the y-axis. The second equation has as its graph a line with slope -2. Thus, the two lines intersect in one point so that the system has exactly one solution and is consistent and independent.

(b) $\quad 2x - 5y = -1$
$\quad\quad -6x + 15y = 1$

(c) $\quad 2x - y + 5 = 0$
$\quad\quad 3y + 6 = 0$

Answers: (a) infinitely many (dependent) (b) no solution (inconsistent) (c) exactly one (consistent and independent)

Note When one (or both) of the equations in a system is of the form $ax = c$, as in Example 1(c), or the form $by = c$, you can find the number of solutions by inspection. Remember that the graphs of these equations are lines parallel to one of the axes so that parallel or coinciding lines will result only if both equations in the system are of the same type.

❸ THE SUBSTITUTION METHOD

Now that we know how to find the number of solutions to a system, we turn our attention to the first of two algebraic methods for solving a system, the **substitution method**. This method, based on the substitution axiom of equality which states that any quantity can be substituted for its equal, is illustrated in the next example.

EXAMPLE 2 THE SUBSTITUTION METHOD

Solve by the substitution method.

$$2x - y = -6$$
$$5x + 2y = 3$$

We can solve either equation for either variable; however, we can avoid fractions by solving the first equation for y.

$$y = 2x + 6$$

Substitute $2x + 6$ for y in the second equation.

$$5x + 2(2x + 6) = 3$$
$$5x + 4x + 12 = 3$$
$$9x + 12 = 3$$
$$9x = -9$$
$$x = -1$$

PRACTICE EXERCISE 2

Solve by the substitution method.

$$3x + 5y = 6$$
$$7x - y = 14$$

Substitute -1 for x in the first equation.

$$2(-1) - y = -6$$
$$-2 - y = -6$$
$$-y = -4$$
$$y = 4$$

Check: Substitute -1 for x and 4 for y in both original equations.

$$2x - y = -6 \qquad 5x + 2y = 3$$
$$2(-1) - (4) \stackrel{?}{=} -6 \qquad 5(-1) + 2(4) \stackrel{?}{=} 3$$
$$-2 - 4 \stackrel{?}{=} -6 \qquad -5 + 8 \stackrel{?}{=} 3$$
$$-6 = -6 \qquad 3 = 3$$

The solution is $(-1, 4)$.

Answer: **(2, 0)**

To Solve a System Using the Substitution Method

1. Solve one of the equations for one of the variables.
2. Substitute that value of the variable in the *remaining* equation.
3. Solve this new equation and substitute the numerical solution into either of the two original equations to find the value of the second variable.
4. Check your solution in *both* original equations.

④ THE ELIMINATION METHOD

Suppose we are given the following system.

$$3x + 5y = -2$$
$$5x + 3y = 2$$

If we use the substitution method in this case, it is impossible to avoid fractions. An alternative method of solving systems such as this, the **elimination method,** is based on *equivalent* systems. Two systems are **equivalent** if they have exactly the same solutions. Recall that if both sides of an equation are multiplied by the same nonzero number, or if the same expression is added to or subtracted from both sides, the resulting equation is equivalent to the original. When these operations are applied to one or both equations in a system, the resulting system is equivalent to the original. Transforming equations in this way is the substance of the elimination method, sometimes called the **addition-subtraction method,** as illustrated in the next example.

EXAMPLE 3   Using the Elimination Method

Solve by the elimination method.

$$3x + 5y = -2$$
$$5x + 3y = 2$$

Multiply the first equation by -5 and the second by 3.

$$-15x - 25y = 10 \qquad \text{-5 times the first equation}$$
$$\underline{15x + 9y = 6} \qquad \text{3 times the second equation}$$
$$-16y = 16 \qquad \text{Add to eliminate } x$$
$$y = -1$$

PRACTICE EXERCISE 3

Solve by the elimination method.

$$2x + 7y = 17$$
$$5x - 8y = -34$$

Substitute -1 for y in the first equation.

$$3x + 5(-1) = -2$$
$$3x - 5 = -2$$
$$3x = 3$$
$$x = 1$$

The solution is $(1, -1)$. Check by substituting $x = 1$ and $y = -1$ in each of the original equations.

Answer: $(-2, 3)$

To Solve a System Using the Elimination Method

1. Apply the multiplication axiom to one or both equations (if necessary) to transform them so that addition or subtraction will eliminate a variable.
2. Eliminate the variable by adding or subtracting.
3. Solve the resulting single-variable equation and substitute this value into one of the original equations and solve.
4. Check your work by substitution in both original equations.

The question "Which method should I use?" is often asked. Actually, it's your choice; use whichever you prefer or the one with which you are most comfortable. However, if one of the equations in the system has a coefficient of 1 or -1 for one of the variables, it is easy to solve for that variable (without introducing fractions), and the substitution method might be preferred in this case. Other things to do that will make solving systems easier are: eliminate decimals and fractions first, and write both equations in the form $ax + by = c$.

EXAMPLE 4   A DEPENDENT SYSTEM

Solve by either method.
$$3x - 2y = 5$$
$$-3x + 2y = -5$$

We might observe that adding directly will eliminate x.

$$3x - 2y = 5$$
$$\underline{-3x + 2y = -5}$$
$$0 = 0$$

Unfortunately, we eliminated both variables and obtained the identity $0 = 0$. Suppose we apply the substitution technique, and solve the first equation for x.

$$3x = 2y + 5$$
$$x = \frac{2}{3}y + \frac{5}{3}$$

Substitute into the second equation.

$$-3\left(\frac{2}{3}y + \frac{5}{3}\right) + 2y = -5$$
$$-2y - 5 + 2y = -5$$
$$0 = 0$$

PRACTICE EXERCISE 4

Solve by either method.
$$4x - 2y = -6$$
$$-2x + y = 3$$

Again we obtain the identity $0 = 0$. Upon closer examination of the original system, we see that the graphs of the equations are coinciding lines [both have slope $\frac{3}{2}$ and y-intercept $\left(0, -\frac{5}{2}\right)$]. As a result, the system is dependent and there are infinitely many solutions. Any pair of numbers that satisfies either equation is a solution to the system. In order to express the solution in such cases, solve one of the equations for y.

$$y = \frac{3}{2}x - \frac{5}{2}$$

Then the solutions to the system are of the form $\left(x, \frac{3}{2}x - \frac{5}{2}\right)$ for x any real number. In particular, $\left(0, -\frac{5}{2}\right)$, $(-1, -4)$, and $\left(2, \frac{1}{2}\right)$ are three solutions found by letting x equal 0, -1, and 2, respectively.

Answer: $(x, 2x + 3)$ for x any real number

EXAMPLE 5   AN INCONSISTENT SYSTEM

Solve by either method.
$$2x - y = 5$$
$$-4x + 2y = -3.$$

Solve the first equation for y and substitute into the second.

$$-4x + 2(2x - 5) = -3 \qquad y = 2x - 5$$
$$-4x + 4x - 10 = -3$$
$$-10 = -3$$

We obtain a contradiction. Similarly, if we multiply the first equation by 2 and add the result to the second, we obtain a contradiction.

$$\begin{aligned} 4x - 2y &= 10 \\ -4x + 2y &= -3 \\ \hline 0 &= 7 \end{aligned}$$

Whenever a contradiction results, the system is inconsistent, the graphs of the equations are parallel lines, and the system has no solution.

PRACTICE EXERCISE 5

Solve by either method.
$$4x - 2y = 3$$
$$-2x + y = 3$$

Answer: no solution

When Solving a System by Substitution or Elimination

1. If an identity results, the system is dependent and has infinitely many solutions. The solutions are of the form (x, y) where x is any real number and y is an expression in terms of x obtained by solving either equation for y.
2. If a contradiction results, the system is inconsistent and has no solution.

6.1   EXERCISES A

1. State whether $(2, -3)$ is a solution to the given system.

   (a) $x + y = -1$
    $3x - y = 9$

   (b) $2x - 4 = 0$
    $y + 3 = 0$

Without solving, in Exercises 2–5, (a) give the number of solutions to the system, (b) tell whether the lines are parallel, coinciding, or intersecting, and (c) state whether the system is inconsistent, dependent, or consistent and independent.

2. $3x - y = 2$
   $x - 3y = 2$

3. $3x + 2y = -5$
   $6x + 4y = 5$

4. $5x - 3y + 2 = 0$
   $-5x + 3y - 2 = 0$

5. $x + 5 = 0$
   $5 + y = 0$

In Exercises 6–8 solve each system by the substitution method.

6. $x + 2y = 7$
   $2x + y = 2$

7. $5x - 2y = 3$
   $-10x + 4y = -6$

8. $3x - y = 2$
   $2y - 8 = 0$

In Exercises 9–11 solve each system by the elimination method.

9. $2x - 5y = 6$
   $4x + 3y = 12$

10. $2s - 11t = 3$
    $-4s + 22t = -1$

11. $0.3x - 0.7y = 1.6$
    $0.5x + 0.6y = 0.9$

In Exercises 12–17 solve each system by either the substitution method or the elimination method, whichever seems more appropriate.

12. $5x - 7y = -2$
    $7x - 5y = 2$

13. $5u - 3v - 2 = 0$
    $-5u + 3v - 1 = 0$

14. $\dfrac{3}{2}x - \dfrac{1}{3}y = \dfrac{9}{20}$
    $\dfrac{3}{4}x + \dfrac{2}{9}y = \dfrac{11}{10}$

15. $0.02x + 1.05y = -1.07$
    $0.1x - 0.6y = 0.5$

16. Solve for x and y assuming that a and b are nonzero constants.
    $ax + by = 1$
    $3ax - by = -5$

17. Find a and b such that $(-1, 3)$ is a solution to the system.
    $ax - by = 11$
    $-2ax - by = 14$

18. Find the value(s) for m for which the system has no solution.
    $x - 4y = m$
    $-2x + 8y = 4$

19. Show that the system has exactly one solution for every real number m.
    $2x + y = m$
    $x - 4y = 3$

20. Find the value(s) for m for which the system has infinitely many solutions.
    $5x + 2y = m$
    $-15x - 6y = 9$

FOR REVIEW

The following exercises from Chapter 1 will help you prepare for the next section. In Exercises 21–22, evaluate each expression when $x = -1$, $y = 3$, and $z = -2$.

21. $2x + 4y - z$

22. $-3x + y + 5z$

Find the numerical value of y in each equation in Exercises 23–24 if $x = 2$ and $z = -3$.

23. $2x - y + 3z = -3$

24. $7x + 2y - z = 17$

ANSWERS: 1. (a) yes (b) yes 2. (a) exactly one (b) intersect (c) consistent and independent 3. (a) none (b) parallel (c) inconsistent 4. (a) infinitely many (b) coincide (c) dependent 5. (a) exactly one (b) intersect (c) consistent and independent 6. $(-1, 4)$ 7. $\left(x, \frac{5}{2}x - \frac{3}{2}\right)$ x any real number 8. $(2, 4)$ 9. $(3, 0)$ 10. no solution 11. $(3, -1)$ 12. $(1, 1)$ 13. no solution 14. $\left(\frac{4}{5}, \frac{9}{4}\right)$ 15. $(-1, -1)$ 16. $\left(-\frac{1}{a}, \frac{2}{b}\right)$ 17. $a = 1; b = -4$ 18. If m is any real number except -2, the system will have no solution. 19. Since the slope of the line corresponding to the first equation is -2 and the slope of the line corresponding to the second equation is $1/4$, the lines intersect regardless of the value of m. The system has exactly one solution. 20. If $m = -3$, the system has infinitely many solutions. 21. 12 22. -4 23. $y = -2$ 24. $y = 0$

6.1 EXERCISES B

1. State whether $(-1, 5)$ is a solution to the given system.

(a) $3x - 2y = -13$
$x + 3y = 16$

(b) $4x - y = -9$
$4x + 4 = 0$

Without solving, in Exercises 2–5, (a) give the number of solutions to the system, (b) tell whether the lines are parallel, coinciding, or intersecting, and (c) state whether the system is inconsistent, dependent, or consistent and independent.

2. $5x + y = 3$
$2x - y = 7$

3. $5x + 5y = 2$
$3x + 3y = -2$

4. $5x + 5y - 5 = 0$
$3x + 3y - 3 = 0$

5. $x + 2 = 0$
$3x - 1 = 0$

In Exercises 6–8 solve each system by the substitution method.

6. $4x - y = -3$
$3x + 5y = 15$

7. $5x - 2y = 3$
$-10x + 4y = -3$

8. $8x + 3y = -1$
$2x + 4 = 0$

In Exercises 9–11 solve each system by the elimination method.

9. $2a - 7b = 3$
$3a + 2b = -8$

10. $3x - 8y = 5$
$-6x + 16y = -10$

11. $\frac{2}{3}x + \frac{1}{9}y = 6$
$\frac{1}{4}x + \frac{3}{4}y = 15$

In Exercises 12–17 solve each system by either the substitution method or the elimination method, whichever seems more appropriate.

12. $3y - 12 = 0$
    $3x + y = 1$

13. $7u - 7v - 7 = 0$
    $-3u + 3v + 3 = 0$

14. $\dfrac{5}{2}x + \dfrac{2}{3}y = 1$
    $\dfrac{3}{4}x - \dfrac{1}{9}y = -\dfrac{5}{2}$

15. $0.3x + 1.2y = 0.3$
    $0.2x - 4.8y = 5.8$

16. Solve for x and y assuming that a and b are nonzero constants.
    $ax - y = 2b$
    $ax + 2y = -b$

17. Find a and b such that $(-1, 3)$ is a solution to the system.
    $3ax + by = -15$
    $ax - 4by = 60$

18. Find the value(s) for m for which the system has no solution.
    $x + 2y = m$
    $-2x - 4y = 3$

19. Show that the system has exactly one solution for every real number m.
    $3x - y = 2$
    $x + 5y = m$

20. Find the value(s) for m for which the system has infinitely many solutions.
    $3x - y = m$
    $-9x + 3y = -6$

FOR REVIEW

The following exercises from Chapter 1 will help you prepare for the next section. In Exercises 21–22, evaluate each expression when $x = -1$, $y = 3$, and $z = -2$.

21. $5x - y + 3z$

22. $-2x + y + 3z$

Find the numerical value of y in each equation in Exercises 23–24 if $x = 2$ and $z = -3$.

23. $3x - 2y + z = -11$

24. $2x + 3y - z = 7$

6.1 EXERCISES C

Solve each system in Exercises 1–6.

1. $2(x + y) - 5(x - y) = 24$
   $3(x + y) + (x - y) = 2$
   [Answer: $(-1, 3)$]

2. $3(x - 2y) - (2x + y) = 14$
   $5(x - 2y) - 2(2x + y) = 24$

3. $\dfrac{1}{x} + \dfrac{1}{y} = -1$
   $\dfrac{3}{x} - \dfrac{6}{y} = 24$
   $\left[\text{Hint: Let } u = \dfrac{1}{x} \text{ and } v = \dfrac{1}{y}.\right]$

4. $\dfrac{4}{x-1} + \dfrac{1}{y+2} = \dfrac{17}{4}$
   $\dfrac{3}{x-1} - \dfrac{2}{y+2} = \dfrac{5}{2}$
   [Answer: $(2, 2)$]

5. $2|x| + 3|y| = 7$
   $5|x| - 3|y| = 7$
   [Answer: $(2, 1), (2, -1), (-2, 1), (-2, -1)$]

6. $7\sqrt{x} + 3\sqrt[3]{y} = 41$
   $2\sqrt{x} - 5\sqrt[3]{y} = 0$

410 CHAPTER 6 SYSTEMS OF EQUATIONS AND INEQUALITIES

7. Find the values of *a* and *b* so that the line with equation $ax + by = 6$ passes through the points with coordinates $(3, 2)$ and $(0, -2)$.
   [Answer: $a = 4, b = -3$]

8. Show that the graph of the quadratic function $f(x) = x^2 - 4x + 3$ intersects the graph of the linear function $g(x) = -3x + 5$ at the points $(2, -1)$ and $(-1, 8)$.

6.2 LINEAR SYSTEMS IN MORE THAN TWO VARIABLES

STUDENT GUIDEPOSTS

1. Linear Systems in Three Variables
2. The Reduction Method
3. Inconsistent Systems
4. Dependent Systems
5. Nonsquare Systems
6. Homogeneous Systems

❶ LINEAR SYSTEMS IN THREE VARIABLES

An equation of the form

$$ax + by + cz = d$$

when *a*, *b*, *c*, and *d* are constant real numbers and *x*, *y*, and *z* are variables, is called a **linear equation in three variables.** Since the graph of a linear equation in three variables is a plane in space, not a line, the term *linear equation* here is a bit misleading (perhaps *first-degree equation* is better).

A solution to a linear equation such as $2x + y - 3z = 3$ is an **ordered triple** of numbers. For example, $(1, -2, -1)$ is a solution since if *x*, *y*, and *z* are replaced with 1, -2, and -1, in that order, a true equation results.

$$2(1) + (-2) - 3(-1) \stackrel{?}{=} 3$$
$$2 - 2 + 3 \stackrel{?}{=} 3$$
$$3 = 3$$

A **system of three linear equations in three variables** (or again simply a system of equations) is a trio of linear equations such as the following.

$$\begin{aligned} x + y - z &= 4 \\ 2x + y + z &= 1 \\ 3x - 2y - z &= 3 \end{aligned}$$

A solution to such a system is an ordered triple that is a solution to *all three* equations. It is easy to verify by substitution that $(1, 1, -2)$ is a solution to the above system.

As with systems of two equations, a system of three equations can have exactly one solution (be **consistent and independent**), no solution (be **inconsistent**), or infinitely many solutions (be **dependent**). Graphically, these possibilities correspond to three planes *A*, *B*, and *C*, intersecting in exactly one point as in Figure 6.3(a), having no single point in all three planes as in Figure 6.3(b), and having infinitely many points in common as in Figure 6.3(c).

(a) Point P is common to all three planes—consistent and independent system

(b) No points are common to all three planes—inconsistent system

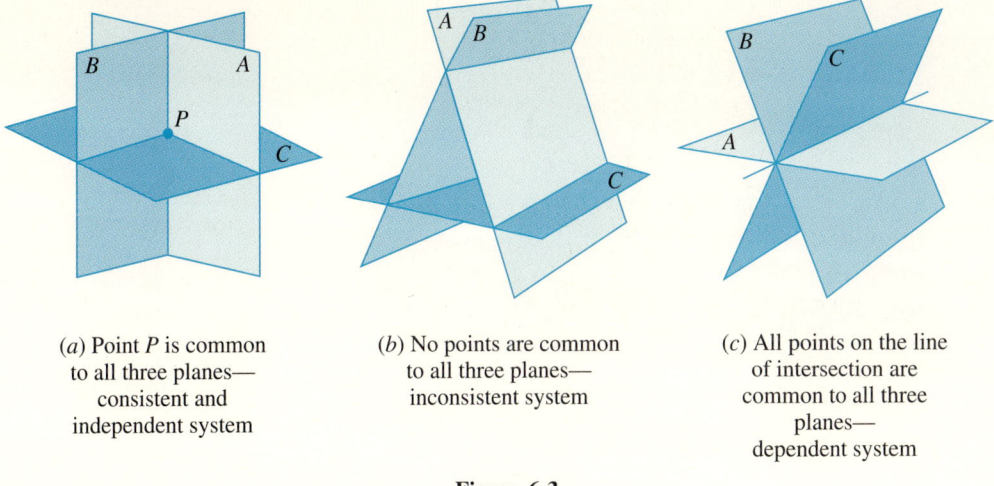

(c) All points on the line of intersection are common to all three planes—dependent system

Figure 6.3

❷ THE REDUCTION METHOD

A system of three linear equations in three variables can be solved algebraically by either substitution or elimination. A third way, the **reduction method,** using both elimination and substitution, transforms the system into a system of two equations in two variables. This technique is perhaps the easiest to learn and the quickest to apply. We summarize below the steps in the reduction method.

> **To Solve a System of Three Equations in Three Variables**
>
> 1. Select any two of the three equations and eliminate a variable.
> 2. Use the equation in the original system that was not used in the first step together with either of the other two equations and eliminate the *same* variable.
> 3. Solve the system of two equations in two variables.
> 4. Substitute the values of the two variables into any of the original equations to obtain the value of the third variable.
> 5. If at any step a contradiction is obtained, the system has no solution.

To solve a system of equations, study the system to decide which of the three variables is the easiest to eliminate. This can save time and effort in later steps. We will illustrate the reduction method by solving the system given earlier in this section.

EXAMPLE 1   LINEAR SYSTEM IN THREE VARIABLES

Solve the system.

(A) $x + y - z = 4$
(B) $2x + y + z = 1$
(C) $3x - 2y - z = 3$

Notice that the multiplication rule is not needed if we eliminate the variable z. Adding equations **(A)** and **(B)** eliminates z and results in an equation in x and y. Then we pair equation **(C)** with **(A)** and subtract. This gives a second equation in x and y.

PRACTICE EXERCISE 1

Solve the system.

$3x + 2y + 3z = 3$
$4x - 5y + 7z = 1$
$2x + 3y - 2z = 6$

$$\begin{aligned}&\textbf{(A)}\quad x+y-z=4\\&\textbf{(B)}\quad 2x+y+z=1\end{aligned}\qquad\begin{aligned}&\textbf{(A)}\quad x+y-z=4\\&\textbf{(C)}\quad 3x-2y-z=3\end{aligned}$$

$$\textbf{(A)}+\textbf{(B)}=\textbf{(D)}\ 3x+2y\ \ =5\qquad\textbf{(A)}-\textbf{(C)}=\textbf{(E)}\ -2x+3y\ \ =1$$

$$\textbf{(D)}\quad 3x+2y=5$$
$$\textbf{(E)}\quad -2x+3y=1$$

We have now reduced the original system to a system of two equations in the two variables x and y. To eliminate x, we multiply **(D)** by 2 and **(E)** by 3, and add the results.

$$\begin{aligned}\textbf{2(D)}\quad 6x+4y&=10\\ \textbf{3(E)}\quad -6x+9y&=\ \ 3\\ \hline 13y&=13\\ y&=\ \ 1\end{aligned}$$

We can now find the value of x by substituting 1 for y in either **(D)** or **(E)**. Suppose we choose **(D)**.

$$3x+2(\mathbf{1})=5$$
$$3x=3$$
$$x=1$$

Finally, to find the value of z, we substitute 1 for x and 1 for y in any of the original equations **(A)**, **(B)**, or **(C)**. Suppose we use **(A)**.

$$(\mathbf{1})+(\mathbf{1})-z=4$$
$$2-z=4$$
$$z=-2$$

Thus, the solution to the system is $(1, 1, -2)$.

Answer: $(2, 0, -1)$

❸ INCONSISTENT SYSTEMS

If a contradiction is obtained at any step of solving a system by the reduction method, the system is inconsistent and has no solution. This is illustrated in the next example.

EXAMPLE 2 AN INCONSISTENT SYSTEM

Solve the system.

$$\begin{aligned}\textbf{(A)}\quad x+y-z&=\ \ 4\\ \textbf{(B)}\quad x-y-z&=\ \ 2\\ \textbf{(C)}\quad 3x-y-3z&=-4\end{aligned}$$

By adding **(A)** and **(B)**, we can eliminate y. Then by pairing **(C)** with **(A)** and adding, y is again eliminated.

$$\begin{aligned}&\textbf{(A)}\quad x+y-z=4\\&\textbf{(B)}\quad x-y-z=2\end{aligned}\qquad\begin{aligned}&\textbf{(A)}\quad x+y-z=\ \ 4\\&\textbf{(C)}\quad 3x-y-3z=-4\end{aligned}$$

$$\textbf{(A)}+\textbf{(B)}=\textbf{(D)}\ 2x-2z=6\qquad\textbf{(A)}+\textbf{(C)}=\textbf{(F)}\ 4x-4z=0$$

$$\tfrac{1}{2}\textbf{(D)}=\textbf{(E)}\ \ x-z=3\qquad \tfrac{1}{4}\textbf{(F)}=\textbf{(G)}\ \ x-z=0$$

$$\textbf{(E)}\quad x-z=3$$
$$\textbf{(G)}\quad x-z=0$$

PRACTICE EXERCISE 2

Solve the system.

$$\begin{aligned}2x-y+3z&=4\\-4x+2y-6z&=1\\8x-3y+5z&=0\end{aligned}$$

Subtracting **(G)** from **(E)** gives the contradiction

$$0 = 3.$$

Thus, the original system has no solution.

Answer: no solution (system is inconsistent)

④ DEPENDENT SYSTEMS

If an identity is obtained in solving a system of three equations, the system has either no solution or infinitely many solutions. Notice that this differs from a system of two equations when an identity *always* means there are infinitely many solutions. However, if an identity is obtained after using *all three equations*, then we know the system is dependent and has infinitely many solutions. When there are infinitely many solutions, they are expressed as for a system of two equations.

EXAMPLE 3   A DEPENDENT SYSTEM

Solve the system.

$$\begin{aligned}\textbf{(A)} \quad x - 3y + 2z &= 6 \\ \textbf{(B)} \quad 4x - 2y + 3z &= 14 \\ \textbf{(C)} \quad 2x + 4y - z &= 2\end{aligned}$$

First eliminate x from **(A)** and **(C)**, and then eliminate x from **(B)** and **(C)**.

$$\begin{array}{rl} 2\textbf{(A)} & 2x - 6y + 4z = 12 \\ \textbf{(C)} & 2x + 4y - z = 2 \\ \hline 2\textbf{(A)} - \textbf{(C)} = \textbf{(D)} & \quad -10y + 5z = 10 \end{array}$$

$$-\frac{1}{5}\textbf{(D)} = \textbf{(E)} \qquad 2y - z = -2$$

$$\begin{array}{rl} \textbf{(B)} & 4x - 2y + 3z = 14 \\ 2\textbf{(C)} & 4x + 8y - 2z = 4 \\ \hline \textbf{(B)} - 2\textbf{(C)} = \textbf{(F)} & \quad -10y + 5z = 10 \end{array}$$

$$-\frac{1}{5}\textbf{(F)} = \textbf{(G)} \qquad 2y - z = -2$$

$$\begin{aligned}\textbf{(E)} \quad 2y - z &= -2 \\ \textbf{(G)} \quad 2y - z &= -2\end{aligned}$$

When **(E)** and **(G)** are subtracted, the result is the identity

$$0 = 0.$$

To express the solution, select one of the two equivalent equations in y and z, say **(E)**, and solve for one of the variables, say z.

$$\begin{aligned} 2y - z &= -2 \\ -z &= -2y - 2 \\ z &= 2y + 2 \end{aligned}$$

PRACTICE EXERCISE 3

Solve the system.

$$\begin{aligned} x - 3y + z &= 4 \\ x + 5y - z &= 2 \\ -2x + 2y - z &= -7 \end{aligned}$$

Substitute this value of z into any of the original equations, say **(A)**, and solve for x in terms of y.

$$x - 3y + 2(2y + 2) = 6 \qquad z = 2y + 2$$
$$x - 3y + 4y + 4 = 6$$
$$x + y = 2$$
$$x = 2 - y$$

Both x and z have now been expressed in terms of y. If y is any real number, then the ordered triple $(2 - y, y, 2y + 2)$ is a solution to the system. Particular solutions can be found by selecting a value of y and evaluating the expressions for x ($x = 2 - y$) and for z ($z = 2y + 2$). Three such solutions are $(2, 0, 2)$, $(1, 1, 4)$, and $(3, -1, 0)$, found by letting $y = 0$, 1, and -1, respectively.

Answer: $(x, 3 - x, 13 - 4x)$, for x any real number (system is dependent)

CAUTION

In Example 3, if we solved equation **(E)** for y instead of z, and substituted into **(B)** to find x, then the form of the solution we obtain would be $\left(-\frac{z}{2} + 3, \frac{z}{2} - 1, z\right)$, for z any real number. The three particular solutions we found would then be generated by letting $z = 2$, 4, and 0. Also, if we had eliminated one of the other variables instead of x in the very first step, the solution triple might have been expressed using x as any number with y and z in terms of x. In other words, there are three ways to represent the solutions when there are infinitely many of them. All three ways yield the same set of solutions.

5 NONSQUARE SYSTEMS

Thus far we have considered systems of *two* equations in *two* variables, and *three* equations in *three* variables. We now expand the discussion to include more general systems.

n × n Systems

If $x_1, x_2, x_3, \ldots, x_n$ represent n variables, where n is a positive integer, and $c, a_1, a_2, a_3, \ldots, a_n$, represent $n + 1$ constant real numbers such that at least one $a_j \neq 0$, then

$$a_1x_1 + a_2x_2 + a_3x_3 + \cdots + a_nx_n = c$$

is a linear equation in n variables. A system of n linear equations in n variables is called an **n × n system** or a **square system of order n.**

In the preceding section we solved 2×2 systems, and in this section we used the reduction method to solve 3×3 systems. The reduction method can be used to solve any $n \times n$ system. We reduce it to an $(n - 1) \times (n - 1)$ system, and continue. For example, a 4×4 system can be reduced to a 3×3 system, then to a 2×2 system.

Systems of equations that have fewer equations than variables, called **nonsquare systems,** have either infinitely many solutions or no solution, and cannot have a unique solution. We illustrate the method for solving such systems in the next example.

6.2 LINEAR SYSTEMS IN MORE THAN TWO VARIABLES

EXAMPLE 4 Nonsquare Systems

Solve the nonsquare systems.

(a) **(A)** $x - 2y + z = 3$
    (B) $x + 4y - z = -7$

By adding **(A)** and **(B)**, we can eliminate z and obtain

$$2x + 2y = -4$$
$$x + y = -2$$
$$y = -x - 2. \quad \text{Solve for } y \text{ in terms of } x$$

Substitute $-x - 2$ for y in either of the original equations—we'll choose **(A)**—and solve for z in terms of x.

$$x - 2(-x - 2) + z = 3$$
$$x + 2x + 4 + z = 3$$
$$z = -3x - 1$$

Thus, the solutions to the system are $(x, -x - 2, -3x - 1)$, for x any real number.

(b) **(A)** $-x + 3y + z = 2$
    (B) $2x - 6y - 2z = -2$

If equation **(A)** is multiplied by 2 and the result added to **(B)**, the contradiction $0 = 2$ is obtained. Thus, the system has no solution.

PRACTICE EXERCISE 4

Solve the nonsquare systems.

(a) $2x + y - z = 5$
    $x - y + 4z = 1$

(b) $5x - y + 2z = 1$
    $-10x + 2y - 4z = 1$

Answers: (a) $(x, 7 - 3x, 2 - x)$, for x any real number   (b) no solution

---

Note Recognizing that a system of two equations in three variables corresponds to two planes in space, it is clear that there must be either infinitely many solutions (the planes coincide or intersect in a line containing infinitely many points) or no solution (the planes are parallel). In no way can the two planes intersect in exactly one point.

6 HOMOGENEOUS SYSTEMS

If all the constant terms in a system of equations are zero, the system is called **homogeneous.** For example,

$$a_1 x_1 + a_2 x_2 + a_3 x_3 = 0$$
$$b_1 x_1 + b_2 x_2 + b_3 x_3 = 0$$
$$c_1 x_1 + c_2 x_2 + c_3 x_3 = 0$$

is a homogeneous 3×3 system. One obvious solution to a homogeneous system such as this is the **trivial solution,** $(0, 0, 0)$. Homogeneous systems often have nontrivial solutions also, and these can be found by the reduction method.

EXAMPLE 5 A Homogeneous System

Solve the system.

(A) $2x + y - z = 0$
(B) $-x + y + z = 0$
(C) $-3x - 3y + z = 0$

PRACTICE EXERCISE 5

Solve the system.

$3x + y - z = 0$
$2x - 3y + z = 0$
$-x + 5y - 2z = 0$

416 CHAPTER 6 SYSTEMS OF EQUATIONS AND INEQUALITIES

First eliminate z from **(A)** and **(B)**, and then eliminate z from **(A)** and **(C)**.

$$\begin{array}{ll} \textbf{(A)} & 2x + y - z = 0 \\ \textbf{(B)} & -x + y + z = 0 \\ \hline \textbf{(A)} + \textbf{(B)} & x + 2y = 0 \end{array} \qquad \begin{array}{ll} \textbf{(A)} & 2x + y - z = 0 \\ \textbf{(C)} & -3x - 3y + z = 0 \\ \hline \textbf{(A)} + \textbf{(C)} & -x - 2y = 0 \end{array}$$

We then obtain the following system.

$$\textbf{(D)} \quad x + 2y = 0$$
$$\textbf{(E)} \quad -x - 2y = 0$$

Adding **(D)** and **(E)** we get $0 = 0$, an identity, so the system has infinitely many solutions. Solve **(D)** for x, $x = -2y$. Substitute $-2y$ for x in **(B)**.

$$-(-2y) + y + z = 0$$
$$3y + z = 0$$
$$z = -3y$$

The solution to the system is:

$$(-2y, y, -3y) \text{ for } y \text{ any real number.}$$

Answer: The only solution is $(0, 0, 0)$.

6.2 EXERCISES A

Solve each system of equations in Exercises 1–12.

1. $\quad x + y + z = 3$
   $-x + 2y - z = 0$
   $\,3x - y + 2z = 2$

2. $x + y + z = 6$
   $x \quad\;\; - z = -2$
   $\quad\; y + 3z = 11$

3. $\;x + 5y - z = 2$
   $4x - y + 3z = 3$
   $8x - 2y + 6z = 7$

4. $\;3x + y + z = 0$
   $-5x + 5y + z = 0$
   $\;x + 2y + z = 0$

5. $2x + y \quad\;\; = 0$
   $\;x - 3y + z = 0$
   $3x + y - z = 0$

6. $\;x - 3y + 2z = -1$
   $4x + 3y + 3z = 6$

7. $\quad x - 3y + 5z = 2$
   $-2x + 6y - 10z = 7$

8. $\dfrac{1}{4}x - \dfrac{1}{3}y - \dfrac{1}{2}z = -2$
   $\dfrac{1}{2}x - \dfrac{1}{2}y + \dfrac{1}{4}z = 2$
   $-\dfrac{1}{4}x + \dfrac{1}{2}y - \dfrac{1}{2}z = -1$

9. $\dfrac{1}{x} - \dfrac{2}{y} - \dfrac{1}{z} = 2$

$\dfrac{2}{x} - \dfrac{1}{y} + \dfrac{1}{z} = 7$

$\dfrac{3}{x} + \dfrac{2}{y} + \dfrac{1}{z} = 2$

[*Hint:* Let $u = 1/x$, $v = 1/y$, and $w = 1/z$.]

10. $x - y + z + w = 2$
$x + y - z + w = 4$
$x + y + z - w = -2$
$x - y - z - w = 0$

11. **ANALYTIC GEOMETRY** Two distinct points determine a unique straight line. In a similar manner, three noncollinear points, no two of which are on the same vertical line, determine a unique parabola with equation $y = ax^2 + bx + c$. By substituting the coordinates of three given points into this equation, the result is a 3×3 system in a, b, and c, the solution to which determines the equation. Find the equation of the parabola determined by the points $(1, 0)$, $(-2, -12)$, and $(2, 0)$.

12. **ANALYTIC GEOMETRY** Three noncollinear points determine a unique circle with equation $x^2 + y^2 + ax + by + c = 0$. Find the equation of the circle determined by the points $(1, 2)$, $(4, -1)$, $(-2, -1)$. Tell what happens if you try to use the same technique when the three points are collinear.

FOR REVIEW

13. Solve the system for x and y assuming that a and b are nonzero constants.

$ax - by = 2$
$2ax + by = 4$

14. Find value(s) for m so that the system is dependent.

$2x - 5y = 3$
$6x - 15y = m$

15. Find value(s) for m so that the system is inconsistent.

$3x + 7y = 2m$
$-6x - 14y = 5$

ANSWERS: 1. $(-1, 1, 3)$ 2. $(1, 2, 3)$ 3. no solution 4. $(x, 2x, -5x)$ x any real number 5. $(0, 0, 0)$ 6. $\left(x, -\dfrac{1}{3}x + 1, 1 - x\right)$ x any real number 7. no solution 8. $(8, 6, 4)$ 9. $\left(1, -\dfrac{1}{2}, \dfrac{1}{3}\right)$ 10. $(1, 0, -1, 2)$ 11. $y = -x^2 + 3x - 2$ 12. $x^2 + y^2 - 2x + 2y - 7 = 0$; the resulting system of equations has no solution (the system is inconsistent). 13. $(2/a, 0)$ 14. $m = 9$ 15. If m is any real number except $-5/4$, the system is inconsistent.

6.2 EXERCISES B

Solve each system of equations in Exercises 1–12.

1. $3x - y + z = 10$
$x + 2y - z = -2$
$-2x + y + z = 0$

2. $3x + 2y + z = 2$
$x - 2y - z = 2$
$2x - y + z = 7$

3. $x - y + z = -8$
   $2x + y + 2z = -1$
   $x + y + z = 2$

4. $x + 2y - 3z = 0$
   $2y + z = 0$
   $x + 4y - 2z = 0$

5. $5x - y + z = 0$
   $2x + y - 2z = 0$
   $x - y - z = 0$

6. $2x - y + 5z = 3$
   $2x + y - z = 1$

7. $5x + y - 3z = 2$
   $-15x - 3y + 9z = 5$

8. $\frac{1}{2}x - \frac{1}{4}y + \frac{1}{3}z = 1$
   $\frac{1}{4}x + \frac{1}{2}y - \frac{1}{2}z = 4$
   $\frac{3}{4}x - \frac{1}{2}y + \frac{2}{3}z = 1$

9. $\frac{1}{x} + \frac{2}{y} - \frac{1}{z} = -3$
   $\frac{2}{x} + \frac{1}{y} + \frac{3}{z} = 12$
   $\frac{1}{x} - \frac{1}{y} + \frac{2}{z} = 9$

10. $5x - y - 3z + w = 0$
    $2x - 3y + z - 4w = 0$
    $-x + 2y + 5z + w = 0$
    $2x + 2y - 3z - w = 0$

11. **ANALYTIC GEOMETRY** Two distinct points determine a unique straight line. In a similar manner, three noncollinear points, no two of which are on the same vertical line, determine a unique parabola with equation $y = ax^2 + bx + c$. By substituting the coordinates of three given points into this equation, the result is a 3×3 system in a, b, and c, the solution to which determines the equation. Find the equation of the parabola determined by the points $(1, 8)$, $(-1, 4)$, and $(3, 20)$.

12. **ANALYTIC GEOMETRY** Three noncollinear points determine a unique circle with equation $x^2 + y^2 + ax + by + c = 0$. Find the equation of the circle determined by the points $(-2, 5)$, $(3, 0)$, $(-2, -5)$. Tell what happens if you try to use the same technique when the three points are collinear.

FOR REVIEW

13. Solve the system for x and y assuming that a and b are nonzero constants.

    $ax + by = 2$
    $2ax - by = 1$

14. Find value(s) for m so that the system is dependent.

    $x + 3y = 2$
    $3x + 9y = m$

15. Find value(s) for m so that the system is inconsistent.

    $2x - 5y = m$
    $-4x + 10y = 8$

6.2 EXERCISES C

Solve each system in Exercises 1–4.

1. $x + y + z - w = 1$
   $2x - y + 3z - w = -2$
   $x - y + 2z - w = 3$
   [Answer: $\left(x, -\frac{1}{2}x - \frac{7}{2}, -x - 5, -\frac{1}{2}x - \frac{19}{2}\right)$ x any real number]

2. $2x - y - z + 3w = -1$
   $x + 5y - 3z + 2w = 7$
   $-4x + 2y + 2z - 6w = 5$

3. $x + 2y = 3$
   $x - 2y = 5$
   $3x + y = 1$
   [Answer: no solution]

4. $x - y = 2$
   $x + y = 0$
   $x - 3y = 4$

MODELING IN STATISTICS In a modeling problem represented by four points (x_1, y_1), (x_2, y_2), (x_3, y_3), and (x_4, y_4), a statistician might use the **least squares regression parabola**, $y = ax^2 + bx + c$, as the curve best fitted to represent this collection of points. The values for a, b, and c are determined by the following system of equations.

$$(\Sigma x^2)a + (\Sigma x)b + 4c = \Sigma y$$
$$(\Sigma x^3)a + (\Sigma x^2)b + (\Sigma x)c = \Sigma xy$$
$$(\Sigma x^4)a + (\Sigma x^3)b + (\Sigma x^2)c = \Sigma x^2 y$$

In this notation, Σx^2 represents $x_1^2 + x_2^2 + x_3^2 + x_4^2$, for example. The symbol Σ is the Greek letter sigma, and it represents "sum" in mathematics. Find the least squares regression parabola for the points given in Exercises 5–6. Can you see that the parabola obtained comes "close" to passing through the given four points?

5. $(-1, 0)$, $(0, 2)$, $(0, 1)$, and $(-2, 0)$
   $\left[\text{Answer: } y = \frac{3}{4}x^2 + \frac{9}{4}x + \frac{3}{2}\right]$

6. $(1, 0)$, $(2, 0)$, $(0, 1)$, and $(0, 2)$

6.3 PROBLEM SOLVING USING SYSTEMS OF EQUATIONS

STUDENT GUIDEPOSTS

1 Applications Using Two Equations **2** Applications Using Three Equations

① APPLICATIONS USING TWO EQUATIONS

While we have been working with the ATTACK method of problem solving, presented in Section 2.2 and used since then, it has probably become clear that the *Translation* step for finding an equation is the most difficult to master. Sometimes things are simplified by using more than one variable and more than one equation; we will use this technique in the present section. Study the examples carefully, and try to pattern your work after them.

It often helps to recognize special types of problems. In a **combination problem** or **mixture problem,** two quantities are combined in different ways. In many of these problems, a *total value* of the combination or mixture must be found. The **total value** of something composed of a number of units, each having the same value, is given by

total value = (value per unit)(number of units).

For example, the total value of 7 pounds of ground beef worth $1.39 per pound is

total value = (value per pound)(number of pounds)
= ($1.39)(7) = $9.73.

In a similar manner, a "total value" might translate to an amount when percents are used. For example, the amount of salt in 5 L of a 25% saline solution is $(0.25)(5) = 1.25$ L, which is the percent per liter times the number of liters. Usually in a combination or mixture problem, we will set up two equations that might be called the *quantity equation* and the *value equation*. With the aid of this general analysis of combination or mixture problems, consider the first example.

EXAMPLE 1 A BUSINESS (MIXTURE) PROBLEM

The owner of a candy store received an order for 50 pounds of a party mix for a large wedding. He plans to combine candy selling for $1.50 per pound with nuts selling for $1.00 per pound, and the mix should sell for $1.20 per pound. How many pounds of each should he use?

Analysis: We recognize this as a mixture problem so we should look for the two equations, the value equation and the quantity equation. Since we are looking for two quantities, select two variables to represent them.

Tabulation: Let c = no. pounds of candy,

n = no. pounds of nuts.

A table sometimes helps to organize the information that is given and that is to be found, and by converting monetary units to cents, we can avoid decimals.

	No. lb	Value/lb	Total value
Candy	c	150	$150c$
Nuts	n	100	$100n$
Mixture	50	120	$(120)(50)$

Translation: The first equation we obtain is the *quantity equation*

(no. lb candy) + (no. lb nuts) = (no. lb mixture)

which translates to

$$c + n = 50.$$

Next we obtain the *value equation*

(value of candy) + (value of nuts) = (value of mix)

which translates to

$$150c + 100n = (120)(50).$$

We have the following system.

$$c + n = 50$$
$$150c + 100n = (120)(50)$$

We simplify the second equation by dividing through by 50; $3c + 2n = 120$. We solve the first for c, $c = 50 - n$, and substitute into the second.

$$3(50 - n) + 2n = 120$$
$$150 - 3n + 2n = 120$$
$$-n = -30$$
$$n = 30$$

Now substitute 30 for n in $c = 50 - n$.

$$c = 50 - 30 = 20$$

PRACTICE EXERCISE 1

Bill Ewing bought 5 shirts (of the same value) and 4 pairs of socks (of the same value) for $87.00. Later in the week he returned to the same store and bought 2 more shirts and 6 more pairs of socks (at the same prices) for $48.00. Find the price of one shirt and one pair of socks.

Approximation: We obtain 30 lb of nuts and 20 lb of candy, and these amounts are reasonable for the problem.

Check: 20 lb + 30 lb = 50 lb, and 20($1.50) + 30($1.00) = $30.00 + $30.00 = $60.00 which is the same as 50($1.20) = $60.00.

Thus, the store owner should mix 20 lb of candy and 30 lb of nuts to make the party mix.

Answer: shirt: $15.00, socks: $3.00

We now solve the applied problem given in the chapter introduction. The solution shown is more abbreviated than the one above since many of the steps are performed mentally.

EXAMPLE 2 A CHEMISTRY PROBLEM

A chemist has two solutions, each containing a certain percentage of acid. If one solution is 5% acid and the other is 15% acid, how much of each should be mixed to obtain 20 L of a solution that is 12% acid?

PRACTICE EXERCISE 2

The radiator in an automobile holds 14 quarts. How much antifreeze should be mixed with a 20% antifreeze solution to obtain a 40% antifreeze mixture that will fill the radiator?

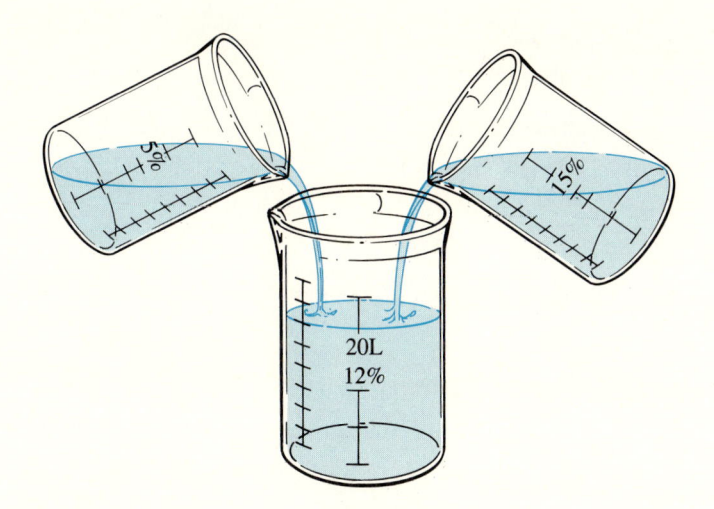

Figure 6.4

Let x = number of liters of the 5% acid solution,
y = number of liters of the 15% acid solution.

The quantity equation is

$$x + y = 20.$$

The value equation equates the amount of acid in the solutions.

$0.05x$ = the amount of acid in x L of the 5% solution
$0.15y$ = the amount of acid in y L of the 15% solution
$(0.12)(20)$ = the amount of acid in 20 L of the 12% mixture

Thus, we have

$$0.05x + 0.15y = (0.12)(20)$$
$$5x + 15y = 240 \quad \text{Clear all decimals}$$
$$x + 3y = 48. \quad \text{Divide both sides by 5}$$

Solve the first equation for x and substitute in the second.

$$20 - y + 3y = 48 \qquad x = 20 - y$$
$$2y = 28$$
$$y = 14$$

Then since $x = 20 - y$, $x = 20 - 14 = 6$. The chemist should mix 6 L of the 5% solution with 14 L of the 15% solution.

Answer: 3.5 qt of antifreeze should be mixed with 10.5 qt of the 20% antifreeze solution.

Many motion-rate problems can be simplified by using a system of equations, as shown in the next example.

EXAMPLE 3 A MOTION-RATE PROBLEM

By going 20 mph for one period of time and then 30 mph for another, a boater traveled from Glen Canyon Dam to the end of Lake Powell, a distance of 180 miles. If he had gone 21 mph throughout the same period of time, he would only have reached Hite Marina, a distance of 147 miles. How many hours did he travel at each speed?

PRACTICE EXERCISE 3

Susan Katz sailed her boat 210 miles upstream from a marina at a constant speed in a time of 7 hours. She returned downstream to the same marina at the same speed in a time of 6 hours. What was the speed of the boat in still water and the speed of the current?

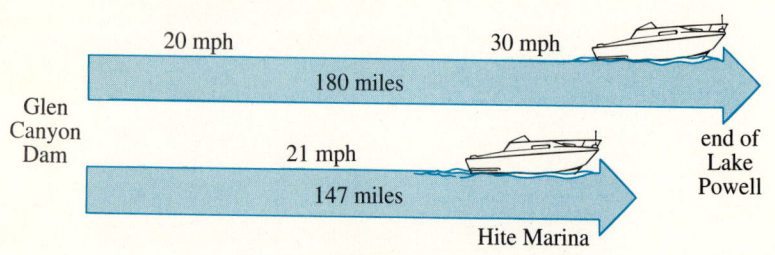

Figure 6.5

Let x = the number of hours traveled at 20 mph,
y = the number of hours traveled at 30 mph.

Since
$$(\text{distance}) = (\text{rate}) \cdot (\text{time})$$

and the total distance traveled is the sum of the two distances traveled at the two rates, the first equation is

$$20x + 30y = 180.$$

At a rate of 21 mph for the total time, $x + y$, he traveled 147 miles. Thus, the second equation is

$$21(x + y) = 147.$$

Simplify the equations before attempting to solve, dividing the first equation by 10 and the second by 7.

$$2x + 3y = 18$$
$$3x + 3y = 21$$

Subtracting the first from the second, we have

$$x = 3.$$

Then
$$2(3) + 3y = 18$$

so that $3y = 12$, or $y = 4$. He traveled 3 hours at 20 mph and 4 hours at 30 mph.

Answer: boat: 32.5 mph, current: 2.5 mph

② APPLICATIONS USING THREE EQUATIONS

The problem in the next example translates to a system of three equations in three variables.

EXAMPLE 4 A Geometry Problem

In a triangle, the largest angle is 70° more than the smallest angle, and the remaining angle is 10° more than three times the smallest angle. Find the measure of each angle.

Let x = the measure of the smallest angle,
y = the measure of the middle angle,
z = the measure of the largest angle.

Since z is 70° more than x,

$$x + 70 = z$$

or
(A) $x - z = -70$.

Since y is 10° more than 3 times x,

$$3x + 10 = y$$

or
(B) $3x - y = -10$.

Finally, the sum of the measures of the angles of a triangle is 180°.

(C) $x + y + z = 180$

Hence, we must solve the following system.

(A) $x + z = -70$
(B) $3x - y = -10$
(C) $x + y + z = 180$

Notice that y is missing in (A). If we add (B) and (C) the result is also an equation with y missing.

(D) $4x + z = 170$ (B) + (C) = (D)

Thus we must solve the following system.

(A) $x - z = -70$
(D) $4x + z = 170$

Add (A) and (D).

$$5x = 100$$
$$x = 20$$

Substitute 20 for x in (A).

$$20 - z = -70$$
$$-z = -90$$
$$z = 90$$

Substitute 20 for x in (B).

$$3(20) - y = -10$$
$$60 - y = -10$$
$$-y = -70$$
$$y = 70$$

The angles have measure 20°, 70°, and 90°.

PRACTICE EXERCISE 4

Samantha has an average of 88 on three tests in accounting. If her first test score was 13 points higher than the second, and the third was 5 points higher than the second, find Sam's three scores.

Answer: 95, 82, 87

In Section 3.7 we saw that many situations which involve two ordered pairs of data can be described by a linear function model. When three ordered pairs of data are known, sometimes a quadratic function model will describe the situation. In Section 6.2, Exercise 11 showed us how to find a quadratic function, whose graph passes through three given noncollinear points, by solving a system of three equations in three variables. This modeling technique, called **curve fitting,** has many applications as shown in the next example.

EXAMPLE 5 A BUSINESS PROBLEM

The owner of a small business made a profit of $200 the first week of the year, $300 the second week, and $500 the third week. When she plotted the points (1, 200), (2, 300), and (3, 500), she felt that a parabola might fit this data. Find the quadratic function that passes through the three points, and use it to predict the profit for the fifth week.

Substitute the coordinates of the three points into the equation $y = ax^2 + bx + c$, and obtain the following system.

(A) $\quad a + b + c = 200$
(B) $\quad 4a + 2b + c = 300$
(C) $\quad 9a + 3b + c = 500$

Subtract (A) from (B) to eliminate c and obtain (D), and subtract (A) from (C) to eliminate c and obtain (E).

(D) $\quad 3a + b = 100$
(E) $\quad 8a + 2b = 300$

Multiply (E) by $-1/2$ and add to (D).

$$\begin{aligned} 3a + b &= 100 \\ -4a - b &= -150 \\ \hline -a &= -50 \\ a &= 50 \end{aligned}$$

Substitute 50 for a in (D).

$$3(50) + b = 100$$
$$b = -50$$

Substitute 50 for a and -50 for b in (A).

$$50 + (-50) + c = 200$$
$$c = 200$$

Thus, the quadratic function that fits these data points is $y = 50x^2 - 50x + 200$. To find the estimated profit in week 5, substitute 5 for x.

$$y = 50(5)^2 - 50(5) + 200$$
$$= 1250 - 250 + 200 = \$1200$$

PRACTICE EXERCISE 5

Repeat Example 5, this time for a business with a profit of $50 the first week, $100 the second, and $300 the third.

Answer: $y = 75x^2 - 175x + 150$; $1150

Note The quadratic model obtained in Example 5 must be used carefully. When $x = 0$, for example, $y = 50(0)^2 - 50(0) + 200 = \$200$. Does this make any sense? How can there be a profit of $200 when no time (weeks) has passed? As with many functional models, be sure to keep in mind the domain of the function. In this case, we must assume that $x \geq 1$. Clearly negative values of x, as well as 0, would not be realistic solutions to the problem.

6.3 EXERCISES A

In Exercises 1–14, solve using a system of equations.

1. **NUMBER** The sum of two numbers is 44 and their difference is 20. Find the numbers.

2. **CONSUMER** Four books (of the same kind) and six pens (of the same kind) cost $9.00. Three books and nine pens also cost $9.00. Find the cost of one book and one pen.

3. **RECREATION** If there were 600 people at a play, the total receipts were $980, and the admission price was $2.00 for adults and $1.00 for children, how many adults and how many children were in attendance?

4. **RETAILING** A candy mix sells for $1.10 per pound. If the mix is composed of two kinds of candy, one worth $0.90 per pound and the other worth $1.50 per pound, how many pounds of each would be in a 60-pound mixture?

5. **COIN** A collection of dimes and quarters is worth $4.20. If the total number of coins in the collection is 30, how many of each are there?

6. **CHEMISTRY** A chemist has one solution that is 25% acid and another that is 50% acid. How much of each should be used to make 25 L of a 40% acid solution?

7. **GEOMETRY** Two angles are supplementary, and one measures 4° more than seven times the measure of the other. Find the angles.

8. **RATE-MOTION** By traveling 40 km/hr for one period of time and then 50 km/hr for another, Mario traveled 370 km. Had he gone 10 km/hr faster throughout, he would have traveled 450 km. How many hours did he travel at each rate?

9. **SPORTS** To stay in good physical condition, a professor wants to jog or play tennis 15 times each month. If he estimates that a total of 26 hr can be spent on these activities each month, each time he jogs he spends 1.2 hr, and each time he plays tennis he spends 2.8 hours, how many times should he pursue each activity during the month to use the entire 26 hr?

10. **AGE** The sum of the ages of Milt, Lew, and Jenny is 53. Jenny is 5 years younger than Lew, and in two years, Milt will be the same age as Lew is now. How old is each?

11. **SPORTS** The total number of seats in a basketball sports arena is 12,000. The arena is divided into three sections, courtside, endzone, and balcony, and there are twice as many balcony seats as courtside seats. For the conference championship game, ticket prices were $10.00 for courtside, $8.00 for balcony, and $7.00 for endzone. If the arena was sold out for the game and the total receipts were $99,000, how many seats were courtside?

12. **CONSUMER** Becky has a collection of nickels, dimes, and quarters with a total value of $4.60. Twice the number of nickels is the same as three times the number of quarters, and the total number of coins is 40. How many of each are there?

13. **INVESTMENT** Larry has $5000 divided into three separate investments. Part of the money is invested in bonds at 8%, part in certificates at 7%, and the rest is in a mutual fund. If the fund does well and earns 6%, the total earnings from all the investments will amount to $345. However, if the fund does not do well, he will lose 3% on this investment, and the total earnings from the three will amount to only $165. How much is invested in each category?

14. **CONSUMER** The cost of operating a home air conditioner can be approximated by a quadratic model as a function of the temperature setting on the thermostat. Through experimentation, Mike Ratliff collected the data in the table below. Find the quadratic function that fits this data and use it to predict the operating cost on a day when the thermostat is set at 100°. Does your answer seem reasonable? Can you think of a situation in which it would be reasonable?

Thermostat setting	Cost/day of operating the air conditioner
70°	$6
80°	$5
90°	$3

FOR REVIEW

Solve the systems in Exercises 15–16.

15. $3x + y - z = 0$
    $x - y + 2z = 0$
    $7x + y = 0$

16. $5x + 2y - z = 2$
    $3x - 3y + z = -1$

Sketch the graph of each linear equation in Exercises 17–19. These problems will help you prepare for the material in the next section.

17. $4x + 2y = -6$

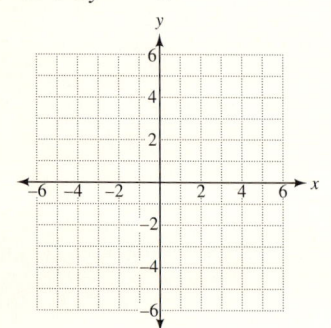

18. $3x + 3y = 0$

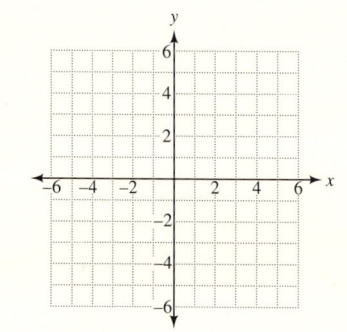

19. $2y - 2 = 0$

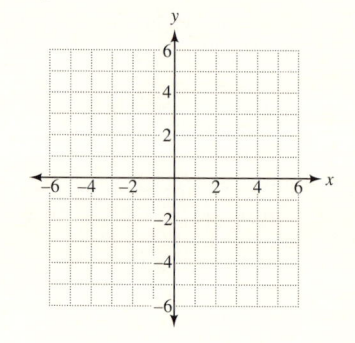

ANSWERS: 1. 12, 32 2. book: $1.50; pen: $0.50 3. 220 children; 380 adults 4. 40 lb of $0.90 candy; 20 lb of $1.50 candy 5. 22 dimes; 8 quarters 6. 10 L of 25% solution; 15 L of 50% solution 7. 22°, 158° 8. 3 hr at 40 km/hr; 5 hr at 50 km/hr 9. jog 10 times; play tennis 5 times 10. Milt is 18; Lew is 20; Jenny is 15. 11. 3000 seats 12. 12 nickels; 20 dimes; 8 quarters 13. $1500 in bonds; $1500 in certificates; $2000 in mutual fund 14. $y = -\frac{1}{200}x^2 + \frac{13}{20}x - 15$; $0.00; if the exterior temperature never exceeded 100° the air conditioner might not turn on at all. 15. $(x, -7x, -4x)$ x any real number 16. $(x, 8x - 1, 21x - 4)$ x any real number

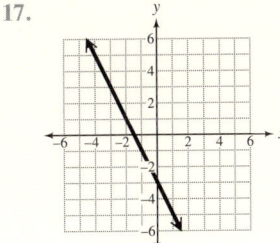

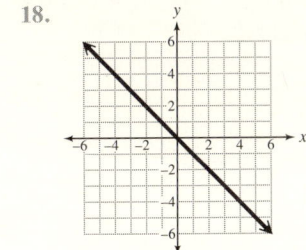

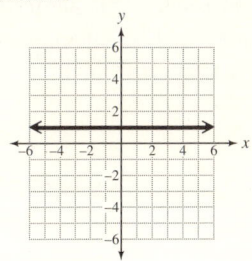

17. 18. 19.

6.3 EXERCISES B

In Exercises 1–14 solve using a system of equations.

1. **AGE** Twice the sum of Sam's and Joe's ages is 40. In 2 years, Sam will be the same age as Joe is now. How old is each?

2. **CONSUMER** Seven pads of paper and five pencils cost $5.20. Two pads and 18 pencils cost $4.80. Find the price of each.

3. **SPORTS** A total of 8340 people paid to see a college basketball game. Tickets were $1.50 for children and $2.50 for adults, and the total receipts were $19,620. How many children and how many adults paid to see the game?

4. **RETAILING** The owner of the Coffee Mill wishes to mix two blends of coffee, one selling for $1.80 per pound and the other for $2.40 per pound, to obtain a 40-pound mixture selling for $2.10 per pound. How many pounds of each must he use?

5. **COIN** Cindy has 40 coins consisting of nickels and dimes. If the total value of the collection is $3.35, find the number of nickels and the number of dimes in the collection.

6. **CHEMISTRY** A lab technician obtains 10 gallons of a 30% saline solution by mixing some 20% solution with some 50% solution. How much of each must she use?

7. **GEOMETRY** In a right triangle, one acute angle measures 6° more than twice the measure of the other. Find each acute angle.

8. **RECREATION** A cruise boat sails 48 miles downstream in 2 hours and returns the 48 miles upstream in 3 hours. Find the speed of the boat and the speed of the stream.

9. **BUSINESS** A shipping crate contains two types of items, one weighing 3 pounds each and the other 0.5 pound each. Suppose it is known that there are 90 items in the crate. If the crate weighs 30 pounds when empty, and now weighs 200 pounds, how many of each type of item are in the crate?

10. **EDUCATION** The average of a student's three scores is 74. If the first is 21 points less than the second, and twice the third is 15 more than the sum of the first two, find all three scores.

11. **RETAILING** On Wednesday, an appliance dealer sold 3 stoves, 4 refrigerators, and 2 washers for a total of $4950. On Thursday she sold 2 stoves, 5 refrigerators, and 1 washer for a total of $4650. On Friday she sold 4 stoves and 2 refrigerators for $3100. If the stoves, refrigerators, and washers were all of the same value, respectively, what was the price of a stove?

12. **COIN** A collection of 100 coins consisting of nickels, dimes, and quarters has a value of $16.00. If the number of nickels plus the number of dimes is equal to the number of quarters, find the number of each.

13. **INVESTMENT** Gail Taggart has $10,000 divided into three separate investments. Part is invested in a mutual fund that earns 8%, part is invested in time certificates that earn 7%, and the rest is invested in a business. If the business does well, it will earn 10% and her total earnings will amount to $790. If the business loses 2%, her total earnings will amount to only $550. How much is invested in each category?

14. **BUSINESS** The operating costs of producing a number of items can be approximated by a quadratic model as a function of the number of items produced. Through experimentation, the owner of a manufacturing firm collected the data in the table below. Find the quadratic function that fits this data, and use it to predict the operating costs on a day when 50 items are produced. What are the production costs when no items are produced? Can you justify these costs?

No. items produced	Daily operating costs
10	$200
20	$400
30	$700

FOR REVIEW

Solve the systems in Exercises 15–16.

15. $2x + y - 3z = 0$
    $x - y + 5z = 0$
    $3x + y + 2z = 0$

16. $7x - 3y + z = 0$
    $-14x + 6y - 2z = 1$

Sketch the graph of each linear equation in Exercises 17–19. These problems will help you prepare for the material in the next section.

17. $2x - y = 4$

18. $x - 4y = 0$

19. $x + 1 = 0$

6.3 EXERCISES C

In Exercises 1–4 solve using a system of equations.

1. **NUTRITION** A lab technician wishes to place the animals in an experiment on a specific diet of 15 grams of protein and 5 grams of fat. She is able to purchase two food mixes, Special Diet, which is 12% protein and 2% fat, and Control K, which is 20% protein and 8% fat. How many grams of each mix (to the nearest tenth of a gram) should she combine to obtain the right diet mixture for her animals?
[Answer: 53.6 g of Control K; 35.7 g of Special Diet]

2. **CHEMISTRY** To maintain the reaction in a laboratory experiment, a chemist must add 19.5 grams of sodium and 2.2 grams of potassium every day. The chemist has Compound A that contains 30% sodium and 4% potassium and Compound B that contains 25% sodium and 2% potassium. How many grams of each should be added each day to maintain the reaction?

3. **MANUFACTURING** Aspen Stoves manufactures three models of woodburning stoves, the Sierra, the San Juan, and the Blue Ridge. Each stove must pass through three stages: cutting, welding, and finishing. The total number of production hours weekly is 195 for cutting, 200 for welding, and 190 for finishing. The number of hours required in each stage for each stove is summarized in the table below. How many of each type of stove must be manufactured weekly for the company to operate at full production capacity?

Stage	Sierra	San Juan	Blue Ridge
Cutting	5	5	2
Welding	4	6	2
Finishing	4	5	3

[Answer: Sierra: 15; San Juan: 20; Blue Ridge: 10]

4. **BUSINESS** The owner of a chemical plant received an order for a special blend of fertilizer that consists of, among other things, 760 pounds of nitrate, 540 pounds of phosphate, and 106 pounds of iron. He has available three mixes with the compositions shown in the table below. How many pounds of each should be blended to fill the order?

	Nitrate (in %)	Phosphate (in %)	Iron (in %)
GroWell	20	15	4
YieldHi	25	20	2
FastGro	30	18	5

6.4 LINEAR SYSTEMS OF INEQUALITIES

STUDENT GUIDEPOSTS

1 Linear Inequalities

2 Systems of Linear Inequalities

1 LINEAR INEQUALITIES

In Chapter 2 we solved linear inequalities in one variable and graphed the solutions on a number line. We now consider *linear inequalities in two variables*.

Linear Inequalities in Two Variables

An inequality in one of the forms

$$ax + by < c, \quad ax + by > c, \quad ax + by \leq c, \quad \text{or} \quad ax + by \geq c,$$

where a, b, and c are constant real numbers, is a **linear inequality in two variables** x and y. A **solution** to a linear inequality in two variables is an ordered pair of numbers, which, when substituted for x and y, makes the inequality true.

For example, $(-1, 0)$ is a solution to the linear inequality $2x + y < -1$ since

$$2(-1) + 0 < -1$$
$$-2 < -1 \quad \text{True}$$

is true. On the other hand, $(4, -3)$ is not a solution since

$$2(4) + (-3) < -1$$
$$8 - 3 < -1$$
$$5 < -1 \quad \text{False}$$

is false.

One way to identify the solution to a linear inequality is to show its graph, that is, the set of all points in the plane whose coordinates solve the inequality. Graphing a linear inequality is not much more difficult than graphing a linear equation. In fact, to begin, we temporarily replace the inequality symbol with an equal sign and graph.

$$ax + by = c,$$

which is a straight line. This **boundary line** divides the plane into two regions called **half-planes** as shown in Figure 6.6. If the inequality is $\leq$ or $\geq$, the points on the boundary line are included with the half-plane, and the region is a **closed half-plane**. If the inequality is $<$ or $>$, the boundary line is not included, and the region is an **open half-plane**. We indicate a closed half-plane with a solid line and an open half-plane with a dashed line.

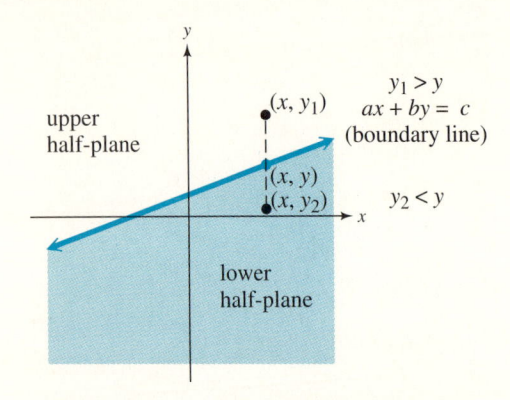

Figure 6.6 Half-Planes

Since $y_1 > y$, the point (x, y_1) in the upper half-plane in Figure 6.6 would satisfy the inequality $ax + by > c$. Similarly, since $y_2 < y$, the point (x, y_2) in the lower half-plane would satisfy the inequality $ax + by < c$. The solutions of a given inequality correspond to all points in exactly *one* of the half-planes determined by the boundary line.

Thus we have the following method.

To Graph an Inequality in Two Variables

1. Graph the boundary line using a solid line if the inequality is $\leq$ or $\geq$ or a dashed line if it is $<$ or $>$.
2. Choose any point not on the boundary line (the point $(0, 0)$ is often selected) and use it as a **test point** by substituting its coordinates into the inequality.
3. Shade the half-plane containing the test point if a true inequality is obtained, or shade the half-plane that does not contain the test point if a false inequality results.

6.4 LINEAR SYSTEMS OF INEQUALITIES   431

| EXAMPLE 1   GRAPHING A LINEAR INEQUALITY | PRACTICE EXERCISE 1 |

Graph $4x + 2y \geq -6$.

First graph the line $4x + 2y = -6$ using a solid line since the inequality is $\geq$. Use the intercepts $(0, -3)$ and $(-3/2, 0)$ to obtain the graph. Now select a test point; $(0, 0)$ is easy to use.

$$4x + 2y \geq -6$$
$$4(0) + 2(0) \geq -6$$
$$0 \geq -6 \quad \text{This is true}$$

Since we obtained a true inequality, shade the half-plane containing $(0, 0)$ to obtain the graph in Figure 6.7.

Graph $3x - 2y \leq 6$.

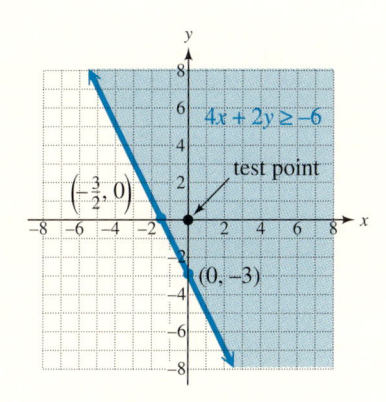

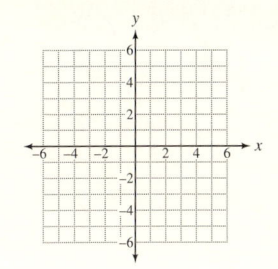

Figure 6.7

Answer: The graph is given at the end of this section.

In view of Example 1, it is easy to see that the graphs of $4x + 2y \leq -6$, $4x + 2y > -6$, and $4x + 2y < -6$ appear as in Figure 6.8(a), (b), and (c), respectively.

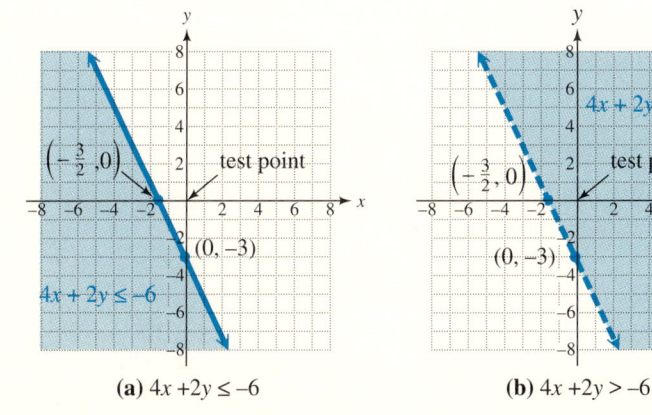

(a) $4x + 2y \leq -6$ (b) $4x + 2y > -6$ (c) $4x + 2y < -6$

Figure 6.8

When $a = 0$ or $b = 0$ in a linear inequality in two variables, the graph of the inequality is an upper or lower half-plane or else a left or right half-plane, respectively.

EXAMPLE 2 SPECIAL LINEAR INEQUALITIES

(a) Graph $2y - 2 < 0$.

First graph the line $2y - 2 = 0$, which simplifies to $y = 1$, using a dashed line. The test point $(0, 0)$ shows that the graph is the lower, open half-plane in Figure 6.9(a). Had we been asked to graph $2y - 2 \geq 0$, the graph is the upper, closed half-plane in Figure 6.9(b).

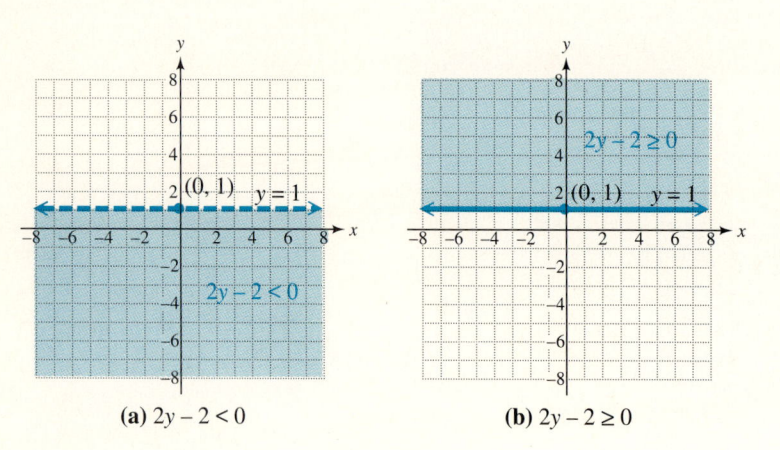

(a) $2y - 2 < 0$ **(b)** $2y - 2 \geq 0$

Figure 6.9

(b) Graph $3x - 3 \geq 0$.

The graph of $3x - 3 \geq 0$, or equivalently, of $x \geq 1$, is given in Figure 6.10(a). Similarly, the graph of $3x - 3 < 0$ is shown in Figure 6.10(b).

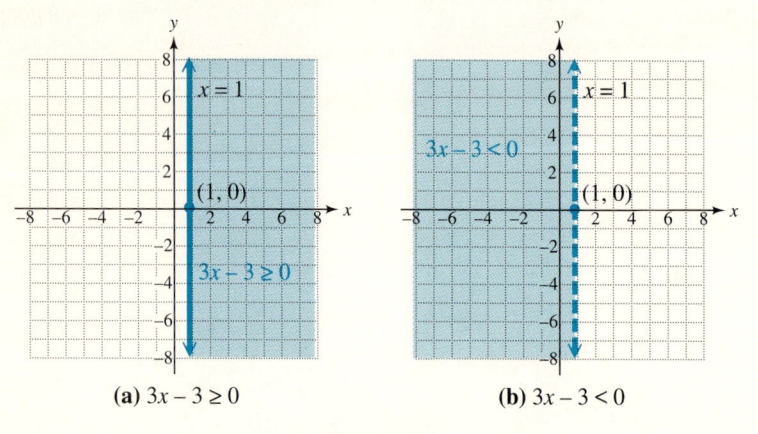

(a) $3x - 3 \geq 0$ **(b)** $3x - 3 < 0$

Figure 6.10

PRACTICE EXERCISE 2

(a) Graph $5y + 2 \geq 0$.

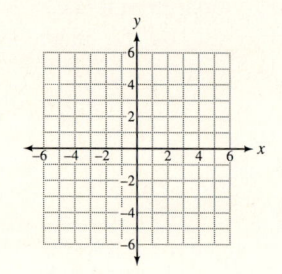

(b) Graph $6x + 3 < 0$.

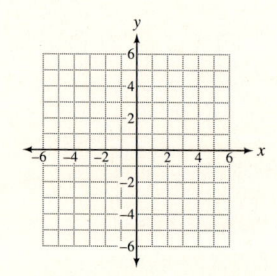

Answer: The graphs are given at the end of this section.

② SYSTEMS OF LINEAR INEQUALITIES

As you might expect from our previous work, a solution to a **system of linear inequalities** is an ordered pair of numbers that solves every inequality in the system. The graph of such a system is the set of points in the plane corresponding to these solutions. This graph is found by sketching the graphs of all the inequalities in the system in the same plane and identifying the region where these graphs overlap or intersect.

EXAMPLE 3 Graphing a System of Inequalities

Graph the system. $2x + 3y < 6$
$-2x + y \geq 1$

Graph $2x + 3y = 6$ using a dashed line and intercepts (0, 2) and (3, 0), and graph $-2x + y = 1$ using a solid line and intercepts (0, 1) and (−1/2, 0). The test point (0, 0) can be used for both inequalities.

$2x + 3y < 6$	$-2x + y \geq 1$
$2(0) + 3(0) < 6$	$-2(0) + 0 \geq 1$
$0 < 6$ Is true	$0 \geq 1$ Is false

The graph of the system is the region formed by the overlap of the open half-plane below the line $2x + 3y = 6$ and the closed half-plane above the line $-2x + y = 1$. See Figure 6.11.

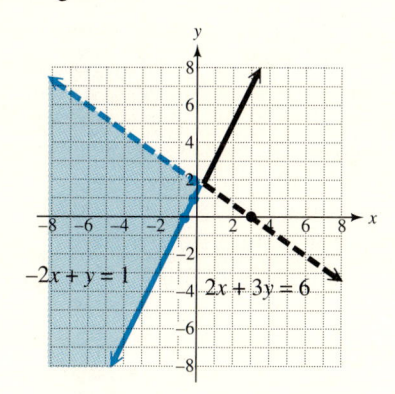

Figure 6.11 Graph of a System of Inequalities

Practice Exercise 3

Graph the system.

$3x - y > 6$
$3x - 3 \geq 0$

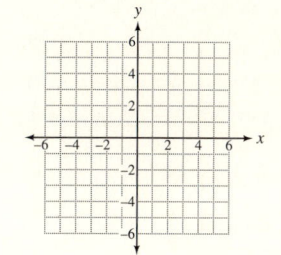

Answer: The graph is given at the end of this section.

The graph of the system in Example 3 is said to be **unbounded** since it extends infinitely far in some directions. Many systems that have **bounded graphs,** enclosed on all sides by line segments, play an important role in a method of problem-solving discussed in the next section.

EXAMPLE 4 A Bounded Graph

Graph the system.
$$x \leq 0$$
$$y \geq 0$$
$$2x - 5y \geq -10$$
$$x - y \geq -3$$

Note that the first two inequalities describe the points on the positive y-axis, the negative x-axis, and in the second quadrant. Turning to the last two inequalities, we graph the lines $2x - 5y = -10$ and $x - y = -3$. The point of intersection of the two lines, $(-5/3, 4/3)$, can be found by solving the system of equations

$$2x - 5y = -10$$
$$x - y = -3.$$

Practice Exercise 4

Graph the system.

$x \geq 0$
$y \geq 0$
$2x + y \leq 4$
$x + 2y \leq 4$

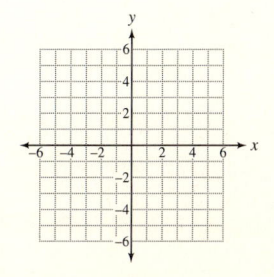

434 CHAPTER 6 SYSTEMS OF EQUATIONS AND INEQUALITIES

The region that satisfies the two inequalities $2x - 5y \geq -10$ and $x - y \geq -3$ is cross-shaded in Figure 6.12(a). The graph of the original system is the bounded portion of the plane in the second quadrant which also satisfies the two inequalities above, as shown in Figure 6.12(b).

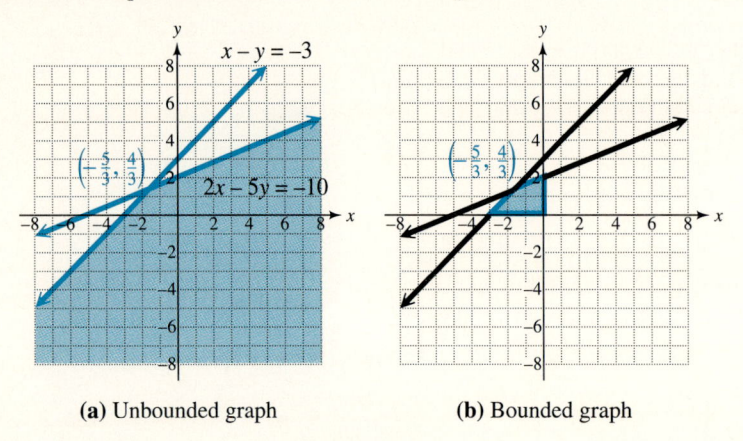

(a) Unbounded graph

(b) Bounded graph

Figure 6.12

Answer: The graph is given at the end of this section.

Many applied problems involving two variables can be described by a system of inequalities. The graph of the system shows all possible solutions to the problem. This is illustrated in the following example and will be considered in greater detail in the next section.

EXAMPLE 5 A BUSINESS PROBLEM

A retail store owner stocks two models of typewriters, the Executive and the Standard. He has discovered that due to demand it is necessary to have at least twice as many Executive models as Standard models. Also, at all times he must have on hand at least 10 Executive models and 5 Standard models. Finally, due to limitations in space, he has room for no more than 30 typewriters at any time. Graph the system of inequalities that describes this information.

Let x = no. Executive models,
 y = no. Standard models.

The phrase "at least twice as many Executive models as Standard models" translates to:

$$x \geq 2y.$$

The phrase "at least 10 Executive models" becomes:

$$x \geq 10.$$

"At least 5 Standard models" translates to:

$$y \geq 5.$$

"Room for no more than 30 typewriters" becomes:

$$x + y \leq 30.$$

PRACTICE EXERCISE 5

The owner of CompuTech has two models of personal computers in stock, Model 20 and Model 30. To maintain an appropriate stock, she has found that at all times at least two of each model must be kept in her inventory. Also, to minimize costs the number of Model 30 computers plus four times the number of Model 20 computers must not exceed 14. Find a system of inequalities, using x for the number of Model 20 and y for the number of Model 30 computers, that describes her inventory, and graph the system.

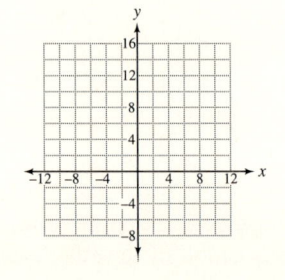

6.4 LINEAR SYSTEMS OF INEQUALITIES   435

The system of inequalities is

$$x \geq 2y$$
$$x \geq 10$$
$$y \geq 5$$
$$x + y \leq 30.$$

The graph of the system is given in Figure 6.13, where the point (20, 10) was located by solving the system

$$x - 2y = 0.$$
$$x + y = 30.$$

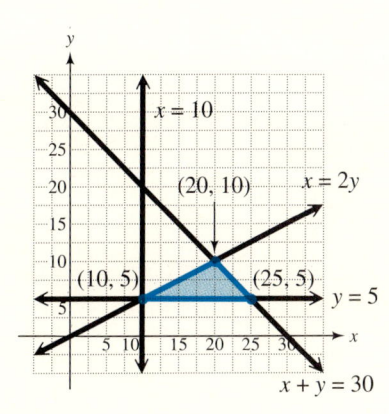

Figure 6.13

Answer: The graph of the system is given at the end of this section.

ANSWERS TO PRACTICE EXERCISES:

1.

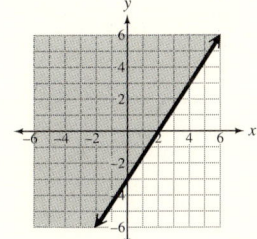

2. (a)

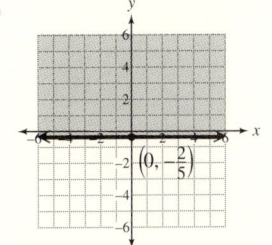

2. (b)

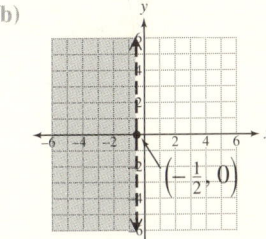

3.

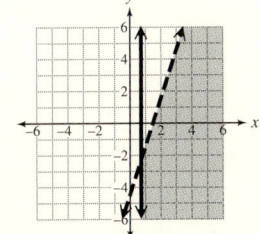

4.

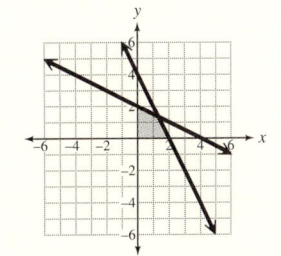

5.

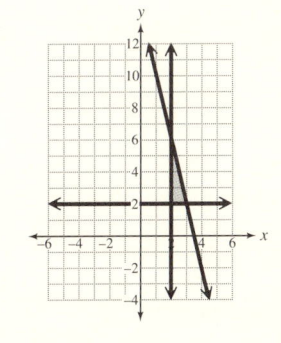

6.4 EXERCISES A

In Exercises 1–6, sketch the graph of each inequality.

1. $3x - 2y \leq 6$

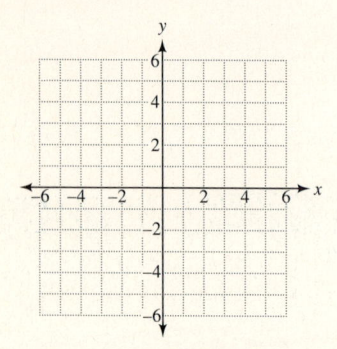

2. $2x + y \geq 3$

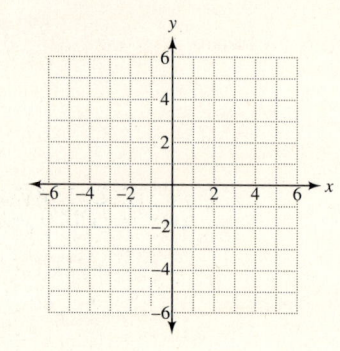

3. $x - y < -2$

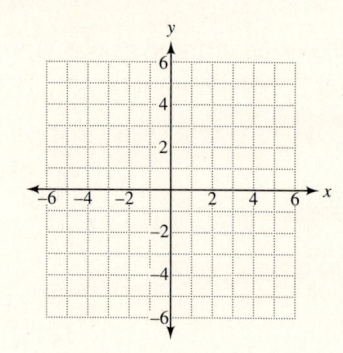

4. $3x + 2y < 0$

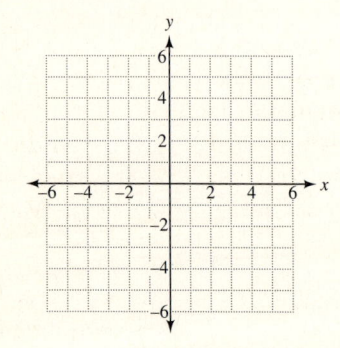

5. $4y - 8 \leq 0$

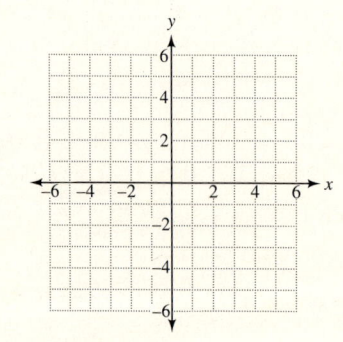

6. $4y + 12 > 0$

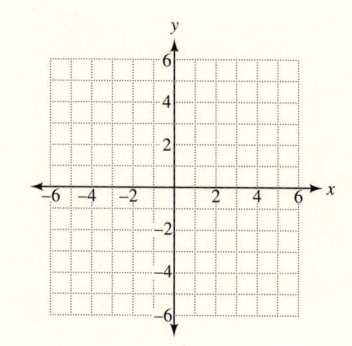

In Exercises 7–12, sketch the graph of each system of inequalities.

7. $2x - 3y \geq 6$
$x + y < -1$

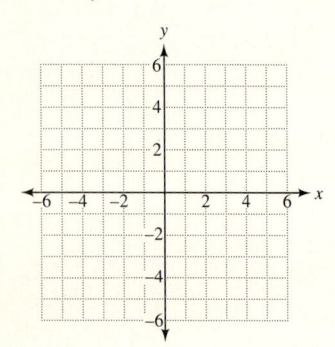

8. $2x - 3y < 6$
$x + y \geq -1$

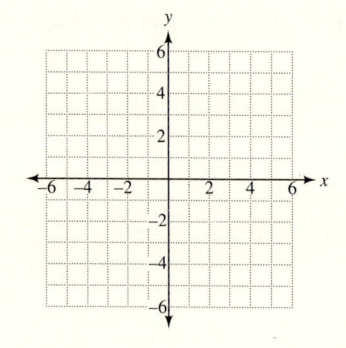

9. $2x - 2y \geq -4$
$x - y < 1$

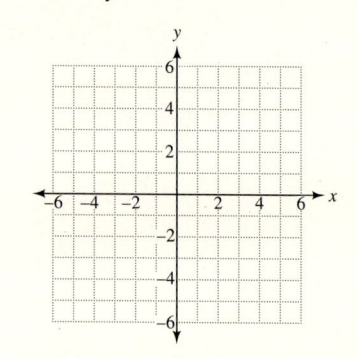

10. $-2x + 3y \leq 6$
$x + 1 > 0$

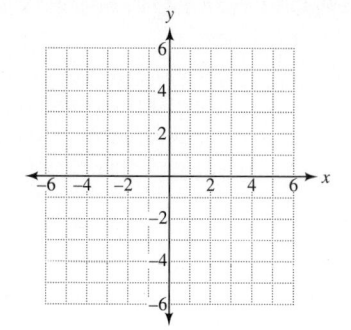

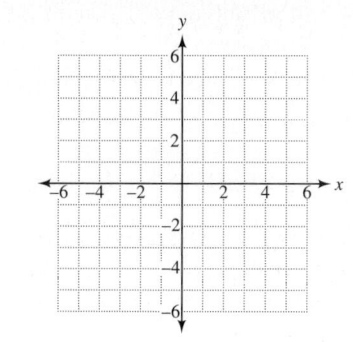

11. $x \geq 0$
$y \geq 0$
$3x + 7y < 21$

12. $x \leq 0$
$y \geq 0$
$2x - y \geq -6$
$x + 2y \leq 2$

In Exercises 13–14, graph the system described by the given information.

13. BUSINESS A dealer sells two models of mobile homes in a particular park, the Princess and the Knight. Due to demand, he must have at least three times as many Princess homes available as Knights. At all times he wants at least 6 Princess homes and 2 Knight homes available and ready for occupancy. The Princess model costs the dealer $30,000, the Knight model costs $20,000, and the dealer wants to keep his inventory costs at $600,000 or less. Form the system of inequalities described by this information and graph the system.

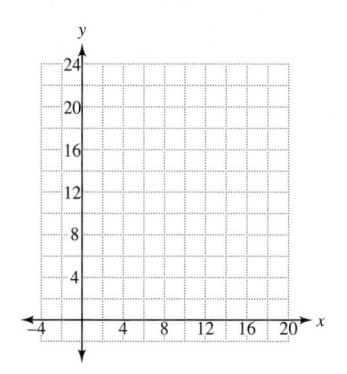

14. MANUFACTURING A manufacturer of patio picnic tables has two models, the Standard and the Deluxe. The table below shows information about production. The manufacture of each type of table cannot use more than the workhours available in each manufacturing stage each week. Form the system of inequalities described by this information and graph the system.

	Standard workhours per table	Deluxe workhours per table	Maximum available workhours per week
Construction stage	3	6	96
Finishing stage	1	4	36

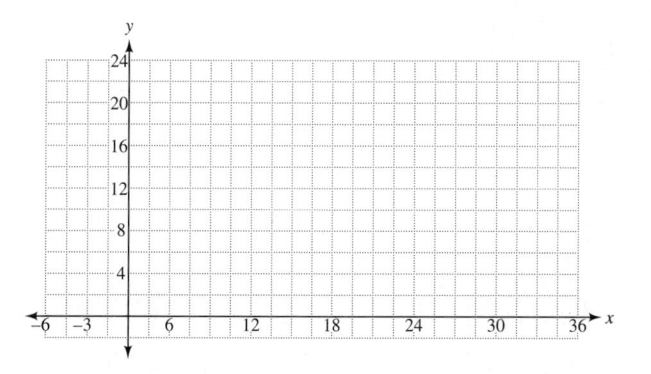

FOR REVIEW

Solve.

15. **CONSUMER** If 2 lb of candy and 3 lb of nuts cost a total of $4.90, and 3 lb of candy and 5 lb of nuts cost a total of $7.90, find the cost of 1 lb of nuts and 1 lb of candy.

16. **GEOMETRY** The smallest angle of a triangle measures 28° less than the largest angle, and the measure of the largest angle less the measure of the middle-sized angle is 2°. Find the measure of each angle.

17. **GEOMETRY** The perimeter of a rectangular pasture is 50 mi and the length is 1 mi more than twice the width. Find the dimensions of the pasture.

18. **CURVE FITTING** Determine the constants a and b for the function $f(x) = ae^x - be^{-x} + 1$ if $f(0) = 2$ and $f(\ln 3) = 4$. What is the function?

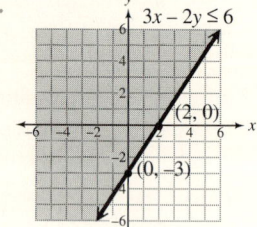

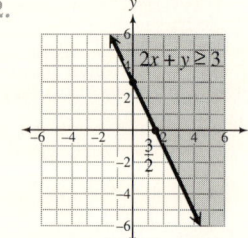

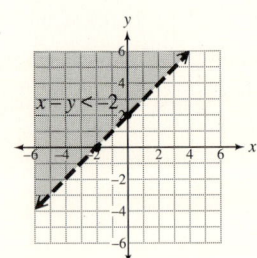

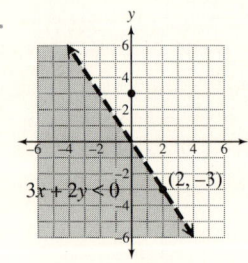

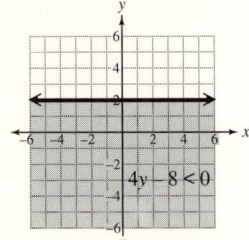

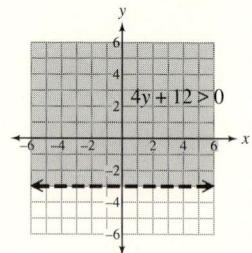

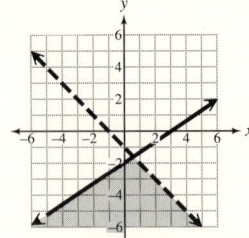

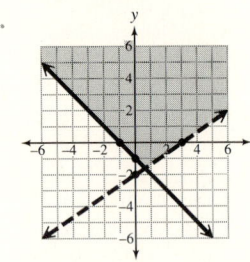

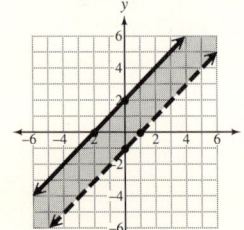

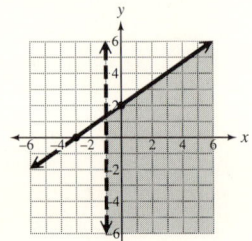

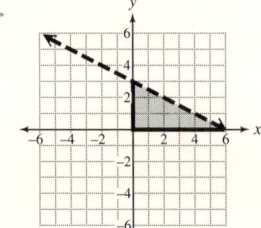

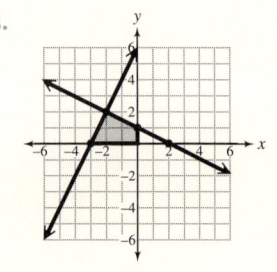

13. Let x = no. of Princess homes,
y = no. of Knight homes.
$$x \geq 6$$
$$y \geq 2$$
$$x - 3y \geq 0$$
$$30{,}000x + 20{,}000y \leq 600{,}000$$

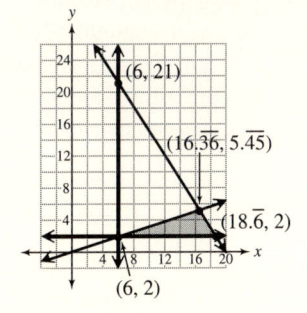

14. Let x = no. of Standard tables,
y = no. of Deluxe Tables.
$$x \geq 0$$
$$y \geq 0$$
$$3x + 6y \leq 96$$
$$x + 4y \leq 36$$

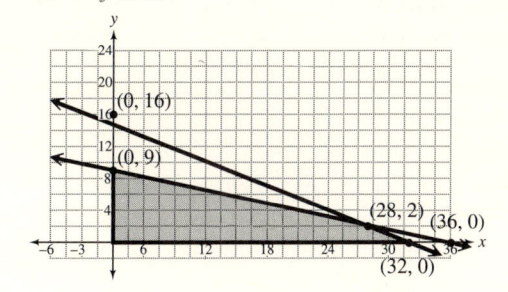

15. candy: $0.80/lb; nuts: $1.10/lb   16. 42°, 68°, 70°   17. 17 mi by 8 mi   18. $a = 1$; $b = 0$; $f(x) = e^x + 1$

6.4 EXERCISES B

In Exercises 1–6, sketch the graph of each inequality.

1. $3x - 2y > 6$
2. $2x + y < 3$
3. $x - y \geq -2$
4. $3x + 2y \geq 0$
5. $3x + 9 > 0$
6. $3x - 15 \leq 0$

In Exercises 7–12, sketch the graph of each system of inequalities.

7. $2x - 3y < 6$
   $x + y \leq -1$

8. $2x - 3y \geq 6$
   $x + y > -1$

9. $2x - 2y < -4$
   $x - y \geq 1$

10. $x - 2 \leq 0$
    $y + 1 > 0$

11. $x \geq 0$
    $y \leq 0$
    $x - y \leq 4$

12. $x \geq 1$
    $y \geq 2$
    $x + 3y \leq 19$
    $3x + 2y \leq 22$

In Exercises 13–14, graph the system described by the given information.

13. **BUSINESS** Ace Electronics has two models of video recorders in stock, the Elite and the Superior. The owner has found that he has to have at least 1 of the Elite model in stock, and at least 1 but no more than 3 of the Superior models in stock. At no time should the total number of recorders exceed 6. Using x for the number of Superior models and y for the number of Elite models, find a system of inequalities that describes this information, and graph the system.

14. **MANUFACTURING** Recreation Furniture manufactures two types of picnic tables, one that seats four and another that seats six. The table below shows the production information. The manufacture of each type of table cannot use more than the workhours available each week in each manufacturing stage. Form the system of inequalities described by this information and graph the system.

	Workhours for four-seat table	Workhours for six-seat table	Maximum workhours per week
Construction	6	8	120
Finishing	1	3	30

440 CHAPTER 6 SYSTEMS OF EQUATIONS AND INEQUALITIES

FOR REVIEW

Solve.

15. **CHEMISTRY** A chemist has one solution that is 5% salt and a second that is 10% salt. How many liters of each should be mixed to obtain 40 L of an 8% salt solution?

16. **INVESTMENT** Boris received an inheritance in the amount of $20,000. He invested the money in three different accounts, one paying 10%, one 11%, and the other 8%, simple interest. If the amount invested at 10% was three times that invested at 8% and his total interest earnings for the year amounted to $1960, how much was invested at each rate?

17. **NUMBER** The sum of the digits of a two-digit number is 5. If the digits are reversed, the resulting number is 9 less than the original. Find the number.

18. **CURVE-FITTING** Find the constants a and b for the function $f(x) = ae^x + be^{-x} + 1$ if $f(0) = 2$ and $f(\ln 2) = 9/2$. What is the function?

6.4 EXERCISES C

Graph each inequality in Exercises 1–3.

1. $y < |x|$
2. $|y| \leq x$ [Hint: Write as a system of two inequalities.]
3. $|x| - |y| \leq 1$
4. $|y| \leq |x|$

Find a system of inequalities whose graph is given in Exercises 5–6.

5.

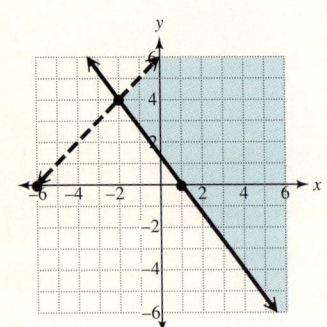

$$\begin{bmatrix} \text{Answer: } 4x + 3y \geq 4 \\ x - y > -6 \end{bmatrix}$$

6.

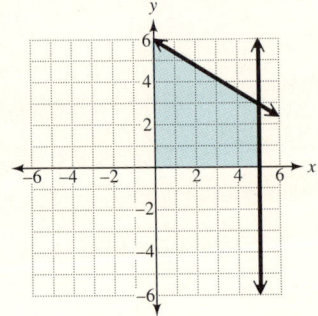

6.5 LINEAR PROGRAMMING

Systems of linear equations and inequalities provide the tools for a method of problem-solving called **linear programming**. Linear programming, developed by George B. Danzig in the 1940's, was first used by the military to aid in allocating supplies during World War II. Today it is widely used to solve problems of interest to the business community. One such problem involves finding the maximum or minimum value of an expression of the form

$$P = ax + by,$$

called an **objective function**, subject to certain limitations, called **constraints**, on the variables x and y. For example, $ax + by$ might denote the profit resulting from "programming" the resources of a production company in such a way that x units of one product and y units of another are produced. The constraints in the problem are given as a system of inequalities which must be graphed.

Suppose we consider the problem of maximizing the objective function $P = 50x + 35y$ subject to the following constraints:

$$x \geq 0$$
$$y \geq 0$$
$$x + y \leq 18$$
$$5x + 3y \leq 60$$

In other words, we want to find the largest value of $50x + 35y$ where x and y satisfy the above system of inequalities. Any pair (x, y) that solves the system of constraints is a **feasible solution** to the problem. The collection of all feasible solutions, shown in Figure 6.14, is the **feasible region.** Any feasible solution that determines a maximum value (or minimum value in other problems) is an **optimal solution** to the problem.

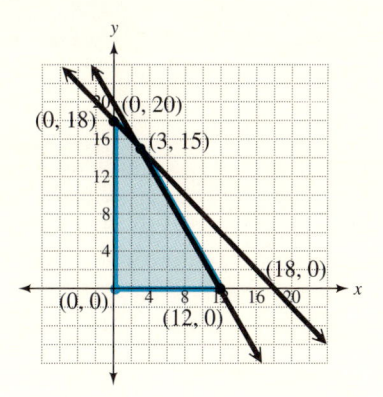

Figure 6.14 Feasible Region

In our work, the feasible region always has a boundary consisting of lines or line segments intersecting in points, each of which is called a **vertex.** The existence of an optimal solution depends on the nature of the feasible region, a consideration beyond the scope of our work. However, under the assumption that an optimal solution will exist, the next theorem tells us where it must be found.

Fundamental Theorem of Linear Programming

If a linear programming problem has an optimal solution, then at least one vertex of the feasible region will provide that solution.

Suppose we return to the problem of finding the maximum value of the objective function $P = 50x + 35y$ subject to the constraints graphed in Figure 6.14. The vertices (plural of vertex) are $(0, 0)$, $(0, 18)$, $(3, 15)$, and $(12, 0)$. From the above theorem, an optimal solution must be found among these pairs. The possibilities are summarized in a table.

Vertex	$50x + 35y$	
(0, 0)	0	
(0, 18)	630	
(3, 15)	675	← Optimal solution
(12, 0)	600	

Thus, considering all possible values of x and y such that (x, y) is in the feasible region, $x = 3$ and $y = 15$ produce the largest value of $50x + 35y$, which is 675.

We can summarize the general procedure in the following algorithm.

> **To Find an Optimal Solution to a Linear Programming Problem**
>
> 1. Graph the system of linear inequalities forming the constraints, and in so doing, identify the feasible region.
> 2. Find the vertices of the feasible region by solving, two at a time, the appropriate linear equations that describe the boundary of the region.
> 3. Complete a table of values for the objective function $P = ax + by$ using all of the vertices.
> 4. If $ax + by$ is to be maximized (minimized), the largest (smallest) value in the table is an optimal solution.

EXAMPLE 1 A BUSINESS PROBLEM

The Mutter Manufacturing Company makes two types of trailer hitches, a standard model and a heavy-duty model. It can produce up to a total of 18 hitches each day using up to 60 total workhours. Experience has shown that it takes 3 workhours to produce one standard hitch, while 5 workhours are needed to make one heavy-duty hitch. If a heavy-duty model returns a profit of $50 when sold, and a standard model returns a profit of $35, how many of each should be made in a day in order to maximize profit?

First, we translate the problem into a system of inequalities and identify the objective function.

Let x = the number of heavy-duty hitches produced daily,
   y = the number of standard hitches produced daily.

In a problem such as this, x and y cannot be negative, so that

$$x \geq 0$$
$$y \geq 0.$$

Up to a total of 18 hitches can be produced daily.

$$x + y \leq 18$$

Since $5x$ and $3y$ are, respectively, the number of workhours required daily to produce the desired number of heavy-duty and standard hitches, we have

$$5x + 3y \leq 60.$$

These four inequalities form the constraints for our problem, and we want to maximize the profit, given by the objective function

$$\text{profit} = 50x + 35y,$$

subject to these constraints.

We now recognize that these conditions are the same as those we analyzed at the beginning of this section so that the maximum profit of $675 will be realized when 3 heavy-duty hitches and 15 standard hitches are produced daily.

PRACTICE EXERCISE 1

Repeat the problem in Example 1, this time assuming that a heavy-duty hitch returns a profit of $30 when sold, and a standard-model hitch returns a profit of $70.

Answer: When 18 standard hitches and no heavy-duty hitches are made, the maximum profit will occur.

EXAMPLE 2 A PRODUCTION PROBLEM

Ponderosa Productions manufactures two different novelty items, Sam the Skunk and Willie the Weasel. For each item, three different phases of production are required, A, B, and C. To make one skunk, phase A takes 1 minute, phase B takes 4 minutes, and phase C takes 6 minutes. To complete one weasel, A takes 3 minutes, B takes 3 minutes, and C takes 1 minute. Due to maintenance conditions, phase A is available only 33 minutes each hour, while phases B and C are each in operation only 42 minutes every hour. The profit on each skunk is $2.00, and the profit on each weasel is $3.00. Find the number of units of each item that should be produced hourly to maximize profit.

Let x = the number of skunks produced each hour,
  y = the number of weasels produced each hour.

Each skunk requires 1 minute in phase A. Thus x skunks require $1x = x$ minutes. Each weasel requires 3 minutes in phase A, so that y weasels require $3y$ minutes. Hence, the total number of minutes per hour that phase A is in operation is $x + 3y$. Since A can be used no more than 33 minutes every hour, we have

$$x + 3y \leq 33.$$

Applying the same type of reasoning to phases B and C,

$$4x + 3y \leq 42$$
$$6x + y \leq 42.$$

These three inequalities, together with the obvious ones, $x \geq 0$ and $y \geq 0$, form the constraints to our problem. The feasible region is sketched in Figure 6.15 where the points (3, 10) and (6, 6) are found, respectively, by solving the systems:

$$\begin{array}{c} x + 3y = 33 \\ 4x + 3y = 42 \end{array} \quad \text{and} \quad \begin{array}{c} 4x + 3y = 42 \\ 6x + y = 42. \end{array}$$

The objective function to be maximized is

$$\text{profit} = 2x + 3y.$$

The possibilities are summarized in the following table.

Vertex	$2x + 3y$
(0, 11)	33
(3, 10)	36 ← Optimal solution
(6, 6)	30
(7, 0)	14
(0, 0)	0

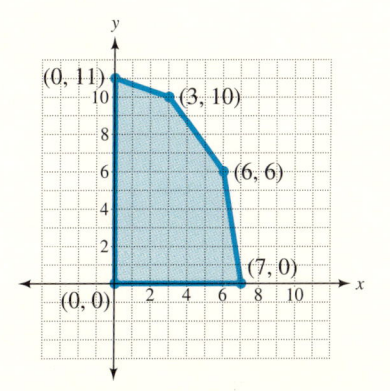

Figure 6.15 Feasible Region

Thus, the maximum profit will be realized when 3 skunks and 10 weasels are produced each hour.

PRACTICE EXERCISE 2

Repeat the problem in Example 2, this time assuming that the profits on each item are reversed.

Answer: When 6 of each are produced, the profit will be maximized.

444 CHAPTER 6 SYSTEMS OF EQUATIONS AND INEQUALITIES

Solving a linear programming problem is simply a matter of translating the given information into a system of inequalities, finding the vertices of the graph of the system, and substituting these values into the objective function. In the two examples above, the feasibility region was bounded. This is not always the case. For example, suppose we find the maximum and minimum values of the objective function $P = 3x + 7y$, subject to the constraints

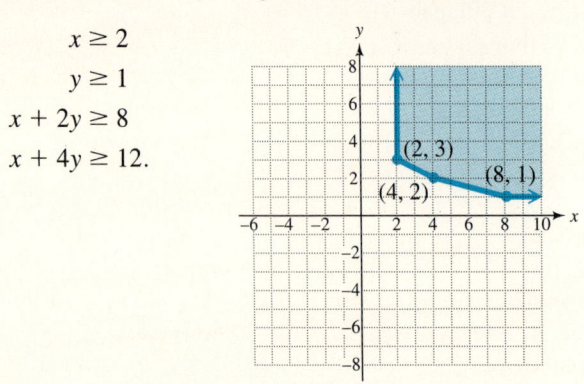

$$x \geq 2$$
$$y \geq 1$$
$$x + 2y \geq 8$$
$$x + 4y \geq 12.$$

Figure 6.16 Unbounded Feasible Region

The feasible region is shown in Figure 6.16 where the points $(2, 3)$, $(4, 2)$, and $(8, 1)$ are found, respectively, by solving the following three systems.

$$\begin{array}{lll} x = 2 & x + 2y = 8 & x + 4y = 12 \\ x + 2y = 8 & x + 4y = 12 & y = 1 \end{array}$$

Notice that in this case, the feasible region is unbounded and that points can be chosen to make $3x + 7y$ as large as we please. For example, the points $(10, 10)$, $(20, 20)$, $(30, 30)$, and so forth, are all in the feasible region. When $3x + 7y$ is evaluated at these points, the value gets larger and larger. Thus, it should be clear that $3x + 7y$ does not assume a maximum value in the region. The minimum value, just as in previous examples, will occur at a vertex.

Vertex	$3x + 7y$
(2, 3)	27
(4, 2)	26 ← Optimal solution
(8, 1)	31

Thus, the minimum value of $3x + 7y$ is 26.

6.5 EXERCISES A

In Exercises 1–2 find the maximum and minimum values of the given objective function, subject to the constraints graphed in the feasible region.

1. $P = x + 5y$

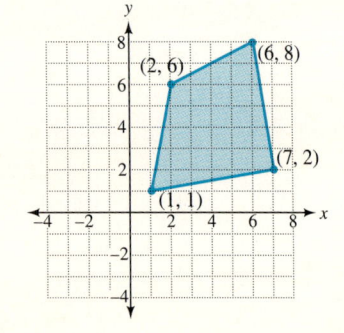

2. $P = 75x + 100y$

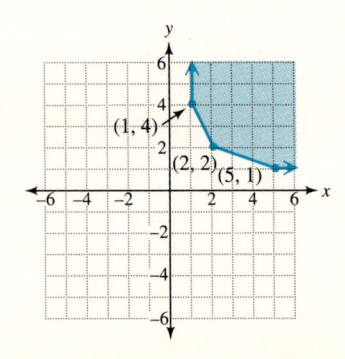

In Exercises 3–4 find the maximum and minimum values of the objective function $P = 10x + 50y$, subject to the given constraints.

3. $x \geq 0$
    $y \geq 0$
    $x + 2y \leq 6$
    $x + y \leq 4$

4. $x \geq 2$
    $2y \geq 1$
    $5x + 6y \geq 28$

Solve each linear programming problem in Exercises 5–8.

5. **BUSINESS** The Dribble-Well basketball manufacturing firm makes a profit of $20 on its True-Shot model and $13 on its True-Bounce model. To meet the demand of wholesalers, the daily production of True-Shots should be between 20 and 100, inclusive, whereas the number of True-Bounce models should be between 10 and 70, inclusive. To maintain quality control, the total number of balls produced each day should not exceed 150. How many of each type should be manufactured daily to maximize profit?

6. **BUSINESS** An oil refinery can produce up to 5000 barrels of oil per day. Two types are produced, type G, used for gasoline, and type H, used for heating oil. At least 1000 and at most 3500 barrels of type G must be produced each day. If there is a profit of $7 a barrel for G and $3 a barrel for H, find the number of barrels of each that should be produced to maximize the daily profit.

7. **AGRICULTURE** A dryland farmer has 100 acres on which he can grow corn and wheat. It costs $5 per acre for seed to plant corn and $8 per acre for seed to plant wheat. Labor and fuel costs amount to $20 per acre for corn and $12 per acre for wheat. If he is fortunate with the weather, he can earn $220 per acre on the corn and $250 per acre on the wheat. If he can spend up to $704 for seed and up to a total of $1640 for labor and fuel, how many acres of each should he plant to maximize his profit?

8. **MANUFACTURING** A company manufactures two products, A and B. Three different machines, X, Y, and Z are used to make each product. In order to make one unit of A, machine X must be used for 2 hours, machine Y for 1 hour, and machine Z for 1 hour. The manufacture of one unit of B requires 1 hour of X, 2 hours of Y, and 1 hour of Z. The profit is $275 on each unit of A and $180 on each unit of B. Machine X can be used at most 18 hours a day, Y can be used at most 20 hours, and Z at most 11 hours daily. How many of each product should be produced daily to maximize profit?

FOR REVIEW

In Exercises 9–10 sketch the graph of each system of inequalities.

9. $2x - 5y \leq 10$
    $x + 2y < -2$

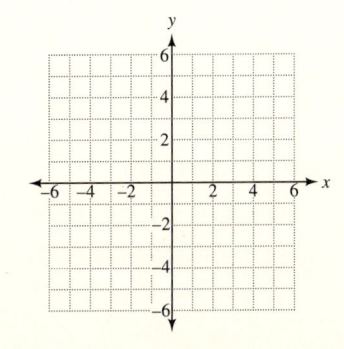

10. $y \geq x$
    $y \geq -x$

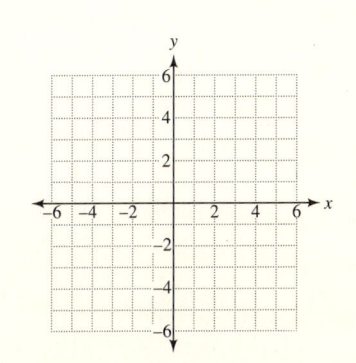

446 CHAPTER 6 SYSTEMS OF EQUATIONS AND INEQUALITIES

11. **SPORTS** An auditorium, to be used for a closed-circuit televised championship boxing match, seats 1000 spectators. Promoters plan to charge $20 for some seats, $10 for others, and hope to make at least $16,000 on the event. If at least 300 seats are to be sold for $10 each, find a system of inequalities that describes this situation and sketch its graph.

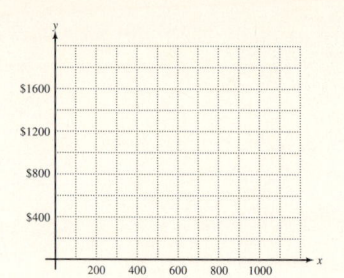

The following exercises will help you prepare for the material presented in the next section.

12. What is the degree of the polynomial $P(x) = (x - 2)(x + 3)^2(x^2 - 5)$?

13. If $2 + 3i$ is a zero of polynomial $P(x)$, with real coefficients, what is another zero of $P(x)$?

14. Suppose $x - 2$ is one factor of $P(x) = x^3 - x^2 - x - 2$. What is another factor?

15. Suppose $f(x) = \frac{1}{x-2} + \frac{2}{x+2}$. Find another representation for f which has only one term.

Solve each system of equations in Exercises 16–17.

16. $A + B = 3$
    $2A - 2B = -2$

17. $A + C = 4$
    $ B - 2C = -4$
    $-A + B + C = 4$

ANSWERS: 1. maximum value: 46; minimum value: 6   2. there is no maximum value; minimum value: 350   3. maximum value: 150; minimum value: 0   4. there is no maximum value; minimum value: 75   5. 100 True-Shot models and 50 True-Bounce models   6. 3500 barrels of type G and 1500 barrels of type H   7. 32 acres of corn and 68 acres of wheat   8. 7 units of A and 4 units of B
9.   10. 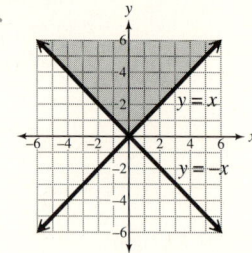   11. Let $x =$ no. of $20 seats,
$y =$ no. of $10 seats.
$x + y \leq 1000$
$20x + 10y \geq 16{,}000$
$x \geq 0$
$y \geq 300$

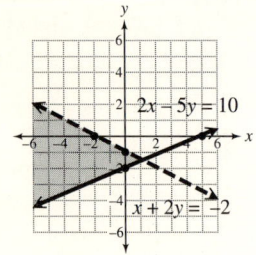

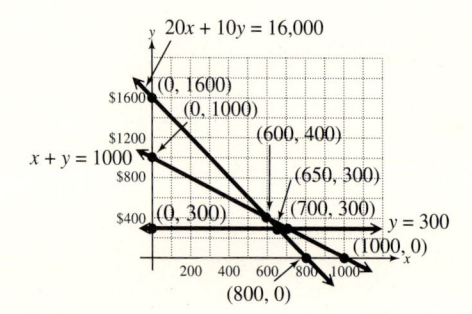

12. 5   13. $2 - 3i$ 14. $x^2 + x + 1$ 15. $f(x) = \frac{3x - 2}{x^2 - 4}$ 16. $A = 1$; $B = 2$ 17. $A = 1$; $B = 2$; $C = 3$

6.5 EXERCISES B

In Exercises 1–2 find the maximum and minimum values of the given objective function, subject to the constraints graphed in the feasible region.

1. $P = 5x + 2y$

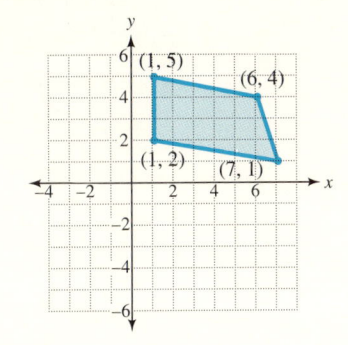

2. $P = 15x + 80y$

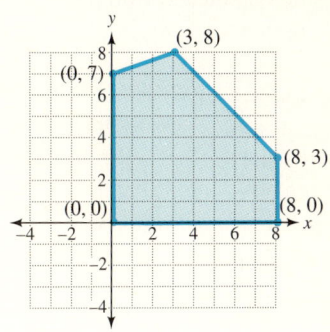

In Exercises 3–4 find the maximum and minimum values of the objective function $P = 10x + 50y$, subject to the given constraints.

3.
$$x \geq 1$$
$$y \geq 1$$
$$x + 5y \leq 26$$
$$3x + 2y \leq 26$$

4.
$$7 \geq x \geq 2$$
$$y \geq 1$$
$$3y - 2x \leq 11$$
$$2y + 3x \leq 29$$

Solve each linear programming problem in Exercises 5–8.

5. BUSINESS The Dribble-Well basketball manufacturing firm makes a profit of $20 on its True-Shot model and $20 on its True-Bounce model. To meet the demand of wholesalers, the daily production of True-Shots should be between 20 and 100, inclusive, whereas the number of True-Bounce models should be between 10 and 70, inclusive. To maintain quality control, the total number of balls produced each day should not exceed 150. How many of each type should be manufactured daily to maximize profit?

6. BUSINESS An oil refinery can produce up to 5000 barrels of oil per day. Two types are produced, type G, used for gasoline, and type H, used for heating oil. At least 1000 and at most 3500 barrels of type G must be produced each day. If there is a profit of $4 a barrel for G and $8 a barrel for H, find the number of barrels of each that should be produced to maximize the daily profit.

7. AGRICULTURE A dryland farmer has 100 acres on which he can grow corn and wheat. It costs $5 per acre for seed to plant corn and $8 per acre for seed to plant wheat. Labor and fuel costs amount to $20 per acre for corn and $12 per acre for wheat. If he is fortunate with the weather, he can earn $270 per acre on the corn and $160 per acre on the wheat. If he can spend up to $704 for seed and up to a total of $1640 for labor and fuel, how many acres of each should he plant to maximize his profit?

8. MANUFACTURING A company manufactures two products, A and B. Three different machines, X, Y, and Z are used to make each product. In order to make one unit of A, machine X must be used for 2 hours, machine Y for 1 hour, and machine Z for 1 hour. The manufacture of one unit of B requires 1 hour of X, 2 hours of Y, and 1 hour of Z. The profit is $200 on each unit of A and $210 on each unit of B. Machine X can be used at most 18 hours a day, Y can be used at most 20 hours, and Z at most 11 hours daily. How many of each product should be produced daily to maximize profit?

FOR REVIEW

In Exercises 9–10 sketch the graph of each system of inequalities.

9.
$$2x - 5y > 10$$
$$x + 2y \geq -2$$

10.
$$y \leq x$$
$$y \leq -x$$

11. **PRODUCTION** Two Wheel Inc. makes two types of racing bikes, the Standard and the Professional models. The table below summarizes the weekly production data. Give the various combinations of bicycles that can be made each week under the limitations on the workhours available in each stage of production.

	Workhours for Standard model	Workhours for Professional model	Maximum workhours per week
Construction	8	12	216
Finishing	2	2	48

The following exercises will help you prepare for the material presented in the next section.

12. What is the degree of the polynomial $P(x) = (x + 1)(x - 2)^3(x^2 + 5)$?

13. If $3 - 5i$ is a zero of polynomial $P(x)$, with real coefficients, what is another zero of $P(x)$?

14. Suppose $x + 1$ is one factor of $P(x) = x^3 + x^2 - 2x - 2$. What is another factor?

15. Suppose $f(x) = \frac{1}{x+1} + \frac{3}{x-1}$. Find another representation for f which has only one term.

Solve each system of equations in Exercises 16–17.

16. $A + B = -1$
    $2A - B = 4$

17. $A - B = -1$
    $B - C = 3$
    $A + B + C = 2$

6.5 EXERCISES C

Solve each linear programming problem.

1. **BUSINESS** A publishing company has two distribution centers, one in California (C), and another in Pennsylvania (P). There are 1500 copies of *College Algebra* stored at C and 1200 copies stored at P. Two schools, Central Technology University (CTU) and Mountain College (MC), order 1000 copies and 750 copies, respectively. To ship from C to CTU, the cost is $0.50 per book; from C to MC, the cost is $0.40 per book; from P to CTU, the cost is $0.45 per book; and from P to MC, the cost is $0.60 per book. How should the order be filled to minimize the total shipping cost? [*Hint:* If x is the number of books shipped to CTU from C, then $1000 - x$ is the number shipped to CTU from P.]

   [Answer: all 1000 books shipped from P to CTU; all 750 books shipped from C to MC]

2. Repeat the problem in Exercise 1 using the costs: $0.30 per book from C to CTU, $0.50 per book from C to MC, $0.35 per book from P to CTU, and $0.45 per book from P to MC.

6.6 PARTIAL FRACTIONS

STUDENT GUIDEPOSTS

1. Conditions for Using Partial Fractions
2. Types of Decomposition of a Rational Function

In Section 1.5 we reviewed addition of rational expressions such as

$$\frac{1}{x-2} + \frac{2}{x+2} = \frac{3x-2}{x^2-4}.$$

In this section, the reverse procedure is considered. Suppose we are given the rational function

$$f(x) = \frac{3x-2}{x^2-4}$$

and want to express $f(x)$ as the sum of two simpler fractions, called **partial fractions.** The partial fraction decomposition of f is, in fact,

$$f(x) = \frac{1}{x-2} + \frac{2}{x+2}.$$

Decompositions like this are important in calculus and other advanced courses. They involve solving a system of linear equations.

1 CONDITIONS FOR USING PARTIAL FRACTIONS

The partial fraction decomposition of a rational function

$$f(x) = \frac{P(x)}{Q(x)}$$

depends on two basic assumptions:

1. The degree of $P(x)$ is less than the degree of $Q(x)$.
2. The polynomial $Q(x)$ can be factored into a product of powers of linear and quadratic factors in such a way that the quadratic factors have no real zeros.

Neither of these two assumptions should seem unreasonable. In fact, if $f(x)$ does not satisfy the first, we could divide $Q(x)$ into $P(x)$ and obtain a fractional remainder that satisfies this requirement. For example,

$$f(x) = \frac{x^3 + 2x + 1}{x^2 - 3}$$

can be written as

$$f(x) = x + \frac{5x+1}{x^2-3}. \quad \text{After dividing by } x^2 - 3$$

In this form we can concentrate on the fractional part of $f(x)$ which does satisfy the first condition. As for the second assumption, we know from Chapter 4 that a polynomial of degree n has n zeros (real or complex) and, as a result, can be factored into n linear factors. Since the complex zeros occur in conjugate pairs, the linear factors corresponding to these pairs can be multiplied to give one quadratic factor with real coefficients.

2 TYPES OF DECOMPOSITION OF A RATIONAL FUNCTION

> **Partial Fraction Decomposition of a Rational Function**
>
> Let $f(x) = \dfrac{P(x)}{Q(x)}$ be a rational function reduced to lowest terms, with the degree of $P(x)$ less than the degree of $Q(x)$.
>
> 1. If $(ax + b)^k$ is a factor of $Q(x)$, for k a natural number, then the partial fraction decomposition of $f(x)$ contains terms of the form
>
> $$\frac{A_1}{(ax + b)} + \frac{A_2}{(ax + b)^2} + \cdots + \frac{A_k}{(ax + b)^k},$$
>
> where $A_1, A_2, \ldots, A_k$ are constant real numbers.
>
> 2. If $(ax^2 + bx + c)^m$ is a factor of $Q(x)$, for m a natural number, then the partial fraction decomposition of $f(x)$ contains terms of the form
>
> $$\frac{B_1 x + C_1}{(ax^2 + bx + c)} + \frac{B_2 x + C_2}{(ax^2 + bx + c)^2} + \cdots + \frac{B_m x + C_m}{(ax^2 + bx + c)^m},$$
>
> where $B_1, B_2, \ldots, B_m$ and $C_1, C_2, \ldots, C_m$ are constant real numbers.

Decomposing a fraction into partial fractions, best illustrated by examples, involves finding the linear and quadratic factors of the denominator, expressing the function in terms of general partial fractions, determining a system of linear equations, and solving the system. The system of equations is determined by using the fact that if two polynomials are equal, the coefficients of like terms are equal. (See Exercise 2 in 4.3 Exercises C.) We begin by finding the decomposition of the rational function given earlier.

EXAMPLE 1   LINEAR FACTORS TO THE FIRST POWER

Find the partial fraction decomposition of

$$f(x) = \frac{3x - 2}{x^2 - 4}.$$

Since the degree of the numerator is 1 and the degree of the denominator is 2, we proceed by factoring the denominator.

$$f(x) = \frac{3x - 2}{(x - 2)(x + 2)}$$

Using the partial fraction decomposition theorem, the decomposition must have two terms $\dfrac{A}{x - 2}$ and $\dfrac{B}{x + 2}$. Thus,

$$\frac{3x - 2}{(x - 2)(x + 2)} = \frac{A}{x - 2} + \frac{B}{x + 2},$$

where A and B are constant real numbers, yet to be determined. Multiply both sides of this equation by the LCD, $(x - 2)(x + 2)$.

$$3x - 2 = A(x + 2) + B(x - 2)$$
$$3x - 2 = Ax + 2A + Bx - 2B$$
$$3x - 2 = (A + B)x + (2A - 2B)$$

PRACTICE EXERCISE 1

Find the partial fraction decomposition of

$$f(x) = \frac{x - 3}{x^2 - 1}.$$

Since the polynomials on both sides are equal, the coefficients of the like terms must be equal. Equating the coefficients of x and equating the constant terms gives the following system of two linear equations in the two variables A and B.

$$A + B = 3$$
$$2A - 2B = -2$$

Solving this system gives $A = 1$ and $B = 2$. Then

$$f(x) = \frac{3x - 2}{x^2 - 4} = \frac{A}{x - 2} + \frac{B}{x + 2}$$

$$= \frac{1}{x - 2} + \frac{2}{x + 2}. \quad \text{Substitute 1 for } A \text{ and 2 for } B$$

Notice that we obtained the sum of fractions given in our introductory remarks.

Answer: $f(x) = \frac{2}{x+1} + \frac{-1}{x-1}$

Note For rational functions such as the one in Example 1 in which the denominator involves only linear factors raised to the first power, there is a quicker way to find A and B. By substituting values for x that make the various factors zero, A and B can be found without solving a system. For example, suppose we begin with

$$3x - 2 = A(x + 2) + B(x - 2).$$

Let $x = -2$: $\quad 3(-2) - 2 = A(-2 + 2) + B(-2 - 2) \quad$ When $x = -2$, A is eliminated
$$-8 = A(0) \quad + B(-4)$$
$$-8 = -4B$$
$$2 = B$$

Let $x = 2$: $\quad 3(2) - 2 = A(2 + 2) + B(2 - 2) \quad$ When $x = 2$, B is eliminated
$$4 = A(4) \quad + B(0)$$
$$4 = 4A$$
$$1 = A$$

EXAMPLE 2 LINEAR FACTORS WITH HIGHER POWERS

Find the partial fraction decomposition of

$$f(x) = \frac{4x^2 - 4x + 4}{(x - 1)^2(x + 1)}.$$

Using the partial fraction decomposition theorem,

$$\frac{4x^2 - 4x + 4}{(x - 1)^2(x + 1)} = \frac{A}{x - 1} + \frac{B}{(x - 1)^2} + \frac{C}{x + 1}.$$

Multiply both sides by the LCD, $(x - 1)^2(x + 1)$.

$$4x^2 - 4x + 4 = A(x - 1)(x + 1) + B(x + 1) + C(x - 1)^2.$$

Clear parentheses and collect like terms.

$$4x^2 - 4x + 4 = (A + C)x^2 + (B - 2C)x + (-A + B + C).$$

PRACTICE EXERCISE 2

Find the partial fraction decomposition of

$$f(x) = \frac{-2x^2 + 11x - 12}{(x - 1)(x - 2)^2}.$$

452 CHAPTER 6 SYSTEMS OF EQUATIONS AND INEQUALITIES

Equate coefficients of like terms to obtain the following system.

$$A \quad + C = 4$$
$$B - 2C = -4$$
$$-A + B + C = 4$$

Solving this system gives $A = 1$, $B = 2$, and $C = 3$. Thus,

$$f(x) = \frac{4x^2 - 4x + 4}{(x-1)^2(x+1)} = \frac{1}{x-1} + \frac{2}{(x-1)^2} + \frac{3}{x+1}.$$

Answer: $f(x) = \frac{1}{x-2} + \frac{2}{(x-2)^2} + \frac{-3}{x-1}$

The next example illustrates what to do when the denominator has one linear and one quadratic factor.

EXAMPLE 3   ONE LINEAR AND ONE QUADRATIC FACTOR

Find the partial fraction decomposition of

$$f(x) = \frac{-7x - 1}{(x^2 + 2)(x - 3)}.$$

The form of the decomposition is

$$\frac{-7x - 1}{(x^2 + 2)(x - 3)} = \frac{Ax + B}{x^2 + 2} + \frac{C}{x - 3}.$$

$-7x - 1 = (Ax + B)(x - 3) + C(x^2 + 2)$ Multiply both sides by the LCD, $(x^2 + 2)(x - 3)$

$-7x - 1 = (A + C)x^2 + (-3A + B)x + (-3B + 2C)$ Clear parentheses and collect like terms

Thus, we obtain the system

$$A \quad\quad + C = 0$$
$$-3A + B \quad\quad = -7$$
$$\quad\quad -3B + 2C = -1,$$

which has solutions $A = 2$, $B = -1$, and $C = -2$. Hence,

$$f(x) = \frac{-7x - 1}{(x^2 + 2)(x - 3)} = \frac{2x - 1}{x^2 + 2} + \frac{-2}{x - 3}.$$

PRACTICE EXERCISE 3

Find the partial fraction decomposition of

$$f(x) = x^2 + x + 10 \ (x^2 + 5)(x - 1)d.$$

Answer: $f(x) = \frac{-x}{x^2 + 5} + \frac{2}{x - 1}$

EXAMPLE 4   DIVIDING FIRST BY THE DENOMINATOR

Find the partial fraction decomposition of

$$f(x) = \frac{x^5 + 5x^3 + x^2 + 5x + 1}{x^4 + 2x^2 + 1}.$$

Since the degree of the numerator exceeds the degree of the denominator, first divide the numerator by the denominator to obtain the following form.

$$f(x) = x + \frac{3x^3 + x^2 + 4x + 1}{x^4 + 2x^2 + 1}$$

PRACTICE EXERCISE 4

Find the partial fraction decomposition of

$$f(x) = \frac{x^5 + 6x^3 + x^2 + 9x + 1}{x^4 + 4x^2 + 4}$$

We can now concentrate on the fraction which satisfies the hypotheses of the decomposition theorem. First factor the denominator.

$$\frac{3x^3 + x^2 + 4x + 1}{x^4 + 2x^2 + 1} = \frac{3x^3 + x^2 + 4x + 1}{(x^2 + 1)^2}$$

Then
$$\frac{3x^3 + x^2 + 4x + 1}{(x^2 + 1)^2} = \frac{Ax + B}{x^2 + 1} + \frac{Cx + D}{(x^2 + 1)^2}.$$

Multiply both sides by the LCD, $(x^2 + 1)^2$, clear parentheses, and collect like terms.

$$3x^3 + x^2 + 4x + 1 = Ax^3 + Bx^2 + (A + C)x + (B + D)$$

Immediately we recognize that $A = 3$ and $B = 1$. Since $A + C = 4$ and $A = 3$, $C = 1$. Also, with $B + D = 1$ and $B = 1$, $D = 0$. Therefore,

$$f(x) = \frac{x^5 + 5x^3 + x^2 + 5x + 1}{x^4 + 2x^2 + 1} = x + \frac{3x + 1}{x^2 + 1} + \frac{x}{(x^2 + 1)^2}.$$

Answer: $f(x) = x + \frac{2x+1}{x^2 + 2} + \frac{x-1}{(x^2 + 2)^2}$

6.6 EXERCISES A

In Exercises 1–6 use constants A, B, C, and D, to express the partial fraction decomposition of each rational function. Do not solve for the constants.

1. $f(x) = \dfrac{3x + 2}{(x - 1)(x + 5)}$

2. $f(x) = \dfrac{6x^2 - 7}{(x + 3)^2(x + 5)}$

3. $f(x) = \dfrac{x^3 - 5}{(x^2 + 2x + 10)^2}$

4. $f(x) = \dfrac{x + 1}{x^3 - 2x^2 - 3x}$

5. $f(x) = \dfrac{x - 3}{x^2(x + 2) - 2x(x + 2) - 3(x + 2)}$

6. $f(x) = \dfrac{x^2 + 5x + 5}{x^4 + 5x^2 + 4}$

In Exercises 7–12 find the real numbers A, B, C, D, and E so that each rational function f(x) has the given fractional decomposition.

7. $f(x) = \dfrac{5x - 1}{x^2 - 1} = \dfrac{A}{x - 1} + \dfrac{B}{x + 1}$

8. $f(x) = \dfrac{1 - x}{(x + 1)^2} = \dfrac{A}{x + 1} + \dfrac{B}{(x + 1)^2}$

9. $f(x) = \dfrac{6x^2 + 1}{(x^2 + 1)(x - 2)} = \dfrac{Ax + B}{x^2 + 1} + \dfrac{C}{x - 2}$

10. $f(x) = \dfrac{x^3 + x^2 + 3x + 1}{(x^2 + 2)^2} = \dfrac{Ax + B}{x^2 + 2} + \dfrac{Cx + D}{(x^2 + 2)^2}$

454 CHAPTER 6 SYSTEMS OF EQUATIONS AND INEQUALITIES

11. $f(x) = \dfrac{-2x^2 + 15x + 13}{(x-2)^2(x+3)} = \dfrac{A}{x-2} + \dfrac{B}{(x-2)^2} + \dfrac{C}{x+3}$

12. $f(x) = \dfrac{4x}{(x^2+1)^2(x-1)} = \dfrac{Ax+B}{x^2+1} + \dfrac{Cx+D}{(x^2+1)^2} + \dfrac{E}{x-1}$

    [*Hint:* After multiplying by the LCD, clearing parentheses, and combining like terms, substitute 1 for x to find E first.]

In Exercises 13–16 find the partial fraction decomposition of each rational function.

13. $f(x) = \dfrac{3x-3}{x^2-9}$

14. $f(x) = \dfrac{6x^2 + 20x + 19}{(x+2)^2(x+1)}$

15. $f(x) = \dfrac{-9x-7}{(x^2+1)(x-5)}$

16. $f(x) = \dfrac{2x^3 - x^2 + 3x - 1}{(x^2+1)^2}$

FOR REVIEW

17. Find the maximum and minimum values of $P = 2x - 5y$ subject to the constraints graphed in the feasible region shown.

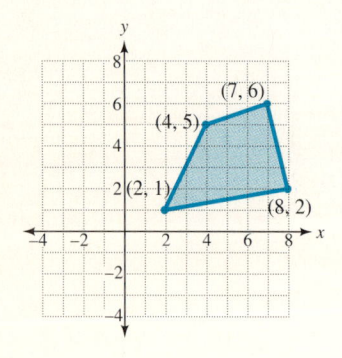

18. **BUSINESS** The Backcourt Manufacturing firm makes a profit of $15 on its Pro-Model tennis racquet and $10 on its Hacker-Model racquet. To meet the wholesale demand, the daily production of the Pro-Model should be between 10 and 50, inclusive, whereas the Hacker-Model should be between 20 and 40, inclusive. To maintain quality control, the total number of racquets manufactured daily should not exceed 80. How many of each type of racquet should be manufactured daily to maximize profits?

ANSWERS: 1. $\dfrac{3x+2}{(x-1)(x+5)} = \dfrac{A}{x-1} + \dfrac{B}{x+5}$ 2. $\dfrac{6x^2-7}{(x+3)^2(x+5)} = \dfrac{A}{x+3} + \dfrac{B}{(x+3)^2} + \dfrac{C}{x+5}$ 3. $\dfrac{x^3-5}{(x^2+2x+10)^2} = \dfrac{Ax+B}{x^2+2x+10} + \dfrac{Cx+D}{(x^2+2x+10)^2}$ 4. $\dfrac{x+1}{x^3-2x^2-3x} = \dfrac{A}{x} + \dfrac{B}{x-3}$ 5. $\dfrac{x-3}{x^2(x+2) - 2x(x+2) - 3(x+2)} = \dfrac{A}{x+2} + \dfrac{B}{x+1}$ 6. $\dfrac{x^2+5x+5}{x^4+5x^2+4} = \dfrac{Ax+B}{x^2+1} + \dfrac{Cx+D}{x^2+4}$ 7. $A = 2$; $B = 3$ 8. $A = -1$; $B = 2$ 9. $A = 1$; $B = 2$; $C = 5$ 10. $A = 1$; $B = 1$; $C = 1$; $D = -1$ 11. $A = 0$; $B = 7$; $C = -2$ 12. $A = -1$; $B = -1$; $C = -2$; $D = 2$; $E = 1$ 13. $f(x) = \dfrac{1}{x-3} + \dfrac{2}{x+3}$ 14. $f(x) = \dfrac{1}{x+2} + \dfrac{-3}{(x+2)^2} + \dfrac{5}{x+1}$ 15. $f(x) = \dfrac{2x+1}{x^2+1} + \dfrac{-2}{x-5}$ 16. $f(x) = \dfrac{2x-1}{x^2+1} + \dfrac{x}{(x^2+1)^2}$ 17. maximum value: 6; minimum value: -17 18. 50 Pro-Model racquets and 30 Hacker-Model racquets

6.6 EXERCISES B

In Exercises 1–6 use constants A, B, C, and D, to express the partial fraction decomposition of each rational function. Do not solve for the constants.

1. $f(x) = \dfrac{4x^2 + 1}{(x - 2)(x + 2)(x - 5)}$

2. $f(x) = \dfrac{x^2 + x + 1}{(x^2 + x + 5)(x + 4)}$

3. $f(x) = \dfrac{x^2 - x + 5}{(x - 3)^2(x^2 - 2x + 7)}$

4. $f(x) = \dfrac{x^2 + x - 4}{x^3 - 3x^2 - 10x}$

5. $f(x) = \dfrac{x + 2}{x^4 - 16}$

6. $f(x) = \dfrac{x^2 - 4x - 5}{x^2(x^2 - 1) - x(x^2 - 1) - 20(x^2 - 1)}$

In Exercises 7–12 find the real numbers A, B, C, D, and E so that each rational function f(x) has the given fractional decomposition.

7. $f(x) = \dfrac{7x + 16}{(x - 2)(x + 4)} = \dfrac{A}{x - 2} + \dfrac{B}{x + 4}$

8. $f(x) = \dfrac{4x - 13}{(x - 3)^2} = \dfrac{A}{x - 3} + \dfrac{B}{(x - 3)^2}$

9. $f(x) = \dfrac{7x^2 + 16x + 17}{(x^2 + x + 1)(x + 4)} = \dfrac{Ax + B}{x^2 + x + 1} + \dfrac{C}{x + 4}$

10. $f(x) = \dfrac{5x^3 - 15x^2 + 26x - 2}{(x^2 - 3x + 5)^2} = \dfrac{Ax + B}{x^2 - 3x + 5} + \dfrac{Cx + D}{(x^2 - 3x + 5)^2}$

11. $f(x) = \dfrac{-3x^3 - 2x^2 + 2x + 19}{(x + 1)^2(x - 2)^2} = \dfrac{A}{x + 1} + \dfrac{B}{(x + 1)^2} + \dfrac{C}{x - 2} + \dfrac{D}{(x - 2)^2}$

12. $f(x) = \dfrac{3x^4 + 6x^3 + 3x^2 - 3x}{(x + 2)^2(x^2 + x + 1)^2} = \dfrac{A}{x + 2} + \dfrac{Bx + C}{x^2 + x + 1} + \dfrac{Dx + E}{(x^2 + x + 1)^2}$

In Exercises 13–16 find the partial fraction decomposition of each rational function.

13. $f(x) = \dfrac{-4x - 1}{(x - 2)(x + 1)}$

14. $f(x) = \dfrac{4x^2 - 8x - 8}{x^3 - 4x}$

15. $f(x) = \dfrac{3x^2 + x + 3}{(2x^2 + 3)(x + 1)}$

16. $f(x) = \dfrac{2x^3 + 7x + 1}{x^4 + 4x^2 + 4}$

FOR REVIEW

17. Find the maximum and minimum values of $P = 10x - 5y$ subject to the given constraints.

$$6 \geq x \geq 0$$
$$y \geq 0$$
$$x + 3y \leq 12$$

18. EDUCATION An exam consists of two types of questions, short-answer (SA) that are worth 3 points each, and essay (E) worth 8 points each. Suppose it takes you 2 minutes for each SA question and 10 minutes for each E question. You are not allowed to answer more than 18 questions, and there is a 60-minute time limit for taking the exam. If you assume all your answers are correct, how many of each type question should you answer to get the best possible score?

6.6 EXERCISES C

In Exercises 1–4 find the partial fraction decomposition of each rational function.

1. $f(x) = \dfrac{7x^2 + 13x + 10}{x^3 + 2x^2 - 4x - 8}$

$\left[\text{Answer: } f(x) = \dfrac{3}{x+2} + \dfrac{-3}{(x+2)^2} + \dfrac{4}{x-2}\right]$

2. $f(x) = \dfrac{5x^2 + 4x + 1}{x^3 - x^2 + 4x - 4}$

3. $f(x) = \dfrac{x^3 - 2x^2 + 7x - 2}{x^2 - 2x + 1}$

$\left[\text{Answer: } f(x) = x + \dfrac{6}{x-1} + \dfrac{4}{(x-1)^2}\right]$

4. $f(x) = \dfrac{2x^3 + x^2 - 22x + 17}{x^2 + x - 12}$

CHAPTER 6 EXTENSION PROJECT

How can a computer be used to solve a system of equations? As a matter of fact, there are several ways. Suppose we concentrate for the moment on a linear system of two equations in two variables,

$$ax + by = c$$
$$dx + ey = f, \quad (1)$$

where a, b, c, d, e, and f are real-number constants. You can solve this system using the elimination method to obtain

$$x = \dfrac{ce - bf}{ae - bd} \quad \text{and} \quad y = \dfrac{af - cd}{ae - bd}.$$

Using these general solutions we can easily write a program for calculating x and y, as shown in PROGRAM I. Also given with this program is a run which solves the system

$$2x + 3y = 5$$
$$x - 5y = -17.$$

Suppose we consider a general 3×3 system,

$$ax + by + cz = d$$
$$ex + fy + gz = h \quad (2)$$
$$ix + jy + kz = l.$$

If we solve this system using the reduction method,

$$x = \dfrac{dfk + bgl + chj - cfl - dgj - bhk}{afk + bgi + cej - cfi - agj - bek}$$

$$y = \dfrac{ahk + dgi + cel - chi - agl - dek}{afk + bgi + cej - cfi - agj - bek}$$

$$z = \dfrac{afl + bhi + dej - dfi - ahj - bel}{afk + bgi + cej - cfi - agj - bek}.$$

With a program similar to PROGRAM I, x, y, and z could be evaluated for any given system. Rather than doing this, however, suppose we look at the **least squares regression parabola** model used to find the parabola that best fits four given points, (x_1, y_1), (x_2, y_2), (x_3, y_3), and (x_4, y_4) discussed in Exercises C of Section 6.2. By combining the method for solving a system of three equations in three variables with the system for obtaining the coefficients in the least squares parabola, we present a program, PROGRAM II, that finds the equation for the least squares regression parabola. Given with the program is a run that finds the equation requested in Exercise 5 (Exercises C, Section 6.2). Compare this with what you obtained then.

THINGS TO DO:

1. Solve the systems in (1) and (2) to verify the solutions given.

2. Use PROGRAM I to solve some of the exercises in Section 6.1. What happens when $ae - bd = 0$?

3. Write a program for solving a general 3×3 system by using parts of PROGRAM II. (We used notation there that will help you.)

4. Use your program in 3 to solve some of the systems in Section 6.2.

5. Find the least squares parabola that fits other pairs of four points and consider its graph in relation to the points.

6. Use PROGRAM I as a model and write a program that will find the vertices, test them, and solve a linear programming problem.

```
                          PROGRAM II
10 FOR M = 1 TO 4
20 INPUT X(M),Y(M)                PROGRAM I
30 NEXT M
40 A=0                10 INPUT A,B,C,D,E,F
50 D=0                20 X = (C*E-B*F)/(A*E-B*D)
60 B=0                30 Y = (A*F-C*D)/(A*E-B*D)
70 E=0                40 PRINT "THE SOLUTION IS: (";X;",";Y:")"
80 H=0                50 END
90 I=0
100 L=0               RUN
110 FOR M = 1 TO 4    ? 2,3,5,1,-5,-17
120 A=A+X(M)^2        THE SOLUTION IS: (-2,3)
130 B=B+X(M)
140 D=D+Y(M)
150 E=E+X(M)^3
160 H=H+X(M)*Y(M)     RUN
170 I=I+X(M)^4        ? -1,0
180 L=L+X(M)^2*Y(M)   ? 0,2
190 NEXT M            ? 0,1
200 C=4               ? -2,0
210 F=A               THE LEAST SQUARES REGRESSION PARABOLA THAT FITS
220 G=B               THE POINT IS: y = ( .75 )x^2 + ( 2.25 )x + ( 1.5 )
230 J=E
240 K=A
250 NX=D*F*K+B*G*L+C*H*J-C*F*L-D*G*J-B*H*K
260 NY=A*H*K+D*G*I+C*E*L-C*H*I-A*G*L-D*E*K
270 NZ=A*F*L+B*H*I+D*E*J-D*F*I-A*H*J-B*E*L
280 AD=A*F*K+B*G*I+C*E*J-C*F*I-A*G*J-B*E*K
290 U=NX/AD
300 V=NY/AD
310 W=NZ/AD
320 PRINT"THE LEAST SQUARES REGRESSION PARABOLA THAT FITS"
330 PRINT "THE POINTS IS: y = (";U;")x^2 + (";V;")x + (";W;")"
340 END
```

CHAPTER 6 REVIEW

KEY TERMS

6.1

1. A **solution** to a **system of two linear equations in two variables**,

$$a_1x + b_1y = c_1$$
$$a_2x + b_2y = c_2,$$

is an ordered pair of numbers that is a solution to *both* equations in the system.

2. If a system has infinitely many solutions (the graphs of the equations are coinciding lines), the system is **dependent**.

3. If a system has no solution (the graphs of the equations are parallel lines), the system is **inconsistent**.

4. If a system has exactly one solution (the graphs of the equations are intersecting lines), the system is **consistent** and **independent**.

5. Two systems are **equivalent** if they have exactly the same solutions.

6.2

1. A solution to a **system of three linear equations in three variables** is an **ordered triple** of numbers that solves each equation in the system.

2. A system of n linear equations in n variables is an **$n \times n$ system** or a **square system of order n**.

3. Systems of equations that have fewer equations than variables are called **nonsquare systems**.

4. If all the constant terms in a system of equations are zero, the system is **homogeneous**. The one obvious solution to a homogeneous system, (0, 0, 0), is the **trivial solution**.

6.4

1. A **linear inequality in two variables** x and y is an inequality of the form $ax + by < c$, $ax + by > c$, $ax + by \leq c$, or $ax + by \geq c$, where a, b, and c are real numbers, a and b not both zero.

2. The graph of the line that results when an inequality symbol is replaced with an equal sign in an inequality, $ax + by = c$, is the **boundary line** of the graph, and it divides the plane into **half-planes.**

3. A **test point** (often (0, 0)) is used to find the half-plane that contains points on the graph of a linear inequality.

4. The graph of a **system of linear inequalities** is the region in the plane common to the graphs of each inequality in the system. If the graph extends infinitely far in some directions, the graph is **unbounded;** otherwise, it is **bounded.**

6.5 **Linear programming** is a technique to find maximum or minimum values of an **objective function** subject to certain **constraints** on the variables given by a system of linear inequalities. Any ordered pair of numbers that solves the system of constraints is a **feasible solution** to the problem. The collection of all feasible solutions is the **feasible region.** Any feasible solution that determines a maximum or minimum value of the objective function is an **optimal solution** to the problem. A **vertex** of the feasible region is a point of intersection of two boundary lines. **The Fundamental Theorem of Linear Programming** states that if a linear programming problem has an optimal solution, then at least one vertex of the feasible region will provide the solution.

6.6 When a rational function is written as a sum of two or more simpler fractions, the function is said to be **decomposed into partial fractions.**

REVIEW EXERCISES

Part I

6.1   **1.** State whether $(2, -3)$ is a solution to the system.

$$3x + y = 3$$
$$-x + 2y = -8$$

2. Solve the following system using the substitution method.

$$3x + y = -1$$
$$-2x - 3y = -11$$

3. Solve the following system using the elimination method.

$$3x + 5y = -7$$
$$-4x + 2y = -8$$

4. Find the value(s) of m so that the system is dependent.

$$2x - 3y = m$$
$$-4x + 6y = 5$$

In Exercises 5–6 solve each system using either substitution or elimination, whichever seems more appropriate.

5. $\quad 3x - 5y = 2$
$\quad -6x + 10y = -1$

6. $\dfrac{3}{x} + \dfrac{1}{y} = 2$

$\dfrac{5}{x} - \dfrac{3}{y} = 8$

6.2   Solve each system in Exercises 7–9.

7. $2x + 3y + 4z = 0$
$\quad x - 5y + 3z = -1$
$\quad 3x + y + z = 5$

8. $x - 2y + 4z = 0$
$\quad x + 2y - 5z = 0$
$\quad 2x \quad\quad - z = 0$

9. $3x + 2y \quad\quad = 12$
$\quad -x + y - 5z = 1$

10. The parabola with equation $y = ax^2 + bx + c$ passes through the points $(1, 4)$, $(-1, 10)$, and $(3, 14)$. Find the values of a, b, and c.

6.3   *In Exercises 11–15 solve using a system of equations.*

11. RATE-MOTION By going 40 mph for one period of time and then 50 mph for another, a man traveled 370 miles. If he had gone 55 mph throughout the same period of time, he would have traveled 440 miles. How long did he travel at each speed?

12. GEOMETRY Two angles are complementary, and one measures 6° less than seven times the measure of the other. Find the measure of each angle.

13. **BUSINESS** How many pounds of candy at $1.60 per lb and how many pounds of candy at $1.20 per lb must be mixed to obtain 80 pounds of a mixture worth $1.36 per lb?

14. **INVESTMENT** Jack Pritchard has $9000 divided into three separate investments. Part is invested in bonds that earn 9%, part is invested in certificates that earn 8%, and part is invested in stocks. If the stocks do well, they will earn 10% and his total earnings will amount to $780. If the market exhibits a downward trend, his stocks will lose 3%, and his total earnings will amount to $520. How much is invested in each category?

15. The polynomial function $P(x) = x^3 + ax + b$ has the property that $P(1) = 4$ and $P(-1) = 12$. Find the values of a and b.

6.4 *In Exercises 16–17 sketch the graph of each system of linear equations.*

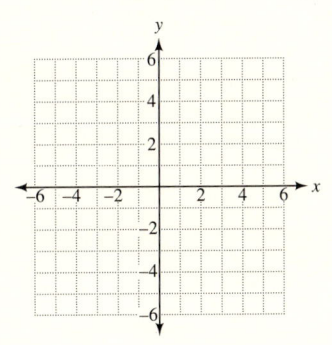

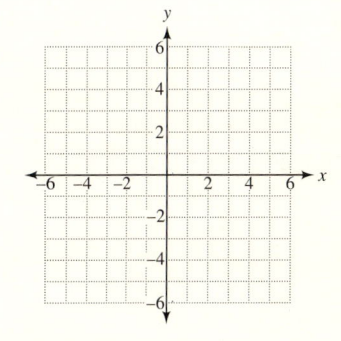

16. $2x + y \leq 2$
    $y - 1 > 0$

17. $x \geq 0$
    $y \geq 0$
    $3x + 2y < 6$

6.5 18. Find the maximum and minimum values of the objective function $P = 18x + 3y$ subject to the constraints graphed in the feasible region shown.

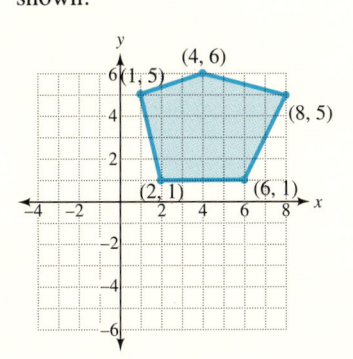

19. Find the maximum and the minimum values of the objective function $P = 6x - 10y$ subject to the following constraints.

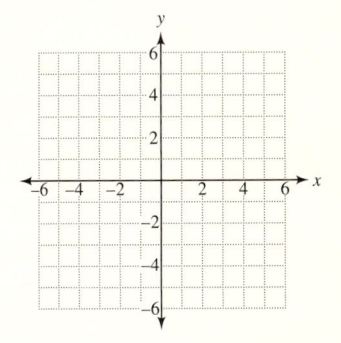

    $1 \leq x \leq 5$
    $y \geq 2$
    $y - 3x \leq 0$
    $2x + 3y \leq 22$

20. **BUSINESS** The owner of a fast-food diner has at most 90 lb of ground beef that can be used to make hamburgers and taco filler daily. Each hamburger contains 1/3 lb of beef and each taco contains 1/6 lb of beef. Suppose profit on hamburgers is 42¢, and profit on tacos is 20¢. Labor costs for making a hamburger amount to 18¢, and the labor costs for making a taco are 6¢. If the owner can pay at most $36 for the labor costs of producing these two items, what number of hamburgers and what number of tacos should be made to maximize his profits? What is the maximum profit?

6.6 21. Find real numbers A and B for which the rational function $f(x)$ has the given partial fraction decomposition.

$$f(x) = \frac{-2x - 5}{(x + 4)^2} = \frac{A}{x + 4} + \frac{B}{(x + 4)^2}$$

22. Find the partial fraction decomposition of the rational function

$$f(x) = \frac{8x^2 - 4x + 16}{(x^2 + 3)(x - 1)}.$$

Part II

In Exercises 23–26 solve using a system of equations.

23. **CHEMISTRY** A chemist has one solution that is 15% acid and another that is 20% acid. How many gallons of each should he mix together to obtain 100 gallons of a solution that is 18% acid?

24. **COIN** A collection of 70 coins consisting of nickels, dimes, and quarters has a value of $8.00. If the number of dimes is twice the number of quarters, how many of each type of coin are in the collection?

25. **GEOMETRY** The smallest angle of a triangle has measure one-third the measure of the middle-sized angle, and the largest angle measures 5° more than the middle-sized angle. Find the measure of each angle.

26. **CONSUMER** Jeff has found that the gasoline costs y of operating his car can be approximated by a quadratic function of the speed x at which the car is driven. Suppose he has collected the data in the table below. Use this data to find the function, and find the cost of operating the car at 65 mph. Does the model make sense when the speed is 0 mph?

Speed	Cost/mile
30 mph	6¢
40 mph	5¢
50 mph	8¢

Solve each system in Exercises 27–30.

27. $2x + 2y = 6$
    $-3x - 3y = -9$

28. $2(x + 3y) + (x + y) = -28$
    $3(x + 3y) - (x + y) = -32$

29. $2x + y - z = 0$
    $-x - y + 5z = 0$
    $3x - 2y + z = 0$

30. $3x - y + z = 4$
    $-6x + 2y - 2z = 4$

31. Find value(s) of m so that the system is inconsistent.

    $2x - 3y = m$
    $-4x + 6y = 4$

32. Find real numbers A, B, C, and D for which the rational function $f(x)$ has the given partial fraction decomposition.

    $$f(x) = \frac{x^3 + 2x^2 + 10x + 7}{(x^2 + x + 7)^2}$$
    $$= \frac{Ax + B}{x^2 + x + 7} + \frac{Cx + D}{(x^2 + x + 7)^2}$$

33. Find the partial fraction decomposition of the rational function

    $$f(x) = \frac{2x^3 - 16x^2 + 28x - 12}{x^2 - 8x + 12}$$

34. Find the maximum and minimum values of the objective function $P = 24x + 5y$ subject to the constraints graphed in the feasible region below.

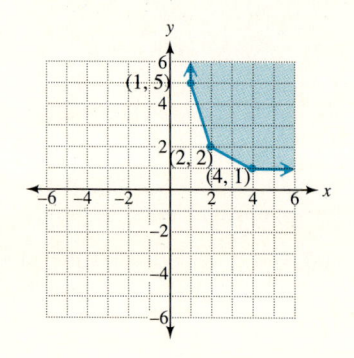

35. Sketch the graph of the system of linear inequalities.

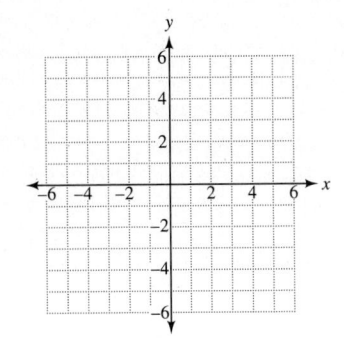

$$1 \leq x \leq 4$$
$$y \geq 1$$
$$3x - 4y \geq -12$$

ANSWERS: 1. yes 2. $(-2, 5)$ 3. $(1, -2)$ 4. $m = -\frac{5}{2}$ 5. no solution 6. $(1, -1)$ 7. $(2, 0, -1)$ 8. $\left(x, \frac{9}{2}x, 2x\right)$ x any real number 9. $(2 - 2z, 3z + 3, z)$ z any real number 10. $a = 2, b = -3, c = 5$ 11. 3 hr at 40 mph; 5 hr at 50 mph 12. 12°, 78° 13. 32 lb of $1.60/lb candy; 48 lb of $1.20/lb candy 14. $2000 in bonds; $5000 in certificates; $2000 in stocks 15. $a = -5, b = 8$ 16.   17.

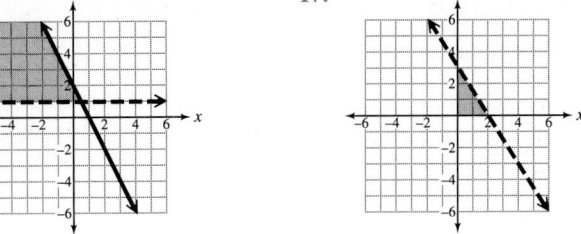

18. maximum value: 159; minimum value: 33 19. maximum value: 10; minimum value: -48 20. 60 hamburgers and 420 tacos; $109.20 21. $A = -2; B = 3$ 22. $f(x) = \frac{3x-1}{x^2+3} + \frac{5}{x-1}$ 23. 40 gal of 15% solution; 60 gal of 20% solution 24. 25 nickels; 30 dimes; 15 quarters 25. 25°, 75°, 80° 26. $y = \frac{1}{50}x^2 - \frac{3}{2}x + 33$; 20¢; no, the cost of gas is then 33¢ per mile, which does not make sense. 27. $(x, 3 - x)$ x any real number 28. $(0, -4)$ 29. $(0, 0, 0)$ 30. no solution 31. For m any real number except -2, the system will be inconsistent. 32. $A = 1; B = 1; C = 2; D = 0$ 33. $f(x) = 2x + \frac{1}{x-2} + \frac{3}{x-6}$ 34. there is no maximum value; minimum value: 49 35.

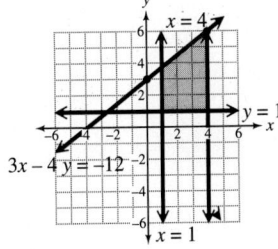

CHAPTER 6 TEST

Solve.

1. $2x - 5y = -16$
   $x + 2y = 19$

2. $2(x + y) - 3(x - y) = -2$
   $3(x + y) + (x - y) = 8$

3. An airplane travels 3000 miles with a tailwind in 5 hours. It travels 3000 miles with a headwind in 6 hours. Find the speed of the plane and the speed of the wind.

4. A grocer mixed candy worth $0.80 per pound with nuts worth $0.70 per pound to obtain 20 pounds of a mixture worth $0.77 per pound. How many pounds of candy did he use?

5. $x - 2y + z = 1$
   $2x + y - 3z = -8$
   $3x - y - 5z = -13$

6. $2x + 3y + 8z = 0$
   $x - 2y - 3z = 0$
   $-x + 5y + 9z = 0$

Solve.

7. A man has $5000 divided into three separate investments. Part is invested in bonds that earn 9%, part is invested in certificates that earn 8%, and part is invested in stocks. If the stocks do well, they will earn 10%, and the total earnings will amount to $450. If the market takes a downward trend, his stocks will lose 3%, and his total earnings will amount to $190. How much is invested in bonds?

1. _____

2. _____

3. _____

4. _____

5. _____

6. _____

7. _____

CHAPTER 6 TEST
CONTINUED

8. Sketch the graph of the system of inequalities.

$$2x + y \geq 4$$
$$x - 1 > 0$$

8.

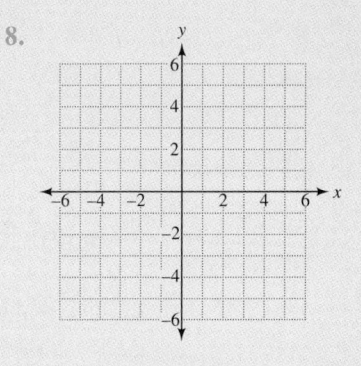

9. The polynomial function $P(x) = x^3 + ax^2 + bx + c$ has the property that $P(1) = 5$, $P(-1) = 9$, $P(2) = 15$. Find the values of a, b, and c.

9. _____

10. Find the maximum and minimum values of the objective function $P = 5x + 10y$ subject to the constraints graphed in the feasible region shown.

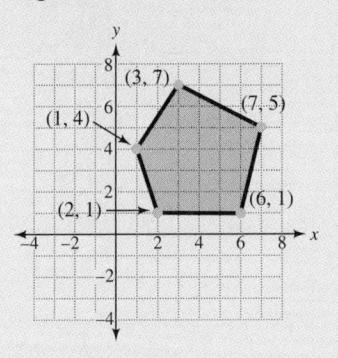

10. _____

11. The Graduation Class Ring Company designs and sells two types of rings, a Valedictorian model and a Salutatorian model. It can make up to a total of 24 rings each day using up to 60 total workhours. It takes 3 hours to make one Valedictorian ring, while 2 workhours are required to make one Salutatorian model. If the profit returned on a Valedictorian model is $30 and the profit on a Salutatorian model is $40, how many of each should be made daily to maximize profit?

11. _____

12. Find the partial fraction decomposition of the given rational function.

$$f(x) = \frac{x + 18}{x^2 + x - 12}$$

12. _____

463

CHAPTER 7

Matrices and Determinants

AN APPLIED PROBLEM IN POLITICS

A candidate for political office in Virginia hired a public relations firm to promote his campaign in three population centers: Richmond, Norfolk, and Roanoke. Potential supporters are to be contacted by telephone, mail, and direct personal contact. The cost of each of these is estimated to be $0.50 by telephone, $0.30 by mail, and $0.75 for a personal contact. The table gives the number of contacts in each category in each area. Use matrices to determine the total cost for the promotion.

	Telephone	Mail	Personal
Richmond	5000	10,000	1000
Norfolk	8000	20,000	6000
Roanoke	3000	5000	700

Analysis

Data given by a table such as the one in the problem can be placed nicely in a matrix, in this case a matrix that is 3×3. A 3×1 matrix consisting of the estimated cost for each type of contact can also be obtained. If these two matrices are multiplied, the result is a 3×1 column vector with entries consisting of the total cost in each of the three cities. The scalar product of $[1 \; 1 \; 1]$ with this matrix will give the total cost for the campaign.

Tabulation

Form the matrix N that gives the number of contacts in each area.

$$N = \begin{bmatrix} \text{Tele} & \text{Mail} & \text{Pers} \\ 5000 & 10{,}000 & 1000 \\ 8000 & 20{,}000 & 6000 \\ 3000 & 5000 & 700 \end{bmatrix} \begin{matrix} \text{Rich} \\ \text{Nor} \\ \text{Roan} \end{matrix}$$

Form the matrix C that gives the estimated cost for each type of contact.

$$C = \begin{bmatrix} 0.50 \\ 0.30 \\ 0.75 \end{bmatrix} \begin{matrix} \text{Telephone} \\ \text{Mail} \\ \text{Personal} \end{matrix}$$

Translation

The total cost is given by

$$T \cdot (NC)$$

where $T = [1 \; 1 \; 1]$.

Matrices and operations on matrices provide streamlined methods for solving a variety of applied problems. With increased use of computers, matrices have become important tools for organizing and manipulating large sets of data. The example at the left, although solvable without matrices, illustrates the power of matrices when more complex problems are involved (see Example 4 in Section 7.2). We begin this chapter with an introduction to matrix theory and show how matrices can be used to solve systems of equations. The presentation concludes with a study of *determinants* and their properties which have numerous applications in mathematics.

7.1 MATRICES

STUDENT GUIDEPOSTS

1. Matrix Terminology
2. Addition of Matrices
3. Subtraction of Matrices
4. Scalar Multiplication
5. Problem Solving with Matrices

1 MATRIX TERMINOLOGY

Rectangular arrays of numbers, or **matrices** (singular, **matrix**), arise in numerous ways. For example, a table of data, such as the one below showing grade distributions in two different mathematics courses, is in effect a matrix.

	A	B	C	D	F
MATH 110	8	10	21	4	5
MATH 130	6	12	18	9	3

Capital letters denote matrices, and the numbers or **elements** of the matrix are enclosed in brackets. The matrix G, given below, abbreviates the distribution of grades in the table.

$$G = \begin{bmatrix} 8 & 10 & 21 & 4 & 5 \\ 6 & 12 & 18 & 9 & 3 \end{bmatrix}$$

The **order** or **dimension** of a matrix with m **rows** (numbers in horizontal lines) and n **columns** (numbers in vertical lines) is $m \times n$, read m by n. The order of matrix G is 2×5. Consider the following matrices.

$$E = \begin{bmatrix} 1 & 2 & -3 \\ 4 & 0 & 5 \end{bmatrix} \quad B = \begin{bmatrix} 1 \\ 7 \\ -3 \end{bmatrix} \quad C = \begin{bmatrix} 2 & 4 \\ -1 & 3 \end{bmatrix} \quad D = [2 \quad 7 \quad -5 \quad 1]$$

The order of E is 2×3, the order of B is 3×1, the order of C is 2×2, and the order of D is 1×4. A matrix of order $n \times n$, such as C, is a **square matrix of order n**. A matrix of order $m \times 1$, such as B, is a **column vector of order m**. A matrix of order $1 \times n$, such as D, is a **row vector of order n**.

A general $m \times n$ matrix is often written in the form

$$A = \begin{bmatrix} a_{11} & a_{12} & a_{13} & \cdots & a_{1n} \\ a_{21} & a_{22} & a_{23} & \cdots & a_{2n} \\ a_{31} & a_{32} & a_{33} & \cdots & a_{3n} \\ \vdots & \vdots & \vdots & & \vdots \\ a_{m1} & a_{m2} & a_{m3} & \cdots & a_{mn} \end{bmatrix}$$

in which the double-subscript notation identifies the row and column containing a particular element. The element a_{ij} is found in the ith row and the jth column. This notation is often abbreviated to

$$A = [a_{ij}]_{m \times n}.$$

Equality of Matrices

Two matrices A and B are **equal**, written $A = B$, if and only if they have the same order and corresponding elements are equal.

466 CHAPTER 7 MATRICES AND DETERMINANTS

Using the notation for a general matrix, if $A = [a_{ij}]_{m \times n}$ and $B = [b_{ij}]_{m \times n}$, $A = B$ if and only if $a_{ij} = b_{ij}$ for every $i = 1, 2, \ldots m$ and $j = 1, 2, \ldots n$.

EXAMPLE 1 EQUALITY OF MATRICES

Consider the following matrices.

$$A = \begin{bmatrix} 2 & 0 & 3^2 \\ (-1)^3 & \sqrt[3]{27} & 1 \end{bmatrix}, \quad B = \begin{bmatrix} \sqrt{4} & 0 & 9 \\ -1 & 3 & 2^0 \end{bmatrix}, \quad C = \begin{bmatrix} 1 & 2 \\ 3 & 4 \end{bmatrix},$$

$$D = \begin{bmatrix} u & v \\ w & x \end{bmatrix}, \quad E = \begin{bmatrix} 1 \\ 2 \\ 3 \\ 4 \end{bmatrix}, \quad F = [1 \ \ 2 \ \ 3 \ \ 4].$$

(a) Since A and B are both of order 2×3 and corresponding elements are equal, $A = B$.

(b) If $C = D$, then u must be 1, v must be 2, w must be 3, and x must be 4.

(c) Although C, the square matrix of order 2, E, the column vector of order 4, and F, the row vector of order 4, all contain the same elements, they are not equal since they have different orders.

PRACTICE EXERCISE 1

Consider the following matrices.

$$U = \begin{bmatrix} 3 & -4 \\ 2 & 5 \end{bmatrix} \quad V = \begin{bmatrix} a & b \\ c & d \end{bmatrix}$$

$$W = [-3 \ \ 5 \ \ 0] \quad X = \begin{bmatrix} -3 \\ 5 \\ 0 \end{bmatrix}$$

(a) If $U = V$, find the values of a, b, c, and d.

(b) Are W and X equal?

(c) What is the order of matrix U?

Answers: (a) $a = 3$, $b = -4$, $c = 2$, $d = 5$ (b) no (c) 2×2

❷ ADDITION OF MATRICES

We can operate on matrices in ways similar to the ways we operate on numbers. The first operation we consider is *addition*. The sum of two matrices is defined just as you would expect it to be. To add two matrices of the same order, we simply add corresponding elements.

> **Addition of Matrices**
>
> Let $A = [a_{ij}]_{m \times n}$ and $B = [b_{ij}]_{m \times n}$ be two matrices. The sum of A and B is the $m \times n$ matrix given by
>
> $$A + B = [a_{ij} + b_{ij}]_{m \times n}.$$
>
> Addition of matrices with different orders is not defined.

EXAMPLE 2 ADDITION OF MATRICES

Consider the following matrices.

$$A = \begin{bmatrix} 1 & 2 \\ -3 & 4 \end{bmatrix} \quad B = \begin{bmatrix} 0 & -3 \\ 2 & 5 \end{bmatrix} \quad C = \begin{bmatrix} -4 & -2 \\ 3 & 1 \end{bmatrix}$$

$$D = \begin{bmatrix} 0 & -2 & 5 \\ 3 & 7 & -1 \end{bmatrix} \quad E = \begin{bmatrix} -5 & 7 & 2 \\ 4 & -1 & 3 \end{bmatrix} \quad F = \begin{bmatrix} 5 & 0 & -1 \\ 2 & -1 & 0 \end{bmatrix}$$

(a) $A + B = \begin{bmatrix} 1 & 2 \\ -3 & 4 \end{bmatrix} + \begin{bmatrix} 0 & -3 \\ 2 & 5 \end{bmatrix} = \begin{bmatrix} 1+0 & 2+(-3) \\ -3+2 & 4+5 \end{bmatrix} = \begin{bmatrix} 1 & -1 \\ -1 & 9 \end{bmatrix}$

(b) $B + A = \begin{bmatrix} 0 & -3 \\ 2 & 5 \end{bmatrix} + \begin{bmatrix} 1 & 2 \\ -3 & 4 \end{bmatrix} = \begin{bmatrix} 0+1 & -3+2 \\ 2+(-3) & 4+5 \end{bmatrix} = \begin{bmatrix} 1 & -1 \\ -1 & 9 \end{bmatrix}$

PRACTICE EXERCISE 2

Use the matrices given in Example 2 to find the following.

(a) $B + C$

(b) $C + B$

3. Which of these matrices are square matrices?

4. Which of these matrices are column vectors?

5. Are A and C equal?

6. What element is in the first row and second column of D?

7. Give the zero matrix with the same dimension as A.

8. Identify the element a_{21} in matrix A.

9. Give the matrix $-A$.

10. Find $A + B$.

11. Find $A + D$.

12. Find $D - E$.

13. Find $B - E$.

14. Find $-2A$.

15. Find $a_{12}B$.

16. Find $2A - 3B$.

17. Find $A + 0$.

18. Find $0 - E$.

In Exercises 19–20 find the values of a, b, and c.

19. $\begin{bmatrix} a & b \\ c & d \end{bmatrix} + \begin{bmatrix} -1 & 3 \\ 0 & 2 \end{bmatrix} = \begin{bmatrix} 6 & -2 \\ 5 & 1 \end{bmatrix}$

20. $\begin{bmatrix} a & 1 \\ 0 & b \end{bmatrix} - \begin{bmatrix} b & -2 \\ 3 & 2a \end{bmatrix} = \begin{bmatrix} -5 & 3 \\ -3 & 8 \end{bmatrix}$

21. **SPORTS** Matrices F and S give the points scored, number of rebounds, and minutes played for the starting five in a two-game basketball tournament.

$$F = \begin{bmatrix} \text{Points} & \text{Rebounds} & \text{Minutes} \\ 21 & 5 & 35 \\ 18 & 9 & 32 \\ 10 & 7 & 28 \\ 8 & 3 & 30 \\ 15 & 12 & 32 \end{bmatrix} \begin{matrix} \text{Hurd} \\ \text{Spencer} \\ \text{Murchison} \\ \text{Payne} \\ \text{Duane} \end{matrix}$$

472 CHAPTER 7 MATRICES AND DETERMINANTS

$$S = \begin{bmatrix} 19 & 4 & 33 \\ 26 & 8 & 30 \\ 8 & 10 & 26 \\ 8 & 5 & 32 \\ 17 & 12 & 28 \end{bmatrix} \begin{matrix} \text{Hurd} \\ \text{Spencer} \\ \text{Murchison} \\ \text{Payne} \\ \text{Duane} \end{matrix}$$

(a) Find the matrix that gives the totals for the tournament in each category.

(b) Find the matrix $\frac{1}{2}(F + S)$ and discuss what it represents.

FOR REVIEW

Solve each system in Exercises 22–23.

22. $3x + 5y = 9$
$2x - 7y = -25$

23. $4x - y = -3$
$- 2y - 5z = -1$
$3x + 2z = -2$

ANSWERS: 1. 2×2 2. 1×4 3. A, B, and C 4. G 5. yes 6. -1 7. $\begin{bmatrix} 0 & 0 \\ 0 & 0 \end{bmatrix}$ 8. 3 9. $\begin{bmatrix} -1 & -2 \\ -3 & -4 \end{bmatrix}$

10. $\begin{bmatrix} 6 & 0 \\ 12 & 0 \end{bmatrix}$ 11. undefined 12. $\begin{bmatrix} 5 & -4 & 4 \\ -3 & -4 & -2 \end{bmatrix}$ 13. undefined 14. $\begin{bmatrix} -2 & -4 \\ -6 & -8 \end{bmatrix}$ 15. $\begin{bmatrix} 10 & -4 \\ 18 & -8 \end{bmatrix}$

16. $\begin{bmatrix} -13 & 10 \\ -21 & 20 \end{bmatrix}$ 17. A 18. $-E$ 19. $a = 7$; $b = -5$; $c = 5$; $d = -1$ 20. $a = -3$; $b = 2$

21. (a) $F + S = \begin{bmatrix} 40 & 9 & 68 \\ 44 & 17 & 62 \\ 18 & 17 & 54 \\ 16 & 8 & 62 \\ 32 & 24 & 60 \end{bmatrix}$ (b) $\frac{1}{2}(F + S) = \begin{bmatrix} 20 & 4.5 & 34 \\ 22 & 8.5 & 31 \\ 9 & 8.5 & 27 \\ 8 & 4 & 31 \\ 16 & 12 & 30 \end{bmatrix}$ The matrix $\frac{1}{2}(F + S)$ represents the average points, rebounds, and minutes played for each player in the tournament. 22. $(-2, 3)$ 23. $(0, 3, -1)$

7.1 EXERCISES B

Exercises 1–18 refer to the following matrices.

$$A = \begin{bmatrix} 1 & 2 \\ 3 & 4 \end{bmatrix} \quad B = \begin{bmatrix} 5 & -2 \\ 9 & -4 \end{bmatrix} \quad C = \begin{bmatrix} 5^0 & \sqrt{4} \\ |-3| & 2^2 \end{bmatrix} \quad D = \begin{bmatrix} 3 & -1 & 4 \\ 2 & 0 & 5 \end{bmatrix}$$

$$E = \begin{bmatrix} -2 & 3 & 0 \\ 5 & 4 & 7 \end{bmatrix} \quad F = [3 \; 2 \; 0 \; -1] \quad G = \begin{bmatrix} 3 \\ 2 \\ 0 \\ -1 \end{bmatrix}$$

1. What is the order of matrix D?

2. What is the order of matrix G?

3. Which of these matrices have order 2×3?

4. Which of these matrices are row vectors?

5. Are F and G equal?

6. What element is in the second row and second column of A?

7. Give the zero matrix with the same dimension as E.

8. Identify the element e_{23} in matrix E.

9. Give the matrix $-D$.

10. Find $D + E$.

11. Find $B + E$.

12. Find $A - B$.

13. Find $A - D$.

14. Find $-4E$.

15. Find $b_{12}A$.

16. Find $3E - 2D$.

17. Find $E + 0$.

18. Find $0 - A$.

In Exercises 19–20 find the values of a, b, c, and d.

19. $\begin{bmatrix} a & b \\ c & d \end{bmatrix} - \begin{bmatrix} 2 & 3 \\ 7 & -1 \end{bmatrix} = \begin{bmatrix} 6 & 0 \\ 4 & 2 \end{bmatrix}$

20. $\begin{bmatrix} 3 & -b \\ 2 & a \end{bmatrix} + \begin{bmatrix} 1 & 3a \\ -2 & -b \end{bmatrix} = \begin{bmatrix} 4 & 7 \\ 0 & 3 \end{bmatrix}$

21. **BUSINESS** Video Rentals Inc. has two stores in Barstow, California. The total number of VHS cassettes, Beta cassettes, and recorders rented during the months of January and February are given in matrices J and F.

$$J = \begin{bmatrix} 2100 & 400 & 55 \\ 3500 & 700 & 60 \end{bmatrix} \begin{matrix} \text{Store 1} \\ \text{Store 2} \end{matrix}$$

(VHS Beta Recorders)

$$F = \begin{bmatrix} 3200 & 750 & 84 \\ 4300 & 900 & 105 \end{bmatrix} \begin{matrix} \text{Store 1} \\ \text{Store 2} \end{matrix}$$

(a) Find the matrix that gives the two-month totals in each category.
(b) Find the matrix $\frac{1}{2}(J + F)$ and discuss what it represents.
(c) Find the matrix that gives the increase in rentals from January to February.

FOR REVIEW

Solve each system of equations in Exercises 22–23.

22. $2x + 6y = 12$
    $-x - 3y = 6$

23. $x + 5y + 2z = -37$
    $3x + 2y - z = -5$
    $-3x - 3y + z = 11$

7.1 EXERCISES C

In Exercises 1–6 use the following general 2 × 2 matrices, real numbers m and n, and properties of real numbers to prove each statement.

$$A = \begin{bmatrix} a_{11} & a_{12} \\ a_{21} & a_{22} \end{bmatrix}, \quad B = \begin{bmatrix} b_{11} & b_{12} \\ b_{21} & b_{22} \end{bmatrix}, \quad C = \begin{bmatrix} c_{11} & c_{12} \\ c_{21} & c_{22} \end{bmatrix}, \quad 0 = \begin{bmatrix} 0 & 0 \\ 0 & 0 \end{bmatrix}.$$

1. $A + (B + C) = (A + B) + C$

2. $A + B = B + A$

3. $m(A + B) = mA + mB$

4. $(m + n)A = mA + nA$

5. $A + 0 = 0 + A = A$

6. $A + (-A) = (-A) + A = 0$

7.2 MATRIX MULTIPLICATION

STUDENT GUIDEPOSTS

1. Scalar (Dot) Products
2. Multiplication of Matrices
3. Properties of Matrix Multiplication
4. Problem Solving with Matrices

1 SCALAR (DOT) PRODUCTS

In this section we will define two kinds of matrix multiplication. These operations might seem strange to you at first, since unlike addition and subtraction, the product of two matrices *is not* found by multiplying corresponding elements nor do the two matrices have to be of the same order. One reason for the definition we use comes from systems of equations; and, as we shall soon see, there are other good reasons from the standpoint of applications of matrices.

We begin by defining the *scalar product* or *dot product* (not to be confused with multiplying a matrix by a scalar introduced in Section 7.1) of a row vector of order n times a column vector of order n. The result is a real number, or scalar, *not* a matrix.

Scalar (Dot) Product of Two Vectors

Let $A = [a_1 \; a_2 \; a_3 \; \ldots \; a_n]$ and $B = \begin{bmatrix} b_1 \\ b_2 \\ b_3 \\ \vdots \\ b_n \end{bmatrix}$.

The **scalar product** of A and B is given by

$$A \cdot B = a_1 b_1 + a_2 b_2 + a_3 b_3 + \cdots + a_n b_n.$$

CAUTION

The dot written between the two vectors (matrices) in the above definition is very important. If it is omitted, the product is interpreted differently and results in a matrix (not a real number). This is defined shortly.

EXAMPLE 1 SCALAR PRODUCT OF TWO VECTORS

Let $A = [3 \; 2 \; 1 \; 5]$ and $B = \begin{bmatrix} 300 \\ 450 \\ 500 \\ 650 \end{bmatrix}$. Find $A \cdot B$.

$$A \cdot B = [3 \; 2 \; 1 \; 5] \cdot \begin{bmatrix} 300 \\ 450 \\ 500 \\ 650 \end{bmatrix} = 3(300) + 2(450) + 1(500) + 5(650)$$

$$= 900 + 900 + 500 + 3250 = 5550$$

PRACTICE EXERCISE 1

Let $A = [1 \; -3 \; 4]$ and $B = \begin{bmatrix} 10 \\ 20 \\ 30 \end{bmatrix}$. Find $A \cdot B$.

Answer: 70

This "row times column" method of multiplication may seem a bit artificial at first, but it has numerous applications. For example, suppose that an appliance dealer sells 3 stereos for $300 each, 2 stereos for $450 each, 1 stereo for $500, and 5 stereos for $650 each. The vector A in Example 1 represents the number of stereos sold at each price, and the vector B represents the price of each type of stereo. The scalar product $A \cdot B$ gives the total revenue produced by selling these stereos, $5550.

❷ MULTIPLICATION OF MATRICES

Keep in mind that the scalar product of a row vector and a column vector is a real number, not a matrix. We now define multiplication of matrices in such a way that the product will be a matrix. Scalar products are used in this definition.

> **Multiplication of Matrices**
>
> Let $A = [a_{ij}]_{m \times n}$ and $B = [b_{ij}]_{n \times p}$ be two matrices with the property that the number of columns of A is the same as the number of rows of B. The **product** of A and B is the $m \times p$ matrix
>
> $$AB = [c_{ij}]_{m \times p},$$
>
> where c_{ij} is the scalar product of the ith row of A and the jth column of B.

The diagram below illustrates the restrictions imposed by the definition of multiplication.

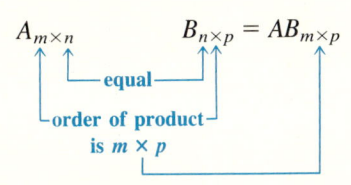

$$A_{m \times n} \quad B_{n \times p} = AB_{m \times p}$$
(equal; order of product is $m \times p$)

EXAMPLE 2 MULTIPLYING MATRICES

Let $A = \begin{bmatrix} 1 & -2 & 3 \\ 4 & 0 & 5 \end{bmatrix}$ and $B = \begin{bmatrix} 2 & -1 \\ -3 & 1 \\ 0 & 4 \end{bmatrix}$. Find AB.

Since it is

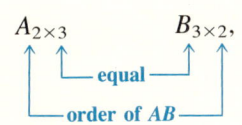

$$A_{2 \times 3} \quad B_{3 \times 2},$$
(equal; order of AB)

the product AB is defined and has order 2×2.

Thus,

$$AB = \begin{bmatrix} [1 \ -2 \ 3] \cdot \begin{bmatrix} 2 \\ -3 \\ 0 \end{bmatrix} & [1 \ -2 \ 3] \cdot \begin{bmatrix} -1 \\ 1 \\ 4 \end{bmatrix} \\ [4 \ 0 \ 5] \cdot \begin{bmatrix} 2 \\ -3 \\ 0 \end{bmatrix} & [4 \ 0 \ 5] \cdot \begin{bmatrix} -1 \\ 1 \\ 4 \end{bmatrix} \end{bmatrix}$$

(row 1 of A · column 1 of B; row 1 of A · column 2 of B; row 2 of A · column 1 of B; row 2 of A · column 2 of B)

PRACTICE EXERCISE 2

Let $A = \begin{bmatrix} 2 & -1 \\ 0 & 3 \end{bmatrix}$ and $B = \begin{bmatrix} 3 & -1 & 0 \\ 4 & 2 & 1 \end{bmatrix}$. Find AB.

$$= \begin{bmatrix} (1)(2) + (-2)(-3) + (3)(0) & (1)(-1) + (-2)(1) + (3)(4) \\ (4)(2) + (0)(-3) + (5)(0) & (4)(-1) + (0)(1) + (5)(4) \end{bmatrix}$$

$$= \begin{bmatrix} 8 & 9 \\ 8 & 16 \end{bmatrix}.$$

Answer: $\begin{bmatrix} 2 & -4 & -1 \\ 12 & 6 & 3 \end{bmatrix}$

Note Keep in mind that matrix multiplication is a row-times-column operation. We take the scalar product of the rows in the first matrix times the columns in the second. When we take the scalar product of the ith row in the first times the jth column in the second, the resulting scalar is the element in the ith row and jth column of the product.

EXAMPLE 3 More Products of Matrices

Find each product.

(a) $\begin{bmatrix} 1 & -2 \\ 0 & -3 \end{bmatrix}_{2 \times 2} \begin{bmatrix} 3 & 0 & 4 \\ -1 & 5 & 0 \end{bmatrix}_{2 \times 3}$

(equal — order of product)

$$= \begin{bmatrix} (1)(3) + (-2)(-1) & (1)(0) + (-2)(5) & (1)(4) + (-2)(0) \\ (0)(3) + (-3)(-1) & (0)(0) + (-3)(5) & (0)(4) + (-3)(0) \end{bmatrix}$$

$$= \begin{bmatrix} 5 & -10 & 4 \\ 3 & -15 & 0 \end{bmatrix}$$

(b) $\begin{bmatrix} 3 & 0 & 4 \\ -1 & 5 & 0 \end{bmatrix}_{2 \times 3} \begin{bmatrix} 1 & -2 \\ 0 & -3 \end{bmatrix}_{2 \times 2}$ This product is not defined.

(not equal)

(c) $[2 \; 0 \; -3]_{1 \times 3} \begin{bmatrix} 1 & 3 & -2 \\ 4 & -1 & 5 \\ 2 & 3 & -5 \end{bmatrix}_{3 \times 3}$

$(2)(1) + (0)(4) + (-3)(2) = 2 + 0 - 6$

$= [2 + 0 - 6 \quad 6 + 0 - 9 \quad -4 + 0 + 15]$

(equal — order of product)

$$= [-4 \quad -3 \quad 11]$$

(d) $\begin{bmatrix} 2 & -1 \\ 3 & 5 \end{bmatrix} \begin{bmatrix} 4 & 0 \\ 1 & -2 \end{bmatrix} = \begin{bmatrix} 8-1 & 0+2 \\ 12+5 & 0-10 \end{bmatrix} = \begin{bmatrix} 7 & 2 \\ 17 & -10 \end{bmatrix}$

(e) $\begin{bmatrix} 4 & 0 \\ 1 & -2 \end{bmatrix} \begin{bmatrix} 2 & -1 \\ 3 & 5 \end{bmatrix} = \begin{bmatrix} 8+0 & -4+0 \\ 2-6 & -1-10 \end{bmatrix} = \begin{bmatrix} 8 & -4 \\ -4 & -11 \end{bmatrix}$

Practice Exercise 3

Find each product.

(a) $\begin{bmatrix} -1 & 0 & 2 \\ 2 & 4 & -1 \\ 3 & -1 & 1 \end{bmatrix} \begin{bmatrix} 1 & -2 \\ 0 & 3 \\ 4 & -1 \end{bmatrix}$

(b) $\begin{bmatrix} 1 & -2 \\ 0 & 3 \\ 4 & -1 \end{bmatrix} \begin{bmatrix} -1 & 0 & 2 \\ 2 & 4 & -1 \\ 3 & -1 & 1 \end{bmatrix}$

(c) $[3 \; -1] \begin{bmatrix} 2 & 0 \\ 4 & -5 \end{bmatrix}$

(d) $\begin{bmatrix} 5 & -1 \\ 3 & 0 \end{bmatrix} \begin{bmatrix} 2 & 4 \\ -1 & 3 \end{bmatrix}$

(e) $\begin{bmatrix} 2 & 4 \\ -1 & 3 \end{bmatrix} \begin{bmatrix} 5 & -1 \\ 3 & 0 \end{bmatrix}$

Answers: (a) $\begin{bmatrix} 7 & 0 \\ -2 & 9 \\ 7 & -10 \end{bmatrix}$

(b) product not defined

(c) $[2 \; 5]$ (d) $\begin{bmatrix} 11 & 17 \\ 6 & 12 \end{bmatrix}$

(e) $\begin{bmatrix} 22 & -2 \\ 4 & 1 \end{bmatrix}$

3 PROPERTIES OF MATRIX MULTIPLICATION

Matrix multiplication is not a commutative operation. That is, if A and B are matrices, then AB does not necessarily equal BA. This is clear if A is of order 2×2 and B is of order 2×3, then AB has order 2×2 but BA is not even defined. (See Example 3(a) and (b).) However, even for square matrices when the products in either order are defined, changing the order of multiplication may result in different products. In Example 3(d) and (e), we saw that

$$\begin{bmatrix} 2 & -1 \\ 3 & 5 \end{bmatrix} \begin{bmatrix} 4 & 0 \\ 1 & -2 \end{bmatrix} \neq \begin{bmatrix} 4 & 0 \\ 1 & -2 \end{bmatrix} \begin{bmatrix} 2 & -1 \\ 3 & 5 \end{bmatrix}.$$

On the other hand, if A is of order $m \times n$, B is of order $n \times p$, and C is of order $p \times q$, it can be shown that

$$(AB)C = A(BC)$$

so that matrix multiplication satisfies the **associative law of multiplication.** Also, for matrices of the appropriate order, the **distributive laws** are true.

$$A(B + C) = AB + AC \quad \text{and} \quad (E + F)D = ED + FD$$

Examples of these properties are considered in the exercises at the end of this section.

Consider the following two products.

$$\begin{bmatrix} 2 & -1 \\ 3 & 5 \end{bmatrix} \begin{bmatrix} 1 & 0 \\ 0 & 1 \end{bmatrix} = \begin{bmatrix} 2+0 & 0-1 \\ 3+0 & 0+5 \end{bmatrix} = \begin{bmatrix} 2 & -1 \\ 3 & 5 \end{bmatrix}$$

$$\begin{bmatrix} 1 & 0 \\ 0 & 1 \end{bmatrix} \begin{bmatrix} 2 & -1 \\ 3 & 5 \end{bmatrix} = \begin{bmatrix} 2+0 & -1+0 \\ 0+3 & 0+5 \end{bmatrix} = \begin{bmatrix} 2 & -1 \\ 3 & 5 \end{bmatrix}$$

Notice that in both cases the product is equal to

$$\begin{bmatrix} 2 & -1 \\ 3 & 5 \end{bmatrix}.$$

The matrix $\begin{bmatrix} 1 & 0 \\ 0 & 1 \end{bmatrix}$

is the 2×2 version of an important group of matrices.

Identity Matrix

The matrix I_n of order $n \times n$ given by

$$I_n = \begin{bmatrix} 1 & 0 & 0 & \cdots & 0 \\ 0 & 1 & 0 & \cdots & 0 \\ 0 & 0 & 1 & \cdots & 0 \\ \vdots & \vdots & \vdots & & \vdots \\ 0 & 0 & 0 & \cdots & 1 \end{bmatrix}$$

with 1's down the **main diagonal** and 0's everywhere else is the **identity matrix of order n.**

Identity matrices play a role in matrix multiplication similar to the role that 1 plays in multiplication of numbers. If $A_{n \times n}$ is an arbitrary matrix, then

$$I_n A_{n \times n} = A_{n \times n} I_n = A_{n \times n}.$$

We have seen that the operations on matrices satisfy many of the properties similar to those for the operations on real numbers. The one exception thus far is the commutative law of multiplication. Another familiar property of real numbers, the zero-product rule, does not carry over to matrix multiplication. Consider the matrices

$$A = \begin{bmatrix} 1 & 4 \\ 1 & 4 \end{bmatrix} \text{ and } B = \begin{bmatrix} 4 & 4 \\ -1 & -1 \end{bmatrix}.$$

Then

$$AB = \begin{bmatrix} 1 & 4 \\ 1 & 4 \end{bmatrix} \begin{bmatrix} 4 & 4 \\ -1 & -1 \end{bmatrix} = \begin{bmatrix} 0 & 0 \\ 0 & 0 \end{bmatrix} = 0.$$

However, neither A nor B is the zero matrix. We see that matrix multiplication does not satisfy the zero-product rule, since AB can equal 0 while neither $A = 0$ nor $B = 0$.

Transpose of a Matrix

The **transpose** of an $m \times n$ matrix A, denoted by A^T, is the $n \times m$ matrix formed by interchanging the rows and columns of A. That is, row 1 of A becomes column 1 of A^T, row 2 of A becomes column 2 of A^T, and so forth.

For example, if $A = \begin{bmatrix} 2 & 3 \\ -1 & 5 \end{bmatrix}$, then $A^T = \begin{bmatrix} 2 & -1 \\ 3 & 5 \end{bmatrix}$, and if $B = \begin{bmatrix} 3 & -1 \\ 0 & 5 \\ 2 & 8 \end{bmatrix}$, then $B^T = \begin{bmatrix} 3 & 0 & 2 \\ -1 & 5 & 8 \end{bmatrix}$. The transpose of a matrix will be used in later sections.

④ PROBLEM SOLVING WITH MATRICES

We conclude this section with the application of matrix multiplication given in the introduction to this chapter.

EXAMPLE 4 A Problem in Politics

A candidate for political office in Virginia hired a public relations firm to promote his campaign in three population centers of the state: Richmond, Norfolk, and Roanoke. Potential supporters are to be contacted by telephone, mail, and direct personal contact. The cost of each of these is estimated to be $0.50 by telephone, $0.30 by mail, and $0.75 for a personal contact. The table below summarizes the desired number of contacts in each category in each area. Use matrices to determine the total cost for the promotion.

	Telephone	Mail	Personal
Richmond	5000	10,000	1000
Norfolk	8000	20,000	6000
Roanoke	3000	5000	700

The matrix

$$C = \begin{bmatrix} 0.50 \\ 0.30 \\ 0.75 \end{bmatrix} \begin{matrix} \text{Telephone} \\ \text{Mail} \\ \text{Personal contact} \end{matrix}$$

gives the estimated cost for each type of contact.

PRACTICE EXERCISE 4

Find the total cost of the campaign in Example 4 if the estimated contact costs are: $0.60 by telephone, $0.25 by mail, and $0.85 for a personal contact.

The matrix

$$N = \begin{bmatrix} \text{Telephone} & \text{Mail} & \text{Personal} \\ 5000 & 10{,}000 & 1000 \\ 8000 & 20{,}000 & 6000 \\ 3000 & 5000 & 700 \end{bmatrix} \begin{matrix} \text{Richmond} \\ \text{Norfolk} \\ \text{Roanoke} \end{matrix}$$

gives the number of contacts desired in each area. The product NC gives the column vector with the total costs in each area.

$$NC = \begin{bmatrix} 5000 & 10{,}000 & 1000 \\ 8000 & 20{,}000 & 6000 \\ 3000 & 5000 & 700 \end{bmatrix} \begin{bmatrix} 0.50 \\ 0.30 \\ 0.75 \end{bmatrix} = \begin{bmatrix} 6250 \\ 14{,}500 \\ 3525 \end{bmatrix} \begin{matrix} \text{Total cost in Richmond} \\ \text{Total cost in Norfolk} \\ \text{Total cost in Roanoke} \end{matrix}$$

The total cost in all three areas can be determined by finding the scalar product of the row vector $T = [1 \ \ 1 \ \ 1]$ with the column vector NC.

$$T \cdot (NC) = [1 \ \ 1 \ \ 1] \cdot \begin{bmatrix} 6250 \\ 14{,}500 \\ 3525 \end{bmatrix} = 24{,}275$$

Thus, the total cost for the campaign is $24,275.

Answer: $24,895

7.2 EXERCISES A

Find the scalar products in Exercises 1–2.

1. $[1 \ \ -2] \cdot \begin{bmatrix} 3 \\ 5 \end{bmatrix}$

2. $[2 \ \ 3 \ \ 5] \cdot \begin{bmatrix} -1 \\ 0 \\ 2 \end{bmatrix}$

In Exercises 3–4 determine m and n so that each product will be defined.

3. $A_{m \times n} B_{3 \times 2} = C_{4 \times 2}$

4. $A_{4 \times m} B_{2 \times n} = C_{4 \times 3}$

Exercises 5–14 refer to the following matrices.

$$A = \begin{bmatrix} 2 & 5 \\ -1 & 3 \end{bmatrix} \quad B = \begin{bmatrix} 0 & -2 \\ 1 & 4 \end{bmatrix} \quad C = \begin{bmatrix} 1 & 1 \\ 1 & 1 \end{bmatrix} \quad D = \begin{bmatrix} 1 & 3 & -2 \\ 4 & 0 & 5 \end{bmatrix}$$

$$E = \begin{bmatrix} 0 & 4 \\ 4 & 0 \end{bmatrix} \quad F = \begin{bmatrix} 3 & 0 \\ -2 & 1 \\ 0 & 5 \end{bmatrix} \quad G = [3 \ \ -2] \quad H = \begin{bmatrix} -2 \\ 4 \\ 1 \end{bmatrix}$$

5. Find AB.

6. Find CD.

7. Find FD.

8. Find DH.

9. Find B^T.

10. Find B^2.

11. Find F^T.

12. Find $(A^T)^T E$.

13. Find $A^T + B^T$.

14. Find $(A + B)^T$.

In Exercises 15–22 use the following matrices to verify each statement.

$$A = \begin{bmatrix} 2 & 1 \\ 0 & 5 \end{bmatrix} \quad B = \begin{bmatrix} 3 & 1 \\ 1 & -2 \end{bmatrix} \quad C = \begin{bmatrix} 0 & 4 \\ -2 & 1 \end{bmatrix}$$

15. $AB \neq BA$

16. $A(BC) = (AB)C$

17. $AI = A$

18. $A + (-A) = 0$

19. $A + 0 = A$

20. $A(B + C) = AB + AC$

21. $(B + C)A = BA + CA$

22. $A(B + C) \neq (B + C)A$

23. SPORTS Matrices F and S give the points scored, number of rebounds, and minutes played for the starting five in a two-game basketball tournament.

$$F = \begin{bmatrix} \text{Points} & \text{Rebounds} & \text{Minutes} \\ 21 & 5 & 35 \\ 18 & 9 & 32 \\ 10 & 7 & 28 \\ 8 & 3 & 30 \\ 15 & 12 & 32 \end{bmatrix} \begin{matrix} \text{Hurd} \\ \text{Spencer} \\ \text{Murchison} \\ \text{Payne} \\ \text{Duane} \end{matrix}$$

$$S = \begin{bmatrix} 19 & 4 & 33 \\ 26 & 8 & 30 \\ 8 & 10 & 26 \\ 8 & 5 & 32 \\ 17 & 12 & 28 \end{bmatrix} \begin{matrix} \text{Hurd} \\ \text{Spencer} \\ \text{Murchison} \\ \text{Payne} \\ \text{Duane} \end{matrix}$$

(a) Find $[1 \ \ 1 \ \ 1 \ \ 1 \ \ 1]F$ and discuss what it represents.

(b) Find [1 1 1 1 1]$(F + S)$ and discuss what it represents.

(c) Find $\frac{1}{2}$[1 1 1 1 1]$(F + S)$ and discuss what it represents.

24. BUSINESS Video Rentals Inc. has two stores in Barstow, California. The total number of VHS cassettes, Beta cassettes, and recorders rented during the months of January and February are given in matrices J and F.

$$J = \begin{matrix} & \text{VHS} & \text{Beta} & \text{Recorders} \\ & \begin{bmatrix} 2100 & 400 & 55 \\ 3500 & 700 & 60 \end{bmatrix} & & \begin{matrix} \text{Store 1} \\ \text{Store 2} \end{matrix} \end{matrix}$$

$$F = \begin{bmatrix} 3200 & 750 & 84 \\ 4300 & 900 & 105 \end{bmatrix} \begin{matrix} \text{Store 1} \\ \text{Store 2} \end{matrix}$$

Suppose that the rental rates charged by both stores are \$3 for VHS, \$4 for Beta, and \$10 for a recorder. Use the matrix

$$C = \begin{bmatrix} 3 \\ 4 \\ 10 \end{bmatrix}$$

to answer the following.

(a) Find JC and interpret the result.

(b) Find FC and interpret the result.

(c) Find $(J + F)C$ and interpret the result.

(d) Find [1 1] $\cdot ((J + F)C)$ and interpret the result.

FOR REVIEW

Use the following matrices in Exercises 25–28.

$$A = \begin{bmatrix} -5 & 0 \\ 2 & 7 \end{bmatrix} \quad B = \begin{bmatrix} 0 & 3 \\ -2 & 1 \end{bmatrix} \quad C = \begin{bmatrix} u & v \\ w & z \end{bmatrix}$$

25. If $A = C$, find the values of u, v, w, and z.

26. Identify the element b_{21} in matrix B.

27. Find $A + 0$.

28. Find $A - B$.

The systems in Exercises 29–30 will be solved in the next section by matrix methods. Find the solution to each system here by techniques given in the preceding chapter.

29. $2x + y = 1$
$x - 3y = 11$

30. $2x - y + z = 1$
$x + 3y - z = -4$
$3x + 2y + z = 0$

ANSWERS: 1. -7 2. 8 3. $m = 4; n = 3$ 4. $m = 2; n = 3$ 5. $\begin{bmatrix} 5 & 16 \\ 3 & 14 \end{bmatrix}$ 6. $\begin{bmatrix} 5 & 3 & 3 \\ 5 & 3 & 3 \end{bmatrix}$ 7. $\begin{bmatrix} 3 & 9 & -6 \\ 2 & -6 & 9 \\ 20 & 0 & 25 \end{bmatrix}$

8. $\begin{bmatrix} 8 \\ -3 \end{bmatrix}$ 9. $\begin{bmatrix} 0 & 1 \\ -2 & 4 \end{bmatrix}$ 10. $\begin{bmatrix} -2 & -8 \\ 4 & 14 \end{bmatrix}$ 11. $\begin{bmatrix} 3 & -2 & 0 \\ 0 & 1 & 5 \end{bmatrix}$ 12. $\begin{bmatrix} 20 & 8 \\ 12 & -4 \end{bmatrix}$ 13. $\begin{bmatrix} 2 & 0 \\ 3 & 7 \end{bmatrix}$ 14. $\begin{bmatrix} 2 & 0 \\ 3 & 7 \end{bmatrix}$

15. $AB = \begin{bmatrix} 7 & 0 \\ 5 & -10 \end{bmatrix} \neq \begin{bmatrix} 6 & 8 \\ 2 & -9 \end{bmatrix} = BA$ 16. $A(BC) = \begin{bmatrix} 0 & 28 \\ 20 & 10 \end{bmatrix} = (AB)C$ 17. $I = \begin{bmatrix} 1 & 0 \\ 0 & 1 \end{bmatrix}$ and $AI = A$

18. $-A$ is the additive inverse of A so their sum is the 2×2 zero matrix. 19. Since in this case 0 represents the 2×2 zero matrix, $A + 0 = A$. 20. $A(B + C) = \begin{bmatrix} 5 & 9 \\ -5 & -5 \end{bmatrix} = AB + AC$ 21. $(B + C)A = \begin{bmatrix} 6 & 28 \\ -2 & -6 \end{bmatrix} = BA + CA$

22. $A(B + C) = \begin{bmatrix} 5 & 9 \\ -5 & -5 \end{bmatrix} \neq \begin{bmatrix} 6 & 28 \\ -2 & -6 \end{bmatrix} = (B + C)A$ 23. (a) $[1 \ 1 \ 1 \ 1 \ 1]F = [72 \ 36 \ 157]$ represents the total points, rebounds, and minutes played by the starters in the first game. (b) $[1 \ 1 \ 1 \ 1 \ 1](F + S) = [150 \ 75 \ 306]$ represents the total points, rebounds, and minutes played by the starters in the tournament.
(c) $\frac{1}{2}[1 \ 1 \ 1 \ 1 \ 1](F + S) = [75 \ 37.5 \ 153]$ represents the average number of points, rebounds, and minutes played by the starters in the tournament. 24. (a) $JC = \begin{bmatrix} 8450 \\ 13,900 \end{bmatrix}$ represents the total revenue in rentals at each store in the month of January. (b) $FC = \begin{bmatrix} 13,440 \\ 17,550 \end{bmatrix}$ represents the total revenue in rentals at each store in the month of February. (c) $(J + F)C = \begin{bmatrix} 21,890 \\ 31,450 \end{bmatrix}$ represents the total revenue in rentals at each store during the two-month period.
(d) $[1, \ 1] \cdot ((J + F)C) = \$53{,}340$ represents the total rental revenue at both stores during the two-month period.
25. $u = -5; v = 0; w = 2; z = 7$ 26. $b_{21} = -2$ 27. A 28. $\begin{bmatrix} -5 & -3 \\ 4 & 6 \end{bmatrix}$ 29. $(2, -3)$ 30. $(-1, 0, 3)$

7.2 EXERCISES B

Find the scalar products in Exercises 1–2.

1. $[-1 \ \ 6] \cdot \begin{bmatrix} 0 \\ -7 \end{bmatrix}$

2. $[0 \ \ 2 \ \ 1 \ \ -3] \cdot \begin{bmatrix} 4 \\ 1 \\ -2 \\ 5 \end{bmatrix}$

In Exercises 3–4 determine m and n so that each product will be defined.

3. $A_{2 \times 3} B_{m \times n} = C_{2 \times 5}$

4. $A_{m \times 4} B_{n \times 2} = C_{5 \times 2}$

Exercises 5–14 refer to the following matrices.

$$A = \begin{bmatrix} 2 & 5 \\ -1 & 3 \end{bmatrix} \quad B = \begin{bmatrix} 0 & -2 \\ 1 & 4 \end{bmatrix} \quad C = \begin{bmatrix} 1 & 1 \\ 1 & 1 \end{bmatrix} \quad D = \begin{bmatrix} 1 & 3 & -2 \\ 4 & 0 & 5 \end{bmatrix}$$

$$E = \begin{bmatrix} 0 & 4 \\ 4 & 0 \end{bmatrix} \quad F = \begin{bmatrix} 3 & 0 \\ -2 & 1 \\ 0 & 5 \end{bmatrix} \quad G = [3 \ \ -2] \quad H = \begin{bmatrix} -2 \\ 4 \\ 1 \end{bmatrix}$$

5. Find AC.

6. Find DE.

7. Find GE.

8. Find A^T.

9. Find A^2.

10. Find D^T.

11. Find AA^T.

12. Find $A(BC)$.

13. Find $B^T + E^T$.

14. Find $(B + E)^T$.

In Exercises 15–22 use the following matrices to verify each statement.

$$A = \begin{bmatrix} 0 & 1 \\ 2 & 3 \end{bmatrix} \quad B = \begin{bmatrix} 4 & 2 \\ 2 & -1 \end{bmatrix} \quad C = \begin{bmatrix} 1 & 0 \\ -1 & 5 \end{bmatrix}$$

15. $AB \neq BA$

16. $A(BC) = (AB)C$

17. $AI = A$

18. $A + (-A) = 0$

19. $A + 0 = A$

20. $A(B + C) = AB + AC$

21. $(B + C)A = BA + CA$

22. $A(B + C) \neq (B + C)A$

23. **BUSINESS** Video Rentals Inc. has two stores in Barstow, California. The total number of VHS cassettes, Beta cassettes, and recorders rented during the months of January and February are given in matrices J and F.

$$J = \begin{matrix} & \text{VHS} & \text{Beta} & \text{Recorders} \\ & \begin{bmatrix} 2100 & 400 & 55 \\ 3500 & 700 & 60 \end{bmatrix} & & \begin{matrix} \text{Store 1} \\ \text{Store 2} \end{matrix} \end{matrix}$$

$$F = \begin{bmatrix} 3200 & 750 & 84 \\ 4300 & 900 & 105 \end{bmatrix} \begin{matrix} \text{Store 1} \\ \text{Store 2} \end{matrix}$$

(a) Find $[1 \ 1]J$ and discuss what it represents.
(b) Find $[1 \ 1](J + F)$ and discuss what it represents.
(c) Find $([1 \ 1](J + F)) \cdot \begin{bmatrix} 1 \\ 1 \\ 1 \end{bmatrix}$ and discuss what it represents.

24. **VETERINARY SCIENCE** A veterinarian blends two brands of dog food into three different mixes. The amounts of protein and fat, measured in grams per ounce, in each of the two brands are given in matrix N. The amounts of each brand, measured in ounces, that he uses in each mix are given in matrix M.

$$N = \begin{matrix} & \text{Brand A} & \text{Brand B} \\ & \begin{bmatrix} 6 & 8 \\ 4 & 3 \end{bmatrix} & \begin{matrix} \text{Protein} \\ \text{Fat} \end{matrix} \end{matrix}$$

$$M = \begin{matrix} & \text{Mix 1} & \text{Mix 2} & \text{Mix 3} \\ & \begin{bmatrix} 12 & 18 & 6 \\ 12 & 6 & 18 \end{bmatrix} & & \begin{matrix} \text{Brand A} \\ \text{Brand B} \end{matrix} \end{matrix}$$

(a) Find the matrix $NM = [c_{ij}]_{2 \times 3}$ and discuss what it represents.
(b) What does the element c_{12} represent?
(c) What does the element c_{21} represent?
(d) Notice that each mix contains 24 ounces of dog food. Find $\frac{1}{24}NM$ and interpret the results.
(e) Suppose that Brand A costs $0.15 per ounce and Brand B costs $0.20 per ounce. Find the matrix that gives total cost for each mix. Which mix is the most expensive?

FOR REVIEW

Use the following matrices in Exercises 25–28.

$$A = \begin{bmatrix} -5 & 0 \\ 2 & 7 \end{bmatrix} \quad B = \begin{bmatrix} 0 & 3 \\ -2 & 1 \end{bmatrix} \quad C = \begin{bmatrix} u & v \\ w & z \end{bmatrix}$$

25. If $B = C$, find the values of u, v, w, and z.

26. Identify the element a_{12} in matrix A.

27. Find $A + B$.

28. Find $3A - 7B$.

1. Interchange the two rows to obtain a 1 in the upper left corner of the matrix.

$$\begin{matrix}(B)\\(A)\end{matrix}\begin{bmatrix}1 & -3 & 11\\ 2 & 1 & 1\end{bmatrix}\quad\text{(A) and (B) interchanged}$$

2. Keep the first row, multiply it by -2, and add the result to the second row.

$$\begin{matrix}(B)\\(C)\end{matrix}\begin{bmatrix}1 & -3 & 11\\ 0 & 7 & -21\end{bmatrix}\leftarrow (C) = -2(B) + (A)$$

3. Multiply the second row by 1/7 to obtain 1 in the second column.

$$\begin{matrix}(B)\\(D)\end{matrix}\begin{bmatrix}1 & -3 & 11\\ 0 & 1 & -3\end{bmatrix}\leftarrow (D) = \tfrac{1}{7}(C)$$

4. To obtain 0 in the second column of the first row, keep the second row, multiply it by 3, and add the result to the first row.

$$\begin{matrix}(E)\\(D)\end{matrix}\begin{bmatrix}1 & 0 & 2\\ 0 & 1 & -3\end{bmatrix}\leftarrow (E) = 3(D) + B$$

Since $x = 2$ and $y = -3$, the solution to the system is $(2, -3)$.

Answer: $(-1, 3)$

For a system of three linear equations in three variables, the elementary row operations are used to transform the augmented matrix into the reduced echelon form

$$\begin{bmatrix}1 & 0 & 0 & p\\ 0 & 1 & 0 & q\\ 0 & 0 & 1 & r\end{bmatrix}$$

which corresponds to the equivalent system

$$x = p$$
$$ y = q$$
$$ z = r.$$

Then the solution to the original system is (p, q, r).

The following diagram shows the best procedure for obtaining 1's and 0's in the augmented matrix of a system of three linear equations. In all cases, once the 1 is obtained in a column, it is used to obtain the two 0's.

1. Obtain this 1 first
2. Obtain these 0's second
3. Obtain this 1 third
4. Obtain these 0's fourth
5. Obtain this 1 fifth
6. Obtain these 0's sixth

$$\begin{bmatrix}1 & 0 & 0 & p\\ 0 & 1 & 0 & q\\ 0 & 0 & 1 & r\end{bmatrix}$$

EXAMPLE 2   THE GAUSSIAN METHOD (3 × 3 SYSTEM)

Solve using the Gaussian method.

$$2x - y + z = 1$$
$$x + 3y - z = -4$$
$$3x + 2y + z = 0$$

PRACTICE EXERCISE 2

Solve using the Gaussian method.

$$x - 2y + z = 0$$
$$2x + y - 3z = 5$$
$$3x - y - z = 5$$

The augmented matrix is

$$\begin{array}{c} \text{(A)} \\ \text{(B)} \\ \text{(C)} \end{array} \begin{bmatrix} 2 & -1 & 1 & 1 \\ 1 & 3 & -1 & -4 \\ 3 & 2 & 1 & 0 \end{bmatrix}.$$

1. To obtain 1 in the upper left corner, interchange the first two rows.

$$\begin{array}{c} \text{(B)} \\ \text{(A)} \\ \text{(C)} \end{array} \begin{bmatrix} 1 & 3 & -1 & -4 \\ 2 & -1 & 1 & 1 \\ 3 & 2 & 1 & 0 \end{bmatrix} \quad \text{(A) and (B) interchanged}$$

2. Keep the first row, multiply it by -2, and add the result to the second row. Then multiply the first row by -3 and add to the third row. By doing so, we obtain two 0's in the first column.

$$\begin{array}{c} \text{(B)} \\ \text{(D)} \\ \text{(E)} \end{array} \begin{bmatrix} 1 & 3 & -1 & -4 \\ 0 & -7 & 3 & 9 \\ 0 & -7 & 4 & 12 \end{bmatrix} \begin{array}{l} \leftarrow \text{(D)} = -2\text{(B)} + \text{(A)} \\ \leftarrow \text{(E)} = -3\text{(B)} + \text{(C)} \end{array}$$

3. Obtain 1 in the second column of the second row by multiplying by $-1/7$.

$$\begin{array}{c} \text{(B)} \\ \text{(F)} \\ \text{(E)} \end{array} \begin{bmatrix} 1 & 3 & -1 & -4 \\ 0 & 1 & -\dfrac{3}{7} & -\dfrac{9}{7} \\ 0 & -7 & 4 & 12 \end{bmatrix} \leftarrow \text{(F)} = -\tfrac{1}{7}\text{(D)}$$

4. Keep the second row, multiply it by -3, and add the result to the first row. Then multiply the second row by 7 and add to the third. Again, in one writing we obtain two 0's in the second column.

$$\begin{array}{c} \text{(G)} \\ \text{(F)} \\ \text{(H)} \end{array} \begin{bmatrix} 1 & 0 & \dfrac{2}{7} & -\dfrac{1}{7} \\ 0 & 1 & -\dfrac{3}{7} & -\dfrac{9}{7} \\ 0 & 0 & 1 & 3 \end{bmatrix} \begin{array}{l} \leftarrow \text{(G)} = -3\text{(F)} + \text{(B)} \\ \\ \leftarrow \text{(H)} = 7\text{(F)} + \text{(E)} \end{array}$$

5. In this case, we obtained a 1 in the third row with no additional effort.

6. Keep the third row, multiply it by 3/7 and add to the second, and finally, multiply the third row by $-2/7$ and add to the first.

$$\begin{array}{c} \text{(J)} \\ \text{(I)} \\ \text{(H)} \end{array} \begin{bmatrix} 1 & 0 & 0 & -1 \\ 0 & 1 & 0 & 0 \\ 0 & 0 & 1 & 3 \end{bmatrix} \begin{array}{l} \leftarrow \text{(J)} = -\tfrac{2}{7}\text{(H)} + \text{(G)} \\ \leftarrow \text{(I)} = \tfrac{3}{7}\text{(H)} + \text{(F)} \end{array}$$

Since $x = -1$, $y = 0$, and $z = 3$, the solution to the system is $(-1, 0, 3)$.

Answer: $(2, 1, 0)$

The real effectiveness of the Gaussian method becomes apparent when systems with a large number of variables are involved. As with 2×2 systems and 3×3 systems, the augmented matrix of higher order systems is transformed into the reduced echelon form

$$\begin{bmatrix} 1 & 0 & 0 & \cdots & 0 & p \\ 0 & 1 & 0 & \cdots & 0 & q \\ 0 & 0 & 1 & \cdots & 0 & r \\ \vdots & \vdots & \vdots & & \vdots & \vdots \\ 0 & 0 & 0 & \cdots & 1 & v \end{bmatrix}$$

by moving from left to right by columns, first obtaining the desired 1 and then the remaining zeros. In fact, it is not difficult to program a computer to carry out these arithmetic operations, obtaining the solution to extremely large systems in a matter of seconds.

EXAMPLE 3 THE GAUSSIAN METHOD (4 × 4 SYSTEM)

Solve using the Gaussian method.

$$w + x + y + z = 2$$
$$w - x + 3y + z = 3$$
$$x - y - z = -1$$
$$2w + x + 3y + z = 0$$

The augmented matrix is

(A) $\begin{bmatrix} 1 & 1 & 1 & 1 & 2 \\ 1 & -1 & 3 & 1 & 3 \\ 0 & 1 & -1 & -1 & -1 \\ 2 & 1 & 3 & 1 & 0 \end{bmatrix}$
(B)
(C)
(D)

1. Since there is already a 1 in the upper left corner, use it to obtain 0's in the second and fourth rows.

(A) $\begin{bmatrix} 1 & 1 & 1 & 1 & 2 \\ 0 & -2 & 2 & 0 & 1 \\ 0 & 1 & -1 & -1 & -1 \\ 0 & -1 & 1 & -1 & -4 \end{bmatrix}$ (E) = −(A) + (B)
(E)
(C)
(F)   (F) = −2(A) + (D)

2. Interchange the second and third rows to obtain a 1 in the desired position.

(A) $\begin{bmatrix} 1 & 1 & 1 & 1 & 2 \\ 0 & 1 & -1 & -1 & -1 \\ 0 & -2 & 2 & 0 & 1 \\ 0 & -1 & 1 & -1 & -4 \end{bmatrix}$ (E) and (C) interchanged
(C)
(E)
(F)

3. Use the 1 in the second row to obtain 0's in the first, third, and fourth rows.

(G) $\begin{bmatrix} 1 & 0 & 2 & 2 & 3 \\ 0 & 1 & -1 & -1 & -1 \\ 0 & 0 & 0 & -2 & -1 \\ 0 & 0 & 0 & -2 & -5 \end{bmatrix}$ (G) = −(C) + (A)
(C)
(H)   (H) = 2(C) + (E)
(I)   (I) = (C) + (F)

4. At this point we might observe that it will be impossible to obtain a 1 in the desired position in the third row. In fact, by replacing the fourth row with −(H) + (I), the matrix becomes

(G) $\begin{bmatrix} 1 & 0 & 2 & 2 & 3 \\ 0 & 1 & -1 & -1 & -1 \\ 0 & 0 & 0 & -2 & -1 \\ 0 & 0 & 0 & 0 & -4 \end{bmatrix}$
(C)
(H)
(J)   (J) = −(H) + (I)

in which the fourth row corresponds to the contradiction $0w + 0x + 0y + 0z = -4$ or $0 = -4$. As was the case with solution methods in the preceding chapter, this indicates that the original system has no solution.

PRACTICE EXERCISE 3

Solve using the Gaussian method.

$$w + x - y + z = 2$$
$$w - x + 2y + z = 4$$
$$w \qquad - y - z = -1$$
$$2w + x \qquad + z = 3$$

Answer: $(1, -1, 0, 2)$

EXAMPLE 4 A DEPENDENT SYSTEM

Solve by the Gaussian method.

$$\begin{aligned} x \quad\quad - z &= -3 \\ -2x - y - 2z &= 2 \\ 3x + y + z &= -5 \end{aligned}$$

The augmented matrix is

$$\begin{matrix}(A)\\(B)\\(C)\end{matrix} \begin{bmatrix} 1 & 0 & -1 & -3 \\ -2 & -1 & -2 & 2 \\ 3 & 1 & 1 & -5 \end{bmatrix}.$$

1. Use the 1 in the first row to obtain 0's in the second and third rows.

$$\begin{matrix}(A)\\(D)\\(E)\end{matrix} \begin{bmatrix} 1 & 0 & -1 & -3 \\ 0 & -1 & -4 & -4 \\ 0 & 1 & 4 & 4 \end{bmatrix} \quad \begin{aligned}(D) &= 2(A) + (B) \\ (E) &= -3(A) + (C)\end{aligned}$$

2. Multiply the second row by -1 to obtain the desired 1, and use it to obtain a 0 in the third row.

$$\begin{matrix}(A)\\(F)\\(G)\end{matrix} \begin{bmatrix} 1 & 0 & -1 & -3 \\ 0 & 1 & 4 & 4 \\ 0 & 0 & 0 & 0 \end{bmatrix} \quad \begin{aligned}(F) &= (-1)(D) \\ (G) &= (-1)(F) + E\end{aligned}$$

3. At this point we obtain the identity $0 = 0$ in the third row. Once this happens it is probably best to write the equations involving the variables and solve by other methods.

$$\begin{aligned} x \quad\quad - z &= -3 \\ y + 4z &= 4 \end{aligned}$$

Then $x = z - 3$ and $y = 4 - 4z$ so the system has infinitely many solutions of the form $(z - 3, 4 - 4z, z)$ for z any real number.

PRACTICE EXERCISE 4

Solve using the Gaussian method.

$$\begin{aligned} x - y + z &= 3 \\ 2x - 2y + 2z &= -1 \\ -x + y - z &= -3 \end{aligned}$$

Answer: no solution (the system is inconsistent)

7.3 EXERCISES A

In Exercises 1–2 assume that the given matrix is the augmented matrix of a system of equations. Follow the indicated steps and write the matrix in reduced echelon form.

1. $\begin{matrix}(A)\\(B)\end{matrix} \begin{bmatrix} 2 & 3 & -4 \\ 1 & -2 & 5 \end{bmatrix}$

   (a) Interchange rows (A) and (B).

   (b) Retain (B) and replace (A) with (C) $= -2(B) + (A)$.

   (c) Replace (C) with (D) $= \frac{1}{7}(C)$.

   (d) Retain (D) and replace (B) with (E) $= 2(D) + (B)$.

2. (A) $\begin{bmatrix} 2 & 1 & -1 & 4 \\ 3 & -1 & 0 & 3 \\ -4 & 2 & 3 & -10 \end{bmatrix}$
   (B)
   (C)

   (a) Replace (A) with (D) = (B) − (A).

   (b) Retain (D), replace (B) with (E) = −3(D) + (B), and replace (C) with (F) = 4(D) + (C).

   (c) Retain (D) and (F) and replace (E) with (G) = (−1)[(E) + (F)].

   (d) Retain (G), replace (D) with (H) = 2(G) + (D), and replace (F) with (I) = 6(G) + (F).

   (e) Retain (H) and (G) and replace (I) with (J) = $-\frac{1}{17}$(I).

   (f) Retain (J), replace (G) with (K) = 4(J) + (G), and replace (H) with (L) = 7(J) + (H).

In Exercises 3–12, solve each system by the Gaussian method.

3. $x + 3y = 1$
   $3x + 7y = 5$

4. $2x - y = 1$
   $-4x + 2y = 1$

5. $4x - 2y = -3$
   $x + y = 3$

6. $5x - 3y = -2$
   $2x + 7y = -9$

7. $3x + 2y = 1$
   $-9x - 5y = -3$

8. $x + 2y = 1$
   $-2x - 4y = -2$

9. $x + y + z = 2$
   $-x - y + 3z = 6$
   $2x + y - z = -1$

10. $x + z = 1$
    $ y + z = 4$
    $2x + y = -3$

11. $2x - y + 3z = 0$
    $x + y - z = 4$
    $ 2y + 5z = -3$

12. $3x - y + 2z = 2$
    $x - z = 1$
    $2x - y + 3z = 1$

FOR REVIEW

Exercises 13–20 refer to the following matrices.

$$A = \begin{bmatrix} 1 & 0 \\ -2 & 5 \end{bmatrix} \quad B = \begin{bmatrix} -2 & 5 \\ 1 & 4 \end{bmatrix} \quad C = \begin{bmatrix} 3 & -1 & 2 \\ 0 & -2 & 4 \end{bmatrix}$$

$$D = \begin{bmatrix} 2 & 1 & -1 \end{bmatrix} \quad E = \begin{bmatrix} 3 \\ 0 \\ 5 \end{bmatrix} \quad F = \begin{bmatrix} 4 & -3 \end{bmatrix}$$

13. What is the order of C?

14. Identify the element c_{13} in matrix C.

15. Give the matrix $-B$.

16. Find $A + B$.

17. Find $2A - 4B$.

18. Find $(AB)^T$.

19. Find AI.

20. Find ED.

21. BUSINESS Suppose that $C = \begin{bmatrix} 200 & 300 & 400 & 550 \end{bmatrix}$ gives the selling prices of four models of television sets, and

$$N = \begin{bmatrix} 5 \\ 3 \\ 4 \\ 2 \end{bmatrix}$$

gives the number of each model sold on Saturday. Find $C \cdot N$ and interpret the result.

ANSWERS: 1. (a) (B) $\begin{bmatrix} 1 & -2 & 5 \\ 2 & 3 & -4 \end{bmatrix}$ (A) (b) (B) (C) $\begin{bmatrix} 1 & -2 & 5 \\ 0 & 7 & -14 \end{bmatrix}$ (c) (B) (D) $\begin{bmatrix} 1 & -2 & 5 \\ 0 & 1 & -2 \end{bmatrix}$ (d) (E) (D) $\begin{bmatrix} 1 & 0 & 1 \\ 0 & 1 & -2 \end{bmatrix}$

2. (a) (B) (C) $\begin{bmatrix} 1 & -2 & 1 & -1 \\ 3 & -1 & 0 & 3 \\ -4 & 2 & 3 & -10 \end{bmatrix}$ (b) (E) (F) $\begin{bmatrix} 1 & -2 & 1 & -1 \\ 0 & 5 & -3 & 6 \\ 0 & -6 & 7 & -14 \end{bmatrix}$ (c) (G) (F) $\begin{bmatrix} 1 & -2 & 1 & -1 \\ 0 & 1 & -4 & 8 \\ 0 & -6 & 7 & -14 \end{bmatrix}$

(d) (G) (I) $\begin{bmatrix} 1 & 0 & -7 & 15 \\ 0 & 1 & -4 & 8 \\ 0 & 0 & -17 & 34 \end{bmatrix}$ (e) (G) (J) $\begin{bmatrix} 1 & 0 & -7 & 15 \\ 0 & 1 & -4 & 8 \\ 0 & 0 & 1 & -2 \end{bmatrix}$ (f) (K) (J) $\begin{bmatrix} 1 & 0 & 0 & 1 \\ 0 & 1 & 0 & 0 \\ 0 & 0 & 1 & -2 \end{bmatrix}$ 3. $(4, -1)$ 4. no solution

5. $\left(\frac{1}{2}, \frac{5}{2}\right)$ 6. $(-1, -1)$ 7. $\left(\frac{1}{3}, 0\right)$ 8. $(1 - 2y, y)$ for y any real number 9. $(1, -1, 2)$ 10. $(-2, 1, 3)$
11. $(2, 1, -1)$ 12. $(x, 5x - 4, x - 1)$ for x any real number 13. 2×3 14. 2

15. $\begin{bmatrix} 2 & -5 \\ -1 & -4 \end{bmatrix}$ 16. $\begin{bmatrix} -1 & 5 \\ -1 & 9 \end{bmatrix}$ 17. $\begin{bmatrix} 10 & -20 \\ -8 & -6 \end{bmatrix}$ 18. $\begin{bmatrix} -2 & 9 \\ 5 & 10 \end{bmatrix}$ 19. A 20. $\begin{bmatrix} 6 & 3 & -3 \\ 0 & 0 & 0 \\ 10 & 5 & -5 \end{bmatrix}$ 21. $C \cdot N = 4600$.

Thus, the gross sales of the four models on Saturday was $4600.

7.3 EXERCISES B

In Exercises 1–2 assume that the given matrix is the augmented matrix of a system of equations. Follow the indicated steps and write the matrix in reduced echelon form.

1. (A) $\begin{bmatrix} 4 & -5 & 5 \\ 1 & 6 & -6 \end{bmatrix}$
   (B)

   (a) Interchange (A) and (B).
   (b) Retain (B) and replace (A) with (C) = −4(B) + (A).
   (c) Replace (C) with (D) = $-\frac{1}{29}$(C).
   (d) Retain (D) and replace (B) with (E) = −6(D) + (B).

2. (A) $\begin{bmatrix} 1 & -2 & 5 & 4 \\ 3 & 1 & 2 & 5 \\ -4 & 3 & -1 & -11 \end{bmatrix}$
   (B)
   (C)

   (a) Retain (A), replace (B) with (D) = −3(A) + (B), and replace (C) with (E) = 4(A) + (C).
   (b) Retain (A) and (E) and replace (D) with (F) = (−1)[2(D) + 3(E)].
   (c) Retain (F), replace (A) with (G) = 2(F) + (A), and replace (E) with (H) = 5(F) + (E).
   (d) Retain (G) and (F) and replace (H) with (I) = $-\frac{1}{136}$(H).
   (e) Retain (I), replace (F) with (J) = 31(I) + (F), and replace (G) with (K) = 57(I) + (G).

In Exercises 3–12, solve each system by the Gaussian method.

3. $4x - y = 3$
   $x + 3y = 4$

4. $3x + 6y = 4$
   $x - 5y = -1$

5. $2x - 3y = 1$
   $-4x + 6y = 2$

6. $3x + y = -6$
   $5x - 2y = -10$

7. $5x - 4y = -1$
   $2x + 8y = 2$

8. $3x - 5y = 2$
   $-9x + 15y = -6$

9. $x - y + z = 3$
   $2x + y + 3z = 3$
   $-x + 2y - 4z = -4$

10. $x - 2y = -1$
    $y + z = 7$
    $x \quad - z = -2$

11. $3x - y + 2z = -2$
    $4x + y + z = 3$
    $7x - 2y - 2z = 9$

12. $4x - y + 2z = 4$
    $3x + y = 5$
    $x - 2y + 2z = -3$

FOR REVIEW

Exercises 13–20 refer to the following matrices.

$$A = \begin{bmatrix} 1 & 0 \\ -2 & 5 \end{bmatrix} \quad B = \begin{bmatrix} -2 & 5 \\ 1 & 4 \end{bmatrix} \quad C = \begin{bmatrix} 3 & -1 & 2 \\ 0 & -2 & 4 \end{bmatrix}$$

$$D = [2 \ \ 1 \ \ -1] \quad E = \begin{bmatrix} 3 \\ 0 \\ 5 \end{bmatrix} \quad F = [4 \ \ -3]$$

13. What is the order of D?

14. Identify the element c_{21} in matrix C.

15. Give the matrix C^T.

16. Find $B - C$.

17. Find $D \cdot E$.

18. Find FA.

19. Find $C + 0$.

20. Give the matrix A^T.

21. **BUSINESS** Suppose that $C = [300 \quad 350 \quad 400 \quad 500]$ gives the selling prices of four models of washing machines, and

$$N = \begin{bmatrix} 2 \\ 4 \\ 3 \\ 6 \end{bmatrix}$$

gives the number of each model sold on Friday. Find $C \cdot N$ and interpret the result.

7.3 EXERCISES C

In Exercises 1–4, solve each system by the Gaussian method.

1. $w + x - y + z = 4$
   $w + 2y - z = -3$
   $w - x + 2z = 5$
   $2w + x + 2y - 3z = -6$
   [Answer: $(1, 0, -1, 2)$]

2. $2w - x - 2y + z = -1$
   $w + x - y = -2$
   $-w - x + 4z = -1$
   $ 2x + y - 3z = 1$

3. $\sqrt{x} + y^2 - \dfrac{1}{z} = 4$
   $2\sqrt{x} - y^2 + \dfrac{1}{z} = 2$
   $-\sqrt{x} - 2y^2 - \dfrac{1}{z} = -3$
   [Answer: $(4, 1, -1)$ or $(4, -1, -1)$]

4. $x^3 - \sqrt{y} + 2^z = 7$
   $2x^3 + 3\sqrt{y} - 2^z = 23$
   $-x^3 + \sqrt{y} + 2^z = -3$

5. The total resistance R in the circuit shown here with resistors R_1 and R_2, connected in parallel, can be obtained using the formula $\frac{1}{R} = \frac{1}{R_1} + \frac{1}{R_2}$. Suppose three resistors A, B, and C are given. When A and B are connected in parallel, the total resistance is $3.\overline{3}$ ohms, when A and C are connected in parallel, the total resistance is 4 ohms, and when B and C are connected in parallel, the total resistance is $6.\overline{6}$ ohms. Write a system of equations that describes this information and solve the system using the Gaussian method to obtain the resistance of each of A, B, and C.

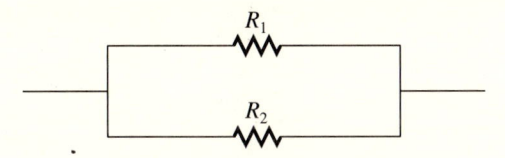

[Answer: A: 5 ohms, B: 10 ohms, C: 20 ohms]

7.4 THE INVERSE OF A SQUARE MATRIX

STUDENT GUIDEPOSTS

1. Solving Matrix Equations
2. The Inverse of a Square Matrix
3. Writing Systems as Matrix Equations
4. Solving Systems Using the Inverse Method
5. Problem Solving Using Inverses

1 SOLVING MATRIX EQUATIONS

In our study of the algebra of matrices, numerous similarities have been observed between matrices and the algebra of real numbers. Properties of real numbers allow us to solve equations involving real-number variables, and properties of matrices enable us to solve matrix equations. For example, consider the matrix equation

$$A + X = B,$$

where $A = \begin{bmatrix} 1 & -3 \\ 2 & 5 \end{bmatrix}$, $X = \begin{bmatrix} w & x \\ y & z \end{bmatrix}$, and $B = \begin{bmatrix} 0 & 2 \\ -4 & 7 \end{bmatrix}$.

Substitute these matrices into the matrix equation and add the two matrices on the left side.

$$\begin{bmatrix} 1 & -3 \\ 2 & 5 \end{bmatrix} + \begin{bmatrix} w & x \\ y & z \end{bmatrix} = \begin{bmatrix} 0 & 2 \\ -4 & 7 \end{bmatrix}$$

$$\begin{bmatrix} 1+w & -3+x \\ 2+y & 5+z \end{bmatrix} = \begin{bmatrix} 0 & 2 \\ -4 & 7 \end{bmatrix}$$

Use the definition of equality of matrices to find w, x, y, and z.

$$\begin{array}{llll} 1 + w = 0 & -3 + x = 2 & 2 + y = -4 & 5 + z = 7 \\ w = -1 & x = 5 & y = -6 & z = 2 \end{array}$$

Thus, we have solved the matrix equation for the variable matrix X obtaining

$$X = \begin{bmatrix} -1 & 5 \\ -6 & 2 \end{bmatrix}.$$

Using the properties of matrices, the equation would have been solved as follows.

$$\begin{array}{ll} A + X = B & \\ -A + (A + X) = -A + B & \text{Add the negative of } A \\ (-A + A) + X = -A + B & \text{Associative law} \\ 0 + X = -A + B & A \text{ and } -A \text{ are additive inverses} \\ X = -A + B & 0 \text{ is the zero matrix} \end{array}$$

We see that obtaining the value for X is merely a matter of performing the matrix operation $-A + B$, or, equivalently, $B - A$, on the constant matrices A and B.

2 THE INVERSE OF A SQUARE MATRIX

The next logical step is to ask how we can solve a matrix equation of the form

$$AX = B.$$

To solve the algebraic equation

$$ax = b,$$

we multiply both sides by the reciprocal or multiplicative inverse of a, $1/a$, or a^{-1}.

$$\begin{aligned} a^{-1}(ax) &= a^{-1}b \\ (a^{-1}a)x &= a^{-1}b \\ 1 \cdot x &= a^{-1}b \\ x &= a^{-1}b \end{aligned}$$

To solve the matrix equation similar to $ax = b$, we need to find a multiplicative inverse of the matrix A. For reasons that will soon become clear, only square matrices can have inverses. A **nonsingular matrix** is a square matrix that has an inverse. A **singular matrix** is a matrix that does not have an inverse. It can be shown that if a matrix A is nonsingular, then the matrix A^{-1} is unique.

The Inverse of a Matrix

Let A be an $n \times n$ square matrix with I the $n \times n$ identity matrix. An $n \times n$ matrix A^{-1} is the **inverse** of A if and only if

$$AA^{-1} = A^{-1}A = I.$$

CAUTION

We read the symbol A^{-1} as "A inverse," and do not use the notation $\frac{1}{A}$ since the operation of division has not been defined for matrices.

Now that we have defined the inverse of a square matrix, the next task is to find that inverse. Suppose we begin with the matrix

$$A = \begin{bmatrix} 3 & 4 \\ 2 & 3 \end{bmatrix}.$$

If A^{-1} exists, then it must be a 2×2 matrix of the form

$$A^{-1} = \begin{bmatrix} a & b \\ c & d \end{bmatrix},$$

and

$$AA^{-1} = \begin{bmatrix} 3 & 4 \\ 2 & 3 \end{bmatrix} \begin{bmatrix} a & b \\ c & d \end{bmatrix} = \begin{bmatrix} 1 & 0 \\ 0 & 1 \end{bmatrix} = I.$$

Multiply and use the definition of equality to obtain two systems of equations.

$$\begin{bmatrix} 3a + 4c & 3b + 4d \\ 2a + 3c & 2b + 3d \end{bmatrix} = \begin{bmatrix} 1 & 0 \\ 0 & 1 \end{bmatrix}$$

$$3a + 4c = 1 \qquad 3b + 4d = 0$$
$$2a + 3c = 0 \qquad 2b + 3d = 1$$

Solving these systems gives $a = 3$, $b = -4$, $c = -2$, and $d = 3$. Thus,

$$A^{-1} = \begin{bmatrix} 3 & -4 \\ -2 & 3 \end{bmatrix}.$$

It is easy to show that this is the inverse since

$$AA^{-1} = \begin{bmatrix} 3 & 4 \\ 2 & 3 \end{bmatrix} \begin{bmatrix} 3 & -4 \\ -2 & 3 \end{bmatrix} = \begin{bmatrix} 1 & 0 \\ 0 & 1 \end{bmatrix} = \begin{bmatrix} 3 & -4 \\ -2 & 3 \end{bmatrix} \begin{bmatrix} 3 & 4 \\ 2 & 3 \end{bmatrix} = A^{-1}A.$$

It was noted earlier that not all matrices have inverses. It can be shown that

$$B = \begin{bmatrix} 1 & 2 \\ 1 & 2 \end{bmatrix}$$

does not have an inverse. If we follow the same procedure used above for matrix A, the two systems of equations in a and c and in b and d are inconsistent and have no solution.

The process of finding the inverse of a matrix, when it exists, can be simplified by reducing a particular augmented matrix to echelon form. This is similar to what was done in the previous section. Consider again the systems obtained when we found A^{-1} above.

$$3a + 4c = 1 \qquad 3b + 4d = 0$$
$$2a + 3c = 0 \qquad 2b + 3d = 1$$

If we solve these systems using the Gaussian method, the augmented matrices are

$$\begin{bmatrix} 3 & 4 & 1 \\ 2 & 3 & 0 \end{bmatrix} \quad \text{and} \quad \begin{bmatrix} 3 & 4 & 0 \\ 2 & 3 & 1 \end{bmatrix}.$$

Since these two matrices are identical in the first two columns, rather than reducing each to echelon form separately, we can speed the process by combining them into a single augmented matrix of the form

$$\begin{matrix} \text{(A)} \\ \text{(B)} \end{matrix} \begin{bmatrix} 3 & 4 & | & 1 & 0 \\ 2 & 3 & | & 0 & 1 \end{bmatrix}.$$

This matrix is of the form $[A|I]$, where A is the original matrix and I the 2×2 identity matrix. By using the elementary row operations on this augmented matrix, we can transform it to the form $[I|A^{-1}]$, using the following steps.

1. To obtain a 1 in the upper left corner, retain (B) and replace (A) with (C) = $(-1)(B) + (A)$.

$$\begin{matrix} \text{(C)} \\ \text{(B)} \end{matrix} \begin{bmatrix} 1 & 1 & | & 1 & -1 \\ 2 & 3 & | & 0 & 1 \end{bmatrix}$$

2. Retain (C) and replace (B) with (D) = $(-2)(C) + (B)$.

$$\begin{matrix} \text{(C)} \\ \text{(D)} \end{matrix} \begin{bmatrix} 1 & 1 & | & 1 & -1 \\ 0 & 1 & | & -2 & 3 \end{bmatrix}$$

3. Retain (D) and replace (C) with (E) = $(-1)(D) + (C)$.

$$\begin{matrix} \text{(E)} \\ \text{(D)} \end{matrix} \begin{bmatrix} 1 & 0 & | & 3 & -4 \\ 0 & 1 & | & -2 & 3 \end{bmatrix}$$

Notice that the matrix $[I|A^{-1}]$ does give the value of A^{-1} obtained before. Although we have illustrated this method for finding A^{-1} with a 2×2 matrix A, it can be used for any $n \times n$ matrices.

To Find the Inverse of a Square Matrix

Let A be an $n \times n$ matrix.

1. Form the augmented matrix $[A|I]$.
2. Use the elementary row operations to reduce $[A|I]$.
3. If A^{-1} exists, the augmented matrix will be reduced to $[I|A^{-1}]$.
4. If A^{-1} does not exist, a row of zeros will be produced to the left of the vertical line.

EXAMPLE 1 THE INVERSE OF A 2 × 2 MATRIX

Find the inverse of $A = \begin{bmatrix} 3 & 4 \\ 1 & 2 \end{bmatrix}$.

Form the augmented matrix $[A|I]$.

$$\begin{matrix} \text{(A)} \\ \text{(B)} \end{matrix} \begin{bmatrix} 3 & 4 & | & 1 & 0 \\ 1 & 2 & | & 0 & 1 \end{bmatrix}$$

PRACTICE EXERCISE 1

Find the inverse of

$$A = \begin{bmatrix} -2 & -1 \\ 3 & 1 \end{bmatrix}.$$

1. To obtain a 1 in the upper left corner, switch rows **(A)** and **(B)**.

$$\begin{matrix} \textbf{(B)} \\ \textbf{(A)} \end{matrix} \begin{bmatrix} 1 & 2 & | & 0 & 1 \\ 3 & 4 & | & 1 & 0 \end{bmatrix}$$

2. Retain **(B)**, replace **(A)** with **(C)** = -3**(B)** + **(A)**.

$$\begin{matrix} \textbf{(B)} \\ \textbf{(C)} \end{matrix} \begin{bmatrix} 1 & 2 & | & 0 & 1 \\ 0 & -2 & | & 1 & -3 \end{bmatrix}$$

3. Retain **(B)**, replace **(C)** with **(D)** = $-\frac{1}{2}$**(C)**.

$$\begin{matrix} \textbf{(B)} \\ \textbf{(D)} \end{matrix} \begin{bmatrix} 1 & 2 & | & 0 & 1 \\ 0 & 1 & | & -\frac{1}{2} & \frac{3}{2} \end{bmatrix}$$

4. Retain **(D)**, replace **(B)** with **(E)** = -2**(D)** + **(B)**.

$$\begin{matrix} \textbf{(E)} \\ \textbf{(D)} \end{matrix} \begin{bmatrix} 1 & 0 & | & 1 & -2 \\ 0 & 1 & | & -\frac{1}{2} & \frac{3}{2} \end{bmatrix}$$

Thus, $A^{-1} = \begin{bmatrix} 1 & -2 \\ -\frac{1}{2} & \frac{3}{2} \end{bmatrix}$.

Check by showing that $AA^{-1} = A^{-1}A = I$.

Answer: $A^{-1} = \begin{bmatrix} 1 & 1 \\ -3 & -2 \end{bmatrix}$

EXAMPLE 2 THE INVERSE OF 3 × 3 MATRIX

Find the inverse of $A = \begin{bmatrix} 1 & 0 & -1 \\ -2 & 1 & 0 \\ 2 & -2 & 3 \end{bmatrix}$.

Form the augmented matrix $[A|I]$.

$$\begin{matrix} \textbf{(A)} \\ \textbf{(B)} \\ \textbf{(C)} \end{matrix} \begin{bmatrix} 1 & 0 & -1 & | & 1 & 0 & 0 \\ -2 & 1 & 0 & | & 0 & 1 & 0 \\ 2 & -2 & 3 & | & 0 & 0 & 1 \end{bmatrix}$$

1. Retain **(A)**, replace **(B)** with **(D)** = 2**(A)** + **(B)**, and replace **(C)** with **(E)** = -2**(A)** + **(C)**.

$$\begin{matrix} \textbf{(A)} \\ \textbf{(D)} \\ \textbf{(E)} \end{matrix} \begin{bmatrix} 1 & 0 & -1 & | & 1 & 0 & 0 \\ 0 & 1 & -2 & | & 2 & 1 & 0 \\ 0 & -2 & 5 & | & -2 & 0 & 1 \end{bmatrix}$$

2. Retain **(D)** and replace **(E)** with **(F)** = 2**(D)** + **(E)**.

$$\begin{matrix} \textbf{(A)} \\ \textbf{(D)} \\ \textbf{(F)} \end{matrix} \begin{bmatrix} 1 & 0 & -1 & | & 1 & 0 & 0 \\ 0 & 1 & -2 & | & 2 & 1 & 0 \\ 0 & 0 & 1 & | & 2 & 2 & 1 \end{bmatrix}$$

3. Retain **(F)**, replace **(D)** with **(G)** = 2**(F)** + **(D)**, and replace **(A)** with **(H)** = **(F)** + **(A)**.

$$\begin{matrix} \textbf{(H)} \\ \textbf{(G)} \\ \textbf{(F)} \end{matrix} \begin{bmatrix} 1 & 0 & 0 & | & 3 & 2 & 1 \\ 0 & 1 & 0 & | & 6 & 5 & 2 \\ 0 & 0 & 1 & | & 2 & 2 & 1 \end{bmatrix}$$

Thus, $A^{-1} = \begin{bmatrix} 3 & 2 & 1 \\ 6 & 5 & 2 \\ 2 & 2 & 1 \end{bmatrix}$. To check, show that $AA^{-1} = A^{-1}A = I$.

PRACTICE EXERCISE 2

Find the inverse of

$$A = \begin{bmatrix} 1 & 1 & -1 \\ 2 & 1 & 1 \\ 3 & -2 & -1 \end{bmatrix}.$$

Answer: $\begin{bmatrix} \frac{1}{13} & \frac{3}{13} & \frac{2}{13} \\ \frac{5}{13} & \frac{2}{13} & -\frac{3}{13} \\ -\frac{7}{13} & \frac{5}{13} & -\frac{1}{13} \end{bmatrix}$

498 CHAPTER 7 MATRICES AND DETERMINANTS

❸ WRITING SYSTEMS AS MATRIX EQUATIONS

Now that we know how to find inverses, suppose we return to the problem of solving the matrix equation

$$AX = B$$

posed at the beginning of this section. If A is nonsingular, then A^{-1} exists and

$$A^{-1}(AX) = A^{-1}B$$
$$(A^{-1}A)X = A^{-1}B$$
$$IX = A^{-1}B$$
$$X = A^{-1}B.$$

Since matrix multiplication is not commutative, $A^{-1}B$ is the correct form for X, not BA^{-1}.

Matrix equations arise in a natural way with systems of linear equations. For example, consider the system

$$2x - y = -4$$
$$3x + 4y = 5.$$

If we let

$$A = \begin{bmatrix} 2 & -1 \\ 3 & 4 \end{bmatrix}, \quad X = \begin{bmatrix} x \\ y \end{bmatrix}, \quad \text{and} \quad B = \begin{bmatrix} -4 \\ 5 \end{bmatrix},$$

then A is the **coefficient matrix** of the system, X is the **variable matrix**, and B is the **constant matrix**. Consider the matrix equation $AX = B$.

$$AX = B$$
$$\begin{bmatrix} 2 & -1 \\ 3 & 4 \end{bmatrix} \begin{bmatrix} x \\ y \end{bmatrix} = \begin{bmatrix} -4 \\ 5 \end{bmatrix}$$
$$\begin{bmatrix} 2x - y \\ 3x + 4y \end{bmatrix} = \begin{bmatrix} -4 \\ 5 \end{bmatrix}$$

By the definition of equality of matrices,

$$2x - y = -4$$
$$\text{and} \quad 3x + 4y = 5.$$

Thus, the matrix equation $AX = B$ is equivalent to the system of two equations in two variables. If A has an inverse A^{-1}, the matrix equation can be solved for X.

$$X = A^{-1}B$$

Suppose we try to find A^{-1}.

(A)
(B) $\begin{bmatrix} 2 & -1 & | & 1 & 0 \\ 3 & 4 & | & 0 & 1 \end{bmatrix}$

(C)
(B) $\begin{bmatrix} 1 & 5 & | & -1 & 1 \\ 3 & 4 & | & 0 & 1 \end{bmatrix}$ (C) = (−1)[(−1)(B) + (A)]

(C)
(D) $\begin{bmatrix} 1 & 5 & | & -1 & 1 \\ 0 & -11 & | & 3 & -2 \end{bmatrix}$ (D) = (−3)(C) + (B)

(C)
(E) $\begin{bmatrix} 1 & 5 & | & -1 & 1 \\ 0 & 1 & | & -\dfrac{3}{11} & \dfrac{2}{11} \end{bmatrix}$ (E) = $-\tfrac{1}{11}$(D)

(F)
(E) $\begin{bmatrix} 1 & 0 & | & \dfrac{4}{11} & \dfrac{1}{11} \\ 0 & 1 & | & -\dfrac{3}{11} & \dfrac{2}{11} \end{bmatrix}$ (F) = (−5)(E) + (C)

Thus, $A^{-1} = \begin{bmatrix} \frac{4}{11} & \frac{1}{11} \\ -\frac{3}{11} & \frac{2}{11} \end{bmatrix}$, which can also be written as $A^{-1} = \frac{1}{11}\begin{bmatrix} 4 & 1 \\ -3 & 2 \end{bmatrix}$, avoiding fractions in the matrix. Then the solution to the matrix equation is

$$X = A^{-1}B = \frac{1}{11}\begin{bmatrix} 4 & 1 \\ -3 & 2 \end{bmatrix}\begin{bmatrix} -4 \\ 5 \end{bmatrix}$$
$$= \frac{1}{11}\begin{bmatrix} -11 \\ 22 \end{bmatrix} = \begin{bmatrix} -1 \\ 2 \end{bmatrix}.$$

Since $X = \begin{bmatrix} x \\ y \end{bmatrix} = \begin{bmatrix} -1 \\ 2 \end{bmatrix}$, we have $x = -1$ and $y = 2$, giving the solution $(-1, 2)$ to the original system.

❹ SOLVING SYSTEMS USING THE INVERSE METHOD

Clearly the system of equations above could be solved more quickly using either the substitution or the elimination method discussed in the previous chapter. However, this technique, called the **inverse method,** is important when used in conjunction with a computer or even a programmable calculator.

EXAMPLE 3 THE INVERSE METHOD (2 × 2 SYSTEM)

Solve the following system using the inverse method.

$$3x + 4y = 9$$
$$x + 2y = 5$$

The system is equivalent to the matrix equation $AX = B$ where

$$A = \begin{bmatrix} 3 & 4 \\ 1 & 2 \end{bmatrix}, \quad X = \begin{bmatrix} x \\ y \end{bmatrix}, \quad \text{and} \quad B = \begin{bmatrix} 9 \\ 5 \end{bmatrix}.$$

In Example 1 we found the inverse of the coefficient matrix A to be

$$A^{-1} = \begin{bmatrix} 1 & -2 \\ -\frac{1}{2} & \frac{3}{2} \end{bmatrix}.$$

The solution to the matrix equation is

$$\begin{bmatrix} x \\ y \end{bmatrix} = X = A^{-1}B = \begin{bmatrix} 1 & -2 \\ -\frac{1}{2} & \frac{3}{2} \end{bmatrix}\begin{bmatrix} 9 \\ 5 \end{bmatrix} = \begin{bmatrix} -1 \\ 3 \end{bmatrix},$$

making $x = -1$ and $y = 3$. Thus, $(-1, 3)$ is the solution to the system.

PRACTICE EXERCISE 3

Solve the system using the inverse method.

$$-2x - y = 0$$
$$3x + y = 2$$

[*Hint:* Use the results of Practice Exercise 1.]

Answer: $(2, -4)$

EXAMPLE 4 THE INVERSE METHOD (3 × 3 SYSTEM)

Solve the following system using the inverse method.

$$x \quad\quad - z = 2$$
$$-2x + y \quad\quad = -8$$
$$2x - 2y + 3z = 13$$

PRACTICE EXERCISE 4

Solve the system using the inverse method.

$$x + y - z = -2$$
$$2x + y + z = 3$$
$$3x - 2y - z = -10$$

The system is equivalent to the matrix equation $AX = B$ where

$$A = \begin{bmatrix} 1 & 0 & -1 \\ -2 & 1 & 0 \\ 2 & -2 & 3 \end{bmatrix}, \quad X = \begin{bmatrix} x \\ y \\ z \end{bmatrix}, \quad \text{and} \quad B = \begin{bmatrix} 2 \\ -8 \\ 13 \end{bmatrix}.$$

In Example 2, we found the inverse matrix

$$A^{-1} = \begin{bmatrix} 3 & 2 & 1 \\ 6 & 5 & 2 \\ 2 & 2 & 1 \end{bmatrix}.$$

The solution to the matrix equation is

$$\begin{bmatrix} x \\ y \\ z \end{bmatrix} = X = A^{-1}B = \begin{bmatrix} 3 & 2 & 1 \\ 6 & 5 & 2 \\ 2 & 2 & 1 \end{bmatrix} \begin{bmatrix} 2 \\ -8 \\ 13 \end{bmatrix} = \begin{bmatrix} 3 \\ -2 \\ 1 \end{bmatrix},$$

making $x = 3$, $y = -2$, and $z = 1$. Thus, $(3, -2, 1)$ is the solution to the system.

[*Hint:* Use the results of Practice Exercise 2.]

Answer: $(-1, 2, 3)$

5 PROBLEM SOLVING USING INVERSES

The inverse method can be a useful tool for solving several systems of equations, all with the same coefficient matrix A. Once the inverse is found, to obtain solutions to the systems, simply multiply the inverse by each of the constant matrices B. This is illustrated in the next example.

EXAMPLE 5 SPORTS PROBLEM SOLVED USING INVERSES

The Olympic Sports Committee schedules four exhibition basketball games between the U.S. Olympic Team and an NBA team. Tickets are to be sold for $5 and $8 with a given total revenue production specified. Assuming that all seats will be sold in every arena, use the information given in the following table to find the number of seats at each price that must be sold for each game.

	Game 1	Game 2	Game 3	Game 4
Total arena seats	9000	12,000	8400	21,000
Revenue produced	$63,000	$81,000	$66,000	$114,000

Let x = the number of tickets sold for $5,
    y = the number of tickets sold for $8.

In effect, we have four systems of equations to solve, each having the same coefficient matrix A. If n is the number of seats in an arena and r is the desired revenue for the game in that arena, then

$$x + y = n$$
$$5x + 8y = r$$

represents the general system which must be solved for the four particular values of n and r given in the table. Consider the matrix equation

$$\begin{bmatrix} 1 & 1 \\ 5 & 8 \end{bmatrix} \begin{bmatrix} x \\ y \end{bmatrix} = \begin{bmatrix} n \\ r \end{bmatrix}.$$

PRACTICE EXERCISE 5

In Example 5, if one additional game, scheduled in a domed stadium seating 35,000, was also sold out with a revenue of $220,000, how many $5 and $8 seats were sold?

The inverse of the coefficient matrix A is

$$A^{-1} = \frac{1}{3}\begin{bmatrix} 8 & -1 \\ -5 & 1 \end{bmatrix}.$$

Then $X = A^{-1}B$ becomes

$$\begin{bmatrix} x \\ y \end{bmatrix} = \frac{1}{3}\begin{bmatrix} 8 & -1 \\ -5 & 1 \end{bmatrix}\begin{bmatrix} n \\ r \end{bmatrix}.$$

To find the number of tickets at each price, substitute the appropriate values for n and r and multiply.

For Game 1: $n = 9000$ and $r = \$63,000$.

$$\begin{bmatrix} x \\ y \end{bmatrix} = \frac{1}{3}\begin{bmatrix} 8 & -1 \\ -5 & 1 \end{bmatrix}\begin{bmatrix} 9000 \\ 63,000 \end{bmatrix} = \frac{1}{3}\begin{bmatrix} 9000 \\ 18,000 \end{bmatrix} = \begin{bmatrix} 3000 \\ 6000 \end{bmatrix}$$

Thus, $x = 3000$ (\$5 seats) and $y = 6000$ (\$8 seats).

For Game 2: $n = 12,000$ and $r = \$81,000$.

$$\begin{bmatrix} x \\ y \end{bmatrix} = \frac{1}{3}\begin{bmatrix} 8 & -1 \\ -5 & 1 \end{bmatrix}\begin{bmatrix} 12,000 \\ 81,000 \end{bmatrix} = \frac{1}{3}\begin{bmatrix} 15,000 \\ 21,000 \end{bmatrix} = \begin{bmatrix} 5000 \\ 7000 \end{bmatrix}$$

Thus, $x = 5000$ (\$5 seats) and $y = 7000$ (\$8 seats).

For Game 3: $n = 8400$ and $r = \$66,000$.

$$\begin{bmatrix} x \\ y \end{bmatrix} = \frac{1}{3}\begin{bmatrix} 8 & -1 \\ -5 & 1 \end{bmatrix}\begin{bmatrix} 8400 \\ 66,000 \end{bmatrix} = \frac{1}{3}\begin{bmatrix} 1200 \\ 24,000 \end{bmatrix} = \begin{bmatrix} 400 \\ 8000 \end{bmatrix}$$

Thus, $x = 400$ (\$5 seats) and $y = 8000$ (\$8 seats).

For Game 4: $n = 21,000$ and $r = \$114,000$.

$$\begin{bmatrix} x \\ y \end{bmatrix} = \frac{1}{3}\begin{bmatrix} 8 & -1 \\ -5 & 1 \end{bmatrix}\begin{bmatrix} 21,000 \\ 114,000 \end{bmatrix} = \frac{1}{3}\begin{bmatrix} 54,000 \\ 9000 \end{bmatrix} = \begin{bmatrix} 18,000 \\ 3000 \end{bmatrix}$$

Thus, $x = 18,000$ (\$5 seats) and $y = 3000$ (\$8 seats.)

Answer: 20,000 \$5 seats, and 15,000 \$8 seats

7.4 EXERCISES A

In Exercises 1–2 show that $B = A^{-1}$ by finding AB and BA.

1. $A = \begin{bmatrix} 2 & 3 \\ -3 & -5 \end{bmatrix}$, $B = \begin{bmatrix} 5 & 3 \\ -3 & -2 \end{bmatrix}$

2. $A = \begin{bmatrix} 1 & 0 & 1 \\ 2 & 1 & 1 \\ 3 & 2 & 2 \end{bmatrix}$, $B = \begin{bmatrix} 0 & 2 & -1 \\ -1 & -1 & 1 \\ 1 & -2 & 1 \end{bmatrix}$

In Exercises 3–8 find the inverse of each matrix.

3. $\begin{bmatrix} 3 & -4 \\ 4 & -5 \end{bmatrix}$

4. $\begin{bmatrix} 2 & -1 \\ 4 & -3 \end{bmatrix}$

5. $\begin{bmatrix} 2 & 2 \\ 2 & 2 \end{bmatrix}$

6. $\begin{bmatrix} 2 & -2 & 3 \\ 1 & 0 & -1 \\ -2 & 1 & 0 \end{bmatrix}$

7. $\begin{bmatrix} 1 & 2 & 0 \\ 0 & 1 & 1 \\ -1 & 1 & 1 \end{bmatrix}$

8. $\begin{bmatrix} 1 & 3 & -1 \\ 3 & 1 & 0 \\ -2 & 0 & 1 \end{bmatrix}$

9. Find the values of x and y.

$$\begin{bmatrix} x \\ y \end{bmatrix} = \begin{bmatrix} 2 & -1 \\ 3 & 5 \end{bmatrix} \begin{bmatrix} -2 \\ 4 \end{bmatrix}$$

10. Find the values of x, y, and z.

$$\begin{bmatrix} x \\ y \\ z \end{bmatrix} = \begin{bmatrix} 2 & -1 & 3 \\ 4 & 0 & 1 \\ 2 & -2 & 0 \end{bmatrix} \begin{bmatrix} 1 \\ -1 \\ 3 \end{bmatrix}$$

Use the inverse method to solve each system in Exercises 11–12. The inverse of the coefficient matrix was found in Exercises 3–4.

11. $3x - 4y = 2$
    $4x - 5y = 3$

12. $2x - y = 11$
    $4x - 3y = 25$

Use the inverse method to solve each system in Exercises 13–15. The inverse of the coefficient matrix was found in Exercises 6–8.

13. $2x - 2y + 3z = -5$
    $x - z = 5$
    $-2x + y = -4$

14. $x + 2y = 4$
    $ y + z = 6$
    $-x + y + z = 2$

15. $x + 3y - z = 13$
    $3x + y = 12$
    $-2x + z = -7$

Use the inverse method to solve each system for the given values of a, b, and c in Exercises 16–17.

16. $2x + y = a$
    $5x + 3y = b$

    (a) $a = 1$, $b = 1$

    (b) $a = 6$, $b = 18$

    (c) $a = -1$, $b = 0$

17. $x - z = a$
    $x + y = b$
    $ y + 2z = c$

    (a) $a = 5$, $b = 2$, $c = -6$

    (b) $a = 0$, $b = 5$, $c = 6$

    (c) $a = -3$, $b = -4$, $c = 0$

18. Suppose that $A = \begin{bmatrix} 1 & 3 \\ 2 & 7 \end{bmatrix}$ and $B = \begin{bmatrix} -3 & 2 \\ 2 & -1 \end{bmatrix}$.

    (a) Find AB.
    (b) Find $(AB)^{-1}$.
    (c) Find A^{-1}.
    (d) Find B^{-1}.
    (e) Find $A^{-1}B^{-1}$.
    (f) Find $B^{-1}A^{-1}$.

    (g) What might be concluded in view of (a)–(f)?

Solve by the inverse method.

19. BIOLOGY A biologist has three groups of animals in an experiment. He wants to prepare a different diet for each group by specifying the amounts of protein and fat. He purchases two food mixes: Supergrow, which contains 30% protein and 6% fat by weight, and Healthy Mix, which contains 20% protein and 2% fat by weight. How many ounces of each food should be combined in order to obtain the mixtures shown in the following table?

	Mix 1	Mix 2	Mix 3
Protein	13.5 ounces	14 ounces	16 ounces
Fat	2.4 ounces	2.6 ounces	3.1 ounces

FOR REVIEW

In Exercises 20–21 solve each system using the Gaussian method.

20. $4x - 5y = -38$
    $2x + 3y = 14$

21. $x - 2y + z = -4$
    $2x \quad - z = 7$
    $-3x + 4y + z = -8$

ANSWERS: 1. Since $AB = BA = I$, $B = A^{-1}$. 2. Since $AB = BA = I$, $B = A^{-1}$. 3. $\begin{bmatrix} -5 & 4 \\ -4 & 3 \end{bmatrix}$ 4. $\begin{bmatrix} \frac{3}{2} & -\frac{1}{2} \\ 2 & -1 \end{bmatrix}$

5. the inverse does not exist 6. $\begin{bmatrix} 1 & 3 & 2 \\ 2 & 6 & 5 \\ 1 & 2 & 2 \end{bmatrix}$ 7. $\begin{bmatrix} 0 & 1 & -1 \\ \frac{1}{2} & -\frac{1}{2} & \frac{1}{2} \\ -\frac{1}{2} & \frac{3}{2} & -\frac{1}{2} \end{bmatrix}$ 8. $\begin{bmatrix} -\frac{1}{10} & \frac{3}{10} & -\frac{1}{10} \\ \frac{3}{10} & \frac{1}{10} & \frac{3}{10} \\ -\frac{1}{5} & \frac{3}{5} & \frac{4}{5} \end{bmatrix}$

9. $x = -8; y = 14$ 10. $x = 12; y = 7; z = 4$ 11. (2, 1) 12. (4, −3) 13. (2, 0, −3) 14. (4, 0, 6) 15. (3, 3, −1) 16. (a) (2, −3) (b) (0, 6) (c) (−3, 5) 17. (a) (2, 0, −3) (b) (1, 4, 1) (c) (−2, −2, 1)
18. (a) $\begin{bmatrix} 3 & -1 \\ 8 & -3 \end{bmatrix}$ (b) $\begin{bmatrix} 3 & -1 \\ 8 & -3 \end{bmatrix}$ (c) $\begin{bmatrix} 7 & -3 \\ -2 & 1 \end{bmatrix}$ (d) $\begin{bmatrix} 1 & 2 \\ 2 & 3 \end{bmatrix}$ (e) $\begin{bmatrix} 1 & 5 \\ 0 & -1 \end{bmatrix}$ (f) $\begin{bmatrix} 3 & -1 \\ 8 & -3 \end{bmatrix}$
(g) It would appear that $(AB)^{-1} = B^{-1}A^{-1}$ and $(AB)^{-1} \neq A^{-1}B^{-1}$. 19. Mix 1: 35 oz of Supergrow and 15 oz of Healthy Mix; Mix 2: 40 oz of Supergrow and 10 oz of Healthy Mix; Mix 3: 50 oz of Supergrow and 5 oz of Healthy Mix 20. (−2, 6) 21. (1, 0, −5)

7.4 EXERCISES B

In Exercises 1–2 show that $B = A^{-1}$ by finding AB and BA.

1. $A = \begin{bmatrix} 1 & 5 \\ 1 & 3 \end{bmatrix}$, $B = \begin{bmatrix} -\frac{3}{2} & \frac{5}{2} \\ \frac{1}{2} & -\frac{1}{2} \end{bmatrix}$

2. $A = \begin{bmatrix} 2 & 3 & -3 \\ 0 & -1 & -1 \\ 1 & 4 & 3 \end{bmatrix}$, $B = \begin{bmatrix} -\frac{1}{4} & \frac{21}{4} & \frac{3}{2} \\ \frac{1}{4} & \frac{9}{4} & -\frac{1}{2} \\ -\frac{1}{4} & \frac{5}{4} & \frac{1}{2} \end{bmatrix}$

In Exercises 3–8 find the inverse of each matrix.

3. $\begin{bmatrix} 1 & -6 \\ -1 & 7 \end{bmatrix}$

4. $\begin{bmatrix} 3 & -3 \\ 2 & -1 \end{bmatrix}$

5. $\begin{bmatrix} 0 & 5 & -1 \\ 2 & 4 & 3 \end{bmatrix}$

6. $\begin{bmatrix} 3 & 0 & 1 \\ 1 & 2 & 4 \\ 1 & 1 & 2 \end{bmatrix}$

7. $\begin{bmatrix} 1 & -1 & 1 \\ 2 & 1 & -1 \\ 1 & -2 & 3 \end{bmatrix}$

8. $\begin{bmatrix} 2 & 1 & -1 \\ 1 & -1 & 0 \\ 0 & 1 & 3 \end{bmatrix}$

9. Find the values of x and y.

$$\begin{bmatrix} x \\ y \end{bmatrix} = \begin{bmatrix} 3 & 0 \\ -4 & 1 \end{bmatrix} \begin{bmatrix} -3 \\ 5 \end{bmatrix}$$

10. Find the values of x, y, and z.

$$\begin{bmatrix} x \\ y \\ z \end{bmatrix} = \begin{bmatrix} -3 & 1 & 5 \\ 0 & 2 & 7 \\ 4 & 2 & -2 \end{bmatrix} \begin{bmatrix} 3 \\ 0 \\ 2 \end{bmatrix}$$

Use the inverse method to solve each system in Exercises 11–12. The inverse of the coefficient matrix was found in Exercises 3–4.

11. $\quad x - 6y = -23$
$\quad\;\; -x + 7y = 26$

12. $\quad 3x - 3y = 12$
$\quad\;\; 2x - y = 6$

Use the inverse method to solve each system in Exercises 13–15. The inverse of the coefficient matrix was found in Exercises 6–8.

13. $3x \quad\quad\; + z = 8$
$\quad\; x + 2y + 4z = -1$
$\quad\; x + y + 2z = 1$

14. $\quad x - y + z = 5$
$\quad 2x + y - z = 11$
$\quad\; x - 2y + 3z = -15$

15. $2x + y - z = 9$
$\quad\; x - y \quad\quad = 6$
$\quad\quad\quad y + 3z = -1$

Use the inverse method to solve each system for the given values of a, b, and c in Exercises 16–17.

16. $7x - 2y = a$
$\;\; 5x - 2y = b$
(a) $a = 22, b = 14$
(b) $a = -7, b = -5$
(c) $a = 18, b = 14$

17. $\quad x + y - z = a$
$\quad 2x \quad\quad + z = b$
$\quad\; x - y \quad\quad = c$
(a) $a = 4, b = 10, c = 6$
(b) $a = 5, b = 7, c = 0$
(c) $a = 0, b = -5, c = 1$

18. Suppose that $A = \begin{bmatrix} 1 & 0 \\ 2 & 3 \end{bmatrix}$ and $B = \begin{bmatrix} -1 & 1 \\ 0 & 2 \end{bmatrix}$.
   (a) Find AB.
   (b) Find $(AB)^{-1}$.
   (c) Find A^{-1}.
   (d) Find B^{-1}.
   (e) Find $A^{-1}B^{-1}$.
   (f) Find $B^{-1}A^{-1}$.
   (g) What might be concluded in view of (a)–(f)?

Solve by the inverse method.

19. **BUSINESS** Ponderosa Stoves sells two models of wood-burning stoves, the Sierra selling for $300 and the Aspen selling for $500. During the four weeks in January, the owner sets a goal of selling the total number of stoves, and earning the weekly revenue, shown in the following table. How many of each model must be sold during each week to reach that goal?

	Week 1	Week 2	Week 3	Week 4
Total stoves sold	12	15	10	20
Total revenue earned	$5400	$7500	$3400	$9000

FOR REVIEW

In Exercises 20–21 solve each system using the Gaussian method.

20. $6x - 2y = 1$
$-3x + y = 1$

21. $x - y + z = -3$
$2x + y\phantom{{}+z} = 0$
$\phantom{2x +{}}2y - z = 4$

7.4 EXERCISES C

Find the inverse of the 4 × 4 matrix given in Exercises 1–2.

1. $\begin{bmatrix} 1 & 0 & 0 & -1 \\ 0 & 2 & 0 & 1 \\ 1 & 0 & 1 & -1 \\ 0 & 1 & -1 & 0 \end{bmatrix}$

2. $\begin{bmatrix} 1 & -1 & 0 & -2 \\ 0 & 1 & 0 & 0 \\ -1 & 0 & 0 & 3 \\ 2 & -1 & 1 & 3 \end{bmatrix}$

$\left[\text{Answer: } \begin{bmatrix} 3 & 1 & -2 & -2 \\ -1 & 0 & 1 & 1 \\ -1 & 0 & 1 & 0 \\ 2 & 1 & -2 & -2 \end{bmatrix} \right]$

3. If $A = \begin{bmatrix} a & b \\ c & d \end{bmatrix}$ is a matrix with the property that $ad - bc \neq 0$, prove that $A^{-1} = \frac{1}{ad - bc} \begin{bmatrix} d & -b \\ -c & a \end{bmatrix}$.

Use the formula for A^{-1} found in Exercise 3 to find the inverse of each matrix in Exercises 4–6.

4. $\begin{bmatrix} 2 & 3 \\ -1 & 5 \end{bmatrix}$

5. $\begin{bmatrix} 2 & -2 \\ 1 & -1 \end{bmatrix}$

6. $\begin{bmatrix} 7 & -2 \\ 3 & 4 \end{bmatrix}$

$\left[\text{Answer: } \frac{1}{13}\begin{bmatrix} 5 & -3 \\ 1 & 2 \end{bmatrix} \right]$

7. If A is a nonsingular $n \times n$ matrix, and if B and C are $n \times n$ matrices such that $AB = AC$, prove that $B = C$. [*Hint:* Consider multiplying both sides of $AB = AC$ on the left by the inverse of A.]

In Exercises 8–9, find conditions on a, b, and c under which A^{-1} exists; then find A^{-1}.

8. $A = \begin{bmatrix} a & 0 \\ 0 & b \end{bmatrix}$

9. $A = \begin{bmatrix} a & 0 & 0 \\ 0 & b & 0 \\ 0 & 0 & c \end{bmatrix}$

$\left[\text{Answer: } ab \neq 0; A^{-1} = \frac{1}{ab}\begin{bmatrix} b & 0 \\ 0 & a \end{bmatrix} \right]$

7.5 DETERMINANTS AND CRAMER'S RULE

STUDENT GUIDEPOSTS

1. The Determinant of a 2 × 2 Matrix
2. Cramer's Rule for Systems of Two Equations
3. Determinants of Matrices of Higher Order
4. Cramer's Rule for Systems of Three Equations

① THE DETERMINANT OF A 2 × 2 MATRIX

Associated with every square matrix is a real number called its *determinant*. There are numerous applications of determinants, the most famous of which is a method of solving systems of n linear equations in n variables called *Cramer's rule*, named after Gabriel Cramer (1704–52). Throughout this section and the next, all matrices we look at will be square matrices. We begin by defining the determinant of a 2 × 2 matrix.

Determinant of a Matrix

Let $A = \begin{bmatrix} a & b \\ c & d \end{bmatrix}$ be a 2 × 2 matrix. The **determinant** of A is given by

$$|A| = \begin{vmatrix} a & b \\ c & d \end{vmatrix} = ad - bc.$$

Note The vertical bars for the determinant $|A|$ should not be confused with the absolute value notation for real numbers. Also, we replace the square brackets of a matrix with vertical bars for the determinant of the matrix. The formula for a 2 × 2 determinant can be remembered from the following diagram.

$$\begin{vmatrix} a & b \\ c & d \end{vmatrix} \qquad \begin{vmatrix} a & b \\ c & d \end{vmatrix}$$
$$ad - bc$$

EXAMPLE 1 EVALUATING 2 × 2 DETERMINANTS

Evaluate the following determinants.

(a) $\begin{vmatrix} 1 & 2 \\ 3 & 4 \end{vmatrix} = (1)(4) - (2)(3) = 4 - 6 = -2$

(b) $\begin{vmatrix} 3 & 9 \\ 2 & 6 \end{vmatrix} = (3)(6) - (9)(2) = 18 - 18 = 0$

(c) $\begin{vmatrix} a_1 & b_1 \\ a_2 & b_2 \end{vmatrix} = a_1 b_2 - a_2 b_1$

PRACTICE EXERCISE 1

Evaluate the following determinants.

(a) $\begin{vmatrix} -1 & 3 \\ 5 & 2 \end{vmatrix}$

(b) $\begin{vmatrix} 6 & 6 \\ 6 & 6 \end{vmatrix}$

(c) $\begin{vmatrix} c_1 & b_1 \\ c_2 & b_2 \end{vmatrix}$

Answers: (a) -17 (b) 0
(c) $c_1 b_2 - c_2 b_1$

---

② CRAMER'S RULE FOR SYSTEMS OF TWO EQUATIONS

To understand how we arrive at the definition of a 2 × 2 determinant, suppose we solve a general system of two linear equations in two variables x and y,

$$a_1 x + b_1 y = c_1$$
$$a_2 x + b_2 y = c_2,$$

where $a_1, b_1, c_1, a_2, b_2,$ and c_2 are real numbers. Multiply the first equation by b_2 and the second by b_1.

$$a_1 b_2 x + b_1 b_2 y = c_1 b_2 \qquad b_2 \text{ times the first equation}$$
$$a_2 b_1 x + b_2 b_1 y = c_2 b_1 \qquad b_1 \text{ times the second equation}$$

Subtract the second from the first to eliminate y.

$$a_1b_2x - a_2b_1x = c_1b_2 - c_2b_1 \qquad b_1b_2y - b_2b_1y = 0 \cdot y = 0$$
$$(a_1b_2 - a_2b_1)x = c_1b_2 - c_2b_1 \qquad \text{Factor out } x$$
$$x = \frac{c_1b_2 - c_2b_1}{a_1b_2 - a_2b_1} \qquad \text{Divide by } a_1b_2 - a_2b_1, \text{ if } a_1b_2 - a_2b_1 \neq 0$$

Similarly, we may first eliminate x and then solve for y.

$$y = \frac{a_1c_2 - a_2c_1}{a_1b_2 - a_2b_1}$$

Consider the coefficient matrix for this system.

$$A = \begin{bmatrix} a_1 & b_1 \\ a_2 & b_2 \end{bmatrix}.$$

The denominator in each fractional expression for x and y is the determinant of the coefficient matrix A,

$$|A| = \begin{vmatrix} a_1 & b_1 \\ a_2 & b_2 \end{vmatrix} = a_1b_2 - a_2b_1.$$

Also, since the numerator of each expression has the appearance of a determinant, we define the matrices A_x and A_y with determinants

$$|A_x| = \begin{vmatrix} c_1 & b_1 \\ c_2 & b_2 \end{vmatrix} = c_1b_2 - c_2b_1, \qquad \text{The numerator for } x$$

and $\quad |A_y| = \begin{vmatrix} a_1 & c_1 \\ a_2 & c_2 \end{vmatrix} = a_1c_2 - a_2c_1.$ $\quad$ The numerator for y

Notice that A_x can be obtained from A by replacing the column of coefficients of x with the constants, and that A_y is obtained in a similar way replacing the coefficients of y with the constants. Then

$$x = \frac{c_1b_2 - c_2b_1}{a_1b_2 - a_2b_1} = \frac{\begin{vmatrix} c_1 & b_1 \\ c_2 & b_2 \end{vmatrix}}{\begin{vmatrix} a_1 & b_1 \\ a_2 & b_2 \end{vmatrix}} = \frac{|A_x|}{|A|},$$

and

$$y = \frac{a_1c_2 - a_2c_1}{a_1b_2 - a_2b_1} = \frac{\begin{vmatrix} a_1 & c_1 \\ a_2 & c_2 \end{vmatrix}}{\begin{vmatrix} a_1 & b_1 \\ a_2 & b_2 \end{vmatrix}} = \frac{|A_y|}{|A|}.$$

We have just proved the following theorem.

Cramer's Rule

The system
$$a_1x + b_1y = c_1$$
$$a_2x + b_2y = c_2$$

has solutions

$$x = \frac{|A_x|}{|A|} \quad \text{and} \quad y = \frac{|A_y|}{|A|}, \quad \text{if } |A| \neq 0,$$

where $\quad |A| = \begin{vmatrix} a_1 & b_1 \\ a_2 & b_2 \end{vmatrix}, \quad |A_x| = \begin{vmatrix} c_1 & b_1 \\ c_2 & b_2 \end{vmatrix}, \quad \text{and} \quad |A_y| = \begin{vmatrix} a_1 & c_1 \\ a_2 & c_2 \end{vmatrix}.$

EXAMPLE 2   Using Cramer's Rule

Solve the system using Cramer's rule.

$$3x - 4y = -5$$
$$5x + y = 7$$

$$|A| = \begin{vmatrix} 3 & -4 \\ 5 & 1 \end{vmatrix} = (3)(1) - (-4)(5) = 23 \qquad \text{Determinant of the coefficient matrix}$$

$$|A_x| = \begin{vmatrix} -5 & -4 \\ 7 & 1 \end{vmatrix} = (-5)(1) - (-4)(7) = 23 \qquad \text{Replace } x\text{-coefficients in } A \text{ with the constants}$$

$$|A_y| = \begin{vmatrix} 3 & -5 \\ 5 & 7 \end{vmatrix} = (3)(7) - (-5)(5) = 46 \qquad \text{Replace } y\text{-coefficients in } A \text{ with the constants}$$

Then $\quad x = \dfrac{|A_x|}{|A|} = \dfrac{23}{23} = 1 \text{ and } y = \dfrac{|A_y|}{|A|} = \dfrac{46}{23} = 2.$

The solution to the system is (1, 2). Check this in both equations.

Practice Exercise 2

Solve the system using Cramer's rule.

$$2x - y = -8$$
$$3x + 7y = 22$$

Answer: $|A| = 17$, $|A_x| = -34$, $|A_y| = 68$, $(-2, 4)$

Note   If the determinant of the coefficient matrix, $|A|$, is zero, Cramer's rule does not apply. In such cases, the system has either no solution (if either $|A_x|$ or $|A_y|$ is not zero) or infinitely many solutions (if $|A_x|$ and $|A_y|$ are both zero), and the solution techniques from Chapter 6 should be used.

❸ DETERMINANTS OF MATRICES OF HIGHER ORDER

To extend Cramer's rule to systems of three equations in three variables, the determinant of a 3×3 matrix must first be defined. Rather than concentrating on the 3×3 case, we will introduce the notions of *minors* and *cofactors* of the elements of a square matrix, and use them to obtain the determinant of an arbitrary $n \times n$ matrix for $n \geq 3$. In fact, as we shall see, a 3×3 determinant is given in terms of three 2×2 determinants, a 4×4 determinant in terms of four 3×3 determinants, and so forth.

The Minor of an Element

Let A be a square matrix of order $n \geq 3$. The **minor** of an element a_{ij}, denoted by M_{ij}, is the determinant of the matrix obtained by deleting the ith row and the jth column of A.

EXAMPLE 3   Finding Minors of Elements

Consider the matrix

$$A = \begin{bmatrix} 1 & 2 & 3 \\ 4 & 5 & 6 \\ 7 & 8 & 9 \end{bmatrix}.$$

Practice Exercise 3

Use matrix A in Example 3 and find the following minors.

(a) M_{23}

(a) Find M_{12}.

M_{12} is the minor of the element $a_{12} = 2$. Eliminate the 1st row and the second column of A and evaluate the determinant of the result.

$$\begin{bmatrix} 1 & 2 & 3 \\ 4 & 5 & 6 \\ 7 & 8 & 9 \end{bmatrix} \qquad M_{12} = \begin{vmatrix} 4 & 6 \\ 7 & 9 \end{vmatrix} = 36 - 42 = -6$$

(b) Find M_{31}.

M_{31} is the minor of $a_{31} = 7$. Eliminate the 3rd row and the 1st column of A.

$$\begin{bmatrix} 1 & 2 & 3 \\ 4 & 5 & 6 \\ 7 & 8 & 9 \end{bmatrix} \qquad M_{31} = \begin{vmatrix} 2 & 3 \\ 5 & 6 \end{vmatrix} = 12 - 15 = -3$$

(b) M_{33}

Answers: (a) -6 (b) -3

A number closely related to the minor of an element is its cofactor.

The Cofactor of an Element

Let A be a square matrix of order $n \geq 3$. The **cofactor** of an element a_{ij}, denoted by A_{ij}, is given by

$$A_{ij} = (-1)^{i+j} M_{ij},$$

where M_{ij} is the minor of a_{ij}.

Note The minor and cofactor of an element differ at most in the sign of the number. For the element a_{ij},

$A_{ij} = M_{ij}$ if $i + j$ is even.   Then $(-1)^{i+j} = +1$
$A_{ij} = -M_{ij}$ if $i + j$ is odd.   Then $(-1)^{i+j} = -1$

The following "checkerboard" array gives the signs applied to corresponding minors to obtain the cofactors for a 3×3 matrix.

$$\begin{bmatrix} + & - & + \\ - & + & - \\ + & - & + \end{bmatrix} \qquad \text{Signs for cofactors}$$

EXAMPLE 4   FINDING COFACTORS OF ELEMENTS

Consider the matrix

$$A = \begin{bmatrix} \overset{+}{1} & \overset{-}{2} & \overset{+}{3} \\ \overset{-}{4} & \overset{+}{5} & \overset{-}{6} \\ \overset{+}{7} & \overset{-}{8} & \overset{+}{9} \end{bmatrix}.$$

(a) Find A_{21}.

A_{21} is the cofactor of the element $a_{21} = 4$. Eliminate the 2nd row and 1st column to determine M_{21}.

$$A_{21} = (-1)^{2+1} M_{21} = (-1)^{2+1} \begin{vmatrix} 2 & 3 \\ 8 & 9 \end{vmatrix} = (-1)^3 [18 - 24] = (-1)(-6) = 6$$

PRACTICE EXERCISE 4

Use matrix A in Example 4 and find the following cofactors.

(a) A_{23}

Since the element $a_{21} = 4$ is in a "−" position in the checkerboard of signs, we could have shortened the work as follows.

$$A_{21} = -M_{21} = -\begin{vmatrix} 2 & 3 \\ 8 & 9 \end{vmatrix} = -(18 - 24) = 6$$

(b) Find A_{13}.

A_{13} is the cofactor of the element $a_{13} = 3$. Since a_{13} is in a "+" position,

$$A_{13} = +M_{13} = +\begin{vmatrix} 4 & 5 \\ 7 & 8 \end{vmatrix} = +[32 - 35] = +[-3] = -3.$$

(b) A_{33}

Answers: (a) 6 (b) −3

The determinant of a square matrix of order $n \geq 3$ can be found using a method called **expansion by cofactors**.

Expansion by Cofactors

Let A be a square matrix of order $n \geq 3$. The determinant of A, $|A|$, is found by adding the products of the elements in *any* row or column and their respective cofactors.

EXAMPLE 5 EXPANDING DETERMINANTS BY COFACTORS

Use expansion by cofactors to find $|A|$ if

$$A = \begin{bmatrix} 1 & 2 & -3 \\ -1 & 0 & 4 \\ 5 & 2 & 3 \end{bmatrix}.$$

To illustrate the method and show that any row or column yields the same result, we find $|A|$ using **(a)** elements in the second column and **(b)** elements in the first row.

(a) $|A| = a_{12}A_{12} + a_{22}A_{22} + a_{32}A_{32}$

$$= (2)(-1)^{1+2}\begin{vmatrix} -1 & 4 \\ 5 & 3 \end{vmatrix} + (0)(-1)^{2+2}\begin{vmatrix} 1 & -3 \\ 5 & 3 \end{vmatrix}$$

$$+ (2)(-1)^{3+2}\begin{vmatrix} 1 & -3 \\ -1 & 4 \end{vmatrix}$$

$$= (-2)(-3 - 20) + (0)(3 - (-15)) + (-2)(4 - 3)$$
$$= (-2)(-23) + 0 + (-2)(1) = 46 - 2 = 44$$

(b) $|A| = a_{11}A_{11} + a_{12}A_{12} + a_{13}A_{13}$

$$= (1)(-1)^{1+1}\begin{vmatrix} 0 & 4 \\ 2 & 3 \end{vmatrix} + (2)(-1)^{1+2}\begin{vmatrix} -1 & 4 \\ 5 & 3 \end{vmatrix}$$

$$+ (-3)(-1)^{1+3}\begin{vmatrix} -1 & 0 \\ 5 & 2 \end{vmatrix}$$

$$= (1)(0 - 8) + (-2)(-3 - 20) + (-3)(-2 - 0)$$
$$= (1)(-8) + (-2)(-23) + (-3)(-2) = -8 + 46 + 6 = 44$$

Thus, we obtain the same result in both expansions.

PRACTICE EXERCISE 5

Consider the matrix.

$$A = \begin{bmatrix} 2 & 1 & 0 \\ -1 & 4 & 0 \\ 3 & 2 & -1 \end{bmatrix}.$$

(a) Find $|A|$ using elements in the third row.

(b) Find $|A|$ using elements in the third column.

(c) Which of these two expansions required less work?

Answers: $|A| = -9$; it was easier to use the third column and take advantage of the 0's.

7.5 DETERMINANTS AND CRAMER'S RULE

Note When expanding determinants, look for rows or columns that have 0 entries. By selecting such rows or columns, you can shorten the work, since a zero makes the corresponding term in the expansion equal to zero.

EXAMPLE 6 A 4 × 4 DETERMINANT

Find

$$|A| = \begin{vmatrix} \overset{+}{-1} & \overset{-}{2} & \overset{+}{0} & \overset{-}{3} \\ \overset{-}{2} & \overset{+}{4} & \overset{-}{3} & \overset{+}{-1} \\ \overset{+}{0} & \overset{-}{1} & \overset{+}{0} & \overset{-}{-2} \\ \overset{-}{-3} & \overset{+}{-1} & \overset{-}{1} & \overset{+}{2} \end{vmatrix}.$$

Suppose we expand along the third row taking advantage of the two zeros. The signs for the cofactors have been placed in color above each element.

$$|A| = 0 + (-1)(-1)\begin{vmatrix} -1 & 0 & 3 \\ 2 & 3 & -1 \\ -3 & 1 & 2 \end{vmatrix} + 0 + (-2)(-1)\begin{vmatrix} -1 & 2 & 0 \\ 2 & 4 & 3 \\ -3 & -1 & 1 \end{vmatrix}$$

$$= 0 + (-1)\left[(-1)\begin{vmatrix} 3 & -1 \\ 1 & 2 \end{vmatrix} + (3)\begin{vmatrix} 2 & 3 \\ -3 & 1 \end{vmatrix}\right] + 0 + (2)\left[(-1)\begin{vmatrix} 4 & 3 \\ -1 & 1 \end{vmatrix} + (-2)\begin{vmatrix} 2 & 3 \\ -3 & 1 \end{vmatrix}\right]$$

Expand along first row   Expand along first row

$$= 0 + (-1)[(-1)(6 + 1) + (3)(2 + 9)] + 0 + (2)[(-1)(4 + 3) + (-2)(2 + 9)]$$
$$= 0 + (-1)[26] + 0 + (2)[-29] = -26 - 58 = -84$$

PRACTICE EXERCISE 6

Find $|A|$ in Example 6 by expanding down the third column.

Answer: You should obtain -84 again.

❹ CRAMER'S RULE FOR SYSTEMS OF THREE EQUATIONS

The definition of the value of a third-order determinant would seem reasonable if we were to solve the following general system of three equations in the three variables x, y, and z.

$$a_1 x + b_1 y + c_1 z = d_1$$
$$a_2 x + b_2 y + c_2 z = d_2$$
$$a_3 x + b_3 y + c_3 z = d_3$$

The solutions to this system have the following form.

$$x = \frac{\begin{vmatrix} d_1 & b_1 & c_1 \\ d_2 & b_2 & c_2 \\ d_3 & b_3 & c_3 \end{vmatrix}}{\begin{vmatrix} a_1 & b_1 & c_1 \\ a_2 & b_2 & c_2 \\ a_3 & b_3 & c_3 \end{vmatrix}}, \quad y = \frac{\begin{vmatrix} a_1 & d_1 & c_1 \\ a_2 & d_2 & c_2 \\ a_3 & d_3 & c_3 \end{vmatrix}}{\begin{vmatrix} a_1 & b_1 & c_1 \\ a_2 & b_2 & c_2 \\ a_3 & b_3 & c_3 \end{vmatrix}}, \quad z = \frac{\begin{vmatrix} a_1 & b_1 & d_1 \\ a_2 & b_2 & d_2 \\ a_3 & b_3 & d_3 \end{vmatrix}}{\begin{vmatrix} a_1 & b_1 & c_1 \\ a_2 & b_2 & c_2 \\ a_3 & b_3 & c_3 \end{vmatrix}}$$

This result is also known as Cramer's rule, this time for a system of three linear equations in three variables. The determinant of the coefficient matrix appears in the denominator of each fraction, and the determinant in each numerator is formed by replacing the coefficients of the respective variable with the constants. As with systems of two equations in two variables, we use the following notation.

$$|A| = \begin{vmatrix} a_1 & b_1 & c_1 \\ a_2 & b_2 & c_2 \\ a_3 & b_3 & c_3 \end{vmatrix} \quad |A_x| = \begin{vmatrix} d_1 & b_1 & c_1 \\ d_2 & b_2 & c_2 \\ d_3 & b_3 & c_3 \end{vmatrix} \quad |A_y| = \begin{vmatrix} a_1 & d_1 & c_1 \\ a_2 & d_2 & c_2 \\ a_3 & d_3 & c_3 \end{vmatrix} \quad |A_z| = \begin{vmatrix} a_1 & b_1 & d_1 \\ a_2 & b_2 & d_2 \\ a_3 & b_3 & d_3 \end{vmatrix}$$

The solutions to the system, using Cramer's rule, are

$$x = \frac{|A_x|}{|A|}, \quad y = \frac{|A_y|}{|A|}, \quad z = \frac{|A_z|}{|A|}.$$

EXAMPLE 7 CRAMER'S RULE (3 × 3 SYSTEM)

Solve the system using Cramer's rule.

$$x + y - z = 4$$
$$2x + y + z = 1$$
$$3x - 2y - z = 3$$

It is easy to verify that the four determinants are

$$|A| = \begin{vmatrix} 1 & 1 & -1 \\ 2 & 1 & 1 \\ 3 & -2 & -1 \end{vmatrix} = 13 \qquad |A_x| = \begin{vmatrix} 4 & 1 & -1 \\ 1 & 1 & 1 \\ 3 & -2 & -1 \end{vmatrix} = 13$$

$$|A_y| = \begin{vmatrix} 1 & 4 & -1 \\ 2 & 1 & 1 \\ 3 & 3 & -1 \end{vmatrix} = 13 \qquad |A_z| = \begin{vmatrix} 1 & 1 & 4 \\ 2 & 1 & 1 \\ 3 & -2 & 3 \end{vmatrix} = -26.$$

Thus,

$$x = \frac{|A_x|}{|A|} = \frac{13}{13} = 1, \; y = \frac{|A_y|}{|A|} = \frac{13}{13} = 1, \; z = \frac{|A_z|}{|A|} = \frac{-26}{13} = -2.$$

PRACTICE EXERCISE 7

Solve the system using Cramer's rule.

$$y + 4z = 6$$
$$3x \quad + z = 7$$
$$5y - z = 9$$

Answer: $|A| = 63$, $|A_x| = 126$, $|A_y| = 126$, $|A_z| = 63$, $(2, 2, 1)$

We were able to solve the system in Example 7 in less time by using the reduction method from Chapter 6. However, if a computer or programmable calculator is used, once the method of evaluating a determinant has been programmed into the device, the coefficients and constants can be entered and solutions to systems are available at the push of a button. Also, there may be occasions when it is not necessary to find solutions for all the variables in a system. Cramer's rule provides a way to find the value of one variable without finding the rest.

7.5 EXERCISES A

Evaluate the determinants in Exercises 1–4.

1. $\begin{vmatrix} 3 & 5 \\ 2 & 4 \end{vmatrix}$

2. $\begin{vmatrix} -1 & 0 \\ 3 & 4 \end{vmatrix}$

3. $\begin{vmatrix} 2 & 5 \\ -3 & -1 \end{vmatrix}$

4. $\begin{vmatrix} -2 & -3 \\ b & a \end{vmatrix}$

Use matrix $A = \begin{bmatrix} 2 & -1 & 3 \\ 1 & -2 & 4 \\ -3 & 5 & -4 \end{bmatrix}$ to answer Exercises 5–8.

5. Find the minor of the element 3.

6. Find the cofactor of the element 3.

7. Find $|A|$ by expanding along the first row.

8. Find $|A|$ by expanding along the first column.

Evaluate the determinants in Exercises 9–11.

9. $\begin{vmatrix} 0 & 2 & -1 \\ 1 & -3 & 0 \\ 0 & 1 & -2 \end{vmatrix}$

10. $\begin{vmatrix} 3 & 1 & 2 \\ 4 & 0 & -2 \\ -2 & 1 & 3 \end{vmatrix}$

11. $\begin{vmatrix} 1 & -2 & 1 \\ 2 & 3 & 2 \\ 3 & 1 & 3 \end{vmatrix}$

In Exercises 12–16 solve each system using Cramer's rule.

12. $x + 2y = -1$
$2x - 3y = 12$

13. $3x + 2y = -3$
$6x - 3y = 8$

14. $4x - y = -3$
$2x - 3y = 1$

15. $3x - y = -11$
$2x + 4z = -2$
$ - 3y + 5z = -1$

16. $2x + y - z = -3$
$x - y + z = 0$
$-3x + 2y - 4z = -3$

Solve for x in Exercises 17–19.

17. $\begin{vmatrix} x & 2 \\ 3 & 1 \end{vmatrix} = 3$

18. $\begin{vmatrix} x & 1 \\ 1 & x \end{vmatrix} = 3$

19. $\begin{vmatrix} x & 0 & 0 \\ 2 & x & -1 \\ -3 & 2 & 1 \end{vmatrix} = 3$

20. Evaluate the 4×4 determinant.

$$\begin{vmatrix} 2 & 3 & 0 & -2 \\ 1 & -2 & -1 & 0 \\ 0 & 1 & 0 & 0 \\ -1 & 4 & -2 & 3 \end{vmatrix}$$

21. (a) Evaluate. $\begin{vmatrix} 0 & 0 \\ 0 & 0 \end{vmatrix}$

(b) Evaluate. $\begin{vmatrix} 0 & 0 & 0 \\ 0 & 0 & 0 \\ 0 & 0 & 0 \end{vmatrix}$

(c) Evaluate. $\begin{vmatrix} 0 & 0 & 0 & 0 \\ 0 & 0 & 0 & 0 \\ 0 & 0 & 0 & 0 \\ 0 & 0 & 0 & 0 \end{vmatrix}$

(d) What is the determinant of any $n \times n$ zero matrix?

22. (a) Evaluate. $\begin{vmatrix} a & 0 \\ 0 & b \end{vmatrix}$

(b) Evaluate. $\begin{vmatrix} a & 0 & 0 \\ 0 & b & 0 \\ 0 & 0 & c \end{vmatrix}$

(c) Evaluate. $\begin{vmatrix} a & 0 & 0 & 0 \\ 0 & b & 0 & 0 \\ 0 & 0 & c & 0 \\ 0 & 0 & 0 & d \end{vmatrix}$

(d) Matrices with determinants shown in (a), (b), and (c) are called **diagonal matrices.** What is the determinant of any diagonal matrix?

23. (a) Evaluate. $\begin{vmatrix} a & 0 \\ b & 0 \end{vmatrix}$

(b) Evaluate. $\begin{vmatrix} a & b & 0 \\ c & d & 0 \\ e & f & 0 \end{vmatrix}$

(c) Evaluate. $\begin{vmatrix} a & b & c & 0 \\ d & e & f & 0 \\ g & h & i & 0 \\ j & k & l & 0 \end{vmatrix}$

(d) What is the determinant of any matrix that has a column of zeros?

24. If $A = \begin{bmatrix} a & b \\ c & d \end{bmatrix}$, find $|A|$ and $|A^T|$ and compare the results.

25. If $A = \begin{bmatrix} a & b \\ c & d \end{bmatrix}$, show that $|A^{-1}| = \frac{1}{|A|}$, provided $|A| \neq 0$.

26. Show that $\begin{vmatrix} a & b \\ na & nb \end{vmatrix} = 0$.

27. Show that $\begin{vmatrix} a & b \\ c & d \end{vmatrix} = -\begin{vmatrix} b & a \\ d & c \end{vmatrix}$.

FOR REVIEW

28. Find the inverse of the matrix.

$$\begin{vmatrix} 2 & -4 \\ -1 & 3 \end{vmatrix}$$

29. Use the inverse method to solve the system. The inverse of the coefficient matrix was found in Exercise 28.

$$2x - 4y = -20$$
$$-x + 3y = 15$$

ANSWERS: 1. 2 2. −4 3. 13 4. −2a + 3b 5. −1 6. −1 7. −19 8. −19 9. 3 10. 6 11. 0 12. (3, −2) 13. ($\frac{1}{3}$, −2) 14. (−1, −1) 15. (−3, 2, 1) 16. (−1, 1, 2) 17. 9 18. ±2 19. 1, −3 20. 0 21. (a) 0 (b) 0 (c) 0 (d) 0 22. (a) ab (b) abc (c) abcd (d) The determinant is the product of the elements along the main diagonal, $a_{11} a_{22} a_{33} \ldots a_{nn}$. 23. (a) 0 (b) 0 (c) 0 (d) 0 24. $|A| = ad - bc = |A^T|$

25. $A^{-1} = \begin{bmatrix} \frac{d}{ad-bc} & \frac{-b}{ad-bc} \\ \frac{-c}{ad-bc} & \frac{a}{ad-bc} \end{bmatrix}$, and $|A^{-1}| = \frac{1}{ad-bc} = \frac{1}{|A|}$. 26. $\begin{bmatrix} a & b \\ na & nb \end{bmatrix} = anb - bna = 0$

27. $\begin{bmatrix} a & b \\ c & d \end{bmatrix} = ad - bc$ and $-\begin{bmatrix} b & a \\ d & c \end{bmatrix} = -[bc - ad] = ad - bc$. 28. $\begin{bmatrix} \frac{3}{2} & 2 \\ \frac{1}{2} & 1 \end{bmatrix}$ 29. (0, 5)

7.5 EXERCISES B

Evaluate the determinants in Exercises 1–4.

1. $\begin{vmatrix} 5 & 7 \\ 6 & 1 \end{vmatrix}$
2. $\begin{vmatrix} 0 & -2 \\ 4 & 8 \end{vmatrix}$
3. $\begin{vmatrix} -2 & -7 \\ -3 & -4 \end{vmatrix}$
4. $\begin{vmatrix} a & b \\ b & a \end{vmatrix}$

Use matrix $A = \begin{bmatrix} 2 & -1 & 3 \\ 1 & -2 & 4 \\ -3 & 5 & -4 \end{bmatrix}$ to answer Exercises 5–8.

5. Find the minor of the element 4.
6. Find the cofactor of the element 4.
7. Find $|A|$ by expanding along the second row.
8. Find $|A|$ by expanding along the second column.

Evaluate the determinants in Exercises 9–11.

9. $\begin{vmatrix} 2 & 0 & -1 \\ 0 & 1 & 0 \\ -1 & 3 & -2 \end{vmatrix}$
10. $\begin{vmatrix} -1 & 0 & 2 \\ 3 & -1 & 1 \\ 1 & -2 & -3 \end{vmatrix}$
11. $\begin{vmatrix} 1 & 1 & 1 \\ -1 & 2 & -2 \\ 3 & 4 & -3 \end{vmatrix}$

In Exercises 12–16 solve each system using Cramer's rule.

12. $2x - y = 8$
    $3x + 5y = -14$

13. $3x + 4y = 1$
    $5x - 2y = 6$

14. $7x - 3y = -6$
    $4x + 2y = 4$

15. $2x + 3z = -8$
    $ y - 4z = 9$
    $x - 3y = -4$

16. $x - 2y - z = -1$
    $2x + y + z = 2$
    $-x - y + 2z = 7$

Solve for x in Exercises 17–19.

17. $\begin{vmatrix} 2 & x \\ 1 & -3 \end{vmatrix} = 5$
18. $\begin{vmatrix} 3 & x \\ x & 1 \end{vmatrix} = -6$
19. $\begin{vmatrix} 0 & 0 & x \\ x & 1 & -3 \\ 3 & -1 & 7 \end{vmatrix} = -10$

20. Evaluate the 4 × 4 determinant.

    $\begin{vmatrix} 2 & -3 & 0 & -1 \\ 0 & -1 & 0 & 1 \\ 5 & 4 & 1 & -2 \\ -2 & 1 & 0 & 2 \end{vmatrix}$

21. (a) Evaluate. $\begin{vmatrix} 1 & 0 \\ 0 & 1 \end{vmatrix}$

    (b) Evaluate. $\begin{vmatrix} 1 & 0 & 0 \\ 0 & 1 & 0 \\ 0 & 0 & 1 \end{vmatrix}$

    (c) Evaluate. $\begin{vmatrix} 1 & 0 & 0 & 0 \\ 0 & 1 & 0 & 0 \\ 0 & 0 & 1 & 0 \\ 0 & 0 & 0 & 1 \end{vmatrix}$

    (d) What is the determinant of any identity matrix?

22. (a) Evaluate. $\begin{vmatrix} a & b \\ 0 & c \end{vmatrix}$

    (b) Evaluate. $\begin{vmatrix} a & b & c \\ 0 & d & e \\ 0 & 0 & f \end{vmatrix}$

(c) Evaluate. $\begin{vmatrix} a & b & c & d \\ 0 & e & f & g \\ 0 & 0 & h & i \\ 0 & 0 & 0 & j \end{vmatrix}$

(d) Matrices with determinants shown in (a), (b), and (c) are called **triangular matrices.** What is the determinant of any triangular matrix?

23. (a) Evaluate. $\begin{vmatrix} a & b \\ 0 & 0 \end{vmatrix}$

(b) Evaluate. $\begin{vmatrix} a & b & c \\ d & e & f \\ 0 & 0 & 0 \end{vmatrix}$

(c) Evaluate. $\begin{vmatrix} a & b & c & d \\ e & f & g & h \\ i & j & k & l \\ 0 & 0 & 0 & 0 \end{vmatrix}$

(d) What is the determinant of any matrix that has a row of zeros?

24. If $A = \begin{vmatrix} a & b & c \\ d & e & f \\ g & 0 & 0 \end{vmatrix}$, find $|A|$ and $|A^T|$ and compare the results.

25. If $A = \begin{bmatrix} a & b \\ c & d \end{bmatrix}$ and $B = \begin{bmatrix} e & f \\ g & h \end{bmatrix}$, show that $|AB| = |A| |B|$.

26. Show that $\begin{vmatrix} a & b \\ c & d \end{vmatrix} = - \begin{vmatrix} c & d \\ a & b \end{vmatrix}$.

27. Show that $\begin{vmatrix} na & nb \\ nc & nd \end{vmatrix} = n^2 \begin{vmatrix} a & b \\ c & d \end{vmatrix}$.

FOR REVIEW

28. Find the inverse of the matrix.

$$\begin{bmatrix} 1 & -1 & 1 \\ 0 & 2 & -1 \\ 2 & 3 & 0 \end{bmatrix}$$

29. Use the inverse method to solve the system. The inverse of the coefficient matrix was found in Exercise 28.

$$\begin{aligned} x - y + z &= 0 \\ 2y - z &= -2 \\ 2x + 3y &= -4 \end{aligned}$$

7.5 EXERCISES C

1. Show that the equation of the line passing through the two points (x_1, y_1) and (x_2, y_2) is given by the equation

$$\begin{vmatrix} x & y & 1 \\ x_1 & y_1 & 1 \\ x_2 & y_2 & 1 \end{vmatrix} = 0.$$

2. Use the equation in Exercise 1 to find the equation of the line passing through the two points $(2, 0)$ and $(-1, 3)$.
   [Answer: $x + y - 2 = 0$]

3. Show that the area of a triangle with vertices (x_1, y_1), (x_2, y_2), and (x_3, y_3) is the absolute value of

$$\frac{1}{2} \begin{vmatrix} x_1 & y_1 & 1 \\ x_2 & y_2 & 1 \\ x_3 & y_3 & 1 \end{vmatrix}.$$

[*Hint:* Make a sketch using three points in quadrant I and show that the area of the triangle can be found by considering the areas of three trapezoids.]

4. Use the expression in Exercise 3 to find the area of the triangle with vertices $(0, 1)$, $(3, 5)$, and $(-1, 2)$.
   [Answer: $\frac{7}{2}$ square units]

5. If A and B are 2×2 matrices such that $|AB| = 0$, must $|A| = 0$ or $|B| = 0$? Explain. [*Hint:* Use Exercises B, 25.]

6. If A and B are 2×2 matrices such that $|AB| = 0$, must $A = 0$ or $B = 0$? Explain.

Let A be an $n \times n$ matrix, I the $n \times n$ identity matrix, and x any real number. Define the function $f(x) = |A - xI|$. In more advanced courses, $f(x)$ is called the **characteristic polynomial** of A, and the zeros of $f(x)$ are **characteristic values** (**eigenvalues**) of A. In Exercises 7–8, find the characteristic polynomial and the characteristic values of the given matrix A.

7. $A = \begin{vmatrix} 2 & 2 \\ 3 & 1 \end{vmatrix}$

8. $A = \begin{vmatrix} 1 & 0 & 0 \\ 0 & 2 & 0 \\ 0 & 0 & 3 \end{vmatrix}$

[Answer: $f(x) = x^2 - 3x - 4$; 4, -1]

[Answer: $f(x) = -x^3 + 6x^2 - 11x + 6$; 1, 2, 3]

7.6 MORE ON DETERMINANTS

STUDENT GUIDEPOSTS

1 Properties of Determinants

2 Using Determinants to Find Inverses (Optional)

1 PROPERTIES OF DETERMINANTS

Some of the exercises in Section 7.5 hinted at a number of properties of determinants. The theorems we present now can be helpful for finding determinants of square matrices of order 3 or more, and they have numerous applications in more advanced courses in matrix theory. Due to difficulties with notation, proofs of these theorems for general $n \times n$ matrices are somewhat involved. As a result, only informal proofs are given using 3×3 matrices. For easy reference, the theorems in this section are assigned numbers.

THEOREM 1 If A is a square matrix of order n, then $|A| = |A^T|$.

In the exercises in Section 7.5, we showed that $|A| = |A^T|$ if A is of order 2. Suppose that A is of order 3 given by

$$A = \begin{bmatrix} a_1 & a_2 & a_3 \\ a_4 & a_5 & a_6 \\ a_7 & a_8 & a_9 \end{bmatrix}.$$

Then

$$A^T = \begin{bmatrix} a_1 & a_4 & a_7 \\ a_2 & a_5 & a_8 \\ a_3 & a_6 & a_9 \end{bmatrix}.$$

Expand $|A|$ along the first row and $|A^T|$ along the first column.

$$|A| = a_1 \begin{vmatrix} a_5 & a_6 \\ a_8 & a_9 \end{vmatrix} - a_2 \begin{vmatrix} a_4 & a_6 \\ a_7 & a_9 \end{vmatrix} + a_3 \begin{vmatrix} a_4 & a_5 \\ a_7 & a_8 \end{vmatrix}$$

$$|A^T| = a_1 \begin{vmatrix} a_5 & a_8 \\ a_6 & a_9 \end{vmatrix} - a_2 \begin{vmatrix} a_4 & a_7 \\ a_6 & a_9 \end{vmatrix} + a_3 \begin{vmatrix} a_4 & a_7 \\ a_5 & a_8 \end{vmatrix}$$

Since the corresponding 2×2 determinants are equal, $|A| = |A^T|$.

518 CHAPTER 7 MATRICES AND DETERMINANTS

Theorem 1 allows us to conclude that any property of determinants involving rows is also true for the columns.

> **THEOREM 2** If A is a square matrix of order n, and B is a matrix formed by multiplying all elements in a row (or column) of A by a constant k, then $|B| = k|A|$.

Suppose A is a general matrix of order 3 given by

$$A = \begin{bmatrix} a_1 & a_2 & a_3 \\ a_4 & a_5 & a_6 \\ a_7 & a_8 & a_9 \end{bmatrix},$$

and B is formed from A by multiplying all elements in row 1 by k.

$$B = \begin{bmatrix} ka_1 & ka_2 & ka_3 \\ a_4 & a_5 & a_6 \\ a_7 & a_8 & a_9 \end{bmatrix}.$$

Find $|B|$ by expanding along the first row.

$$|B| = ka_1 \begin{vmatrix} a_5 & a_6 \\ a_8 & a_9 \end{vmatrix} - ka_2 \begin{vmatrix} a_4 & a_6 \\ a_7 & a_9 \end{vmatrix} + ka_3 \begin{vmatrix} a_4 & a_5 \\ a_7 & a_8 \end{vmatrix}$$

$$= k\left(a_1 \begin{vmatrix} a_5 & a_6 \\ a_8 & a_9 \end{vmatrix} - a_2 \begin{vmatrix} a_4 & a_6 \\ a_7 & a_9 \end{vmatrix} + a_3 \begin{vmatrix} a_4 & a_5 \\ a_7 & a_8 \end{vmatrix} \right)$$

$$= k(|A|) = k|A|$$

A similar argument will prove the theorem for k multiplied by any other row or column.

EXAMPLE 1 PROPERTIES OF DETERMINANTS

Given that

$$A = \begin{bmatrix} 1 & 2 & -3 \\ -1 & 0 & 4 \\ 5 & 2 & 3 \end{bmatrix}$$

with $|A| = 44$, find $|B|$ if

$$B = \begin{bmatrix} 3 & 6 & -9 \\ -1 & 0 & 4 \\ 5 & 2 & 3 \end{bmatrix}.$$

Since the elements in row 1 of B are 3 times the elements in row 1 of A,

$$|B| = 3|A| = 3(44) = 132.$$

PRACTICE EXERCISE 1

Using matrix A in Example 1, with $|A| = 44$, find $|C|$ if

$$C = \begin{bmatrix} 1 & -1 & 5 \\ 2 & 0 & 2 \\ -3 & 4 & 3 \end{bmatrix}.$$

Answer: Since $C = A^T$, by Theorem 1, $|C| = 44$.

Theorem 2 can also be used in reverse order to factor out a number common to all the elements in any row or column. For instance, consider the matrix B given in Example 1. If we knew that $|B| = 132$, then

$$|B| = 3|A| = 132$$

would imply that $|A| = \frac{132}{3} = 44$.

THEOREM 3 If A is a square matrix of order n, and every element in a row (or column) of A is 0, then $|A| = 0$.

This theorem follows immediately from Theorem 2 by factoring out the common factor 0 from the elements in the row (or column) of $|A|$ that are all 0.

EXAMPLE 2 PROPERTIES OF DETERMINANTS

Given that

$$A = \begin{bmatrix} -3 & 0 & 5 \\ 2 & 0 & -7 \\ 4 & 0 & 8 \end{bmatrix},$$

find $|A|$.

Since the elements in the second column of A are all 0, by Theorem 3, $|A| = 0$. This could also be seen directly by expanding $|A|$ along the second column.

PRACTICE EXERCISE 2

Given that

$$B = \begin{bmatrix} 6 & -4 & 8 \\ 1 & 3 & -5 \\ 0 & 0 & 0 \end{bmatrix},$$

find $|B|$.

Answer: $|B| = 0$ by Theorem 3, since all elements in row 3 are 0.

THEOREM 4 If A is a square matrix of order n, and B is a matrix formed by interchanging two rows (or two columns) of A, then $|B| = -|A|$.

This theorem was verified for 2×2 determinants in the exercises of the previous section. The proof for 3×3 determinants is essentially the same, but the notation is somewhat involved.

EXAMPLE 3 PROPERTIES OF DETERMINANTS

Given that

$$A = \begin{bmatrix} 1 & 2 & -3 \\ -1 & 0 & 4 \\ 5 & 2 & 3 \end{bmatrix}$$

with $|A| = 44$, show that $|B| = -44$ if

$$B = \begin{bmatrix} 1 & 2 & -3 \\ 5 & 2 & 3 \\ -1 & 0 & 4 \end{bmatrix}.$$

Since B is formed from A by interchanging the second and third rows, $|B| = -|A| = -44$.

PRACTICE EXERCISE 3

Using matrix A in Example 3 with $|A| = 44$, find $|C|$ if

$$C = \begin{bmatrix} 2 & 1 & -3 \\ 0 & -1 & 4 \\ 2 & 5 & 3 \end{bmatrix}.$$

Answer: Since C is formed from A by interchanging the first two columns, $|C| = -|A| = -44$.

THEOREM 5 Let A be a square matrix of order n. If the corresponding elements in two rows (or columns) of A are equal, then $|A| = 0$.

Let A be an $n \times n$ matrix with two equal rows (or columns). If these two rows (or columns) are interchanged, the resulting matrix is the same as A. But by Theorem 4, the determinant of the new matrix, which is $|A|$, must be equal to the negative of the determinant of the original matrix, $-|A|$. Thus,

$$|A| = -|A|$$
$$2|A| = 0$$
$$|A| = 0.$$

EXAMPLE 4 PROPERTIES OF DETERMINANTS

Find

$$\begin{vmatrix} 1 & 2 & 2 \\ 3 & 2 & 2 \\ -7 & 2 & 2 \end{vmatrix}.$$

Since the second and third columns are the same, by Theorem 5 the determinant is 0.

PRACTICE EXERCISE 4

Find

$$\begin{vmatrix} -3 & -3 & -3 \\ 1 & 4 & -6 \\ -3 & -3 & -3 \end{vmatrix}.$$

Answer: Since the first and third rows are the same, the determinant is 0.

THEOREM 6 Let A be a square matrix of order n. If all the elements in one row (or column) of A are a multiple of the elements in another row (or column) of A, then $|A| = 0$.

Suppose

$$A = \begin{bmatrix} ka_1 & ka_2 & ka_3 \\ a_1 & a_2 & a_3 \\ a_4 & a_5 & a_6 \end{bmatrix},$$

in which the elements of row 1 are all k-multiples of the elements in row 2. Then by Theorem 2,

$$|A| = k \begin{vmatrix} a_1 & a_2 & a_3 \\ a_1 & a_2 & a_3 \\ a_4 & a_5 & a_6 \end{vmatrix}.$$

Since two rows of this determinant are equal, by Theorem 5 the determinant is 0. Thus,

$$|A| = k(0) = 0.$$

EXAMPLE 5 PROPERTIES OF DETERMINANTS

Find

$$\begin{vmatrix} 1 & -5 & 3 \\ -1 & 5 & -2 \\ 3 & -15 & 7 \end{vmatrix}.$$

Since the elements in the second column are -5 times the corresponding elements in the first column, by Theorem 6 the determinant is 0.

PRACTICE EXERCISE 5

Find

$$\begin{vmatrix} 1 & -2 & 0 \\ 3 & 7 & -4 \\ 3 & -6 & 0 \end{vmatrix}.$$

Answer: Since the elements in the third row are 3 times those in the first row, the determinant is 0.

> **THEOREM 7** If A is a square matrix of order n, and B is a matrix formed by replacing a row (or column) of A by the sum of that row (or column) and a constant multiple of another row (or column) then $|B| = |A|$.

Suppose that A is of order 3 given by

$$A = \begin{bmatrix} a_1 & a_2 & a_3 \\ a_4 & a_5 & a_6 \\ a_7 & a_8 & a_9 \end{bmatrix}, \text{ and } B = \begin{bmatrix} a_1 + ka_4 & a_2 + ka_5 & a_3 + ka_6 \\ a_4 & a_5 & a_6 \\ a_7 & a_8 & a_9 \end{bmatrix}.$$

Expand $|B|$ along the first row.

$$|B| = (a_1 + ka_4) \begin{vmatrix} a_5 & a_6 \\ a_8 & a_9 \end{vmatrix} - (a_2 + ka_5) \begin{vmatrix} a_4 & a_6 \\ a_7 & a_9 \end{vmatrix} + (a_3 + ka_6) \begin{vmatrix} a_4 & a_5 \\ a_7 & a_8 \end{vmatrix}$$

$$= a_1 \begin{vmatrix} a_5 & a_6 \\ a_8 & a_9 \end{vmatrix} - a_2 \begin{vmatrix} a_4 & a_6 \\ a_7 & a_9 \end{vmatrix} + a_3 \begin{vmatrix} a_4 & a_5 \\ a_7 & a_8 \end{vmatrix}$$

$$+ k \left(a_4 \begin{vmatrix} a_5 & a_6 \\ a_8 & a_9 \end{vmatrix} - a_5 \begin{vmatrix} a_4 & a_6 \\ a_7 & a_9 \end{vmatrix} + a_6 \begin{vmatrix} a_4 & a_5 \\ a_7 & a_8 \end{vmatrix} \right)$$

$$= \begin{vmatrix} a_1 & a_2 & a_3 \\ a_4 & a_5 & a_6 \\ a_7 & a_8 & a_9 \end{vmatrix} + k \begin{vmatrix} a_4 & a_5 & a_6 \\ a_4 & a_5 & a_6 \\ a_7 & a_8 & a_9 \end{vmatrix}$$

$$= |A| \quad + k \quad (0) \qquad \text{Two rows are equal so the determinant is 0}$$

$$= |A|$$

Notice the similarity between the Gaussian method for solving linear systems and the method indicated in Theorem 7. We use Theorem 7 to transform a determinant into equal determinants with 0 entries in all but one position in a row or column. By expanding along this row or column, an $n \times n$ determinant can be found by computing a single $(n-1) \times (n-1)$ determinant. The next example will clarify the process.

EXAMPLE 6 PROPERTIES OF DETERMINANTS

Use Theorem 7 to find

$$\begin{vmatrix} 2 & 3 & -1 \\ 4 & -2 & 1 \\ -5 & -1 & 3 \end{vmatrix}.$$

When using Theorem 7 it is helpful to key on an element equal to 1, if possible. Suppose we use 1 in the second row and third column and force the remaining elements in the third column (-1 and 3) to be 0.

$$\begin{vmatrix} 2 & 3 & -1 \\ 4 & -2 & 1 \\ -5 & -1 & 3 \end{vmatrix} = \begin{vmatrix} 6 & 1 & 0 \\ 4 & -2 & 1 \\ -17 & 5 & 0 \end{vmatrix} \begin{matrix} \leftarrow \text{Replaced with (1)(Row 2) + (Row 1)} \\ \\ \leftarrow \text{Replaced with } (-3)(\text{Row 2}) + (\text{Row 3}) \end{matrix}$$

$$= (-1) \begin{vmatrix} 6 & 1 \\ -17 & 5 \end{vmatrix} \quad \text{Expanding down the third column}$$

$$= (-1)(30 + 17) = -47$$

PRACTICE EXERCISE 6

Use Theorem 7 to find

$$\begin{vmatrix} 2 & 1 & 1 \\ 1 & -1 & 1 \\ 1 & -2 & -1 \end{vmatrix}.$$

Answer: 7

2 USING DETERMINANTS TO FIND INVERSES (OPTIONAL)

In Section 7.4 we found inverses of square matrices starting with the augmented matrix $[A|I]$ and changing it to the form $[I|A^{-1}]$ using the elementary row operations. Another way to find inverses using determinants and cofactors is given below, but the verification is beyond the level of our work in this text.

> ### To Find the Inverse of a Square Matrix
>
> Let A be an $n \times n$ square matrix, $n \geq 2$, with *nonzero* determinant $|A|$. (If $|A| = 0$, A has no inverse.)
>
> 1. Form **cof A**, the matrix of cofactors of A, by replacing every element of A with its cofactor.
> 2. Form the transpose of cof A, $(\text{cof } A)^T$.
> 3. The inverse of A, A^{-1}, is the matrix formed by taking the scalar product of $\frac{1}{|A|}$ and $(\text{cof } A)^T$. That is,
>
> $$A^{-1} = \frac{1}{|A|}(\text{cof } A)^T$$

EXAMPLE 7 FINDING THE INVERSE OF A 3 × 3 MATRIX

Find A^{-1} if $A = \begin{bmatrix} 1 & 1 & -1 \\ 2 & 1 & 1 \\ 3 & -2 & -1 \end{bmatrix}$.

We first find $|A|$, because if $|A| = 0$, A^{-1} does not exist. Expand along the first row.

$$|A| = (1)(-1)^{1+1}\begin{vmatrix} 1 & 1 \\ -2 & -1 \end{vmatrix} + (1)(-1)^{1+2}\begin{vmatrix} 2 & 1 \\ 3 & -1 \end{vmatrix} + (-1)(-1)^{1+3}\begin{vmatrix} 2 & 1 \\ 3 & -2 \end{vmatrix}$$

$$= (1)(-1 + 2) + (-1)(-2 - 3) + (-1)(-4 - 3)$$

$$= (1)(1) + (-1)(-5) + (-1)(-7) = 1 + 5 + 7 = 13$$

Since $|A| = 13 \neq 0$, we find the matrix of cofactors.

$$\text{cof } A = \begin{bmatrix} A_{11} & A_{12} & A_{13} \\ A_{21} & A_{22} & A_{23} \\ A_{31} & A_{32} & A_{33} \end{bmatrix} = \begin{bmatrix} +\begin{vmatrix} 1 & 1 \\ -2 & -1 \end{vmatrix} & -\begin{vmatrix} 2 & 1 \\ 3 & -1 \end{vmatrix} & +\begin{vmatrix} 2 & 1 \\ 3 & -2 \end{vmatrix} \\ -\begin{vmatrix} 1 & -1 \\ -2 & -1 \end{vmatrix} & +\begin{vmatrix} 1 & -1 \\ 3 & -1 \end{vmatrix} & -\begin{vmatrix} 1 & 1 \\ 3 & -2 \end{vmatrix} \\ +\begin{vmatrix} 1 & -1 \\ 1 & 1 \end{vmatrix} & -\begin{vmatrix} 1 & -1 \\ 2 & 1 \end{vmatrix} & +\begin{vmatrix} 1 & 1 \\ 2 & 1 \end{vmatrix} \end{bmatrix} = \begin{bmatrix} 1 & 5 & -7 \\ 3 & 2 & 5 \\ 2 & -3 & -1 \end{bmatrix}$$

Then

$$(\text{cof } A)^T = \begin{bmatrix} 1 & 3 & 2 \\ 5 & 2 & -3 \\ -7 & 5 & -1 \end{bmatrix} \text{ and } A^{-1} = \frac{1}{|A|}(\text{cof } A)^T = \frac{1}{13}\begin{bmatrix} 1 & 3 & 2 \\ 5 & 2 & -3 \\ -7 & 5 & -1 \end{bmatrix}.$$

Since $A^{-1}A$ must be I, we can check our work.

$$A^{-1}A = \frac{1}{13}\begin{bmatrix} 1 & 3 & 2 \\ 5 & 2 & -3 \\ -7 & 5 & -1 \end{bmatrix}\begin{bmatrix} 1 & 1 & -1 \\ 2 & 1 & 1 \\ 3 & -2 & -1 \end{bmatrix} = \frac{1}{13}\begin{bmatrix} 13 & 0 & 0 \\ 0 & 13 & 0 \\ 0 & 0 & 13 \end{bmatrix} = \begin{bmatrix} 1 & 0 & 0 \\ 0 & 1 & 0 \\ 0 & 0 & 1 \end{bmatrix} = I$$

PRACTICE EXERCISE 7

Find A^{-1} if

$$A = \begin{bmatrix} 1 & 0 & -1 \\ -2 & 1 & 0 \\ 2 & -2 & 3 \end{bmatrix}.$$

Answer: $\begin{bmatrix} 3 & 2 & 1 \\ 6 & 5 & 2 \\ 2 & 2 & 1 \end{bmatrix}$

Compare this with the result in Example 2 of Section 7.4.

If we define the determinant of a 1×1 matrix to be the number itself, the above method will also apply to finding determinants of 2×2 matrices whose cofactors are of order 1×1. For example, if $A = [4]$ and $B = [-2]$ are 1×1 matrices, define $|A| = 4$ and $|B| = -2$.

EXAMPLE 8 Finding the Inverse of a 2 × 2 Matrix

Find A^{-1} if
$$A = \begin{bmatrix} 1 & 2 \\ 3 & 4 \end{bmatrix}.$$

Since $|A| = 4 - 6 = -2 \neq 0$, A^{-1} exists.

$$\text{cof } A = \begin{bmatrix} A_{11} & A_{12} \\ A_{21} & A_{22} \end{bmatrix} = \begin{bmatrix} +|[4]| & -|[3]| \\ -|[2]| & +|[1]| \end{bmatrix} = \begin{bmatrix} 4 & -3 \\ -2 & 1 \end{bmatrix}$$

$$(\text{cof } A)^T = \begin{bmatrix} 4 & -2 \\ -3 & 1 \end{bmatrix} \quad \text{and} \quad A^{-1} = \frac{1}{|A|}(\text{cof } A)^T = \frac{1}{-2}\begin{bmatrix} 4 & -2 \\ -3 & 1 \end{bmatrix}$$

Check: $A^{-1}A = \frac{1}{-2}\begin{bmatrix} 4 & -2 \\ -3 & 1 \end{bmatrix}\begin{bmatrix} 1 & 2 \\ 3 & 4 \end{bmatrix} = -\frac{1}{2}\begin{bmatrix} -2 & 0 \\ 0 & -2 \end{bmatrix} = \begin{bmatrix} 1 & 0 \\ 0 & 1 \end{bmatrix} = I.$

PRACTICE EXERCISE 8

Find A^{-1} if
$$A = \begin{bmatrix} -2 & -1 \\ 3 & 1 \end{bmatrix}$$

Answer: $\begin{bmatrix} 1 & 1 \\ -3 & -2 \end{bmatrix}$
Compare this with the result in Practice Exercise 1, Section 7.4.

7.6 EXERCISES A

In Exercises 1–9, give the theorem in this section that justifies each statement. Do not evaluate the determinants.

1. $\begin{vmatrix} 1 & 3 \\ -1 & 5 \end{vmatrix} = \begin{vmatrix} 1 & -1 \\ 3 & 5 \end{vmatrix}$

2. $\begin{vmatrix} 0 & 7 \\ 0 & -8 \end{vmatrix} = 0$

3. $\begin{vmatrix} 4 & -4 \\ -1 & 3 \end{vmatrix} = 4\begin{vmatrix} 1 & -1 \\ -1 & 3 \end{vmatrix}$

4. $\begin{vmatrix} -5 & 0 \\ 1 & 6 \end{vmatrix} = -\begin{vmatrix} 0 & -5 \\ 6 & 1 \end{vmatrix}$

5. $\begin{vmatrix} 1 & 1 & -6 \\ 2 & 2 & 7 \\ 3 & 3 & -8 \end{vmatrix} = 0$

6. $\begin{vmatrix} 1 & 3 \\ 4 & -5 \end{vmatrix} = \begin{vmatrix} 1 & 3 \\ 4-4 & -5-12 \end{vmatrix}$

7. $\begin{vmatrix} -2 & -6 \\ 1 & 3 \end{vmatrix} = 0$

8. $3\begin{vmatrix} -1 & 2 & 0 \\ 2 & -1 & 5 \\ -1 & 2 & 0 \end{vmatrix} = \begin{vmatrix} -1 & 2 & 0 \\ 6 & -3 & 15 \\ -1 & 2 & 0 \end{vmatrix}$

9. $\begin{vmatrix} 5 & 0 & 7 \\ -6 & 0 & 5 \\ 1 & 0 & -4 \end{vmatrix} = 0$

In Exercises 10–11 the determinant on the left was transformed to the determinant on the right using Theorem 7. Find the value of x.

10. $\begin{vmatrix} 1 & 4 \\ -3 & 5 \end{vmatrix} = \begin{vmatrix} 1 & 0 \\ -3 & x \end{vmatrix}$

11. $\begin{vmatrix} 1 & -1 & 3 \\ 0 & 2 & -2 \\ 4 & -3 & 5 \end{vmatrix} = \begin{vmatrix} 1 & -1 & 3 \\ 0 & 2 & -2 \\ x & 0 & -4 \end{vmatrix}$

524 CHAPTER 7 MATRICES AND DETERMINANTS

12. Use Theorem 7 to introduce two zeros in row 1 of the determinant.

$$\begin{vmatrix} 2 & 1 & -5 \\ -3 & 4 & -2 \\ 1 & 3 & -1 \end{vmatrix}$$

13. Use Theorem 7 to introduce two zeros in column 2 of the determinant.

$$\begin{vmatrix} 2 & -3 & 1 \\ -1 & -2 & 3 \\ 4 & 1 & -2 \end{vmatrix}$$

In Exercises 14–17 use Theorem 7 to introduce two zeros in a row or column of each determinant and evaluate the determinant.

14. $\begin{vmatrix} 0 & 1 & 2 \\ -1 & 2 & 5 \\ 2 & -3 & 1 \end{vmatrix}$
15. $\begin{vmatrix} 2 & -1 & 5 \\ 3 & 1 & -2 \\ 4 & -3 & 2 \end{vmatrix}$
16. $\begin{vmatrix} 4 & -5 & 7 \\ 2 & 3 & -3 \\ -6 & -2 & -4 \end{vmatrix}$
17. $\begin{vmatrix} 2 & 3 & 0 & -2 \\ 1 & 4 & -1 & 0 \\ 3 & -2 & 2 & 5 \\ -2 & 0 & 1 & -3 \end{vmatrix}$

Use the optional method to find the inverse of each matrix in Exercises 18–21.

18. $\begin{bmatrix} 2 & 1 \\ 5 & -3 \end{bmatrix}$
19. $\begin{bmatrix} 2 & 2 \\ 2 & 2 \end{bmatrix}$
20. $\begin{bmatrix} 1 & 2 & 0 \\ 0 & 1 & 1 \\ -1 & 1 & 1 \end{bmatrix}$
21. $\begin{bmatrix} 1 & 3 & -1 \\ 3 & 1 & 0 \\ -2 & 0 & 1 \end{bmatrix}$

FOR REVIEW

22. Solve the system using Cramer's rule. You may wish to use Theorem 7 when evaluating determinants.

$$\begin{aligned} 2x \quad - z &= -6 \\ 3y + 5z &= 29 \\ x - y + z &= 0 \end{aligned}$$

ANSWERS: 1. Theorem 1 2. Theorem 3 3. Theorem 2 4. Theorem 4 5. Theorem 5 6. Theorem 7
7. Theorem 6 8. Theorem 2 9. Theorem 3 10. $x = 17$ 11. $x = 1$ 12. $\begin{vmatrix} 0 & 1 & 0 \\ -11 & 4 & 18 \\ -5 & 3 & 14 \end{vmatrix}$ 13. $\begin{vmatrix} 14 & 0 & -5 \\ 7 & 0 & -1 \\ 4 & 1 & -2 \end{vmatrix}$ 14. 9
15. -59 16. -104 17. -98 18. $-\frac{1}{11}\begin{bmatrix} -3 & -1 \\ -5 & 2 \end{bmatrix}$ 19. does not exist since the determinant is zero
20. $-\frac{1}{2}\begin{bmatrix} 0 & -2 & 2 \\ -1 & 1 & -1 \\ 1 & -3 & 1 \end{bmatrix}$ 21. $-\frac{1}{10}\begin{bmatrix} 1 & -3 & 1 \\ -3 & -1 & -3 \\ 2 & -6 & -8 \end{bmatrix}$ 22. $(-1, 3, 4)$

7.6 EXERCISES B

In Exercises 1–9, give the theorem in this section that justifies each statement. Do not evaluate the determinants.

1. $\begin{vmatrix} 0 & 2 \\ -5 & 4 \end{vmatrix} = \begin{vmatrix} 0 & -5 \\ 2 & 4 \end{vmatrix}$
2. $\begin{vmatrix} 6 & -2 \\ 0 & 0 \end{vmatrix} = 0$
3. $\begin{vmatrix} -2 & 1 \\ 0 & 5 \end{vmatrix} = \frac{1}{2}\begin{vmatrix} -4 & 1 \\ 0 & 5 \end{vmatrix}$

4. $\begin{vmatrix} -3 & 2 \\ 6 & 7 \end{vmatrix} = -\begin{vmatrix} 6 & 7 \\ -3 & 2 \end{vmatrix}$
5. $\begin{vmatrix} -1 & 0 & 5 \\ 7 & 4 & -3 \\ -1 & 0 & 5 \end{vmatrix} = 0$

6. $\begin{vmatrix} 7 & 5 \\ 1 & -2 \end{vmatrix} = \begin{vmatrix} 7 & 5 + 14 \\ 1 & -2 + 2 \end{vmatrix}$
7. $\begin{vmatrix} -4 & 12 \\ 1 & -3 \end{vmatrix} = 0$

8. $\begin{vmatrix} 5 & 0 & 1 \\ -3 & 2 & 6 \\ 1 & 4 & -2 \end{vmatrix} = -\begin{vmatrix} 1 & 0 & 5 \\ 6 & 2 & -3 \\ -2 & 4 & 1 \end{vmatrix}$
9. $\begin{vmatrix} 1 & -3+3 & 0 \\ 2 & -4+6 & 1 \\ -3 & 2-9 & -1 \end{vmatrix} = \begin{vmatrix} 1 & -3 & 0 \\ 2 & -4 & 1 \\ -3 & 2 & -1 \end{vmatrix}$

In Exercises 10–11, the determinant on the left was transformed to the determinant on the right using Theorem 7. Find the value of x.

10. $\begin{vmatrix} 2 & 6 \\ -1 & 3 \end{vmatrix} = \begin{vmatrix} 0 & x \\ -1 & 3 \end{vmatrix}$

11. $\begin{vmatrix} 0 & 2 & 4 \\ 1 & -3 & -1 \\ 2 & 5 & -2 \end{vmatrix} = \begin{vmatrix} 0 & 2 & 4 \\ 1 & -3 & -1 \\ x & 11 & 0 \end{vmatrix}$

12. Use Theorem 7 to introduce two zeros in row 1 of the determinant.

$\begin{vmatrix} 2 & -3 & 1 \\ -1 & -2 & 3 \\ 4 & 1 & -2 \end{vmatrix}$

13. Use Theorem 7 to introduce two zeros in column 2 of the determinant.

$\begin{vmatrix} 2 & 1 & -5 \\ -3 & 4 & -2 \\ 1 & 3 & -1 \end{vmatrix}$

In Exercises 14–17, use Theorem 7 to introduce two zeros in a row or column of each determinant and evaluate the determinant.

14. $\begin{vmatrix} 1 & 0 & 3 \\ -2 & 4 & -1 \\ 2 & 1 & 5 \end{vmatrix}$

15. $\begin{vmatrix} 2 & -3 & 5 \\ -1 & -2 & 4 \\ 6 & 3 & 1 \end{vmatrix}$

16. $\begin{vmatrix} 2 & 5 & -2 \\ -3 & 6 & 3 \\ -5 & 2 & 4 \end{vmatrix}$

17. $\begin{vmatrix} -3 & 1 & 0 & 5 \\ 2 & -4 & -1 & 3 \\ -1 & 0 & -2 & 2 \\ 3 & -1 & 4 & -2 \end{vmatrix}$

Use the optional method to find the inverse of each matrix in Exercises 18–21.

18. $\begin{bmatrix} 3 & 0 \\ 2 & -1 \end{bmatrix}$

19. $\begin{bmatrix} 0 & 5 \\ 0 & -7 \end{bmatrix}$

20. $\begin{bmatrix} 2 & -2 & 3 \\ 1 & 0 & -1 \\ -2 & 1 & 0 \end{bmatrix}$

21. $\begin{bmatrix} 1 & 4 & 1 \\ 2 & 0 & 2 \\ 3 & 5 & 3 \end{bmatrix}$

FOR REVIEW

22. Solve the system using Cramer's rule. You may wish to use Theorem 7 when evaluating determinants.

$$\begin{aligned} 3x - 5y + 7z &= 4 \\ 2x + 3y - 4z &= -6 \\ 4x - 7y - 8z &= -12 \end{aligned}$$

7.6 EXERCISES C

1. Show without evaluating that $(-1, 3)$ and $(2, 7)$ are solutions to the equation

$\begin{vmatrix} x & -2 & y \\ -1 & -2 & 3 \\ 2 & -2 & 7 \end{vmatrix} = 0.$

[*Hint:* Use Theorem 5.]

2. Show without evaluating that $(4, 1)$ and $(-2, 3)$ are solutions to the equation

$\begin{vmatrix} -1 & -1 & -1 \\ x & 4 & -2 \\ y & 1 & 3 \end{vmatrix} = 0.$

3. In 7.5 EXERCISES C we showed that the area of a triangle with vertices (x_1, y_1), (x_2, y_2), and (x_3, y_3) is the absolute value of

$\dfrac{1}{2} \begin{vmatrix} x_1 & y_1 & 1 \\ x_2 & y_2 & 1 \\ x_3 & y_3 & 1 \end{vmatrix}.$

What can be said about the three points if this determinant is zero?

4. Consider two lines with equations $ax + by = e$ and $cx + dy = f$. Prove that the lines are parallel when

$\begin{vmatrix} a & b \\ c & d \end{vmatrix} = 0$, but $\begin{vmatrix} a & e \\ c & f \end{vmatrix} \neq 0$ or $\begin{vmatrix} e & b \\ f & d \end{vmatrix} \neq 0.$

CHAPTER 7 EXTENSION PROJECT

How does a computer perform operations on matrices? Some forms of BASIC, usually on large mainframe computers, have the capability of reading specific matrix commands using the MAT format. Other forms, usually found on personal computers, do not have this feature. Nevertheless, it is easy to write simple programs, such as PROGRAM I below, which perform basic operations.

Dimension statements are used to give the order or dimension of the matrices under consideration. PROGRAM I will add two 2×2 matrices and multiply a matrix by a scalar. For the example run below,

$$A = \begin{bmatrix} 1 & 2 \\ 3 & 4 \end{bmatrix}, \quad B = \begin{bmatrix} -2 & 3 \\ -4 & 5 \end{bmatrix}, \quad \text{and } k = 5.$$

PROGRAM I

```
10  DIM A(2,2),B(2,2),C(2,2),D(2,2)
20  PRINT"THIS PROGRAM ADDS MATRICES A AND B"
30  PRINT"AND MULTIPLIES A BY SCALAR k"
40  PRINT "INPUT MATRIX A"
50  FOR I = 1 TO 2
60  FOR J = 1 TO 2
70  INPUT A(I,J)
80  NEXT J
90  NEXT I
100 PRINT "INPUT MATRIX B"
110 FOR I = 1 to 2
120 FOR J = 1 to 2
130 INPUT B(I,J)
140 NEXT J
150 NEXT I
160 PRINT "INPUT SCALAR k"
170 INPUT K
180 FOR I = 1 TO 2
190 FOR J = 1 TO 2
200 C(I,J) = A(I,J)+B(I,J)
210 D(I,J) = K*A(I,J)
220 NEXT J
230 NEXT I
240 PRINT "A"
250 PRINT A(1,1),A(1,2)
260 PRINT A(2,1),A(2,2)
270 PRINT "B"
280 PRINT B(1,1),B(1,2)
290 PRINT B(2,1),B(2,2)
300 PRINT "A+B"
310 PRINT C(1,1),C(1,2)
320 PRINT C(2,1),C(2,2)
330 PRINT "kA"
340 PRINT D(1,1),D(1,2)
350 PRINT D(2,1),D(2,2)
360 END
```

```
RUN
THIS PROGRAM ADDS MATRICES A AND B
AND MULTIPLIES A BY SCALAR k
INPUT MATRIX A
? 1
? 2
? 3
? 4
INPUT MATRIX B
? -2
? 3
? -4
? 5
INPUT SCALAR k
? 5
A
1           2
3           4
B
-2          3
-4          5
A+B
-1          5
-1          9
kA
5           10
15          20
```

In the Chapter 6 Extension Project we showed one way that a computer can be used to solve a system of equations. You have probably discovered by now that what was given there was an adaptation of Cramer's rule. The inverse method of solving systems can also be used. BASIC PROGAM II, given below, solves a 2×2 system of equations. The attached run solves the system

$$2x + y = 1$$
$$3x - 2y = 12$$

written as the matrix equation $AX = B$, with

$$A = \begin{bmatrix} 2 & 1 \\ 3 & -2 \end{bmatrix}, \quad B = \begin{bmatrix} 1 \\ 12 \end{bmatrix}, \quad \text{and } X = \begin{bmatrix} x \\ y \end{bmatrix}.$$

```
                          PROGRAM II
10  DIM A(2,2),B(2,1),C(2,2),X(2,1)
20  PRINT"THIS PROGRAM SOLVES A 2X2 SYSTEM"
30  PRINT"AS A MATRIX EQUATION AX = B"            RUN
40  PRINT"INPUT COEFFICIENT MATRIX A"             THIS PROGRAM SOLVES A 2X2 SYSTEM
50  FOR I = 1 TO 2                                AS A MATRIX EQUATION AX = B
60  FOR J = 1 TO 2                                INPUT COEFFICIENT MATRIX A
70  INPUT A(I,J)                                  ? 2
80  NEXT J                                        ? 1
90  NEXT I                                        ? 3
100 PRINT "INPUT CONSTANT MATRIX B"               ? -2
110 FOR I = 1 to 2                                INPUT CONSTANT MATRIX B
120 INPUT B(I,1)                                  ? 1
130 NEXT I                                        ? 12
140 D = A(1,1)*A(2,2)-A(2,1)*A(1,2)               MATRIX A    IS:
150 C(1,1) = A(2,2)/D                             2                   1
160 C(1,2) = -A(1,2)/D                            3                  -2
170 C(2,1) = -A(2,1)/D                            MATRIX B IS:
180 C(2,2) = A(1,1)/D                             1
190 PRINT "MATRIX A IS:"                          12
200 PRINT A(1,1),A(1,2)                           MATRIX A^-1 IS:
210 PRINT A(2,1),A(2,2)                           .2857143            .1428571
220 PRINT "MATRIX B IS:"                          .4285715           -.2857143
230 PRINT B(1,1)                                  MATRIX X IS:
240 PRINT B(2,1)                                  2
250 PRINT "MATRIX A^-1 IS:"                       -3
260 PRINT C(1,1),C(1,2)                           THUS, THE SOLUTION IS: ( 2 ,-3 )
270 PRINT C(2,1),C(2,2)
280 X(1,1) = C(1,1)*B(1,1) + C(1,2)*B(2,1)
290 X(2,1) = C(2,1)*B(1,1) + C(2,2)*B(2,1)
300 PRINT "MATRIX X IS:"
310 PRINT X(1,1)
320 PRINT X(2,1)
330 PRINT "THUS, THE SOLUTION IS: (";X(1,1);",";X(2,1);")"
```

A computer can also be programmed to solve a system using the Gauss row-reduction method. The program is easier to write using four subroutines: one for identifying a nonzero element of any row, one for switching two rows, one for multiplying a row by a constant and adding to another row, and one for multiplying a row by a constant.

THINGS TO DO:

1. Use PROGRAM I to find sums and scalar multiples of various matrices.
2. Modify PROGRAM I to find sums and scalar multiples of 3×3 matrices, 2×3 matrices, and 3×2 matrices.
3. Solve several systems in Section 6.1 using PROGRAM II.
4. Can you modify PROGRAM II to solve 3×3 systems? Begin by writing a program that will find the inverse of a 3×3 matrix. You might use the alternate method for finding inverses presented in Section 7.6.
5. Write a BASIC program that will solve a 2×2 system using the Gaussian method. Use your program to solve some of the exercises in Section 6.1.

CHAPTER 7 REVIEW

KEY TERMS

7.1
1. A **matrix** is a rectangular array of numbers or **elements.**
2. The **order** or **dimension** of a matrix with m rows and n columns is $m \times n$.
3. A matrix of order $n \times n$ is a **square matrix**; a matrix of order $m \times 1$ is a **column vector of order m**; and a matrix of order $1 \times n$ is a **row vector of order n**.
4. Two matrices are **equal** if they have the same order and corresponding elements are equal.
5. If $A = [a_{ij}]$ and $B = [b_{ij}]$ are two matrices of the same order, the **sum** of A and B is the matrix $A + B = [a_{ij} + b_{ij}]$.
6. A matrix with 0's for all elements is a **zero matrix.**
7. If $A = [a_{ij}]$ is a matrix, the **negative** or **additive inverse** of A is the matrix $-A = [-a_{ij}]$.
8. If $A = [a_{ij}]$ and $B = [b_{ij}]$ are two matrices of the same order, the **difference** of A and B is the matrix $A - B = [a_{ij} - b_{ij}]$.
9. If $A = [a_{ij}]$ is a matrix and k is a real number, the **scalar multiple** of A by k is the matrix $kA = [ka_{ij}]$.

7.2
1. If A is a row vector of order n and B is a column vector of order n, the **scalar product** or **dot product** of A and B is the real number $A \cdot B = a_1 b_1 + a_2 b_2 + \cdots + a_n b_n$.
2. Let $A = [a_{ij}]$ be of order $m \times n$ and $B = [b_{ij}]$ be of order $n \times p$. The **product** of A and B is the $m \times p$ matrix $AB = [c_{ij}]$ where c_{ij} is the scalar product of the ith row of A and the jth column of B.
3. The **identity matrix of order n,** denoted by I_n, is the $n \times n$ matrix with 1's down the **main diagonal** and 0's everywhere else.
4. If A is an $m \times n$ matrix, the $n \times m$ matrix, A^T, formed by interchanging the rows and columns of A is the **transpose** of A.

7.3 The **Gaussian method** is a method used to solve a system of equations by transforming the **augmented matrix** of the system into **reduced echelon form** using the **elementary row operations.**

7.4
1. Let A be an $n \times n$ matrix. An $n \times n$ matrix A^{-1} with the property that $AA^{-1} = A^{-1}A = I_n$ is the **inverse of A.**
2. A **nonsingular matrix** is a square matrix that has an inverse. A **singular matrix** is a matrix with no inverse.
3. A system of equations can be written as a matrix equation $AX = B$, where A is the **coefficient matrix,** X is the **variable matrix,** and B is the **constant matrix.**
4. The **inverse method** for solving a system of equations uses A^{-1}, the inverse of the coefficient matrix A, by multiplying the inverse times the constant matrix.

7.5
1. The **determinant** of a 2×2 matrix is a real number given by
$$\begin{vmatrix} a & b \\ c & d \end{vmatrix} = ad - bc.$$
2. **Cramer's rule** is a method used to solve a system of equations using determinants.
3. Let A be a square matrix of order $n \geq 3$. The **minor** of the element a_{ij}, denoted by M_{ij}, is the determinant of the matrix obtained by deleting the ith row and jth column of A. The **cofactor** of a_{ij} is given by $A_{ij} = (-1)^{i+j} M_{ij}$.
4. **Expansion by cofactors** is a method for finding the determinant of a square matrix of order $n \geq 3$ by adding the products of the elements in any row or column and their respective cofactors.

REVIEW EXERCISES

Part I

7.1– Exercises 1–14 refer to the following matrices.
7.2

$$A = \begin{bmatrix} -1 & 2 \\ 3 & 4 \end{bmatrix} \qquad B = \begin{bmatrix} 0 & -3 \\ 1 & 5 \end{bmatrix} \qquad C = \begin{bmatrix} -2 & 4 & 5 \\ 0 & -1 & 2 \end{bmatrix}$$

$$D = \begin{bmatrix} 1 & -3 & 4 \end{bmatrix} \qquad E = \begin{bmatrix} 2 \\ -1 \\ 0 \end{bmatrix} \qquad F = \begin{bmatrix} 2 & 0 & -1 \\ 4 & 1 & 3 \\ -2 & -1 & 5 \end{bmatrix}$$

1. What is the order of matrix C?
2. Give the zero matrix with the same order as A.
3. Identify the element a_{21} in matrix A.
4. Give the matrix $-B$.
5. Give the matrix C^T.
6. Find $A + B$.
7. Find $A - B$.
8. Find $-3C$.
9. Find $2A - 3B$.
10. Find $D \cdot E$.
11. Find AC.
12. Find CA.
13. Find CF.
14. Find $|C|$.

15. **BUSINESS** Island Rentals has two car rental agencies on the island of Jamaica. The total numbers of compact, intermediate, and luxury cars rented during the months of July and August are given in matrices J and A.

$$J = \begin{bmatrix} 200 & 180 & 20 \\ 300 & 120 & 40 \end{bmatrix} \begin{matrix} \text{Agency 1} \\ \text{Agency 2} \end{matrix}$$

with columns labeled Compact, Intermediate, Luxury.

$$A = \begin{bmatrix} 150 & 120 & 30 \\ 220 & 100 & 10 \end{bmatrix} \begin{matrix} \text{Agency 1} \\ \text{Agency 2} \end{matrix}$$

Suppose that the average revenue per rental is $80 for a compact car, $120 for an intermediate car, and $150 for a luxury car. The revenue matrix for such rentals is given by

$$C = \begin{bmatrix} 80 \\ 120 \\ 150 \end{bmatrix}.$$

(a) Find the matrix that gives the two-month totals in each category.
(b) Find the matrix $\frac{1}{2}(J + A)$ and discuss what it represents.
(c) Find $[1 \quad 1]J$ and discuss what it represents.
(d) Find $[1 \quad 1](J + A)$ and discuss what it represents.
(e) Find $([1 \quad 1](J + A)) \cdot \begin{bmatrix} 1 \\ 1 \\ 1 \end{bmatrix}$ and discuss what it represents.
(f) Find JC and interpret the result.
(g) Find $(J + A)C$ and interpret the result.
(h) Find $[1 \quad 1] \cdot ((J + A)C)$ and interpret the result.

7.3 In Exercises 16–17, solve each system using the Gaussian method.

16. $x - 3y = 4$
    $4x + 5y = -1$

17. $x - 2y + 3z = 7$
    $-x + 3y + 2z = 8$
    $3x - 4y - z = -9$

7.4 In Exercises 18–19, find the inverse of each matrix.

18. $\begin{bmatrix} -5 & -2 \\ 3 & 1 \end{bmatrix}$

19. $\begin{bmatrix} 3 & 0 & 1 \\ 1 & -1 & 0 \\ 0 & 1 & 2 \end{bmatrix}$

In Exercises 20–21, use the inverse method to solve each system. The inverse of the coefficient matrix was found in Exercises 18–19.

20. $-5x - 2y = 7$
    $3x + y = -5$

21. $3x + z = 9$
    $x - y = 4$
    $ y + 2z = 4$

22. **BUSINESS** The Fun and Fitness Center is conducting a three-month membership drive. The annual individual membership fee is $300 and the annual family membership fee is $400. During the three months, the owners have set a goal of selling the total number of memberships and earning the annual revenue shown in the following table. How many of each type of membership must be sold during each month in order to reach their goal?

	January	February	March
Total memberships	30	50	65
Annual revenue produced	$10,000	$17,000	$23,000

7.5 *Evaluate the determinants in Exercises 23–24.*

23. $\begin{vmatrix} -3 & 5 \\ 1 & 7 \end{vmatrix}$

24. $\begin{vmatrix} 2 & -1 & 3 \\ -2 & 2 & 4 \\ 1 & -3 & 5 \end{vmatrix}$

25. Find the cofactor of the element 4 in the matrix whose determinant is given in Exercise 24.

In Exercises 26–27 solve each system using Cramer's rule.

26. $-2x + y = 9$
    $3x + 4y = 14$

27. $3x - y + z = -12$
    $-2x + y + 3z = 9$
    $x - y - 2z = -6$

Solve for x in Exercises 28–29.

28. $\begin{vmatrix} 2 & x \\ x & 5 \end{vmatrix} = 6$

29. $\begin{vmatrix} x & 1 & 3 \\ 0 & 2 & x \\ 0 & 1 & -1 \end{vmatrix} = -3$

7.6 *In Exercises 30–33 state a theorem that justifies each statement. Do not evaluate the determinants.*

30. $\begin{vmatrix} 0 & 0 \\ 5 & -7 \end{vmatrix} = 0$

31. $\begin{vmatrix} 3 & -2 & 3 \\ 5 & 0 & 5 \\ 1 & 6 & 1 \end{vmatrix} = 0$

32. $\begin{vmatrix} 2 & -3 \\ 1 & 7 \end{vmatrix} = - \begin{vmatrix} -3 & 2 \\ 7 & 1 \end{vmatrix}$

33. $\begin{vmatrix} 3 & -3 \\ 0 & 4 \end{vmatrix} = 3 \begin{vmatrix} 1 & -1 \\ 0 & 4 \end{vmatrix}$

34. If the determinant $\begin{vmatrix} 0 & 1 \\ x & 5 \end{vmatrix}$ was obtained from $\begin{vmatrix} 2 & 1 \\ -3 & 5 \end{vmatrix}$ using Theorem 7 of Section 7.6, find the value of x.

35. Evaluate the following determinant by introducing two zeros in column 2.

$$\begin{vmatrix} -2 & 5 & 2 \\ 4 & 1 & 7 \\ 1 & -3 & -1 \end{vmatrix}$$

36. Suppose that $A = \begin{bmatrix} a_1 & a_2 & 0 & 0 \\ a_3 & a_4 & 0 & 0 \\ 0 & 0 & b_1 & b_2 \\ 0 & 0 & b_3 & b_4 \end{bmatrix}$. Show that $|A| = \begin{vmatrix} a_1 & a_2 \\ a_3 & a_4 \end{vmatrix} \begin{vmatrix} b_1 & b_2 \\ b_3 & b_4 \end{vmatrix}$.

Part II

37. **BUSINESS** Recreation Enterprises manufactures two types of picnic tables, the standard model and the deluxe model. Costs are $20 for labor and $25 for materials to make one standard model, and $40 for labor and $80 for materials to make one deluxe model. During each week, the total resources available for labor and material costs fluctuate due to employee vacations and variations in delivery schedules. The owner has determined that during the next 3 weeks, his available resources are limited according to the schedule in the table below. Use the inverse method to determine the number of tables of each type he should schedule for construction during each week.

	Week 1	Week 2	Week 3
Labor	$1800	$1400	$2200
Materials	$3300	$2200	$3950

38. Find A^{-1} if $A = \begin{bmatrix} 2 & -1 \\ 3 & 2 \end{bmatrix}$.

39. Find A^{-1} if $A = \begin{bmatrix} 2 & -1 & 1 \\ 1 & 0 & 2 \\ 0 & 1 & -5 \end{bmatrix}$.

Write the systems in Exercises 40–41 as matrix equations and solve using the inverse method.

40. $2x - y = 11$
    $3x + 2y = -1$
    [*Hint:* See Exercise 38.]

41. $2x - y + z = 3$
    $x + 2z = 0$
    $ y - 5z = 5$
    [*Hint:* See Exercise 39.]

Solve each system in Exercises 42–43 using the Gaussian method.

42. $2x - y = 11$
    $3x + 2y = -1$

43. $2x - y + z = 3$
    $x + 2z = 0$
    $ y - 5z = 5$

Use the following matrices for Exercises 44–64.

$$A = \begin{bmatrix} 2 & 3 \\ -1 & 4 \end{bmatrix} \quad B = \begin{bmatrix} 1 & 3 & 0 \\ -2 & 5 & 3 \end{bmatrix} \quad C = \begin{bmatrix} 2 \\ 1 \end{bmatrix} \quad D = \begin{bmatrix} 3 & 4 \end{bmatrix}$$

$$0 = \begin{bmatrix} 0 & 0 \\ 0 & 0 \end{bmatrix} \quad G = \begin{bmatrix} 3 & -2 & 4 \\ 1 & -4 & 0 \end{bmatrix} \quad H = \begin{bmatrix} x \\ y \end{bmatrix} \quad I = \begin{bmatrix} 1 & 0 \\ 0 & 1 \end{bmatrix}$$

44. Find $A + 0$.
45. Find $B + G$.
46. Find $2G - B$.
47. Find AI.
48. Find IA.
49. Find AC.
50. Find CA.
51. Find AH.
52. Find AB.
53. Find BA.
54. Find DC.
55. Find CD.
56. Find A^T.
57. Find B^T.
58. Find $|A|$.
59. Find $D \cdot C$.
60. Find A^{-1}.
61. Find AA^{-1}.
62. Find $|B|$.
63. Find B^{-1}.
64. If $H = C$, find x and y.

65. Solve for *x only* using Cramer's rule.

$$x - 2y + 4z = 1$$
$$x + y + z = 4$$
$$2x - y + z = 1$$

ANSWERS: 1. 2×3 2. $\begin{bmatrix} 0 & 0 \\ 0 & 0 \end{bmatrix}$ 3. 3 4. $\begin{bmatrix} 0 & 3 \\ -1 & -5 \end{bmatrix}$ 5. $\begin{bmatrix} -2 & 0 \\ 4 & -1 \\ 5 & 2 \end{bmatrix}$ 6. $\begin{bmatrix} -1 & -1 \\ 4 & 9 \end{bmatrix}$ 7. $\begin{bmatrix} -1 & 5 \\ 2 & -1 \end{bmatrix}$
8. $\begin{bmatrix} 6 & -12 & -15 \\ 0 & 3 & -6 \end{bmatrix}$ 9. $\begin{bmatrix} -2 & 13 \\ 3 & -7 \end{bmatrix}$ 10. 5 11. $\begin{bmatrix} 2 & -6 & -1 \\ -6 & 8 & 23 \end{bmatrix}$ 12. undefined 13. $\begin{bmatrix} 2 & -1 & 39 \\ -8 & -3 & 7 \end{bmatrix}$
14. undefined 15. (a) $J + A = \begin{bmatrix} 350 & 300 & 50 \\ 520 & 220 & 50 \end{bmatrix}$ (b) $\frac{1}{2}(J + A) = \begin{bmatrix} 175 & 150 & 25 \\ 260 & 110 & 25 \end{bmatrix}$ represents the average number of rentals per agency in each category over the two-month period. (c) $[1 \;\; 1]J = [500 \;\; 300 \;\; 60]$ represents the total number of rentals in each category at both agencies during July. (d) $[1 \;\; 1](J + A) = [870 \;\; 520 \;\; 100]$ represents the total number of rentals in each category at both agencies during the two-month period. (e) $([1 \;\; 1](J + A)) \cdot \begin{bmatrix} 1 \\ 1 \\ 1 \end{bmatrix} =$ 1490 represents the total number of rentals in every category at both agencies during the two-month period. (f) $JC = \begin{bmatrix} 40{,}600 \\ 44{,}400 \end{bmatrix}$ represents the total revenue produced by each agency during July. (g) $(J + A)C = \begin{bmatrix} 71{,}500 \\ 75{,}500 \end{bmatrix}$ represents the total revenue produced by each agency during the two-month period. (h) $[1 \;\; 1] \cdot ((J + A)C) = \$147{,}000$ represents the total revenue produced by both agencies during the two-month period. 16. $(1, -1)$ 17. $(-2, 0, 3)$
18. $\begin{bmatrix} 1 & 2 \\ -3 & -5 \end{bmatrix}$ 19. $\begin{bmatrix} \frac{2}{5} & -\frac{1}{5} & -\frac{1}{5} \\ \frac{2}{5} & -\frac{6}{5} & -\frac{1}{5} \\ -\frac{1}{5} & \frac{3}{5} & \frac{3}{5} \end{bmatrix}$ 20. $(-3, 4)$ 21. $(2, -2, 3)$ 22. January: 20 individual and 10 family memberships; February: 30 individual and 20 family memberships; March: 30 individual and 35 family memberships
23. -26 24. 42 25. 5 26. $(-2, 5)$ 27. $(-3, 3, 0)$ 28. ± 2 29. $-3, 1$ 30. Since a row is all zeros, the determinant is 0. 31. Since two columns are equal, the determinant is 0. 32. Since two columns were interchanged, the determinant is negated. 33. When a row is multiplied by 3, the determinant is multiplied by 3. 34. $x = -13$ 35. -11
36. $|A| = \begin{vmatrix} a_1 & a_2 & 0 & 0 \\ a_3 & a_4 & 0 & 0 \\ 0 & 0 & b_1 & b_2 \\ 0 & 0 & b_3 & b_4 \end{vmatrix} = a_1 \begin{vmatrix} a_4 & 0 & 0 \\ 0 & b_1 & b_2 \\ 0 & b_3 & b_4 \end{vmatrix} - a_3 \begin{vmatrix} a_2 & 0 & 0 \\ 0 & b_1 & b_2 \\ 0 & b_3 & b_4 \end{vmatrix} = a_1 a_4 \begin{vmatrix} b_1 & b_2 \\ b_3 & b_4 \end{vmatrix} - a_3 a_2 \begin{vmatrix} b_1 & b_2 \\ b_3 & b_4 \end{vmatrix} =$
$(a_1 a_4 - a_3 a_2) \begin{vmatrix} b_1 & b_2 \\ b_3 & b_4 \end{vmatrix} = \begin{vmatrix} a_1 & a_2 \\ a_3 & a_4 \end{vmatrix} \begin{vmatrix} b_1 & b_2 \\ b_3 & b_4 \end{vmatrix}$ 37. Week 1: 20 standard models and 35 deluxe models; Week 2: 40 standard models and 15 deluxe models; Week 3: 30 standard models and 40 deluxe models 38. $\begin{bmatrix} \frac{2}{7} & \frac{1}{7} \\ -\frac{3}{7} & \frac{2}{7} \end{bmatrix}$
39. $\begin{bmatrix} \frac{1}{4} & \frac{1}{2} & \frac{1}{4} \\ -\frac{5}{8} & \frac{5}{4} & \frac{3}{8} \\ -\frac{1}{8} & \frac{1}{4} & -\frac{1}{8} \end{bmatrix}$ 40. $(3, -5)$ 41. $(2, 0, -1)$ 42. $(3, -5)$ 43. $(2, 0, -1)$ 44. A 45. $\begin{bmatrix} 4 & 1 & 4 \\ -1 & 1 & 3 \end{bmatrix}$
46. $\begin{bmatrix} 5 & -7 & 8 \\ 4 & -13 & -3 \end{bmatrix}$ 47. A 48. A 49. $\begin{bmatrix} 7 \\ 2 \end{bmatrix}$ 50. not defined 51. $\begin{bmatrix} 2x + 3y \\ -x + 4y \end{bmatrix}$ 52. $\begin{bmatrix} -4 & 21 & 9 \\ -9 & 17 & 12 \end{bmatrix}$ 53. not defined 54. $[10]$ 55. $\begin{bmatrix} 6 & 8 \\ 3 & 4 \end{bmatrix}$ 56. $\begin{bmatrix} 2 & -1 \\ 3 & 4 \end{bmatrix}$ 57. $\begin{bmatrix} 1 & -2 \\ 3 & 5 \\ 0 & 3 \end{bmatrix}$ 58. 11 59. 10 60. $\begin{bmatrix} \frac{4}{11} & -\frac{3}{11} \\ \frac{1}{11} & \frac{2}{11} \end{bmatrix}$ 61. I 62. not defined
63. not defined 64. $x = 2; y = 1$ 65. $|A| = -12; |A_x| = -12; x = 1$

CHAPTER 7 TEST

1. Find $2A - 3B$ if $A = \begin{bmatrix} -2 & 0 \\ 4 & 1 \end{bmatrix}$ and $B = \begin{bmatrix} 3 & -1 \\ 0 & 5 \end{bmatrix}$.

   1. _____

2. Find the scalar product.

   $$[-1 \quad 3 \quad 5] \cdot \begin{bmatrix} 2 \\ 0 \\ -1 \end{bmatrix}$$

   2. _____

3. Find AB if $A = \begin{bmatrix} -1 & 3 \\ 4 & 2 \end{bmatrix}$ and $B = \begin{bmatrix} -2 & 0 \\ -1 & 5 \end{bmatrix}$.

   3. _____

4. Mutter's Menswear has two stores in Denver. The total number of suits, pants, and shirts sold during the month of January is given in J.

   4. _____

   $$J = \begin{bmatrix} \text{suits} & \text{pants} & \text{shirts} \\ 27 & 48 & 74 \\ 35 & 43 & 61 \end{bmatrix} \begin{matrix} \text{Store 1} \\ \text{Store 2} \end{matrix}$$

   Matrix C represents the selling price of each item. That is, c_{11} is the price of one suit, c_{21} is the price of one pair of pants, and c_{31} is the price of one shirt (all in dollars).

   $$C = \begin{bmatrix} 100 \\ 20 \\ 15 \end{bmatrix}$$

   Find $[1 \quad 1] \cdot (JC)$, and interpret the result.

5. Solve the system using the Gaussian method.

   $$\begin{aligned} 3x + y &= 1 \\ -x + 2y &= -5 \end{aligned}$$

   5. _____

6. Solve the system using the Gaussian method.

   $$\begin{aligned} 2x - y + 3z &= 10 \\ -x - y + 4z &= 17 \\ 5x + 3y - 2z &= -13 \end{aligned}$$

   6. _____

CHAPTER 7 TEST
CONTINUED

7. If $A = \begin{bmatrix} 1 & 3 \\ 2 & -1 \end{bmatrix}$ find A^{-1}.

7. _____

8. Solve the system using the inverse method.
   [*Hint:* Use Problem 7.]

$$x + 3y = 1$$
$$2x - y = -5$$

8. _____

9. If $A = \begin{bmatrix} 1 & 1 & 1 \\ 2 & 0 & 1 \\ 0 & 1 & -1 \end{bmatrix}$ find A^{-1}.

9. _____

10. Solve the system using the inverse method.
    [*Hint:* Use Problem 9.]

$$x + y + z = 1$$
$$2x + z = 0$$
$$ y - z = -2$$

10. _____

11. Solve the system using Cramer's Rule.

$$2x - 3y = 1$$
$$5x - 4y = 20$$

11. _____

12. Evaluate the determinant by first introducing two zeros in a row or column.

$$\begin{vmatrix} 3 & -4 & -3 \\ -1 & 5 & 6 \\ 2 & -2 & 4 \end{vmatrix}$$

12. _____

13. Without evaluating, explain why $|A| = 0$ if

$$A = \begin{bmatrix} -3 & 2 & 1 \\ 0 & 0 & 0 \\ -5 & -7 & 8 \end{bmatrix}.$$

13. _____

534

CHAPTER 8

Topics in Analytic Geometry

AN APPLIED PROBLEM IN ARCHITECTURE

A domed ceiling is a parabolic surface. For the best lighting on the floor, a light source is to be placed at the focus of the surface. If 10.0 m down from the top of the dome the ceiling is 15.0 m wide, find the best location for the light source.

Analysis

To solve this problem we begin by looking at a cross-section of the ceiling: a parabola. Make a sketch of the parabola placing the vertex at the origin of a coordinate system. Using the standard form of the equation of a parabola opening down with the vertex at the origin, we can locate the focus of the parabola. The y-coordinate of the focus will give us the location of the light source.

Tabulation

Sketch the cross-section of the ceiling and identify the two points on the curve given by the fact that 10.0 m down the ceiling is 15.0 m wide.

In this chapter we consider properties and graphs of a general *second-degree equation in two variables*. Since most of these graphs can be obtained by intersecting a cone with a plane, the curves are called *conic sections*. Conic sections have many applications in science, navigation, astronomy, and chemistry, as well as in architecture, like the one at the left (see Example 5 in Section 8.4).

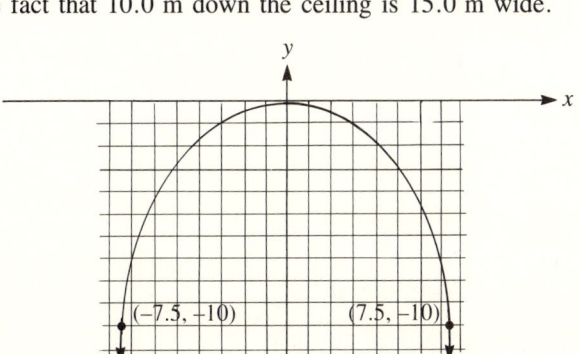

Translation

The standard form for the equation of this parabola is $x^2 = 4py$. Substitute 7.5 for x and -10 for y to find p.

8.1 THE CIRCLE

STUDENT GUIDEPOSTS

1. Conic Sections
2. Equations of a Circle

1 CONIC SECTIONS

The study of **analytic geometry** involves describing geometric figures with algebraic equations. Our work here will require us to solve two types of problems:

1. Given a figure, find the equation that describes it.
2. Given an equation, graph the corresponding figure in a coordinate system.

Our primary interest is in curves called **conic sections** or simply **conics**. The name comes from the fact that these curves can be obtained by intersecting a right circular cone with a plane. We will geometrically describe the four conics, the circle, ellipse, hyperbola, and parabola, and derive an equation for each that fits the description.

Another approach to the study of conics is to describe them as particular cases of a general **second-degree equation** or **quadratic equation** in two variables

$$Ax^2 + Bxy + Cy^2 + Dx + Ey + F = 0,$$

where at least one of A, B, or C is not zero. In Chapter 3 we studied the parabola as the graph of the quadratic function $y = ax^2 + bx + c$, which you can see is a particular case of the general quadratic equation in two variables.

The study of conics dates back to Menaechmus, a geometer in Plato's academy. The Greeks probably used these curves primarily to solve construction problems, but the application of conics continues to grow in modern times. The reflecting properties of a parabolic surface allow it to be used in telescopes, radar units, navigational systems, and headlights on an automobile. Planets and satellites follow elliptical orbits, and atomic particles travel in hyperbolic paths.

Figure 8.1 shows how the circle, ellipse, hyperbola, and parabola are obtained by intersecting a double-napped right circular cone with a plane.

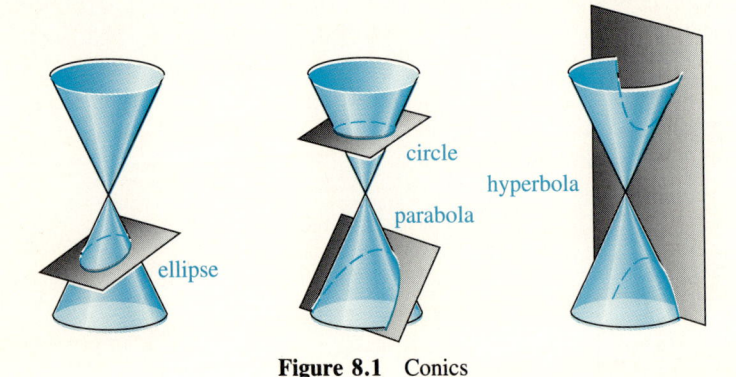

Figure 8.1 Conics

Actually, the circle is a special case of the ellipse in which the plane is perpendicular to the axis of the cone. Notice that none of the planes in Figure 8.1 passes through the vertex of the cone. When the intersecting plane does pass through the vertex, as shown in Figure 8.2, we obtain the **degenerate conic sections**. Our discussion is centered primarily on the conics in Figure 8.1.

8.1 THE CIRCLE 537

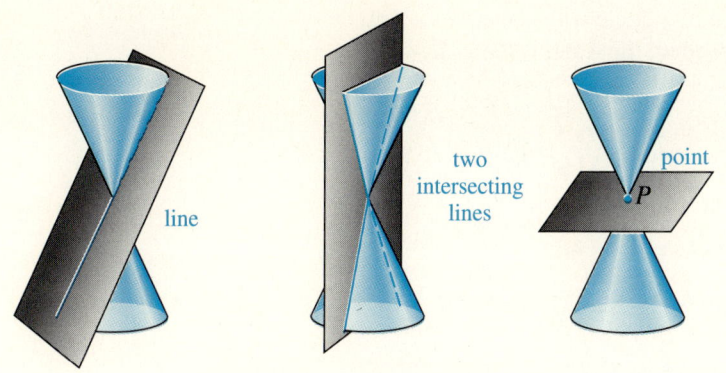

Figure 8.2 Degenerate Conics

❷ EQUATIONS OF A CIRCLE

The first conic section to be discussed is the *circle*. We define a circle geometrically and then derive an equation.

The Circle

Let (h, k) be any point in the plane and r be any positive real number. The collection of all points in the plane that are r units from (h, k) is a **circle** with **center** (h, k) and **radius** r.

The distance formula

$$d = \sqrt{(x_1 - x_2)^2 + (y_1 - y_2)^2},$$

which gives the distance between two points with coordinates (x_1, y_1) and (x_2, y_2), is used to derive the equation of a circle. Let (x, y) be an arbitrary point on the circle with radius r and center at (h, k). See Figure 8.3.

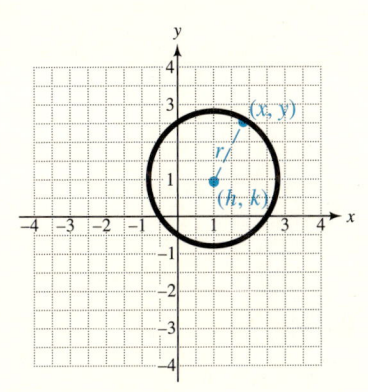

Figure 8.3 Circle

Since every point (x, y) is r units from (h, k),

$$r = \sqrt{(x - h)^2 + (y - k)^2}$$
$$r^2 = (x - h)^2 + (y - k)^2. \quad \text{Square both sides}$$

Standard Form of the Equation of a Circle

The **standard form** of the equation of the circle with center (h, k) and radius r is

$$(x - h)^2 + (y - k)^2 = r^2$$

When $(h, k) = (0, 0)$, then the equation of a circle with center at the origin is $x^2 + y^2 = r^2$. If the standard form is expanded, the result is the *general form* of the equation of a circle.

$$(x - h)^2 + (y - k)^2 = r^2$$
$$x^2 - 2hx + h^2 + y^2 - 2ky + k^2 = r^2$$
$$x^2 + y^2 - 2hx - 2ky + (h^2 + k^2 - r^2) = 0$$

This is a second-degree equation with **general form**

$$x^2 + y^2 + Dx + Ey + F = 0.$$

EXAMPLE 1 EQUATIONS OF A CIRCLE

Find the standard and general forms of the equation of a circle with center $(2, -1)$ and radius 3.

Substitute $h = 2$, $k = -1$, and $r = 3$ in the standard form.

$$(x - h)^2 + (y - k)^2 = r^2$$
$(x - 2)^2 + (y - (-1))^2 = 3^2$ Watch the sign on k
$(x - 2)^2 + (y + 1)^2 = 9$ Standard form

Expanding,

$$x^2 - 4x + 4 + y^2 + 2y + 1 = 9$$
$x^2 + y^2 - 4x + 2y - 4 = 0.$ General form

PRACTICE EXERCISE 1

Find the standard and general forms of the equation of a circle with center $(7, 0)$ and radius $\sqrt{13}$.

Answer: $(x - 7)^2 + y^2 = 13$;
$x^2 + y^2 - 14x + 36 = 0$

As Example 1 illustrates, obtaining the general form from the standard form is simply a matter of clearing the parentheses and collecting like terms. However, to find the standard form from the general form, we need to complete the squares in both x and y. This is shown in the next example.

EXAMPLE 2 FINDING THE STANDARD FORM

Find the standard form of the equation $x^2 + y^2 - 6x + 2y - 6 = 0$, and then graph the circle.

First write the x terms together and the y terms together, and place the constant term on the right, leaving space to write in numbers to complete both squares.

$$x^2 - 6x \quad + y^2 + 2y \quad = 6$$

To complete the square on x, add the square of one-half the coefficient of x, namely $(-3)^2$ or 9. Similarly, add $(1)^2$ or 1 to complete the square on y. We have added 9 and 1 to the left side of the equation and $9 + 1$ or 10 to the right side.

$$x^2 - 6x + 9 + y^2 + 2y + 1 = 6 + 10$$
$$(x - 3)^2 + (y + 1)^2 = 16$$
$$(x - 3)^2 + (y + 1)^2 = 4^2$$

Since $x - h = x - 3$, $h = 3$. Also $y - k = y + 1 = y - (-1)$ implies that $k = -1$. Thus, the center of the circle is $(3, -1)$ and the radius is 4. See Figure 8.4.

PRACTICE EXERCISE 2

Find the standard form of the equation $x^2 + y^2 + 4x - 6y + 4 = 0$, and then graph the circle.

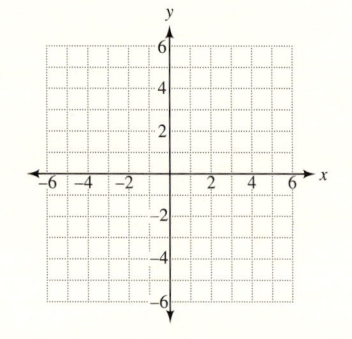

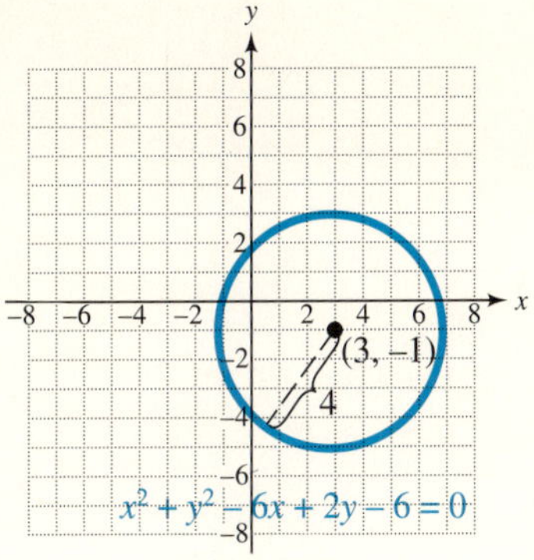

8.1 THE CIRCLE

Figure 8.4

Answer: $(x + 2)^2 + (y - 3)^2 = 9$
The circle has center $(-2, 3)$ and radius 3.

Generally the quadratic equation in two variables of the form

$$Ax^2 + Cy^2 + Dx + Ey + F = 0,$$

where $A = C \neq 0$, can be written in the standard form of the equation of a circle by dividing by A and completing the squares. For example,

$3x^2 + 3y^2 + 12x - 24y - 15 = 0$
$x^2 + y^2 + 4x - 8y - 5 = 0$ Divide by 3
$x^2 + 4x + 4 + y^2 - 8y + 16 = 5 + 20$ Complete the squares
$(x + 2)^2 + (y - 4)^2 = 5^2.$ Standard form

CAUTION

There are equations that appear to be equations of a circle that are actually degenerate conics. For example, when the squares are completed in $x^2 + y^2 - 4x + 6y + 13 = 0$, we obtain

$$(x - 2)^2 + (y + 3)^2 = 0.$$

Since $(2, -3)$ is the only solution to this equation, the graph is the single point $(2, -3)$. This can be thought of as a point circle. There are no points that satisfy the equation $2x^2 + 2y^2 + 4x - 2y + 3 = 0$ since by completing the squares we obtain

$$(x + 1)^2 + \left(y - \frac{1}{2}\right)^2 = -\frac{1}{4}.$$

Clearly the left side of this equation is nonnegative and thus the equation has no real solutions.

We conclude this section with an application of the circle.

EXAMPLE 3   AN ENGINEERING PROBLEM

A highway tunnel is to be designed in the shape of a semicircle with diameter 12.0 m. Find an equation for the semicircle and determine the vertical clearance at a point 4.0 m from the centerline.

Set a coordinate system with the centerline of the roadway at $(0, 0)$ as shown in Figure 8.5. The equation of the circle of which the semicircle

PRACTICE EXERCISE 3

Can a truck with an oversized load 6 m wide and 6 m high pass through the tunnel in Example 3?

is the top half becomes

$$(x-0)^2 + (y-0)^2 = 6^2$$
$$x^2 + y^2 = 36.$$

Solving for y,

$$y^2 = 36 - x^2$$
$$y = \pm\sqrt{36 - x^2}.$$

Using only the positive value of y for the top half of the circle,

$$y = \sqrt{36 - x^2}. \quad \text{Equation of semicircle}$$

The vertical clearance 4.0 m from the centerline can be found by letting $x = 4.0$ m.

$$y = \sqrt{36 - (4.0)^2} = \sqrt{36 - 16} = \sqrt{20} \approx 4.5$$

Thus, the vertical clearance is approximately 4.5 m.

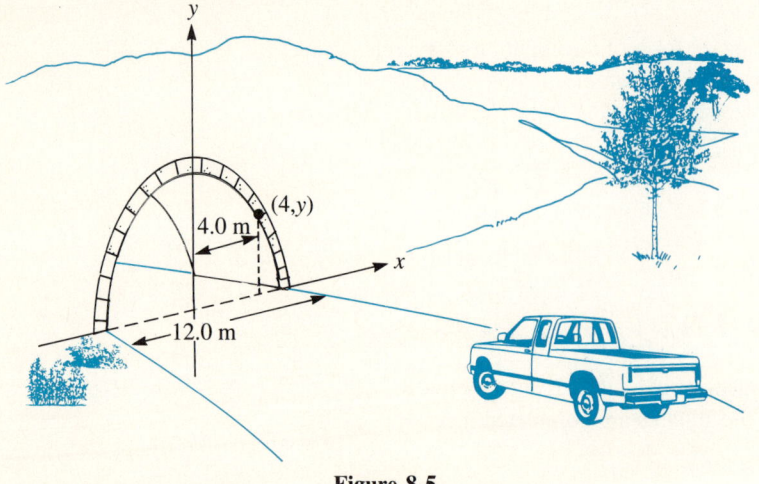

Figure 8.5

Answer: No; the vertical clearance 3 m from the centerline is only about 5.2 m.

8.1 EXERCISES A

Find the standard and general forms of the equation of each circle described in Exercises 1–3.

1. Center $(-3, 2)$, radius 1

2. Center $\left(\dfrac{1}{4}, -1\right)$, radius 3

3. Center $(0, 0)$, radius 1

Find the standard form of the equation of each circle described in Exercises 4–9.

4. Center at $(0, 0)$, passing through $(0, 5)$

5. Center at $(1, -5)$, passing through $(7, 3)$

6. Center at $(4, -1)$, radius $\sqrt{3}$

7. Center at $(6, 8)$, tangent to the x-axis

8. Endpoints of a diameter at $(1, -4)$ and $(5, 2)$

9. Endpoints of a diameter at $(-7, 8)$ and $(4, 7)$

In Exercises 10–13 graph each circle.

10. $x^2 + y^2 = 16$

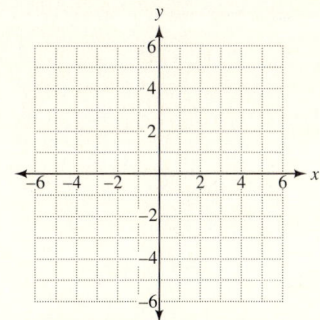

11. $(x - 1)^2 + (y + 4)^2 = 4$

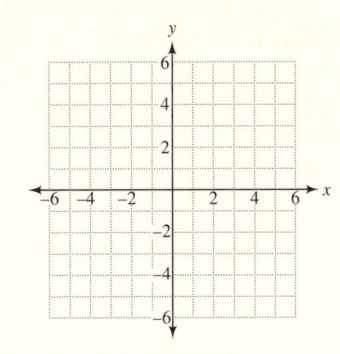

12. $x^2 + y^2 + 4y = 5$

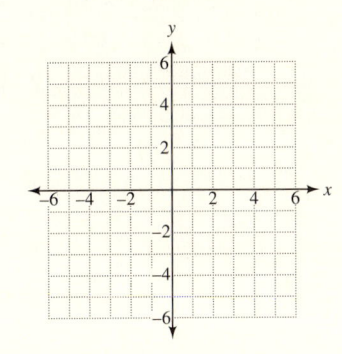

13. $x^2 + y^2 - 6x - 8y + 21 = 0$

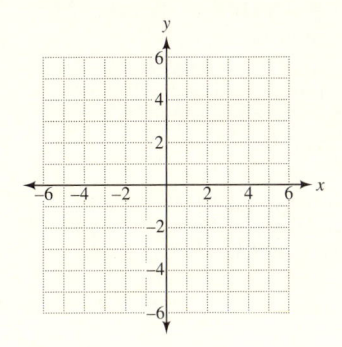

In Exercises 14–16, give the standard form of each equation.

14. $x^2 + y^2 - 4x - 2y + 4 = 0$

15. $4x^2 + 4y^2 - 4x + 24y - 63 = 0$

16. $9x^2 + 9y^2 + 6x - 6y - 142 = 0$

17. ENGINEERING A canal with cross-section a semicircle is 10 ft deep at the center. Find an equation for the semicircle and use it to find the depth 4 ft from the edge.

18. ENGINEERING A pipe is circular with cross-sectional inside radius 20 cm. Is it possible to insert a board that is 25 cm thick and 31 cm wide into the pipe? See the figure below.

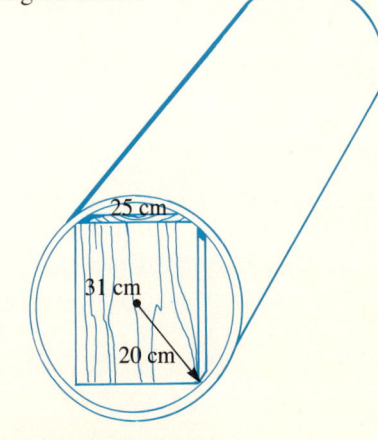

19. What is the graph of the second-degree equation $x^2 + y^2 + 1 = 0$? Can you see why this degenerate conic is sometimes called an *imaginary circle?*

20. What is the graph of the second-degree equation $x^2 - y^2 = 0$? [*Hint:* Factor and use the zero-product rule.]

ANSWERS: 1. $(x + 3)^2 + (y - 2)^2 = 1^2$; $x^2 + y^2 + 6x - 4y + 12 = 0$ 2. $\left(x - \frac{1}{4}\right)^2 + (y + 1)^2 = 3^2$; $16x^2 + 16y^2 - 8x + 32y - 127 = 0$ 3. $(x - 0)^2 + (y - 0)^2 = 1^2$ or $x^2 + y^2 = 1$; $x^2 + y^2 - 1 = 0$ 4. $x^2 + y^2 = 25$ 5. $(x - 1)^2 + (y + 5)^2 = 100$ 6. $(x - 4)^2 + (y + 1)^2 = 3$ 7. $(x - 6)^2 + (y - 8)^2 = 64$ 8. $(x - 3)^2 + (y + 1)^2 = 13$ 9. $\left(x + \frac{3}{2}\right)^2 + \left(y - \frac{15}{2}\right)^2 = \frac{61}{2}$ 10.–13. [graphs] 14. $(x - 2)^2 + (y - 1)^2 = 1^2$ 15. $\left(x - \frac{1}{2}\right)^2 + (y + 3)^2 = 5^2$ 16. $\left(x + \frac{1}{3}\right)^2 + \left(y - \frac{1}{3}\right)^2 = 4^2$ 17. $y = -\sqrt{100 - x^2}$, 8 ft 18. yes 19. There is no graph. Since the equation is similar to the standard form of the equation of a circle, this degenerate conic might be called an imaginary circle. 20. The graph is two intersecting lines with equations $y = x$ and $y = -x$.

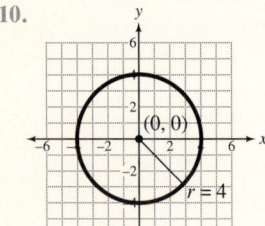

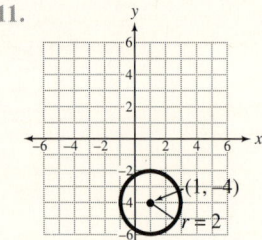

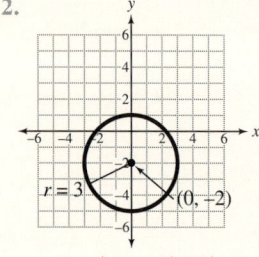

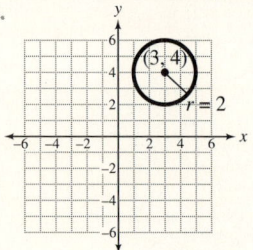

8.1 EXERCISES B

Find the standard and general forms of the equation of each circle described in Exercises 1–3.

1. Center $(5, -3)$, radius 2
2. Center $\left(-\frac{1}{2}, 0\right)$, radius 1
3. Center $\left(0, \frac{1}{5}\right)$, radius 4

Find the standard form of the equation of each circle described in Exercises 4–9.

4. Center at $(0, 0)$, passing through $(6, 8)$
5. Center at $(-1, -2)$, passing through $(4, -3)$
6. Center at $(-3, 5)$, radius $\sqrt{6}$
7. Center at $(-5, 3)$, tangent to the *y*-axis
8. Endpoints of a diameter at $(0, 6)$ and $(6, 0)$
9. Endpoints of a diameter at $(3, -2)$ and $(10, -9)$

In Exercises 10–13 graph each circle.

10. $x^2 + y^2 = 25$
11. $(x - 2)^2 + (y + 3)^2 = 16$
12. $x^2 + y^2 - 4x - 5 = 0$
13. $x^2 + y^2 + 10x + 2y + 10 = 0$

In Exercises 14–16, give the standard form of each equation.

14. $x^2 + y^2 + 2x - 10y + 17 = 0$
15. $16x^2 + 16y^2 - 8x - 8y - 62 = 0$
16. $3x^2 + 3y^2 - 18x + 6y + 30 = 0$

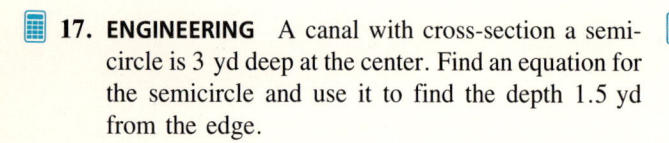

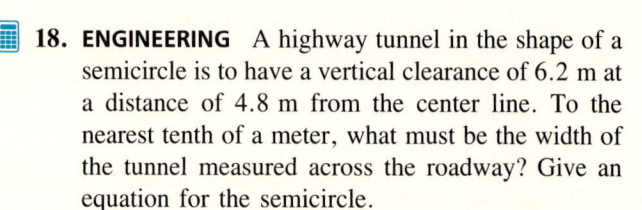

17. **ENGINEERING** A canal with cross-section a semicircle is 3 yd deep at the center. Find an equation for the semicircle and use it to find the depth 1.5 yd from the edge.

18. **ENGINEERING** A highway tunnel in the shape of a semicircle is to have a vertical clearance of 6.2 m at a distance of 4.8 m from the center line. To the nearest tenth of a meter, what must be the width of the tunnel measured across the roadway? Give an equation for the semicircle.

19. What is the graph of the second-degree equation $x^2 + y^2 = 0$? Can you see why this degenerate conic is sometimes called a *point circle*?

20. What is the graph of the second-degree equation $x^2 - 2xy + y^2 - x + y = 0$? [*Hint:* The left side can be factored into $(y - x)(y - x + 1)$. Use the zero-product rule.]

8.1 EXERCISES C

For Exercises 1–2 a **chord** *of a circle is defined to be a line segment joining two distinct points on the circle.*

1. Prove that a radius perpendicular to a chord bisects the chord.

2. Suppose that $(3, -4)$ is the midpoint of a chord of the circle $x^2 + y^2 = 100$.
 (a) Find the equation of the line that contains the chord.
 (b) Find the length of the chord.

3. The equation $x^2 + y^2 + cx + cy = 0$ describes a **family of circles** with each value of c representing a different circle. Describe this family in words.

4. Find the equation of the circle in the family $x^2 + y^2 - 4x + 2y + c(x^2 + y^2 - 2y - 4) = 0$ whose center is on the line $2x + 4y = 1$.
 [Answer: $x^2 + y^2 - 3x + y - 1 = 0$]

8.2 THE ELLIPSE

STUDENT GUIDEPOSTS

1. Definition of an Ellipse
2. Equation of an Ellipse
3. Translation of Axes
4. General Form of an Ellipse

1 DEFINITION OF AN ELLIPSE

A circle is the set of points in a plane whose distance from *one* fixed point is a constant. If a string is attached to *two* fixed pins on a piece of paper and drawn taut with a pencil, the oval-shaped curve traced by moving the pencil around the pins while keeping the string taut is an *ellipse*. See Figure 8.6. Notice that since the string does not change length in this process, the sum of the distances from the fixed points to each point on the curve remains constant. This idea leads to the formal definition of an ellipse.

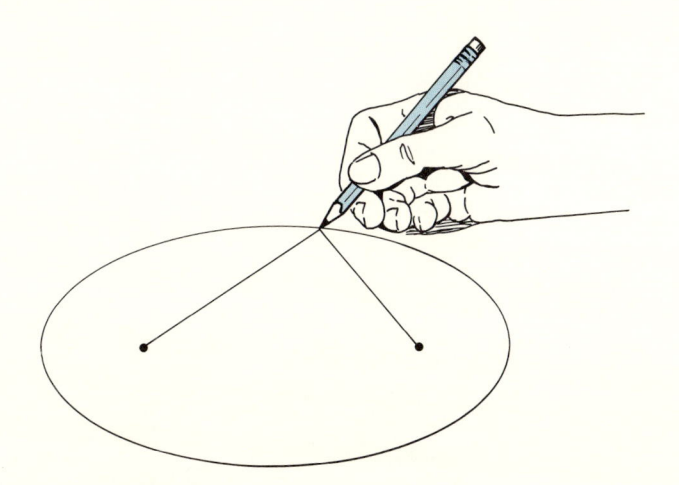

Figure 8.6 Construction of an Ellipse

The Ellipse

Let F_1 and F_2 be two fixed points in a plane. An **ellipse** with **foci** (plural of **focus**) F_1 and F_2 is the collection of all points in the plane the sum of whose distances from F_1 and F_2 is a constant. The midpoint of the line segment joining F_1 and F_2 is the **center** of the ellipse.

2 EQUATION OF AN ELLIPSE

The equation of an ellipse is developed by considering an ellipse with center at the origin as shown in Figure 8.7. The points $(-a, 0)$ and $(a, 0)$ are the **vertices** of the ellipse. The line segment joining the vertices which passes through the foci and has length $2a$ is the **major axis.** The line segment through the center of the ellipse with endpoints $(0, -b)$ and $(0, b)$ is the **minor axis.** Notice that the minor axis has length $2b$ and the distance between the foci, $F_1(-c, 0)$ and $F_2(c, 0)$ is $2c$.

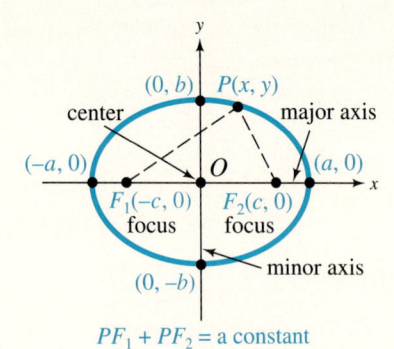

Figure 8.7 Ellipse

If $P(x, y)$ is any point on the ellipse, then by definition

$$PF_1 + PF_2 = \text{a constant.}$$

If point P is at $(a, 0)$, then $PF_1 = a + c$ and $PF_2 = a - c$ so that

$$PF_1 + PF_2 = (a + c) + (a - c) = 2a.$$

Thus, the sum of the distances must be $2a$. Using the point $P(x, y)$, by the distance formula

$$PF_1 = \sqrt{(x + c)^2 + (y - 0)^2} = \sqrt{(x + c)^2 + y^2}$$

and

$$PF_2 = \sqrt{(x - c)^2 + (y - 0)^2} = \sqrt{(x - c)^2 + y^2}.$$

Thus, $\qquad PF_1 + PF_2 = \sqrt{(x + c)^2 + y^2} + \sqrt{(x - c)^2 + y^2} = 2a.$

Isolate one radical and square both sides.

$$\sqrt{(x + c)^2 + y^2} = 2a - \sqrt{(x - c)^2 + y^2}$$
$$(x + c)^2 + y^2 = 4a^2 - 4a\sqrt{(x - c)^2 + y^2} + (x - c)^2 + y^2$$
$$x^2 + 2xc + c^2 + y^2 = 4a^2 - 4a\sqrt{(x - c)^2 + y^2} + x^2 - 2xc + c^2 + y^2$$
$$4xc - 4a^2 = -4a\sqrt{(x - c)^2 + y^2}$$
$$xc - a^2 = -a\sqrt{(x - c)^2 + y^2}$$

Square both sides again.

$$x^2c^2 - 2xca^2 + a^4 = a^2[(x - c)^2 + y^2]$$
$$= a^2[x^2 - 2xc + c^2 + y^2]$$
$$= a^2x^2 - 2a^2xc + a^2c^2 + a^2y^2$$
$$x^2(c^2 - a^2) - a^2y^2 = -a^2(a^2 - c^2)$$

Multiply both sides by -1.

$$x^2(a^2 - c^2) + a^2y^2 = a^2(a^2 - c^2)$$

If point P is at $(0, b)$, as in Figure 8.8, $PF_1 = PF_2 = a$ (since $PF_1 + PF_2 = 2a$ and $PF_1 = PF_2$). Since $PO = b$ and $F_2O = c$, by the Pythagorean theorem,

$$a^2 = b^2 + c^2 \quad \text{so that} \quad a^2 - c^2 = b^2.$$

Substitute b^2 for $a^2 - c^2$ in the equation above, then divide through by a^2b^2.

$$x^2b^2 + a^2y^2 = a^2b^2$$

$$\frac{x^2b^2}{a^2b^2} + \frac{a^2y^2}{a^2b^2} = \frac{a^2b^2}{a^2b^2}$$

$$\frac{x^2}{a^2} + \frac{y^2}{b^2} = 1$$

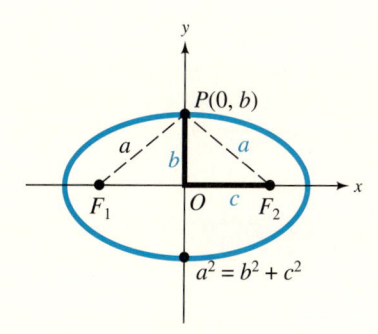

Figure 8.8

This equation is the **standard form** of the equation of an ellipse centered at the origin with x-intercepts $(a, 0)$ and $(-a, 0)$ and y-intercepts $(0, b)$ and $(0, -b)$. Remember that the foci are on the x-axis at $(-c, 0)$ and $(c, 0)$ and that $a > b$.

Ellipse with Horizontal Major Axis

The equation

$$\frac{x^2}{a^2} + \frac{y^2}{b^2} = 1,$$

where $a > b$, is an ellipse with graph shown in Figure 8.9. a, b, and c are related by $b^2 = a^2 - c^2$.

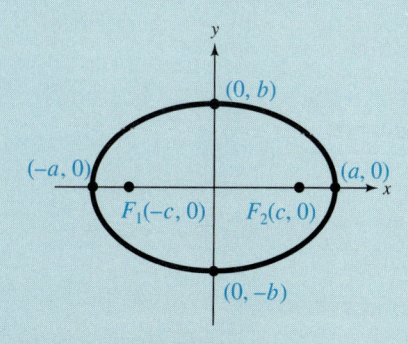

Figure 8.9

EXAMPLE 1 ELLIPSE WITH HORIZONTAL MAJOR AXIS

Find the equation of the ellipse with foci at $(-\sqrt{5}, 0)$ and $(\sqrt{5}, 0)$ and x-intercepts $(-3, 0)$ and $(3, 0)$.

Since the foci are on the x-axis, we know $a = 3$. Thus,

$$b^2 = a^2 - c^2 = (3)^2 - (\sqrt{5})^2 = 4.$$
$$b = 2$$

Substitute $a = 3$ and $b = 2$ in the standard form.

$$\frac{x^2}{3^2} + \frac{y^2}{2^2} = 1$$

$$\frac{x^2}{9} + \frac{y^2}{4} = 1$$

PRACTICE EXERCISE 1

Find the equation of the ellipse with foci $(-4, 0)$ and $(4, 0)$ and x-intercepts $(-6, 0)$ and $(6, 0)$.

Answer: $\frac{x^2}{36} + \frac{y^2}{20} = 1$

An ellipse may have its major axis on the y-axis with endpoints $(0, -a)$ and $(0, a)$. The foci are at $(0, -c)$ and $(0, c)$; endpoints of the minor axis $(-b, 0)$ and $(b, 0)$.

Ellipse with Vertical Major Axis

The equation

$$\frac{x^2}{b^2} + \frac{y^2}{a^2} = 1,$$

where $a > b$, is an ellipse with graph shown in Figure 8.10. a, b, and c are related by $b^2 = a^2 - c^2$.

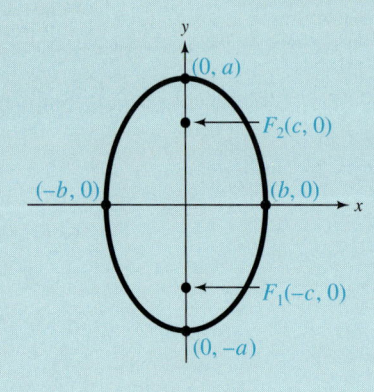

Figure 8.10

EXAMPLE 2 GRAPHING AN ELLIPSE

Graph the equation

$$25x^2 + 16y^2 = 400.$$

Divide by 400.

$$\frac{25x^2}{400} + \frac{16y^2}{400} = 1$$

$$\frac{x^2}{16} + \frac{y^2}{25} = 1 \qquad \text{Standard form}$$

PRACTICE EXERCISE 2

Graph the equation $x^2 + 9y^2 = 36$.

Observe that since $25 > 16$ this is the standard form of an ellipse with vertical major axis. Also, $a^2 = 25$, or $a = 5$, and $b^2 = 16$ which means $b = 4$. The x-intercepts are $(-4, 0)$ and $(4, 0)$ while the y-intercepts are $(0, -5)$ and $(0, 5)$.

Another way to find the x-intercepts is to set $y = 0$ in the original equation and solve for x.

$$\frac{x^2}{16} + \frac{0^2}{25} = 1$$

$$\frac{x^2}{16} = 1$$

$$x^2 = 16$$

$$x = \pm 4$$

Thus, the x-intercepts are $(4, 0)$ and $(-4, 0)$. Similarly, by setting $x = 0$, we obtain the y-intercepts $(0, -5)$ and $(0, 5)$. The graph of this ellipse is given in Figure 8.11.

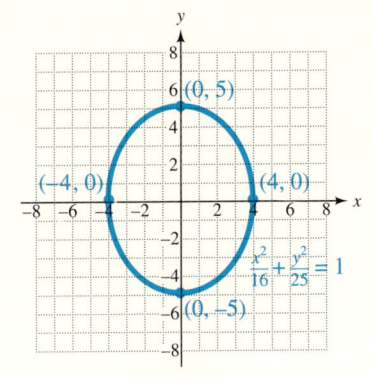

Figure 8.11

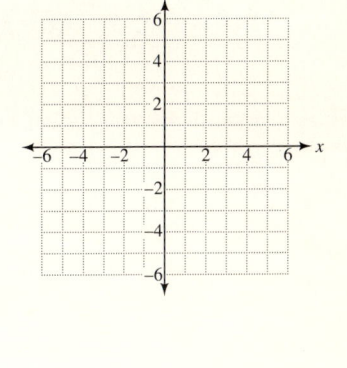

Answer: The graph is the ellipse with horizontal major axis and intercepts $(\pm 6, 0)$ and $(0, \pm 2)$.

The next example gives one application of ellipses and shows how a circle can be used to approximate an ellipse in certain situations.

EXAMPLE 3   An Aerospace Problem

A communications company has determined that the most economical orbit for one of its space labs is an ellipse which has a maximum distance above the surface of the earth of 200 miles and a minimum distance of 100 miles. Use 4000 miles as the radius of the earth and determine the equation of the ellipse. Estimate the distance traveled by the space lab in one revolution.

The coordinate system in Figure 8.12 has been set to have the maximum of the orbit occur on the x-axis. The minimum is then on the y-axis. Thus, $a = 4200$ and $b = 4100$. The equation is

$$\frac{x^2}{(4200)^2} + \frac{y^2}{(4100)^2} = 1.$$

Notice that for this ellipse a and b are relatively the same size. If $a = b$, the equation would be a circle. With this information we have an easy way to estimate the distance traveled by the space lab in one revolution. Calculate the circumference of a circle with radius $r = \frac{a+b}{2}$.

Practice Exercise 3

At what rate, in feet per second, is the space lab in Example 3 traveling if it makes one revolution every 24 hours?

$$C = 2\pi r = 2\pi\left(\frac{4200 + 4100}{2}\right)$$
$$\approx 26{,}100 \text{ mi}$$

Thus, the space lab travels approximately 26,100 mi in each revolution.

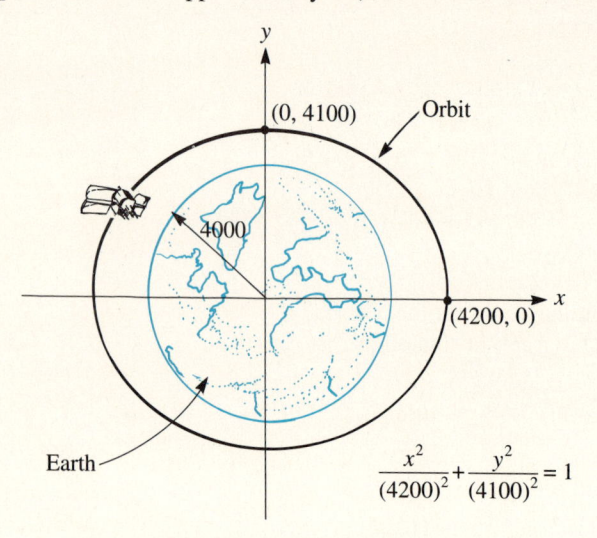

Figure 8.12 Space Lab Orbit

Answer: 1595 ft/sec

③ TRANSLATION OF AXES

To discuss an ellipse with center at any point (h, k), we use *translation of axes*. Consider the ellipse in Figure 8.13 with center at (h, k) in an xy-coordinate system. Now establish a new $x'y'$-coordinate system with origin at (h, k) in the xy-coordinate system. Notice in Figure 8.13 that the coordinates of a point on the ellipse, $P(x, y) = P(x', y')$, satisfy

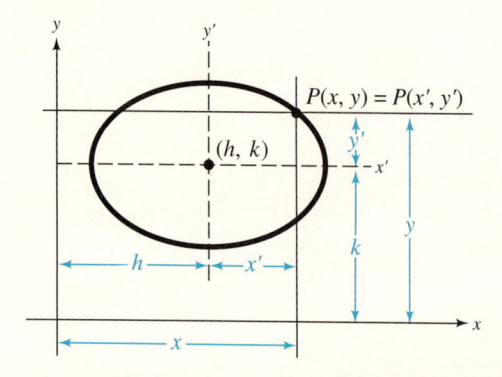

$$x = x' + h$$
$$y = y' + k$$
$$\text{or} \quad x' = x - h$$
$$y' = y - k.$$

Figure 8.13 Translation of Axes

These are transformation equations which hold when a **translation of axes** takes place. This means a new coordinate system is established in such a way that the x'-axis is parallel to the x-axis and the y'-axis is parallel to the y-axis.

The equation for the ellipse in Figure 8.13 is of the form

$$\frac{x'^2}{a^2} + \frac{y'^2}{b^2} = 1$$

in the $x'y'$-coordinate system. Using $x' = x - h$ and $y' = y - k$, we have the following.

Standard Forms of the Equation of an Ellipse

The **standard form** of the equation of an ellipse with center (h, k) and horizontal major axis is

$$\frac{(x - h)^2}{a^2} + \frac{(y - k)^2}{b^2} = 1, \text{ with } a > b.$$

The **standard form** of the equation of an ellipse with center at (h, k) and vertical major axis is

$$\frac{(x - h)^2}{b^2} + \frac{(y - k)^2}{a^2} = 1, \text{ with } a > b.$$

EXAMPLE 4 IDENTIFYING AND GRAPHING ELLIPSES

Find the standard form and graph the ellipse that has $b = 4$ and foci at $(-1, 3)$ and $(5, 3)$.

The center of the ellipse is the midpoint of the line segment joining the foci.

$$h = \frac{-1 + 5}{2} = 2 \quad \text{and} \quad k = \frac{3 + 3}{2} = 3$$

Thus $(h, k) = (2, 3)$. Also $2c = 5 - (-1) = 6$, so $c = 3$.

$$a^2 = b^2 + c^2 = (4)^2 + (3)^2 = 25$$

Therefore, $a = 5$ and the standard form of the ellipse is

$$\frac{(x - 2)^2}{25} + \frac{(y - 3)^2}{16} = 1.$$

The graph intersects the line $y = 3$ at points $a = 5$ units on each side of the center $(2, 3)$, and the line $x = 2$ at points $b = 4$ units above and below the center $(2, 3)$. See Figure 8.14.

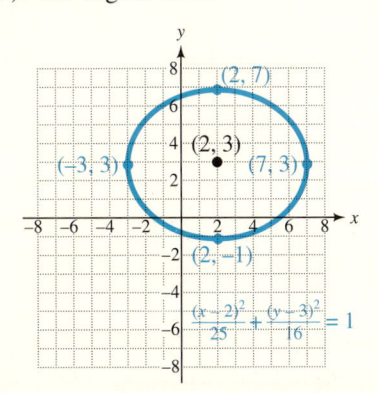

Figure 8.14

PRACTICE EXERCISE 4

Find the standard form and graph the ellipse that has $a = 5$ and foci at $(1, -1)$ and $(1, 5)$.

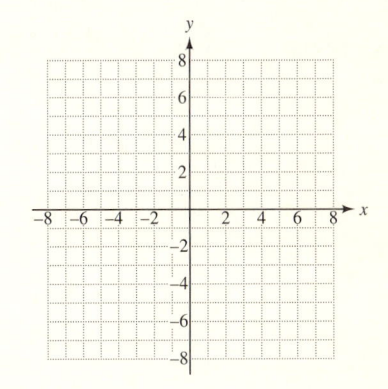

Answer: $\frac{(x - 1)^2}{16} + \frac{(y - 2)^2}{25} = 1$
The ellipse has vertical major axis and passes through the points $(1, 7)$, $(-3, 2)$, $(1, -3)$, and $(5, 2)$.

④ GENERAL FORM OF AN ELLIPSE

If the parentheses are cleared in the standard form, we obtain the **general form of the equation of an ellipse.**

$$Ax^2 + Cy^2 + Dx + Ey + F = 0 \qquad A \neq C, A \text{ and } C \text{ same signs}$$

For the equation to represent an ellipse, A and C must have the same signs and must not be equal. In order to convert from general form to standard form, proceed as with the circle and complete the squares.

EXAMPLE 5   FINDING THE STANDARD FORM

Write $36x^2 + 64y^2 + 108x - 128y - 431 = 0$ in standard form.

$36x^2 + 108x \quad + 64y^2 - 128y \quad = 431$ Rewrite

$36(x^2 + 3x \quad) + 64(y^2 - 2y \quad) = 431$ Factor out coefficients of squared terms

$36\left(x^2 + 3x + \dfrac{9}{4}\right) + 64(y^2 - 2y + 1) = 431 + 36\left(\dfrac{9}{4}\right) + 64(1)$ Complete the squares

Do not just add 9/4 and 1 to the right side since they are multiplied on the left side by 36 and 64, respectively.

$$36\left(x + \dfrac{3}{2}\right)^2 + 64(y - 1)^2 = 576$$

$$\dfrac{36\left(x + \dfrac{3}{2}\right)^2}{576} + \dfrac{64(y - 1)^2}{576} = \dfrac{576}{576} \qquad \text{Divide by 576}$$

$$\dfrac{\left(x + \dfrac{3}{2}\right)^2}{16} + \dfrac{(y - 1)^2}{9} = 1$$

The equation is an ellipse with center at $\left(-\dfrac{3}{2}, 1\right)$, $a = 4$, and $b = 3$.

PRACTICE EXERCISE 5

Write $4x^2 + y^2 + 40x - 12y + 100 = 0$ in standard form.

Answer: $\dfrac{(x + 5)^2}{9} + \dfrac{(y - 6)^2}{36} = 1$

8.2 EXERCISES A

In Exercises 1–5 find the equation of each ellipse with center at the origin.

1. Intercepts $(\pm 7, 0)$ and $(0, \pm 6)$

2. Foci $(\pm 3, 0)$ and x-intercepts $(\pm 5, 0)$

3. Foci $(0, \pm 3)$ and x-intercepts $(\pm 4, 0)$

4. Length of major axis 12 and y-intercepts $(0, \pm 5)$

5. Major axis intercepts $(\pm 10, 0)$ and passing through $(5, \sqrt{3})$
   [*Hint:* Use the equation in standard form to find b.]

In Exercises 6–8 find the equation of each ellipse.

6. Center at $(5, -2)$, $a = 4$, and foci $(3, -2)$ and $(7, -2)$

7. Foci $(\pm 4, 3)$ and y-intercepts $(0, 0)$ and $(0, 6)$

8. Foci $(7, -1)$ and $(7, -7)$ and $a = 8$

Graph each ellipse in Exercises 9–12.

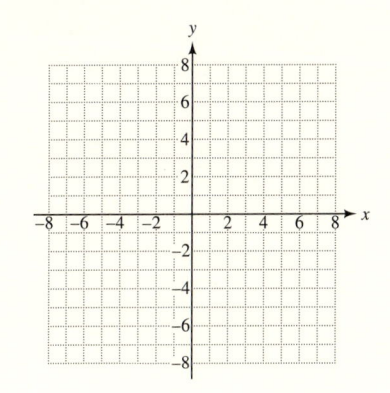

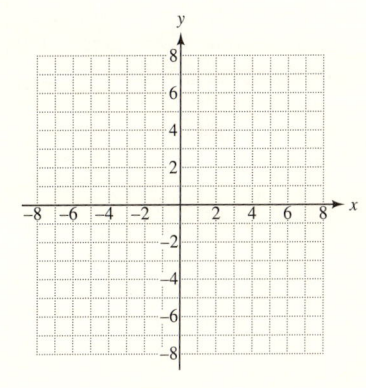

9. $\dfrac{x^2}{49} + \dfrac{y^2}{25} = 1$

10. $\dfrac{(x-2)^2}{16} + \dfrac{(y+2)^2}{4} = 1$

11. $4x^2 + 25y^2 - 24x + 50y - 39 = 0$

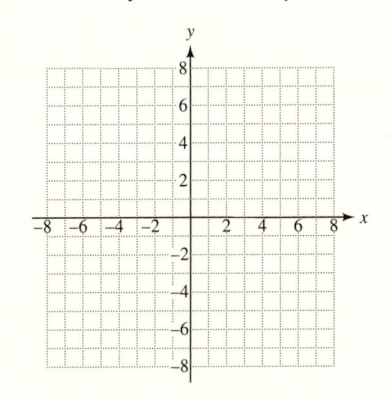

12. $16x^2 + 4y^2 + 32x - 20y + 5 = 0$

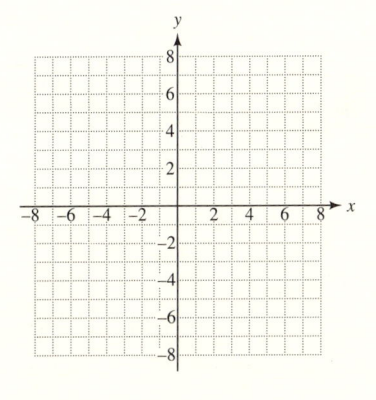

In Exercises 13–15 write each equation in the form $\dfrac{(x-h)^2}{m} + \dfrac{(y-k)^2}{n} = 1$.

13. $8x^2 + 7y^2 - 56 = 0$

14. $9x^2 + 50y^2 + 54x - 300y + 306 = 0$

15. $4x^2 + 9y^2 + 16x + 88 = 0$

16. ENGINEERING An elliptical riding path is to be built on a rectangular piece of property that measures 3 mi by 6 mi. Find an equation for the ellipse if the path is to touch the center of the property line on all sides.

17. AEROSPACE A satellite is to be put into an elliptical orbit around one of the moons of a planet. The moon is a sphere with radius 900 km. Give an equation for the ellipse if the distance of the satellite from the surface of the moon varies from 200 km to 300 km. Estimate the distance traveled by the satellite in one revolution.

18. ENGINEERING A bridge over a canal is a semiellipse. The equation of the ellipse is $\frac{x^2}{16} + \frac{y^2}{9} = 1$ for a coordinate system centered at the middle of the canal and horizontal x-axis across the canal. Solve the equation for y and determine the appropriate vertical clearance for boats 3.0 m from the edge of the canal.

FOR REVIEW

19. Find the equation of the circle with center $(2, -5)$ and passing through $(4, -1)$.

20. Write the equation in standard form.
$$x^2 + y^2 + 4x - 16y + 59 = 0$$

21. ENGINEERING A highway tunnel in the shape of a semicircle has a diameter measuring 30 ft. Find the equation of this semicircle using a coordinate system with the origin at the center line of the roadway. What is the vertical clearance 5 ft from the edge of the tunnel?

ANSWERS: 1. $\frac{x^2}{49} + \frac{y^2}{36} = 1$ 2. $\frac{x^2}{25} + \frac{y^2}{16} = 1$ 3. $\frac{x^2}{16} + \frac{y^2}{25} = 1$ 4. $\frac{x^2}{36} + \frac{y^2}{25} = 1$ 5. $\frac{x^2}{100} + \frac{y^2}{4} = 1$ 6. $\frac{(x-5)^2}{16} + \frac{(y+2)^2}{12} = 1$
7. $\frac{x^2}{25} + \frac{(y-3)^2}{9} = 1$ 8. $\frac{(x-7)^2}{55} + \frac{(y+4)^2}{64} = 1$
9. [graph: ellipse with intercepts $(\pm 7, 0)$ and $(0, \pm 5)$]
10. [graph: circle centered at $(2, -2)$, with $x = 2$, $y = -2$]
11. [graph: circle centered at $(3, -1)$, with $x = 3$, $y = -1$]
12. [graph: ellipse centered at $(-1, 5/2)$, with $x = -1$, $y = 5/2$]

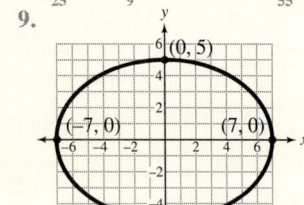

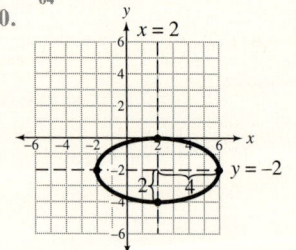

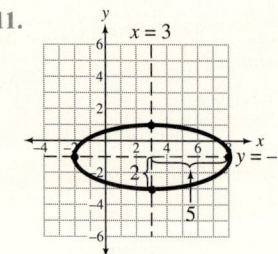

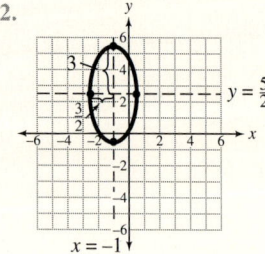

13. $\frac{x^2}{7} + \frac{y^2}{8} = 1$ 14. $\frac{(x+3)^2}{25} + \frac{(y-3)^2}{9/2} = 1$ 15. not possible, there is no graph 16. $\frac{x^2}{9} + \frac{4y^2}{9} = 1$

17. $\frac{x^2}{(1200)^2} + \frac{y^2}{(1100)^2} = 1$; approximately 7200 km 18. $y = \frac{3}{4}\sqrt{16 - x^2}$; 2.9 m 19. $(x-2)^2 + (y+5)^2 = 20$
20. $(x+2)^2 + (y-8)^2 = 9$ 21. $y = \sqrt{225 - x^2}$; 11.2 ft

8.2 EXERCISES B

In Exercises 1–5 find the equation of each ellipse with center at the origin.

1. Intercepts $(\pm 6, 0)$ and $(0, \pm 7)$

2. Foci $(\pm \sqrt{7}, 0)$ and y-intercepts $(0, \pm 4)$

3. Foci $(0, \pm \sqrt{7})$ and y-intercepts $(0, \pm \sqrt{11})$

4. Length of minor axis 6 and foci $(0, \pm 6)$

5. Minor axis intercepts $(\pm 2, 0)$ and passing through $(\sqrt{2}, 2\sqrt{2})$

In Exercises 6–8 find the equation of each ellipse.

6. Center at $(-2, 5)$, $b = 4$, and foci $(-2, 3)$ and $(-2, 7)$

7. Foci $(5, \pm 3)$ and x-intercepts $(4, 0)$ and $(6, 0)$

8. Foci $(-2, -3)$ and $(4, -3)$ and $b = 7$

Graph each ellipse in Exercises 9–12.

9. $\dfrac{x^2}{4} + \dfrac{y^2}{25} = 1$

10. $\dfrac{(x+3)^2}{49} + \dfrac{(y-1)^2}{16} = 1$

11. $4x^2 + 9y^2 - 16x + 36y + 16 = 0$

12. $36x^2 + 25y^2 - 36x - 50y - 191 = 0$

In Exercises 13–15 write each equation in the form $\dfrac{(x-h)^2}{m} + \dfrac{(y-k)^2}{n} = 1$.

13. $10x^2 + 12y^2 - 60 = 0$

14. $16x^2 + 21y^2 - 128x + 210y + 753 = 0$

15. $100x^2 + 4y^2 - 300x + 16y + 241 = 0$

16. **ENGINEERING** A railroad tunnel is in the shape of a semiellipse with horizontal major axis. The height of the tunnel at the center is 30 ft and the vertical clearance must be 25 ft at a point 20 ft from the center. Find an equation for the ellipse.

17. **CARPENTRY** A rectangular board is m by n units where $m > n$. To construct the largest elliptical tabletop the foci of the ellipse must be determined. How far from the short side of the board will the foci be located?

18. A string is connected to the foci in Exercise 17 and pulled taut by a pencil in order to draw the ellipse. What length string must be used?

FOR REVIEW

19. Find the equation of the circle with endpoints of a diameter at $(-3, 2)$ and $(6, 1)$.

20. Write the equation in standard form.
$$36x^2 + 36y^2 - 36x + 24y - 131 = 0$$

21. **ENGINEERING** A canal with cross-section a semicircle is to have a depth of 3 ft at a distance of 4 ft from the center of the waterline in a full canal. Find the equation of the semicircle. What is the maximum depth of the water in the canal?

8.2 EXERCISES C

Find the equation of each set of points described.

1. The ellipse with center at the origin, horizontal major axis, $a = 5b$, and passing through $(7, 2)$.

2. The set of points obtained when each abscissa of the points on $x^2 + y^2 = 16$ is divided by 2.

3. The distance of each point from the line $x = 8$ is twice its distance from the point $(2, 0)$.
   [Answer: $3x^2 + 4y^2 - 48 = 0$]

4. The distance of each point from the line $y = 6$ is twice its distance from the point $(0, 3)$.

8.3 THE HYPERBOLA

STUDENT GUIDEPOSTS

1. Definition of a Hyperbola
2. Equation of a Hyperbola
3. Asymptotes of a Hyperbola
4. Translation of Axes
5. General Form of a Hyperbola
6. Alternate Form of a Hyperbola

1 DEFINITION OF A HYPERBOLA

A *hyperbola* is defined much like an ellipse. The distinction between the two definitions is that the hyperbola is the *difference* of distances instead of the *sum*. One of the more interesting occurrences of a hyperbola is related to the sonic boom of a jet airplane. When a jet breaks the sound barrier, the sonic wave is in the shape of a cone which intersects the surface of the earth in a curve that is roughly hyperbolic as shown in Figure 8.15. Many modern buildings also have cross-sectional structures in the shape of a hyperbola.

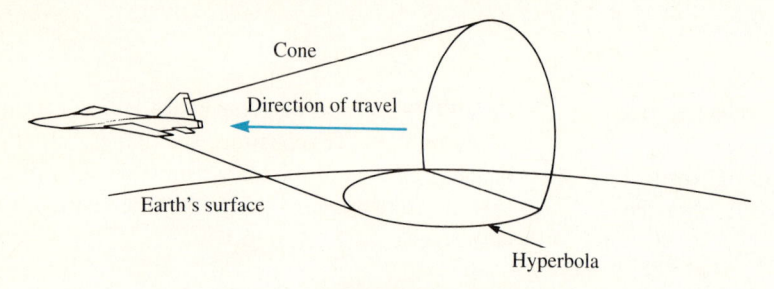

Figure 8.15

The Hyperbola

Let F_1 and F_2 be two fixed points in a plane. A **hyperbola** with **foci** F_1 and F_2 is the collection of all points in the plane the difference of whose distances from F_1 and F_2 is a constant. The midpoint of the line segment joining F_1 and F_2 is the **center** of the hyperbola.

2 EQUATION OF A HYPERBOLA

The equation of a hyperbola is developed by considering the graph given in Figure 8.16.

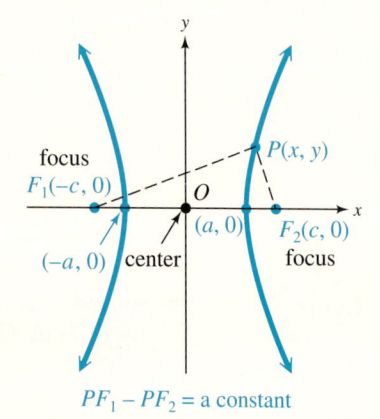

$PF_1 - PF_2 =$ a constant

Figure 8.16 Hyperbola

It can be shown that the difference in the distances is the constant $2a$, and thus

$$PF_1 - PF_2 = 2a.$$

If a point on the other side or **branch** of the hyperbola is chosen, we write

$$PF_2 - PF_1 = 2a.$$

In either case it is true that

$$|PF_1 - PF_2| = 2a.$$

Using this relation and techniques similar to those employed for the ellipse, we can derive the standard form of the equation of a hyperbola. Note that $a < c$ for the hyperbola, and we define b using

$$b^2 = c^2 - a^2.$$

The points $(a, 0)$ and $(-a, 0)$ are the **vertices** (plural of **vertex**) and the line segment joining the vertices is the **transverse axis** of the hyperbola. The line through $(0, -b)$ and $(0, b)$ is the **conjugate axis.**

Hyperbola with Horizontal Transverse Axis

The equation

$$\frac{x^2}{a^2} - \frac{y^2}{b^2} = 1$$

has as its graph a hyperbola with center at $(0, 0)$, foci at $(\pm c, 0)$, and vertices at $(\pm a, 0)$. a, b, and c are related by $b^2 = c^2 - a^2$.

EXAMPLE 1   Hyperbola (Horizontal Transverse Axis)

Find the equation of a hyperbola with foci $(\pm 5, 0)$ and $b = 2$.
Since $c = 5$ and $b = 5$ we can determine a.

$$a^2 = c^2 - b^2$$
$$a^2 = 5^2 - 2^2 = 21$$

Thus, the vertices are $(\sqrt{21}, 0)$ and $(-\sqrt{21}, 0)$ and the equation is

$$\frac{x^2}{21} - \frac{y^2}{4} = 1.$$

Practice Exercise 1

Find the equation of a hyperbola with foci $(\pm 3, 0)$ and $a = 2$.

Answer: $\frac{x^2}{4} - \frac{y^2}{5} = 1$

Note Graphing a hyperbola requires a different approach from that for the ellipse. Notice that the hyperbola has only two intercepts. Setting $y = 0$, we show that the vertices are the x-intercepts.

$$\frac{x^2}{a^2} - \frac{0}{b^2} = 1$$
$$x = \pm a$$

However, if we set $x = 0$, the equation $-\frac{y^2}{b^2} = 1$ has no solution, and there are no y-intercepts.

3 ASYMPTOTES OF A HYPERBOLA

To gain more information for graphing, solve

$$\frac{x^2}{a^2} - \frac{y^2}{b^2} = 1$$

for y.

$$\frac{y^2}{b^2} = \frac{x^2}{a^2} - 1 = \frac{x^2 - a^2}{a^2} \qquad \text{Common denominator is } a^2$$

$$y^2 = \frac{b^2}{a^2}(x^2 - a^2) \qquad \text{Multiply by } b^2$$

$$y = \pm \frac{b}{a}\sqrt{x^2 - a^2} \qquad \text{Take square root of both sides}$$

For $|x|$ very large $x^2 - a^2 \approx x^2$ and the hyperbola will be close to the lines

$$y = \frac{b}{a}x \quad \text{and} \quad y = -\frac{b}{a}x.$$

We call these lines **asymptotes**.

To graph a hyperbola, draw the asymptotes as shown in Figure 8.17. The best procedure is to construct a rectangle with sides parallel to the axes and passing through the points $(a, 0)$, $(-a, 0)$, $(0, b)$, and $(0, -b)$. The asymptotes will then pass through the corners of this rectangle. Now plot the vertices $(a, 0)$ and $(-a, 0)$. The hyperbola passes through the vertices and approaches the asymptotes.

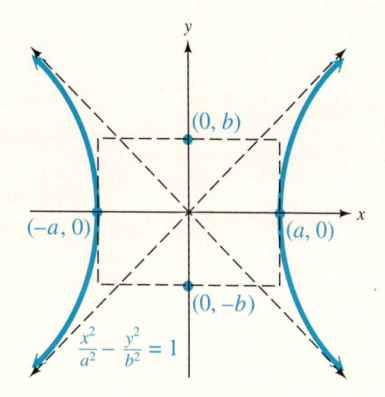

Figure 8.17 Hyperbola (Horizontal Transverse Axis)

EXAMPLE 2 GRAPHING A HYPERBOLA

Graph the hyperbola with equation $\frac{x^2}{4} - \frac{y^2}{9} = 1$.

Since the equation can be written $\frac{x^2}{2^2} - \frac{y^2}{3^2} = 1$, $a = 2$ and $b = 3$. (Note that a need not be greater than b as was the case with the ellipse.) Now construct the rectangle with sides through the points $(2, 0)$, $(-2, 0)$, $(0, 3)$, and $(0, -3)$. Draw the asymptotes through the corners of the rectangle. The hyperbola passes through the vertices $(2, 0)$ and $(-2, 0)$ and approaches the asymptotes as in Figure 8.18.

PRACTICE EXERCISE 3

Graph the hyperbola with equation

$$\frac{x^2}{16} - \frac{y^2}{4} = 1.$$

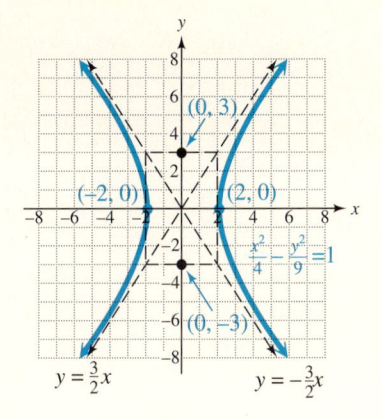

Figure 8.18

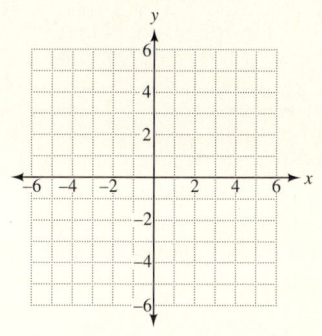

Answer: The hyperbola opens left and right with x-intercepts $(\pm 4, 0)$ and asymptotes the diagonals of the rectangle through $(\pm 4, 0)$ and $(0, \pm 2)$.

Only hyperbolas with transverse axis along the x-axis and conjugate axis along the y-axis have been discussed in the preceding paragraphs. If the y^2-term is positive and the x^2-term negative in the standard form, the hyperbola will have transverse axis along the y-axis. The standard form of the equation of these hyperbolas is given in the next theorem.

Hyperbola with Vertical Transverse Axis

The equation

$$\frac{y^2}{a^2} - \frac{x^2}{b^2} = 1$$

has as its graph a hyperbola with center at $(0, 0)$, foci at $(0, \pm c)$, and vertices at $(0, \pm a)$. a, b, and c are related by $b^2 = c^2 - a^2$.

The asymptotes for these hyperbolas are $y = \pm \dfrac{a}{b} x$, but the procedure for constructing them using the diagonals of the related rectangle is exactly the same. See Figure 8.19.

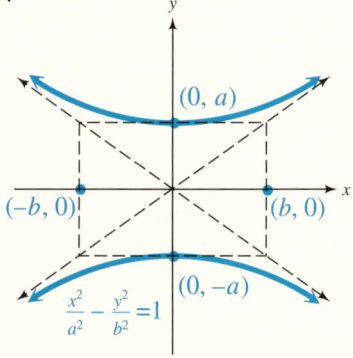

Figure 8.19 Hyperbola (Vertical Transverse Axis)

EXAMPLE 3 HYPERBOLA (VERTICAL TRANSVERSE AXIS)

Graph the hyperbola with foci at $(0, \pm\sqrt{13})$ and vertices $(0, \pm 3)$.
  Knowing $c = \sqrt{13}$ and $a = 3$, find b.

$$b^2 = c^2 - a^2 = (\sqrt{13})^2 - (3)^2 = 13 - 9 = 4$$
$$b = 2$$

PRACTICE EXERCISE 3

Graph the hyperbola with foci $(0, \pm\sqrt{29})$ and vertices $(0, \pm 2)$.

The equation of the hyperbola is

$$\frac{y^2}{9} - \frac{x^2}{4} = 1.$$

Construct a rectangle through $(2, 0)$, $(-2, 0)$, $(0, 3)$, and $(0, -3)$ and draw the asymptotes through the corners, $(2, 3)$, $(-2, 3)$, $(-2, -3)$, and $(2, -3)$. The hyperbola, shown in Figure 8.20, passes through vertices $(0, \pm 3)$ and approaches the asymptotes.

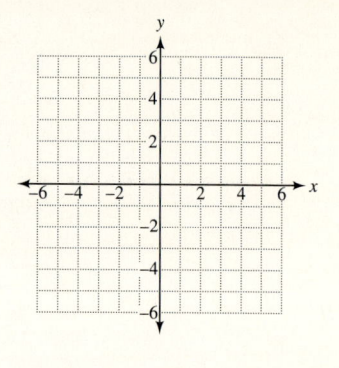

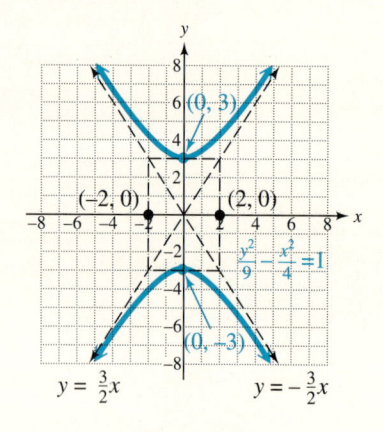

Figure 8.20

Answer: The equation is $\frac{y^2}{4} - \frac{x^2}{25} = 1$, and the graph opens up and down through intercepts $(0, \pm 2)$ with asymptotes $y = \pm \frac{2}{5} x$.

4 TRANSLATION OF AXES

Using an argument similar to that for ellipses, we can translate axes and discuss hyperbolas centered at points other than the origin and obtain the two *standard forms*.

Standard Forms of the Equation of a Hyperbola

The **standard form** of the equation of a hyperbola with center (h, k) and horizontal transverse axis is

$$\frac{(x - h)^2}{a^2} - \frac{(y - k)^2}{b^2} = 1.$$

The **standard form** of the equation of a hyperbola with center (h, k) and vertical transverse axis is

$$\frac{(y - k)^2}{a^2} - \frac{(x - h)^2}{b^2} = 1.$$

EXAMPLE 4 GRAPHING A HYPERBOLA

Graph the hyperbola

$$\frac{(x + 1)^2}{9} - \frac{(y - 2)^2}{16} = 1.$$

PRACTICE EXERCISE 4

Graph the hyperbola

$$\frac{(y - 2)^2}{16} - \frac{(x + 1)^2}{9} = 1.$$

Since $$\frac{(x-(-1))^2}{3^2} - \frac{(y-2)^2}{4^2} = 1,$$

$(h, k) = (-1, 2)$, $a = 3$, and $b = 4$. First locate the center, $(-1, 2)$, and then sketch the lines $x = -1$ and $y = 2$. Think of these lines as new coordinate axes, with the given hyperbola centered at the origin of this system. We use $a = 3$ and $b = 4$, relative to these new axes, to sketch the rectangle and the asymptotes. Since the hyperbola has a horizontal transverse axis, we know it passes through the points 3 units right and left of $(-1, 2)$, that is, through $(2, 2)$ and $(-4, 2)$. Using this information we can sketch the graph in Figure 8.21.

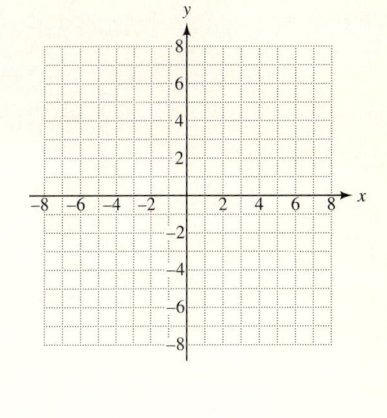

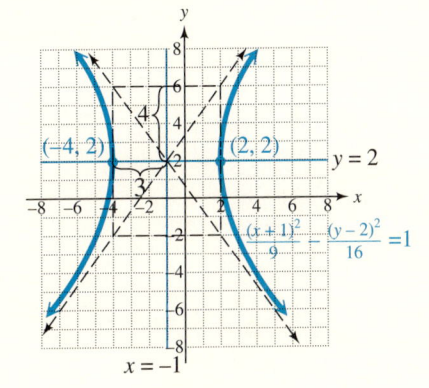

Figure 8.21

Answer: The graph opens up and down through $(-1, 6)$ and $(-1, -2)$ and with the same asymptotes as shown in Figure 8.21.

5 GENERAL FORM OF A HYPERBOLA

As was the case with ellipses, if the standard form of the equation of a hyperbola is expanded, the result is the *general form of the hyperbola*. For example, expanding the equation in Example 4 gives the general form

$$16x^2 - 9y^2 + 32x + 36y - 164 = 0.$$

All of the hyperbolas discussed in this section can be written in the **general form**

$$Ax^2 + Cy^2 + Dx + Ey + F = 0,$$

where A and C have opposite signs. If the general form is given, complete the squares in both x and y to find the standard form as shown in the next example.

EXAMPLE 5 FINDING THE STANDARD FORM

Write the equation $25x^2 - 9y^2 - 100x - 54y - 206 = 0$ in standard form.

$$25x^2 - 100x \quad - 9y^2 - 54y \quad = 206$$
$$25(x^2 - 4x \quad) - 9(y^2 + 6y \quad) = 206$$
$$25(x^2 - 4x + 4) - 9(y^2 + 6y + 9) = 206 + 100 - 81$$
$$25(x-2)^2 - 9(y+3)^2 = 225$$
$$\frac{(x-2)^2}{9} - \frac{(y+3)^2}{25} = 1$$
$$\frac{(x-2)^2}{3^2} - \frac{(y+3)^2}{5^2} = 1$$

PRACTICE EXERCISE 5

Write the equation $-4x^2 + y^2 - 40x - 2y - 115 = 0$ in standard form.

Answer: $\frac{(y-1)^2}{16} - \frac{(x+5)^2}{4} = 1$

6 ALTERNATE FORM OF A HYPERBOLA

The hyperbolas we have studied thus far open left and right or up and down. Considering the most general quadratic equation

$$Ax^2 + Bxy + Cy^2 + Dx + Ey + F = 0$$

leads to another type of equation with a hyperbola for its graph having the axes as asymptotes. If $A = 0$, $C = 0$, $D = 0$, and $E = 0$, the resulting equation has the form $Bxy + F = 0$, which can be simplified to $xy = c$, with c a constant real number. When $c > 0$, the graph is a hyperbola with the axes as asymptotes and the two branches in quadrants I and III. If $c < 0$, the graph is a hyperbola with the axes as asymptotes and the two branches in quadrants II and IV.

EXAMPLE 6 GRAPHING A HYPERBOLA (ALTERNATE FORM)

Graph $2xy - 6 = 0$.

Simplifying, we have $xy = 3$, which is graphed by plotting points in Figure 8.22. The graph is a hyperbola with the coordinate axes as asymptotes. Observe that the graph appears to be "rotated" with respect to the graphs we have studied previously.

x	y
1	3
2	3/2
3	1
−1	−3
−2	−3/2
−3	−1

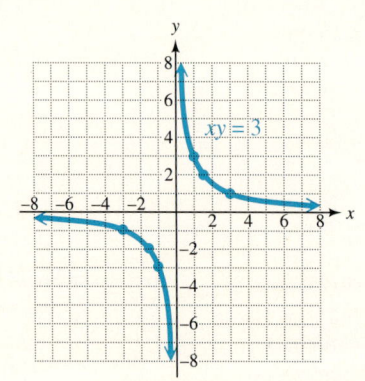

Figure 8.22

PRACTICE EXERCISE 6

Graph $2xy + 6 = 0$.

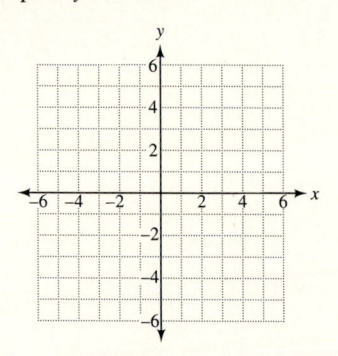

Answer: The graph is similar to the one in Figure 8.22, but has the two branches in quadrants II and IV.

8.3 EXERCISES A

Find the standard form of the equation of each hyperbola in Exercises 1–6.

1. Foci at $(\pm 5, 0)$ and vertices at $(\pm 3, 0)$

2. $a = b = 3$, transverse axis along the y-axis, and center at the origin

3. Foci $(\pm 7, 0)$ and length of transverse axis is 10

4. Vertices $(0, \pm 6)$ and asymptotes $y = \pm \dfrac{3}{4} x$

5. Center at $(-3, 2)$, foci $(-3, 8)$ and $(-3, -4)$, and vertices $(-3, 6)$ and $(-3, -2)$

6. Vertices $(\pm 4, 0)$ and passing through $(8, 6)$

−10.0 m the width is 15.0 m, the points (−7.5, −10) and (7.5, −10) are on the parabola. Use (7.5, −10) in the equation to find p.

$$x^2 = 4py$$
$$(7.5)^2 = 4p(-10)$$
$$p \approx -1.4$$

Thus, the light should be placed 1.4 m down from the top of the dome.

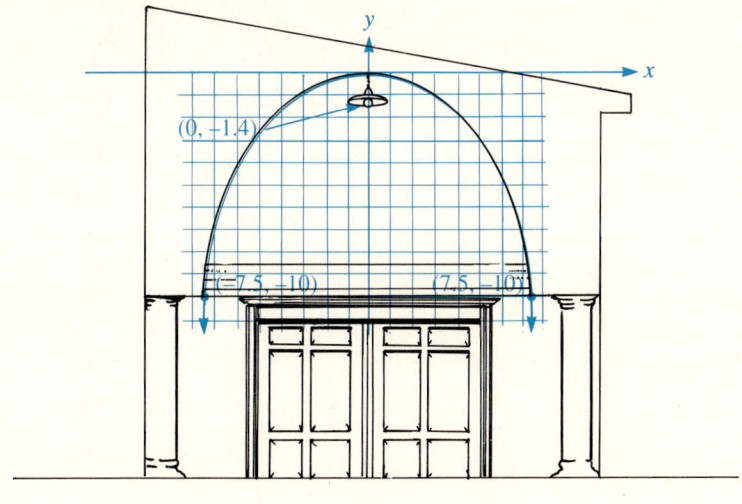

Figure 8.28

Answer: 3.1 m down from the top of the dome

⑤ GENERAL FORM OF A PARABOLA

If the equation of a parabola is expanded, the *general form* of the equation results. For example, consider the equation

$$(y + 1)^2 = 5(x - 3).$$

Expanding, we have

$$y^2 + 2y + 1 = 5x - 15$$
$$y^2 - 5x + 2y + 16 = 0.$$

The **general form of a parabola** can always be written in one of the two forms $Ax^2 + Dx + Ey + F = 0$ ($A, E \neq 0$) or $Cy^2 + Dx + Ey + F = 0$ ($C, D \neq 0$).

To convert from general form to standard form, complete the square on the squared variable.

EXAMPLE 6 FINDING THE STANDARD FORM

Write the equation $y^2 - 2x - 2y - 5 = 0$ in standard form and give the vertex, focus, and directrix. Graph the equation.

Complete the square on y.

$$\begin{aligned} y^2 - 2y &= 2x + 5 & &\text{Isolate the } y\text{-terms} \\ y^2 - 2y + 1 &= 2x + 5 + 1 & &\text{Add 1 to both sides} \\ (y - 1)^2 &= 2x + 6 \\ (y - 1)^2 &= 2(x + 3) \end{aligned}$$

PRACTICE EXERCISE 6

Write the equation $x^2 - 4x + 4y + 12 = 0$ in standard form and give the vertex, focus, and directrix. Graph the equation.

The parabola opens to the right and the vertex is $(-3, 1)$. Since $4p = 2$, $p = \frac{1}{2}$. Thus, the focus is $\frac{1}{2}$ unit to the right of $(-3, 1)$. The focus is

$$(h + p, k) = \left(-3 + \frac{1}{2}, 1\right) = \left(-\frac{5}{2}, 1\right).$$

The directrix is the vertical line $x = h - p = -3 - \frac{1}{2} = -\frac{7}{2}$. Plot the additional points $(-1, -1)$ and $(-1, 3)$ to graph the parabola. See Figure 8.29.

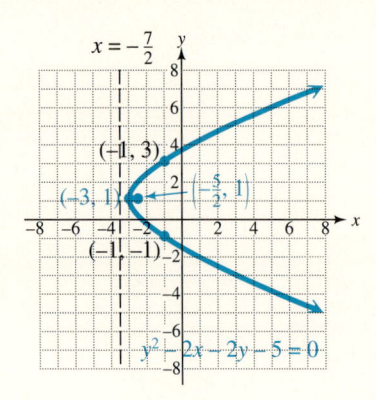

Figure 8.29

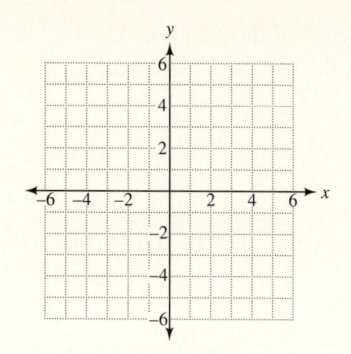

Answer: $(x - 2)^2 = -4(y + 2)$; vertex: $(2, -2)$; focus: $(2, -3)$; directrix: $y = -1$; the graph opens down and passes through $(0, -3)$ and $(4, -3)$

8.4 EXERCISES A

In Exercises 1–6, give the equation of each parabola.

1. Vertex at the origin and focus $\left(0, \frac{3}{2}\right)$

2. Vertex at $(0, 0)$ and focus $\left(\frac{7}{2}, 0\right)$

3. Directrix $x = -3$ and focus $(3, 0)$

4. Directrix $y = -1$ and focus $(4, 5)$

5. Vertex $(4, -6)$, axis of symmetry $x = 4$, and passing through $(0, -8)$

6. Vertex $(8, -7)$, axis of symmetry parallel to the x-axis, and passing through $(6, -8)$

Sketch the graph of each parabola in Exercises 7–10.

7. $y^2 = 2x$

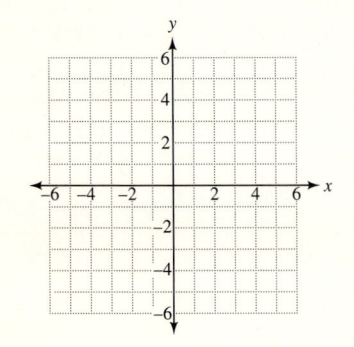

8. $(x - 2)^2 = -(y + 1)$

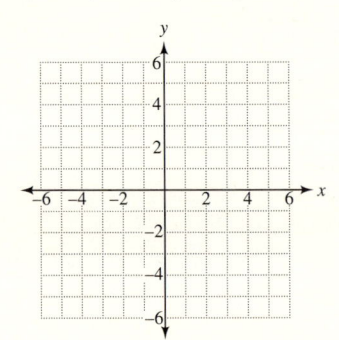

9. $x^2 - 4y + 8 = 0$

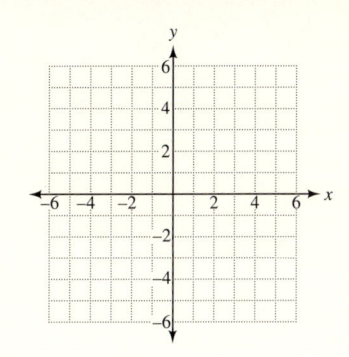

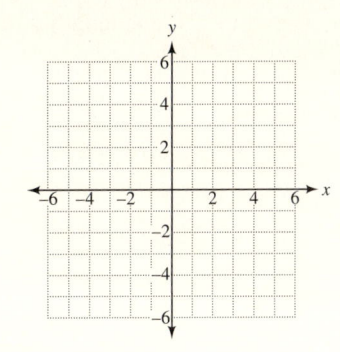

10. $y^2 + 4x + y - \dfrac{47}{4} = 0$

In Exercises 11–13, write each equation in the standard form
$$(y - k)^2 = 4p(x - h) \quad or \quad (x - h)^2 = 4p(y - k)$$
and give the vertex, focus, and directrix.

11. $y^2 - 12x + 24 = 0$

12. $x^2 + 8x + 9y + 25 = 0$

13. $4x^2 - 12x - 2y - 3 = 0$

In Exercises 14–17, complete the squares to determine if the equation represents a circle, ellipse, hyperbola, parabola, or degenerate conic.

14. $5x^2 - 2y^2 + x - 7 = 0$

15. $-x^2 - y^2 + x + y + 12 = 0$

16. $4x^2 - 49y^2 + 8x + 4 = 0$

17. $20x^2 + 10y^2 - 5x + 15y - 75 = 0$

18. ARCHITECTURE A modern building has an entry in the shape of a parabolic arch which is 32.5 ft high and 24.2 ft wide at the base. Find an equation for the parabola (see the figure) if the vertex is put at the origin of the coordinate system.

19. ENGINEERING A tunnel is in the shape of a parabola. The maximum height is 12.8 m and it is 10.2 m wide at the base. What is the vertical clearance 1.5 m from the edge of the tunnel?

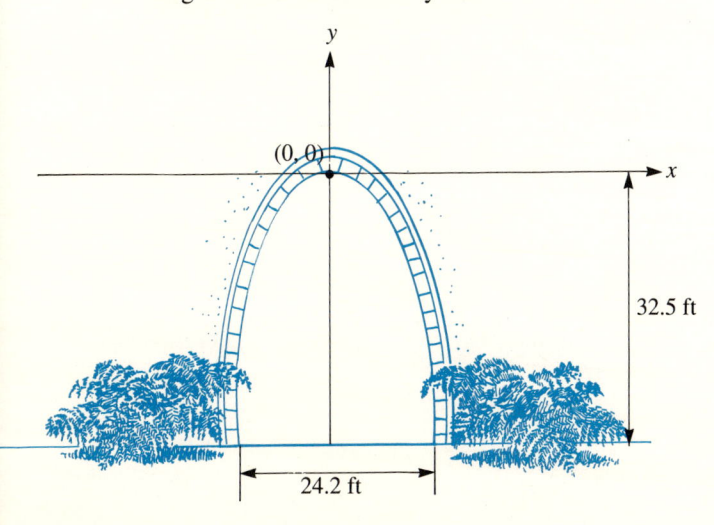

572 CHAPTER 8 TOPICS IN ANALYTIC GEOMETRY

FOR REVIEW

20. Give the equation of the hyperbola with center $(4, -2)$, foci $(4, -7)$ and $(4, 3)$, and length of the transverse axis 6.

21. Write the equation in standard form and identify the graph.

$$6x^2 - 5y^2 - 24x - 30y - 51 = 0$$

In Exercises 22–24, sketch the graph of each equation. These problems will assist you in the material presented in the next section.

22. $(x - 2)^2 + y^2 = 4$

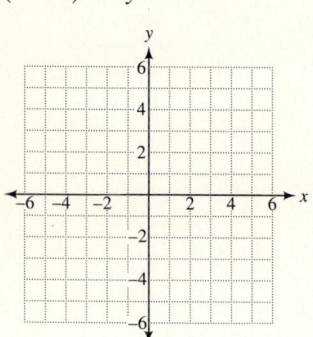

23. $y = \log_2 x$

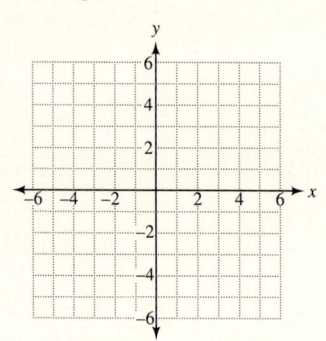

24. $y = |x - 1| + 1$

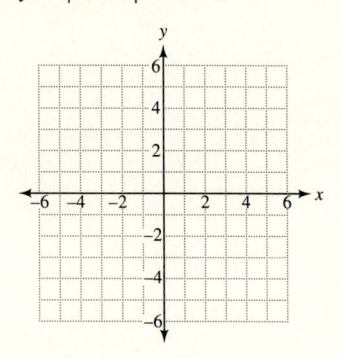

ANSWERS: 1. $x^2 = 6y$ 2. $y^2 = 14x$ 3. $y^2 = 12x$ 4. $(x - 4)^2 = 12(y - 2)$ 5. $(x - 4)^2 = -8(y + 6)$ 6. $(y + 7)^2 = -\frac{1}{2}(x - 8)$ 7. 8. 9. 10.

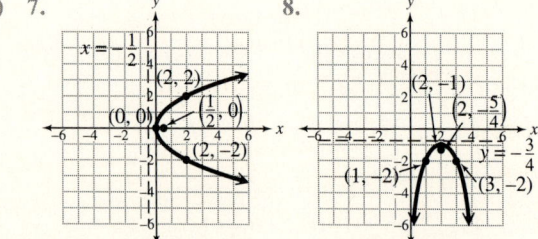

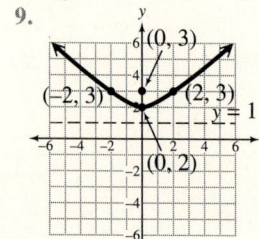

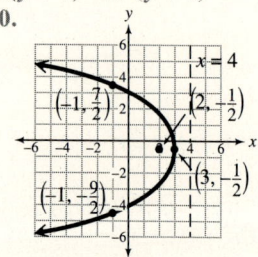

11. $y^2 = 12(x - 2)$; $(2, 0)$; $(5, 0)$; $x = -1$ 12. $(x + 4)^2 = -9(y + 1)$; $(-4, -1)$; $(-4, -\frac{13}{4})$; $y = \frac{5}{4}$ 13. $\left(x - \frac{3}{2}\right)^2 = \frac{1}{2}(y + 6)$; $\left(\frac{3}{2}, -6\right)$; $\left(\frac{3}{2}, -\frac{47}{8}\right)$; $y = -\frac{49}{8}$ 14. hyperbola 15. circle 16. degenerate conic (graph is two intersecting lines) 17. ellipse 18. $x^2 = -4.5y$ 19. 6.4 m 20. $\frac{(y + 2)^2}{9} - \frac{(x - 4)^2}{16} = 1$ 21. $\frac{(x - 2)^2}{5} - \frac{(y + 3)^2}{6} = 1$; hyperbola

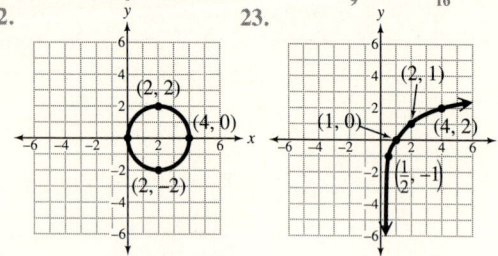

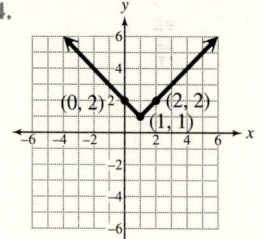

22. 23. 24.

8.4 EXERCISES B

In Exercises 1–6, give the equation of each parabola.

1. Vertex at $(0, 0)$ and focus $(-3, 0)$

2. Vertex at the origin and focus $(0, -1)$

3. Directrix $y = 5$ and focus $(0, -5)$

4. Directrix $x = 8$ and focus $(-2, 6)$

5. Vertex $(-5, -1)$, axis of symmetry $y = -1$, and passing through $(-2, -4)$

6. Vertex $\left(\frac{1}{2}, 1\right)$, axis of symmetry parallel to the y-axis, and passing through $\left(-\frac{9}{2}, 6\right)$

Sketch the graph of each parabola in Exercises 7–10.

7. $x^2 = -4y$

8. $(y - 1)^2 = 8(x + 3)$

9. $x^2 - 4x - 2y = 0$

10. $y^2 + 4x - 6y + 15 = 0$

In Exercises 11–13, write each equation in the standard form

$$(y - k)^2 = 4p(x - h) \quad \text{or} \quad (x - h)^2 = 4p(y - k)$$

and give the vertex, focus, and directrix.

11. $x^2 - 10x - 6y + 25 = 0$

12. $y^2 + 5x - 16y + 74 = 0$

13. $25y^2 - 75x - 20y - 11 = 0$

In Exercises 14–17, complete the squares to determine if the equation represents a circle, ellipse, hyperbola, parabola, or degenerate conic.

14. $-3x^2 + 2x - y + 8 = 0$

15. $4x^2 + 4y^2 - x + 3y - 15 = 0$

16. $64x^2 + 25y^2 - 200y + 400 = 0$

17. $36x^2 - 100y^2 - 36x - 400y - 1291 = 0$

18. ASTRONOMY A radiotelescope has a parabolic surface. If it is 1.2 m deep and 8.4 m wide, how far is the focus from the vertex? See the figure.

19. AGRICULTURE A cross-section of an irrigation canal is a parabola. If the surface of the water is 34.6 ft wide and the canal is 22.0 ft deep at the center, how deep is it 5.0 ft from the edge?

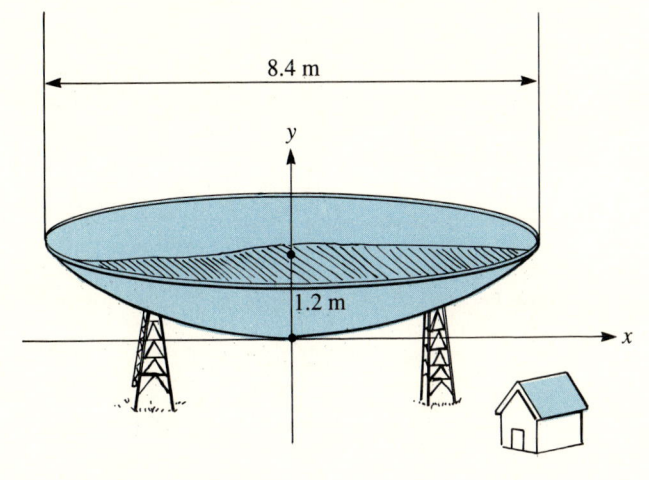

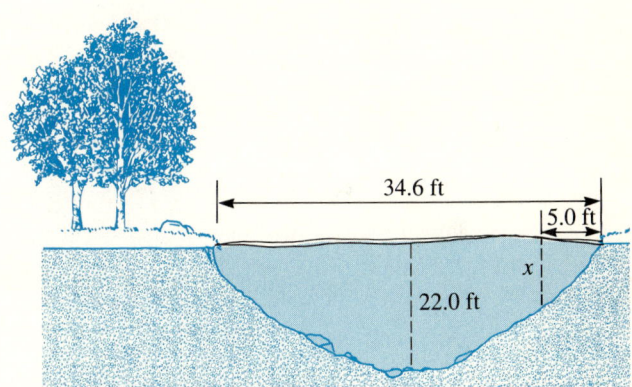

FOR REVIEW

20. Give the equation of the hyperbola with vertices $(\pm 5, 0)$ and passing through $(10, 6)$.

21. Write the equation in standard form and identify the graph.

$$10y^2 - 8x^2 - 144x + 100y - 478 = 0$$

In Exercises 22–24, sketch the graph of each equation. These problems will assist you in the material presented in the next section.

22. $\dfrac{x^2}{4} + \dfrac{y^2}{9} = 1$

23. $\dfrac{x^2}{4} - \dfrac{y^2}{9} = 1$

24. $y = e^x$

8.4 EXERCISES C

1. Construct the graph of the parabola $y^2 = 4px$ ($p > 0$) and graph its directrix. Construct a line through the focus parallel to the directrix. From the points of intersection of this line and the parabola, draw line segments to the point of intersection of the axis of the parabola and the directrix. Prove that these line segments are perpendicular.

2. Find the equation of the set of points each of which has distance from $(3, -4)$ that is 4 units less than its distance from the line $x + 5 = 0$.
[Answer: $y^2 + 8y - 8x + 24 = 0$]

3. Find the equation of the parabola with vertex at the origin, axis the x-axis, and passing through the point (r, s).

4. Use the definition to find the equation of the parabola with directrix $x + 2y - 1 = 0$ and focus at the origin. [*Hint:* The distance from a point (x_1, y_1) to a line $ax + by + c = 0$ is given by $\dfrac{|ax_1 + by_1 + c|}{\sqrt{a^2 + b^2}}$.]

8.5 NONLINEAR SYSTEMS

STUDENT GUIDEPOSTS

1. Interpreting Nonlinear Systems Graphically
2. The Substitution Method
3. The Elimination Method
4. More Complex Second-Degree Systems
5. Systems Involving Other Functions
6. Graphing Systems of Nonlinear Inequalities

1 INTERPRETING NONLINEAR SYSTEMS GRAPHICALLY

When we studied systems of two linear equations, the graphs of the equations gave useful information about the solutions. The same is true for **nonlinear systems**. A solution to a nonlinear system such as

$$x^2 + y^2 = 25$$
$$x + y = 1$$

is an ordered pair of numbers that satisfies both equations. The points of intersection of the two graphs correspond to real-number solutions. In the system above, the graph of the first equation is a circle, and the graph of the second is a straight line. In general, the graphs of a circle and a line can be related in one of three different ways, as shown in Figure 8.30.

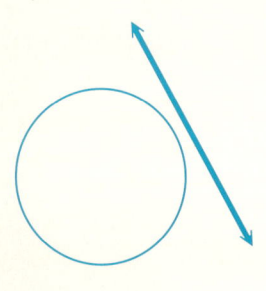

no points of intersection
no real solutions

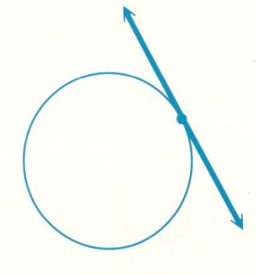

one point of intersection
one real solution

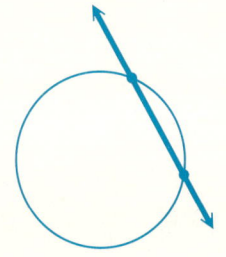

two points of intersection
two real solutions

Figure 8.30 Intersecting a Circle and a Line

If we graph both of the above equations in the same coordinate system, as in Figure 8.31, the points of intersection appear to have coordinates $(-3, 4)$ and $(4, -3)$. By substitution, it is easy to show that these pairs of numbers satisfy both equations, hence are solutions to the system.

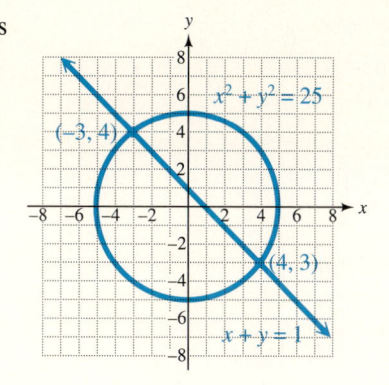

Figure 8.31

② THE SUBSTITUTION METHOD

It is generally better to solve systems algebraically rather than graphically. When one equation is a first-degree equation, solve it for one of the variables and substitute the result into the other equation. Suppose we use the *substitution method* to solve the system given above. Solve $x + y = 1$ for y and substitute in $x^2 + y^2 = 25$.

$$x^2 + (1 - x)^2 = 25 \quad y = 1 - x$$
$$x^2 + 1 - 2x + x^2 = 25$$
$$2x^2 - 2x - 24 = 0$$
$$x^2 - x - 12 = 0 \quad \text{Divide out 2}$$
$$(x - 4)(x + 3) = 0$$

Setting each factor equal to zero gives the following.

$$x - 4 = 0 \quad \text{or} \quad x + 3 = 0$$
$$x = 4 \quad\quad\quad x = -3$$
$$y = 1 - x \quad\quad y = 1 - x$$
$$= 1 - 4 = -3 \quad = 1 - (-3) = 4$$

The solutions are $(4, -3)$ and $(-3, 4)$, obtained previously by graphing.

EXAMPLE 1 USING THE SUBSTITUTION METHOD

Solve. $xy = -5$
$x - y = 2$

The first equation is a hyperbola, and the second equation is a line. Three possibilities for the graphs are shown in Figure 8.32.

PRACTICE EXERCISE 1

Solve. $x^2 + y^2 = 2$
$xy = 1$

[*Hint:* Solve the second equation for y and substitute into the first.]

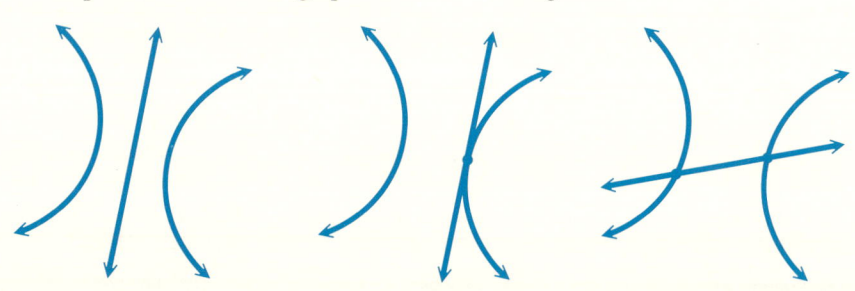

no points of intersection
no real solutions

one point of intersection
one real solution

two points of intersection
two real solutions

Figure 8.32 Intersecting a Hyperbola and a Line

$$x = y + 2 \qquad \text{Solve } x - y = 2 \text{ for } x$$
$$(y + 2)y = -5 \qquad \text{Substitute } y + 2 \text{ for } x \text{ in } xy = -5$$
$$y^2 + 2y = -5$$
$$y^2 + 2y + 5 = 0$$
$$y = \frac{-2 \pm \sqrt{4 - 4(1)(5)}}{2}$$
$$= \frac{-2 \pm \sqrt{-16}}{2}$$
$$= \frac{-2 \pm 4i}{2} = -1 \pm 2i$$

In this case there are no real solutions; the two graphs do not intersect. However, the complex-number solutions are $(1 + 2i, -1 + 2i)$ and $(1 - 2i, -1 - 2i)$.

Answer: $(1, 1), (-1, -1)$

If we look at the solutions to a system of two second-degree equations graphically, several possibilities exist. For example, the system in Practice Exercise 1 has as graphs a circle and a hyperbola, which intersect in only two points, $(1, 1)$ and $(-1, -1)$. Some of the many possibilities for systems with two second-degree equations are shown in Figure 8.33 using a hyperbola and an ellipse.

no points of intersection
no real solutions

one point of intersection
one real solution

two points of intersection
two real solutions

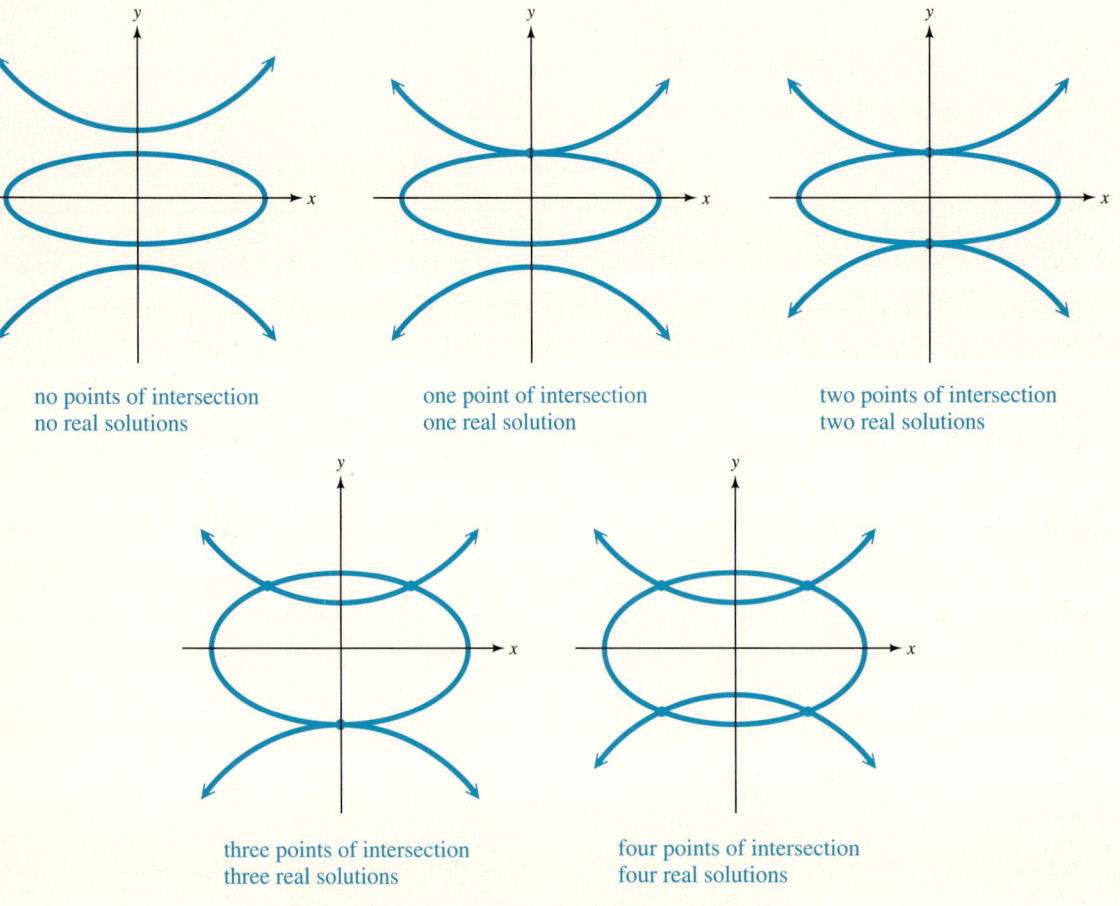

three points of intersection
three real solutions

four points of intersection
four real solutions

Figure 8.33 Intersecting a Hyperbola and an Ellipse

③ THE ELIMINATION METHOD

Some systems of two second-degree equations can be solved using substitution as in Practice Exercise 1. Others require an *elimination method* similar to the one we used for linear systems. We illustrate this technique in the next example.

EXAMPLE 2 Using the Elimination Method

Solve. $x^2 + y^2 = 14$
$x^2 - y^2 = 4$

Add to eliminate y.　　$2x^2 = 18$
$x^2 = 9$
$x = \pm 3$

Substitute 3 for x in the first equation.

$$(3)^2 + y^2 = 14$$
$$9 + y^2 = 14$$
$$y^2 = 5$$
$$y = \pm\sqrt{5}$$

Two of the solutions are $(3, \sqrt{5})$ and $(3, -\sqrt{5})$. When -3 is substituted for x in the first equation we obtain $y = \pm\sqrt{5}$. Thus, there are four solutions: $(3, \sqrt{5})$, $(3, -\sqrt{5})$, $(-3, \sqrt{5})$, and $(-3, -\sqrt{5})$. Check these in both equations.

PRACTICE EXERCISE 2

Solve.

$5x^2 - 2y^2 = 2$
$2x^2 + 3y^2 = 35$

Answer: $(2, 3)$, $(2, -3)$, $(-2, 3)$, and $(-2, -3)$

④ MORE COMPLEX SECOND-DEGREE SYSTEMS

For some second-degree systems, substitution or elimination may not be applicable. Sometimes it is possible to transform the system to an equation with constant term equal to zero, then factor, and use the zero-product rule. This technique is illustrated in the next example.

EXAMPLE 3 A More Complex System

Solve. $x^2 + 15xy + 9y^2 = -5$
$7x^2 + 9xy + 27y^2 = 25$

Multiply the first equation by 5 and add the result to the second equation. The resulting equation has constant term zero.

$$12x^2 + 84xy + 72y^2 = 0$$
$$x^2 + 7xy + 6y^2 = 0 \quad \text{Divide by 12}$$
$$(x + 6y)(x + y) = 0 \quad \text{Factor}$$
$$x = -6y \quad \text{or} \quad x = -y$$

$(-6y)^2 + 15(-6y)y + 9y^2 = -5$　　$(-y)^2 + 15(-y)y + 9y^2 = -5$　　Substitute for x in the first equation
$36y^2 - 90y^2 + 9y^2 = -5$　　$y^2 - 15y^2 + 9y^2 = -5$
$-45y^2 = -5$　　$-5y^2 = -5$
$y^2 = \dfrac{1}{9}$　　$y^2 = 1$
$y = \pm\dfrac{1}{3}$　　$y = \pm 1$

PRACTICE EXERCISE 3

Solve.

$x^2 + 2xy - 7y^2 = 4$
$x^2 - 4xy - y^2 = -2$

$$x = -6y = -6\left(\frac{1}{3}\right) = -2 \qquad x = -y = -(+1) = -1$$

$$x = -6y = -6\left(-\frac{1}{3}\right) = 2 \qquad x = -y = -(-1) = 1$$

Substitute for y to find the numerical value of x

The solutions are $\left(-2, \frac{1}{3}\right), \left(2, -\frac{1}{3}\right), (-1, 1),$ and $(1, -1)$.

Answer: $\left(\frac{3\sqrt{2}}{2}, \frac{\sqrt{2}}{2}\right), \left(-\frac{3\sqrt{2}}{2}, -\frac{\sqrt{2}}{2}\right),$ $\left(-\frac{i\sqrt{2}}{2}, \frac{i\sqrt{2}}{2}\right), \left(\frac{i\sqrt{2}}{2}, -\frac{i\sqrt{2}}{2}\right)$

5 SYSTEMS INVOLVING OTHER FUNCTIONS

A nonlinear system may involve functions other than second-degree equations, such as logarithmic or exponential functions. Sometimes these can be solved in a manner similar to the ones we have studied.

EXAMPLE 4 A NONLINEAR SYSTEM (EXPONENTIALS)

Solve. $2e^x + y = 4$
$e^x + 3y = 7$

Multiply the first equation by 3 and subtract the second to obtain

$$5e^x = 5$$
$$e^x = 1.$$

Then $x = 0$, and substituting in the first equation,

$$2e^0 + y = 4$$
$$2 + y = 4$$
$$y = 2.$$

The solution is $(0, 2)$. This system could also be solved by substitution. Solve the first equation for y ($y = 4 - 2e^x$) and substitute in the second to obtain $-5e^x = -5$, which also reduces to $e^x = 1$.

PRACTICE EXERCISE 4

Solve.

$$y + \ln x = 3$$
$$y - \ln x^2 = 0$$

Answer: $(e, 2)$

Note For some systems of nonlinear inequalities, it is very difficult, if not impossible, to find solutions using algebraic techniques. These systems are often studied graphically using a graphing calculator or graphing software on a computer. Solutions can be approximated by this technique. For more information on this subject, see the Chapter 8 Extension Project.

6 GRAPHING SYSTEMS OF NONLINEAR INEQUALITIES

The technique for graphing nonlinear systems of inequalities parallels that for graphing linear systems of inequalities. We can temporarily replace the inequality symbol with an equal sign and graph the resulting equation with a solid ($\leq$ or $\geq$) or dashed ($<$ or $>$) curve. Test points selected from the regions in the plane determined by the boundary curve are used to identify the solution to the inequality. The region common to all solutions is the graph of the system. The four graphs in Figure 8.34 illustrate some of the possible nonlinear inequalities that might be found in a system.

8.5 NONLINEAR SYSTEMS 579

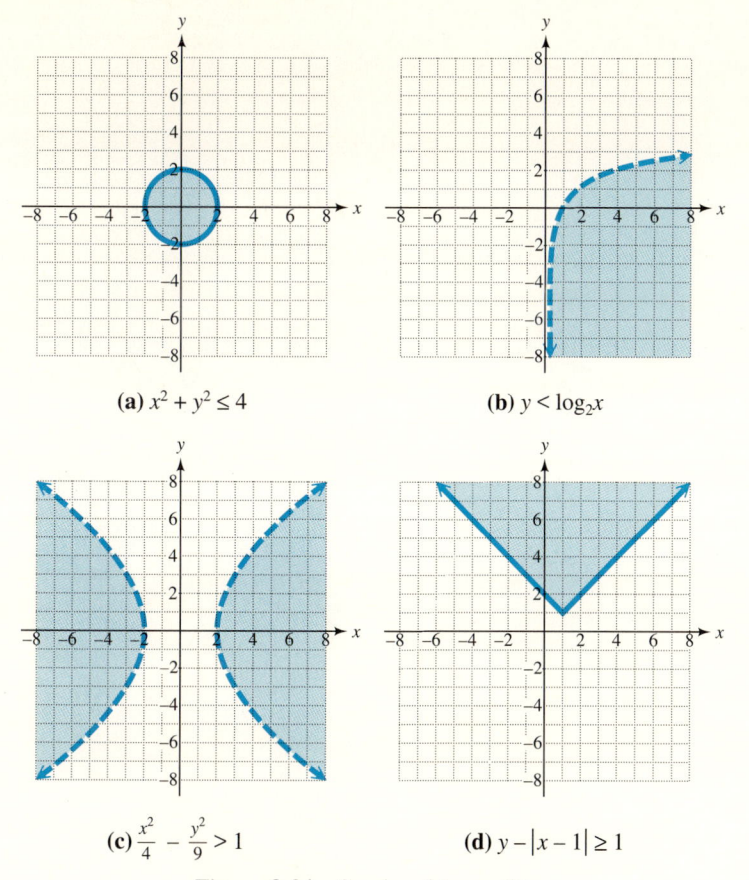

(a) $x^2 + y^2 \leq 4$

(b) $y < \log_2 x$

(c) $\dfrac{x^2}{4} - \dfrac{y^2}{9} > 1$

(d) $y - |x - 1| \geq 1$

Figure 8.34 Graphs of Inequalities

EXAMPLE 5 A Nonlinear System of Inequalities

Graph the system. $\quad y \geq x^2 - 1$
$\qquad\qquad\qquad\quad y < x + 1$

Replace the inequality symbols with equal signs. $y = x^2 - 1$ is the equation of a parabola. The test point $(0, 0)$ gives the true inequality $0 \geq -1$ and thus the graph of $y \geq x^2 - 1$ is all points on or "inside" the parabola. Use $(0, 0)$ in $y < x + 1$ to get the true inequality $0 < 1$. This shows that the second inequality has as its graph all points "below" the line $y = x + 1$. Putting this information together, we can obtain the graph of the system in Figure 8.35.

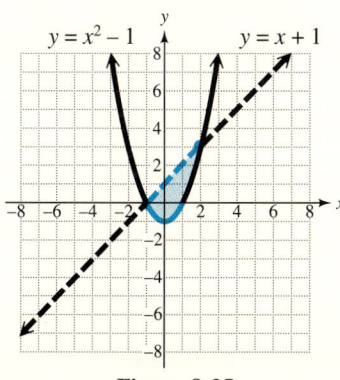

Figure 8.35

PRACTICE EXERCISE 5

Graph the system.

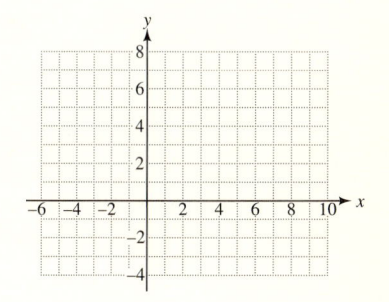

$y \geq |x - 5|$
$y < \log_2 x$

Answer: The graph is given in Figure 8.36.

Figure 8.36

8.5 EXERCISES A

Solve each system in Exercises 1–10.

1. $x^2 + y^2 = 10$
   $x - y = 2$

2. $x + y = 3$
   $x^2 - y^2 = 3$

3. $5x^2 + xy - y^2 = -1$
   $y - 2x = 1$

4. $x^2 + 3y^2 = 37$
   $2x^2 - y^2 = 46$

5. $x^2 + y^2 = 5$
   $xy = 2$

6. $x^2 + 2xy + y^2 = 9$
   $x^2 - 2xy - y^2 = 9$

7. $x^2 + 2xy + 2y^2 = 10$
   $2x^2 + xy + 22y^2 = 50$

8. $10^x + y = 11$
   $10^x + 2y = 12$

9. $y + \log_4 (x + 3) = 6$
   $y - \log_4 x = 5$

10. $|y| + x = 7$
    $2|y| - x = 5$

Graph each system in Exercises 11–16.

11. $y \geq x^2$
    $y \leq x + 2$

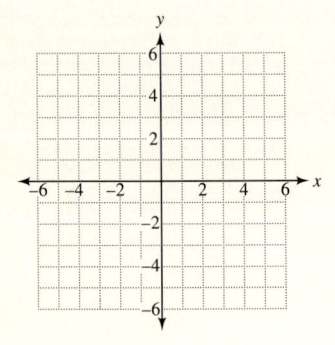

12. $y - |x| \geq 0$
    $y + |x| < 4$

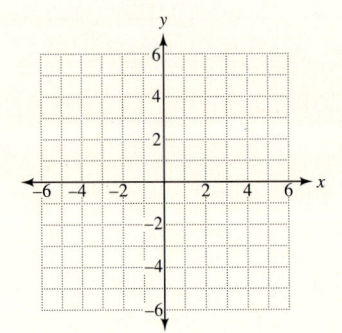

13. $y - 2^x \geq 0$
    $y - x \geq 2$

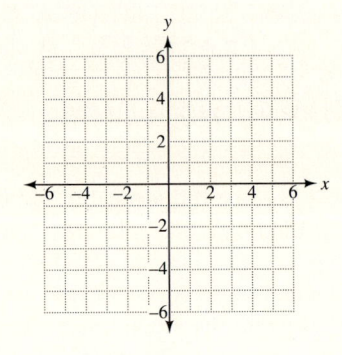

14. $y + |x - 2| < 0$
    $x - |y + 1| > 0$

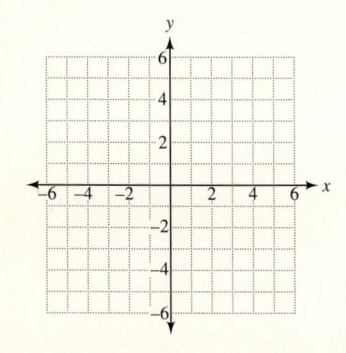

15. $y \geq 0$
    $y \leq \sqrt{x - 1}$
    $y > x - 3$

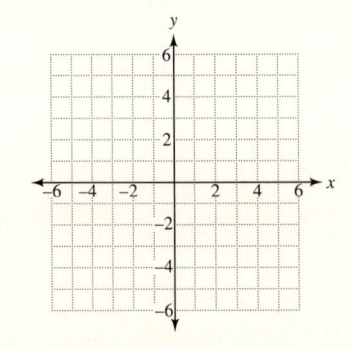

16. $x^2 + y^2 > 4$
    $x^2 + y^2 < 16$

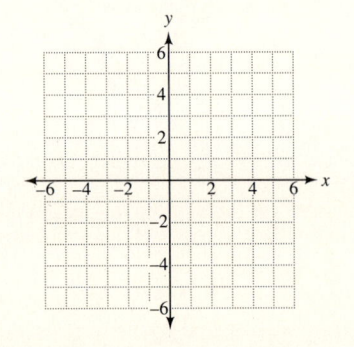

17. GEOMETRY Find the dimensions of a rectangular garden with perimeter 20 yd and area 24 yd².

18. ECONOMICS The weekly demand equation for a certain product is given by $xy = 100$, where x is the number of people willing to pay y dollars for the product. The supply equation for the product is given by $x^2 - xy = 44$. The **market equilibrium point** occurs at a point where the supply curve intersects the demand curve. Find the market equilibrium point for this product.

19. BUSINESS Outdoor Industries manufactures and sells gas barbeque grills. If x represents the number of grills made and sold each week, the total weekly cost is given by $C = 50x + 1000$, and the total weekly revenue produced is given by $R = 100x - 0.2x^2$. Find the break-even values of x, the values of x for which the weekly cost and revenue are equal.

FOR REVIEW

20. Give the vertex, focus, directrix, and line of symmetry of the parabola with equation $(x - 1)^2 = -16(y + 3)$.

Exercises 21–22, reviewing functional notation, will help you prepare for the material in Chapter 9. Find $f(1)$, $f(3)$, and $f(10)$ for each function.

21. $f(n) = 2n + 1$

22. $f(n) = (-1)^n n^2$

ANSWERS: 1. (3, 1), (−1, −3) 2. (2, 1) 3. (0, 1), (1, 3) 4. (5, 2), (5, −2), (−5, 2), (−5, −2) 5. (2, 1), (−2, −1), (1, 2), (−1, −2) 6. (3, 0), (3, −6), (−3, 0), (−3, 6) 7. (4, −1), (−4, 1), ($\sqrt{2}$, $\sqrt{2}$), (−$\sqrt{2}$, −$\sqrt{2}$) 8. (1, 1) 9. (1, 5) 10. (3, 4), (3, −4)
11. 12. 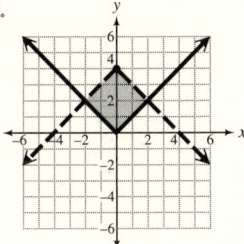 13. 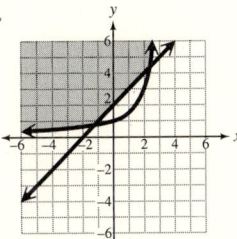 14. 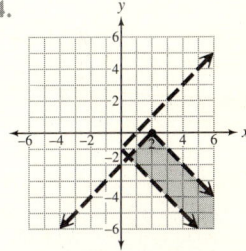
15. 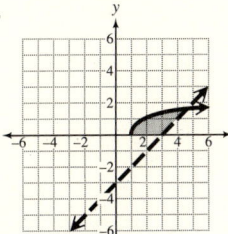 16. 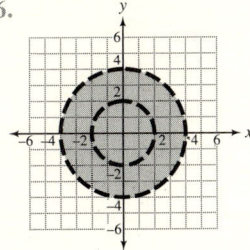 17. 6 yd by 4 yd 18. 12 people; approximately $8.33 19. approximately 22 grills and 228 grills 20. vertex: (1, −3); focus: (1, −7); directrix: $y = 1$; line of symmetry: $x = 1$ 21. 3; 7; 21 22. −1; −9; 100

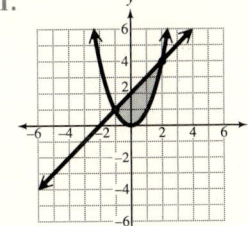

8.5 EXERCISES B

Solve each system in Exercises 1–10.

1. $2x^2 + y^2 = 9$
$y - 2x = 3$

2. $xy = 12$
$2x - y = -2$

3. $x^2 + 2y^2 = 8$
$x^2 - 2y^2 = 0$

4. $x^2 + 2y^2 = 8$
$2x^2 - y^2 = 1$

5. $x^2 + y^2 = 29$
   $xy = 10$

6. $3xy - y^2 = -13$
   $2xy + y^2 = -2$

7. $2x^2 - 5xy + 2y^2 = 20$
   $8x^2 - 4xy + y^2 = 20$

8. $2^x - y = 2$
   $2^x + y = 6$

9. $\log_3 x - 2y = 2$
   $\log_3 x^2 + 2y = 1$

10. $|x| + |y| = 5$
    $2|x| - |y| = 1$

Graph each system in Exercises 11–16.

11. $y + x^2 \leq 0$
    $x + y > -2$

12. $\log_2 x - y > 0$
    $2x - y \leq 4$

13. $x^2 + 4y^2 \leq 16$
    $y^2 - x \leq 0$

14. $x^2 + y^2 \leq 16$
    $(x-2)^2 + y^2 \geq 4$

15. $x \leq 0$
    $y \geq 0$
    $x + y^2 < 1$

16. $x^2 + y^2 < 9$
    $x^2 - y^2 > 1$

17. **ENGINEERING** A rectangular solar heating surface must be 8 inches longer than it is wide. If the area of the surface must be 180 in^2, find the dimensions of the surface.

18. **NUMBER** The sum of the squares of the digits of a positive two-digit number is 100, and the tens digit is 2 less than the units digit. Find the number.

19. **BUSINESS** Outdoor Industries manufactures and sells snack tables. If x represents the number of tables made and sold each week, the total weekly cost is given by $C = 10x + 80$, and the total weekly revenue produced is given by $R = 25x - 0.2x^2$. Find the break-even values of x, the values of x for which the weekly cost and revenue are equal.

FOR REVIEW

20. Give the vertex, focus, directrix, and line of symmetry of the parabola with equation $(y - 1)^2 = 16(x + 3)$.

Exercises 21–22, reviewing functional notation, will help you prepare for the material in Chapter 9. Find $f(1)$, $f(3)$, and $f(10)$ for each function.

21. $f(n) = n^2 - 1$

22. $f(n) = \dfrac{n+1}{n}$

8.5 EXERCISES C

Find all real solutions to each system of equations.

1. $x^2 + y^2 + z^2 = 14$
   $xz = -6$
   $x - z = 5$

2. $xy + z = -2$
   $x - yz = 10$
   $x^2 + yz = -10$

3. $2x^2 + y - z = 2$
   $x - y^2 + z = -1$
   $x + y = 0$

4. $x^2 + y^2 + z^4 = 5$
   $x^2 + 4y + z^4 = 0$
   $3x^2 - 2y^2 + 3z = 10$
   [Answers: $(2, -1, 0)$; $(-2, -1, 0)$; $(\sqrt{3}, -1, 1)$; $(-\sqrt{3}, -1, 1)$]

CHAPTER 8 EXTENSION PROJECT

Many nonlinear systems cannot be solved algebraically. Instead, we can approximate solutions graphically using a graphics calculator or graphing software on a personal computer. For example, consider the following system.

$$y - \ln x = 0$$
$$x^2 + y^2 = 4$$

Suppose we graph these two equations in the same coordinate system as shown in Figure 8.37.

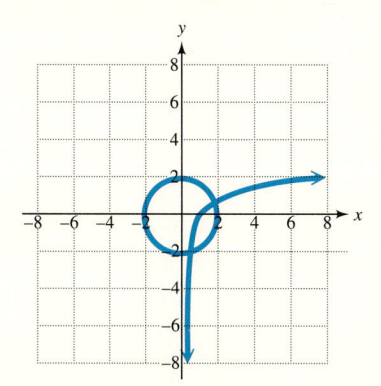

Figure 8.37

From the graphs we can see that there are two solutions to the system, one in quadrant I and another in quadrant IV. Let us attempt to approximate the solution in quadrant I. It appears that the solution has x-coordinate in the interval $[1, 2]$. By changing the scale on our graph, we can "zoom in" and graph the system on this interval as shown in Figure 8.38.

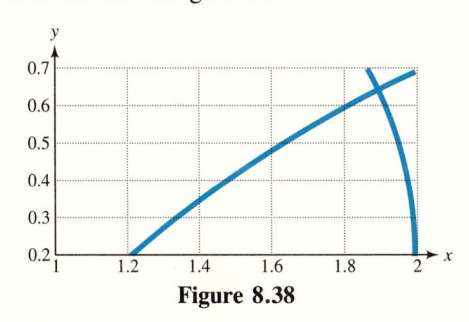

Figure 8.38

It now appears that the x-coordinate of the point of intersection is between 1.8 and 2.0. As a result, we "zoom in" once more to plot the graphs on $[1.8, 2]$ as shown in Figure 8.39.

Figure 8.39

Using this graph we can determine that the x-coordinate of the point is somewhere between 1.88 and 1.90. Closing in on this interval, we graph the system once more as shown in Figure 8.40. This process could be continued until we have found any degree of accuracy desired.

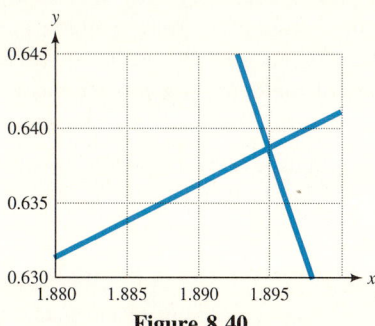

Figure 8.40

At this point it would appear that the point of intersection has coordinates approximated by (1.895, 0.639). Suppose we try these values. Substituting into the first equation gives

$$0.639 - \ln(1.895) =$$
$$0.639 - 0.639218838 \approx -0.000218838 \approx 0$$

and into the second equation gives

$$(1.895)^2 + (0.639)^2 =$$
$$3.591025 + 0.408321 \approx 3.999346 \approx 4.$$

Thus, (1.895, 0.639) is a fairly accurate approximation to the point of intersection of the curves, and to the solution in the first quadrant to the system.

THINGS TO DO:

1. Approximate the second solution, in quadrant IV, to the system given above.

2. Approximate the solution, in quadrant I, to the system
$$x^2 + y^2 = 1$$
$$x^2 - y = 0.$$

The actual value of this solution can be found using the substitution method. Compare this value with your approximation.

3. Approximate the solution to various other systems that involve logarithmic, exponential, and absolute value functions with second-degree equations.

4. Solving the system
$$y - \ln x = 0$$
$$x^2 + y^2 = 4$$
is equivalent to finding the solutions to the equation $x^2 + (\ln x)^2 = 4$. Explain why this is true, and use this technique to solve several other systems.

CHAPTER 8 REVIEW

KEY TERMS

8.1 1. **Analytic geometry** is an area of mathematics that involves describing geometric figures with algebraic equations.

2. **Conic sections,** or simply **conics,** are curves formed by intersecting a plane and a right circular cone, but not through the vertex of the cone. If the plane passes through the vertex, the result is a **degenerate conic.**

3. A **second-degree equation,** or **quadratic equation in two variables,** is an equation of the form

$$Ax^2 + Bxy + Cy^2 + Dx + Ey + F = 0.$$

4. Let (h, k) be a point in the plane with r any positive real number. The set of all points in the plane that are r units from (h, k) is a **circle** with **center** (h, k) and **radius** r.

8.2 Let F_1 and F_2 be two points in a plane. An **ellipse** with **foci** (plural of **focus**) F_1 and F_2 is the collection of all points in the plane the sum of whose distances from F_1 and F_2 is a constant. The midpoint of the line segment joining F_1 and F_2 is the **center** of the ellipse, the line segment through the foci is the **major axis** of the ellipse, and the line perpendicular to the major axis through the center of the ellipse is the **minor axis** of the ellipse.

8.3 Let F_1 and F_2 be two points in a plane. A **hyperbola** with **foci** F_1 and F_2 is the collection of all points in the plane the difference of whose distances from F_1 and F_2 is a constant. The midpoint of the line segment joining F_1 and F_2 is the **center** of the hyperbola, the intercepts of the hyperbola are the **vertices** of the hyperbola, the line segment joining the vertices is the **transverse axis** of the hyperbola, and the line perpendicular to the transverse axis through the center is the **conjugate axis.** The lines approached by the hyperbola for large values of $|x|$ are the **asymptotes** of the hyperbola.

8.4 Let F be a fixed point and L a fixed line in the plane. A **parabola** with **focus** F and **directrix** L is the set of all points in the plane equidistant from F and L. The line through the focus perpendicular to the directrix is the **axis of symmetry** of the parabola, and the point of intersection of the parabola and the axis of symmetry is the **vertex** of the parabola.

8.5 A system of two or more nonlinear equations in two variables is a **nonlinear system of equations.**

REVIEW EXERCISES

Part I

8.1 1. Find the equation of the circle with center $(-3, \frac{2}{3})$ and radius 4.

2. Sketch the graph. $x^2 + y^2 = 9$

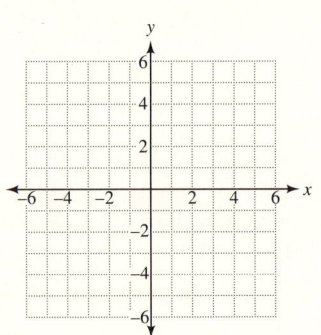

3. Complete the squares to determine if the equation represents a circle, ellipse, hyperbola, parabola, or degenerate conic.

$$3x^3 + 3y^2 - 6x + 12y - 40 = 0$$

4. **ENGINEERING** A canal with cross-section a semicircle is 15.0 ft deep at the center. Find the equation of the semicircle by placing a coordinate system with origin at the center of the water level in a full canal. What is the depth of the water in the canal 3.5 ft from the edge of the canal?

8.2 5. Find the equation of the ellipse centered at the origin with intercepts (± 3, 0) and (0, ± 7).

6. Sketch the graph.
$$\frac{x^2}{25} + \frac{y^2}{4} = 1$$

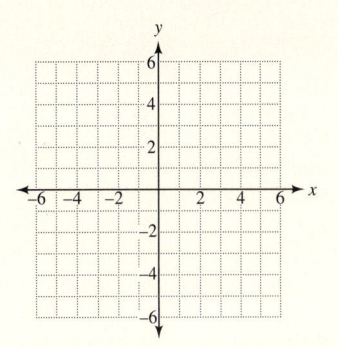

7. Complete the squares to determine if the equation represents a circle, ellipse, hyperbola, parabola, or degenerate conic.
$$3x^2 + 2y^2 - 6x + 12y - 60 = 0$$

8. **ENGINEERING** A highway tunnel is in the shape of a semiellipse with maximum height 12.8 m and width 16.2 m. Find the vertical clearance 4.4 m from the edge of the tunnel.

8.3 9. Find the equation of the hyperbola with foci (± 4, 0) and length of the transverse axis 6.

10. Sketch the graph.
$$y^2 - x^2 = 4$$

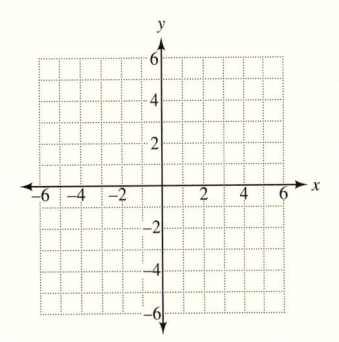

11. Complete the squares to determine if the equation represents a circle, ellipse, hyperbola, parabola, or degenerate conic.
$$3x^2 - 3y^2 - 6x + 12y - 80 = 0$$

12. **ARCHITECTURE** The roof of a basketball arena is in the shape of the hyperbola $y^2 - x^2 = 1600$, where x and y are in feet. Refer to the figure and determine the height h of the outside walls.

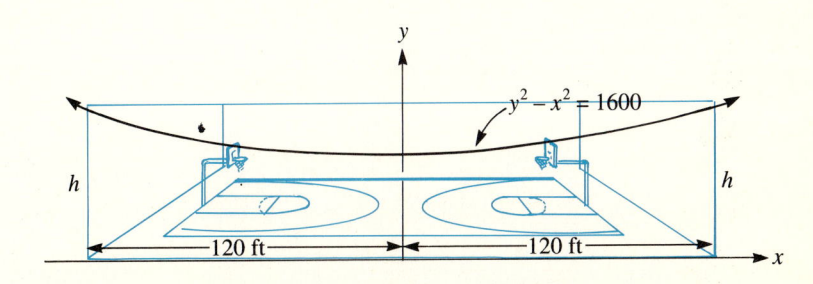

8.4 13. Find the equation of the parabola with directrix $x = -3$ and focus $(5, 2)$.

14. Sketch the graph.
$$x^2 + 2y = 0$$

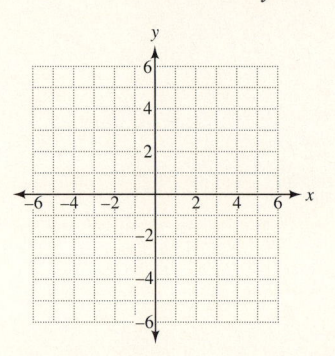

15. Complete the square to determine if the equation represents a circle, ellipse, hyperbola, parabola, or degenerate conic.
$$3x^2 - 6x + 12y - 48 = 0$$

16. **ARCHITECTURE** A domed ceiling is to be constructed with a parabolic vertical cross-section. Plans call for a 16-ft beam across the dome at a point 6 ft below the top. If the highest point of the dome is 20 ft from the floor and the origin of a coordinate system is put on the floor below the high point, find the equation of the parabola.

8.5 Solve each system in Exercises 17–18.

17. $x^2 + y^2 = 5$
    $xy = 2$

18. $2x^2 - 6xy + 12y^2 = 8$
    $x^2 - xy + 6y^2 = 8$

Graph each system in Exercises 19–20.

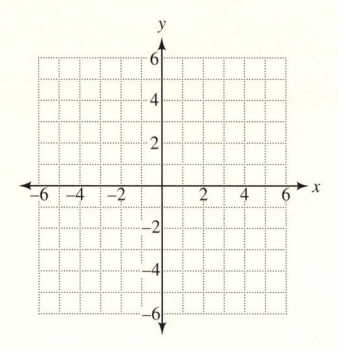

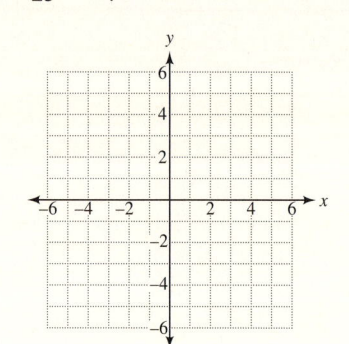

19. $x \leq 4 - y^2$
    $x > |y| - 2$

20. $x^2 + y^2 \geq 4$
    $\dfrac{x^2}{25} + \dfrac{y^2}{4} \leq 1$

Part II

Sketch the graph of each conic in Exercises 21–24.

21. $x^2 + y^2 - 4x + 8y + 11 = 0$

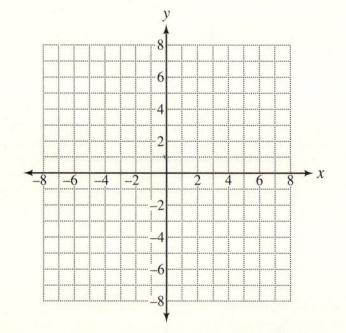

22. $4x^2 + y^2 + 4x - 6y - 6 = 0$

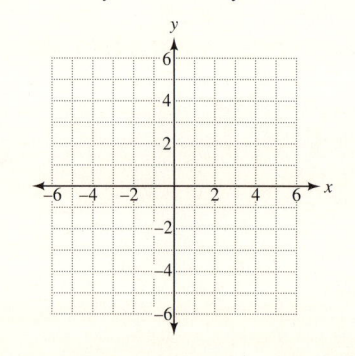

23. $4x^2 - y^2 + 16x - 2y - 21 = 0$

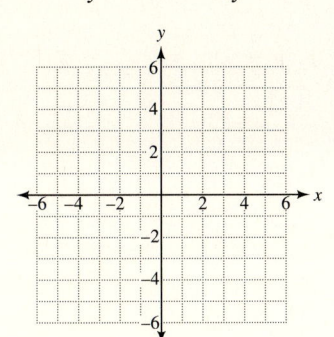

24. $y^2 + 6x - 4y + 22 = 0$

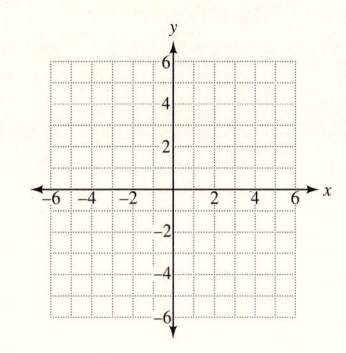

Solve each system in Exercises 25–30.

25. $2x^2 + y^2 = 3$
$x - y = 2$

26. $x^2 + y^2 = 20$
$xy = 8$

27. $3x^2 + xy - 6y^2 = 8$
$x^2 + 2xy - 4y^2 = 4$

28. $3^x + y = 10$
$3^x - y = 8$

29. $x^2 + y^2 = 169$
$x + y = 17$

30. $x^2 + y^2 = 25$
$x^2 - y^2 = 25$

Find the equation of each conic in Exercises 31–34.

31. Circle with endpoints of a diameter at $(-2, -3)$ and $(6, 7)$

32. Ellipse with foci $(2, -5)$ and $(2, -1)$, and $b = 1$

33. Hyperbola with vertices $(0, \pm 5)$ and passing through $(3, 10)$

34. Parabola with vertex $(1, -2)$, axis of symmetry $x = 1$, and passing through $(5, 2)$

Graph each system in Exercises 35–37.

35. $x^2 + y^2 \leq 16$
$y > |x|$

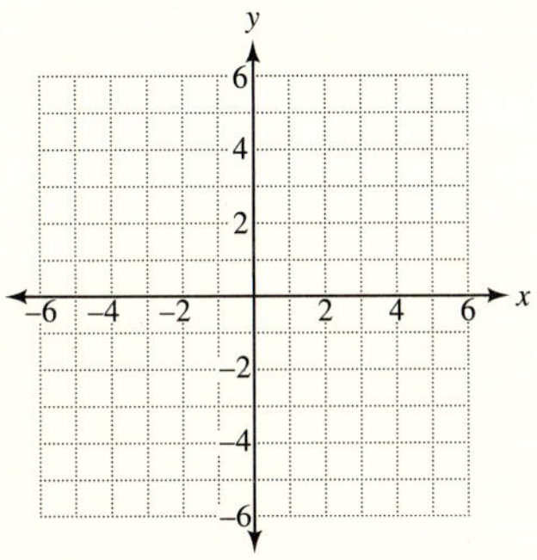

36. $y - 3^x \geq 0$
$3x - 4y \geq -9$

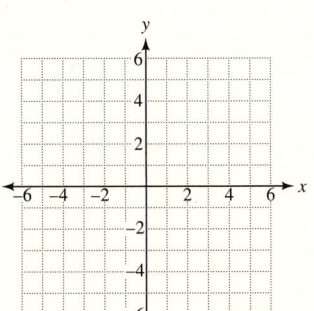

37. $|y| - x \leq 0$
$x^2 + y^2 < 16$

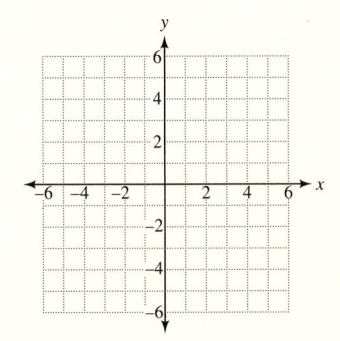

38. ENGINEERING A footbridge over a creek is in the shape of a parabola. If the span of the bridge is 32 ft and the arch rises 4 ft above the bank of the creek, find the equation of the parabola assuming that its vertex is at the origin and it opens down.

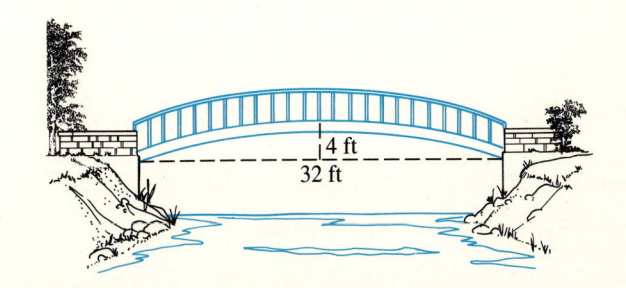

588 CHAPTER 8 TOPICS IN ANALYTIC GEOMETRY

In Exercises 39–40, complete the squares to determine if the equation represents a circle, ellipse, hyperbola, parabola, or degenerate conic.

39. $3x^2 + 3y^2 - 6x - 6y + 6 = 0$

40. $9x^2 - 4y^2 - 18x - 16y - 43 = 0$

ANSWERS:
1. $(x + 3)^2 + \left(y - \frac{2}{3}\right)^2 = 16$
2. [graph]
3. circle
4. $y = -\sqrt{225 - x^2}$; 9.6 ft
5. $\frac{x^2}{9} + \frac{y^2}{49} = 1$
6. [graph]
7. ellipse
8. 11.4 m
9. $\frac{x^2}{9} - \frac{y^2}{7} = 1$
10. [graph]
11. hyperbola
12. 126.5 ft
13. $(y - 2)^2 = 16(x - 1)$
14. [graph]
15. parabola
16. $x^2 = -\frac{32}{3}(y - 20)$
17. $(2, 1), (-2, -1), (1, 2), (-1, -2)$
18. $\left(\sqrt{6}, \frac{\sqrt{6}}{3}\right), \left(-\sqrt{6}, -\frac{\sqrt{6}}{3}\right), (2, 1), (-2, -1)$
19. [graph]
20. [graph]
21. [graph]
22. [graph]
23. [graph]
24. [graph]
25. $\left(\frac{1}{3}, -\frac{5}{3}\right), (1, -1)$
26. $(2, 4), (-2, -4), (4, 2), (-4, -2)$
27. $(2, 1), (-2, -1), (2i, 2i), (-2i, -2i)$
28. $(2, 1)$
29. $(5, 12), (12, 5)$
30. $(5, 0), (-5, 0)$
31. $(x - 2)^2 + (y - 2)^2 = 41$
32. $(x - 2)^2 + \frac{(y + 3)^2}{5} = 1$
33. $\frac{y^2}{25} - \frac{x^2}{3} = 1$
34. $(x - 1)^2 = 4(y + 2)$
35. [graph]
36. [graph]
37. [graph]
38. $y = -\frac{1}{64}x^2$
39. degenerate conic (a point circle)
40. hyperbola

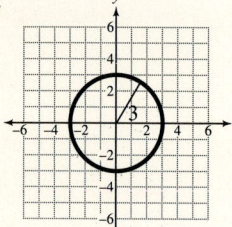

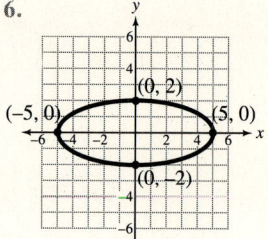

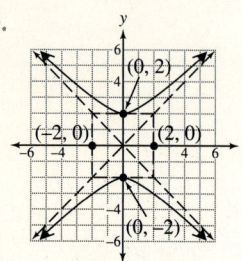

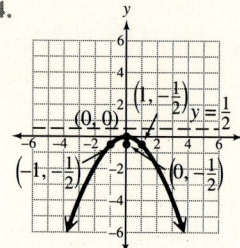

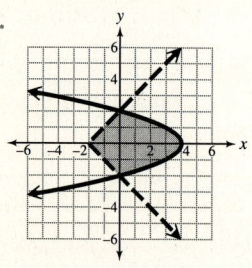

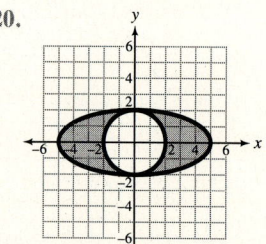

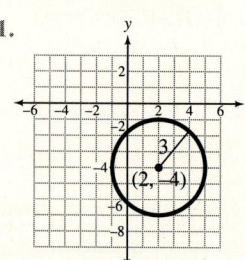

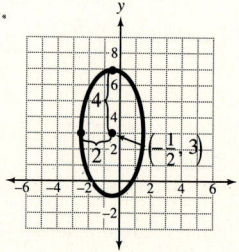

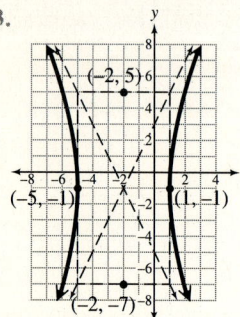

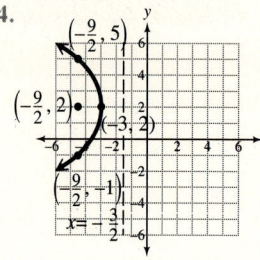

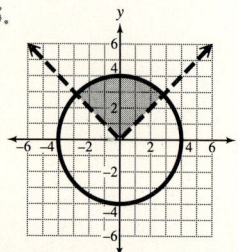

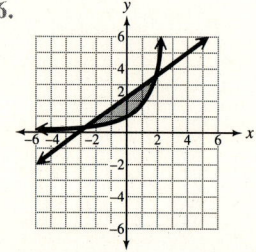

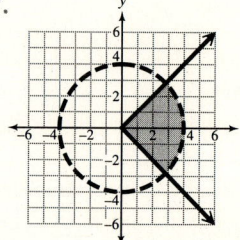

CHAPTER 8 TEST

1. Find the standard form of the equation of a circle with center $(-2, 4)$ and radius 5.

2. Graph $x^2 + y^2 - 2x + 4y - 4 = 0$.

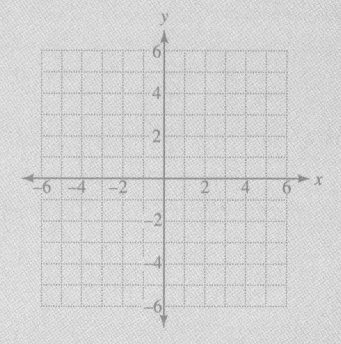

3. Find the equation of the ellipse with $a = 3$ and foci $(2, -1)$ and $(6, -1)$.

4. Graph $\dfrac{(x-1)^2}{25} + \dfrac{(y+2)^2}{4} = 1$.

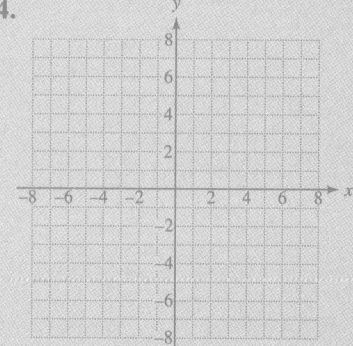

5. Find the standard form of the equation of the hyperbola with general form $8y^2 - x^2 - 16y - 4x - 4 = 0$.

6. Graph $\dfrac{(x-1)^2}{25} - \dfrac{(y+2)^2}{4} = 1$.

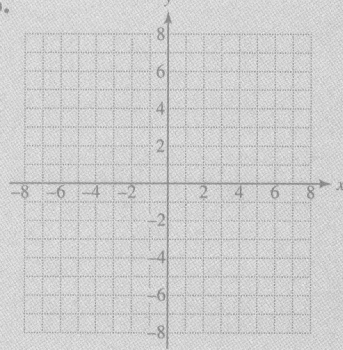

CHAPTER 8 TEST
CONTINUED

7. Find the standard form of the equation of the parabola with vertex (4, −5), axis of symmetry parallel to the *x*-axis, and passing through (0, −7).

   7. _____

8. The entryway to a building is in the shape of a parabolic arch that is 40 ft high (in the center) and 18 ft wide at the base (total width). What is the vertical clearance 2.5 ft from the center of the base of the arch (to the nearest tenth of a foot)?

   8. _____

9. Solve the system.

$$x^2 + y^2 = 81$$
$$x^2 - y^2 = 81$$

   9. _____

10. Solve the system.

$$y + e^x = 2$$
$$y - e^x = 0$$

    10. _____

11. Graph the system.

$$x^2 + y^2 \leq 4$$
$$y \geq x^2$$

    11.

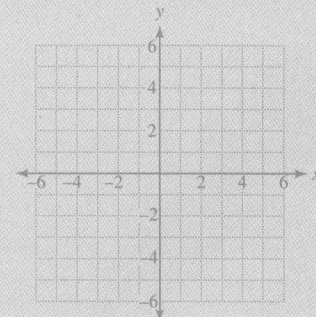

12. Graph the system.

$$y \geq 2^x$$
$$y \leq x^2 + 3$$

    12.

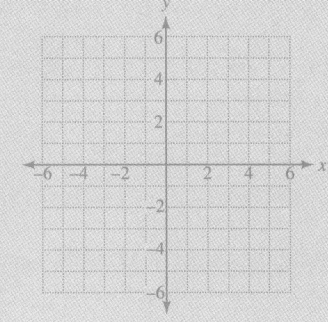

CHAPTER 9

Sequences, Series, and Probability

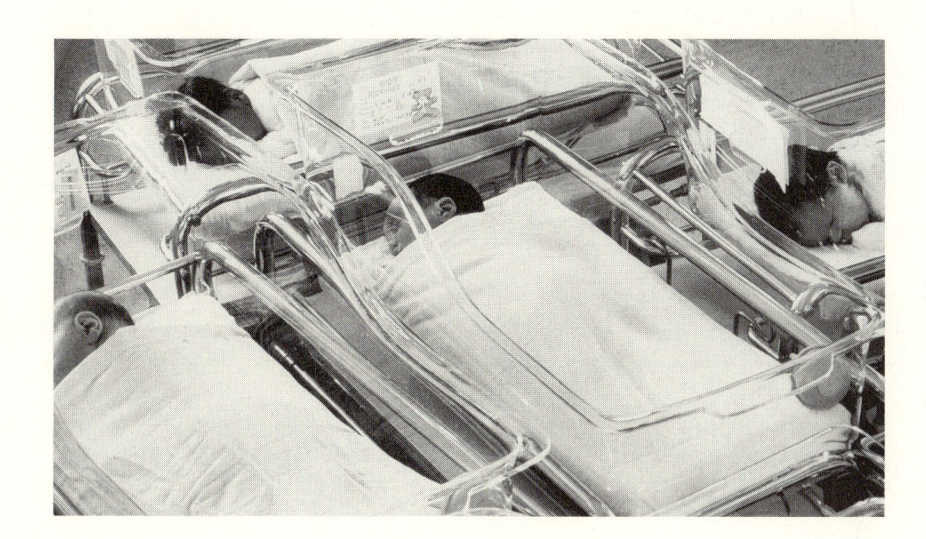

AN APPLIED PROBLEM IN BIOLOGY

A couple plans to have two children. Assuming that it is equally likely that a boy or a girl is born, what is the probability that the couple will have two girls? What is the probability of two boys? What is the probability of one boy and one girl?

Analysis
We can find the probability of an event if we know the sample space for the particular experiment. First we list the elements in the sample space, then identify the outcomes that are in each particular event, then use the formula for finding the probability of each event.

Tabulation
The sample space for this experiment is

$$S = \{gg, gb, bg, bb\}$$

where gg means girl-then-girl, gb means girl-then-boy, bg means boy-then-girl, and bb means boy-then-boy.

Translation
Substitute into the formulas:

$$P(gg) = \frac{n(gg)}{n(S)} \qquad P(bb) = \frac{n(bb)}{n(S)}$$

$$P(1 \text{ boy and } 1 \text{ girl}) = P(gb) + P(bg)$$

The topics in this chapter not only are of interest in college algebra but give a foundation for study in more advanced mathematics and provide a variety of interesting applications in such areas as banking, demography, physics, education, recreation, and biology. The application given at the left is solved completely in Example 5 in Section 9.8. We begin with general sequences and series, then shift to two special types, *arithmetic* and *geometric*, and introduce the idea of mathematical induction. Permutations, combinations, and counting techniques follow, with a discussion of the binomial theorem. A brief introduction to probability concludes the chapter.

9.1 SEQUENCES AND SERIES

> **STUDENT GUIDEPOSTS**
> 1. Sequences
> 2. Recursive Definition
> 3. Series
> 4. Properties of Series

1 SEQUENCES

Informally we think of a sequence as a collection of numbers arranged in a particular order. This means there is a first number, a second number, a third number, and so forth. Thus, each number in the sequence corresponds to a natural number. This suggests the following formal definition.

> **Infinite Sequence**
>
> An **infinite sequence** is a function with domain the set of natural numbers.

As an example, consider the function defined by

$$f(n) = 2n. \quad n = 1, 2, 3, 4, \ldots$$

The three dots mean that the pattern continues. Instead of using the usual function notation we usually write

$$x_n = 2n. \quad n = 1, 2, 3, 4, \ldots$$

That is, a letter with a subscript, such as x_n, is used instead of $f(n)$. For the sequence defined by $x_n = 2n$,

$$\begin{aligned} x_1 &= 2(1) = 2 \\ x_2 &= 2(2) = 4 \\ x_3 &= 2(3) = 6 \\ x_4 &= 2(4) = 8 \\ &\vdots \\ x_n &= 2(n) = 2n \\ &\vdots \end{aligned}$$

A sequence is frequently defined by indicating its range. Thus, the sequence above can be written

$$2, 4, 6, 8, \ldots, 2n, \ldots.$$

Each number is called a **term** of the sequence, and the terms are written in order of increasing n. The term $2n$ is called the ***n*th term** or the **general term.** Note the use of the three dots after the general term to indicate an infinite sequence. We will also study sequences which are finite.

> **Finite Sequence**
>
> A **finite sequence** with m terms is a function with domain the set of natural numbers from 1 to m.

For example,

$$2, 4, 6, 8$$

indicates a sequence with four terms, and

$$3, 6, 9, \ldots, 3n$$

is a sequence with n terms. The three dots used in this example indicate the presence of the terms between 9 and $3n$.

The following examples illustrate the use of sequence notation.

EXAMPLE 1 CALCULATING TERMS OF A SEQUENCE

Find the first four terms and the seventh term of each sequence. Assume the domain in each case is the set of natural numbers.

(a) $x_n = \dfrac{1}{n+1}$. Use $n = 1, 2, 3, 4$, and 7 in order.

$$x_1 = \frac{1}{1+1} = \frac{1}{2}$$
$$x_2 = \frac{1}{2+1} = \frac{1}{3}$$
$$x_3 = \frac{1}{3+1} = \frac{1}{4}$$
$$x_4 = \frac{1}{4+1} = \frac{1}{5}$$
$$x_7 = \frac{1}{7+1} = \frac{1}{8}$$

The complete sequence can be written

$$\frac{1}{2}, \frac{1}{3}, \frac{1}{4}, \frac{1}{5}, \ldots, \frac{1}{n+1}, \ldots$$

(b) $a_n = 2^n - 1$.

$$a_1 = 2^1 - 1 = 2 - 1 = 1$$
$$a_2 = 2^2 - 1 = 4 - 1 = 3$$
$$a_3 = 2^3 - 1 = 8 - 1 = 7$$
$$a_4 = 2^4 - 1 = 16 - 1 = 15$$
$$a_7 = 2^7 - 1 = 128 - 1 = 127$$

The infinite sequence can be written

$$1, 3, 7, 15, \ldots, 2^n - 1, \ldots$$

(c) $b_n = (-1)^n(n^2 + 1)$

$$b_1 = (-1)^1(1^2 + 1) = (-1)(2) = -2$$
$$b_2 = (-1)^2(2^2 + 1) = (1)(5) = 5$$
$$b_3 = (-1)^3(3^2 + 1) = (-1)(10) = -10$$
$$b_4 = (-1)^4(4^2 + 1) = (1)(17) = 17$$
$$b_7 = (-1)^7(7^2 + 1) = (-1)(50) = -50$$

Notice that the factor $(-1)^n$ changes the sign on each term. When n is odd, $(-1)^n = -1$, and when n is even, $(-1)^n = 1$.

PRACTICE EXERCISE 1

Find the second, fourth, fifth, and seventh terms of each sequence.

(a) $x_n = 2n^2 + 3$

(b) $a_n = \dfrac{n-2}{n+2}$

(c) $b_n = (-1)^{n+1}(3n - 5)$

Answers: (a) 11, 35, 53, 101
(b) $0, \frac{1}{3}, \frac{3}{7}, \frac{5}{9}$ (c) $-1, -7,$ 10, 16

In Example 1 we were given the general or nth term of a sequence and were asked to find several of the terms. If we are given the first few terms of a sequence, we may be able to discover a formula for the general term.

EXAMPLE 2 Finding A General Term

Find a formula for a_n given the first few terms of the sequence.

(a) 1, 3, 5, 7, . . .

Here each of the terms is an odd integer; in fact, they are consecutive odd integers. Since even integers can be written $2n$, odd integers can be written as $2n + 1$ or $2n - 1$. In order for the sequence to start with $n = 1$, we try $2n - 1$.

$$a_1 = 2(1) - 1 = 1$$
$$a_2 = 2(2) - 1 = 3$$
$$a_3 = 2(3) - 1 = 5$$
$$a_4 = 2(4) - 1 = 7$$

Thus, $a_n = 2n - 1$ gives the first few terms and is the required general term.

(b) $-2, 5, -10, 17, \ldots$

First we notice that the terms alternate in signs. This tells us that the general term has a factor of $(-1)^n$. After some study we also notice that the absolute value of each term is one larger than the perfect squares 1, 4, 9, and 16. This suggests that $a_n = (-1)^n(n^2 + 1)$. Check.

Practice Exercise 2

Find a formula for a_n.

(a) 7, 12, 17, 22, . . .

(b) 3, -6, 9, -12, . . .

Answers: (a) $a_n = 5n + 2$
(b) $a_n = (-1)^{n+1}3n$

2 RECURSIVE DEFINITION

Consider the following way of defining a sequence. Let $x_1 = 5$ and $x_{n+1} = 3x_n$. Then

$$x_1 = 5$$
$$x_2 = x_{1+1} = 3x_1 = 15 \quad n = 1 \text{ and } n + 1 = 2$$
$$x_3 = x_{2+1} = 3x_2 = 45 \quad n = 2 \text{ and } n + 1 = 3$$
$$x_4 = x_{3+1} = 3x_3 = 135 \quad n = 3 \text{ and } n + 1 = 4$$
$$\vdots$$
$$x_{n+1} = 3x_n$$
$$\vdots$$

With x_1 given and x_{n+1} defined in terms of x_n, we can give the whole sequence. This method of defining a sequence is called a **recursive definition**. Recursive techniques are used extensively with computers.

EXAMPLE 3 Using the Recursive Definition

Let $a_1 = -2$ and $a_n = 2a_{n-1} + 3$ for $n \geq 2$. Find $a_2, a_3, a_4,$ and a_5. Notice that in this recursive formula a_n is defined in terms of a_{n-1} for $n \geq 2$.

Practice Exercise 3

Let $a_{n+1} = 5 - a_n$ and $a_1 = 15$. Find $a_2, a_4, a_5,$ and a_7.

$$a_2 = 2a_{2-1} + 3 = 2a_1 + 3 = 2(-2) + 3 = -1$$
$$a_3 = 2a_{3-1} + 3 = 2a_2 + 3 = 2(-1) + 3 = 1$$
$$a_4 = 2a_{4-1} + 3 = 2a_3 + 3 = 2(1) + 3 = 5$$
$$a_5 = 2a_{5-1} + 3 = 2a_4 + 3 = 2(5) + 3 = 13$$

Answers: $-10, -10, 15, 15$

3 SERIES

Associated with each sequence is a **series,** the indicated sum of the terms of the sequence. For example, associated with the sequence

$$1, 3, 5, 7, 9, 11, 13$$

is the series

$$1 + 3 + 5 + 7 + 9 + 11 + 13.$$

Notice that terms of a sequence are separated by commas, but the word *series* means that terms are added.

The Greek letter Σ (sigma), used as a **summation symbol,** shortens writing in a series.

Summation Notation

The sum of the sequence $a_1, a_2, a_3, \ldots, a_n$ can be written

$$\sum_{k=1}^{n} a_k = a_1 + a_2 + a_3 + \cdots + a_n,$$

where k is the **index** on the summation, 1 is the **lower limit,** and n is the **upper limit of summation.**

For example, if $a_k = 2k$ and $n = 7$, we write

$$\sum_{k=1}^{n} a_k = \sum_{k=1}^{7} 2k$$
$$= 2(1) + 2(2) + 2(3) + 2(4) + 2(5) + 2(6) + 2(7)$$
$$= 2 + 4 + 6 + 8 + 10 + 12 + 14 = 56.$$

EXAMPLE 4 Using the Summation Symbol

Give the value of each series.

(a) $\displaystyle\sum_{k=1}^{5} (k^2 - k) = (1^2 - 1) + (2^2 - 2) + (3^2 - 3) + (4^2 - 4) + (5^2 - 5)$

$$= 0 + 2 + 6 + 12 + 20 = 40$$

(b) $\displaystyle\sum_{k=2}^{6} (-1)^k \sqrt{k} = (-1)^2\sqrt{2} + (-1)^3\sqrt{3} + (-1)^4\sqrt{4} + (-1)^5\sqrt{5} + (-1)^6\sqrt{6}$

$$= (1)\sqrt{2} + (-1)\sqrt{3} + (1)(2) + (-1)\sqrt{5} + (1)\sqrt{6}$$
$$= \sqrt{2} - \sqrt{3} + 2 - \sqrt{5} + \sqrt{6}$$

PRACTICE EXERCISE 4

Give the value of each series.

(a) $\displaystyle\sum_{k=1}^{7} (5k - 6)$

(b) $\displaystyle\sum_{k=3}^{5} (-1)^k (k^2 + 2)$

Answers: (a) 98 (b) -20

Notice in Example 4(b) that the lower limit of the summation is 2. In general the lower limit can be any whole number less than or equal to the upper limit. Thus,

$$\sum_{k=3}^{8} a_k = a_3 + a_4 + a_5 + a_6 + a_7 + a_8.$$

If a_k is a constant, that is $a_k = c$ for all k, then

$$a_1 + a_2 + a_3 + \cdots + a_n = c + c + c + \cdots + c$$
$$= nc.$$

This gives the following theorem.

Sum of Constants

If $a_k = c$ for all k, then

$$\sum_{k=1}^{n} a_k = nc.$$

Thus, if $a_k = 5$ for $k = 1, 2, \ldots, 8$, then

$$\sum_{k=1}^{8} a_k = \sum_{k=1}^{8} 5 = 8(5) = 40.$$

❹ PROPERTIES OF SERIES

Other properties can be discovered which involve sums of terms, differences of terms, and a constant times each term of a series.

Properties of Summation Notation

If $a_1, a_2, a_3, \ldots, a_n$ and $b_1, b_2, b_3, \ldots, b_n$ are sequences and c is a constant, then

1. $\displaystyle\sum_{k=1}^{n} (a_k + b_k) = \sum_{k=1}^{n} a_k + \sum_{k=1}^{n} b_k,$

2. $\displaystyle\sum_{k=1}^{n} (a_k - b_k) = \sum_{k=1}^{n} a_k - \sum_{k=1}^{n} b_k,$

3. $\displaystyle\sum_{k=1}^{n} c\, a_k = c \sum_{k=1}^{n} a_k.$

The proof of the parts of this theorem involves the repeated use of the commutative and associative laws as well as the distributive law. For example,

$$\sum_{k=1}^{n} (a_k + b_k) = (a_1 + b_1) + (a_2 + b_2) + (a_3 + b_3) + \cdots + (a_n + b_n)$$
$$= (a_1 + a_2 + a_3 + \cdots + a_n) + (b_1 + b_2 + b_3 + \cdots + b_n)$$
$$= \sum_{k=1}^{n} a_k + \sum_{k=1}^{n} b_k.$$

9.1 EXERCISES A

In Exercises 1–4, give the first five terms, the eighth term, and the twelfth term of each sequence.

1. $a_n = 4n$

2. $x_n = \left(-\dfrac{1}{2}\right)^n$

3. $b_n = (-1)^n(n^2 + 5)$

4. $x_n = (-1)^n + (-1)^{n+1}$

Determine a formula for a_n given the first few terms of the sequence in Exercises 5–8.

5. 2, 6, 10, 14, ...

6. 1, −4, 9, −16, ...

7. $\dfrac{1}{2}, \dfrac{2}{3}, \dfrac{3}{4}, \dfrac{4}{5}, \ldots$

8. log 3, log 6, log 9, log 12, ...

In Exercises 9–11, find the second, third, fourth, and fifth terms of each sequence.

9. $a_1 = 3;\; a_{n+1} = 4a_n$

10. $b_1 = 8;\; b_n = -3b_{n-1}$

11. $x_1 = \dfrac{2}{3};\; x_{n+1} = 9x_n + 5$

Write out each series in Exercises 12–14.

12. $\displaystyle\sum_{k=1}^{5} x_k$

13. $\displaystyle\sum_{m=1}^{4} \dfrac{1}{2m+1}$

14. $\displaystyle\sum_{k=0}^{6} \dfrac{(-1)^{k+1}}{k+1}$

In Exercises 15–18, write each series using sigma summation notation.

15. $1 + 2 + 3 + 4 + \cdots + n$

16. $1 + \dfrac{1}{\sqrt{2}} + \dfrac{1}{\sqrt{3}} + \cdots + \dfrac{1}{\sqrt{n}}$

17. $2 + 4 + 6 + 8$

Evaluate each series in Exercises 18–23.

18. $\displaystyle\sum_{k=1}^{6} (2k-1)$

19. $\displaystyle\sum_{i=0}^{3} (i^2 + 1)$

20. $\displaystyle\sum_{m=1}^{5} [1 + (-1)^m]m^2$

21. $\displaystyle\sum_{k=1}^{50} 3$

22. $\displaystyle\sum_{k=1}^{4} (k+1)(k-2)$

23. $\displaystyle\sum_{m=1}^{5} \left(\dfrac{1}{m} - \dfrac{1}{m+1}\right)$

24. Compare the series $\displaystyle\sum_{k=2}^{5} (k+1)(k+2)$ and $\displaystyle\sum_{k=1}^{4} (k+2)(k+3)$.

25. Find the approximate value of a_{30} by repeated use of the square root key on a calculator if $a_1 = 6$ and $a_{n+1} = \sqrt{a_n}$.

26. Use your calculator to find a_{1000} if $a_n = \left(1 + \frac{1}{n}\right)^n$.

27. Use your calculator to find $1 + 1 + \frac{1}{2} + \frac{1}{6} + \frac{1}{24} + \frac{1}{120}$. Compare the results with Exercise 26.

Let S_n denote the sum of the first n terms of a sequence. In Exercises 28–29 use a calculator to find S_5 to four decimal places.

28. $a_n = n + \sqrt{n}$

29. $b_1 = 0.5;\ b_{n+1} = 2b_n + b_n^2$

In Exercises 30–31 find a simplified formula for S_n, the sum of the first n terms.

30. $a_k = \dfrac{1}{k} - \dfrac{1}{k+1}$

31. $b_k = \sqrt{k+1} - \sqrt{k}$

If $\sum\limits_{k=1}^{n} k = \dfrac{n(n+1)}{2}$ and $\sum\limits_{k=1}^{n} k^2 = \dfrac{n(n+1)(2n+1)}{6}$, give each sum in Exercises 32–34.

32. $\sum\limits_{k=1}^{6} k$

33. $\sum\limits_{k=1}^{n} (k^2 + k)$

34. $\sum\limits_{k=1}^{n} (2k^2 + 3k)$

35. Prove that $\sum\limits_{k=1}^{n} (a_k - b_k) = \sum\limits_{k=1}^{n} a_k - \sum\limits_{k=1}^{n} b_k$.

36. STATISTICS The arithmetic mean $\bar{x}$ of a collection of data is given by $\bar{x} = \dfrac{1}{n} \sum\limits_{k=1}^{n} x_k$. Use this series and the series used in Exercises 32–34 to find $\bar{x}$ for the data $1, 2, 3, \ldots, n$.

In Exercises 37–38 use the well-known Fibonacci sequence, defined by

$$a_1 = a_2 = 1 \quad \text{and} \quad a_{k+1} = a_k + a_{k-1}.$$

37. Give the first eight terms of the sequence.

38. If $r_k = \dfrac{a_{k+1}}{a_k}$, find r_8 to three decimal places. As k increases the sequence r_k gives better and better approximations for the *golden ratio*, the ratio of the sides of a rectangle thought to be most pleasing to the eye.

FOR REVIEW

In Exercises 39–40, determine the type of conic section by completing the square.

39. $5x^2 + 10y^2 - 10x + 20y = 0$

40. $5x^2 - 10y^2 - 10x + 20y = 0$

ANSWERS: 1. 4, 8, 12, 16, 20; 32; 48 2. $-\frac{1}{2}, \frac{1}{4}, -\frac{1}{8}, \frac{1}{16}, -\frac{1}{32}; \frac{1}{256}; \frac{1}{4096}$ 3. $-6, 9, -14, 21, -30; 69; 149$ 4. $0, 0, 0, 0, 0; 0; 0$ 5. $4n - 2$ 6. $(-1)^{n+1} n^2$ 7. $\frac{n}{n+1}$ 8. $\log 3n$ 9. $12, 48, 192, 768$ 10. $-24, 72, -216, 648$ 11. $11, 104, 941, 8474$ 12. $x_1 + x_2 + x_3 + x_4 + x_5$ 13. $\frac{1}{3} + \frac{1}{5} + \frac{1}{7} + \frac{1}{9}$ 14. $-1 + \frac{1}{2} - \frac{1}{3} + \frac{1}{4} - \frac{1}{5} + \frac{1}{6} - \frac{1}{7}$ 15. $\sum\limits_{k=1}^{n} k$ 16. $\sum\limits_{k=1}^{n} \frac{1}{\sqrt{k}}$ 17. $\sum\limits_{k=1}^{4} 2k$ 18. 36 19. 18 20. 40 21. 150 22. 12 23. $\frac{5}{6}$ 24. each has value 104 25. 1 26. 2.7169 27. 2.7167 28. 23.3823 29. 687.6060 30. $1 - \frac{1}{n+1}$ 31. $\sqrt{n+1} - 1$ 32. 21 33. $\frac{n(n+1)(n+2)}{3}$ 34. $\frac{n(n+1)(4n+11)}{6}$ 35. proof 36. $\frac{n+1}{2}$ 37. $1, 1, 2, 3, 5, 8, 13, 21$ 38. 1.619 39. ellipse 40. hyperbola

9.1 EXERCISES B

In Exercises 1–4, give the first five terms, the eighth term, and the twelfth term of each sequence.

1. $x_n = 3n + 1$
2. $x_n = (n-1)(n+2)(n-3)$
3. $a_n = (-1)^{n+1} 2^n$
4. $b_n = [1 + (-1)^n] \dfrac{n+1}{n}$

Determine a formula for a_n given the first few terms of the sequence in Exercises 5–8.

5. $8, 11, 14, 17, \ldots$
6. $-3, 9, -27, 81, \ldots$
7. $-\dfrac{2}{3}, -\dfrac{1}{4}, 0, \dfrac{1}{6}, \ldots$
8. $1, \sqrt{3}, \sqrt{5}, \sqrt{7}, \ldots$

In Exercises 9–11, find the second, third, fourth, and fifth terms of each sequence.

9. $x_1 = -2;\ x_{n+1} = x_n + 3$
10. $a_1 = \dfrac{1}{2};\ a_n = 2a_{n-1} - 3$
11. $b_1 = 2;\ b_{n+1} = b_n^2$

Write out each series in Exercises 12–14.

12. $\sum_{k=0}^{4} a_k$
13. $\sum_{m=0}^{5} \pi m$
14. $\sum_{k=1}^{4} \dfrac{(-1)^k}{3k}$

In Exercises 15–18, write each series using sigma summation notation.

15. $\dfrac{1}{2} + \dfrac{2}{3} + \dfrac{3}{4} + \cdots + \dfrac{n}{n+1}$
16. $1 + \dfrac{1}{4} + \dfrac{1}{9} + \cdots + \dfrac{1}{n^2}$
17. $-1 - \dfrac{1}{2} - \dfrac{1}{3} - \dfrac{1}{4} - \dfrac{1}{5}$

Evaluate each series in Exercises 18–23.

18. $\sum_{k=1}^{4} (k-1)^2$
19. $\sum_{k=3}^{4} (-1)^{2k} 2^{-k}$
20. $\sum_{m=1}^{3} [1 + (-1)^m] \dfrac{2m}{m+1}$
21. $\sum_{j=4}^{24} 100$
22. $\sum_{k=1}^{3} \dfrac{k+1}{k+2}$
23. $\sum_{m=1}^{4} \left(\dfrac{1}{m^2} - \dfrac{1}{(m+1)^2} \right)$

24. Compare the series $\sum_{m=3}^{5} \dfrac{m-2}{m+1}$ and $\sum_{k=1}^{3} \dfrac{k}{k+3}$.

25. Find the approximate value of a_{30} by repeated use of the square root key on a calculator if $a_1 = 8$ and $a_n = \sqrt{a_{n-1}}$.

26. Use your calculator to find a_{10000} if $a_n = \left(1 + \dfrac{2}{n}\right)^{n/2}$.

27. Use your calculator to find $1 + 1 + \dfrac{1}{2} + \dfrac{1}{6} + \dfrac{1}{24} + \dfrac{1}{120} + \dfrac{1}{720}$. Compare the results with Exercise 26.

Let S_n denote the sum of the first n terms of a sequence. In Exercises 28–29 use a calculator to find S_5 to four decimal places.

28. $a_n = \sqrt{n+1} - 2$
29. $b_1 = 0.1;\ b_{n+1} = 5b_n - b_n^2$

In Exercises 30–31 find a simplified formula for S_n, the sum of the first n terms.

30. $a_k = \dfrac{1}{k^2} - \dfrac{1}{(k+1)^2}$

31. $b_k = \sqrt{k+2} - \sqrt{k+1}$

If $\sum_{k=1}^{n} k = \dfrac{n(n+1)}{2}$ and $\sum_{k=1}^{n} k^2 = \dfrac{n(n+1)(2n+1)}{6}$, give each sum in Exercises 32–34.

32. $\sum_{k=1}^{6} k^2$

33. $\sum_{k=1}^{n} (k - k^2)$

34. $\sum_{k=1}^{n} \left(\dfrac{1}{2} - k\right)$

35. Prove that $\sum_{k=1}^{n} ca_k = c \sum_{k=1}^{n} a_k$.

36. STATISTICS The arithmetic mean $\bar{x}$ of a collection of data is given by $\bar{x} = \dfrac{1}{n} \sum_{k=1}^{n} x_k$. Use this series and the series used in Exercises 32–34 to find $\bar{x}$ for the data $1, 4, 9, \ldots, n^2$.

The slope m and y-intercept b in the least squares regression line $y = mx + b$ which best approximates a collection of n points (x_k, y_k) can be determined by solving the following system of equations.

$$nb + \left(\sum_{k=1}^{n} x_k\right) m = \sum_{k=1}^{n} y_k$$

$$\left(\sum_{k=1}^{n} x_k\right) b + \left(\sum_{k=1}^{n} x_k^2\right) m = \sum_{k=1}^{n} x_k y_k$$

Use this system to find the regression line for the data given in Exercises 37–38.

37. $(-1, 2), (1, -1), (5, -4)$

38. $(-4, -1), (-2, 2), (0, 1), (3, 5)$

FOR REVIEW

In Exercises 39–40, determine the type of conic section by completing the square.

39. $3x^2 - 6x + 4y = 4$

40. $8x^2 + 8y^2 + 16x + 16y - 64 = 0$

9.1 EXERCISES C

1. Prove that $(k+1)^2 - k^2 = 2k + 1$.

2. Prove that $\sum_{k=1}^{n} [(k+1)^2 - k^2] = (n+1)^2 - 1 = n^2 + 2n$.

3. Use the results of Exercises 1–2 and $\sum_{k=1}^{n} 1 = n$ to prove that $\sum_{k=1}^{n} k = \dfrac{n(n+1)}{2}$.

4. Use $(k+1)^3 - k^3 = 3k^2 + 3k + 1$, $\sum_{k=1}^{n} [(k+1)^3 - k^3] = (n+1)^3 - 1$, and $\sum_{k=1}^{n} 1 = n$ to prove that

$\sum_{k=1}^{n} k^2 = \dfrac{n(n+1)(2n+1)}{6}$.

9.2 ARITHMETIC SEQUENCES AND SERIES

STUDENT GUIDEPOSTS

1. Formula for a_n
2. Formulas for S_n
3. Summary of Formulas
4. Arithmetic Means

There are several special types of sequences which have a wide range of applications. The first of the two types that we will consider is illustrated by the following:

$$1,\ 4,\ 7,\ 10,\ 13,\ 16,\ \ldots\ .$$

Notice that $4 - 1 = 3$, $7 - 4 = 3$, $10 - 7 = 3$, $13 - 10 = 3$, and $16 - 13 = 3$. That is, the successive terms of the sequence differ by a constant.

Arithmetic Sequence

An **arithmetic sequence (progression)** is a sequence in which the successive terms differ by some constant d, called the **common difference.**

1 FORMULA FOR a_n

Since $a_n - a_{n-1} = d$ for all n, we have the recursive formula

$$a_n = a_{n-1} + d.$$

For arithmetic sequences we can develop a formula for a_n in terms of a_1, n, and d. Consider the following:

1st term $= a_1 = a_1 + \mathbf{0}d$
2nd term $= a_2 = a_1 + d = a_1 + \mathbf{1}d$
3rd term $= a_3 = a_2 + d = (a_1 + d) + d = a_1 + \mathbf{2}d$
4th term $= a_4 = a_3 + d = (a_1 + 2d) + d = a_1 + \mathbf{3}d$
5th term $= a_5 = a_4 + d = (a_1 + 3d) + d = a_1 + \mathbf{4}d$

Notice that the nth term a_n is the first term a_1 plus $n - 1$ times the common difference d. Thus,

$$a_n = a_1 + (n-1)d.$$

EXAMPLE 1   FINDING TERMS OF A SEQUENCE

Find the eighth and the twelfth terms of the arithmetic sequence 2, 7, 12, 17, 22, ...
We have $a_1 = 2$ and $d = 5$. Thus,

$$a_8 = a_1 + (8-1)d = 2 + (8-1)5$$
$$= 2 + 35 = 37$$

and $a_{12} = a_1 + (12-1)d = 2 + (11)5$
$$= 2 + 55 = 57.$$

Notice that the n in a_n is the same as the n in $(n - 1)$.

PRACTICE EXERCISE 1

Find the seventh and tenth terms of the sequence -2, 4, 10, 16, 22, ...

Answers: 34, 52

EXAMPLE 2 Finding an Arithmetic Sequence

Find x so that $x + 3$, $2x + 8$, and $4x + 15$ form a three-term arithmetic sequence in the given order. Also, give the sequence.

We use the fact that the difference between successive terms is equal to the common difference d.

$$(2x + 8) - (x + 3) = d \quad \text{and} \quad (4x + 15) - (2x + 8) = d$$

Set both expressions for d equal to each other.

$$(2x + 8) - (x + 3) = (4x + 15) - (2x + 8)$$
$$x + 5 = 2x + 7$$
$$-2 = x$$

Then $x + 3 = -2 + 3 = 1$, $2x + 8 = 2(-2) + 8 = 4$, $4x + 15 = 4(-2) + 15 = 7$, so that the desired arithmetic sequence is 1, 4, 7.

PRACTICE EXERCISE 2

Find x so that $2x$, $x + 3$, and $3x - 3$ form a three-term arithmetic sequence in the given order. Give the sequence.

Answers: $x = 3$; 6, 6, 6

❷ FORMULAS FOR S_n

The sum of the first n terms in an arithmetic sequence can also be determined by a formula. Let S_n denote the sum of the first n terms of an arithmetic sequence. Then we have the series

$$S_n = a_1 + a_2 + a_3 + a_4 + \cdots + a_n$$
$$= a_1 + a_1 + d + a_1 + 2d + a_1 + 3d + \cdots + a_1 + (n - 1)d.$$

Reversing the order of addition, we obtain

$$S_n = a_n + a_{n-1} + a_{n-2} + \cdots + a_1$$
$$= a_1 + (n - 1)d + a_1 + (n - 2)d + a_1 + (n - 3)d + \cdots + a_1.$$

Add corresponding terms in both representations of S_n.

$$S_n = a_1 \qquad\qquad + a_1 + \quad d + a_1 + \quad 2d + \cdots + a_1 + (n - 1)d$$
$$S_n = a_1 + (n - 1)d + a_1 + (n - 2)d + a_1 + (n - 3)d + \cdots + a_1$$
$$\overline{2S_n = 2a_1 + (n - 1)d + 2a_1 + (n - 1)d + 2a_1 + (n - 1)d + \cdots + 2a_1 + (n - 1)d}$$
$$2S_n = n[2a_1 + (n - 1)d] \qquad \text{There are } n \text{ terms of the form } 2a_1 + (n - 1)d$$

$$S_n = \frac{n}{2}[2a_1 + (n - 1)d]$$

EXAMPLE 3 Finding a_n and S_n

Find the fifteenth term and the sum of the first fifteen terms of the arithmetic sequence $-2, 1, 4, 7, 10, \ldots$.

We have $a_1 = -2$ and $d = 3$, and we must calculate a_{15} ($n = 15$) and S_{15}.

$$a_{15} = a_1 + (15 - 1)d = -2 + (14)(3) = -2 + 42 = 40$$
$$S_{15} = \frac{n}{2}[2a_1 + (n - 1)d] = \frac{15}{2}[2(-2) + (15 - 1)3]$$
$$= \frac{15}{2}[-4 + (14)3] = \frac{15}{2}[-4 + 42] = \frac{15}{2}[38] = 285$$

PRACTICE EXERCISE 3

Find a_{12} and S_{12} for the sequence 15, 11, 7, 3, $-1, \ldots$.

Answers: $a_{12} = -29$; $S_{12} = -84$

An alternate form for S_n is easily derived from

$$S_n = \frac{n}{2}[2a_1 + (n-1)d].$$

We write $2a_1$ as $a_1 + a_1$ and observe that $a_1 + (n-1)d = a_n$.

$$S_n = \frac{n}{2}[a_1 + \underbrace{a_1 + (n-1)d}_{a_n}]$$

$$S_n = \frac{n}{2}[a_1 + a_n]$$

In the preceding example, since we had already calculated $a_{15} = 40$, it would have been easier to substitute into the new formula for S_n.

$$S_{15} = \frac{15}{2}[a_1 + a_{15}] = \frac{15}{2}[-2 + 40] = \frac{15}{2}[38] = 285$$

❸ SUMMARY OF FORMULAS

The following theorem summarizes the formulas that have been derived for arithmetic sequences.

Formulas for Arithmetic Sequences

For an arithmetic sequence the general or nth term is given by

$$a_n = a_1 + (n-1)d.$$

The sum of the first n terms is given by

$$S_n = \frac{n}{2}[a_1 + a_n] \quad \text{or} \quad S_n = \frac{n}{2}[2a_1 + (n-1)d].$$

EXAMPLE 4   Finding d, Terms, and a Sum

Suppose that the fifteenth term of an arithmetic sequence is 71, and that the twenty-first term is 101. Find a_1, d, the first five terms, and the sum of the first five terms.

We have

$$a_{15} = 71 = a_1 + (15-1)d = a_1 + 14d$$

and

$$a_{21} = 101 = a_1 + (21-1)d = a_1 + 20d.$$

To find a_1 and d, we must solve the following system.

$$a_1 + 14d = 71$$
$$a_1 + 20d = 101$$

Subtract the first equation from the second.

$$6d = 30$$
$$d = 5$$

Then substitute this value for d in the first equation.

$$a_1 + 14 \cdot 5 = 71$$
$$a_1 = 71 - 70 = 1$$

PRACTICE EXERCISE 4

If $a_{11} = 20$ and $S_9 = 72$, find d, the first five terms, and S_{20}.

The first five terms are 1, 6, 11, 16, 21, and

$$S_5 = \frac{5}{2}[a_1 + a_5] = \frac{5}{2}[1 + 21] = \frac{5}{2}[22] = 55.$$

Answers: $d = 2$; 0, 2, 4, 6, 8; $S_{20} = 380$

There are many applications of arithmetic sequences. Consider the following depreciation problem.

EXAMPLE 5 APPLICATION TO CONSTRUCTION

An earth mover is purchased for $260,000. Assume that it depreciates 7.0% the first year, 6.5% the second year, 6.0% the third year, and continues in the same manner for ten years. If all depreciations apply to the original cost, what is the value of the earth mover in ten years?

We calculate the sum of the depreciations over the ten years with $a_1 = 7.0$, $a_2 = 6.5$, $a_3 = 6.0$, . . . and $d = -0.5$.

$$S_{10} = \frac{10}{2}[2(7.0) + (10-1)(-0.5)]$$

$$= 5[14.0 - 4.5] = 47.5$$

Thus, the total percentage depreciation in ten years is 47.5%. This means that the equipment is worth 52.5% of its original value.

$$0.525(260,000) = 136,500$$

The earth mover is worth $136,500 in ten years.

PRACTICE EXERCISE 5

The enrollment at a university is 10,300. If the increases for the coming years are 3%, 5%, 7%, 9%, . . . , what will be the enrollment in 8 years if all increases are based on the 10,300?

Answer: 18,540

④ ARITHMETIC MEANS

We now consider a property of arithmetic sequences that is a generalization of the average or arithmetic mean of two numbers. Notice that since the average $\frac{a+b}{2}$ is midway between the numbers a and b on a number line,

$$a, \frac{a+b}{2}, b$$

is an arithmetic sequence. We define **arithmetic means** to be all the terms of an arithmetic sequence between a_1 and a_n.

EXAMPLE 6 DETERMINING ARITHMETIC MEANS

Insert four arithmetic means between the numbers 8 and -7.

Since four means must be inserted between 8 and -7, there will be six terms in the sequence, with $a_1 = 8$ and $a_6 = -7$. Use $a_n = a_1 + (n-1)d$ with $n = 6$, $a_1 = 8$, and $a_6 = -7$ to find d.

$$a_6 = a_1 + (6-1)d$$
$$-7 = 8 + 5d$$
$$-3 = d$$

Since $8 + (-3) = 5$, $5 + (-3) = 2$, $2 + (-3) = -1$, and $-1 + (-3) = -4$, the four arithmetic means between 8 and -7 are 5, 2, -1, and -4.

PRACTICE EXERCISE 6

Insert five arithmetic means between 0.5 and 9.5.

Answers: 2.0, 3.5, 5.0, 6.5, 8.0

9.2 EXERCISES A

In Exercises 1–3, tell if each sequence is arithmetic. If it is, give the common difference d.

1. 4, 8, 12, 16, . . .
2. 9, −1, −11, −21, . . .
3. 1, −1, 1, −1, . . .

Find the first six terms of each arithmetic sequence in Exercises 4–6.

4. $a_1 = 2$ and $d = 7$
5. $a_1 = -2$ and $a_2 = 5$
6. $a_1 = \sqrt{3}$ and $d = 4\sqrt{3}$

7. Find x so that x, $x + 4$, and $2x$ form a three-term arithmetic sequence in the given order. Give the sequence.

In Exercises 8–10, find the indicated sum of the arithmetic sequence.

8. $-8, -1, 6, 13, \ldots$; S_{10}
9. $a_4 = 6$ and $a_8 = 26$; S_{12}
10. $\sum_{k=1}^{6}(k+2)$

In Exercises 11–16, some of the numbers n, a_1, a_n, d, and S_n are given. Find the missing ones.

11. $a_1 = 2$, $n = 17$, $d = 3$
12. $a_n = 27$, $S_n = 63$, $a_1 = -9$
13. $a_1 = \dfrac{5}{3}$, $d = \dfrac{1}{6}$, $n = 12$
14. $a_{15} = 4$, $S_{15} = 30$
15. $a_1 = -9$, $a_7 = 21$
16. $a_1 = \log 7$, $d = \log 49$, $n = 5$

17. Insert six arithmetic means between 11 and 32.

18. How many integers between 39 and 146 are divisible by 5?

19. Find the sum of all the even integers between 1 and 201.

20. Prove that the sum of the sequence 2, 4, 6, . . . , 2n is $n^2 + n$.

Solve.

21. **CONSUMER** A new car costs $8400. Assume that it depreciates 2.1% the first year, 1.8% the second year, 1.5% the third year, and continues in the same manner for twelve years. If all depreciations apply to the original cost, what is the value of the car (rounded to the nearest dollar) in twelve years?

22. **RECREATION** A theater has 40 rows with 20 seats in the first row, 23 in the second row, 26 in the third row, and so forth. How many seats are in the theater?

23. **BANKING** If a woman puts $1 in a bank on the first day of September, $2 on the second day of September, $3 on the third day, and so forth, how much money will be in the bank at the end of the month?

24. **PHYSICS** A rock is dropped from the top of a tall cliff and falls 16 ft during the first second, 48 ft during the second, 80 ft during the third, and so on. How many feet does the rock fall during the eighth second?

FOR REVIEW

In Exercises 25–26, give the fourth and fifth terms of each general sequence.

25. $x_n = (-1)^{n+1} 3^{-n}$

26. $x_1 = 6$ and $x_{k+1} = 1 - x_k$

27. Write $\dfrac{1}{2} + \dfrac{2}{5} + \dfrac{3}{10} + \cdots + \dfrac{n}{n^2 + 1}$ in sigma summation notation.

28. Evaluate. $\displaystyle\sum_{k=2}^{4} \dfrac{1}{3k - 2}$

ANSWERS: 1. yes; 4 2. yes; −10 3. no 4. 2, 9, 16, 23, 30, 37 5. −2, 5, 12, 19, 26, 33 6. $\sqrt{3}$, $5\sqrt{3}$, $9\sqrt{3}$, $13\sqrt{3}$, $17\sqrt{3}$, $21\sqrt{3}$ 7. $x = 8$; 8, 12, 16 8. 235 9. 222 10. 33 11. $a_{17} = 50$; $S_{17} = 442$ 12. $n = 7$; $d = 6$ 13. $a_{12} = \frac{7}{2}$; $S_{12} = 31$ 14. $n = 15$; $d = \frac{2}{7}$; $a_1 = 0$ 15. $n = 7$; $d = 5$; $S_7 = 42$ 16. $a_5 = 9 \log 7$; $S_5 = 25 \log 7$ 17. 14, 17, 20, 23, 26, 29 18. 22 19. 10,100 20. proof 21. $7946 22. 3140 23. $465 24. 240 ft 25. $x_4 = -\frac{1}{81}$; $x_5 = \frac{1}{243}$ 26. $x_4 = -5$; $x_5 = 6$ 27. $\displaystyle\sum_{k=1}^{n} \dfrac{k}{k^2 + 1}$ 28. $\dfrac{69}{140}$

9.2 EXERCISES B

In Exercises 1–3, tell if each sequence is arithmetic. If it is, give the common difference d.

1. 4, 8, 10, 12, . . .

2. $\dfrac{2}{3}, 2, \dfrac{10}{3}, \dfrac{14}{3}, \ldots$

3. log 2, log 4, log 8, log 16, . . .

Find the first six terms of each arithmetic sequence in Exercises 4–6.

4. $a_1 = 8$ and $d = -5$

5. $a_1 = 2$ and $a_6 = -13$

6. $a_1 = \ln 10$ and $a_3 = \ln 1000$

7. Find y so that $2y$, $3y + 7$, and $5y + 1$ form a three-term arithmetic sequence in the given order. Give the sequence.

In Exercises 8–10 find the indicated sum of the arithmetic sequence.

8. $a_1 = 3$ and $a_{10} = 57$; S_{10}

9. $a_3 = 9$ and $a_5 = -1$; S_{12}

10. $\displaystyle\sum_{k=1}^{30} (1 - 2k)$

In Exercises 11–16, some of the numbers n, a_1, a_n, d, and S_n are given. Find the missing ones.

11. $a_1 = 24$, $a_n = 3$, $n = 8$

12. $a_1 = -40$, $S_{21} = 210$

13. $a_1 = -7$, $d = 8$, $S_n = 225$

14. $a_1 = 10$, $a_n = -8$, $S_n = 7$

15. $a_1 = \dfrac{3}{4}$, $a_5 = -\dfrac{1}{4}$

16. $a_1 = \ln 3$, $a_4 = \ln 81$

17. Insert eight arithmetic means between 47 and 11.

18. How many integers between −92 and 261 are divisible by 7?

19. Find the sum of all the integers divisible by 5 from 25 through 350.

20. Prove that the sum of the first n odd natural numbers is n^2.

Solve.

21. **BUSINESS** A man earned $3500 the first year he worked. If he received a raise of $750 at the end of each year for 20 years, what was his salary during his twenty-first year of work? How much income did he have during the first 21 years of work?

22. **COINS** A collection of nickels is arranged in a triangular array with 15 coins in the base row, 14 in the next, 13 in the next, and so forth. Find the value of the collection.

23. **FORESTRY** There are 190 logs to be stacked in such a way that one log will be on top, 2 logs in the second row, 3 logs in the third, and so on. How many logs should be put in the row on the bottom of the stack?

24. **DEMOGRAPHY** The population of a town is decreasing by 500 inhabitants each year. If its population at the beginning of 1980 was 20,135, what was its population at the beginning of 1990?

FOR REVIEW

In Exercises 25–26, give the third and sixth terms of each general sequence.

25. $x_n = (-1)^{n+1} 2^{n+1}$

26. $x_1 = -2$ and $x_{k+1} = 1 + 2x_k$

27. Write $\dfrac{1}{3} + \dfrac{4}{6} + \dfrac{9}{11} + \cdots + \dfrac{n^2}{n^2 + 2}$ in sigma summation notation.

28. Evaluate. $\displaystyle\sum_{k=3}^{6} \dfrac{k}{2k - 3}$

9.2 EXERCISES C

1. Use a formula from this section to prove that
$$\sum_{k=1}^{n} k = \frac{n(n+1)}{2}.$$

2. Let $a_k = 4k - 1$. Show that $a_1, a_2, a_3, \ldots$ is an arithmetic sequence and derive a formula for the sum of the first n terms.

3. Let $f(x) = mx + b$. Prove that if $a_n = f(n)$ then $a_1, a_2, a_3, \ldots$ is an arithmetic sequence.

4. Assume that $a_{k+1} = \dfrac{a_k}{1 + a_k}$. If $b_k = \dfrac{1}{a_k}$, prove that $b_1, b_2, b_3, \ldots$ is an arithmetic sequence.

9.3 GEOMETRIC SEQUENCES

STUDENT GUIDEPOSTS

1. Formula for a_n
2. Formulas for S_n
3. Summary of Formulas

The second type of special sequence that we consider is illustrated by

$$1, \frac{1}{2}, \frac{1}{4}, \frac{1}{8}, \frac{1}{16}, \frac{1}{32}, \ldots$$

For this sequence

$$\frac{a_2}{a_1} = \frac{1}{2}, \quad \frac{a_3}{a_2} = \frac{1}{2}, \quad \frac{a_4}{a_3} = \frac{1}{2}, \quad \frac{a_5}{a_4} = \frac{1}{2}, \quad \frac{a_6}{a_5} = \frac{1}{2}, \quad \text{and so on.}$$

Geometric Sequence

A **geometric sequence (progression)** is a sequence with the property that

$$\frac{a_n}{a_{n-1}} = r$$

for all n. The number r is called the **common ratio.**

1 FORMULA FOR a_n

Since $\frac{a_n}{a_{n-1}} = r$, then $a_n = a_{n-1}r$. We can use this relation to derive a formula for the nth or general term of a geometric sequence.

$$\text{1st term} = a_1 = a_1 r^0$$
$$\text{2nd term} = a_2 = a_1 \cdot r = a_1 r^1$$
$$\text{3rd term} = a_3 = a_2 \cdot r = (a_1 r) \cdot r = a_1 r^2$$
$$\text{4th term} = a_4 = a_3 \cdot r = (a_1 r^2) \cdot r = a_1 r^3$$
$$\text{5th term} = a_5 = a_4 \cdot r = (a_1 r^3) \cdot r = a_1 r^4$$

Notice that the nth term a_n is equal to the first term a_1 times r^{n-1}. Thus,

$$a_n = a_1 r^{n-1}.$$

EXAMPLE 1 FINDING TERMS OF A SEQUENCE

Find the seventh and tenth terms of the geometric sequence

$$-2, \ 1, \ -\frac{1}{2}, \ \frac{1}{4}, \ -\frac{1}{8}, \ \ldots$$

We have $a_1 = -2$ and $r = -1/2$.

$$a_7 = a_1 r^{7-1} = (-2)\left(-\frac{1}{2}\right)^6 = -\frac{1}{32}$$

$$a_{10} = a_1 r^{10-1} = (-2)\left(-\frac{1}{2}\right)^9 = \frac{1}{256}$$

Notice that the n in a_n is the same as the n in the exponent $n-1$.

PRACTICE EXERCISE 1

Find the fifth and ninth terms of the sequence $1, -3, 9, -27, \ldots$

Answers: 81, 6561

EXAMPLE 2 FINDING A GEOMETRIC SEQUENCE

Find x so that $x - 3$, $x - 1$, and $2x + 1$ form a three-term geometric sequence in the given order. Also, give the sequence.

Use the fact that the ratio of successive terms is equal to the common ratio r.

$$\frac{x-1}{x-3} = \frac{2x+1}{x-1}$$

$$(x-1)(x-1) = (x-3)(2x+1)$$

$$x^2 - 2x + 1 = 2x^2 - 5x - 3$$

$$0 = x^2 - 3x - 4$$

$$0 = (x+1)(x-4)$$

$$x + 1 = 0 \quad \text{or} \quad x - 4 = 0$$

$$x = -1 \qquad\qquad x = 4$$

If $x = -1$:

$x - 3 = -1 - 3 = -4$
$x - 1 = -1 - 1 = -2$
$2x + 1 = 2(-1) + 1 = -1;$

the sequence is

$-4, \ -2, \ -1 \ \left(r = \dfrac{1}{2}\right)$.

If $x = 4$:

$x - 3 = 4 - 3 = 1$
$x - 1 = 4 - 1 = 3$
$2x + 1 = 2(4) + 1 = 9;$

the sequence is

$1, \ 3, \ 9 \ (r = 3)$.

PRACTICE EXERCISE 2

Find y and the sequence if $3y + 2$, $y + 2$, and y form a three-term geometric sequence in the given order.

Answers: $y = 2$; 8, 4, 2 or $y = -1$; $-1, 1, -1$

② FORMULAS FOR S_n

We now derive a formula for calculating the sum of the first n terms of a geometric sequence. Let S_n denote the sum of the first n terms. The series is

$$S_n = a_1 + a_1 r + a_1 r^2 + a_1 r^3 + \cdots + a_1 r^{n-2} + a_1 r^{n-1}.$$

Multiply by r.

$$rS_n = a_1 r + a_1 r^2 + a_1 r^3 + \cdots + a_1 r^{n-2} + a_1 r^{n-1} + a_1 r^n$$

Subtract.

$$S_n = a_1 + a_1 r + a_1 r^2 + a_1 r^3 + \cdots + a_1 r^{n-2} + a_1 r^{n-1}$$
$$-rS_n = \quad\quad -a_1 r - a_1 r^2 - a_1 r^3 - \cdots - a_1 r^{n-2} - a_1 r^{n-1} - a_1 r^n$$
$$S_n - rS_n = a_1 \qquad\qquad\qquad\qquad\qquad\qquad\qquad\qquad\quad - a_1 r^n$$
$$S_n(1 - r) = a_1 - a_1 r^n$$
$$S_n = \frac{a_1 - a_1 r^n}{1 - r}$$

EXAMPLE 3 FINDING a_n AND S_n

Find the ninth term and the sum of the first nine terms of the following geometric sequence.

$$\frac{2}{5}, \ \frac{1}{5}, \ \frac{1}{10}, \ \frac{1}{20}, \ \frac{1}{40}, \ \ldots$$

PRACTICE EXERCISE 3

Find a_{11} and S_{11} for the sequence 3, -6, 12, $-24, \ldots$.

610 CHAPTER 9 SEQUENCES, SERIES, AND PROBABILITY

We have $a_1 = 2/5$, $r = 1/2$, and we must calculate a_9 ($n = 9$) and S_9.

$$a_9 = a_1 r^{9-1} = \left(\frac{2}{5}\right)\left(\frac{1}{2}\right)^8 = \left(\frac{2}{5}\right)\left(\frac{1}{256}\right) = \frac{1}{640}$$

$$S_9 = \frac{a_1 - a_1 r^9}{1 - r} = \frac{\frac{2}{5} - \frac{2}{5}\left(\frac{1}{2}\right)^9}{1 - \frac{1}{2}} = \frac{\left(\frac{2}{5}\right)\left[1 - \left(\frac{1}{2}\right)^9\right]}{\frac{1}{2}} = \frac{\frac{2}{5}\left(1 - \frac{1}{512}\right)}{\frac{1}{2}}$$

$$= \frac{2}{5}\left(\frac{511}{512}\right)\left(\frac{2}{1}\right) = \frac{511}{640}$$

Answers: $a_{11} = 3072$, $S_{11} = 2049$

An alternate form for S_n is easily derived from

$$S_n = \frac{a_1 - a_1 r^n}{1 - r}$$

by writing $a_1 r^n = r(a_1 r^{n-1}) = r a_n$.

$$S_n = \frac{a_1 - r a_n}{1 - r}$$

In the preceding example, since we had already calculated $a_9 = 1/640$, it would have been easier to substitute into the new formula for S_n.

$$S_9 = \frac{a_1 - r a_9}{1 - r} = \frac{\frac{2}{5} - \left(\frac{1}{2}\right)\left(\frac{1}{640}\right)}{1 - \frac{1}{2}} = \frac{\frac{2}{5} - \frac{1}{1280}}{\frac{1}{2}} = \frac{\frac{512}{1280} - \frac{1}{1280}}{\frac{1}{2}} = \frac{511}{640}$$

③ SUMMARY OF FORMULAS

A summary of the formulas derived for geometric sequences is given in the following theorem.

> **Formulas for Geometric Sequences**
>
> For a geometric sequence the general or nth term is given by
>
> $$a_n = a_1 r^{n-1}.$$
>
> The sum of the first n terms is given by
>
> $$S_n = \frac{a_1 - r a_n}{1 - r} \quad \text{or} \quad S_n = \frac{a_1 - a_1 r^n}{1 - r}.$$

EXAMPLE 4 FINDING r, A TERM, AND A SUM

Suppose that the third term of a geometric sequence is 27 and the fifth term is 243. Find a_1, r, and S_5.

$$a_3 = a_1 r^{3-1} = a_1 r^2 = 27 \quad \text{and} \quad a_5 = a_1 r^{5-1} = a_1 r^4 = 243$$

PRACTICE EXERCISE 4

If in a geometric sequence $r = 4$ and $S_4 = 425$, find a_1, a_6, and S_6.

Solve the first equation for a_1, $a_1 = 27/r^2$, and substitute into the second.

$$a_1 r^4 = \frac{27}{r^2} \cdot r^4 = 27r^2 = 243$$

$$r^2 = 9$$

$$r = \pm\sqrt{9} = \pm 3$$

We obtain two different solutions since there are two values for r.

If $r = 3$: $a_1 r^2 = 27$
$a_1(3)^2 = 27$
$a_1 = 3$

If $r = -3$: $a_1 r^2 = 27$
$a_1(-3)^2 = 27$
$a_1 = 3$

The first sequence is 3, 9, 27, 81, 243, . . .

The second sequence is 3, −9, 27, −81, 243, . . .

$$S_5 = \frac{a_1 - ra_5}{1 - r}$$

$$= \frac{3 - 3 \cdot 243}{1 - 3}$$

$$= \frac{3(1 - 243)}{-2}$$

$$= 363$$

$$S_5 = \frac{a_1 - ra_5}{1 - r}$$

$$= \frac{3 - (-3)(243)}{1 - (-3)}$$

$$= \frac{3(1 + 243)}{4}$$

$$= 183$$

The sum in the first case, where all signs are positive, is greater than the second sum, which includes negative terms.

Answers: $a_1 = 5$, $a_6 = 5120$, and $S_6 = 6825$

EXAMPLE 5   A BANKING PROBLEM

Gloria Bell borrows $1000.00 at 12% interest compounded annually. If she repays the loan in full at the end of three years, how much does she pay?

At beginning of first year
$a_1 = 1000$

At beginning of second year (or end of first year)
$a_2 = 1000 + 0.12(1000)$
$= [1 + 0.12](1000)$
$= (1.12)(1000)$

At beginning of third year (or end of second year)
$a_3 = (1.12)(1000)$
$+ (0.12)(1.12)(1000)$
$= [1 + 0.12](1.12)(1000)$
$= (1.12)^2(1000)$

To obtain the next term of the sequence, multiply the preceding term by 1.12. Thus,

$$a_n = a_1(1.12)^{n-1}.$$

To find n we must count the amount borrowed as the first term in the sequence

$$1000, \ 1000(1.12), \ 1000(1.12)^2, \ 1000(1.12)^3.$$

There are four terms ($n = 4$) in the sequence for the three-year period. Thus,

$$a_4 = 1000(1.12)^{4-1} = 1000(1.12)^3 \approx 1404.93.$$

She has to pay $1404.93 at the end of the three years.

PRACTICE EXERCISE 5

At the beginning of each year for three years Mike Horn deposited $1000 in an account paying 8% compounded annually. What was the value of the account at the end of the third year?

Answer: $3506.11

9.3 EXERCISES A

Find the first six terms of each geometric sequence in Exercises 1–3.

1. $a_1 = 4$ and $r = 2$
2. $a_1 = -16$ and $r = -\dfrac{1}{2}$
3. $a_1 = \sqrt{2}$ and $r = -\sqrt{2}$

4. Find x so that $x + 7$, $x - 3$, and $x - 8$ form a three-term geometric sequence in the given order. Give the sequence.

In Exercises 5–10, find the indicated sum of each geometric sequence.

5. $\dfrac{1}{6}, \dfrac{1}{12}, \dfrac{1}{24}, \ldots; S_9$
6. $4, 24, 144, \ldots; S_6$
7. $a_3 = \dfrac{1}{6}$ and $a_5 = \dfrac{1}{24}; S_5$

8. $a_2 = 10$ and $a_4 = 40; S_8$
9. $\displaystyle\sum_{k=1}^{7} \left(\dfrac{5}{6}\right)^k$
10. $\displaystyle\sum_{n=2}^{5} 4^n$

In Exercises 11–13, some of the numbers n, a_1, a_n, r, and S_n are given. Find the missing ones.

11. $a_1 = 2, n = 6, r = 2$
12. $r = \dfrac{1}{2}, a_9 = 1$
13. $a_7 = \dfrac{1}{5}, r = \dfrac{1}{5}$

14. The terms between a_1 and a_n of a geometric sequence are called the **geometric means** of a_1 and a_n. Insert four geometric means between 5 and -160.

Solve.

15. Find all possible values of r if $a_1 = 3$ and $S_3 = 9$.

16. **CONSUMER** A new car costing $6400 depreciates 20% of its value each year. How much is the car worth at the end of six years?

17. **BANKING** Lori Wade borrows $2000.00 at 11% interest compounded annually. If she pays off the loan in full at the end of four years, how much does she pay?

18. **ECONOMICS** Louis was offered a job for the month of June (thirty days), and was told he would be paid $0.01 at the end of the first day, $0.02 at the end of the second day, $0.04 at the end of the third day, and so forth, doubling each previous day's salary. However, Louis refused the job thinking that the pay was inferior. Would you take the job? Why?

19. **PHYSICS** The tip of a pendulum sweeps out an arc of 20 cm on its first pass. If on each succeeding pass the distance traveled is 4/5 of the distance of the preceding pass, how far has it traveled by the end of the fourth arc?

20. **PHYSICS** A ball is dropped from a height of 12.0 feet. If on each rebound it rises to a height of 3/4 the distance from which it fell, how far (up and down) will the ball have traveled when it hits the ground for the eighth time?

21. **INVESTMENT** If Wade Wright invests $1000 at the first of each year from 1990 through 1999 and interest is 8% compounded annually, how much will he have in the account at the start of the next century?

FOR REVIEW

22. Find the first five terms of the arithmetic sequence with $a_1 = 7$ and $d = -4$.

23. Find the sum of the first nine terms of the arithmetic sequence $-\frac{1}{2}, 0, \frac{1}{2}, 1, \frac{3}{2}, \ldots$.

24. Insert four arithmetic means between 12 and -13.

25. **RECREATION** A theater has 50 rows with 15 seats in the first row, 17 in the second row, 19 in the third row, and so forth. How many seats are in the theater?

ANSWERS: 1. 4, 8, 16, 32, 64, 128 2. $-16, 8, -4, 2, -1, \frac{1}{2}$ 3. $\sqrt{2}, -2, 2\sqrt{2}, -4, 4\sqrt{2}, -8$ 4. $x = 13$; 20, 10, 5 5. $\frac{511}{1536}$ 6. 37,324 7. $\frac{11}{24}$ or $\frac{31}{24}$ 8. $1275(r = 2)$ or $425(r = -2)$ 9. 3.6046 10. 1360 11. $a_6 = 64$; $S_6 = 126$ 12. $n = 9$; $a_1 = 256$; $S_9 = 511$ 13. $n = 7$; $a_1 = 3125$; $S_7 = \frac{19531}{5}$ 14. $-10, 20, -40, 80$ 15. $1, -2$ 16. $1678 17. $3036.14 18. yes, if you would work for approximately $10,700,000 for the month 19. $\frac{1476}{25}$ cm, approximately 59.0 cm 20. 74.4 ft 21. $15,645.49 22. 7, 3, $-1, -5, -9$ 23. $\frac{27}{2}$ 24. 7, 2, $-3, -8$ 25. 3200

9.3 EXERCISES B

Find the first six terms of each geometric sequence in Exercises 1–3.

1. $a_1 = -\frac{1}{8}$ and $r = -2$

2. $a_1 = 64$ and $a_6 = \frac{1}{16}$

3. $a_1 = \sqrt{3}$ and $r = 3\sqrt{3}$

4. Find y so that $2y + 5$, $y + 7$, and $3y - 7$ form a three-term geometric sequence in the given order. Give the sequence.

In Exercises 5–10, find the indicated sum of each geometric sequence.

5. $\frac{1}{4}, \frac{1}{12}, \frac{1}{36}, \ldots$; S_9

6. $-4, 20, -100, \ldots$; S_5

7. $a_1 = -\frac{1}{9}$ and $a_6 = -27$; S_6

8. $a_3 = 15$ and $a_5 = 60$; S_6

9. $\sum_{k=1}^{5} \left(\frac{3}{5}\right)^k$

10. $\sum_{n=2}^{6} 5^n$

In Exercises 11–13, some of the numbers n, a_1, a_n, r, and S_n are given. Find the missing ones.

11. $a_1 = 1$, $r = -2$, $a_n = 64$

12. $a_1 = 2$, $a_n = 32$, $S_n = 62$

13. $r = -2$, $S_n = -63$, $a_n = -96$

14. The terms between a_1 and a_n of a geometric sequence are called the **geometric means** of a_1 and a_n. Insert four geometric means between -8 and $\frac{1}{4}$.

Solve.

15. Find all possible values of r if $a_3 = 32$ and $S_2 = 10$.

16. **CONSUMER** A new car costing $5800 depreciates 25% of its value each year. How much is the car worth at the end of five years?

17. **BANKING** Rafael Mendez borrows $1000.00 at 14% interest compounded annually. If he pays off the loan in full at the end of five years, how much does he pay?

18. **DEMOGRAPHY** The population of a town is increasing by 20% each year. If its present population is 1250, what will be its population seven years from now?

19. **PHYSICS** A ping-pong ball dropped from a height of 32 ft always rebounds 1/4 of the distance of the previous fall. What distance does it rebound the seventh time?

20. **PHYSICS** A ball is dropped from a height of 18.0 feet. If on each rebound it rises to a height of 2/3 the distance from which it fell, how far (up and down) will the ball have traveled when it hits the ground for the sixth time?

21. **INVESTMENT** If Randee Wire invests $1000 at the first of each year from 1990 through 1999 and interest is 12% compounded annually, how much will she have in the account at the start of the next century?

FOR REVIEW

22. Find the first five terms of the arithmetic sequence with $a_1 = -12$ and $a_6 = 23$.

23. Find the sum of the first nine terms of the arithmetic sequence if $a_1 = -18$ and $a_9 = 38$.

24. Find the sum of the integers divisible by 12 between -111 and 25.

25. **DEMOGRAPHY** The population of a town is increasing by 700 inhabitants each year. If its present population is 1250, what will be its population seven years from now?

9.3 EXERCISES C

1. If a, b, c is a geometric sequence, then b is the geometric mean between a and c. Prove that if all the terms are positive, then $b = \sqrt{ac}$.

2. Consider the geometric sequence $1, \frac{1}{2}, \frac{1}{4}, \frac{1}{8}, \ldots$. Without making an exact calculation, determine an approximate value for S_{100}. [Answer: 2]

3. Let $a_1, a_2, a_3, \ldots$ be a geometric sequence. Prove that $\sqrt{a_1}, \sqrt{a_2}, \sqrt{a_3}, \ldots$ is a geometric sequence.

4. Let $a_1, a_2, a_3, \ldots$ be an arithmetic sequence and b a positive real number. Prove that $b^{a_1}, b^{a_2}, b^{a_3}, \ldots$ is a geometric sequence.

9.4 INFINITE GEOMETRIC SEQUENCES AND SERIES

STUDENT GUIDEPOSTS

1. Infinite Geometric Sequences
2. Sum of an Infinite Sequence
3. Converting Repeating Decimals to Fractions

1 INFINITE GEOMETRIC SEQUENCES

In order to understand better the concept of infinite sequence, consider the following physical problem. A radioactive substance changes to other elements as time passes. The time required for one-half of the substance to change (decay) is called the half-life. Suppose the half-life of one radioactive substance is one day. Then each day one-half of the substance present at the beginning of the day will decay.

If we start with 1 gram, after one day 0.5 grams will have decayed. Since the second day starts with 0.5 grams, one-half of that or 0.25 grams will decay the second day. The number of grams that decay each day defines a geometric sequence with $r = 0.5$.

Grams that decay first day: $a_1 = 0.5 = (0.5)^1$
Grams that decay second day: $a_2 = 0.25 = (0.5)^2$
Grams that decay third day: $a_3 = 0.125 = (0.5)^3$
$\vdots$
Grams that decay nth day: $a_n = (0.5)^n$

If we use a calculator to find the sum of the first n terms of the sequence for $n = 1, 2, 3, 4, 5, \ldots, 12$, we obtain the following numbers rounded to four decimal places.

$S_1 = 0.5000 \quad S_7 = 0.9922$
$S_2 = 0.7500 \quad S_8 = 0.9961$
$S_3 = 0.8750 \quad S_9 = 0.9980$
$S_4 = 0.9375 \quad S_{10} = 0.9990$
$S_5 = 0.9688 \quad S_{11} = 0.9995$
$S_6 = 0.9844 \quad S_{12} = 0.9998$

The S_n are getting close to 1.0000 as n increases, but that is what we would expect since in the physical problem we started with 1 gram of substance. However, even for large n the value of S_n will never be exactly 1 since each day only one-half of the substance decays.

2 SUM OF AN INFINITE SEQUENCE

If we could add up all the terms of the infinite sequence,

$$0.5, (0.5)^2, (0.5)^3, (0.5)^4, \ldots, (0.5)^n, \ldots,$$

we would then expect to obtain the exact value of 1. Using our normal adding process we cannot actually do this, but we can define a way to determine infinite sums such as the one in the radioactive decay problem. If we consider the sequence

$$S_1, S_2, S_3, S_4, \ldots, S_n, \ldots,$$

which is the **sequence of partial sums** of the given sequence

$$a_1, a_2, a_3, a_4, \ldots, a_n, \ldots,$$

and the S_n's approach some fixed number S as n gets larger and larger, then S is called the **sum of the geometric sequence**. We denote this as

$$S = a_1 + a_2 + a_3 + a_4 + \cdots = \sum_{n=1}^{\infty} a_n.$$

For the example of radioactive decay,

$$S = 0.5 + (0.5)^2 + (0.5)^3 + (0.5)^4 + \cdots = \sum_{n=1}^{\infty} (0.5)^n = 1.$$

In the definition of the sum of an infinite sequence, we require the S_n to approach a fixed number S in order to have a sum. For the geometric sequence

$$10, 10, 10, 10, \ldots$$

where $r = 1$, the sequence of partial sums is

$$10, 20, 30, 40, 50, 60, \ldots,$$

and the S_n's get larger and larger and do not approach a fixed number. By looking at the formula for S_n given in Section 9.3, we can show that a geometric sequence has a sum for $|r| < 1$ and does not for $|r| \geq 1$. Suppose we write the formula as follows:

$$S_n = \frac{a_1 - a_1 r^n}{1 - r} = \frac{a_1}{1 - r} - \frac{a_1}{1 - r} r^n.$$

For the example with $r = 0.5$,

$$S_n = \frac{a_1}{1 - r} - \frac{a_1}{1 - r}(0.5)^n.$$

As n becomes large, $(0.5)^n$ becomes small and $\frac{a_1}{1-r}(0.5)^n$ approaches zero. Thus the sum is approximately $\frac{a_1}{1-r}$. In fact, for any r such that $-1 < r < 1$ or $|r| < 1$, as n increases r^n goes to zero, and we have the following result.

Sum of Infinite Geometric Sequence

An infinite geometric sequence, with first term a_1 and common ratio r satisfying $|r| < 1$, has a sum given by

$$S = \frac{a_1}{1 - r}.$$

For example, the geometric sequence

$$5, \frac{5}{3}, \frac{5}{9}, \frac{5}{27}, \ldots$$

with $r = \frac{1}{3}$, has sum

$$S = \frac{a_1}{1 - r} = \frac{5}{1 - \frac{1}{3}} = \frac{5}{\frac{2}{3}} = \frac{15}{2}.$$

We have already stated that for $r = 1$, there is no sum of an infinite geometric sequence, but what about $|r| > 1$? In this case the r^n in the formula for S_n

becomes large as *n* increases and the infinite sequence has no sum. For example, the sequence

$$3, 9, 27, 81, \ldots, 3^n, \ldots$$

with $r = 3$, has no sum.

EXAMPLE 1 USING THE SUM FORMULA

Find the first five terms of an infinite geometric sequence with $a_1 = 5/2$ and $S = 5$.

Substitute into the formula.

$$S = \frac{a_1}{1-r}$$

$$5 = \frac{\frac{5}{2}}{1-r}$$

$$5(1-r) = \frac{5}{2}$$

$$5 - 5r = \frac{5}{2}$$

$$r = \frac{1}{2}$$

Since $|r| < 1$, there is an infinite sequence with the conditions stated and the first five terms are

$$\frac{5}{2}, \frac{5}{4}, \frac{5}{8}, \frac{5}{16}, \frac{5}{32}.$$

PRACTICE EXERCISE 1

Find the first five terms of the infinite geometric sequence with $a_1 = 8$ and $S = 2/3$.

Answer: no solution; $r = -11$ and $|-11| > 1$

CAUTION

In Example 1 if *r* had been such that $|r| \geq 1$, there would have been no sequence satisfying the given properties. See Practice Exercise 1.

③ CONVERTING REPEATING DECIMALS TO FRACTIONS

A repeating decimal can be written as an infinite geometric sequence, and we can use the formula for *S* to change the decimal to a fraction.

EXAMPLE 2 CONVERTING DECIMALS TO FRACTIONS

Convert $2.3\overline{4}$ to a fraction.

First write $2.3\overline{4}$ as follows:

$$2.3\overline{4} = 2.3 + 0.04 + 0.004 + 0.0004 + \cdots$$

$$= \frac{23}{10} + (0.04 + 0.004 + 0.0004 + \cdots).$$

PRACTICE EXERCISE 2

Convert $5.\overline{63}$ to a fraction.

618 CHAPTER 9 SEQUENCES, SERIES, AND PROBABILITY

The series $(0.04 + 0.004 + 0.0004 + \cdots)$ is an infinite geometric series with $a_1 = 0.04$ and $r = 0.1$.

$$S = \frac{a_1}{1-r} = \frac{0.04}{1-0.1} = \frac{0.04}{0.9} = \frac{4}{90}$$

$$2.3\overline{4} = \frac{23}{10} + \frac{4}{90} = \frac{207}{90} + \frac{4}{90} = \frac{211}{90}$$

Answer: $\frac{62}{11}$

| **EXAMPLE 3** APPLICATION TO PHYSICS | **PRACTICE EXERCISE 3** |

A ball rebounds 4/5 as far as it falls. If the ball is dropped from a height of 40 ft, how far does it travel (up and down) before coming to rest?

Use an infinite geometric series to approximate the total distance traveled. Make a sketch like the one in Figure 9.1 describing the situation.

The tip of a pendulum sweeps out an arc of 84 cm on the first pass. If on each succeeding pass the distance traveled is 7/9 of the preceding pass, how far will it travel before coming to rest?

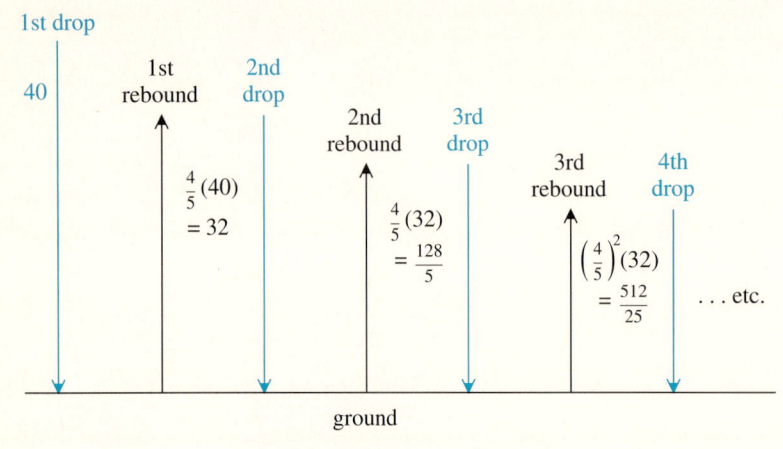

Figure 9.1 Rebounding Ball

We need to find the following sum.

$$40 + 32 + 32 + \frac{128}{5} + \frac{128}{5} + \frac{512}{25} + \frac{512}{25} + \cdots$$

$$= 40 + 2\left[32 + \frac{128}{5} + \frac{512}{25} + \cdots\right]$$

Since there are two of each number after the first drop, we simply double the sum of the "ups" from that point on to obtain the sum of the "ups" and "downs." Then the formula is applied only to the series in the brackets,

$$32 + \frac{128}{5} + \frac{512}{25} + \cdots,$$

with $a_1 = 32$ and $r = 4/5$.

$$S = \frac{a_1}{1-r} = \frac{32}{1 - \frac{4}{5}} = \frac{32}{\frac{1}{5}} = 5 \cdot 32 = 160$$

Then $40 + 2[32 + 128/5 + 512/25 + \cdots] = 40 + 2[160] = 40 + 320 = 360$. The total distance traveled by the ball is 360 ft.

Answer: 378 cm

9.4 EXERCISES A

In Exercises 1–3, state whether each sequence has a sum.

1. 100, 10, 1, 0.1, ...
2. $\frac{1}{64}, \frac{1}{16}, \frac{1}{4}, 1, \ldots$
3. 125, −75, 45, −27, ...

In Exercises 4–9, find the sum of each infinite geometric sequence.

4. $\frac{1}{2}, \frac{1}{4}, \frac{1}{8}, \frac{1}{16}, \ldots$
5. $\frac{1}{16}, \frac{1}{8}, \frac{1}{4}, \frac{1}{2}, \ldots$
6. $-14, 8, -\frac{32}{7}, \frac{128}{49}, \ldots$

7. 15, 1.5, 0.15, 0.015, ...
8. −64, 48, −36, 27, ...
9. $8, 4\sqrt{2}, 4, 2\sqrt{2}, \ldots$

In Exercises 10–12, find the indicated sum.

10. $\sum_{k=1}^{\infty} \left(\frac{3}{4}\right)^k$
11. $\sum_{k=1}^{\infty} 4^{-k}$
12. $\sum_{n=1}^{\infty} (0.2)^n$

In Exercises 13–15, find the first six terms of the geometric sequence.

13. $a_1 = 25$ and $S = 125$
14. $a_1 = 4$ and $S = -7$
15. $a_1 = -4$ and $S = -5$

In Exercises 16–19, convert each decimal to a fraction.

16. $0.\overline{3}$
17. $0.\overline{21}$
18. $0.\overline{123}$
19. $2.1\overline{5}$

20. Prove that 0.999 ... = 1.

Solve.

21. **RECREATION** A child on a swing traverses an arc of 22 ft. Each pass thereafter, she traverses an arc that is 5/7 the length of the previous arc. How far does she travel before coming to rest?

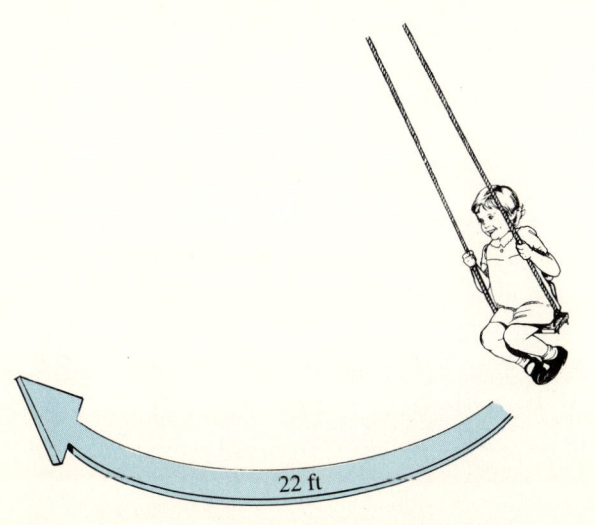

22. **PHYSICS** A ball dropped from a height of 40 ft always rebounds 1/2 the length of the preceding fall. What is the total distance traveled (up and down) if the ball is allowed to continue bouncing until it comes to rest? How might you approximate the total distance traveled when the ball hits the ground for the hundredth time?

23. ECONOMICS Eddie Petrowski receives $5000 on the day of his birth, and 3/5 as much on each birthday as on the previous one. Approximately how much will Eddie receive in his lifetime (assuming he lives a long life)?

24. GEOMETRY A square has area 64 in². A second square is constructed by connecting in order the midpoints of the sides of the first square, a third by connecting in order the midpoints of the sides of the second square, and so forth. (See the figure.) Calculate the sum of the areas of all these squares.

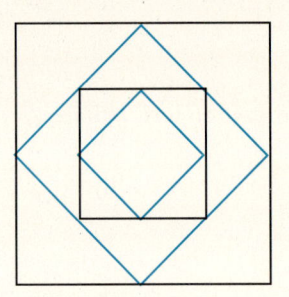

25. ECONOMICS An appliance manufacturing company builds a plant in a small country and will have a payroll of $1,000,000 per year. It is estimated that 80% of this money will be spent in the country, and of this amount 80% will be spent again in the country. This process, the **economic multiplier effect**, should continue without stopping. What will be the total amount of spending generated by any one year of company salaries?

FOR REVIEW

26. Find the first four terms of the geometric sequence with $a_1 = -5$ and $r = 0.08$.

27. Find a_6 for the sequence $-54, 36, -24, 16, \ldots$.

28. Find S_6 for the sequence $-54, 36, -24, 16, \ldots$.

29. If $a_2 = -5$ and $a_5 = 0.04$, find r and S_6.

30. Insert three geometric means between 250 and 0.025.

ANSWERS: 1. yes 2. no 3. yes 4. 1 5. no sum, $|r| = 2 > 1$ 6. $-\frac{98}{11}$ 7. $\frac{50}{3}$ 8. $-\frac{256}{7}$ 9. $16 + 8\sqrt{2}$ 10. 3 11. $\frac{1}{3}$ 12. $\frac{1}{4}$ 13. 25, 20, 16, $\frac{64}{5}$, $\frac{256}{25}$, $\frac{1024}{125}$ 14. no sequence, $|r| = \frac{11}{7} > 1$ 15. $-4, -\frac{4}{5}, -\frac{4}{25}, -\frac{4}{125}, -\frac{4}{625}, -\frac{4}{3125}$ 16. $\frac{1}{3}$ 17. $\frac{7}{33}$ 18. $\frac{41}{333}$ 19. $\frac{97}{45}$ 21. 77 ft 22. 120 ft; use $S = 120$ ft for an approximation 23. $12,500 24. 128 in² 25. $5,000,000 26. $-5, -0.4, -0.032, -0.00256$ 27. $\frac{64}{9}$ 28. $-\frac{266}{9}$ 29. $r = -0.2$; $S_6 = 20.832$ 30. 25, 2.5, 0.25 or $-25, 2.5, -0.25$

9.4 EXERCISES B

In Exercises 1–3, state whether each sequence has a sum.

1. 25, 5, 1, 0.2, . . .
2. $\dfrac{1}{25}, \dfrac{1}{5}, 1, 5, \ldots$
3. 64, −80, 100, −125, . . .

In Exercises 4–9, find the sum of each infinite geometric sequence.

4. $3, 1, \dfrac{1}{3}, \dfrac{1}{9}, \ldots$
5. $\dfrac{1}{81}, \dfrac{1}{27}, \dfrac{1}{9}, \dfrac{1}{3}, \ldots$
6. $2, \sqrt{2}, 1, \dfrac{\sqrt{2}}{2}, \dfrac{1}{2}, \ldots$
7. 40, 28, 19.6, 13.72, . . .
8. 250, −100, 40, −16, . . .
9. $27, 9\sqrt{3}, 9, 3\sqrt{3}, \ldots$

In Exercises 10–12, find the indicated sum.

10. $\displaystyle\sum_{k=1}^{\infty} \left(\dfrac{2}{9}\right)^k$
11. $\displaystyle\sum_{k=1}^{\infty} 3^{-k}$
12. $\displaystyle\sum_{n=1}^{\infty} (-0.6)^n$

In Exercises 13–15, find the first six terms of the geometric sequence.

13. $a_1 = -16$ and $S = -64$
14. $a_1 = 30$ and $S = 20$
15. $a_1 = 3.5$ and $S = -7$

In Exercises 16–19, convert each decimal to a fraction.

16. $0.\overline{8}$
17. $0.\overline{36}$
18. $2.\overline{6}$
19. $4.2\overline{3}$

20. Prove that $4.\overline{9} = 5$.

Solve.

21. **PHYSICS** The tip of a pendulum moves back and forth so that it sweeps out an arc 12 inches in length, and on each succeeding pass, the length of the arc traveled is 7/8 of the length of the preceding pass. What is the total distance traveled by the tip of the pendulum?

22. **PHYSICS** A ball dropped from a height of 15 m always rebounds 3/5 the length of the preceding fall. What is the total distance traveled (up and down) if the ball is allowed to continue bouncing until it comes to rest? How might you approximate the total distance traveled when the ball hits the ground for the two hundredth time?

23. **ECONOMICS** Krista Sekeres receives $1000 the month of her eighteenth birthday, and 0.8 as much each month thereafter as on the previous one. Approximately how much will Krista receive in her lifetime (assuming she lives a long life)?

24. **GEOMETRY** A square has area 100 cm². A second square is constructed by drawing lines at one-third the length and at one-third the width. This is also done for the new square and the process is continued indefinitely (see the figure). Calculate the sum of the areas of all these squares.

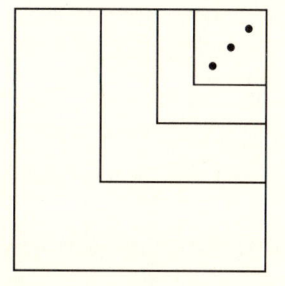

25. ECONOMICS A furniture manufacturing company builds a plant in a small town and will have a payroll of $4,000,000 per year. It is estimated that 75% of this money will be spent in the town, and of this amount 75% will be spent again in the town. This process, the **economic multiplier effect,** should continue without stopping. What will be the total amount of spending generated by any one year of company salaries?

FOR REVIEW

26. Find the first four terms of the geometric sequence with $a_1 = 0.06$ and $r = -10$.

27. Find a_6 for the sequence 32, -80, 200, -500, ...

28. Find S_6 for the sequence 32, -80, 200, -500, ...

29. If $a_3 = 125$ and $a_6 = 1$, find r and S_6.

30. Insert three geometric means between 81 and 256.

9.4 EXERCISES C

1. If $a_n = \frac{1}{2^{n-1}}$, find the sum of all the terms of $a_1, a_2, a_3, \ldots$.

2. If $a_n = \frac{1}{2^{n-1}}$, find the sum of all the terms of $b_1, b_2, b_3, \ldots$ where $b_n = \sqrt{a_n}$. [Answer: $2 + \sqrt{2}$]

3. Consider the arithmetic sequence $a_1, a_2, a_3, \ldots$ where $a_n = 3n + 1$. Find the sum of all the terms of $b_1, b_2, b_3, \ldots$ where $b_n = \left(\frac{1}{2}\right)^{a_n}$.

4. Let $\ln a_1, \ln a_2, \ln a_3, \ldots$ be an arithmetic sequence with common difference $d = -2$ and $a_1 = e$. Find the sum of all the terms of $a_1, a_2, a_3, \ldots$.

9.5 MATHEMATICAL INDUCTION

STUDENT GUIDEPOSTS

1 Method of Proof

2 Principle of Mathematical Induction

1 METHOD OF PROOF

Certain statements or formulas which involve positive integers, such as the sum formulas for arithmetic and geometric sequences, can be proved by the method of **mathematical induction.** This process is used when each positive integer gives rise to a particular statement which must be verified. For example, we might wish to show that the sum of the first n positive integers is equal to $\frac{1}{2}n(n + 1)$, that is,

$$1 + 2 + 3 + \cdots + n = \frac{1}{2}n(n + 1).$$

We can verify by substitution that the formula holds for the first few positive integers.

Substitution	Left side	Right side
$n = 1$	$1 = 1$	$\frac{1}{2}(1)(1 + 1) = \frac{1}{2} \cdot 2 = 1$
$n = 2$	$1 + 2 = 3$	$\frac{1}{2}(2)(2 + 1) = \frac{1}{2} \cdot 6 = 3$
$n = 3$	$1 + 2 + 3 = 6$	$\frac{1}{2}(3)(3 + 1) = \frac{1}{2} \cdot 12 = 6$
$n = 4$	$1 + 2 + 3 + 4 = 10$	$\frac{1}{2}(4)(4 + 1) = \frac{1}{2} \cdot 20 = 10$
$n = 5$	$1 + 2 + 3 + 4 + 5 = 15$	$\frac{1}{2}(5)(5 + 1) = \frac{1}{2} \cdot 30 = 15$

Although we know that the formula is true for the first five positive integers, we have not established it for every possible positive integer. In many situations a statement may be true for the first few positive integers, but not true for all integers. For example, 1, 2, 3, 4, 5, and 6 are all divisors of 60. However, to conclude that all integers are divisors of 60 would be incorrect since, for example, 7 is not such a divisor.

In general, the method of proof by mathematical induction involves three parts.

1. *Verification* of truth for one particular value. Usually we show that the statement is true when $n = 1$.

2. *Induction,* or extension of truth from one particular value to the next. At this step, we use the **induction assumption:** we assume the statement is true when $n = k$. Then we must prove it is also true when $n = k + 1$ (the next integer after k).

3. *Conclusion* of the truth of the statement for all positive integers. Having shown that the result is true for a particular value of n (say $n = 1$), by the induction step we assumed that it is true for any value of n ($n = k$). Then we proved it is true for the next value of n ($n = k + 1$). We conclude that the result is true for any choice of positive integer n.

❷ PRINCIPLE OF MATHEMATICAL INDUCTION

A proof by mathematical induction could be compared to the child's game of tipping over dominoes. The dominoes can be placed so that if one falls over ($n = k$), the next one will also fall over ($n = k + 1$). Thus, if the first one is tipped ($n = 1$), the entire collection of dominoes will fall.

Principle of Mathematical Induction

Let $S(n)$ be a statement involving the positive integer n. Then $S(n)$ is true for every positive integer provided:

1. $S(n)$ is true when $n = 1$.   ($S(1)$ is true.)
2. Under the assumption that $S(n)$ is true when $n = k$, we can show that $S(n)$ is true when $n = k + 1$.   (If $S(k)$ is true, then $S(k + 1)$ is true.)

EXAMPLE 1 Proving a Sum Formula

Show that $$S(n): \quad 1 + 2 + 3 + \cdots + n = \frac{1}{2}n(n+1)$$

is true for every positive integer n.

Verification: Clearly $S(n)$ is true when $n = 1$, since in this case it becomes

$$S(1): \quad 1 = \frac{1}{2} \cdot 1 \cdot (1 + 1) = 1.$$

Induction: Assume that the statement is true when $n = k$.

$$S(k): \quad 1 + 2 + 3 + \cdots + k = \frac{1}{2} \cdot k \cdot (k + 1) \qquad \text{Induction assumption}$$

We must show that the statement is true when $n = k + 1$. That is, we must show that

$$S(k+1): \quad 1 + 2 + 3 + \cdots + k + (k + 1) = \frac{1}{2} \cdot (k + 1)[(k + 1) + 1]$$

is true. The left side of this equation can be rewritten as

$$[1 + 2 + 3 + \cdots + k] + (k + 1).$$

Substitute $(1/2) \cdot k \cdot (k + 1)$ for $[1 + 2 + 3 + \cdots + k]$, since we have assumed the equation is true for k.

$$\begin{aligned}
[1 + 2 + 3 + \cdots + k] + (k + 1) &= \left[\frac{1}{2} \cdot k \cdot (k + 1)\right] + (k + 1) \\
&= \left(\frac{1}{2} \cdot k + 1\right)(k + 1) \qquad \text{Factor out } k + 1 \\
&= \left(\frac{1}{2}k + \frac{1}{2} \cdot 2\right)(k + 1) \qquad 1 = \frac{1}{2} \cdot 2 \\
&= \frac{1}{2}(k + 2)(k + 1) \qquad \text{Factor out } \frac{1}{2} \\
&= \frac{1}{2}(k + 1)(k + 2) \qquad \text{Commute} \\
&= \frac{1}{2}(k + 1)[(k + 1) + 1]
\end{aligned}$$

Thus, using the induction assumption, we have shown that

$$S(k+1): \quad 1 + 2 + 3 + \cdots + k + (k + 1) = \frac{1}{2}(k + 1)[(k + 1) + 1]$$

is true; that is, that $S(n)$ is true when $n = k + 1$.

Conclusion: Since $S(n)$ is true when $n = 1$ (by verification), and since if we assume it is true when $n = k$ we can prove it is true when $n = k + 1$, then we know $S(n)$ is true for every positive integer n.

Practice Exercise 1

Prove that $1 + 4 + 7 + \cdots + (3n - 2) = \frac{n(3n-1)}{2}$ is true for every positive integer n.

Answer: Pattern proof after Example 1.

Usually, in a proof by mathematical induction, the conclusion is shortened somewhat as illustrated in the next example.

EXAMPLE 2   PROVING A SUM FORMULA

Show that $\quad S(n): \quad 2 + 4 + 6 + \cdots + 2n = n(n+1)$

is true for every positive integer n.

Verification: When $n = 1$, the left side is 2 (there is only one term on the left) and the right side is $1(1+1) = 2$. Thus, $S(n)$ is true when $n = 1$.

Induction: Assume that $S(n)$ is true when $n = k$. That is, assume that

$$S(k): \quad 2 + 4 + 6 + \cdots + 2k = k(k+1) \qquad \text{Induction assumption}$$

is true. Then we must show that $S(n)$ is true when $n = k + 1$. That is, we must show that

$$S(k+1): \quad 2 + 4 + 6 + \cdots + 2(k+1) = (k+1)[(k+1) + 1]$$

is true. But the left side of this equation is really

$$[2 + 4 + 6 + \cdots + 2k] + 2(k+1)$$

which becomes $\quad [k(k+1)] + 2(k+1)$

when we substitute $k(k+1)$ for $(2 + 4 + 6 + \cdots + 2k)$. Factoring out $(k+1)$, we have

$$(k+2)(k+1)$$

which is equal to $\quad (k+1)[(k+1) + 1].$

Thus, using the induction assumption, we have shown that

$$2 + 4 + 6 + \cdots + 2(k+1) = (k+1)[(k+1) + 1].$$

Conclusion: By the principle of mathematical induction, $S(n)$ is true for every positive integer n.

PRACTICE EXERCISE 2

Prove that $3 + 3^2 + 3^3 + \cdots + 3^n = \frac{3}{2}(3^n - 1)$ is true for every positive integer n.

Answer: Pattern proof after Example 2.

EXAMPLE 3   PROVING A LAW OF EXPONENTS

Show that $\quad S(n): \quad \left(\dfrac{a}{b}\right)^n = \dfrac{a^n}{b^n}$

is true for every positive integer n.

Verification: When $n = 1$,

$$\left(\frac{a}{b}\right)^1 = \frac{a}{b} = \frac{a^1}{b^1}.$$

Thus, $S(n)$ is true when $n = 1$.

Induction: Assume that $S(n)$ is true when $n = k$. Then,

$$S(k): \quad \left(\frac{a}{b}\right)^k = \frac{a^k}{b^k}. \qquad \text{Induction assumption}$$

PRACTICE EXERCISE 3

Prove that $(ab)^n = a^n b^n$ is true for every positive integer n.

We must now show that $S(n)$ is true when $n = k + 1$.

$$\left(\frac{a}{b}\right)^{k+1} = \left(\frac{a}{b}\right)^k \left(\frac{a}{b}\right)$$

$$= \frac{a^k}{b^k} \cdot \frac{a}{b} \quad \text{Since } \left(\frac{a}{b}\right)^k = \frac{a^k}{b^k}$$

$$= \frac{a^k \cdot a}{b^k \cdot b} = \frac{a^{k+1}}{b^{k+1}}$$

Conclusion: By the principle of mathematical induction, $S(n)$ is true for every positive integer n.

Answer: Pattern proof after Example 3.

EXAMPLE 4 PROVING AN INEQUALITY

Show that
$$S(n): \quad n < 3^n$$
is true for every positive integer n.

Verification: Clearly, when $n = 1$, $1 < 3^1 = 3$ so that $S(1)$ is true.

Induction: Assume that $S(n)$ is true when $n = k$. That is, assume that

$$S(k): \quad k < 3^k \quad \text{Induction assumption}$$

is true. Then we must show that $S(n)$ is true when $n = k + 1$. That is, we must show that

$$S(k+1): \quad k + 1 < 3^{k+1}$$

is true. Since $k < 3^k$ we know that

$$3 \cdot k < 3 \cdot 3^k \quad \text{Multiply both sides by 3}$$
$$3k < 3^{k+1}.$$

But since k is a positive integer,

$$1 \leq k$$

so that
$$k + 1 \leq k + k \quad \text{Add } k \text{ to both sides}$$
$$k + 1 \leq 2k.$$

Also,
$$2 < 3$$
$$2k < 3k. \quad \text{Multiply by } k$$

Combining these facts, we have

$$k + 1 \leq 2k < 3k < 3^{k+1}.$$

Hence, $S(n)$ is true when $n = k + 1$.

Conclusion: By the principle of mathematical induction, $S(n)$ is true for every positive integer n.

PRACTICE EXERCISE 4

Prove that $3^n < 3^{n+1}$ is true for every positive integer n.

Answer: Pattern proof after Example 4.

Note In all of our examples the statements have been true for all $n \geq 1$. Some statements are not true for $n = 1$ but are true for $n \geq m$ for some $m > 1$. For example, $2^n > 10$ is true for all $n \geq 4$. In cases like this we show $S(m)$ to be true and consider the induction assumption for $k \geq m$.

9.5 EXERCISES A

In Exercises 1–3, the statements are not true for all n. Show that $S(1)$, $S(2)$, and $S(3)$ are true, then find the first positive integer n for which $S(n)$ is false.

1. $n < 5$
2. n divides 420
3. $\dfrac{|n - 10|}{n - 10} = -1$

In Exercises 4–6, write out the statements for $S(1)$, $S(k)$, and $S(k + 1)$.

4. $S(n)$: $1 + 3 + 5 + \cdots + (2n - 1) = n^2$
5. $S(n)$: $n < 2^n$

6. $S(n)$: 3 divides $n^3 - n + 3$

In Exercises 7–9 prove that each statement in Exercises 4–6 is true for all positive integers n.

7. $1 + 3 + 5 + \cdots + (2n - 1) = n^2$
8. $n < 2^n$

9. 3 divides $n^3 - n + 3$

In Exercises 10–16 prove that each statement is true for all positive integers n.

10. $5 + 10 + 15 + \cdots + 5n = \dfrac{5}{2}n(n + 1)$
11. $1^2 + 2^2 + 3^2 + \cdots + n^2 = \dfrac{n}{6}(n + 1)(2n + 1)$

12. $2 \le 2^n$
13. $1 + \dfrac{1}{2} + \dfrac{1}{2^2} + \cdots + \dfrac{1}{2^{n-1}} = \dfrac{2^n - 1}{2^{n-1}}$

14. $\left(1 + \dfrac{1}{1}\right)\left(1 + \dfrac{1}{2}\right) \cdots \left(1 + \dfrac{1}{n}\right) = n + 1$
15. $1 + 2n < 3^n$, $n \ge 2$

16. 2 divides $n^2 + n$

628 CHAPTER 9 SEQUENCES, SERIES, AND PROBABILITY

FOR REVIEW

17. Convert $3.6\overline{25}$ to a fraction.

18. PHYSICS The tip of a pendulum moves back and forth in such a way that it sweeps out an arc 18 inches in length, and on each succeeding pass, the length of the arc traveled is 7/9 of the length of the preceding pass. What is the total distance traveled by the tip of the pendulum?

ANSWERS: Due to the nature of these exercises, only partial proofs are presented. **1.** $n = 5$ **2.** $n = 8$ **3.** $n = 10$
4. $S(1): 1 = 1^2$; $S(k): 1 + 3 + 5 + \cdots + (2k - 1) = k^2$; $S(k + 1): 1 + 3 + 5 + \cdots + (2k - 1) + [2(k + 1) - 1] = (k + 1)^2$
5. $S(1): 1 < 2^1$; $S(k): k < 2^k$; $S(k + 1): k + 1 < 2^{k+1}$ **6.** $S(1): 3$ divides $1^3 - 1 + 3 = 3$; $S(k): 3$ divides $k^3 - k + 3$; $S(k + 1): 3$ divides $(k + 1)^3 - (k + 1) + 3$

7. *Verification:* Since $1 = 1^2$, $S(n)$ is true when $n = 1$.

Induction: Assuming that $\quad S(k): \quad 1 + 3 + 5 + \cdots + (2k - 1) = k^2$

we must show that $\quad S(k + 1): \quad 1 + 3 + 5 + \cdots + (2(k + 1) - 1) = (k + 1)^2$.

But $\quad \underbrace{1 + 3 + 5 + \cdots + (2k - 1)} + (2(k + 1) - 1)$

$\quad = \quad k^2 \quad + (2k + 2 - 1) \quad$ By assumption

$\quad = \quad k^2 + 2k + 1$

$\quad = \quad (k + 1)^2 \quad$ Factoring

Conclusion: $S(n)$ is true for every positive integer n by the principle of mathematical induction.

8. *Verification:* $1 < 2^1$ so that $S(n)$ is true when $n = 1$.

Induction: Assuming that $k < 2^k$

we must show that $k + 1 < 2^{k+1}$.
Since $k < 2k$

$\quad 2k < 2 \cdot 2^k \quad$ Multiply by 2
$\quad 2k < 2^{k+1}$.

But since $\quad 1 \leq k$

$\quad k + 1 \leq k + k \quad$ Add k

$\quad k + 1 \leq 2k$.

Thus, $\quad k + 1 \leq 2k < 2^{k+1}$

9. Assuming 3 divides $k^3 - k + 3$, write $S(k + 1)$ in terms of $S(k)$.

$(k + 1)^3 - (k + 1) + 3 = k^3 + 3k^2 + 3k + 1 - k - 1 + 3$
$= k^3 - k + 3 + 3k^2 + 3k$
$= (k^3 - k + 3) + 3(k^2 + k)$

Thus, 3 divides $(k + 1)^3 - (k + 1) + 3$.

10. $5 + 10 + 15 + \cdots + 5k + 5(k + 1) = \frac{5}{2}k(k + 1) + 5(k + 1)$

$= (k + 1)\left(\frac{5}{2}k + 5\right)$

$= \frac{5}{2}(k + 1)[(k + 1) + 1]$

11. *Induction:* Assuming that $1^2 + 2^2 + 3^3 + \cdots + k^2 = \frac{k}{6}(k+1)(2k+1)$ we must show that

$$1^2 + 2^2 + \cdots + k^2 + (k+1)^2 = \frac{k+1}{6}((k+1)+1)(2(k+1)+1).$$

$$1^2 + 2^2 + \cdots + k^2 + (k+1)^2 = \frac{k}{6}(k+1)(2k+1) + (k+1)^2$$

$$= (k+1)\left[\frac{k}{6}(2k+1) + (k+1)\right] = \frac{1}{6}(k+1)[k(2k+1) + 6(k+1)]$$

$$= \frac{(k+1)}{6}[2k^2 + k + 6k + 6] = \frac{k+1}{6}[2k^2 + 7k + 6] = \frac{k+1}{6}(k+2)(2k+3)$$

$$= \frac{k+1}{6}((k+1)+1)(2(k+1)+1)$$

12. *Induction:* Assuming that $2 \le 2^k$ we know that
$2 \cdot 2 \le 2 \cdot 2^k = 2^{k+1}$. Since $2 \le 2 \cdot 2 = 4$, $2 \le 2^{k+1}$

13. $1 + \frac{1}{2} + \frac{1}{2^2} + \cdots + \frac{1}{2^{k-1}} + \frac{1}{2^k} = \frac{2^k - 1}{2^{k-1}} + \frac{1}{2^k} = \frac{2(2^k - 1)}{2 \cdot 2^{k-1}} + \frac{1}{2^k} = \frac{2^{k+1} - 2 + 1}{2^k} = \frac{2^{k+1} - 1}{2^{(k+1)-1}}$

14. Note that $(k+1)\left(1 + \frac{1}{k+1}\right)$ simplifies to the right side of S when $n = k+1$. 15. $S(2)$: $1 + 2(2) < 3^2$ or $5 < 9$. Assume $1 + 2k < 3^k$. $1 + 2(k+1) = 1 + 2k + 2 < 3^k + 2 < 3^k + 3^k < (3)3^k = 3^{k+1}$ 16. Assume 2 divides $k^2 + k$. $(k+1)^2 + (k+1) = k^2 + 2k + 1 + k + 1 = k^2 + k + 2(k+1)$ Thus, 2 divides $(k+1)^2 + (k+1)$. 17. $\frac{3589}{990}$ 18. 81 in

9.5 EXERCISES B

In Exercises 1–3, the statements are not true for all n. Show that S(1), S(2), and S(3) are true, then find the first positive integer n for which S(n) is false.

1. $n > n^2 - 100$
2. $n^2 - n + 5$ is prime
3. $n^3 - 6n^2 + 11n - 6 = 0$

In Exercises 4–6, write out the statements for S(1), S(k), and S(k + 1).

4. $S(n)$: $3 + 6 + 9 + \cdots + 3n = \frac{3}{2}n(n+1)$
5. $S(n)$: $n < n + 1$
6. $S(n)$: 5 divides $6^n - 1$

In Exercises 7–9, prove that each statement in Exercises 4–6 is true for all positive integers n.

7. $3 + 6 + 9 + \cdots + 3n = \frac{3}{2}n(n+1)$
8. $n < n + 1$
9. 5 divides $6^n - 1$

In Exercises 10–16, prove that each statement is true for all positive integers n.

10. $1 + 5 + 9 + \cdots + (4n - 3) = n(2n - 1)$
11. $1^3 + 2^3 + 3^3 + \cdots + n^3 = \frac{1}{4}n^2(n+1)^2$
12. $2n \le 2^n$
13. $2 + 2^2 + 2^3 + \cdots + 2^n = 2^{n+1} - 2$
14. $\frac{1}{1 \cdot 2} + \frac{1}{2 \cdot 3} + \frac{1}{3 \cdot 4} + \cdots + \frac{1}{n(n+1)} = \frac{n}{n+1}$
15. $n + 7 < n^2$, $n \ge 4$
16. $x - 1$ divides $x^{2n} - 1$ if $x \ne 1$

FOR REVIEW

17. Convert $1.\overline{135}$ to a fraction.

18. ECONOMICS A government puts $10,000,000 into the economy of a country. It is estimated that 90% of the money will be spent in the country, and of that amount 90% will be spent again. If this process continues without stopping, what will be the total amount of spending generated by the $10,000,000?

9.5 EXERCISES C

1. Use mathematical induction to prove the sum formula for arithmetic sequences.

$$a_1 + (a_1 + d) + (a_1 + 2d) + \cdots + [a_1 + (n-1)d] = \frac{n}{2}[2a_1 + (n-1)d]$$

2. Use mathematical induction to prove the sum formula for geometric sequences.

$$a_1 + a_1 r + a_1 r^2 + \cdots + a_1 r^{n-1} = \frac{a_1 - a_1 r^n}{1 - r}$$

3. Use mathematical induction to prove that $a - b$ divides $a^n - b^n$. [*Hint:* $a^{k+1} - b^{k+1} = (a^k - b^k)a + (a - b)b^k$]

4. Use mathematical induction to prove that $a + b$ divides $a^{2n-1} + b^{2n-1}$.

9.6 PERMUTATIONS AND COMBINATIONS

STUDENT GUIDEPOSTS

1. Counting Techniques
2. Permutations
3. Factorial Notation
4. Formula for $_nP_r$
5. Distinguishable Permutations
6. Circular Permutations
7. Combinations
8. Formula for $_nC_r$

1 COUNTING TECHNIQUES

The techniques for counting the number of ways that a collection of objects or acts can be arranged, ordered, combined, chosen, or occur in succession come under the heading of **combinatorial algebra.** As an example consider a student who wishes to take three courses. She must choose one math course from algebra (A) or trigonometry (T), one English course from technical writing (W), literature (L), or rhetoric (R), and one science course from chemistry (C), geology (G), or physics (P). The **tree diagram** in Figure 9.2 will help count the number of ways the student can select the three courses.

Each branch of the tree represents one possible selection of courses. For example A, L, G means algebra, literature, and geology were chosen. Notice that 18 choices can be made. But $18 = 2 \cdot 3 \cdot 3$ is the product of the number of math courses, the number of English courses, and the number of science courses. This example leads to the **fundamental counting principle.**

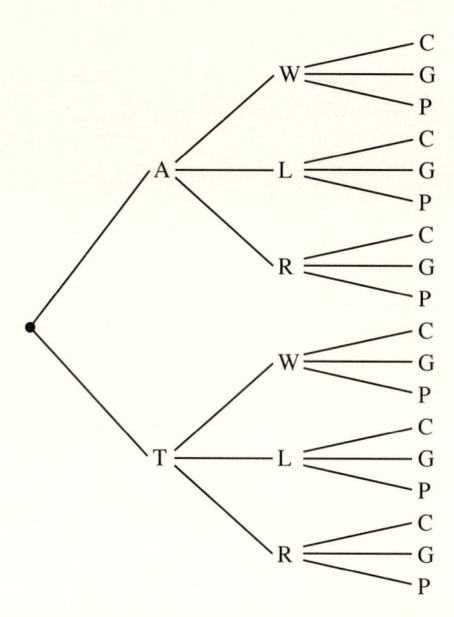

Figure 9.2 Counting

Fundamental Counting Principle

If event E_1 can occur in m_1 ways, E_2 in m_2 ways, E_3 in m_3 ways, etc., then the total number of ways $E_1, E_2, \ldots, E_k$ can occur is

$$m_1 m_2 \ldots m_k.$$

EXAMPLE 1   USING THE COUNTING PRINCIPLE

Consider making three-digit numbers from the digits 1, 2, 3, 4, 5, 6, and 7.

(a) How many ways can this be done if repetition of digits is allowed?

Since there are three digits to select, fill three blanks with the number of ways each digit can be selected. Since digits may be repeated, each of the three blanks will contain a 7.

$$\underline{7} \quad \underline{7} \quad \underline{7}$$

Thus, there are $7 \cdot 7 \cdot 7 = 343$ different three-digit numbers possible when repetition is allowed.

(b) How many ways are there when repetition is not allowed?

There are seven choices for the first slot, only six choices for the second (without repetition, the digit used first is not available now), and only five choices remaining for the third slot.

$$\underline{7} \quad \underline{6} \quad \underline{5}$$

Thus, there are $7 \cdot 6 \cdot 5 = 210$ possible different three-digit numbers when repetition is not allowed.

PRACTICE EXERCISE 1

Consider making four-digit numbers from 1, 2, 3, 4, 5, 6, and 7.

(a) How many ways are there if repetition is allowed?

(b) How many ways are there if repetition is not allowed?

Answers: (a) 2401   (b) 840

EXAMPLE 2 A COUNTING PRINCIPLE APPLICATION

License plates are to be made by using three letters followed by a three-digit number. How many such license plates are possible?

Since there are twenty-six letters and ten numerals (0, 1, 2, . . . , 9) and repetitions are allowed, the number of such plates can be derived by filling slots

$$\underline{26}\ \underline{26}\ \underline{26}\ \underline{10}\ \underline{10}\ \underline{10}$$

to obtain $26^3 \cdot 10^3 = 17{,}576{,}000$ license plates.

PRACTICE EXERCISE 2

How many possible license plates are there with one letter followed by five digits?

Answer: 2,600,000

❷ PERMUTATIONS

We now concentrate on the problem of choosing a collection of objects from a given set of objects.

> **Permutation**
> A **permutation** is any arrangement of a collection of objects in a particular order.

Consider the permutations (arrangements) of the letters A, B, and C.

$$\text{ABC, ACB, BAC, BCA, CAB, CBA}$$

There are six permutations of three objects. This number can be determined by filling three slots. There are 3 choices for the first position, 2 for the second, and 1 for the third.

$$\underline{3} \cdot \underline{2} \cdot \underline{1} = 6$$

The number of permutations of 5 distinct objects is

$$\underline{5} \cdot \underline{4} \cdot \underline{3} \cdot \underline{2} \cdot \underline{1} = 120.$$

The number of permutations of 8 distinct objects is

$$\underline{8} \cdot \underline{7} \cdot \underline{6} \cdot \underline{5} \cdot \underline{4} \cdot \underline{3} \cdot \underline{2} \cdot \underline{1} = 40{,}320.$$

In general, the number of permutations of n distinct objects is

$$n(n-1)(n-2)(n-3) \cdots 2 \cdot 1.$$

❸ FACTORIAL NOTATION

Since products of successive integers from a given integer down to 1 occur so often in our work with permutations, we adopt the **factorial notation**

$$n! = n(n-1)(n-2)(n-3) \cdots 2 \cdot 1 \quad \text{(read ``}n\text{ factorial'')}.$$

Thus, $3! = 3 \cdot 2 \cdot 1 = 6$, $5! = 5 \cdot 4 \cdot 3 \cdot 2 \cdot 1 = 120$, and $8! = 8 \cdot 7 \cdot 6 \cdot 5 \cdot 4 \cdot 3 \cdot 2 \cdot 1 = 40{,}320$. Notice that $1! = 1$.

The phrase "the number of permutations of n objects using all n of the objects" or briefly "the number of permutations of n objects" occurs repeatedly, so we use $_nP_n$ to symbolize this phrase.

> **Permutations of n Objects**
>
> The number of permutations of n objects, ${}_nP_n$, is given by
> $${}_nP_n = n! = n(n-1)(n-2)(n-3) \cdots 2 \cdot 1.$$

❹ FORMULA FOR ${}_nP_r$

Given a collection of objects, we may wish to consider all arrangements of subcollections of a particular size. For example, consider the permutations of the letters A, B, C, D, and E, taken two at a time.

$$\begin{array}{ccccc}
AB & BA & CA & DA & EA \\
AC & BC & CB & DB & EB \\
AD & BD & CD & DC & EC \\
AE & BE & CE & DE & ED
\end{array}$$

In the array, there are twenty permutations of five distinct objects using two of them at a time. This number could have been obtained using the slot-filling technique since there are 5 choices for the first position and 4 for the second.

$$\underline{5} \cdot \underline{4} = 20$$

In general, we are concerned with finding "the number of permutations of n objects using r of the objects at a time." This phrase is symbolized by ${}_nP_r$. Consider filling r slots. There are n choices for the first slot, $n-1$ for the second, $n-2$ for the third, and $n-(r-1)$ for the rth slot.

$$\underline{\frac{n}{1}} \quad \underline{\frac{n-1}{2}} \quad \underline{\frac{n-2}{3}} \quad \underline{\frac{n-3}{4}} \cdots \underline{\frac{n-(r-1)}{r}}$$

Thus $\quad {}_nP_r = n(n-1)(n-2)(n-3) \cdots [n-(r-1)].$

Suppose we multiply on the right by $(n-r)!/(n-r)!$, which is 1.

$$\begin{aligned}
{}_nP_r &= n(n-1)(n-2)(n-3) \cdots [n-(r-1)] \frac{(n-r)!}{(n-r)!} \\
&= \frac{n(n-1)(n-2)(n-3) \cdots [n-(r-1)](n-r)!}{(n-r)!} \\
&= \frac{n!}{(n-r)!}
\end{aligned}$$

This proves the following theorem.

> **Permutations of n Objects Using r**
>
> The number of permutations of n objects using r of the objects at a time ($r < n$) is given by
> $${}_nP_r = \frac{n!}{(n-r)!}.$$

In the previous example, $n = 5$ and $r = 2$ so, using the formula,

$${}_5P_2 = \frac{5!}{(5-2)!} = \frac{5!}{3!} = \frac{5 \cdot 4 \cdot \cancel{3} \cdot \cancel{2} \cdot \cancel{1}}{\cancel{3} \cdot \cancel{2} \cdot \cancel{1}} = 5 \cdot 4 = 20.$$

634 CHAPTER 9 SEQUENCES, SERIES, AND PROBABILITY

We do not evaluate the factorials in the numerator and denominator, we just indicate the products and cancel common factors. Since $5! = 5 \cdot 4 \cdot 3! = 5 \cdot 4 \cdot 3 \cdot 2 \cdot 1$, we could have abbreviated further.

$$_5P_2 = \frac{5!}{(5-2)!} = \frac{5!}{3!} = \frac{5 \cdot 4 \cdot \cancel{3!}}{\cancel{3!}} = 5 \cdot 4 = 20$$

In addition, if we define $0! = 1$, we can use the formula

$$_nP_r = \frac{n!}{(n-r)!}$$

in the case when $n = r$ [$n - r = 0$ so $(n - r)! = 0! = 1$], and the result is the formula for $_nP_n$.

EXAMPLE 3 Calculating $_nP_r$

(a) $_5P_5 = \dfrac{5!}{(5-5)!} = \dfrac{5!}{0!} = \dfrac{5!}{1} = 5! = 5 \cdot 4 \cdot 3 \cdot 2 \cdot 1 = 120$

(b) $_{10}P_4 = \dfrac{10!}{(10-4)!} = \dfrac{10!}{6!} = \dfrac{10 \cdot 9 \cdot 8 \cdot 7 \cdot \cancel{6!}}{\cancel{6!}} = 10 \cdot 9 \cdot 8 \cdot 7 = 5040$

(c) $_5P_0 = \dfrac{5!}{(5-0)!} = \dfrac{5!}{5!} = 1$

PRACTICE EXERCISE 3

(a) $_7P_7 =$

(b) $_9P_6 =$

(c) $_7P_0 =$

Answers: (a) 5040 (b) 60,480 (c) 1

EXAMPLE 4 Application of $_nP_r$

How many ways can a president, vice president, and secretary be chosen for a club with ten members?

Since order is important in the selection, this is a permutation problem.

$$_{10}P_3 = \frac{10!}{(10-3)!} = \frac{10 \cdot 9 \cdot 8 \cdot \cancel{7!}}{\cancel{7!}} = 720$$

PRACTICE EXERCISE 4

How many ways can seven different prizes be given to three different people?

Answer: 210

---

5 DISTINGUISHABLE PERMUTATIONS

Consider the number of arrangements of the seven letters in the word ARIZONA. Since two letters are the same, there are not 7! different arrangements. Each possible arrangement has an identical twin formed by interchanging the two A's. Since the two A's can be arranged 2! ways, divide 7! by 2! to obtain the number of distinguishable arrangements. This illustrates the following theorem.

Distinguishable Permutations

The number of distinguishable permutations of n objects of which n_1 are the same, n_2 are the same, n_3 are the same, and so forth, is

$$\frac{n!}{n_1! n_2! n_3! \cdots}.$$

EXAMPLE 5 FINDING DISTINGUISHABLE PERMUTATIONS

Determine the number of distinguishable permutations of the letters in MISSISSIPPI.

In this case there are eleven letters: one M, four I's, four S's, and two P's. That is, $n_1 = 1$, $n_2 = 4$, $n_3 = 4$, and $n_4 = 2$. The number of distinguishable permutations is

$$\frac{n!}{n_1!n_2!n_3!n_4!} = \frac{11!}{1!4!4!2!} = \frac{11 \cdot 10 \cdot 9 \cdot 8 \cdot 7 \cdot 6 \cdot 5 \cdot 4!}{1 \cdot 4 \cdot 3 \cdot 2 \cdot 1 \cdot 4! \cdot 1 \cdot 2}$$

$$= \frac{11 \cdot 10 \cdot 9 \cdot 8 \cdot 7 \cdot 6 \cdot 5}{4 \cdot 3 \cdot 2 \cdot 2} = 34{,}650.$$

PRACTICE EXERCISE 5

Find the number of distinguishable permutations of the letters in the word ELEMENT.

Answer: 840

6 CIRCULAR PERMUTATIONS

We have been arranging objects in a straight line, and have seen that four objects can be arranged in a straight line in $_4P_4 = 4! = 24$ different ways. Suppose we arranged these same four objects in a symmetric circular pattern. For example, let us arrange A, B, C, and D around a circle. One such arrangement is shown in Figure 9.3, and three others in Figure 9.4.

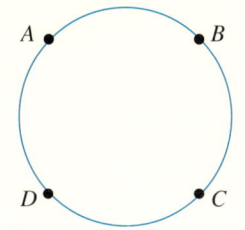

Figure 9.3 Circular Permutations

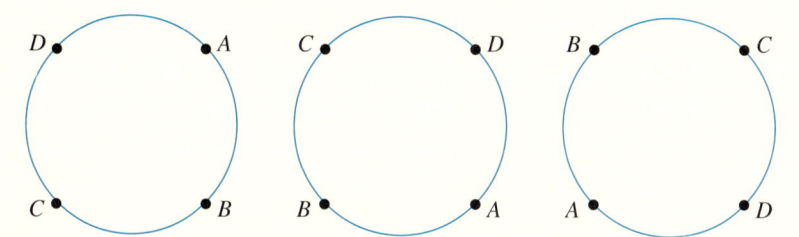

Figure 9.4 Circular Permutations

Are the arrangements in Figure 9.4 different or distinguishable from the one in Figure 9.3? If we ignore the positions of the letters and consider only their relationship to each other, we see that these four arrangements are the same. That is, we can begin at A and move clockwise around any of the circles to get the same arrangement, A B C D and then back to A again. Thus, the four different arrangements ABCD, BCDA, CDAB, and DABC are not distinguishable in a circular arrangement. In general, if there are x distinct circular arrangements of four objects, there would be $4 \cdot x$ arrangements of these objects along a straight line. But since the number of arrangements along a straight line is 4!, we have $4 \cdot x = 4!$ or

$$x = \frac{4!}{4} = \frac{4 \cdot 3!}{4} = 3!$$

This is a special case of the following theorem.

> **Circular Permutations**
>
> The number of distinguishable **circular permutations** (arrangements evenly around a circle) of n objects is $(n - 1)!$.

EXAMPLE 6 DETERMINING CIRCULAR PERMUTATIONS

In how many ways can six people be seated around a circular table?
In this case $n = 6$, so that

$$(n - 1)! = (6 - 1)! = 5! = 120$$

is the number of possible circular permutations of the six people.

PRACTICE EXERCISE 6

In how many ways could nine different plates be placed around a circular table?

Answer: 40,320

7 COMBINATIONS

We have been discussing permutations where it is necessary to consider the order of selection or arrangement. Now consider selecting r objects from n objects without arranging them in any particular order.

> **Combination**
>
> A **combination** is any collection of objects. Order of the selection is not important.

8 FORMULA FOR $_nC_r$

Consider the $_4P_3 = 24$ permutations of the letters A, B, C, and D, using three of them at a time.

ABC	ABD	ACD,	BCD
ACB	ADB	ADC	BDC
BAC	BAD	CAD	CBD
BCA	BDA	CDA	CDB
CAB	DAB	DAC	DBC
CBA	DBA	DCA	DCB

Notice that the six permutations in the first column all involve the same three letters A, B, and C, so they are counted as a single combination (order is not counted for a combination). Similarly, the six permutations listed in the second, third, and fourth columns give rise to the single combinations

ABD,   ACD,   BCD.

As a result, there are only four combinations of the four letters A, B, C, and D using three of them at a time. In general, there are fewer combinations of a group of objects than there are permutations of the group.

We use $_nC_r$ to represent the number of combinations of n objects using r of them at a time. In the preceding example, we found that $_4C_3 = 4$. Since each combination of three objects can be resolved into $6 = 3!$ permutations of the three objects, we see that

$$_4P_3 = (3!) \cdot {_4C_3} = 3! \cdot 4 = 6 \cdot 4 = 24$$

or
$$_4C_3 = \frac{_4P_3}{3!} = \frac{\frac{4!}{(4-3)!}}{3!} = \frac{4!}{(4-3)!3!}$$

which is a special case of the next theorem.

> **Combinations of n Objects Using r of Them**
>
> The number of combinations of n objects using r of the objects at a time ($r \leq n$) is given by
> $$_nC_r = \frac{n!}{(n-r)!r!}.$$

EXAMPLE 7 Calculating $_nC_r$

(a) $_5C_3 = \dfrac{5!}{(5-3)!3!} = \dfrac{5!}{2!3!} = \dfrac{5 \cdot \cancel{4}^{2} \cdot \cancel{3!}}{1 \cdot \cancel{2} \cdot \cancel{3!}} = 10$

(b) $_5C_5 = \dfrac{5!}{(5-5)!5!} = \dfrac{\cancel{5!}}{0!\cancel{5!}} = \dfrac{1}{1} = 1$

(c) $_5C_0 = \dfrac{5!}{(5-0)!0!} = \dfrac{\cancel{5!}}{\cancel{5!}0!} = \dfrac{1}{1} = 1$

Practice Exercise 7

(a) $_7C_4 =$

(b) $_7C_7 =$

(c) $_7C_0 =$

Answers: (a) 35 (b) 1 (c) 1

Parts (b) and (c) of Example 7 do make sense. If we have five objects, there is one way to choose all five of them (simply take them all), and one way to choose none of them (simply do not take them). In fact, there is nothing special about five in these examples;

$$_nC_n = 1 \quad \text{and} \quad _nC_0 = 1$$

for any natural number n.

EXAMPLE 8 Application of $_nC_r$

How many five-card poker hands consisting of three kings and two cards that are not kings are possible in a fifty-two card deck?

There are

$$_4C_3 = \frac{4!}{1!3!} = 4$$

ways to get three of the four kings and

$$_{48}C_2 = \frac{48!}{46!2!} = \frac{\cancel{48}^{24} \cdot 47 \cdot \cancel{46!}}{\cancel{46!} \cdot \cancel{2}} = 1128$$

ways to have two of the forty-eight cards that are not kings. Thus, in total, there are

$$_4C_3 \cdot {_{48}C_2} = 4 \cdot 1128 = 4512$$

ways to have the prescribed hand.

Practice Exercise 8

How many five-card poker hands are possible with 2 aces, 2 kings, and a card that is not an ace or a king?

Answer: 1584

9.6 EXERCISES A

Evaluate $_nP_r$ and $_nC_r$ in Exercises 1–4.

1. $_7P_3$

2. $_{10}P_2$

3. $_9C_0$

4. $_{10}C_7$

Solve.

5. Suppose there are five roads connecting town A to town B and three roads connecting town B to town C. In how many ways can a person travel from A to C via B?

6. **MANUFACTURING** License plates are to be made using two letters followed by four digits. How many plates can be made if repetition of both letters and digits is allowed? How many plates can be made if repetition of letters is possible, but repetition of digits is not allowed? How many plates can be made if neither letters nor digits can be repeated?

7. **MILITARY** A signal is made by placing three flags, one above the other, on a flagpole. If there are eight different flags available, how many possible signals can be flown?

8. In how many ways can the letters in the word SHELF be arranged using all five letters?

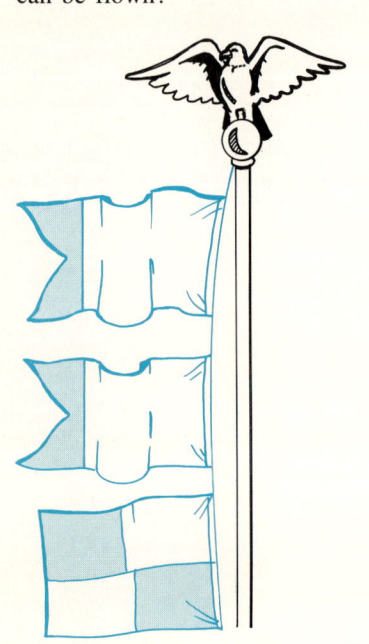

9. LAW ENFORCEMENT In how many ways can a police department arrange eight suspects in a lineup?

10. SPORTS In how many ways can the manager of a baseball team arrange his batting order under the following conditions?
  (a) Any player can bat in any position. [*Hint:* There are nine players on a baseball team.]
  (b) The pitcher must bat last.
  (c) The pitcher must bat last, and the centerfielder must bat in the clean-up position (fourth).
  (d) The pitcher must bat last, and the three outfielders must bat in the first three positions.

11. How many distinguishable permutations of the letters in the given word are possible?
  (a) SHEEP
  (b) LETTER
  (c) TENNESSEE
  (d) AARDVARK

12. MILITARY How many distinct signals can be made on a vertical flagpole with fourteen flags if five of them are white, four are red, three are green, and two are black?

13. In how many ways can eight people be arranged around a circular table?

14. MILITARY In how many ways can a detail of 5 soldiers be chosen from a group of 14?

15. MILITARY In how many ways can a detail of 5 soldiers be chosen from a group of 14 and be presented with 5 distinct awards?

16. EDUCATION In how many ways can a committee of 5 be chosen from 8 seniors and 10 juniors under the following conditions?
  (a) The committee must have exactly 4 seniors.
  (b) The committee must have at least 4 seniors. [*Hint:* Find the number of ways of having 4 seniors and 1 junior and add to this the number of ways of having 5 seniors.]
  (c) The committee must have exactly 3 juniors.
  (d) The committee must have at least 3 juniors.

640 CHAPTER 9 SEQUENCES, SERIES, AND PROBABILITY

17. **GEOMETRY** Points on the same straight line are **collinear.** How many lines are determined by seven points, no three of which are collinear? [*Hint:* How many points determine a line?]

18. **EDUCATION** A student must answer 7 of the first 10 questions on a test. On the next 5 questions, he must answer 4. In how many ways can this be done?

19. **RECREATION** How many 5-card poker hands dealt from a 52-card deck are possible if they must consist of the following cards?

   (a) 5 hearts

   (b) 4 aces and a seven

   (c) 3 aces and two cards that are not aces

   (d) 3 hearts and 2 clubs

   (e) 3 face cards and 2 sevens

20. **CONSUMER** Fred plans to purchase a new car. He has narrowed the choices to one of 3 models, one of 4 colors, and one of 5 upholstery colors. How many possible combinations are still in the running?

21. Give an example to show that $(n!)(m!) \neq (n \cdot m)!$.

22. Prove that $_nC_r = {_nC_{n-r}}$.

FOR REVIEW

In Exercises 23–24 use mathematical induction to prove the statement.

23. $3 + 3^2 + 3^3 + \cdots + 3^n = \dfrac{3}{2}(3^n - 1)$

24. 6 divides $7^n - 1$

ANSWERS: 1. 210 2. 90 3. 1 4. 120 5. 15 6. 6,760,000; 3,407,040; 3,276,000 7. 336 8. 120 9. 40,320 10. (a) 362,880 (b) 40,320 (c) 5040 (d) 720 11. (a) 60 (b) 180 (c) 3780 (d) 3360 12. 2,522,520 13. 5040 14. 2002 15. 240,240 16. (a) 700 (b) 756 (c) 3360 (d) 5292 17. 21 18. 600 19. (a) 1287 (b) 4 (c) 4512 (d) 22,308 (e) 1320 20. 60 21. $n = 2$ and $m = 3$ is one example 22. use the formula for $_nC_{n-r}$ 23. Induction: $3 + 3^2 + 3^3 + \cdots + 3^k + 3^{k+1} = \frac{3}{2}(3^k - 1) + 3^{k+1} = \frac{1}{2}(3^{k+1} - 3) + \frac{2}{2}3^{k+1} = \frac{3}{2}(3^{k+1} - 1)$ 24. $7^{k+1} - 1 = 7(7^k - 1) + 6$

9.6 EXERCISES B

Evaluate $_nP_r$ and $_nC_r$ in Exercises 1–4.

1. $_6P_6$

2. $_8P_0$

3. $_{10}C_3$

4. $_{20}C_{19}$

Solve.

5. How many three-digit symbols can be formed from the digits 0, 1, 2, 3, 4, 5, 6, 7, 8, 9, if repetition of digits is allowed? If repetition of digits is not allowed?

6. **BUSINESS** The Spear Shirt Company has four basic designs for shirts, in either short or long sleeves, with eight different color patterns. How many different shirts must be made to insure that one of each possible kind is made?

7. **EDUCATION** In how many ways can three freshmen and four sophomores line up at the registrar's office if no other restrictions are imposed? If the four sophomores insist on standing first in line? If the sophomores and freshmen must alternate in line? If a sophomore must stand first and no other restrictions are imposed? If the sophomores insist on standing together and the freshmen insist on standing together?

8. In how many ways can the letters in the word SHELF be arranged using four of the letters?

9. **MUSIC** A musician plans to perform nine selections. In how many ways can she arrange the program?

10. A penny, a nickel, a dime, and a quarter are to be arranged in a straight line.
    (a) How many ways can this be done?
    (b) If we distinguish between heads and tails, in how many ways can this be done?

11. In how many distinct ways can $x^4 y^3 z^5$ be expressed without exponents?

12. **RECREATION** If Judith Crist, one of the leading movie critics, has viewed 500 movies this year, in how many ways could she list, in order of preference, the ten best movies she has seen? [*Hint:* You need not evaluate your answer.]

13. In how many ways could the twelve knights of the round table be seated at King Arthur's round table?

14. **EDUCATION** On a test, a student must answer 8 out of 10 questions. In how many ways can she choose the 8 questions?

15. How many ways can 8 people be chosen and arranged in a straight line if there are 10 people to choose from?

16. **SPORTS** A group of 16 students are to play a tennis tournament with 4 four-person teams.
    (a) In how many ways can the first team be formed?
    (b) In how many ways can the second team be formed once the first is formed?
    (c) In how many ways can the third team be formed once the first two are formed?
    (d) In how many ways can the final team be made once the first three are formed?

17. **GEOMETRY** Three noncollinear points determine a circle. How many circles are determined by 5 such points?

18. How many different sums of money can be obtained using any 3 of 5 coins consisting of a penny, a nickel, a dime, a quarter, and a half-dollar?

19. **RECREATION** How many 5-card poker hands dealt from a 52-card deck are possible if they must consist of the following cards?
    (a) 5 black cards
    (b) 4 aces and a joker
    (c) an ace and 4 cards that are not aces
    (d) an ace, a queen, a king, and two cards which are not any of these three
    (e) 2 face cards, 2 aces, and an eight

20. **EDUCATION** A professor plans to give three A's, six B's, ten C's, five D's, and two F's to his class of twenty-six students. In how many ways can he do this?

21. Give an example to show that $n! + m! \neq (n + m)!$.

22. Prove that $_nP_0 = 1$, $_nC_0 = 1$, and $_nC_n = 1$.

FOR REVIEW

In Exercises 23–24 use mathematical induction to prove the statement.

23. $2 + 7 + 12 + \cdots + (5n - 3) = \frac{1}{2}n(5n - 1)$

24. 9 divides $10^{n+1} + 3 \cdot 10^n + 5$

9.6 EXERCISES C

Simplify each factorial expression in Exercises 1–2.

1. $\dfrac{(n + 1)!(n - 3)!}{(n - 1)!n!}$

2. $\dfrac{(n - k)!(n - k + 2)!}{(n - k - 1)!(n - k + 3)!}$

Determine all positive integers n for which each equation in Exercises 3–4 is true.

3. $_nP_3 = {}_{n-1}P_4$

4. $_nC_6 = 7 {}_{n-2}C_4$

9.7 THE BINOMIAL THEOREM

STUDENT GUIDEPOSTS

- **1** Expansion of $(a + b)^n$
- **2** Pascal's Triangle
- **3** Binomial Theorem
- **4** Proof of Binomial Theorem

1 EXPANSION OF $(a + b)^n$

Another application of combinations is in the expansion of $(a + b)^n$ where $a + b$ is any binomial and n is a nonnegative integer. First let us expand several such expressions, and search for a pattern.

$(a + b)^0 =$ $\qquad$ 1
$(a + b)^1 =$ $\qquad$ $a + b$
$(a + b)^2 =$ $\qquad$ $a^2 + 2ab + b^2$
$(a + b)^3 =$ $\qquad$ $a^3 + 3a^2b + 3ab^2 + b^3$
$(a + b)^4 =$ $\qquad$ $a^4 + 4a^3b + 6a^2b^2 + 4ab^3 + b^4$
$(a + b)^5 =$ $\qquad$ $a^5 + 5a^4b + 10a^3b^2 + 10a^2b^3 + 5ab^4 + b^5$
$(a + b)^6 =$ $\qquad$ $a^6 + 6a^5b + 15a^4b^2 + 20a^3b^3 + 15a^2b^4 + 6ab^5 + b^6$

In each case we observe the following:

1. There are always $n + 1$ terms in the expansion.
2. The exponents on a start with n and decrease to 0.
3. The exponents on b start with 0 and increase to n.
4. The sum of the exponents in each term is always n.

❷ PASCAL'S TRIANGLE

To discover the pattern of the numerical coefficients of each term, write the coefficients in the same arrangement as in the preceding expansions.

```
Row 0              1
Row 1             1  1
Row 2            1  2  1
Row 3           1  3  3  1
Row 4          1  4  6  4  1
Row 5         1  5  10  10  5  1
Row 6        1  6  15  20  15  6  1
```

This triangular array is known as **Pascal's triangle.** The row number corresponds to the exponent n in the expansion of $(a + b)^n$. The numbers in any row, other than the first and the last which are always 1, can be found by adding the two numbers immediately above and to the left and right of it. For example, as indicated, 15 is $5 + 10$. Thus, Pascal's triangle gives us one way to determine the coefficients in the expansion of a given binomial.

EXAMPLE 1   USING PASCAL'S TRIANGLE

Expand $(x + 2y)^5$.

In this example, $n = 5$, $a = x$, and $b = 2y$. Row 5 of Pascal's triangle has the following numbers for the coefficients.

$$1 \quad 5 \quad 10 \quad 10 \quad 5 \quad 1$$

We thus substitute $a = x$ and $b = 2y$ into

$$a^5 + \mathbf{5}a^4b + \mathbf{10}a^3b^2 + \mathbf{10}a^2b^3 + \mathbf{5}ab^4 + b^5$$

to obtain $x^5 + 5x^4(2y) + 10x^3(2y)^2 + 10x^2(2y)^3 + 5x(2y)^4 + (2y)^5$
$= x^5 + 5x^4(2y) + 10x^3(4y^2) + 10x^2(8y^3) + 5x(16y^4) + 32y^5$
$= x^5 + 10x^4y + 40x^3y^2 + 80x^2y^3 + 80xy^4 + 32y^5$.

PRACTICE EXERCISE 1

Expand $(2x - y)^5$.

Answer: $32x^5 - 80x^4y + 80x^3y^2 - 40x^2y^3 + 10xy^4 - y^5$

CAUTION

Make sure that you evaluate $b^3 = (2y)^3$ as $2^3 y^3 = 8y^3$ and *not* $2y^3$. This is perhaps the most common mistake made with this expansion technique.

EXAMPLE 2 Using Pascal's Triangle

Expand $(2y - 3)^4$.

Express $(2y - 3)^4$ as $(2y + (-3))^4$; then $n = 4$, $a = 2y$, and $b = -3$. In Row 4 of Pascal's triangle, we find the following coefficients.

$$1 \quad 4 \quad 6 \quad 4 \quad 1$$

Substituting $a = 2y$ and $b = -3$ into

$$a^4 + 4a^3 b + 6a^2 b^2 + 4ab^3 + b^4,$$

we have $(2y)^4 + 4(2y)^3(-3) + 6(2y)^2(-3)^2 + 4(2y)(-3)^3 + (-3)^4$

$= 16y^4 + 4(8y^3)(-3) + 6(4y^2)(9) + 4(2y)(-27) + 81$

$= 16y^4 - 96y^3 + 216y^2 - 216y + 81.$

PRACTICE EXERCISE 2

Expand $(3x + 2)^4$.

Answer: $81x^4 + 216x^3 + 216x^2 + 96x + 16$

When n is large, or when we are interested in finding only one particular term in a binomial expansion, it is helpful if we make one further observation relative to the binomial coefficients in Pascal's triangle. It is easy to verify that Pascal's triangle is in fact an array of combinations.

Row 0	1
Row 1	$_1C_0 \quad _1C_1$
Row 2	$_2C_0 \quad _2C_1 \quad _2C_2$
Row 3	$_3C_0 \quad _3C_1 \quad _3C_2 \quad _3C_3$
Row 4	$_4C_0 \quad _4C_1 \quad _4C_2 \quad _4C_3 \quad _4C_4$
Row 5	$_5C_0 \quad _5C_1 \quad _5C_2 \quad _5C_3 \quad _5C_4 \quad _5C_5$
Row 6	$_6C_0 \quad _6C_1 \quad _6C_2 \quad _6C_3 \quad _6C_4 \quad _6C_5 \quad _6C_6$

For example,

$$_6C_1 = \frac{6!}{(6-1)!1!} = \frac{6!}{5!1!} = 6, \quad _5C_3 = \frac{5!}{2!3!} = 10, \quad _4C_2 = \frac{4!}{2!2!} = 6.$$

As a result, the entries in any row n ($n \geq 1$) can be found by evaluating $_nC_r$ where r has values $0, 1, 2, \ldots, n$.

③ BINOMIAL THEOREM

We now state the binomial theorem using a new notation for $_nC_r$ which is commonly used in this context.

$$\binom{n}{r} = {_nC_r}$$

Thus $\binom{5}{3} = {_5C_3} = \frac{5!}{2!3!} = 10$ and $\binom{7}{0} = {_7C_0} = \frac{7!}{0!7!} = 1$.

Binomial Theorem

For any binomial $a + b$ and any positive integer n,

$$(a+b)^n = \binom{n}{0}a^n + \binom{n}{1}a^{n-1}b^1 + \binom{n}{2}a^{n-2}b^2 + \binom{n}{3}a^{n-3}b^3 + \cdots + \binom{n}{n}b^n.$$

Using the sigma summation notation,

$$(a+b)^n = \sum_{r=0}^{n} \binom{n}{r} a^{n-r} b^r.$$

We call $\binom{n}{r}$ the **binomial coefficient** of the $(r+1)$th term of the expansion.

$$(r+1)\text{th term} = \binom{n}{r} a^{n-r} b^r$$

CAUTION

Notice that since r varies from 0 to n, a particular term is called the $(r+1)$th term. For example, when $r = 0$ we have the $(0+1)$th or 1st term. The 3rd term is the $(2+1)$th term which is the $(r+1)$th term for $r = 2$. Thus, if you are asked for a particular term, identify n, a, and b, and simply remember that *r is 1 less than the number of the needed term*.

❹ PROOF OF BINOMIAL THEOREM

The binomial theorem is proved using mathematical induction. Remember that we must show that $S(1)$ is true, and then prove that if $S(k)$ is true then $S(k+1)$ also is true.

First show that $S(1)$ is true, that is, the binomial theorem holds for $n = 1$.

$$\sum_{r=0}^{1} \binom{1}{r} a^{1-r} b^r = \binom{1}{0} a^{1-0} b^0 + \binom{1}{1} a^{1-1} b^1 = a + b = (a+b)^1$$

Thus, the theorem is true for $n = 1$. That is,

$$(a+b)^1 = \sum_{r=0}^{1} \binom{1}{r} a^{1-r} b^r.$$

Now assume that $(a+b)^k = \sum_{r=0}^{k} \binom{k}{r} a^{k-r} b^r$, and prove that $(a+b)^{k+1} = \sum_{r=0}^{k+1} \binom{k+1}{r} a^{k+1-r} b^r$. To do this start with the assumed equality and multiply both sides by $a + b$.

$$(a+b)^k = \sum_{r=0}^{k} \binom{k}{r} a^{k-r} b^r \qquad \text{Assumed true}$$

$$(a+b)^k (a+b) = \left[\sum_{r=0}^{k} \binom{k}{r} a^{k-r} b^r \right] (a+b)$$

$$(a + b)^{k+1} = \sum_{r=0}^{k} \binom{k}{r} a^{k-r} b^r a + \sum_{r=0}^{k} \binom{k}{r} a^{k-r} b^r b$$

$$= \sum_{r=0}^{k} \binom{k}{r} a^{k+1-r} b^r + \sum_{r=0}^{k} \binom{k}{r} a^{k-r} b^{r+1}.$$

When these sums are expanded and like terms are collected, the following series results.

$$(a + b)^{k+1} = \binom{k}{0} a^{k+1} + \left[\binom{k}{0} + \binom{k}{1}\right] a^k b + \left[\binom{k}{1} + \binom{k}{2}\right] a^{k-1} b^2 + \cdots$$
$$+ \left[\binom{k}{k-1} + \binom{k}{k}\right] a b^k + \binom{k}{k} b^{k+1}.$$

In the exercises for this section you are asked to prove that

$$\binom{k}{0} = \binom{k+1}{0}, \quad \binom{k}{k} = \binom{k+1}{k+1}, \quad \text{and} \quad \binom{k}{r-1} + \binom{k}{r} = \binom{k+1}{r}.$$

Using these equations we have

$$(a + b)^{k+1} = \binom{k+1}{0} a^{k+1} + \binom{k+1}{1} a^k b + \binom{k+1}{2} a^{k-1} b^2 + \cdots$$

$$+ \binom{k+1}{k} a b^k + \binom{k+1}{k+1} b^k = \sum_{r=0}^{k+1} \binom{k+1}{r} a^{k+1-r} b^r.$$

Thus, by mathematical induction, the binomial theorem is established for all positive integers n.

EXAMPLE 3 Using the Binomial Theorem

(a) Find the 5th term in the expansion of $(x + 2y)^7$.

In this example $n = 7$, $a = x$, $b = 2y$, and since we want the 5th term, r will be 4.

$$\binom{n}{r} a^{n-r} b^r = \binom{7}{4} (x)^{7-4} (2y)^4$$

$$= \frac{7!}{3!4!} x^3 (16y^4)$$

$$= 35x^3 (16y^4) = 560x^3 y^4$$

(b) Find the 7th term in the expansion of $(a^2 - 3y)^{10}$.

We have $n = 10$, $a = a^2$, $b = (-3y)$ [be careful to write $(a^2 - 3y)^{10}$ as $(a^2 + (-3y))^{10}$], and $r = 6$ (1 less than 7).

$$\binom{n}{r} a^{n-r} b^r = \binom{10}{6} (a^2)^{10-6} (-3y)^6$$

$$= \frac{10!}{4!6!} (a^2)^4 (729 y^6)$$

$$= 210 a^8 (729 y^6) = 153{,}090 a^8 y^6$$

PRACTICE EXERCISE 3

(a) Find the third term in the expansion of $(x + 2y)^7$.

(b) Find the 4th term in the expansion of $(a^2 - 3y)^{10}$.

Answers: (a) $84x^5 y^2$
(b) $-3240 a^{14} y^3$

9.7 EXERCISES A

In Exercises 1–3 evaluate each binomial coefficient.

1. $\binom{4}{2}$
2. $\binom{5}{0}$
3. $\binom{10}{8}$

Expand each binomial in Exercises 4–9.

4. $(x + y)^5$
5. $(3a - 1)^5$
6. $(3x - y)^4$
7. $(u^2 + v^2)^6$
8. $(a + a^{-1})^7$
9. $(x^{1/2} - y^{1/2})^4$

In Exercises 10–15 find the indicated term in each binomial expansion.

10. 3rd; $(x + 2)^5$
11. 4th; $(x + 2y)^5$
12. 4th; $(x + y^2)^5$
13. 6th; $(3a - b)^7$
14. middle; $(4a - 2)^6$
15. 6th; $(a - a^{-1})^8$

In Exercises 16–17 find the value of the real or complex number raised to a power.

16. $(1 + 1)^7$
17. $(1 + i)^5$

18. Find the term containing x^5 in $(3x - 5y)^6$.

19. Use the first four terms of $(1 - 0.1)^6$ to approximate $(0.9)^6$. Compare your result with a calculator value.

20. Prove that $\binom{n}{0} = \binom{n+1}{0}$ and $\binom{n}{n} = \binom{n+1}{n+1}$.

FOR REVIEW

Solve.

21. **SPORTS** How many ways can the five members of a basketball team be introduced to the crowd if the center must be introduced first?

22. **MANUFACTURING** How many ways can 4 light bulbs be selected from 24 light bulbs for testing?

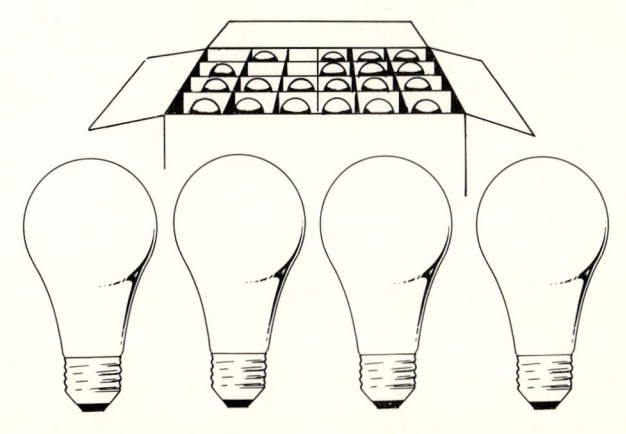

648 CHAPTER 9 SEQUENCES, SERIES, AND PROBABILITY

ANSWERS: 1. 6 2. 1 3. 45 4. $x^5 + 5x^4y + 10x^3y^2 + 10x^2y^3 + 5xy^4 + y^5$ 5. $243a^5 - 405a^4 + 270a^3 - 90a^2 + 15a - 1$ 6. $81x^4 - 108x^3y + 54x^2y^2 - 12xy^3 + y^4$ 7. $u^{12} + 6u^{10}v^2 + 15u^8v^4 + 20u^6v^6 + 15u^4v^8 + 6u^2v^{10} + v^{12}$ 8. $a^7 + 7a^5 + 21a^3 + 35a + 35a^{-1} + 21a^{-3} + 7a^{-5} + a^{-7}$ 9. $x^2 - 4x^{3/2}y^{1/2} + 6xy - 4x^{1/2}y^{3/2} + y^2$ 10. $40x^3$ 11. $80x^2y^3$ 12. $10x^2y^6$ 13. $-189a^2b^5$ 14. $-10,240a^3$ 15. $-56a^{-2}$ 16. 128 17. $-4 - 4i$ 18. $-7290x^5y$ 19. 0.53 20. Use the formula to show that each of the four binomial coefficients is 1. 21. 24 22. 10,626

9.7 EXERCISES B

In Exercises 1–3 evaluate each binomial coefficient.

1. $\binom{6}{4}$
2. $\binom{5}{5}$
3. $\binom{12}{7}$

Expand each binomial in Exercises 4–9.

4. $(x - y)^5$
5. $(x^2 - 2)^4$
6. $(x + 3y)^4$
7. $(u^2 - v^2)^6$
8. $(a^{-2} - a^2)^7$
9. $(x - 1)^9$

In Exercises 10–15 find the indicated term in each binomial expansion.

10. 5th; $(x - 2)^5$
11. 4th; $(x - 2y)^5$
12. 4th; $(x - y^2)^5$
13. 3rd; $(4a - 2)^6$
14. middle; $(2a - 3b)^8$
15. 4th; $(1 - \sqrt{x})^7$

In Exercises 16–17 find the value of the real or complex number raised to a power.

16. $(1 + \sqrt{2})^3$
17. $(1 - 3i)^4$

18. Find the term containing y^5 in $(5x - 2y)^6$.

19. Use the first four terms of $(1 + 0.1)^6$ to approximate $(1.1)^6$. Compare your results with a calculator value.

20. Prove that $\binom{n}{r-1} + \binom{n}{r} = \binom{n+1}{r}$.

FOR REVIEW

Solve.

21. How many distinguishable permutations of the letters in the word ALABAMA are possible?

22. **EDUCATION** How many ways can 4 essays be selected from 24 to be given first, second, third, and fourth place awards?

9.7 EXERCISES C

1. Find the sum of each row of Pascal's triangle through row 6. From the pattern, what is the sum of the numbers in row n?

2. Prove that the sum of the binomial coefficients in the expansion of $(a + b)^n$ is 2^n. [*Hint:* Let $a = 1 = b$.]

3. Prove that $\sum_{k=0}^{n} (-1)^k \binom{n}{k} = 0$.

4. Solve. $\sum_{k=0}^{6} \binom{6}{k} x^{6-k}(-1)^k = 64$ [Answers: 3, -1]

9.8 PROBABILITY

STUDENT GUIDEPOSTS

1. Language of Probability
2. Definition of Probability
3. Probability of *A* or *B*
4. Complementary Events
5. Experimental Probability

1 LANGUAGE OF PROBABILITY

Probability originated with problems related to games of chance, but today it has grown to include applications in such areas as genetics, insurance, physics, social science, and medicine.

Determining the likelihood of a particular event is the substance of probability. To make this precise, we need a variety of terms. We use the term **experiment** to mean the performing of some activity that has more than one possible outcome, such as flipping a coin, drawing a card, or rolling dice. When an experiment is performed, the set of all possible **outcomes** or results is called the **sample space** of the experiment. It is assumed that any particular outcome is just as likely to occur as any other, that is, the outcomes are **equally likely.** Any subset of the sample space is called an **event.** Elements in a specified event are **favorable outcomes** or **successes,** while outcomes not in the event are **unfavorable outcomes** or **failures.**

The experiment of flipping two fair coins has sample space

$$S = \{hh, ht, th, tt\},$$

where *hh* corresponds to the outcome of obtaining a head on each coin, and *ht* corresponds to the outcome of obtaining a head on the first coin and a tail on the second. Note that *ht* and *th* are different. This is clear if one coin is a dime, and the other a quarter, for then certainly there are two ways to obtain one head and one tail. The event

$$E = \{tt\}$$

is the event of obtaining two tails, while

$$F = \{ht, th\}$$

is the event of obtaining one head and one tail.

The experiment of rolling two unloaded dice has a sample space consisting of thirty-six pairs of numbers, each number varying from 1 through 6. This is because we have insisted that each outcome be as likely as any other. We cannot use the possible sums of numbers on the two dice, that is, $\{2, 3, 4, 5, 6, 7, 8, 9, 10, 11, 12\}$, to represent the sample space since the sum of 7 can be obtained in six ways

$$(1, 6), \quad (2, 5), \quad (3, 4), \quad (4, 3), \quad (5, 2), \quad (6, 1)$$

whereas the sum of 2 is less likely since it can only be obtained in one way, $(1, 1)$.

Observe that $(2, 5)$ and $(5, 2)$ are different outcomes. This is perhaps easier to understand if we think of the dice as being of different colors, one red and the other white. A 2 with the red die and a 5 with the white is clearly a different outcome than a 5 with the red and a 2 with the white.

We show the sample space for this experiment as follows.

$$F \begin{cases} (1, 1) & (1, 2) & \mathbf{(1, 3)} & (1, 4) & (1, 5) & (1, 6) \\ (2, 1) & \mathbf{(2, 2)} & (2, 3) & (2, 4) & (2, 5) & \mathbf{(2, 6)} \\ \mathbf{(3, 1)} & (3, 2) & (3, 3) & (3, 4) & \mathbf{(3, 5)} & (3, 6) \\ (4, 1) & (4, 2) & (4, 3) & \mathbf{(4, 4)} & (4, 5) & (4, 6) \\ (5, 1) & (5, 2) & \mathbf{(5, 3)} & (5, 4) & (5, 5) & (5, 6) \\ (6, 1) & \mathbf{(6, 2)} & (6, 3) & (6, 4) & (6, 5) & (6, 6) \end{cases} E$$

The event E of obtaining a sum of 8 is

$$E = \{(6, 2), (5, 3), (4, 4), (3, 5), (2, 6)\}$$

while the event F of obtaining a sum of 4 is

$$F = \{(3, 1), (2, 2), (1, 3)\}.$$

The pairs that give a particular sum are all found along a diagonal from lower left to upper right.

❷ DEFINITION OF PROBABILITY

We are now able to give a definition of probability along with several important properties. We use $n(S)$, read "n of S," to represent the number of elements in set S, and $n(E)$ for the number of elements in E.

Probability

Let S be the sample space of an experiment, and E an event (E is a subset of S). The **probability that E will occur**, or simply the **probability of E**, denoted by $P(E)$, is given by

$$P(E) = \frac{n(E)}{n(S)} = \frac{\text{the number of favorable outcomes}}{\text{the total number of outcomes}}.$$

Since E is a subset of S, $0 \le n(E) \le n(S)$, so that dividing through by $n(S)$, we obtain

$$\frac{0}{n(S)} \le \frac{n(E)}{n(S)} \le \frac{n(S)}{n(S)}$$

or
$$0 \le P(E) \le 1.$$

Hence the probability of an event is always a number between 0 and 1, inclusive. If $P(E) = 0$, we say that E is **impossible** (E cannot occur). If $P(E) = 1$, we say that E is **certain** (E must occur). If E and F are two events such that $P(E) < P(F)$, we say that E is **less likely to occur** than F, or that F is **more likely to occur** than E. If $P(E) = P(F)$, the events are **equally likely**.

EXAMPLE 1   PROBABILITIES WITH COINS

The sample space for the experiment of flipping two fair coins is $S = \{hh, ht, th, tt\}$.

(a) If E is the event of obtaining two heads, what is $P(E)$?

Since $E = \{hh\}$,

$$P(E) = \frac{n(E)}{n(S)} = \frac{1}{4}. \quad \text{Remember that } hh \text{ is one outcome}$$

PRACTICE EXERCISE 1

Use the sample space in Example 1.

(a) What is the probability of two tails?

(b) If F is the event of obtaining at least one head, find $P(F)$.
$$F = \{hh, ht, th\}.$$
Thus,
$$P(F) = \frac{n(F)}{n(S)} = \frac{3}{4}.$$

(b) What is the probability of no tails?

Answers: (a) $\frac{1}{4}$ (b) $\frac{1}{4}$

EXAMPLE 2   PROBABILITIES WITH DICE

Consider the sample space for the experiment of rolling two dice.

(a) Determine the probability of rolling a sum of 8.

The sample space was given earlier in this section.
$$E = \{(6, 2), (5, 3), (4, 4), (3, 5), (2, 6)\}$$
Thus $P(E) = \dfrac{n(E)}{n(S)} = \dfrac{5}{36}.$

(b) Determine the probability of rolling a sum of 3.

From the array of outcomes given in this section, the number of ways of obtaining a three is 2. Thus,
$$P(3) = \frac{2}{36} = \frac{1}{18}.$$

PRACTICE EXERCISE 2

Use the sample space for rolling two dice.

(a) What is the probability of rolling a sum of 10?

(b) What is the probability of rolling a sum of 5?

Answers: (a) $\frac{1}{12}$ (b) $\frac{1}{9}$

Note The probabilities in Example 2 do not mean that if the dice are thrown 36 times exactly 5 sums of 8 and 2 sums of 3 will be obtained. It does mean that if the dice are thrown hundreds of times it is likely that about 5/36 of the tosses will result in a sum of 8, and about 1/18 of the tosses will give a sum of 3.

EXAMPLE 3   PROBABILITIES IN CARD PLAYING

Find the probability of drawing a five-card hand consisting of 3 aces and two cards that are not aces from a deck of 52 cards.

Let E represent the event. There are $_4C_3$ ways of drawing 3 aces, $_{48}C_2$ ways of drawing two non-aces, and $_{52}C_5$ total ways of drawing 5 cards. Thus,
$$P(E) = \frac{_4C_3 \cdot {_{48}C_2}}{_{52}C_5} = \frac{\dfrac{4!}{3!1!}\dfrac{48!}{46!2!}}{\dfrac{52!}{5!47!}} = \frac{94}{54145} \approx 0.0017.$$

PRACTICE EXERCISE 3

What is the probability of drawing a five-card hand of 3 aces and 2 kings?

Answer: $\frac{1}{108,290} \approx 0.0000092$

③ PROBABILITY OF A OR B

When two or more events are joined by the word *or*, we say that the result is the **disjunction** of the events. If A is the event of rolling a 5 with one die, and B is the event of rolling a 3 with one die, then A or B is the event of rolling a 3 or a 5 with one die. In this case, it is clear that

$$P(A \text{ or } B) = P(A) + P(B)$$
$$\frac{2}{6} = \frac{1}{6} + \frac{1}{6}.$$

The **conjunction** of two or more events is formed by joining them with the word *and*. In the above, it is impossible for both A and B to occur on one trial. When this happens, we say that A and B are **mutually exclusive,** and observe that the probability that both A and B occur is 0, that is,

$$P(A \text{ and } B) = 0.$$

Thus, the event (A and B) is impossible.

Suppose that A is the event of drawing an ace and B is the event of drawing a spade in the experiment of drawing one card from a well-shuffled deck of 52 cards. Then (A or B) is the event of drawing an ace or a spade on one draw of a card. Since there are 13 spades and 4 aces, we are tempted to add to obtain 17 as the number of outcomes favorable to (A or B). However, the ace of spades has then been counted twice (both as a spade and as an ace) so that 17 is 1 too many for the number of favorable outcomes. Thus, $P(A \text{ or } B) = 16/52$. This same result can be obtained by the following formula.

$$P(A \text{ or } B) = P(A) + P(B) - P(A \text{ and } B)$$
$$= \frac{4}{52} + \frac{13}{52} - \frac{1}{52} \quad P(A \text{ and } B) \text{ is } \tfrac{1}{52}$$
$$= \frac{16}{52}$$

When A and B are mutually exclusive $P(A \text{ and } B) = 0$ so that the formula still applies. The above examples are specific cases of the next theorem.

Probability of A or B

Let A and B be events. Then

$$P(A \text{ or } B) = P(A) + P(B) - P(A \text{ and } B).$$

If A and B are mutually exclusive, $P(A \text{ and } B) = 0$ and the formula reduces to

$$P(A \text{ or } B) = P(A) + P(B).$$

EXAMPLE 4 PROBABILITIES OF TWO EVENTS

If one card is drawn from a well-shuffled deck of 52 cards, find the probability of drawing the following cards.

(a) An ace or a king

$$P(\text{ace or king}) = P(\text{ace}) + P(\text{king}) - P(\text{ace and king})$$
$$= \frac{4}{52} + \frac{4}{52} - \frac{0}{52} = \frac{8}{52} = \frac{2}{13}$$

PRACTICE EXERCISE 4

Use the sample space of Example 4 to find each probability.

(a) a jack or a queen

(b) A face card or a heart

$$P(\text{face or heart}) = P(\text{face}) + P(\text{heart}) - P(\text{face and heart})$$
$$= \frac{12}{52} + \frac{13}{52} - \frac{3}{52} \quad \text{3 heart face cards}$$
$$= \frac{22}{52} = \frac{11}{26}$$

(c) An ace and a jack

$P(\text{ace and jack}) = 0$ since the two events are mutually exclusive.

(b) a ten or a black card

(c) a five and a seven

Answers: (a) $\frac{2}{13}$ (b) $\frac{7}{13}$ (c) 0

We now solve the applied problem in the introduction to this chapter.

EXAMPLE 5 APPLICATION TO BIOLOGY

A couple plans to have two children. Assuming that it is equally likely that a boy or a girl will be born, what is the probability that the couple will have two girls? What is the probability of two boys? What is the probability of one boy and one girl?

The sample space for this experiment is

$$S = \{gg, gb, bg, bb\},$$

where gb means girl-then-boy and bg means boy-then-girl. The probability of two girls is

$$P(gg) = \frac{n(gg)}{n(S)} = \frac{1}{4}.$$

The probability of two boys is the same.

$$P(bb) = \frac{n(bb)}{n(S)} = \frac{1}{4}$$

However, the probability of one boy and one girl is

$$P(1 \text{ boy and } 1 \text{ girl}) = P(gb) + P(bg) = \frac{1}{4} + \frac{1}{4} = \frac{1}{2}.$$

PRACTICE EXERCISE 5

If a couple plans to have three children, what is the probability of having one girl and two boys? What is the probability of having no more than one boy?

Answers: $\frac{3}{8}, \frac{1}{2}$

❹ COMPLEMENTARY EVENTS

If E is an event and F is the event described as not E, then E and F are **complementary events**. Since either E or F must be true, $P(E \text{ or } F) = 1$. Also, $P(E \text{ and } F) = 0$. Thus,

$$P(E \text{ or } F) = P(E) + P(F) = 1$$
$$\text{or} \quad P(F) = 1 - P(E).$$

Complementary Events

If E is an event and F is the event that E will not occur, then E and F are complementary and

$$P(F) = 1 - P(E).$$

EXAMPLE 6   APPLICATION FROM MEDICINE

During an outbreak of both measles and mumps, health officials warned parents to have children vaccinated. If they did not, the probability of a child contracting measles was 0.012, the probability of contracting the mumps was 0.031, and the probability of contracting both was 0.005.

(a) Find the probability of a child who was not vaccinated contracting the measles or the mumps.

$$P(\text{Me or Mu}) = P(\text{Me}) + P(\text{Mu}) - P(\text{Me and Mu})$$
$$= 0.012 + 0.031 - 0.005$$
$$= 0.038$$

(b) Find the probability of a child not contracting either of the diseases.

Let F be the event of contracting neither of the diseases.

$$P(F) = 1 - P(\text{Me or Mu})$$
$$= 1 - 0.038 = 0.962$$

PRACTICE EXERCISE 6

Use the data in Example 6.

(a) What is the probability of a child not contracting both measles and mumps?

(b) Find the probability of a child not contracting the measles and not contracting the mumps.

Answers: (a) 0.995   (b) 0.962

5   EXPERIMENTAL PROBABILITY

How did the health officials in Example 6 come up with the probabilities that are given? They most likely used experimental data. During similar outbreaks of measles and mumps in previous years, they could have selected representative groups of children, counted them, and counted the number that contracted measles and the number that contracted mumps. This data could then have been used to calculate empirically the probabilities. For example, if 485 children were in a group and 6 of them contracted measles, the probability of contracting measles was

$$P(\text{Me}) = \frac{6}{485} \approx 0.0123711 \approx 0.012.$$

This type of probability is used extensively in all the sciences, in engineering, and in business. Two areas which are visible to most people are weather forecasting and life insurance. Also, many practical experiments are carried out to discover critical information. For example, an architectural firm might study the heights of people who are likely to use a building in order to improve the design of doorways and elevators.

Suppose a representative sample of 60 potential users of a building is selected and their heights determined. To make the data easier to interpret, the heights could be divided into groups called **class intervals.** We might use 66–68 inches as one interval, 68–70 inches for another, and so forth. The number of heights in each class interval is called the **class frequency.** The table below summarizes the data collected.

Height in inches (class intervals)	Number of people (class frequency)
66–68	4
68–70	12
70–72	24
72–74	18
74–76	2

Suppose we assume that a person whose height is exactly 68 inches is in the class interval 68–70 and not in 66–68, and similarly for persons whose height is 70 inches, 72 inches, and 74 inches. Then each of the sixty people is in exactly one class interval. A special type of bar graph, called a **histogram,** can be used to picture the data in the table. The width of each bar in Figure 9.5 represents the range of values in that class interval, and the length or height of the bar corresponds to the class frequency.

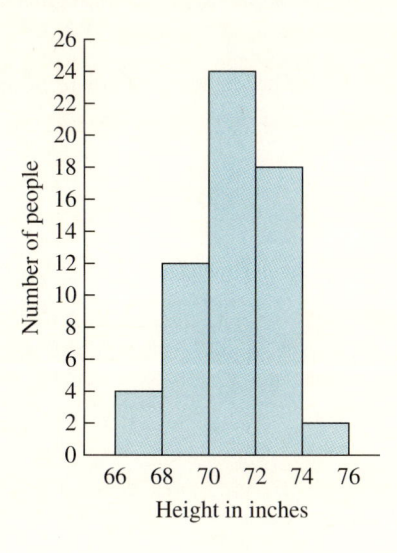

Figure 9.5 Histogram

EXAMPLE 7 Using Data for Probabilities

Use the data presented in the histogram in Figure 9.5 to find the following probabilities.

(a) What is the probability that a user of the proposed building will be 70 inches or taller?

We add the class frequencies for the class intervals 70–72, 72–74, and 74–76.

$$24 + 18 + 2 = 44$$

Since there are 60 people in the sample space, the probability of someone being 70 inches or taller is

$$\frac{44}{60} = \frac{11}{15} \approx 0.73.$$

(b) What is the probability of a person's height being 66–68 or 70–72 or 74–76?

This time we add probabilities.

$$\frac{4}{60} + \frac{24}{60} + \frac{2}{60} = \frac{28}{60} = \frac{7}{15} \approx 0.47$$

Practice Exercise 7

Use the data presented in the histogram in Figure 9.5.

(a) What is the probability that a person is shorter than 72 inches?

(b) What is the probability that a person is at least 68 inches but less than 74 inches?

Answers: (a) $\frac{2}{3}$ (b) $\frac{9}{10}$

9.8 EXERCISES A

In Exercises 1–3 answer yes if the number could be a probability and no if it could not.

1. 0
2. $-\dfrac{1}{2}$
3. 0.00001

RECREATION *Let $S = \{1, 2, 3, 4, 5, 6\}$ be the sample space of rolling one unloaded die. Give the probability of each event in Exercises 4–6.*

4. Rolling a 5
5. Rolling a number greater than 0
6. Rolling a number $n \geq 4$

RECREATION *Let S be the sample space for the experiment of rolling two unloaded dice. Find the probability of each event in Exercises 7–9.*

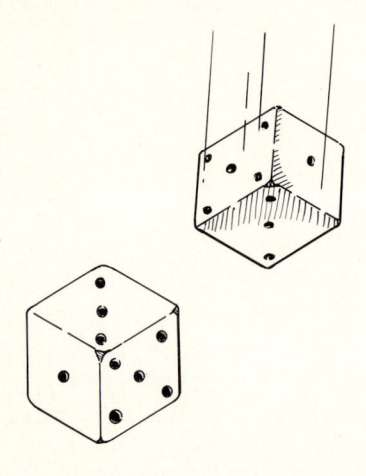

7. Rolling a total of 7
8. Rolling a total of 7 or 11
9. Rolling a total greater than 3

RECREATION *In the game of craps, on the first roll a player wins with a total of 7 or 11 and loses with a total of 2, 3, or 12. Any other total becomes his point, and he must continue to roll until he rolls his point and wins, or rolls a 7 and loses. Find the probability of each event in Exercises 10–11.*

10. Winning on the first roll
11. Rolling a total other than a winning or losing total on the first roll

RECREATION *A roulette wheel contains 38 slots numbered 00, 0, 1, 2, 3, 4, . . . , 35, 36. Eighteen of the slots numbered 1 through 36 are colored black and the rest are colored red. The 00 and 0 slots are colored green and are called house numbers. The wheel is spun and a ball is rolled around the rim in the opposite direction. Eventually the ball falls into a slot. Find the probability of each event in Exercises 12–15.*

12. Ball falls into the number 3 slot
13. Ball falls into a red slot

14. Ball falls into a blue slot

15. Ball falls into an odd-numbered slot

RECREATION *A card is drawn from a well-shuffled deck of 52 cards. In Exercises 16–22 find the probability of each event.*

16. Drawing an ace

17. Drawing a black card

18. Drawing the queen of hearts

19. Drawing a 7 or a queen

20. Drawing a 7 or a diamond

21. Drawing a face card or a heart

22. Drawing a king and a heart

RECREATION *If 5 cards are dealt from a well-shuffled deck of 52 cards, find the probability of the hands in Exercises 23–25.*

23. 5 hearts

24. 4 aces and a card that is not an ace

25. 3 queens and 2 jacks

26. RECREATION If the probability of winning a game is 4/7, what is the probability of losing if there are no ties?

27. A child picks up a telephone and dials seven digits at random; what is the probability that he dials his own number?

28. GENETICS It has been determined that the number of possible genetic combinations that a child of one couple can have is 2^{48}. Assuming that Mr. and Mrs. Levin already have one child, what is the probability that a second child will have the same genetic makeup?

EDUCATION *The bar graph shows possible majors in the College of Arts and Science at a large university. In Exercises 29–32, find the following probabilities relating to students in the College of Arts and Science.*

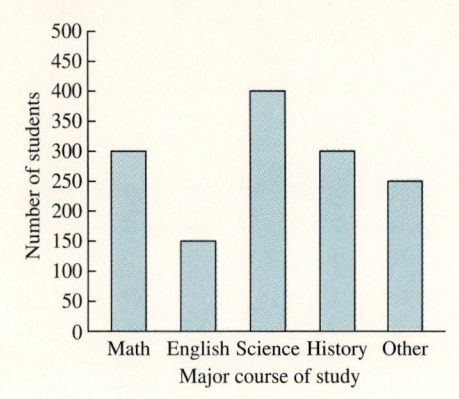

29. What is the probability that a particular student is a math major or a science major?

30. What is the probability that a student is not a history major?

31. What is the probability that a student's major is in math, English, or history?

32. What is the probability that a student's major is not in math, is not in science, and is not in history?

EDUCATION *The number of years of teaching experience of the faculty of a university is shown in the histogram. Use this data to solve Exercises 33–36.*

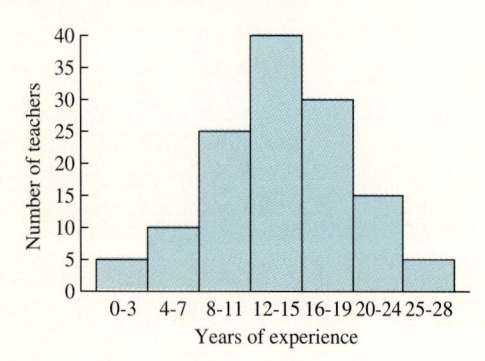

33. What is the probability that a particular professor has less than 4 or more than 24 years of experience?

34. What is the probability that a professor has 8 or more years of experience?

35. What is the probability that a professor does not have at least 8 years of experience?

36. What is the probability that a professor has 4–7 or 20–24 years of experience?

FOR REVIEW

37. Use the binomial theorem to expand $(2x - 5y)^4$.

38. Find the 5th term of $(x^2 - 2y)^7$.

ANSWERS: 1. yes 2. no 3. yes 4. $\frac{1}{6}$ 5. 1 6. $\frac{1}{2}$ 7. $\frac{1}{6}$ 8. $\frac{2}{9}$ 9. $\frac{11}{12}$ 10. $\frac{2}{9}$ 11. $\frac{2}{3}$ 12. $\frac{1}{38}$ 13. $\frac{9}{19}$ 14. 0 15. $\frac{9}{19}$ 16. $\frac{1}{13}$ 17. $\frac{1}{2}$ 18. $\frac{1}{52}$ 19. $\frac{2}{13}$ 20. $\frac{4}{13}$ 21. $\frac{11}{26}$ 22. $\frac{1}{52}$ 23. $\frac{33}{66,640}$ 24. $\frac{1}{54,145}$ 25. $\frac{1}{108,290}$ 26. $\frac{3}{7}$ 27. 10^{-7} 28. 2^{-48} 29. 0.50 30. 0.79 31. 0.54 32. 0.29 33. 0.08 34. 0.88 35. 0.12 36. 0.19 37. $16x^4 - 160x^3y + 600x^2y^2 - 1000xy^3 + 625y^4$ 38. $560x^6y^4$

9.8 EXERCISES B

In Exercises 1–3 answer yes if the number could be a probability and no if it could not.

1. $\frac{2}{3}$
2. 1
3. $\frac{5}{4}$

RECREATION *Let $S = \{1, 2, 3, 4, 5, 6\}$ be the sample space of rolling one unloaded die. Give the probability of each event in Exercises 4–6.*

4. Rolling a number less than 3
5. Rolling a multiple of 3
6. Rolling a number $n < 1$

RECREATION *Let S be the sample space for the experiment of rolling two unloaded dice. Find the probability of each event in Exercises 7–9.*

7. Rolling a total of 11
8. Rolling a total of 5 or 2
9. Rolling a total less than or equal to 11

RECREATION *In the game of craps, on the first roll a player wins with a total of 7 or 11 and loses with a total of 2, 3, or 12. Any other total becomes his point, and he must continue to roll until he rolls his point and wins, or rolls a 7 and loses. Find the probability of the event in Exercise 10 and answer the question in Exercise 11.*

10. Losing on the first roll
11. Of all the points 4, 5, 6, 8, 9, 10, which two are the most likely to be rolled? Least likely to be rolled?

RECREATION *A roulette wheel contains 38 slots numbered 00, 0, 1, 2, 3, 4, . . . , 35, 36. Eighteen of the slots numbered 1 through 36 are colored black and the rest are colored red. The 00 and 0 slots are colored green and are called house numbers. The wheel is spun and a ball is rolled around the rim in the opposite direction. Eventually the ball falls into a slot. Find the probability of each event in Exercises 12–15.*

12. Ball falls into a black slot
13. Ball falls into a green slot
14. Ball falls into a black or red slot
15. Ball falls into the number 0 slot

RECREATION *A card is drawn from a well-shuffled deck of 52 cards. In Exercises 16–22 find the probability of each event.*

16. Drawing a heart
17. Drawing a face card
18. Drawing a joker
19. Drawing a 7 or a black card
20. Drawing a face card or a black card
21. Drawing a face card or a 3
22. Drawing a face card and a heart

RECREATION *If 5 cards are dealt from a well-shuffled deck of 52 cards, find the probability of the hands in Exercises 23–25.*

23. 5 black cards
24. 4 aces and a joker

25. 2 aces, 2 kings, and a card that is not an ace or a king

26. **SPORTS** If the probability of a horse losing a race is 9/11, what is the probability that the horse will win?

27. Claude has flipped a fair coin fifty times and it has fallen heads each time. He conjectures that on the fifty-first flip the likelihood of a tail will certainly increase. Do you agree? What is the probability of a tail on the fifty-first flip?

28. **GENETICS** A couple plans to have three children.
    (a) Give the sample space representing these children according to sex.
    (b) What is the probability of having three boys?
    (c) What is the probability of having three children of the same sex?
    (d) What is the probability of two girls and one boy?
    (e) What is the probability of at least two boys?

ADVERTISING *From research about media advertising strategies on campuses, a company got the results shown in the bar graph. Use this data in Exercises 29–32.*

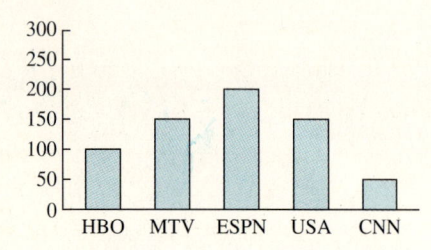

29. What is the probability that a student prefers HBO or MTV?

30. What is the probability that a student prefers anything besides CNN?

31. What is the probability that a student prefers MTV, ESPN, or USA?

32. Which is greater, the probability of a student preferring HBO or CNN or the probability of a student preferring MTV or USA?

ECONOMICS *The histogram gives the fuel economy of automobiles. Assume that* 10 *mpg is in* 10–15 *and not in* 5–10 *and the same for* 5, 15, 20, 25, 30, *and* 35. *Use this data in Exercises 33–36.*

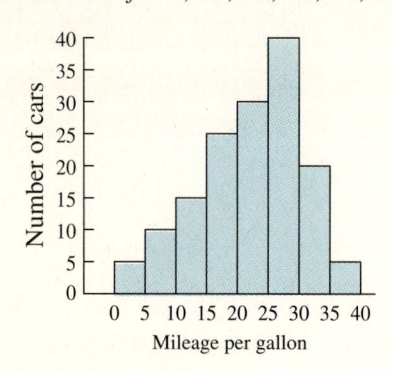

33. What is the probability of a car getting 30 mpg or more?

34. What is the probability of a car getting less than 10 mpg?

35. What is the probability of a car getting at least 15 mpg but less than 30 mpg?

36. What is the probability that a car is not in the 10–30 mpg range?

FOR REVIEW

37. Find the value of the complex number $(2 - i)^5$.

38. Find the term with no x in $(x^{-1} + x)^6$.

9.8 EXERCISES C

1. If a 5-card hand is dealt from a standard 52-card deck, what is the probability of getting a *royal flush* which is ace-king-queen-jack-ten all of the same suit?

    $\left[\text{Answer: } \dfrac{1}{649{,}740} \approx 0.0000015\right]$

2. If a 5-card hand is dealt from a standard 52-card deck, what is the probability of getting *four of a kind* which is a hand such as two-two-two-two-six or ace-ace-ace-ace-ten?

3. If the given number of dice are thrown, what is the probability of them all coming up four?
    (a) 1 (b) 2 (c) 3 (d) 4 (e) n

4. If n dice are thrown, what is the probability of them all coming up the same number?

CHAPTER 9 EXTENSION PROJECT

In Section 9.8 one way that we found probabilities was by dividing the number of favorable outcomes in an experiment by the total number of outcomes. For example, we could determine empirically if a die is fair by rolling it several hundred times and recording each time the numbers (1, 2, 3, 4, 5, 6) came up. Suppose someone has rolled a die 600 times and recorded the results in the bar graph below.

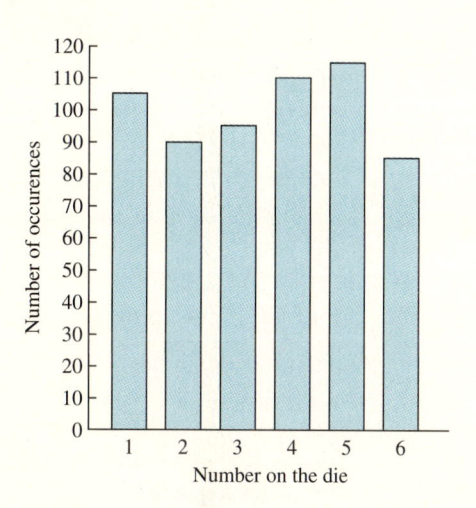

We now read the graph and calculate the probabilities in the following table. Of course, we would expect each probability to be approximately $1/6 \approx 0.167$.

Number	Frequency	Probability	Difference from 0.167
1	105	0.175	0.008
2	90	0.150	−0.017
3	95	0.158	−0.009
4	110	0.183	0.016
5	115	0.192	0.025
6	85	0.142	−0.025

The probabilities are close to 0.167, but in order for you to really decide about the fairness of a die, you should carry out several experiments on the same die and equivalent experiments on several others. You will be asked to do this in "Things to Do."

The study of the frequency of use of the letters of our alphabet provides us with another experiment. It has been shown that the letter *e* occurs most frequently in our writing and the letter *z* the least frequently. Suppose that we select a page in a book, count the total number of letters, and count the number of times the letter *e*, *a*, *s*, and *n* occur. In one experiment we counted a total of 2452 letters and constructed the following table.

Letter	Frequency	Probability
e	321	0.13
a	166	0.07
s	152	0.06
n	190	0.08

The table shows that if a letter is selected at random on the page, the probability is about 0.13 that it will be an *e*. If we do other experiments confirming this number, we can assume that approximately 13% of the letters on a typical page will be the letter *e*. Likewise we would expect approximately 7% to be *a*, 6% to be *s*, and 8% to be *n*.

THINGS TO DO:

1. Select a die, collect data on several hundred rolls, and calculate the probabilities. Perform this experiment again on the same die and perform experiments on several other dice. Use this data to determine how close the probability of each number on a fair die is to 1/6 and which of the dice used in your experiments are fair.

662 CHAPTER 9 SEQUENCES, SERIES, AND PROBABILITY

2. Carry out the letter-counting experiment and compare your results with the probabilities obtained above. Obtain data and probabilities for other letters. To stress the idea of probability, devise a way to select letters at random from a page and see if 13% of the ones selected are the letter *e*.

3. Study the literature on probability and statistics in order to refine your methods of data collection and your calculations of probabilities. In particular, can a computer be used in the collection of data as well as the calculation in an experiment such as the one involving letter counting?

4. Study newspapers and periodicals and watch news reports to observe the variety of ways that this type of mathematics is used in our society.

CHAPTER 9 REVIEW

KEY TERMS

9.1
1. An **infinite sequence** is a function with domain the set of natural numbers.
2. A **finite sequence** with m terms is a function with domain the set of natural numbers from 1 to m.
3. A sequence is determined by a **recursive definition** if x_1 is given and x_{n+1} is defined in terms of x_n.
4. A **series** is the indicated sum of the terms of a sequence.
5. The Greek letter Σ is the **summation symbol**. In $\sum_{k=1}^{n} a_k$ the letter k is the **index** on the summation, 1 is the **lower limit**, and n is the **upper limit of summation**.

9.2
1. An **arithmetic sequence** is a sequence in which the successive terms differ by some constant d, called the **common difference**.
2. The **nth term a_n** of an arithmetic sequence is given by
$$a_n = a_1 + (n - 1)d.$$
3. The **sum of the first n terms S_n** of an arithmetic sequence is given by
$$S_n = \frac{n}{2}[2a_1 + (n-1)d] = \frac{n}{2}[a_1 + a_n].$$
4. The **arithmetic means** are all the terms between a_1 and a_n in an arithmetic sequence.

9.3
1. A **geometric sequence** is a sequence with the property that
$$\frac{a_n}{a_{n-1}} = r$$
for all n. The number r is the **common ratio**.

2. The **nth term a_n** of a geometric sequence is given by
$$a_n = a_1 r^{n-1}.$$
3. The **sum of the first n terms S_n** of a geometric sequence is given by
$$S_n = \frac{a_1 - ra_n}{1 - r} = \frac{a_1 - a_1 r^n}{1 - r}.$$

9.4
1. The number $S_n = a_1 + a_2 + a_3 + \cdots + a_n$ is the **nth partial sum** of a sequence.
2. If the S_n's approach some fixed number S as n gets larger and larger, then S is called the **sum of the geometric sequence.** S is given by $S = \frac{a_1}{1 - r}$ if $|r| < 1$.

9.5
1. The method of **mathematical induction** is used to prove statements or formulas which involve positive integers.
2. The **induction assumption** is that a statement is true for a given positive integer k.

9.6
1. The techniques for counting the number of ways that a collection of objects or acts can be arranged, ordered, combined, chosen, or occur in succession come under the heading of **combinatorial algebra.**
2. The **factorial notation** is given by
$$n! = n(n-1)(n-2)(n-3) \cdots 2 \cdot 1.$$
3. A **permutation** is any arrangement of a collection of objects in a particular order.
$$_nP_r = \frac{n!}{(n-r)!}$$
4. A **combination** is any collection of objects. The order of selection is not important.
$$_nC_r = \frac{n!}{(n-r)!r!}$$

9.7
1. The triangular array of the coefficients in binomial expansions is known as **Pascal's triangle**.
2. The **binomial coefficient** is $\binom{n}{r} = {}_nC_r$.
3. The **binomial theorem** states that

$$(a+b)^n = \sum_{r=1}^{n} \binom{n}{r} a^{n-r} b^r.$$

9.8
1. An **experiment** is the performing of an activity that has several possible outcomes. The set of all possible **outcomes** or results is the **sample space** of the experiment. It is assumed that any particular outcome is just as likely to occur as any other, that is, the outcomes are **equally likely**. Any subset of the sample space is called an **event**. Elements in a specified event are **favorable outcomes** or **successes**, while outcomes not in the event are **unfavorable outcomes** or **failures**.
2. The **probability** of event E, denoted by $P(E)$, is the number of favorable outcomes divided by the total number of outcomes.
3. If $P(E) = 0$, we say E is **impossible**. If $P(E) = 1$, we say that E is **certain**. If $P(E) < P(F)$, we say that E is **less likely to occur** than F, or that F is **more likely to occur** than E. If $P(E) = P(F)$, the events are **equally likely**.
4. When two or more events are joined by the word *or*, we say that the result is the **disjunction** of the events. The **conjunction** of two or more events is formed by joining them with the word *and*.
5. If it is impossible for both A and B to occur on one trial, we say that they are **mutually exclusive**.
6. If E is an event and F is the event described as not E, then E and F are **complementary events**.
7. A **histogram** is a bar graph in which the width of each bar represents the range of values in that **class interval**, and the length or height of the bar corresponds to the **class frequency**.

REVIEW EXERCISES

Part I

9.1 *In Exercises 1–2 the nth term of a sequence is given. Find the first five terms and the eighth term of each.*

1. $a_n = \dfrac{n^2 - 1}{n}$

2. $x_n = \dfrac{(-1)^n}{3n + 1}$

3. Write $\sum_{k=0}^{3} \sqrt{k^2 + 1}$ without using sigma summation notation.

4. Write $\dfrac{1}{2} + \dfrac{4}{3} + \dfrac{9}{4} + \dfrac{16}{5} + \cdots + \dfrac{n^2}{n+1}$ using sigma summation notation.

5. Give the second and third terms of the sequence if $a_1 = -3$ and $a_{n+1} = 1 - 2a_n$.

6. Find the formula for the nth term of 3, 6, 11, 18, 27, ...

Evaluate each series in Exercises 7–8.

7. $\sum_{k=1}^{5} (-1)^{k+1} 2^k$

8. $\sum_{m=3}^{6} \dfrac{m+5}{m-1}$

9.2
9. Write the first six terms of an arithmetic sequence with $a_1 = -9$ and $d = 4$, and find S_{12}.

10. For an arithmetic sequence, $a_1 = 5$, $a_n = 19$, and $S_n = 96$. Find n and d.

11. Insert three arithmetic means between 17 and 5.

12. Find x so that $x + 1$, $2x + 3$, $4x + 1$ forms a three-term arithmetic sequence. Also, give the sequence.

13. CONSUMER A new car costing $9000 depreciates 24% the first year, 20% the second year, 16% the third year and continues in that manner for five years. If all depreciations apply to the original cost, what is the value of the car in five years?

14. A collection of dimes is arranged in a triangular array with 20 coins in the base row, 19 in the next, 18 in the next, and so forth. Find the value of the collection.

9.3

15. Write the first six terms of a geometric sequence with $a_1 = \frac{1}{6}$ and $r = 2$, and find S_6.

16. Find the sum of the first eight terms of the geometric sequence with $a_1 = \frac{1}{27}$ and $a_8 = 81$.

17. Insert four geometric means between -12 and $\frac{3}{8}$.

18. Find a positive number x so that $3x - 1$, $x + 3$, $x - 2$ forms a three-term geometric sequence in the given order. Also, give the sequence.

19. BUSINESS Michael Healy borrows $1500 at 9% interest compounded annually. If he pays off the loan in full at the end of five years, how much does Michael pay?

20. PHYSICS A ball is dropped from a height of 30.0 ft. If, on each rebound, it rises to a height of 4/5 the distance from which it fell, how far does it rise on the fifth rebound, and how far up and down has it traveled when it hits the ground for the sixth time?

9.4

21. Find the sum of the infinite geometric sequence $8, -2, 1/2, -1/8, \ldots$.

22. Find the first five terms of an infinite geometric sequence with $a_1 = 30$ and $S = 36$.

In Exercises 23–24 convert each decimal to a fraction.

23. $1.\overline{2}$

24. $5.\overline{15}$

25. RECREATION A child on a swing traverses an arc of 8 m. Each pass thereafter, he traverses an arc that is 8/9 the length of the previous arc. How far does he travel before coming to rest?

26. PHYSICS A ball dropped from a height of 27 ft always rebounds 2/3 of the height of the previous drop. How far does it travel (up and down) before coming to rest?

9.5 *In Exercises 27–28 use the principle of mathematical induction to prove that each statement is true for every positive integer n.*

27. $S(n)$: $4 + 8 + 12 + \cdots + 4n = 2n(n + 1)$

28. $S(n)$: $1 \cdot 2 + 2 \cdot 3 + 3 \cdot 4 + \cdots + n(n + 1) = \frac{1}{3}n(n + 1)(n + 2)$

9.6 *Evaluate each permutation or combination in Exercises 29–34.*

29. $_3P_2$

30. $_3C_2$

31. $_5P_5$

32. $_5C_5$

33. $_8P_0$

34. $_8C_0$

35. How many serial numbers formed by a letter followed by a three-digit numeral are possible under the following conditions?
(a) No other restrictions are imposed.
(b) The letter must not be an A and repetition of digits is not allowed.
(c) The letter must be an A, B, or C and the first digit cannot be a zero.

36. In how many ways can four officers be chosen from an organization consisting of thirteen members?

37. In how many ways can nine people be placed in a line?

38. In how many ways can nine people be seated around a circular table?

39. How many distinguishable permutations of the letters in the word FLAGSTAFF are there?

40. GEOMETRY How many lines are determined by six points, no three of which are collinear?

41. RECREATION In how many ways can a picnic committee of three men and two women be selected from eight men and eleven women?

42. In how many ways can a committee of four be chosen from an organization consisting of thirteen members?

9.7 **43.** Expand $(3a - z)^5$.

44. Expand $(y^{-1} + y)^4$.

45. Find the 4th term in the binomial expansion of $(2x - y^2)^6$.

46. Find the 5th term in the binomial expansion of $(a - 3b)^7$.

9.8 *One marble is selected, sight unseen, from a bag containing 8 red and 7 black marbles. In Exercises 47–50, determine the probability of the event.*

47. Selecting a red marble

48. Selecting a black marble

49. Selecting a blue marble

50. Selecting a marble that is not blue

RECREATION *A single card is drawn from a well-shuffled deck of 52 cards. In Exercises 51–56, give the probability of the event.*

51. Drawing a spade

52. Drawing a face card

53. Drawing the jack of hearts

54. Drawing a red ace

55. Drawing a ten or a queen

56. Drawing a ten and a queen

57. RECREATION If the probability of winning first prize in a contest is 10^{-4}, what is the probability of not winning?

58. EDUCATION Given that the probability of passing English is 0.92, the probability of passing math is 0.85, and the probability of passing them both is 0.78, find the probability of passing English or math. What is the probability of failing both?

A survey in a community gave the preference of food indicated in the bar graph. Read the graph and use the data to find the probabilities in Exercises 59–60.

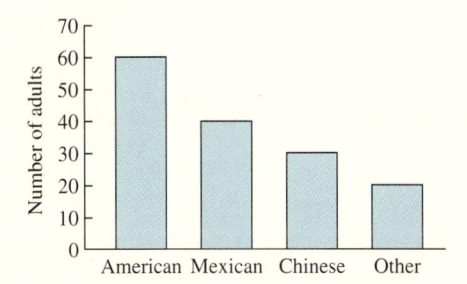

59. If a person is selected at random from the community, what is the probability that they prefer food other than American, Mexican, or Chinese?

60. What is the probability that a person prefers Mexican or Chinese food?

Part II

61. Find the sum of the infinite geometric sequence 16, -8, 4, ...

62. For the general sequence 4, 9, 14, 19, . . . , find the formula for the general term.

63. RECREATION How many 5-card poker hands consisting of 3 aces and 2 eights are possible with an ordinary 52-card deck? What is the probability of being dealt such a hand?

64. Find the sum of the integers divisible by 4 between -25 and 125.

666 CHAPTER 9 SEQUENCES, SERIES, AND PROBABILITY

65. If $x_1 = \frac{1}{3}$ and $x_n = 9x_{n-1} - 5$, give the second and third terms of the sequence.

66. **EDUCATION** In how many ways can a student choose 10 out of 15 problems on a quiz?

RECREATION *A single card is drawn from a well-shuffled deck of 52 cards. In Exercises 67–72, determine the probability of the event.*

67. Drawing a black card

68. Drawing a card that is not a face card

69. Drawing a joker

70. Drawing a heart or a club

71. Drawing a red card or an ace

72. Drawing a face card or a club

73. **DEMOGRAPHY** The population of a town is decreasing by 400 inhabitants each year. If its population is presently 8500, what will its population be in nine years?

Evaluate each permutation or combination in Exercises 74–76.

74. $_6C_4$

75. $_6P_4$

76. $_{12}C_{10}$

77. Find the 4th term of $(x^2 - y^2)^6$.

78. In a geometric sequence $a_1 = 2$ and $a_4 = 16$. Find S_5.

79. Convert $0.\overline{78}$ to a fraction.

80. Use mathematical induction to prove that 8 divides $(9^n - 1)$ for every positive integer n.

ANSWERS: 1. $0, \frac{3}{2}, \frac{8}{3}, \frac{15}{4}, \frac{24}{5}, \frac{63}{8}$ 2. $-\frac{1}{4}, \frac{1}{7}, -\frac{1}{10}, \frac{1}{13}, -\frac{1}{16}; \frac{1}{25}$ 3. $1 + \sqrt{2} + \sqrt{5} + \sqrt{10}$ 4. $\sum_{k=1}^{n} \frac{k^2}{k+1}$ 5. $7, -13$ 6. $a_n = n^2 + 2$ 7. 22 8. $\frac{117}{10}$ 9. $-9, -5, -1, 3, 7, 11; S_{12} = 156$ 10. $n = 8, d = 2$ 11. 14, 11, 8 12. $x = 4$; 5, 11, 17 13. $1800 14. $21 15. $\frac{1}{6}, \frac{1}{3}, \frac{2}{3}, \frac{4}{3}, \frac{8}{3}, \frac{16}{3}; S_6 = \frac{21}{2}$ 16. $\frac{3280}{27}$ 17. $6, -3, \frac{3}{2}, -\frac{3}{4}$ 18. $x = 7$; 20, 10, 5 19. $2307.94 20. 9.8 ft; 191.4 ft 21. $\frac{32}{5}$ 22. $30, 5, \frac{5}{6}, \frac{5}{36}, \frac{5}{216}$ 23. $\frac{11}{9}$ 24. $\frac{170}{33}$ 25. 72 m 26. 135 ft 27. $4 + 8 + 12 + \cdots + 4k + 4(k+1) = 2k(k+1) + 4(k+1) = 2(k+1)(k+2) = 2(k+1)[(k+1)+1]$ 28. $1 \cdot 2 + 2 \cdot 3 + 3 \cdot 4 + \cdots + k(k+1) + (k+1)(k+2) = \frac{1}{3}k(k+1)(k+2) + (k+1)(k+2) = \frac{1}{3}(k+1)[(k+1)+1][(k+2)+1]$ 29. 6 30. 3 31. 120 32. 1 33. 1 34. 1 35. (a) 26,000 (b) 18,000 (c) 2700 36. 17,160 37. 362,880 38. 40,320 39. 30,240 40. 15 41. 3080 42. 715 43. $243a^5 - 405a^4z + 270a^3z^2 - 90a^2z^3 + 15az^4 - z^5$ 44. $y^{-4} + 4y^{-2} + 6 + 4y^2 + y^4$ 45. $-160x^3y^6$ 46. $2835a^3b^4$ 47. $\frac{8}{15}$ 48. $\frac{7}{15}$ 49. 0 50. 1 51. $\frac{1}{4}$ 52. $\frac{3}{13}$ 53. $\frac{1}{52}$ 54. $\frac{1}{26}$ 55. $\frac{2}{13}$ 56. 0 57. $\frac{9999}{10,000}$ 58. 0.99; 0.01 59. $\frac{2}{15}$ 60. $\frac{7}{15}$ 61. $\frac{32}{3}$ 62. $a_n = 5n - 1$ 63. 24; $\frac{1}{108,290}$ 64. 1900 65. $-2, -23$ 66. 3003 67. $\frac{1}{2}$ 68. $\frac{10}{13}$ 69. 0 70. $\frac{1}{2}$ 71. $\frac{7}{13}$ 72. $\frac{11}{26}$ 73. 4900 74. 15 75. 360 76. 66 77. $-20x^6y^6$ 78. 62 79. $\frac{26}{33}$ 80. $9^{k+1} - 1 = 9 \cdot 9^k - 9 + 8 = 9(9^k - 1) + 8$

CHAPTER 9 TEST

1. Give the sixth term of the sequence.
$$a_n = (-1)^{n+1}(2n - 1)$$

2. Find the second and third terms of the sequence.
$$a_1 = 5;\ a_{n+1} = 2a_n - 4$$

3. Use $\sum_{k=1}^{n} k = \frac{n(n+1)}{2}$ and $\sum_{k=1}^{n} k^2 = \frac{n(n+1)(2n+1)}{6}$ to determine the sum.
$$\sum_{k=1}^{4} (k + k^2)$$

4. Some of the numbers for an arithmetic sequence, n, a_1, a_n, d, and S_n are given. Find the missing ones.
$$a_{10} = 24,\ S_{10} = 55$$

5. Insert four arithmetic means between -5 and 10.

6. Logs are to be stacked in such a way that there is one log on top, 2 logs on the second row, 3 logs on the third, and so forth. How many logs should be put on the bottom row if there are 120 logs to be stacked?

7. Some of the numbers for a geometric sequence, n, a_1, a_n, r, and S_n are given. Find the missing ones.
$$a_1 = 3,\ a_n = -\frac{1}{729},\ S_n = \frac{1640}{729}$$

8. Find the sum of the infinite geometric sequence.
$$27,\ 18,\ 12,\ 8,\ \ldots$$

9. Kathy borrowed $10,000 at 12% interest compounded annually. If the loan was paid off (principal and interest) at the end of five years, how much did she pay?

CHAPTER 9 TEST
CONTINUED

10. Use mathematical induction to prove that the statement is true for all positive integers n.

$$4 \text{ divides } 5^n - 1$$

11. Evaluate $_5P_2$.

12. How many three-digit numbers can be formed from the digits 1, 3, 5, 7, 9?

13. How many ways can a committee of three be selected from a club with twelve members?

14. How many distinguishable permutations of the letters in the word CRITICS are possible?

15. Find the 6th term in the binomial expansion of $(2x + y)^7$.

16. A single card is drawn from a deck of 52 cards. What is the probability that it is a spade or a face card?

17. If the probability of rain is $\frac{3}{5}$, what is the probability that it will not rain?

FINAL REVIEW EXERCISES

CHAPTER 1

1. Give the set of integers.

2. What do we call numbers formed by taking quotients of integers? (Division by zero is excluded.)

3. What property is illustrated by $6(3 + x) = 18 + 6x$?

4. The product or quotient of two numbers with opposite signs is always what kind of number?

5. A trinomial has how many terms?

6. Factor $a^3 - b^3$.

7. If k is even, evaluate $\sqrt[k]{a^k}$.

8. If k is odd, evaluate $\sqrt[k]{a^k}$.

Perform the indicated operations in Exercises 9–15.

9. $-3 + (-2)$

10. $(-8) - (-5)$

11. $\left(\dfrac{3}{4}\right)\left(-\dfrac{16}{3}\right)$

12. $\left(-\dfrac{7}{6}\right) \div \left(\dfrac{14}{3}\right)$

13. $|(-2)(-5) - (-6)|$

14. $2x - (-3x + 5)$

15. $(8x^2y^3 - 6xy^2 + 2xy) + (-4xy^2 - 3xy + 5) - (6x^2y^3 + 2xy^2 - 7)$

Simplify.

16. $\left(\dfrac{3^0 x^{-6}}{2y^3}\right)^{-2}$

17. $\left(\dfrac{4x^2 y^{-3}}{x^{-4} y^3}\right)^{-1}$

18. Express 0.0000159 in scientific notation.

Multiply the polynomials in Exercises 19–22.

19. $(8x - 3y)(5x + 4y)$

20. $(3x - 7y)^2$

21. $(5a - 4b)(5a + 4b)$

22. $(2a + 9b)^2$

Factor the polynomials in Exercises 23–28.

23. $x^2 - 6xy + 8y^2$

24. $8x^2 - 2y^4$

25. $5a^2 + 13ab - 6b^2$

26. $-6a^2 + 36ab - 54b^2$

27. $125u^3 + 27v^3$

28. $24u^3 - 81v^3$

In Exercises 29–34, simplify and rationalize denominators when necessary.

29. $\sqrt{125}$

30. $\sqrt[4]{32x^6 y^4}$

31. $5\sqrt{75} - 3\sqrt{48}$

32. $\sqrt[3]{\dfrac{48u^5 v}{3uv^2}}$

33. $\left(\dfrac{8^{2/3} a^{-3/2} b^3}{3a^{-5/2} b^{1/2}}\right)^{-2}$

34. $\dfrac{\sqrt{27} - \sqrt{5}}{\sqrt{3} + \sqrt{5}}$

Perform the indicated operations in Exercises 35–38.

35. $\dfrac{x^2 - 3x + 2}{x^2 - 2x + 1} \cdot \dfrac{x^2 - x}{x^2 - 4}$

36. $\dfrac{4x^2 + 4x + 1}{2x^2 - 9x - 5} \div \dfrac{x^2 + 5x + 25}{x^3 - 125}$

A-1

37. $\dfrac{x+5}{x^2+7x+10} - \dfrac{x-2}{x^2+5x+6}$

38. $\dfrac{x+\dfrac{x}{y}}{x-\dfrac{x}{y}}$

CHAPTER 2

39. When solving an inequality, if we obtain an inequality such as $5 > 7$, the original inequality has how many solutions?

40. When solving $|x| = a$, $a > 0$, what two equations do we solve?

41. What do we call an equation stating that two ratios are equal?

42. To multiply or divide both sides of an inequality by a negative number, what must be done?

Solve each equation in Exercises 43–46.

43. $x - (3 - 2x) = -2x + 7$

44. $\sqrt[3]{x-2} + 3 = 2$

45. $\sqrt{x-2} - \sqrt{x+5} = -1$

46. $\dfrac{2}{x-3} - \dfrac{3}{x-2} = \dfrac{5}{x^2-5x+6}$

47. **CONSUMER** After receiving a 7% raise, Pam now earns $40,660 a year. What was her former salary?

48. **WORK** Henry can do a job in six days and Joe can do the same job in ten days. How long would it take to do the job if they worked together?

49. **RATE-MOTION** An airplane can fly 300 km with the wind and 200 km against the wind in the same period of time. If the speed of the wind is 100 km/hr, what is the speed of the plane in still air?

In Exercises 50–51, solve and graph each inequality on a number line.

50. $|2x - 3| < 5$

51. $|3 - 2x| \geq 2$

52. Solve $\dfrac{1}{a} + \dfrac{1}{b} = \dfrac{1}{c}$ for b.

53. Find the mean proportional between 6 and 24.

54. If the methods of factoring or taking roots fail when you are solving a quadratic equation, what is the next best method to use?

55. For the imaginary number i, what is i^2?

56. What is the conjugate of $9 + 5i$?

Perform the indicated operations in Exercises 57–60.

57. $(5 - 7i) - (-3 - 6i)$

58. $(3 + 4i)(8 - 7i)$

59. $\dfrac{5 - 2i}{-3 + 4i}$

60. $\dfrac{1}{2 + 3i}$

Solve each equation in Exercises 61–66.

61. $3y^2 + y - 10 = 0$

62. $2y^2 + 4y = 3$

63. $(u^2 + 5)^2 - 7(u^2 + 5) + 12 = 0$

64. $\dfrac{2u}{u-2} = -\dfrac{8}{u^2-4} + \dfrac{u}{u+2}$

65. $\sqrt{4x+1} - \sqrt{x-2} = 3$

66. $\dfrac{1}{x+4} + \dfrac{1}{x} = \dfrac{1}{5}$

67. **CONSUMER** A group of men plan to share equally in the $240 rental of a cabin in the mountains. At the last minute two men decide not to go, and this raises the share of each remaining man by $4. How many men were planning to go initially?

68. **NAVIGATION** Two boats leave the same island at the same time cruising at right angles to each other. In two hours they are 50 miles apart. If the speed of one boat is 5 mph faster than the speed of the other, what is the speed of each boat?

CHAPTER 3

69. Find the intercepts and graph $2x + 3y = 12$.

70. Find the vertex, x-intercepts, and graph of $f(x) = -x^2 + 4x + 5$.

71. Find the general form of the equation of the line passing through $(4, -1)$ and $(6, 3)$.

72. Find the general form of the equation of the line passing through $(5, -2)$ and perpendicular to $2x + 5y = 7$.

73. Find the slope and y-intercept of the line with equation $3x + 7y - 2 = 0$.

74. Find the distance between the points with coordinates $(-2, 5)$ and $(6, 1)$.

75. Find the midpoint of the line segment joining $(7, 4)$ and $(-2, 3)$.

76. Which of the following are functions?
    (a) $1 \to 4$, $2 \to 5$, $3 \to 5$
    (b) $4 \to 1$, $5 \to 2$, $5 \to 3$

77. Which of the following are graphs of functions?
    (a)
    (b)

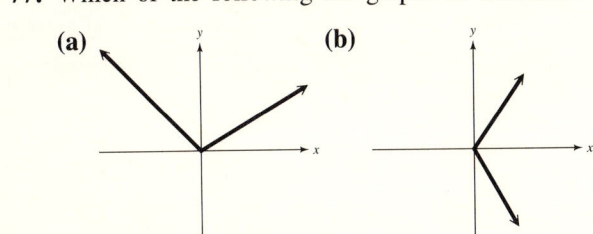

78. For what values of x is $f(x) = -x^2$ increasing? Decreasing?

79. Determine the symmetry of the following relations.
    (a) $x^2 + y^2 = 4$ (b) $xy = 6$

80. Determine the inverse of each of the following functions.
    (a) $f(x) = \dfrac{3}{2}x + 2$ (b) $g(x) = x^2 - 5$, $x \geq 0$

81. If y varies directly as the square of x and inversely as the cube root of z, find the equation of variation if y is 5 when $x = 4$ and $z = 27$.

82. Juan collected the data in the table showing the relationship between the daily cost of operating a machine and the number of units produced by it. Find a quadratic function that fits this data and use it to predict the cost of operation on a day when 10 units are produced.

No. units produced	Daily costs
0	$20
1	$21
3	$29

CHAPTER 4

83. If $P(x)$ is a polynomial and $P(b) = 0$, what is b called relative to the polynomial $P(x)$?

84. What is the remainder when $P(x)$ is divided by $x + 5$?

85. What is the name of the theorem used to answer Exercise 84?

86. If $P(-8) = 0$, give one factor of $P(x)$.

87. What is the name of the theorem used to answer Exercise 86?

88. If p/q is a solution to the polynomial equation $5x^3 - 4x^2 + 3x - 6 = 0$, then p is a factor of what number?

89. Counting multiplicities, a polynomial of degree n has exactly how many zeros?

90. If r is a zero of $Q(x)$ in rational function $f(x) = P(x)/Q(x)$, then $x = r$ is what kind of asymptote of the graph of $y = f(x)$?

91. If $4 - 3i$ is a zero of polynomial $P(x)$, with real coefficients, give one other zero of $P(x)$.

92. If 3 and $1 - 3\sqrt{2}$ are zeros of polynomial $P(x)$, with rational coefficients, what is the least degree $P(x)$ can have?

93. Divide $P(x) = 4x^4 - 2x^3 + x - 9$ by $x - 2$ using synthetic division and give the value of $P(2)$.

94. Use Descartes' rule of signs to find the possible number of negative, positive, and nonreal solutions to $-x^4 + 2x^3 - 3x^2 + x - 7 = 0$.

95. Use the upper and lower bound test to find the smallest positive integer upper bound and the largest negative integer lower bound for the real solutions to $4x^3 - 16x^2 + 11x + 10 = 0$.

96. Find all solutions of $2x^3 + 7x^2 + 2x - 6 = 0$.

97. Graph $P(x) = x^3 + x^2 - 2x$, and give **(a)** the maximum number of turning points, **(b)** the x-intercepts, **(c)** the eventual direction of the graph to the right, and **(d)** the eventual direction of the graph to the left.

98. Graph $f(x) = \frac{1}{x^2 + 2x + 1}$, and give the **(a)** vertical asymptotes, **(b)** horizontal asymptotes, **(c)** oblique asymptotes, **(d)** x-intercepts, **(e)** y-intercept, and **(f)** symmetries of the graph.

Solve and graph each inequality in Exercises 99–100.

99. $2x^2 + 9x - 5 < 0$

100. $\dfrac{3x - 1}{2x - 3} \geq 0$

CHAPTER 5

101. Write $x = a^y$ as an equivalent logarithmic equation.

102. For any base a, evaluate $\log_a 1$.

103. For any base a, evaluate $\log_a a$.

104. The graph of $y = \log_a x$ can be obtained from the graph of $y = a^x$ by reflecting it across what line?

105. Relative to the equation $\log n = x$, what is n called?

106. If a and b are bases and $x > 0$, simplify $\dfrac{\log_a x}{\log_a b}$.

Graph each function in Exercises 107–109.

107. $f(x) = 3^x$

108. $f(x) = \log_3 x$

109. $f(x) = e^{x+1}$

Solve for x in Exercises 110–111.

110. $8^{2/3} = x$

111. $\log_4 \frac{1}{64} = x$

112. Expand $\log_a \frac{x^2}{(y^2z)^3}$ using the properties of logarithms.

113. Express $2 \log_a xyz - \frac{1}{3} \log_a xy^2 + 4 \log_a z$ as a single logarithm and simplify.

Use a calculator to find the value of n in each expression in Exercises 114–119.

114. $n = \log 3.421$

115. $\log n = 0.3271$

116. $n = \ln 0.00351$

117. $\ln n = 2.5378$

118. $n = e^{3.2564}$

119. $\log_5 482$

Solve each equation in Exercises 120–123.

120. $8^{3x+2} = 4^{x-1}$

121. $\log_2 (2x + 2) - \log_2 (x - 2) = 2$

122. $2^x = 5^{x-1}$

123. $\log (\log x) = 0$

124. DEMOGRAPHY Use the equation $y = ce^{kt}$ to predict the population of Dry, Nevada, in the year 2000 if there were 120 people living there in 1970 and 180 people in 1990.

125. BANKING How many compounding periods will it take to change $300 into $1000 if the interest rate is 12% compounded quarterly?

126. AUDIOLOGY Use the sound equation $D = 10 \log \frac{S}{S_0}$, where $S_0 = 10^{-12}$ watt/m^2, to find the decibel level of a noise with intensity measured at 0.45 watt/m^2.

127. GEOLOGY Use the Richter scale formula $M = \log \frac{A}{A_0}$ to find the magnitude of an earthquake that was 1.8×10^7 times more powerful than a reference level zero quake.

CHAPTER 6

128. When the slopes of the lines in a system of two linear equations are equal and the y-intercepts are also equal, how many solutions does the system have?

129. When the slopes of the lines in a system of two linear equations are unequal, the system has how many solutions?

130. When solving a system of equations, if we obtain an equation such as $5 = 0$, how many solutions does the system have?

Solve each system in Exercises 131–134.

131. $3x - 2y = -12$
$x + 4y = 10$

132. $5x + 3y = 5$
$2x - 4y = 28$

133. $x + 3y + 2z = -3$
$3x + 2y - 3z = 13$
$-4x - 3y + 3z = -14$

134. $2x + y - z = 5$
$x - 4y + z = 4$

135. GEOMETRY The second angle of a triangle measures three times the first. The third angle has measure one-fourth the difference of the measures of the first two. Find the measure of each.

Graph each system in Exercises 136–137.

136. $x + y > 5$
$2x - y \leq 4$

137. $x \geq 0$
$y \geq 0$
$3y < -2x + 6$

138. Find the maximum and minimum values of the objective function $P = 15x + 4y$ subject to the constraints graphed in the feasible region shown.

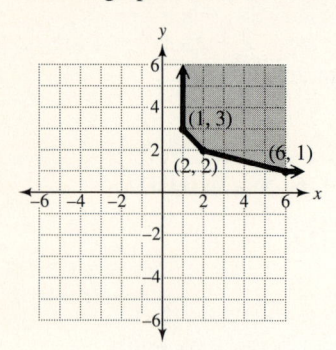

139. Acme Electronics makes two types of radar detectors, models A and B. It can produce up to a total of 50 detectors each day using up to 200 total work-hours. It takes 5 hours to make one model A and 3 hours to make one model B. One model A returns a profit of $60 when sold, and one model B returns a profit of $40. How many of each should be made daily to maximize profit?

In Exercises 140–141, find the partial fraction decomposition of the given rational function.

140. $f(x) = \dfrac{7x - 1}{x^2 + x - 2}$

141. $f(x) = \dfrac{7x^2 - 6x + 5}{(x - 3)(x^2 + 1)}$

CHAPTER 7

142. What is the order of the given matrix?

$$\begin{bmatrix} 2 & 3 & 5 \\ -1 & 5 & 2 \end{bmatrix}$$

143. What is an $n \times n$ matrix called?

144. Can two matrices be added when they have different dimensions?

145. Give the identity matrix of order 3.

146. If the rows and columns of an $m \times n$ matrix are interchanged, the resulting $n \times m$ matrix is called what relative to the original matrix?

147. The matrix A^{-1} having the property that $A^{-1}A = AA^{-1} = I$ is called what relative to A?

148. Give the coefficient matrix of the system

$$x + y = 3$$
$$3x - 2y = 1.$$

149. Give the augmented matrix of the system

$$x + y = 3$$
$$3x - 2y = 1.$$

Perform the indicated operations in Exercises 150–155.

150. $\begin{bmatrix} 2 & 3 \\ -1 & 5 \end{bmatrix} \begin{bmatrix} 4 & -1 & 3 \\ 0 & 2 & -1 \end{bmatrix}$

151. $\begin{bmatrix} 4 & -1 & 3 \\ 0 & 2 & -1 \end{bmatrix} \begin{bmatrix} 2 & 3 \\ -1 & 5 \end{bmatrix}$

152. $\begin{bmatrix} 4 & 0 & -2 \\ 0 & -3 & 7 \end{bmatrix} + \begin{bmatrix} -2 & 0 & 1 \\ 4 & -1 & 5 \end{bmatrix}$

153. If $A = \begin{bmatrix} 3 & 2 & 1 \\ -1 & 7 & 8 \end{bmatrix}$ find A^T.

154. Find A^{-1} if $A = \begin{bmatrix} 3 & -1 \\ 2 & 1 \end{bmatrix}$.

155. Find A^{-1} if $A = \begin{bmatrix} 1 & 0 & 2 \\ 2 & -1 & 1 \\ 0 & 1 & -5 \end{bmatrix}$.

156. Write the system as a matrix equation and solve using the inverse method. [*Hint:* See Exercise 154.]

$$3x - y = -2$$
$$2x + y = -3$$

157. Write the system as a matrix equation and solve using the inverse method. [*Hint:* See Exercise 155.]

$$x + 2z = 4$$
$$2x - y + z = 1$$
$$ y - 5z = -9$$

Solve using the Gaussian method in Exercises 158–159.

158. $3x - y = -2$
$2x + y = -3$

159. $x + 2z = 4$
$2x - y + z = 1$
$ y - 5z = -9$

160. Matrix J represents the number of radios, television sets, and VCR's sold in a store during July. Matrix A represents the number of these items sold in the store in August. Matrix C represents the selling price of each of the three items sold in these two months. Find $C \cdot (J + A)$ and interpret the result.

$$C = [100 \quad 350 \quad 325], \ J = \begin{bmatrix} 3 \\ 2 \\ 4 \end{bmatrix}, \text{ and } A = \begin{bmatrix} 5 \\ 1 \\ 3 \end{bmatrix}.$$

Evaluate the determinants in Exercises 161–163.

161. $\begin{vmatrix} a & b \\ c & d \end{vmatrix}$

162. $\begin{vmatrix} 2 & -5 \\ -3 & 4 \end{vmatrix}$

163. $\begin{vmatrix} 1 & -2 & 3 \\ 2 & 7 & 4 \\ 4 & 5 & 3 \end{vmatrix}$

Solve using Cramer's rule in Exercises 164–165.

164. $2x - 5y = -2$
$-3x + 4y = 10$

165. $x - 2y + 3z = 9$
$2x + 7y + 4z = -6$
$4x + 5y + 3z = 1$

Without evaluating the determinants in Exercises 166–167, tell why each statement is true.

166. $\begin{vmatrix} 0 & 3 \\ 0 & 8 \end{vmatrix} = 0$

167. $\begin{vmatrix} 1 & 2 & -3 \\ 2 & 0 & 5 \\ 1 & 2 & -3 \end{vmatrix} = 0$

168. Solve for x. $\begin{vmatrix} x & 0 & 0 \\ 0 & x & 1 \\ 0 & 1 & 1 \end{vmatrix} = 2$

CHAPTER 8

169. Since circles, ellipses, parabolas, and hyperbolas can be obtained by intersecting a plane with a cone, they are often called what?

170. The collection of all points in a plane, the difference of whose distances from two different points is constant, is called what?

171. The collection of all points in a plane which are located a constant distance from a fixed point in that plane is called what?

172. The collection of all points in a plane, the sum of whose distances from two different fixed points is constant, is called what?

173. The graph of $(x - 2)^2 + (y + 1)^2 = 1$ is what kind of curve?

174. The graph of $(x - 2)^2 - (y + 1)^2 = 1$ is what kind of curve?

175. Give the vertex of the parabola with standard-form equation $(y + 2)^2 = 8(x - 3)$.

176. The equation $2x^2 + y^2 - 4x + 2y - 30 = 0$ has as its graph what kind of curve?

177. The equation $xy = -7$ has as its graph what kind of curve?

178. The equation $2x^2 - 3y^2 - 4x + 2y - 35 = 0$ has as its graph what kind of curve?

179. Find the standard form of the equation of a circle centered at $(3, -1/2)$ with radius $r = 2$.

180. Find the standard form of the equation of an ellipse centered at the origin with intercepts $(\pm 3, 0)$ and $(0, \pm 4)$.

Graph the conics in Exercises 181–183.

181. $4x^2 + 9y^2 - 8x + 36y + 4 = 0$

182. $9x^2 - 4y^2 - 18x - 16y - 43 = 0$

183. $xy = -6$

Solve the systems in Exercises 184–186.

184. $x^2 + y^2 = 13$
$x + y = 5$

185. $4x^2 + y^2 = 17$
$x^2 - 3y^2 = 1$

186. $4^x + y = 1$
$4^x - y = 1$

187. ENGINEERING A drainage ditch has a cross-section in the shape of a semicircle, with the maximum water depth 7.5 ft. What is the depth of the water 2.0 ft from the edge of the ditch?

Graph each nonlinear system in Exercises 188–189.

188. $x^2 + y^2 \leq 25$
$x^2 - y^2 \geq 4$

189. $y \leq \ln x$
$(x - 1)^2 + y^2 \leq 4$

CHAPTER 9

190. What is a sequence called in which each term after the first is obtained by adding a fixed number to the preceding term?

191. What is a sequence called in which each term after the first is obtained by multiplying the preceding term by a fixed number?

192. The formula for calculating the sum of an infinite geometric sequence applies only when the common ratio r satisfies what inequality?

193. Write $\sum_{k=1}^{3} (k^3 - 5)$ without sigma summation notation.

194. Write $\frac{1}{2} + \frac{8}{3} + \frac{27}{4} + \cdots + \frac{n^3}{n+1} + \cdots$ using sigma summation notation.

195. Given the sequence 7, 4, 1, −2, . . . , state whether it is arithmetic or geometric, and find a_8 and S_8.

196. Given the sequence 8, −4, 2, −1, 1/2, . . . , state whether it is arithmetic or geometric, and find a_7 and S_7.

197. A collection of dimes is arranged in a triangular array with 30 coins in the base row, 29 in the next, 28 in the next, and so forth. Find the value of the collection.

198. BANKING Peter borrows $1200.00 at 10% interest compounded annually. If he pays off the loan in full (principal and interest) at the end of 5 years, how much does he pay?

199. RECREATION A child on a swing traverses an arc of 20 feet. Each pass thereafter, he traverses an arc that is 9/10 the length of the previous arc. How far does he travel before coming to rest?

200. Convert $3.\overline{72}$ to a fraction and check your work by division.

201. Use the principle of mathematical induction to prove that $S(n)$ is true for every positive integer n.

$S(n)$: $7 + 14 + 21 + \cdots + 7n = \frac{7}{2}n(n + 1)$

202. When an experiment is performed, what is the set of all possible outcomes called?

203. What is a subset of the sample space in an experiment called?

204. When working with permutations, is order important?

205. Evaluate.
(a) $_3P_2$ (b) $_3C_2$ (c) $_5P_5$ (d) $_5C_5$

206. If a single card is drawn from a well-shuffled deck of 52 cards, what is the probability of drawing the following?
(a) A red card
(b) A spade
(c) A three
(d) A heart or a face card
(e) A heart and a face card
(f) A black card or an ace

207. If E and F are complementary events and $P(E) = 5/6$, what is $P(F)$?

208. How many five-digit numbers can be formed from the digits 0, 1, 2, 3, 4, 5, 6, 7 if zero cannot come first and no digit may be repeated?

209. How many ways can a president, vice president, and secretary be chosen from the members of a 10-member club?

210. How many 5-card poker hands consisting of 3 kings and 2 jacks are possible with an ordinary 52-card deck? What is the probability of being dealt such a hand?

211. How many ways can six people be placed in a line? Around a circle?

212. How many distinguishable permutations of the letters in the word CALCULUS are there?

213. If A and B are mutually exclusive, $P(A) = 2/5$, and $P(B) = 1/3$, find $P(A$ or $B)$.

214. If the probability of passing history is 4/5, of passing math is 2/3, and of passing both is 8/15, what is the probability of passing math or history?

215. Find the fourth term in the expansion of $(2x - y^2)^6$.

ANSWERS: 1. $\{\ldots, -2, -1, 0, 1, 2, \ldots\}$ 2. rational numbers 3. distributive law 4. negative 5. three
6. $(a - b)(a^2 + ab + b^2)$ 7. $|a|$ 8. a 9. -5 10. -3 11. -4 12. $-1/4$ 13. 16 14. $5x - 5$ 15. $2x^2y^3 - 12xy^2 - xy + 12$ 16. $4x^{12}y^6$ 17. $\frac{y^6}{4x^6}$ 18. 1.59×10^{-5} 19. $40x^2 + 17xy - 12y^2$ 20. $9x^2 - 42xy + 49y^2$ 21. $25a^2 - 16b^2$
22. $4a^2 + 36ab + 81b^2$ 23. $(x - 2y)(x - 4y)$ 24. $2(2x - y^2)(2x + y^2)$ 25. $(5a - 2b)(a + 3b)$ 26. $-6(a - 3b)^2$
27. $(5u + 3v)(25u^2 - 15uv + 9v^2)$ 28. $3(2u - 3v)(4u^2 + 6uv + 9v^2)$ 29. $5\sqrt{5}$ 30. $2xy\sqrt[4]{2x^2}$ 31. $13\sqrt{3}$ 32. $\frac{2u\sqrt[3]{2uv^2}}{v}$
33. $\frac{9}{16a^2b^5}$ 34. $2\sqrt{15} - 7$ 35. $\frac{x}{x+2}$ 36. $2x + 1$ 37. $\frac{5}{(x+2)(x+3)}$ 38. $\frac{y+1}{y-1}$ 39. none 40. $x = a$ and $x = -a$
41. proportion 42. reverse the inequality 43. 2 44. 1 45. 11 46. 0 47. $38,000 48. $\frac{15}{4}$ days 49. 500 km/hr
50. $-1 < x < 4$; 51. $x \leq 1/2$ or $x \geq 5/2$;
52. $b = \frac{ac}{a-c}$ 53. ± 12 54. the quadratic formula 55. -1 56. $9 - 5i$ 57. $8 - i$ 58. $52 + 11i$ 59. $\frac{-23 - 14i}{25}$
60. $\frac{2 - 3i}{13}$ 61. $5/3, -2$ 62. $\frac{-2 \pm \sqrt{10}}{2}$ 63. $\pm i, \pm i\sqrt{2}$ 64. -4 65. 2, 6 66. $3 \pm \sqrt{29}$ 67. 12 68. 15 mph, 20 mph
69. $(0, 4), (6, 0)$ 70. vertex: $(2, 9)$; $(5, 0), (-1, 0)$

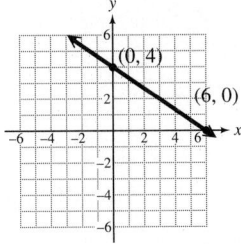

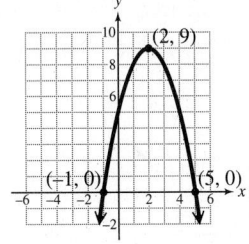

71. $2x - y - 9 = 0$ 72. $5x - 2y - 29 = 0$ 73. $-3/7, (0, 2/7)$ 74. $4\sqrt{5}$ 75. $\left(\frac{5}{2}, \frac{7}{2}\right)$ 76. (a) is a function (b) is *not* a function 77. (a) is the graph of a function (b) is *not* the graph of a function 78. $x \leq 0; x \geq 0$ 79. (a) x-axis, y-axis, origin (b) origin 80. (a) $f^{-1}(x) = \frac{2x - 4}{3}$ (b) $g^{-1}(x) = \sqrt{x + 5}$ 81. $y = \frac{15x^2}{16\sqrt[3]{z}}$ 82. $f(x) = x^2 + 20$; $120
83. zero 84. $P(-5)$ 85. remainder theorem 86. $x + 8$ 87. factor theorem 88. -6 89. n 90. vertical asymptote
91. $4 + 3i$ 92. three 93. 41 94. either 0, 0, 4; 0, 2, 2; or 0, 4, 0 95. lower bound: -1; upper bound: 4
96. $-3/2, -1 \pm \sqrt{3}$ 97. (a) 2 (b) $(-2, 0), (0, 0), (1, 0)$ (c) up (d) down

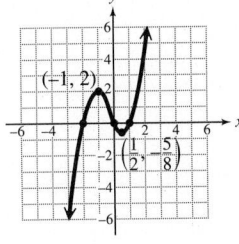

98. (a) $x = -1$ (b) $y = 0$ (c) none (d) none (e) $(0, 1)$ (f) none
99. $-5 < x < 1/2$
100. $x \leq 1/3$ or $x > 3/2$

101. $\log_a x = y$ 102. 0   103. 1   104. $y = x$ 105. antilogarithm   106. $\log_b x$ 107.
108.   109.

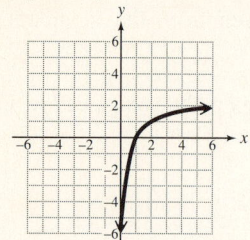

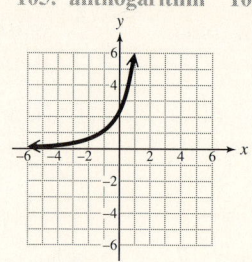

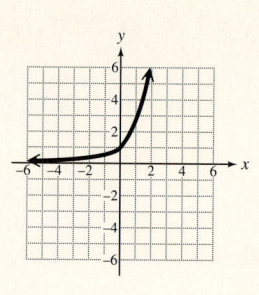

110. 4   111. −3   112. $2\log_a x - 6\log_a y - 3\log_a z$ 113. $\log_a x^{5/3}y^{4/3}z^6$ 114. 0.5342   115. 2.12   116. −5.6521
117. 12.7   118. 26.0   119. 3.8386   120. −8/7   121. 5   122. 1.76   123. 10   124. 220   125. 41   126. 116.5 decibels
127. 7.3   128. infinitely many   129. exactly one   130. none   131. (−2, 3)   132. (4, −5)   133. (−1, 2, −4)
134. $(x, x - 3, 3x - 8)$ x any real number   135. 40°, 120°, 20°   136.
137.

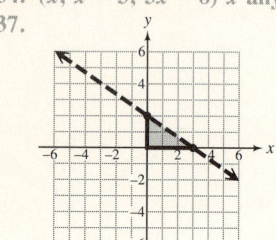

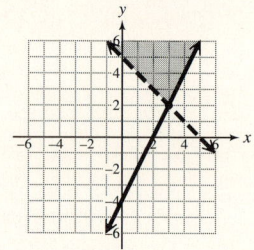

138. maximum value: none; minimum value: 27   139. 25 of each model returns the maximum profit   140. $f(x) = \frac{2}{x-1} + \frac{5}{x+2}$ 141. $f(x) = \frac{2x}{x^2+1} + \frac{5}{x-3}$ 142. 2×3
143. square matrix   144. no

145. $\begin{bmatrix} 1 & 0 & 0 \\ 0 & 1 & 0 \\ 0 & 0 & 1 \end{bmatrix}$ 146. transpose of A 147. inverse of A 148. $\begin{bmatrix} 1 & 1 \\ 3 & -2 \end{bmatrix}$ 149. $\begin{bmatrix} 1 & 1 & 3 \\ 3 & -2 & 1 \end{bmatrix}$ 150. $\begin{bmatrix} 8 & 4 & 3 \\ -4 & 11 & -8 \end{bmatrix}$

151. not defined   152. $\begin{bmatrix} 2 & 0 & -1 \\ 4 & -4 & 12 \end{bmatrix}$ 153. $\begin{bmatrix} 3 & -1 \\ 2 & 7 \\ 1 & 8 \end{bmatrix}$ 154. $\frac{1}{5}\begin{bmatrix} 1 & 1 \\ -2 & 3 \end{bmatrix}$ 155. $\frac{1}{8}\begin{bmatrix} 4 & 2 & 2 \\ 10 & -5 & 3 \\ 2 & -1 & -1 \end{bmatrix}$

156. (−1, −1)   157. (0, 1, 2)   158. (−1, −1)   159. (0, 1, 2)   160. $4125; this represents the total revenue received on these three items during the two-month period.   161. $ad - bc$ 162. −7   163. −73   164. (−6, −2)
165. (2, −2, 1)   166. The determinant is zero because a column is all zeros.   167. The determinant is zero because two rows are the same.   168. $x = 2$ and $x = -1$ 169. conic sections or conics   170. hyperbola   171. circle
172. ellipse   173. circle   174. hyperbola   175. (3, −2)   176. ellipse   177. hyperbola   178. hyperbola
179. $(x - 3)^2 + \left(y + \frac{1}{2}\right)^2 = 4$ 180. $\frac{x^2}{9} + \frac{y^2}{16} = 1$ 181.   182.
183.

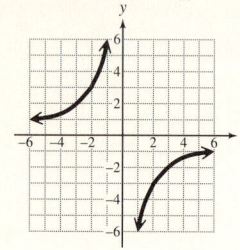

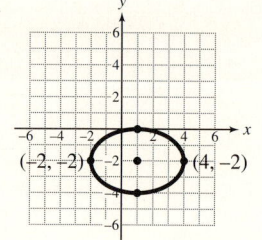

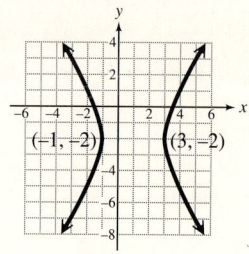

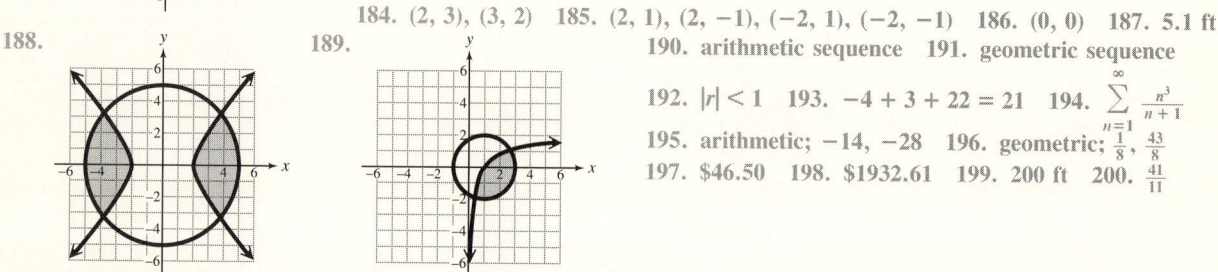

184. (2, 3), (3, 2)   185. (2, 1), (2, −1), (−2, 1), (−2, −1)   186. (0, 0)   187. 5.1 ft
188.   189.   190. arithmetic sequence   191. geometric sequence
192. $|r| < 1$ 193. $-4 + 3 + 22 = 21$ 194. $\sum_{n=1}^{\infty} \frac{n^3}{n+1}$
195. arithmetic; −14, −28   196. geometric; $\frac{1}{8}, \frac{43}{8}$
197. $46.50   198. $1932.61   199. 200 ft   200. $\frac{41}{11}$

201. $S(1)$: $\frac{7}{2}(1)(1 + 1) = 7$ is true. Assume $S(k)$: $7 + 14 + 21 + \cdots + 7k = \frac{7}{2}k(k + 1)$ is true. $S(k + 1)$: $7 + 14 + 21 + \cdots + 7k + 7(k + 1) = \frac{7}{2}k(k + 1) + 7(k + 1) = \frac{7}{2}[k(k + 1) + 2(k + 1)] = \frac{7}{2}[(k + 1)(k + 2)] = \frac{7}{2}(k + 1)(k + 1 + 1)$
Thus, the statement is true for all positive integers n.   202. sample space   203. event   204. yes   205. (a) 6   (b) 3
(c) 120   (d) 1   206. (a) 1/2   (b) 1/4   (c) 1/13   (d) 11/26   (e) 3/52   (f) 7/13   207. 1/6   208. 5880   209. 720
210. 24; $\frac{1}{108,290}$ 211. 720; 120   212. 5040   213. 11/15   214. 14/15   215. $-160x^3y^6$

ANSWERS TO CHAPTER TESTS

Chapter 1 Test

1. false 2. false 3. -8 4. 20 5. $\frac{a^8}{16b^{12}}$ 6. $\frac{y^3}{8x^9}$ 7. $\$2.21 \times 10^4$ 8. $-2a^2b^2 + 5ab - 10$ 9. $4x^4 - 12x^2y + 9y^2$
10. $2x^2 - 5xy - 3y^2$ 11. $a^2 + 4ab + 2b^2$ 12. $(2a - 3b)(2a + b)$ 13. $(u - 2v)(u + 2v)$ 14. $2(x + 5)(x^2 - 5x + 25)$
15. $(a - 2b + 2)(a + 2b)$ 16. $\$130$ 17. $\frac{x+6}{x+5}$ 18. $\frac{x^2 + y^2}{(x-y)^2(x+y)}$ 19. $\frac{2a^2 + b^2}{a(a-b)(a+b)}$ 20. $\frac{a^2(b-1)}{b^2(a-1)}$ 21. $2ab^2\sqrt[3]{4b}$ 22. $\frac{xy\sqrt{x}}{3}$
23. $\frac{\sqrt[3]{2xy^2}}{y}$ 24. $\frac{a + 2\sqrt{a} + 1}{a - 1}$ 25. $m\sqrt[12]{m}$ 26. $\sqrt[8]{(a+b)^7}$ 27. 0.477 28. 1.5 sec

Chapter 2 Test

1. $\frac{2}{3}$ 2. $1, -\frac{5}{3}$ 3. $a = \frac{w}{z-y}$ 4. 12 min 5. 300 km/hr 6. 72 ft, 32 ft 7. $-3 \pm \sqrt{10}$ 8. $5, -3$ 9. $3, -4$ 10. 20 shares 11. 30 mph, 40 mph 12. $\frac{3 \pm i\sqrt{7}}{4}$ 13. $\frac{5}{13} - \frac{14}{13}i$ 14. two complex (nonreal)
15. $x \geq 2$ 16. $(-\infty, 5)$ or $[8, \infty)$ 17. $(-15, 7)$
18. $(-\infty, -3)$ or $(11/3, \infty)$ 19. $x = 0$ 20. 5.6 sec

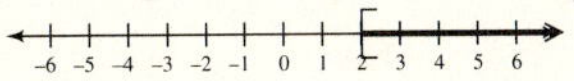

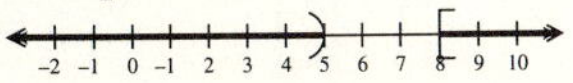

Chapter 3 Test

1. 5 2. $(-1/2, 5)$ 3. $4x + 3y - 13 = 0$ 4. $1/2; (0, -2)$ 5. $x + 2y + 3 = 0$ 6. $\$120,000$ 7. all reals $x \neq 1/3$
8. $\{7, 9, 11\}$ 9. is a function 10. $(-\infty, \infty)$ 11. right 1 up 2 12. odd 13. y-axis only 14. 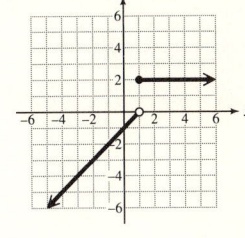
15. $18x^2 - 60x + 49$ 16. one-to-one 17. $f^{-1}(x) = \sqrt{x} - 1$ 18.

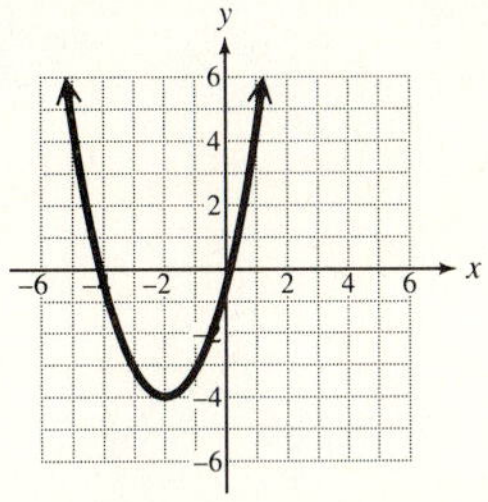

19. -144; 12 and -12 20. (a) $C(n) = 20n + 1200$ (b) $P(n) = 10n - 1200$ (c) 120 units 21. 220 in^3

Chapter 4 Test

1. $Q: x^3 - x + 1; R: 0$ 2. yes 3. 26 4. yes 5. $4 - 3i$ 6. $x^4 - 8x^3 + 23x^2 - 30x + 18$ 7. either 0, 0, 6 or 0, 2, 4
8. upper bound: 2; lower bound: -2 9. $1, -2, -1/2$ 10. 20 in by 10 in by 7 in 11.
12. $x = 1, x = -2$ 13. $y = 0$ 14. $y = x + 7$ 15. y-axis 16.
17. $(-\infty, -5]$ or $(-1, 2]$

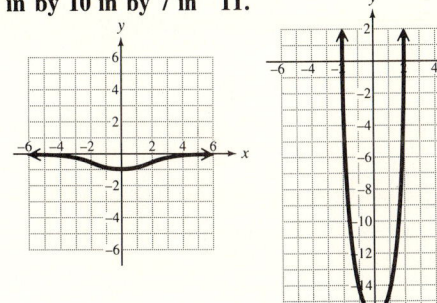

Chapter 5 Test

1. $x = 8$ 2. $a = -2$ 3. 1.2 4. 45,255 5. 1.7964 6. $\log_a x + 2\log_a y - \frac{1}{3}\log_a z$ 7. $\log_a \frac{y^3\sqrt{x}}{z}$ 8. 3.7574
9. 10. 12 11. -1.87 12. 10 13. 13.6 14. 166°F 15. $\$8042.19$ 16. 94 decibels
17. 46 yr

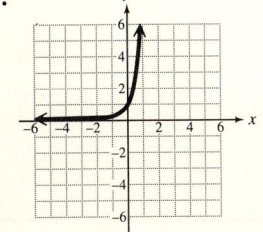

A-11

Chapter 6 Test

1. (7, 6) 2. (2, 0) 3. plane: 550 mph; wind: 50 mph 4. 14 lb 5. (−1, 0, 2) 6. (−z, −2z, z)z any real number
7. $1000 8.   9. $a = 2; b = −3; c = 5$ 10. maximum value: 85; minimum value: 20
11. valedictorian: 0; salutatorian: 24 12. $f(x) = \frac{3}{x-3} + \frac{-2}{x+4}$

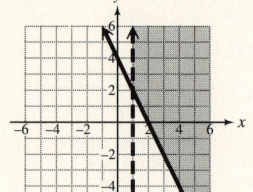

Chapter 7 Test

1. $\begin{bmatrix} -13 & 3 \\ 8 & -13 \end{bmatrix}$ 2. −7 3. $\begin{bmatrix} -1 & 15 \\ -10 & 10 \end{bmatrix}$
4. $10,045; this represents the total revenue received on these items in the two stores in January
5. (1, −2) 6. (−1, 0, 4)
7. $-\frac{1}{7}\begin{bmatrix} -1 & -3 \\ -2 & 1 \end{bmatrix}$ 8. (−2, 1) 9. $\frac{1}{3}\begin{bmatrix} -1 & 2 & 1 \\ 2 & -1 & 1 \\ 2 & -1 & -2 \end{bmatrix}$ 10. (−1, 0, 2) 11. (8, 5) 12. 56
13. because one row is all zeros

Chapter 8 Test

1. $(x + 2)^2 + (y − 4)^2 = 25$ 2.   3. $\frac{(x-4)^2}{9} + \frac{(y+1)^2}{5} = 1$ 4.

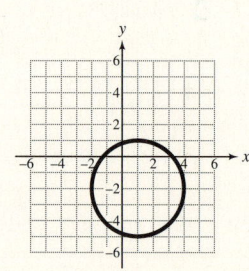

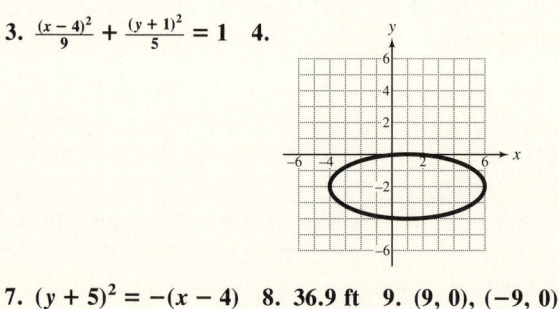

5. $\frac{(y-1)^2}{1} - \frac{(x+2)^2}{8} = 1$ 6. 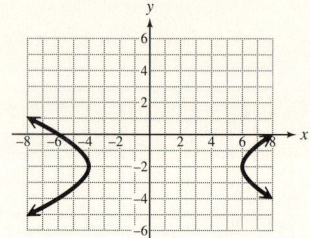   7. $(y + 5)^2 = −(x − 4)$ 8. 36.9 ft 9. (9, 0), (−9, 0)

10. (0, 1) 11.   12.

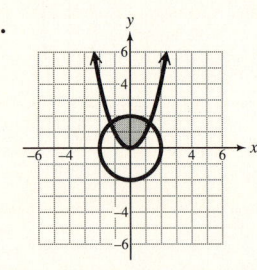

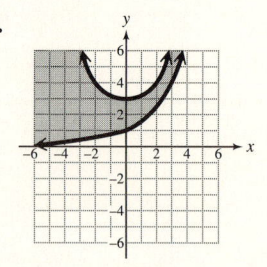

Chapter 9 Test

1. −11 2. 6, 8 3. 40 4. $a_1 = −13; d = \frac{37}{9}$ 5. −2, 1, 4, 7 6. 15 7. $n = 8; r = -\frac{1}{3}$ 8. 81 9. $17,623.42
10. Use the fact that $5^{n+1} − 1 = 5^{n+1} − 5 + 5 − 1 = 5(5^n − 1) + 4$. Since 4 divides $5^n − 1$ (by the induction hypothesis) and 4 divides 4, 4 divides both terms; thus divides the expression. 11. 20 12. 125 13. 220 14. 1260 15. $84x^2y^5$
16. $\frac{11}{26}$ 17. $\frac{2}{5}$

APPENDIX

LOGARITHMIC TABLES AND INTERPOLATION

STUDENT GUIDEPOSTS

1. Finding Logarithms Using the Table of Common Logarithms
2. Finding Antilogarithms Using the Table of Common Logarithms
3. Finding Logarithms and Antilogarithms Using Linear Interpolation

1 FINDING LOGARITHMS USING THE TABLE OF COMMON LOGARITHMS

In the past few years, scientific calculators have all but replaced logarithm tables. However, when a calculator is unavailable, a table giving common logarithms can be used. Such a table is limited in accuracy to three significant digits (four if interpolation is used).

If n is any positive number, n can be written in scientific notation as

$$n = m \times 10^c,$$

where $1 \leq m < 10$ and c is an integer. Using the product rule,

$$\log n = \log(m \times 10^c) = \log m + \log 10^c = \log m + c.$$

Log m, the **mantissa** of log n, is a decimal greater than or equal to 0 and less than 1. The **characteristic** c of log n is an integer. Since any positive number can be written in scientific notation, the table with logarithms of numbers between 1.00 and 9.99 in increments of 0.01, gives us logarithms of any number with three significant digits.

EXAMPLE 1 FINDING LOGARITHMS

(a) Find log 1.23.

Since $1.23 = 1.23 \times 10^0$,

$$\log 1.23 = \log(1.23 \times 10^0)$$
$$= \log 1.23 + 0. \quad \text{The characteristic is 0}$$

The left-hand column of Table 1 shows the numbers 1.0 through 9.9, while the numbers 0 through 9 lead each of the other columns. To find log 1.23, we look down the left column to 1.2, read across to the column headed 3, and find the decimal, .0899, the mantissa of log 1.23. Thus,

$$\log 1.23 = .0899 + 0 = 0.0899.$$

PRACTICE EXERCISE 1

(a) Find log 5.26.

A-13

A-14 APPENDIX

n	0	1	2	3	4	5	6	7	8	9
1.0	.0000	.0043	.0086	.0128	.0170	.0212	.0253	.0294	.0334	.0374
1.1	.0414	.0453	.0492	.0531	.0569	.0607	.0645	.0682	.0719	.0755
1.2	.0792	.0828	.0864	.0899	.0934	.0969	.1004	.1038	.1072	.1106
1.3	.1139	.1173	.1206	.1239	.1271	.1303	.1335	.1367	.1399	.1430
1.4	.1461	.1492	.1523	.1553	.1584	.1614	.1644	.1673	.1703	.1732

(b) Find log 123.

Since $123 = 1.23 \times 10^2$,

$$\begin{aligned} \log 123 &= \log (1.23 \times 10^2) \\ &= \log 1.23 + \log 10^2 \\ &= \log 1.23 + 2 \\ &= .0899 + 2 = 2.0899. \end{aligned}$$

(c) Find log 0.000123.

Since $0.000123 = 1.23 \times 10^{-4}$,

$$\begin{aligned} \log (0.000123) &= \log (1.23 \times 10^{-4}) \\ &= \log 1.23 + \log 10^{-4} \\ &= .0899 + (-4) \\ &= -3.9101. \end{aligned}$$

Note We could have left log 0.000123 in the form $0.0899 - 4$ rather than simplifying it to -3.9101. Actually, for use of a table this form is preferred since it shows the positive mantissa (0.0899) and the characteristic (-4).

❷ FINDING ANTILOGARITHMS USING THE TABLE OF COMMON LOGARITHMS

Up to now we have illustrated how to find a logarithm of a given number. Now we show how to find a number when its logarithm is given, that is, how to find an **antilogarithm (antilog).** For use of the table, the given logarithm must be in standard form showing the positive mantissa and the integer characteristic.

EXAMPLE 2 FINDING ANTILOGARITHMS

(a) Find the antilog of 2.7679.

In effect we are given that

$$\log n = 2.7679,$$

and we are asked to find n. Since $\log n = 2.7679 = 0.7679 + 2$, the mantissa is .7679 and the characteristic is 2. In the body of Table 1 find .7679 in the row headed by 5.8 under the column headed by 6. Thus,

$$\begin{aligned} n &= 5.86 \times 10^2 \\ &= 586. \end{aligned}$$

(b) Find log 5260.

(c) Find log 0.00526.

Answers: (a) 0.7210 (b) 3.7210
(c) $0.7210 + (-3) = -2.2790$

PRACTICE EXERCISE 2

(a) Find the antilog of 1.4786.

(b) Find the antilog of -2.2321.

We must first express
$$\log n = -2.2321$$
in standard form by adding and subtracting 3.
$$\begin{array}{r} 3.0000 - 3 \\ -2.2321 \\ \hline 0.7679 - 3 \end{array}$$

Thus, the mantissa is .7679 and the characteristic is -3. From Table 1,
$$n = 5.86 \times 10^{-3}$$
$$= 0.00586.$$

(b) Find the antilog of -2.0278.

Answers: (a) 30.1 (b) 0.00938

3 FINDING LOGARITHMS AND ANTILOGARITHMS USING LINEAR INTERPOLATION

To find the common logarithm of a number with four significant digits, such as 5138, we can use a process called **linear interpolation.** The word *linear* appears because the process uses a straight line to approximate the portion of the graph of $y = \log x$ between two values in the table. Consider Figure 1.

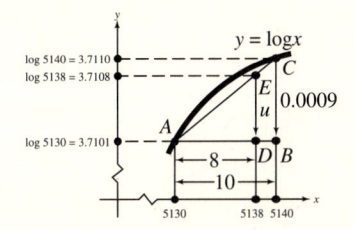

Figure 1 Interpolation

Noting that triangles *ABC* and *ADE* are similar, we conclude that
$$\frac{8}{10} = \frac{u}{0.0009}.$$

Thus,
$$u = \frac{8}{10}(0.0009) \approx 0.0007.$$

To obtain the approximation for log 5138, add 0.0007 to log 5130.
$$\log 5138 \approx \log 5130 + 0.0007 = 3.7101 + 0.0007 = 3.7108.$$

EXAMPLE 3 LINEAR INTERPOLATION (LOGARITHMS)

Use linear interpolation to find log 0.006412. The characteristic is -3 so we concentrate on finding the mantissa for log 6.412.

PRACTICE EXERCISE 3

Use linear interpolation to find log 21.58.

A-16 APPENDIX

```
           x           log x
         ┌─6.410      ┌─0.8069─┐
   ┌0.002┤            │        ├─u
0.010┤   └─6.412      ?────────┘      ┐
   └     ─6.420       ─0.8075         ├─0.0006
```

$$\frac{0.002}{0.010} = \frac{2}{10} = \frac{u}{0.0006}$$

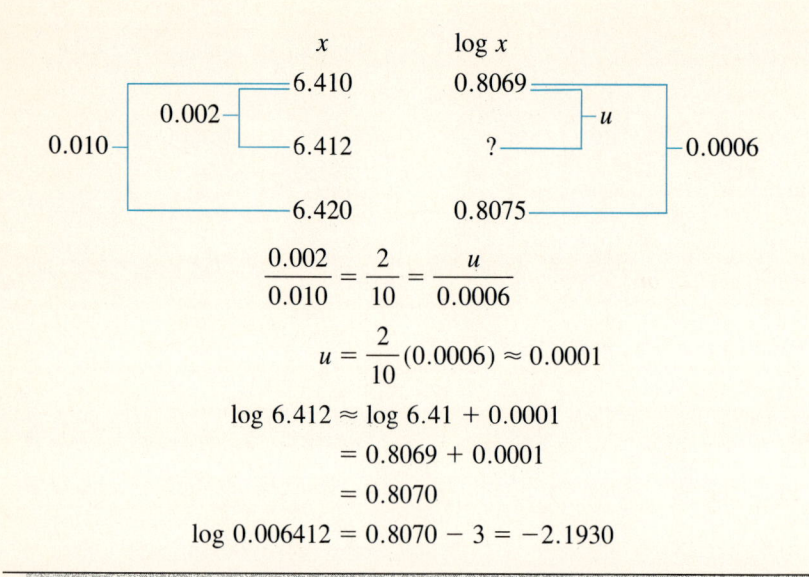

$$u = \frac{2}{10}(0.0006) \approx 0.0001$$

$$\log 6.412 \approx \log 6.41 + 0.0001$$
$$= 0.8069 + 0.0001$$
$$= 0.8070$$
$$\log 0.006412 = 0.8070 - 3 = -2.1930$$

Answer: 1.3341

If a calculation requires finding the antilog of a logarithm whose mantissa is not in Table 1, linear interpolation can be used to approximate the four-digit answer.

EXAMPLE 4 LINEAR INTERPOLATION (ANTILOGS)

Find the antilog of 2.4705.

The characteristic 2 tells us where to place the decimal point in the final answer. Thus we can concentrate on 0.4705. In Table 1, 0.4705 is between log 2.95 = 0.4698 and log 2.96 = 0.4713.

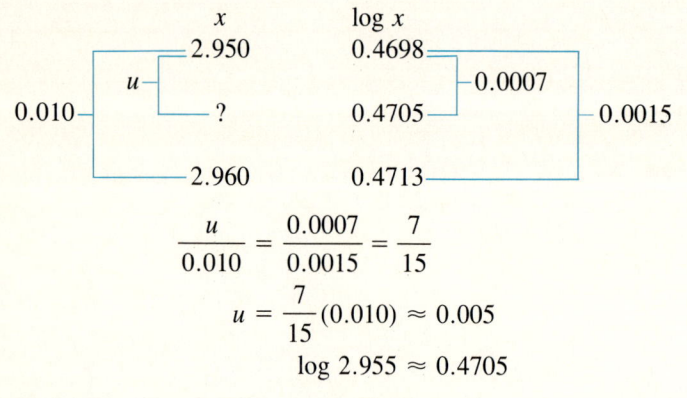

```
           x           log x
         ┌─2.950      ┌─0.4698─┐
   ┌u────┤            │        ├─0.0007   ┐
0.010┤   └─?          └─0.4705─┘          ├─0.0015
   └     ─2.960       ─0.4713             ┘
```

$$\frac{u}{0.010} = \frac{0.0007}{0.0015} = \frac{7}{15}$$

$$u = \frac{7}{15}(0.010) \approx 0.005$$

$$\log 2.955 \approx 0.4705$$

Thus, the antilog of 2.4705 is approximately

$$2.955 \times 10^2 = 295.5.$$

PRACTICE EXERCISE 4

Find the antilog of 1.5746.

Answer: 37.55

APPENDIX EXERCISES A

In Exercises 1–6 use Table 1 to find the common logarithm of each number.

1. 4.68   **2.** 46.8   **3.** 0.0468

4. 0.000279	**5.** 279	**6.** 2.79

In Exercises 7–12 use Table 1 to find the antilogarithm of each number.

7. 0.7364	**8.** 0.9827	**9.** 3.5855
10. 4.7896	**11.** −2.3799	**12.** −3.2692

In Exercises 13–18 use Table 1 and linear interpolation to approximate the logarithm of each number.

13. 3.278	**14.** 6.157	**15.** 437.6
16. 249.7	**17.** 0.003972	**18.** 0.0004256

In Exercises 19–24 use Table 1 and linear interpolation to approximate the antilogarithm of each number.

19. 1.6974	**20.** 1.2390	**21.** 0.5409 − 2
22. 2.6754	**23.** −3.1155	**24.** −2.8118

ANSWERS: 1. 0.6702 2. 1.6702 3. 0.6702 − 2 = −1.3298 4. 0.4456 − 4 = −3.5544 5. 2.4456 6. 0.4456
7. 5.45 8. 9.61 9. 3850 10. 61,600 11. 0.00417 12. 0.000538 13. 0.5156 14. 0.7894 15. 2.6411 16. 2.3974
17. 0.5990 − 3 = −2.4010 18. 0.6290 − 4 = −3.3710 19. 49.82 20. 17.34 21. 0.03475 22. 473.6 23. 0.0007665
24. 0.001542

APPENDIX EXERCISES B

In Exercises 1–6 use Table 1 to find the common logarithm of each number.

1. 7.64	**2.** 76.4	**3.** 0.00764
4. 0.00465	**5.** 465	**6.** 4.65

In Exercises 7–12 use Table 1 to find the antilogarithm of each number.

7. 0.5809	**8.** 0.8733	**9.** 2.7007
10. 6.8739	**11.** −1.6737	**12.** −3.3072

In Exercises 13–18 use Table 1 and linear interpolation to approximate the logarithm of each number.

13. 2.724	**14.** 5.334	**15.** 259.7
16. 8769	**17.** 0.0004421	**18.** 0.07395

In Exercises 19–24 use Table 1 and linear interpolation to approximate the antilogarithm of each number.

19. 1.7342	**20.** 2.2862	**21.** 0.9607 − 3
22. 4.8760	**23.** −2.3589	**24.** −3.7565

TABLE OF COMMON LOGARITHMS

n	0	1	2	3	4	5	6	7	8	9
1.0	.0000	.0043	.0086	.0128	.0170	.0212	.0253	.0294	.0334	.0374
1.1	.0414	.0453	.0492	.0531	.0569	.0607	.0645	.0682	.0719	.0755
1.2	.0792	.0828	.0864	.0899	.0934	.0969	.1004	.1038	.1072	.1106
1.3	.1139	.1173	.1206	.1239	.1271	.1303	.1335	.1367	.1399	.1430
1.4	.1461	.1492	.1523	.1553	.1584	.1614	.1644	.1673	.1703	.1732
1.5	.1761	.1790	.1818	.1847	.1875	.1903	.1931	.1959	.1987	.2014
1.6	.2041	.2068	.2095	.2122	.2148	.2175	.2201	.2227	.2253	.2279
1.7	.2304	.2330	.2355	.2380	.2405	.2430	.2455	.2480	.2504	.2529
1.8	.2553	.2577	.2601	.2625	.2648	.2672	.2695	.2718	.2742	.2765
1.9	.2788	.2810	.2833	.2856	.2878	.2900	.2923	.2945	.2967	.2989
2.0	.3010	.3032	.3054	.3075	.3096	.3118	.3139	.3160	.3181	.3201
2.1	.3222	.3243	.3263	.3284	.3304	.3324	.3345	.3365	.3385	.3404
2.2	.3424	.3444	.3464	.3483	.3502	.3522	.3541	.3560	.3579	.3598
2.3	.3617	.3636	.3655	.3674	.3692	.3711	.3729	.3747	.3766	.3784
2.4	.3802	.3820	.3838	.3856	.3874	.3892	.3909	.3927	.3945	.3962
2.5	.3979	.3997	.4014	.4031	.4048	.4065	.4082	.4099	.4116	.4133
2.6	.4150	.4166	.4183	.4200	.4216	.4232	.4249	.4265	.4281	.4298
2.7	.4314	.4330	.4346	.4362	.4378	.4393	.4409	.4425	.4440	.4456
2.8	.4472	.4487	.4502	.4518	.4533	.4548	.4564	.4579	.4594	.4609
2.9	.4624	.4639	.4654	.4669	.4683	.4698	.4713	.4728	.4742	.4757
3.0	.4771	.4786	.4800	.4814	.4829	.4843	.4857	.4871	.4886	.4900
3.1	.4914	.4928	.4942	.4955	.4969	.4983	.4997	.5011	.5024	.5038
3.2	.5051	.5065	.5079	.5092	.5105	.5119	.5132	.5145	.5159	.5172
3.3	.5185	.5198	.5211	.5224	.5237	.5250	.5263	.5276	.5289	.5302
3.4	.5315	.5328	.5340	.5353	.5366	.5378	.5391	.5403	.5416	.5428
3.5	.5441	.5453	.5465	.5478	.5490	.5502	.5514	.5527	.5539	.5551
3.6	.5563	.5575	.5587	.5599	.5611	.5623	.5635	.5647	.5658	.5670
3.7	.5682	.5694	.5705	.5717	.5729	.5740	.5752	.5763	.5775	.5786
3.8	.5798	.5809	.5821	.5832	.5843	.5855	.5866	.5877	.5888	.5899
3.9	.5911	.5922	.5933	.5944	.5955	.5966	.5977	.5988	.5999	.6010
4.0	.6021	.6031	.6042	.6053	.6064	.6075	.6085	.6096	.6107	.6117
4.1	.6128	.6138	.6149	.6160	.6170	.6180	.6191	.6201	.6212	.6222
4.2	.6232	.6243	.6253	.6263	.6274	.6284	.6294	.6304	.6314	.6325
4.3	.6335	.6345	.6355	.6365	.6375	.6385	.6395	.6405	.6415	.6425
4.4	.6435	.6444	.6454	.6464	.6474	.6484	.6493	.6503	.6513	.6522
4.5	.6532	.6542	.6551	.6561	.6571	.6580	.6590	.6599	.6609	.6618
4.6	.6628	.6637	.6646	.6656	.6665	.6675	.6684	.6693	.6702	.6712
4.7	.6721	.6730	.6739	.6749	.6758	.6767	.6776	.6785	.6794	.6803
4.8	.6812	.6821	.6830	.6839	.6848	.6857	.6866	.6875	.6884	.6893
4.9	.6902	.6911	.6920	.6928	.6937	.6946	.6955	.6964	.6972	.6981
5.0	.6990	.6998	.7007	.7016	.7024	.7033	.7042	.7050	.7059	.7067
5.1	.7076	.7084	.7093	.7101	.7110	.7118	.7126	.7135	.7143	.7152
5.2	.7160	.7168	.7177	.7185	.7193	.7202	.7210	.7218	.7226	.7235
5.3	.7243	.7251	.7259	.7267	.7275	.7284	.7292	.7300	.7308	.7316
5.4	.7324	.7332	.7340	.7348	.7356	.7364	.7372	.7380	.7388	.7396
n	0	1	2	3	4	5	6	7	8	9

LOGARITHMIC TABLES AND INTERPOLATION   A-19

n	0	1	2	3	4	5	6	7	8	9
5.5	.7404	.7412	.7419	.7427	.7435	.7443	.7451	.7459	.7466	.7474
5.6	.7482	.7490	.7497	.7505	.7513	.7520	.7528	.7536	.7543	.7551
5.7	.7559	.7566	.7574	.7582	.7589	.7597	.7604	.7612	.7619	.7627
5.8	.7634	.7642	.7649	.7657	.7664	.7672	.7679	.7686	.7694	.7701
5.9	.7709	.7716	.7723	.7731	.7738	.7745	.7752	.7760	.7767	.7774
6.0	.7782	.7789	.7796	.7803	.7810	.7818	.7825	.7832	.7839	.7846
6.1	.7853	.7860	.7868	.7875	.7882	.7889	.7896	.7903	.7910	.7917
6.2	.7924	.7931	.7938	.7945	.7952	.7959	.7966	.7973	.7980	.7987
6.3	.7993	.8000	.8007	.8014	.8021	.8028	.8035	.8041	.8048	.8055
6.4	.8062	.8069	.8075	.8082	.8089	.8096	.8102	.8109	.8116	.8122
6.5	.8129	.8136	.8142	.8149	.8156	.8162	.8169	.8176	.8182	.8189
6.6	.8195	.8202	.8209	.8215	.8222	.8228	.8235	.8241	.8248	.8254
6.7	.8261	.8267	.8274	.8280	.8287	.8293	.8299	.8306	.8312	.8319
6.8	.8325	.8331	.8338	.8344	.8351	.8357	.8363	.8370	.8376	.8382
6.9	.8388	.8395	.8401	.8407	.8414	.8420	.8426	.8432	.8439	.8445
7.0	.8451	.8457	.8463	.8470	.8476	.8482	.8488	.8494	.8500	.8506
7.1	.8513	.8519	.8525	.8531	.8537	.8543	.8549	.8555	.8561	.8567
7.2	.8573	.8579	.8585	.8591	.8597	.8603	.8609	.8615	.8621	.8627
7.3	.8633	.8639	.8645	.8651	.8657	.8663	.8669	.8675	.8681	.8686
7.4	.8692	.8698	.8704	.8710	.8716	.8722	.8727	.8733	.8739	.8745
7.5	.8751	.8756	.8762	.8768	.8774	.8779	.8785	.8791	.8797	.8802
7.6	.8808	.8814	.8820	.8825	.8831	.8837	.8842	.8848	.8854	.8859
7.7	.8865	.8871	.8876	.8882	.8887	.8893	.8899	.8904	.8910	.8915
7.8	.8921	.8927	.8932	.8938	.8943	.8949	.8954	.8960	.8965	.8971
7.9	.8976	.8982	.8987	.8993	.8998	.9004	.9009	.9015	.9020	.9025
8.0	.9031	.9036	.9042	.9047	.9053	.9058	.9063	.9069	.9074	.9079
8.1	.9085	.9090	.9096	.9101	.9106	.9112	.9117	.9122	.9128	.9133
8.2	.9138	.9143	.9149	.9154	.9159	.9165	.9170	.9175	.9180	.9186
8.3	.9191	.9196	.9201	.9206	.9212	.9217	.9222	.9227	.9232	.9238
8.4	.9243	.9248	.9253	.9258	.9263	.9269	.9274	.9279	.9284	.9289
8.5	.9294	.9299	.9304	.9309	.9315	.9320	.9325	.9330	.9335	.9340
8.6	.9345	.9350	.9355	.9360	.9365	.9370	.9375	.9380	.9385	.9390
8.7	.9395	.9400	.9405	.9410	.9415	.9420	.9425	.9430	.9435	.9440
8.8	.9445	.9450	.9455	.9460	.9465	.9469	.9474	.9479	.9484	.9489
8.9	.9494	.9499	.9504	.9509	.9513	.9518	.9523	.9528	.9533	.9538
9.0	.9542	.9547	.9552	.9557	.9562	.9566	.9571	.9576	.9581	.9586
9.1	.9590	.9595	.9600	.9605	.9609	.9614	.9619	.9624	.9628	.9633
9.2	.9638	.9643	.9647	.9652	.9657	.9661	.9666	.9671	.9675	.9680
9.3	.9685	.9689	.9694	.9699	.9703	.9708	.9713	.9717	.9722	.9727
9.4	.9731	.9736	.9741	.9745	.9750	.9754	.9759	.9763	.9768	.9773
9.5	.9777	.9782	.9786	.9791	.9795	.9800	.9805	.9809	.9814	.9818
9.6	.9823	.9827	.9832	.9836	.9841	.9845	.9850	.9854	.9859	.9863
9.7	.9868	.9872	.9877	.9881	.9886	.9890	.9894	.9899	.9903	.9908
9.8	.9912	.9917	.9921	.9926	.9930	.9934	.9939	.9943	.9948	.9952
9.9	.9956	.9961	.9965	.9969	.9974	.9978	.9983	.9987	.9991	.9996
n	0	1	2	3	4	5	6	7	8	9

INDEX

Abscissa, 175
Absolute value, 5, 9
  of complex numbers, 119
  equations with, 91, 92
  functions, 205
  inequalities with, 160, 161
Addition, 6
  associative property of, 7
  commutative property of, 7, 467
  and distributive property, 7
  of fractional expressions, 47, 48
  of matrices, 466–68
  of polynomials, 25
  of signed numbers, 6
Addition-subtraction method. *See* Elimination method.
Additive identity, 7
Additive inverse, 7
Algebra, combinatorial, 630
Algebraic expressions, 22, 42
  evaluation of, 22
Algebraic fractions, 42, 43
Amortizing, 384
Analytic geometry, 535–79
Antilogarithms, 366, 367
Applications
  aerospace, 547
  agriculture, 71, 233
  archaeology, 387
  architecture, 568
  audiology, 388
  banking, 383, 611
  biology, 653
  business, 156, 199, 234, 240, 247, 375, 424, 434, 442, 443, 469, 500
  chemistry, 351, 386, 421
  construction, 604
  consumer, 143, 144, 351, 383, 384, 420
  engineering, 191, 248, 291, 314, 539, 540
  environmental science, 220
  geology, 387
  geometry, 101, 106, 142, 423
  language of, 99
  using logarithms, 351, 375, 383–88
  manufacturing, 314
  medicine, 654
  mixture problems, 419, 420
  number, 99, 141
  using percent, 100, 101, 156, 423, 604
  physics, 63, 93, 247, 248, 326, 375, 618
  population, 385, 604
  rate-motion, 103–5, 145–47, 233, 422
  ratio and proportion, 105–7
  recreation, 50
  using systems of equations, 401–16, 419–24
  variation, 245–47
  work, 102, 103, 145
Arithmetic means, 604
Arithmetic sequence (progression), 601
  formulas for, 603
Associative property
  of addition, 7
  of multiplication, 7
Asymptotes, 307–11
  horizontal, 308, 310, 311
  of a hyperbola, 556–58
  oblique, 308, 311
  vertical, 308–11
ATTACK strategy, 98, 141, 239, 375, 419,
Augmented matrix, 484
Axiom, 4

Base, 13, 339, 365
  e, 364, 367, 368
  exponential function, 374
  logarithmic function, 367, 368
BASIC, 526
Bel, 388
Bell, Alexander Graham, 388
Binomial coefficient, 645
Binomial theorem, 642–46
Binomials, 24, 32, 33
Bounds for roots, 286, 287
Bracketing, method of, 300, 301

Canceling (dividing out), 44
Cartesian coordinate system, 175
Center
  of a circle, 537
  of an ellipse, 544
  of a hyperbola, 554
Characteristic of logarithms, 681
Circle, 536–40
  center, 537
  equations of, 418, 537
  radius, 537
Circular permutations, 635, 636
Class frequency, 654
Class intervals, 654
Closed half-plane, 430
Closed interval, 157

Closure property, 7
Coefficient matrix, 484, 498
Coefficients of terms, 22, 643
  complex, 117
  leading, 264
  real, 264, 281, 295
Cofactors, 508, 509
  expansion by, 510
Collinear points, 193
Column vector, 465
Combinations, 636
  of n objects taken r at a time, 637
Combinatorial algebra, 630
  combinations, 636
  permutations, 630–37
Combined variation, 248
Commission, 100
Common difference, 601
Common logarithms, 364, 365
Common ratio, 608
Commutative property
  of addition, 7
  of multiplication, 7, 8
Completing the square, 124–26
Complex fractions, 48–51
Complex numbers, 114
  absolute value of, 119
  conjugates of, 116
  equality of, 114
  operations on, 115–17
  reciprocals of, 117
  solving equations, 117, 118
Composite functions, 219
Compound inequalities, 157–59
Compound interest, 382, 383, 392
  continuous, 393
Conditional equation, 85
Conic sections, 536
  degenerate, 536, 537
Conjugate axis of a hyperbola, 555
Conjugates, 61
  of complex numbers, 280
Constant, 22
Constant functions, 197, 205
Constant of variation (proportionality), 245
Contradiction, 85
Coordinate plane. *See* Cartesian coordinate system.
Counting techniques, 630
Cramer, Gabriel, 506
Cramer's rule, 506, 507
Cross-product equation, 106
Cube roots, 56
Cubes, sum and difference of, 38
Curve fitting, 424

Decibel, 388
Decimals, 617
Decreasing functions, 205
Degenerate conics, 536, 537
Degree of a polynomial, 24, 264, 295, 309
Denominators, 42
  least common, 46
  rationalizing, 61
Dependent system, 405, 410, 413
Dependent variable, 200
Descartes, René, 175, 282
Descartes' rule of signs, 282
Determinants, 506
  nonzero, 522
  properties of, 517–21
  second-order, 506
  and solving systems, 498–501
  third-order, 508, 517
Diagonal matrix, 514
Difference of cubes, 37
Difference of squares, 26, 37
Digits, significant, 18
Dimension of a matrix, 465
Direct variation, 245
Directrix of parabola, 564, 565
Discriminant, 128, 129
Distance formula, 103, 177, 238, 537
Distinguishable permutations, 634, 635
Distributive property, 7, 25, 33, 477
Division, 7
  algorithm, 272, 273
  of complex numbers, 117
  of fractional expressions, 45
  of polynomials, 27, 28
  of radicals, 59, 60
  of signed numbers, 7
  synthetic, 266–69
  by zero, 7
Domain, 196, 198, 199, 592
Double root, 123, 279

e, the number, 364, 367, 368
Earnings (effective annual rate), 392, 393
Earthquake
  intensity, 387
  magnitude, 387
Echelon form, reduced, 485
Element, 2
  of a matrix, 465
  of a set, 2
Elementary row operations, 485
Elimination method, 404–6, 577

A-20

Index A-21

Ellipse, 543–50
  equations of, 544–50
  foci, 544
  major axis, 544
  minor axis, 544
  translation of axes, 548
  vertices, 544
Equality
  addition property, 8
  of complex numbers, 114
  of fractions, 42, 43
  of matrices, 465, 485
  multiplication property, 8
  of real numbers, 4
  reflexive property, 5
  substitution property, 5, 8
  symmetric property, 5, 8
  theorems, 8
  transitive property, 4, 5, 8
  trichotomy, 4
Equations
  with absolute value, 91, 92
  of circles, 418, 537
  conditional, 85
  contradiction, 85
  cross-product, 106
  cubic, 264
  definition of, 85
  of ellipses, 544–50
  equivalent, 85
  exponential, 339–41, 374
  fractional, 89–91, 137, 138
  of hyperbolas, 554–60
  identity, 85
  linear, 85, 183–91, 401–16
  literal, 92, 93
  logarithmic, 339–42, 376, 377
  matrix, 493–501
  of parabolas, 228, 229, 564–70
  polynomial, 265
  quadratic, 121, 122, 536
  quadratic in form, 135
  quartic, 264
  radical, 87–89, 138, 139
  regression, 252
  root of, 85, 123, 124, 279, 280
  solutions of, 85, 86, 121–28, 265
  systems of, 401–16, 498–501
Equivalent equations, 85, 339
Equivalent expressions, 43
Equivalent inequalities, 154
Euler, Leonhard, 368
Even functions, 209
Exponential equations, 339–41, 374
Exponential functions, 345–48, 368, 578
  inverse of, 350
Exponential notation, 13, 69
Exponents, 13–15, 340, 341
  approximating, 347, 348
  integer, 15
  irrational, 345, 346
  properties of, 13, 341
  rational, 68–71
  zero, 14
Extraneous roots, 87
Extrapolation, 239
Extremes, 105

Factor, 13
Factor theorem, 274, 275
Factorial notation, 632
Factoring
  binomials, 32, 33
  difference of cubes, 38
  difference of squares, 38
  and distributive property, 33
  by grouping, 33, 36
  perfect square, 37, 38
  polynomials, 32–38
  solving equations by, 121–23
  special formulas, 37
  sum of cubes, 38
  trinomials, 33
Fibonacci, Leonardo, 166
Fibonacci sequence, 166
Finite sequences, 592
First-degree equation. *See* Linear equations.
Focus
  of an ellipse, 544
  of a hyperbola, 554
  of a parabola, 564, 565
FOIL method, 26, 33, 34
Fractional equations, 89–91, 137, 138
Fractions
  complex, 48–51
  equivalent, 42, 43
  evaluating, 42–44
  fundamental principle of, 43, 44
  partial, 449–53
  reciprocals of, 45
  reducing, 44
  simplifying, 44, 49, 50
Functions, 196–201
  absolute value, 205
  algebra of, 218
  composite, 219
  constant, 197, 205
  decomposition, partial fraction, 450–53
  decreasing, 205
  domains of, 196, 198, 199
  even, 209
  exponential, 345–48
  graphing, 175
  greatest integer, 211
  horizontal line test, 221
  horizontal shifts, 207
  identity, 197
  increasing, 205
  inverse, 220–24
  linear, 197
  logarithmic, 348–50
  maximum values of, 232
  minimum values of, 232
  noncontinuous, 211
  notation, 197, 265
  odd, 209
  one-to-one, 221
  polynomial, 295–301
  quadratic, 197, 228–34, 418, 424
  standard form, 229
  ranges of, 196
  rational, 307–15, 450, 451
  symmetry tests, 210, 211
  vertical line test, 204
Fundamental counting principle, 630, 631

Fundamental theorem of algebra, 278–80
Fundamental theorem of linear programming, 441

Gauss, Karl Fredrich, 484
Gaussian method, 484–89
General term, 592, 603, 610
Geometric means, 612, 614
Geometric sequence, 608–11
  infinite, 615–18
  sum of, 616
Golden ratio, 166
Golden rectangle, 166
Graphs
  of absolute value, 205
  boundary line, 430
  bounded, 433
  of circles, 537–39
  of ellipses, 546–50
  of exponential functions, 346, 369
  of first-degree equations, 156
  of functions, 175, 228–32, 295–301, 346–50
  horizontal line test, 221
  horizontal shift, 207
  of hyperbolas, 555–60
  of inequalities, 156, 157, 430–35
    bounded, 433
    half-planes, 430
    linear in two variables, 429–35
    nonlinear, 578, 579
    test point, 430
    unbounded, 433
  of linear equations, 156, 410, 411
  of logarithmic functions, 348–51
  of parabolas, 228–32, 566–70
  plot points on, 175, 176
  of polynomial functions, 295–301
  of rational functions, 311–14
  reflection of, 208
  scatter diagram, 252
  of second-degree equations, 175–77
  of sets, 176
  of systems, 410, 411
  tests for symmetry, 210, 211
  transformations, 207
  turning points of, 297–300
  unbounded, 433
  vertical line test, 204
  vertical shift, 206, 207
Greater than (>), 4
Greater than or equal (≥), 9
Greatest common factor, 33
Greatest integer function, 211
Growth constant, 385

Half-life, 386, 615
Half-open interval, 158
Half-plane, 430
Histogram, 655
Homogeneous systems, 415, 416
Horizontal asymptote, 308, 310, 311
Horizontal line, 184, 185

Horizontal shift, 207
Hyperbolas, 554–60
  alternate form, 560
  asymptote, 556–58
  equations of, 554–60

i, 113
Identity, 85
Identity functions, 197
Identity matrix, 477
Imaginary numbers, 113, 114
Inconsistent systems, 406, 410, 412
Increasing functions, 205
Independent system, 410
Independent variable, 200
Index of a radical, 57, 60
Index of summation, 595
Induction assumption, 623
Induction, mathematical, 622–26
Inequalities
  with absolute value, 160, 161
  compound, 157
  equivalent, 154
  graphing, 156, 157, 430–35
  linear, 154, 429–35
  polynomial, 323, 324
  quadratic, 319–23
  rational, 324–26
  solution of, 154–56, 319, 429, 430
  systems of, 432–35
Infinite interval, 158
Infinite sequences, 592, 615–18
Infinite series, 615–18
Infinity, 158
Initial height, 147
Initial velocity, 146
Integers, 2
  as exponents, 15
  negative, 2
  positive, 2
Intercepts, 183–85
Interest rates, 143, 382, 383, 392, 393
Intermediate value theorem, 300
Interpolation, 683, 684
Interval
  closed, 157
  half-open, 158
  halving, 330
  infinite, 158
  open, 157
  notation, 157
Inverse
  additive, 7
  functions, 220–24, 348–51
  multiplicative, 7, 45, 494
  of square matrix, 494–97
  variation, 247
Irrational numbers, 3
Irrational roots of polynomials, 281

Joint variation, 248

k (growth constant), 385
kth root, 57

Least common denominator, 46
  in fractional equations, 89
Least-square criterion, 252
Leonardo of Pisa, 166

Index

Less than (<), 4
Less than or equal (≤), 9
Like terms, 24
Linear equations
  applications, 423, 424
  general form of, 85, 183
  graphs of, 156
  point-slope form, 187, 188
  slope-intercept form, 188, 189
  solving, 86, 401
  system of three, in three variables, 410–16
  system of two, in two variables, 401–6
  two-intercept form, 195
  two-point form, 193
Linear functions, 197
Linear inequalities, 154
  in two variables, 429–32
Linear interpolation, 683, 684
Linear models, 239
Linear programming, 440–44
  constraints, 440
  feasible region, 441
  feasible solution, 441
  fundamental theorem of, 441
  objective function, 440
  optimal solution, 441
  vertex, 441
Lines
  horizontal, 185
  parallel, 189
  perpendicular, 189
  point-slope equation of, 187
  slope of, 185–87, 402
  slope-intercept equation of, 188, 189
  two-intercept form, 195
  two-point equation of, 193
  vertical, 185
Literal equations, 92, 93
Logarithmic equations, 376, 377
Logarithmic functions, 348–50, 578
Logarithms, 339–42
  anti-, 366
  in applied problems, 351, 375, 383–88
  approximating, 359
  base, 339, 340, 365
  base conversion, 359, 360, 365, 366
  basic rule of, 355
  characteristic of, 681
  common, 364, 365
  definition of, 339, 340
  mantissa of, 681
  natural, 364, 367–69
  power rule, 357, 358
  product rule, 356
  properties of, 342
  quotient rule, 356, 357
  tables, of common, 686
  using tables of, 682

Major axis of ellipse, 544
Malthus, Thomas, 385
Malthusian model, 385
Mantissa, 681
Mathematical induction, 622–26
  principle of, 623
Mathematical models, 238–41

Matrices, 465–523
  addition of, 466
  additive identity, 467
  additive inverse, 467
  associative property, 467, 477
  augmented, 484
  coefficient, 484, 498
  cofactors, 508, 509
    expansion by, 510
  column vector, 465
  commutative property of addition, 467
  constant, 498
  Cramer's rule, 506
  determinant of, 506, 517–21
  diagonal, 514
  dimension of, 465
  distributive property of, 477
  echelon form of, 485
  elementary row operations, 485
  elements of, 465
  equal, 465, 485
  and Gaussian method, 484–89
  identity, 477
  inverse, 494–97
  inverse method, 499, 526
  main diagonal, 477
  minors, 508
  multiplication of, 475, 476
  negative of, 467
  nonsingular, 495
  order of, 465
  row vector, 465
  scalar multiplication, 468
  scalar (dot) product, 474, 522
  singular, 495
  solving equations, 494–501
  square, 465, 508
  subtraction of, 468
  and systems of equations, 498–501
  transpose of, 478
  triangular, 516
  variable, 498
  zero, 467
Mean, arithmetic, 604
Mean proportional, 106
Means, 105
Midpoint formula, 178, 179
Minor, 508
Minor axis of an ellipse, 544
Monomials, 24
Motion problems, 103–5, 145–47, 233, 422
Multiple roots, 279
Multiplication
  associative property of, 7
  commutative property of, 7
  of complex numbers, 116
  and distributive property, 7
  of fractional expressions, 45
  of matrices, 475, 476
  of polynomials, 25–27
  of radicals, 58
  of signed numbers, 6
  by zero, 6, 8
Multiplicative identity, 7
Multiplicative inverse, 7, 45, 494

Natural exponential function, 367, 368
Natural logarithms, 367–69
Natural numbers, 2, 592
Negative exponent, 14, 15
Negative integers, 2, 7, 9
Nonlinear systems of equations, 574–79
Nonsquare systems, 414, 415
Number line, 4
Numbers, 2
  complex, 114
  counting, 2
  irrational, 3
  natural, 2, 592
  rational, 2
  real, 3
  whole, 2

Oblique asymptote, 308, 311
Odd functions, 209
One-to-one function, 221
Open half-plane, 430
Open interval, 157
Order of matrix, 465
Order of operations, 22
Ordered pairs, 175, 200, 401
Ordered triples, 410
Ordinate, 175
Origin, 4, 175

Pairs, ordered, 175, 200, 401
Parabola, 228, 229, 564–70
  axis of symmetry, 565
  directrix, 564, 565
  equations of, 418, 565, 569
  general form, 569
  least squares regression, 419, 456
  line of symmetry, 230
  standard form of, 229, 566
  translation of axes, 567
Parallel lines, 189
Partial fractions, 449–53
  decomposition, 450
Pascal's triangle, 643, 644
Perfect square, 26, 38
Permutations, 632–37
  circular arrangements, 635, 636
  distinguishable, 634, 635
  of n objects, r at a time, 633
Perpendicular lines, 189
Pi (π), 3
Plotting points, 175, 176
Point-slope form, 187, 188
Polynomial equations, 265
Polynomial functions, 295–301
Polynomial inequalities, 323, 324
Polynomials, 23, 24, 264
  addition of, 25
  binomials, 24
    factoring, 32, 33
  bounds of zeros, 286, 287
  degree of, 24, 264, 309
  division of, 27, 28
  factoring, 32–38
  FOIL method, 26, 33, 34
  leading coefficient of, 264
  monomials, 24
  multiplication of, 25–27
  number of zeros, 266, 279, 301

rational roots, 288–92
  with real coefficients, 264
  solutions, 266
  subtraction of, 25
  synthetic division of, 266–69
  trinomials, 24
    factoring, 33–36
  variation in signs, 282
  zero, 264, 273
  zeros of, 266, 330
Population growth, 385
Positive integers, 2
Power rule
  of exponents, 13, 69
  of logarithms, 357
Power theorem, 87
Principal, 382
Principal kth root, 59
Principal square root, 57
Probability, 649–55
  complementary events, 653
  conjunction, 652
  disjunction, 652
  equally likely, 649, 650
  event, 649
  experiment, 649
  favorable outcome, 649
  mutually exclusive, 652
  sample space, 649
  unfavorable outcome, 649
Product formulas, 26
Product rule
  of exponents, 13
  of logarithms, 356
  for radicals, 58
Progressions
  arithmetic, 601
  geometric, 608–11
Proportions, 105
Pythagorean theorem, 142

Quadrants, 176
Quadratic equations, 121–31, 536
  and fractional equations, 137, 138
  general form, 122
  graphs of, 175–77
  reducible to quadratic, 135
  solutions, 123–31
    approximating solutions, 131
    by completing the square, 124–26
    using complex numbers, 117, 118
    equations from, 130
    by factoring, 121–23
    product of, 130
    properties of, 128–30
    by quadratic formula, 126–28
    sums of, 130
    by taking roots, 123, 124
    types of, 128
Quadratic formula, 126–28
Quadratic functions, 197, 228–34
  standard form, 229
Quadratic inequalities, 319–23
Quadratic models, 241
Quotient rule
  of exponents, 13
  of logarithms, 356, 357
  for radicals, 59

Radical equations, 87–89, 138, 139
Radical expressions, 56–63
  to exponential notation, 69
  similar, 60
Radicals ($\sqrt{\ }$), 57
  simplest form, 62
  simplifying, 58–63
Radicand, 57, 60
Radioactive decay, 386
Radius of a circle, 537
Range, 196, 592
Ratio, 105
  and proportion problems, 105–7
Rational exponents, 68–71
Rational expressions, 42–51
  inequalities, 324–26
Rational functions, 307–15
Rational numbers, 2
Rational root theorem, 288
Rationalizing the denominator, 61
Real number system, 4
Real numbers, 3, 6, 281
  axioms of, 7
  operation on, 6, 7
Reciprocals, 7, 494
  of complex numbers, 117
  of fractional expressions, 45
Rectangular coordinates. *See* Cartesian coordinate system.
Recursive definition, 594
Reduction method, 411
Relations, 196, 200
Remainder theorem, 273, 274
Repeating decimals, 2, 617, 618
Richter, Charles, 387
Richter scale, 387
Right triangle, 177, 178
Root, kth, 57
Roots, 279, 280, 288
  approximating, 300, 301
  complex, 280, 281
  extraneous, 87
  irrational, 281
  of multiplicity two, 279
  rational, 288
  real, 286, 287
Row vector, 465

Sample space, 649
Scalar product of matrices, 474, 522
Scatter diagram, 252
Scientific notation, 16–18
  on calculator, 17, 18
Second-degree equations, 121, 122, 536
  See also Quadratic equations.
Sequence, 592–94
  arithmetic, 601, 603
  finite, 592
  general term, 592–94
  geometric, 608–11
  infinite, 592, 615–18
  of partial sums, 615
  recursive definition, 594
Series, 595, 596
  finite, 595, 596
  infinite, 615–18
Sets, 2
  graphing, 176, 177
Shift of graph, 206, 207
Significant digits, 18
Slope, 185–87, 402
  of parallel lines, 189
  of perpendicular lines, 189
Slope-intercept form, 188, 189
Sound intensity, 388
Square matrix, 465
Square roots, 56
  principal, 57
Square systems, 414
Squares
  difference of, 38
  perfect, 26, 38
Substitution method, 403, 404, 575, 576
Subtraction, 7
  of complex numbers, 115
  of fractional expressions, 47, 48
  of matrices, 468
  of polynomials, 25
  of signed numbers, 7
Sum of constants, 596
Sum of cubes, 38
Sum of infinite geometric series, 609, 610, 615
Summation
  index on the, 595
  lower limits of, 595

properties, 596
  symbol, 595
  upper limits of, 595
Supplementary angles, 110
Symmetric property, 5, 8
Symmetry, 210
  line of, 230, 298
  tests for, 210, 211
Synthetic division, 266–69
Systems of equations, 401–16
  Cramer's rule, 506, 507, 511, 512
  elimination method, 404–6
  equivalent, 404
  Gaussian method, 484–89
  graphs of, 410, 411
  homogeneous, 415, 416
  inverse method, 499
  of linear equations, 401–16
  $n \times n$ systems, 414, 415
  nonlinear, 574–79
  nonsquare, 414, 415
  reduction method, 411
  solution of, 401–16, 419–24
  substitution method, 403, 404
  in three variables, 410–16
  in two variables, 401–6
  types of
    consistent (intersect), 402, 404, 405, 410
    dependent (coincide), 402, 405, 410
    inconsistent (parallel), 402, 406, 410
Systems of inequalities, 432–35

Terminating decimals, 2
Terms, 22
  degree of, 24
  reducing to lowest, 44
  of a sequence, 592–94
Test point, 430
Tests for symmetry, 210, 211
Theorem, 8
Transitive property, 4, 5, 8
Translation of axes, 548, 558, 567
Transverse axis, 555
Tree diagram, 630
Triangles
  isosceles, 110
  right, 177, 178

Trichotomy property, 4
Trinomials, 24
  factoring, 33–36
Trivial solution, 415

Variables, 22, 200, 339
Variation, 245–48
  of signs, 282
Velocity, 146
Verification, 623
Vertex, 229
  of an ellipse, 544
  formula for, 229
  of a hyperbola, 555
  of a parabola, 229, 565
Vertical asymptotes, 308–11
Vertical line test, 204
Vertical shift, 206, 207

Whole numbers, 2
Work problems, 102, 103, 145

x-axis, 175
x-coordinate, 175
x-intercept, 183, 230
x-y coordinate system. *See* Cartesian coordinate system.

y-axis, 175
y-coordinate, 175
y-intercept, 183

Zero, 2, 7, 9
  division by, 7, 42
  as an exponent, 14
  multiplication by, 6, 8
  properties of, 8
  theorems, 8
Zero matrix, 467
Zero polynomial, 264, 273
Zero of a polynomial, 266, 279, 301
Zero of a rational function, 311
Zero-product rule, 122, 478, 577